STATE of KNOWLEDGE
of SOIL BIODIVERSITY

Status, challenges and potentialities

Report
2020

Food and Agriculture Organization of the United Nations
Rome, 2020

Required citation
FAO, ITPS, GSBI, SCBD and EC. 2020. *State of knowledge of soil biodiversity - Status, challenges and potentialities*, Report 2020. Rome, FAO.
https://doi.org/10.4060/cb1928en

The designations employed and the presentation of material in this information product do not imply the expression of any opinion whatsoever on the part of the Food and Agriculture Organization of the United Nations (FAO), Global Soil Biodiversity Initiative (GSBI), Secretariat of the Convention of Biological (SCBD) or European Commission (EC) concerning the legal or development status of any country, territory, city or area or of its authorities, or concerning the delimitation of its frontiers or boundaries. The mention of specific companies or products of manufacturers, whether or not these have been patented, does not imply that these have been endorsed or recommended by FAO, GSBI, SCBD or EC in preference to others of a similar nature that are not mentioned.
The views expressed in this information product are those of the author(s) and do not necessarily reflect the views or policies of FAO, GSBI, SCBD or EC.

ISBN 978-92-5-133582-6 [FAO]

© FAO, 2020

Some rights reserved. This work is made available under the Creative Commons Attribution-NonCommercial-ShareAlike 3.0 IGO licence (CC BY-NC-SA 3.0 IGO; https://creativecommons.org/licenses/by-nc-sa/3.0/igo/legalcode/legalcode).

Under the terms of this licence, this work may be copied, redistributed and adapted for non-commercial purposes, provided that the work is appropriately cited. In any use of this work, there should be no suggestion that FAO endorses any specific organization, products or services. The use of the FAO logo is not permitted. If the work is adapted, then it must be licensed under the same or equivalent Creative Commons licence. If a translation of this work is created, it must include the following disclaimer along with the required citation: "This translation was not created by the Food and Agriculture Organization of the United Nations (FAO). FAO is not responsible for the content or accuracy of this translation. The original [Language] edition shall be the authoritative edition."

Disputes arising under the licence that cannot be settled amicably will be resolved by mediation and arbitration as described in Article 8 of the licence except as otherwise provided herein. The applicable mediation rules will be the mediation rules of the World Intellectual Property Organization http://www.wipo.int/amc/en/mediation/rules and any arbitration will be conducted in accordance with the Arbitration Rules of the United Nations Commission on International Trade Law (UNCITRAL).

Third-party materials. Users wishing to reuse material from this work that is attributed to a third party, such as tables, figures or images, are responsible for determining whether permission is needed for that reuse and for obtaining permission from the copyright holder. The risk of claims resulting from infringement of any third-party-owned component in the work rests solely with the user.

Sales, rights and licensing. FAO information products are available on the FAO website (www.fao.org/publications) and can be purchased through publications-sales@fao.org. Requests for commercial use should be submitted via: www.fao.org/contact-us/licence-request. Queries regarding rights and licensing should be submitted to: copyright@fao.org.

Cover: ©FAO/Matteo Sala

CONTENTS

Contributors — XVI

Foreword — XXI

Acknowledgements — XXIII

Abbreviations and Acronyms — XXIV

Chapter 1
Introduction — 1

    1.1 | A growing awareness of the importance of soil biodiversity — 2

    1.2 | Structure of the report — 3

Chapter 2
Global diversity and distribution of soil biodiversity — 7

    2.1 | What is soil biodiversity? — 7

    2.2 | Soil communities — 8

    2.3 | Biodiversity in the soil — 13

    2.3.1 | Bacteria and Archaea — 14

    2.3.2 | Fungi — 19

    2.3.3 | Viruses — 28

    2.3.4 | Algae — 30

    2.3.5 | Protists — 37

    2.3.6 | Nematodes — 44

    2.3.7 | Mites — 48

    2.3.8 | Springtails — 56

    2.3.9 | Earthworms — 62

    2.3.10 | Isopods, millipedes, insects and spiders — 68

    2.3.11 | Termites — 80

    2.3.12 | Ants — 85

    2.3.13 | Soil vertebrates — 92

    2.4 | Spatial and temporal distribution of soil biodiversity — 98

    2.4.1 | Spatial patterns of soil organisms — 98

2.4.2 | Temporal patterns of soil organisms	100
2.4.3 | Below-ground distribution does not follow above-ground patterns	102
2.4.4 | Interactions among soil organisms	104
2.5 | Novel technologies – from indicators to monitoring	109

Chapter 3
Contributions of soil biodiversity to ecosystem functions and services — **115**

3.1 | Defining soil functions	115
3.1.1 | The soil food web approach	119
3.1.2 | The multifunctionality of soil biodiversity	122
3.2 | Defining soil ecosystem services	124
3.3 | Soil biodiversity supports the *Sustainable Development Goals*	127
3.4 | Provisioning ecosystem services	131
3.4.1 | Nutrient cycling	131
3.4.2 | Food production	134
3.4.3 | Raw materials	135
3.4.4 | Clean water	136
3.4.5 | Soil biodiversity and human health	139
3.5 | Regulating ecosystem services	147
3.5.1 | Climate regulation	147
3.5.2 | Soil carbon cycles	153
3.5.3 | Carbon sequestration and storage	157
3.5.4 | Soil formation and erosion prevention	160
3.5.5 | Regulation of water flow	161
3.5.6 | Wastewater treatment	162
3.5.7 | Invasive species	163
3.5.8 | Biodiversity regulation and biocontrol	166
3.5.9 | Genetic diversity	172
3.6 | Cultural services	173
3.6.1 | Spiritual experience and sense of place	173

3.6.2	Aesthetic appreciation and inspiration for culture, art and design	175
3.6.3	Cultural heritage, knowledge and education	180
3.6.4	Recreation and mental and physical health	180
3.7	Economic value of soil biodiversity	182
3.7.1	The challenge of valuing soil organisms	185

Chapter 4
Threats to soil biodiversity - global and regional trends — 191

4.1	Introduction	191
4.2	Threats to soil biodiversity	198
4.2.1	Deforestation	198
4.2.2	Urbanization	200
4.2.3	Agricultural intensification	202
4.2.4	Loss of soil organic matter and soil organic carbon	205
4.2.5	Soil compaction and sealing	210
4.2.6	Soil acidification and nutrient imbalances	213
4.2.7	Pollution	216
4.2.8	Salinization and sodification	221
4.2.9	Fire	224
4.2.10	Erosion and landslides	226
4.2.11	Climate change	232
4.2.12	Invasive species	235
4.3	Regional status of threats to soil biodiversity	237
4.3.1	Sub-Saharan Africa	239
4.3.2	Asia	241
4.3.3	Europe	244
4.3.4	Latin America and the Caribbean	247
4.3.5	North America	251
4.3.6	South West Pacific	253
4.4	Global synthesis	259

Chapter 5
Responses and opportunities   263

5.1 | Introduction to the management of soil biodiversity   263

5.2 | Ecosystem restoration: starting from the ground and leveraging soil biodiversity for sustainability   267

5.2.1 | Land use and land degradation   267

5.2.2 | Novel whole-ecosystem approaches for soil restoration and soil management   272

5.3 | Potential of soil biodiversity in the fight against soil pollution   277

5.3.1 | Use of soil biodiversity as a soil pollution remediation tool   277

5.3.2 | Knowledge gaps and environmental risk assessment   284

5.4 | Solutions to specific soil biodiversity-related problems   291

5.4.1 | Current and future applications of soil biodiversity research on sustainable food production   291

5.4.2 | Microbiome-based approaches to improve plant production   292

5.4.3 | Soil biodiversity as a farmer's tool   294

5.4.4 | Indigenous knowledge related to soil organisms   297

5.5 | Regional examples of novel approaches and applications   298

5.5.1 | Agricultural green development in China   298

5.5.2 | Australia   300

5.5.3 | Latin America   302

5.5.4 | The United States of America   302

5.6 | Future Risks   304

5.7 | Education, mainstreaming and policy   304

Chapter 6
State of soil biodiversity at the national level   311

6.1 | Assessment of soil biodiversity   312

6.2 | Research, capacity development and awareness raising   316

6.3 | Mainstreaming: policies, programmes, regulations and governmental frameworks   318

6.4 | Analysis of the main gaps, barriers and opportunities in the conservation and sustainable use of soil biodiversity   319

Chapter 7
Conclusions and way forward — **323**

    7.1 | What is soil biodiversity and how is it organized? — 323

    7.2 | Status of knowledge on soil biodiversity — 324

    7.3 | Soil biodiversity potential — 325

    7.3.1 | Provision of ecosystem services — 325

    7.3.2 | Food security, nutrition and human health — 326

    7.3.3 | Environmental remediation — 327

    7.3.4 | Climate change — 327

    7.4 | Challenges and gaps — 328

    7.5 | The way forward — 333

Annex I
Country Responses to the Soil Biodiversity Survey — **339**

    1 | Asia — 340

    1.1 | Assessment of soil biodiversity — 340

    1.1.1 | Contributions of soil biodiversity to ecosystems services — 340

    1.1.2 | National assessments — 341

    1.1.3 | Practical applications of soil biodiversity — 341

    1.1.4 | Major practices negatively impacting soil biodiversity — 343

    1.1.5 | Invasive alien species (IAS) — 343

    1.1.6 | Monitoring soil biodiversity — 343

    1.1.7 | Indicators used to evaluate soil biodiversity — 345

    1.2 | Research, capacity development and awareness raising — 345

    1.3 | Mainstreaming: policies, programmes, regulations and governmental frameworks — 347

    1.4 | Analysis of the main gaps, barriers and opportunities in the conservation and sustainable use of soil biodiversity — 348

    2 | Europe and Eurasia — 349

    2.1 | Assessment of soil biodiversity — 349

    2.1.1 | Contributions of soil biodiversity to ecosystems services — 349

    2.1.2 | National assessments — 354

2.1.3 | Practical applications of soil biodiversity	357
2.1.4 | Major practices negatively impacting soil biodiversity	360
2.1.5 | Invasive alien species (IAS)	361
2.1.6 | Monitoring soil biodiversity	362
2.1.7 | Indicators used to evaluate soil biodiversity	368
2.2 | Research, capacity development and awareness raising	369
2.3 | Mainstreaming: policies, programmes, regulations and governmental frameworks	379
2.4 | Analysis of the main gaps, barriers and opportunities in the conservation and sustainable use of soil biodiversity	384
3 | Latin America and the Caribbean (LAC)	386
3.1 | Assessment of soil biodiversity	386
3.1.1 | Contributions of soil biodiversity to ecosystems services	386
3.1.2 | National assessments	389
3.1.3 | Practical applications of soil biodiversity	390
3.1.4 | Major practices negatively impacting soil biodiversity	394
3.1.5 | Invasive alien species (IAS)	394
3.1.6 | Monitoring soil biodiversity	395
3.1.7 | Indicators used to evaluate soil biodiversity	397
3.2 | Research, capacity development and awareness raising	398
3.3 | Mainstreaming: policies, programmes, regulations and governmental frameworks	401
3.4 | Analysis of the main gaps, barriers and opportunities in the conservation and sustainable use of soil biodiversity	406
4 | Near East and North Africa (NENA)	407
4.1 | Assessment of soil biodiversity	407
4.1.1 | Contributions of soil biodiversity to ecosystems services	407
4.1.2 | National assessments	408
4.1.3 | Practical applications of soil biodiversity	408
4.1.4 | Major practices negatively impacting soil biodiversity	409
4.1.5 | Indicators used to evaluate soil biodiversity	409

4.2 | Research, capacity development and awareness raising	409
4.3 | Mainstreaming: policies, programmes, regulations and governmental frameworks	410
4.4 | Analysis of the main gaps, barriers and opportunities in the conservation and sustainable use of soil biodiversity	410
5 | North America	411
5.1 | Assessment of soil biodiversity	411
5.1.1 | Contributions of soil biodiversity to ecosystems services	411
5.1.2 | National assessments	418
5.1.3 | Practical applications of soil biodiversity	419
5.1.4 | Major practices negatively impacting soil biodiversity	420
5.1.5 | Invasive alien species (IAS)	421
5.1.6 | Monitoring soil biodiversity	422
5.1.7 | Indicators used to evaluate soil biodiversity	423
5.2 | Research, capacity development and awareness raising	423
5.3 | Mainstreaming: policies, programmes, regulations and governmental frameworks	424
5.4 | Analysis of the main gaps, barriers and opportunities in the conservation and sustainable use of soil biodiversity	425
6 | South West Pacific	427
6.1 | Assessment of soil biodiversity	427
6.1.1 | Contributions of soil biodiversity to ecosystems services	427
6.1.2 | National assessments	427
6.1.3 | Practical applications of soil biodiversity	427
6.1.4 | Major practices negatively impacting soil biodiversity	428
6.1.5 | Monitoring soil biodiversity	428
6.1.6 | Indicators used to evaluate soil biodiversity	428
6.2 | Research, capacity development and awareness raising	429
6.3 | Mainstreaming: policies, programmes, regulations and governmental frameworks	429
6.4 | Analysis of the main gaps, barriers and opportunities in the conservation and sustainable use of soil biodiversity	430
7 | Sub-Saharan Africa (SSA)	430

7.1	Assessment of soil biodiversity	430
7.1.1	Contributions of soil biodiversity to ecosystems services	430
7.1.2	National assessments	430
7.1.3	Practical applications of soil biodiversity	431
7.1.4	Major practices negatively impacting soil biodiversity	431
7.1.5	Invasive alien species (IAS)	431
7.1.6	Monitoring soil biodiversity	432
7.1.7	Indicators used to evaluate soil biodiversity	432
7.2	Research, capacity development and awareness raising	433
7.3	Mainstreaming: policies, programmes, regulations and governmental frameworks	433
7.4	Analysis of the main gaps, barriers and opportunities in the conservation and sustainable use of soil biodiversity	434

Annex II
National Survey on Status of Soil Biodiversity: Knowledge, Challenges and Opportunities — 437

I	Assessment	439
II	Research, capacity development and awareness raising	442
III	Mainstreaming: policies, regulations and governmental frameworks	442
IV	Analysis of gaps and opportunities	443

Annex III
List of countries that responded to the survey — 445

References — 447

BOXES

Terminology	7
2.2 | Size-structured soil communities	9
2.3.1 | The Rhizobium and nitrogen fixation	16
2.3.1.1 | They are the oldest known evidence of life on Earth!	17
2.3.2 | Arbuscular mycorrhizal fungi	24
2.3.2.1 | Natural soil cover	25
2.3.2.2 | Laying the groundwork	27
2.3.4 | Microfossils, a reference to the past	36
2.3.5 | Testate amoebae	40
2.3.5.1 | Soil-borne pathogen	42
2.3.9 | Micromorphology of biological activity in soils	66
2.4.1 | Pedodiversity	99
2.5 | The bait-lamina test	112
3.3.1 | Soil Biodiversity and the SDGs	130
3.5.1.1 | Climate Change and Soil Biodiversity	149
3.5.1.2 | Nitrification inhibitors	152
3.5.3.1 | Keeping the carbon in the soil	159
3.5.3.2 | Biochar as C-sequestration	159
3.5.8.1 | Anaerobic soil disinfection (ASD) - an alternative fumigation technique	169
3.5.8.2 | Biological control and crops	169
4.2.4.1 | How do SOC fractions influence soil biodiversity?	208
4.2.7.1 | Main Sources of contaminants that impact soil biodiversity	218
5.3.2.1 | Microplastics	290
5.4.3.1 | Purposes of on-farm use of soil microorganisms	296
5.5.2.1	301

FIGURES

2.2.1 | From viruses to mammals	11
2.3 | Soil Biodiversity	13
2.4.4.1 | Interactions between organisms through the soil food web	105
3.1.1.1 | Diagram of a soil food web	120
3.2.1 | Relationship between soil biodiversity and ecosystem services	125
3.3.1 | Healthy soils, a prerequisite to achieve the SDGs	129
3.4.1.1 | Main drivers for the alteration of global cycles	133
3.5.2.1 | Role of soil biodiversity in aggregate stability and carbon storage	156
3.5.8.1 | Synergistic and antagonistic effects of soil microbes and plants	170
3.5.8.2 | Biological control	171
3.6.2.1 | Soil biodiversity and inspiration for art	175
4.1.1 | Major anthropogenic threats to soil biodiversity	193
4.2.2.1 | Collembolan Ecomorphological Index	202
4.2.5.1 | Soil compaction	212
4.2.8.1 | Salinization	222
4.2.10.1 | Satellite images of natural and anthropogenic-enhanced water erosion	230
4.2.10.2 | Satellite images of soil erosion	231
4.3.1 | Terrestrial Ecoregions of the World used for the assessment of regional threats	238
4.3.3.1 | Importance of threats to soil biodiversity in Europe	245
4.4.1 | Estimated levels of current potential threats to soil biodiversity worldwide	260
5.1.1 | Management of soil biodiversity	267
5.2.1.1 | Soil biodiversity as a tool for nature-based solutions	271
5.3.1.1 | Biorremediation	283
6.1.2 | Soil biodiversity and ecosystem services perception (Survey Question 2.1, Annex II)	313
6.1.3 | Major practices that have a negative impact on soil biodiversity (Survey Question: 2.8, Annex II)	314

6.1.4 | Indicators commonly used for soil biodiversity evaluation (Survey Question 2.7)	316
7.1 | Soil Biodiversity map	325

PHOTOS

2.3.1 | Actinobacteria	18
2.3.2a | Mycorrhizal spore	22
2.3.2b | Aspergillus spore	22
2.3.2c | Saprophytic fungus	23
2.3.2d | Basidiomycete fungus	23
2.3.2e | Basidiomycete fungus	24
2.3.3 | Bacteriophage virus	30
2.3.4a | Yellow-green algae	33
2.3.4b | Yellow-green algae	33
2.3.4c | Yellow-green algae	34
2.3.4d | Eustigmatophyte algae	34
2.3.4e | Green algae	35
2.3.4f | Algae cultures	35
2.3.5b | Free-living protist	39
2.3.5c | Ciliated protist	39
2.3.6a | Nematode	46
2.3.6b | Nematode	46
2.3.6c | Nematode	47
2.3.6d | Enchytraeid	47
2.3.6e | Springtail carrying nematodes	48
2.3.7a | Oribatid mite	53
2.3.7b | Opilioacarid mite	54
2.3.7c | Predatory mite	54

2.3.7d | Mite	55
2.3.7e | Predatory mite	55
2.3.8a | Springtail	59
2.3.8b | Springtails	59
2.3.8c | Springtail	60
2.3.8d | Springtail	60
2.3.8e | Springtail	61
2.3.8f | Springtail	61
2.3.8g | Springtail	62
2.3.9a | Earthworm	65
2.3.9b | Flatworm	65
2.3.9c | Predacious flatworm	65
2.3.10a | Centipede	73
2.3.10b | Millipede	73
2.3.10c | Dung beetle	74
2.3.10d | Beetle	74
2.3.10e | Millipede	75
2.3.10f | Millipede	75
2.3.10g | Pseudoscorpion	76
2.3.10h | Spider	76
2.3.10i | Wolf spider	77
2.3.10l | Woodlouse	77
2.3.10m | Woodlouse	78
2.3.10n | Woodlouse	78
2.3.10o | Woodlouse	79
2.3.10p | Woodlouse	79
2.3.11a | Termites	82
2.3.11b | Termites	82

2.3.11c | Worker termite					83

2.3.11d | Termite nest					83

2.3.11e | Termite mound					84

2.3.11f | Termite mound					84

2.3.12a | Ant					87

2.3.12b | Jack Jumper ant					87

2.3.12c | Ant					88

2.3.12d | Leaf-cutter ant					88

2.3.12e | Carpenter ant					89

2.3.12f | Ants					89

2.3.12g | Carpenter ant					90

2.3.12h | Workers carpenter ants					90

2.3.13a | Pocket gopher					95

2.3.13b | Mole					95

2.3.13c | Gopher					96

2.3.13d | Prairie dog					97

CONTRIBUTORS

All names listed here are presented in alphabetic order.

General coordination: Ronald Vargas Rojas *(FAO-GSP)*

Managing Editors:
Initial: Kelly S. Ramirez *(FAO-GSP)*, Monica Kobayashi *(FAO-GSP)*
Current: Rosa Cuevas Corona *(FAO-GSP)*, Vinisa Saynes Santillan *(FAO-GSP)*

Editorial Board: Arwyn Jones *(EU-JRC)*, Caridad Canales *(CBD)*, David Cooper *(CBD)*, Diana H. Wall *(GSBI)*, Edmundo Barrios *(FAO)*, Luca Montanarella *(EU-JRC)*, Peter C. de Ruiter *(ITPS)*, Rosa Maria Poch *(ITPS)* and Wim H. van der Putten *(GSBI)*.

Coordinating Lead Authors: Anne Winding *(Denmark)*, Brajesh K. Singh *(Australia)*, Elizabeth Bach *(United States of America)*, George Brown *(Brazil)*, Junling Zhang, *(China)*, Miguel Cooper *(Brazil)*, Patrice Dion *(Canada)*, Pauline Mele *(Australia)*, Nico Eisenhauer *(Germany)*, Sergio Peña-Neira *(Chile)*, Zöe Lindo *(Canada)*.

Contributing Authors

Chapter 1. Introduction
LEAD AUTHORS:
   Kate Scow, *United States of America*
   Richard D. Bardgett, *United Kingdom*
CONTRIBUTING AUTHORS:
   Dan Pennock, *Canada*
   Ronald Vargas Rojas, *(FAO-GSP)*

Chapter 2. Global diversity & distribution of soil biodiversity
LEAD AUTHORS:
   Brajesh K. Singh, *Australia*
   Nico Eisenhauer, *Germany*
   Zöe Lindo, *Canada*
CONTRIBUTING AUTHORS:
   Amanda Koltz, *United States of America*
   Apolline Auclerc, *France*
   Asma Asemaninejad, *Iran*
   Carlos A. Guerra, *Germany*
   Clément Duckert, *Switzerland*
   Cristiano Nicosia, *Italy*
   David Andrés Donoso Vargas, *Ecuador*
   David. J Eldridge, *Australia*
   Eleonora Egidi, *Australia*
   Emiru Birhane, *Ethiopia*
   Eric Blanchart, *France*
   Fatima Maria de Souza Moreira, *Brazil*
   Jerome Tondoh, *Ivory Coast*
   Joanne B. Emerson, *United States of America*
   Johan van Hoogen, *Switzerland*
   Konstantin B. Gongalsky, *Russian Federation*
   Liesje Mommer, *The Netherlands*
   Manuel Delgado-Baquerizo, *Spain*
   Maria Minor, *New Zealand*
   Maria Tsiafouli, *Greece*
   Marie Josiane Meupia, *Cameroon*
   Marla Schwarzfeld, *Canada*
   Nataliya Rybalka, *Germany*
   Pankaj Trivedi, *United States of America*
   Paul B.L. George, *Wales*
   Rosa Cuevas Corona *(FAO-GSP)*
   Rosa Poch, *Spain*
   Stefan Geisen, *Germany/The Netherlands*
   Stephanie D Jurburg, *Germany*
   Tanel Vahter, *Estonia*
   Thom Kuyper, *The Netherlands*
   Werner Kratz, *Germany*
   Yolanda Cantón Castilla, *Spain*
   Yuan Zeng, *United States of America*

Chapter 3. Contributions of soil biodiversity to ecosystem functions and services

LEAD AUTHORS:

Anne Winding, *Denmark*
Elizabeth Bach, *United States of America*
Pauline Mele, *Australia*

CONTRIBUTING AUTHORS:

Ana Paula Dias Turetta, *Brazil*
Andrew Robertson, *United States of America*
Apolline Auclerc, *France*
Astghik Pepoyan, *Armenia*
Bertha Cecilia Garcia Cienfuegos,
Cairo Robb, Charlene Janion-Scheepers, *South Africa*
Christian Davies, *United States of America*
Cristina Abbate, *Italy*
Daphne Miller, *United States of America*
Erin Cameron, *Canada*
Guénola Pérès, *France*
Jason Hong, *United States of America*
Jianming Xu, *China*
Kate Lajtha, *United States of America*
Kate Scow, *United States of America*
Laura Aldrich-Wolfe, *United States of America*
Linda Blackall, *Australia*
Maria J.I. Briones, *Spain*
Marianne Zandersen, *Denmark*
Marijana Kapovic Solomun, *Bosnia and Herzegovina*
Mette Termansen, *Denmark*
Michael Schloter, *Germany*
Pascal Jouquet, *France*
Rolf Altenburger, *Germany*
Susan Crow, *United States of America*
Tania Runge, *Germany*
Timothy Cavagnaro, *Australia*

Chapter 4. Threats to soil biodiversity - global and regional trends

LEAD AUTHORS:

George Brown, *Brazil*
Miguel Cooper, *Brazil*
Monica Kobayashi, *FAO-GSP*

CONTRIBUTING AUTHORS:

Alberto Orgiazzi, *Italy*
Anahí Domínguez, *Argentina*
Ana Paula Dias Turetta, *Brazil*
André L.C. Franco, *Brazil*
Andrey S. Zaitsev, *Germany*
Anne Winding, *Denmark*
Bente Foereid, *Norway*
Brajesh K. Singh, *Australia*
Carlos Guerra, *Germany*
Claudia Rojas, *Chile*
David Spurgeon, *United Kingdom*
Ece Aksoy, *Turkey*
Fátima Maria Moreira, *Brazil*
Francisco Bautista, *Mexico*
Jianming Xu, *China*
Johannes Rousk, *Sweden*
José Camilo Bedano, *Argentina*
Joseph D. Bagyaraj, *India*
Krishna Saxena, *India*
Laura Fernanda Simões da Silva, *Brazil*
Leho Tedersoo, *Estonia*
Loren Byrne, *United States of America*
Mac A. Callaham, *United States of America*
Madhu Choudhary, *India*
M. Fernanda Aller, *Spain*
Manuel Delgado-Baquerizo, *Spain*
Maria Fuensanta García Orenes, *Spain*
Maria Tsiafouli, *Greece*
Marie de Graaf, *United States of America*
Miranda Hart, *Canada*
Moses Thuita, *Kenya*
Nancy Karanja, *Kenya*
Nathalie Fromin, *France*
Nico Eisenhauer, *Germany*
Nobuhiro Kaneko, *Japan*
Pauline Mele, *Australia*
Pilar Andres Pastor, *Spain*
Raul Ochoa-Hueso, *Spain*
Roman Kuperman, *United States of America*
Stephen Ichami, *Kenya*
Steven J. Fonte, *United States of America*
Vinisa Saynes Santillan, *(FAO-GSP)*
Yunuen Tapia Torres, *Mexico*

Chapter 5. Responses and opportunities
LEAD AUTHORS:
- Junling Zhang, *China*
- Patrice Dion, *Canada*
- Sergio Peña Neira, *Chile*

CONTRIBUTING AUTHORS:
- Adriano Sofo, *Italy*
- Aidan Keith, *United Kingdom*
- Ana Paula Turetta,
- Angela Sessitsch, *Austria*
- Brajesh Singh, *Australia*
- Carlos Garbisu, *Spain*
- Catriona Macdonald, *Australia*
- Céline Pelosi, *France*
- Cornelis A.M. van Gestel, *The Netherlands*
- Dana Elhottová, *Czech Republic*
- E.R. Jasper Wubs, *The Netherlands*
- Fatima Maria de Souza Moreira, *Brazil*
- Fusuo Zhang, *China*
- Ilze Vircava, *Latvia*
- Jianbo Shen, *China*
- Katarina Hedlund, *Sweden*
- Marcel van der Heijden, *Switzerland*
- Mark Brady, *Sweden*
- Matteo Mancini, *Italy*
- Matthew Wallenstein, *United States of America*
- Mattias Rillig, *Germany*
- Nancy Karanja, *Kenya*
- Peter Mortimer, *China*
- Pilar Andres Pastor, *Spain*
- Rubén Borge, *Spain*
- Ruth Schmidt, *Austria*
- Sarah Evans, *United States of America*
- Sheila Okoth, *Kenya*
- Svein Solberg, *Norway*
- Thibaud Decaens, *France*
- Uffe Nielsen, *Australia*
- Vinisa Saynes Santillan, *(FAO-GSP)*
- Warshi Shamila Dandeniya, *Sri Lanka*

Chapter 6. State of soil biodiversity at the national level
LEAD AUTHORS:
- Monica Kobayashi *(FAO-GSP)*

Chapter 7. Conclusions and way forward
LEAD AUTHORS:
- Rosa Cuevas Corona *(FAO-GSP)*
- Ronald Vargas Rojas *(FAO-GSP)*

Reviewers
Intergovernmental Technical Panel on Soils (ITPS)
- Adalberto Benavides Mendoza, *Mexico*
- Attia Rafla, *Tunisia*
- Costanza Calzolari, *Italy*
- Megan Balks, *New Zealand*

Food and Agriculture Organization of the United Nations (FAO)
- Dafydd Pilling
- Irene Hoffmann
- Julie Belanger
- Natalia Rodriguez Eugenio

Global Soil Biodiversity Initiative (GBSI)
- André L.C. Franco
- Carl J. Wepking
- Monica Farfan
- Valerie Behan-Pelletier

Convention of Biological Diversity (CBD)
- Julie Anne Botzas-Coluni

Countries (Chapter 6)
- Aleksi Lehtonen, *Finland*
- Ana Olga Ospina, Colombia Anna Benedetti, *Italy*
- Barbara M. DeRosa-Joynt, *United Sates of America*
- Bianca Moebius-Clune, *United Sates of America*
- Bob Turnock, *Canada*

David Knaebel, *United Sates of America*
Dennis Chessman, *United Sates of America*
Elena Starchenko, *Ukraine*
Jaap Bloem, *The Netherlands*
Jacques Tavares, *Cabo Verde*
Jefé Leão Ribeiro, *Brazil*
Jose Luis Berroteran, *Venezuela*
Juan Aciego Pietri, *Venezuela*
Maria Rivero Santos, *Colombia*
Muhammad Manhal Alzoubi, *Syria*
Pitayakon Limtong, *Thailand*
Raphaël Ngadi Litadi, *Gabon*
Skye Wills, *United Sates of America*
Urwana Querrec, *France*

Edition and publication
Dan Pennock
Hugo Bourhis *(FAO-GSP)*
Isabelle Verbeke *(FAO-GSP)*
Lea Pennock

Art direction
Matteo Sala *(FAO-GSP)*

Photo credits
Abdoulaye Fofana Fall
Aisha Ildarovna Usmanova
Alessandra Arrázola
Andrés Costa
Andy Murray
Anna Wolak
Antonietta La Terza
Christian Omuto
Christian Thine
Christopher Marley
Clément Duckert
Cristiano Nicosia
Daizy Bharti
David Elliott
Daniel Castro Torres

Dmitry Zagumyonnyi
Duur Aanen
Ehteram Sadat Mirghasemi
Emilio Rodríguez
Eric Palevsky
Eye of Science
European Space Agency (ESA)
Felicity Crotty
François-Xavier
Gary Bauchan
Guillermo Peralta
Irina Petrova
Joanne Emerson
Jose Raul Román
Joseph Mowery
Juliane Filser
Koos Boomsma
Lucia Nunez
Lynn Carta
Manuel Vergara
Marton Jolankai
Matteo Sala
Megha Bhaskaran
Nataliya Rybalka
Olga Kozyreva
Oscar Franken
Pavel Krásenský
Pilatluck Lioroongcharoen
Ronald Ochoa
Ronald Vargas
Rosa Poch
Santosh Kumar
Sergii Dymchenko
Steven Fonte
Vinisa Saynes
Werner Kratz
Winston Bess
Yolanda Cantón
Yuan Zeng

© Andy Murray

FOREWORD

Our well-being and the livelihoods of human societies are highly dependent on biodiversity and the ecosystem services it provides. It is essential that we understand these links and the consequences of biodiversity loss for the various global challenges we currently face, including food insecurity and malnutrition, climate change, poverty and diseases. The Agenda 2030 for Sustainable Development sets out a transformative approach to achieve socio-economic development while conserving the environment.

There is increasing attention on the importance of biodiversity for food security and nutrition, especially above-ground biodiversity such as plants and animals. However, less attention is being paid to the biodiversity beneath our feet, soil biodiversity. Yet, the rich diversity of soil organisms drives many processes that produce food, regenerate soil or purify water.

In 2002, the Conference of the Parties (COP) to the Convention on Biological Diversity (CBD) decided at its 6th meeting to establish an International Initiative for the Conservation and Sustainable Use of Soil Biodiversity and since then, the Food and Agriculture Organization of the United Nations (FAO) has been facilitating this initiative. In 2012, FAO members established the Global Soil Partnership to promote sustainable soil management and increase attention to this hidden resource. The Status of the World's Soil Resources (FAO, 2015) concluded that the loss of soil biodiversity is considered one of the main global threats to soils in many regions of the world.

The 14th Conference of the Parties invited FAO, in collaboration with other organizations, to consider the preparation of a report on the state of knowledge on soil biodiversity covering its current status, challenges and potentialities. This report is the result of an inclusive process involving 300 scientists from around the world under the auspices of the FAO's Global Soil Partnership and its Intergovernmental Technical Panel on Soils, the Convention on Biological Diversity, the Global Soil Biodiversity Initiative and the European Commission. The report presents the state of knowledge on soil biodiversity, the threats to it, the solutions that soil biodiversity can provide to problems in different fields, including agriculture, environmental conservation, climate change adaptation and mitigation, nutrition, medicine and pharmaceuticals, remediation of polluted sites, and many others.

The report will make a valuable contribution to raising awareness of the importance of soil biodiversity and highlighting its role in finding solutions to today's global threats; it is a cross-cutting topic at the heart of the alignment of several international policy frameworks, including the Sustainable Development Goals (SDGs) and multilateral environmental agreements. Furthermore, soil biodiversity and the ecosystem services it provides will be critical to the success of the recently declared UN Decade on Ecosystem Restoration (2021-2030) and the upcoming Post-2020 Global Biodiversity Framework.

Soil biodiversity could constitute, if an enabling environment is built, a real nature-based solution to most of the problems humanity is facing today, from the field to the global scale. Therefore efforts to conserve and protect biodiversity should include the vast array of soil organisms that make up more than 25% of the total biodiversity of our planet.

FAO Director-General

QU Dongyu

Executive Secretary of CBD

Elizabeth Maruma Mrema

© Christopher Marley

ACKNOWLEDGEMENTS

The report *State of Knowledge of Soil Biodiversity - Status, Challenges and Potentialities* was made possible thanks to the commitment and expertise of hundreds of individuals, and the collaboration and support of many governments, institutions and partners.

We would like to express our sincere gratitude to the world's leading soil biodiversity scientists and experts who volunteered their time, passion and dedication to the writing of this report. Specifically, the Editorial Board, Lead Authors, Contributing Authors, Reviewers and the Editorial team for their invaluable contributions to the report. Furthermore, we would like to thank all the photographers, scientists and artists who shared with us amazing photographs and arts to ensure that this report reflects the beauty of soil biodiversity.

Further, we thank the many universities, institutions, and Governments that have supported the participation of their scientific experts in this major work.

Finally, we (FAO-GSP) recognize our vital partners, including the ITPS, the Convention on Biological Diversity, the Global Soil Biodiversity Initiative and the Joint Research Centre for their contribution to this report, along with the European Commission, the Russian Federation and the Swiss Confederation who financially supported the development and publication of this report.

ABBREVIATIONS AND ACRONYMS

AAFC | Agriculture and Agri-Food Canada
ABMI | Alberta Biodiversity Monitoring Institute
ACC | 1-aminocyclopropane-1-carboxylic acid
ACIAR | Australian Centre for International Agricultural Research
ADEME | Agence de l'Environnement et de la Maîtrise de l'Énergie
AFICUP | Plan Nacional de Agricultura Familiar, Indígena, Campesina, Urbana y Periurbana
AFM | Arbuscular Fungi Mycorrhiza
AFNOR | Association Française de Normalisation
AFOLU | Forestry and Other Land Use sector
AFRI | Agriculture and Food Research Initiative
AGD | Agriculture Green Development
AHDB | Agriculture and Horticulture Development Board
AIDS | Acquired Immunodeficiency Syndrome
AM | Arbuscular mycorrhizas
AMF | Arbuscular Mycorrhizal Fungi
AnaEE | Analisys and Experimentations on Ecosystems
ANPE | Peruvian National Association of Ecological Producers
ANR | French Research Agency
ARS | Agricultural Research Service
ASD | Anaerobic soil disinfestation
AUREA | Groupe européen dédié à l'économie circulaire- European group dedicated to the circular economy
BASE | Biomes of Australian Soil Environment
BOF | bio-organic fertilizer
BSI | (Biologial fertility Index)
Bt | Bacillus thuringiensis
C | Carbon
°C | degree Celsius
C_4 grasses | C_4 photosynthesis grasses.
Ca | Calcium
Ca^{2+} | Calcium ions
CAD | Canadian Dollars
CanSIS | Canadian Soil Information System
CAP | Common Agricultural policy
CARDI | Caribbean Agricultural Research and Development Institute
CBB | Calvin-Benson-Bassham
CEC | Caution Exchange Capacity
CEI | Collembolan Ecomorphological Index
CELOS | Center for Agricultural Research in Suriname
CEN | European Committee for Standardization
CH_4 | Methane
CIARA | Training and Innovation Foundation for Rural Development
CLM | Community Land Model
CNRS | Centre National de la Recherche Scientifique
CO | carbon monoxide
CO_2 | Carbon Dioxide
CBD | Convention on Biological Diversity
Cd | cadmium
CEDAW | Convention on the Elimination of All Forms of Discrimination against Women
CONAFOR | Comisión Nacional Forestal
COP | Conference of the Parties
Cr | chromium
CREA | Council for Agricultural Research and Economics
Cu | cooper
CUE | carbon utilization efficiency
DAISIE | Delivering Alien Invasives Species Inventories for Europe
DEB | Dynamic Energy Budget
DFG | German Science Foundation
DNA | Deoxyribonucleic Acid
EC | European Commission
ECLAC | Economic Commission for Latin America and the Caribbean
ECM | Ectomycorrhiza
EcoFINDERS | Ecological Function and Biodiversity Indicators in European Soils
EM | Energetic Material
EM CL-20 | China Lake compound 20
EPAMB | Estratégia e Plano de Ação Nacionais para a Biodiversidade (National Biodiversity Program)
ES | ecological stoichiometry
ESP | European Soil Partnership

EJP Soil | European Joint Programme
ESB | Ecosystem Services and Biodiversity
ESB | German Environmental Specimen Bank
EPS | extracellular polymeric substances
ETH-Zurich | Eidgenössische Technische Hochschule Zürich
EU | European Union
EUdaphobase | European Soil-Biology Data Warehouse for Soil Protection
EUGRIS | European Groundwater and Contaminated Land Remediation Information System
EUR | Euro
FAO | Food and Agriculture Organization of the United Nations
Fe | iron
FERA | Federal Emergency Relief Administration
FinBIF | Finnish Biodiversity Information Facility
FIRA | Fideicomisos Instituidos en Relación con la Agricultura
FISBo BGR | FachInformationsSystem Bodenkunde
GAP | Good Agricultural Practices
GAP | Gender Action Plan
GCSAR | General Commission for Scientific Agricultural Research
GESSOL | Gestion and sol "Fonctions environnementales et GEStion du patrimoine SOL"
GHG | Greenhouse Gases
GIS | Geographic Information System
GIZ | Deutsche Gesellschaft für Internationale Zusammenarbeit (German Agency for International Cooperation)
GMEP | Global Marine Environment Protection
GMOs | Genetically Modified Organisms
GRH | growth rate hypothesis
GSBI | Global Soil Biodiversity Initiative
GSOC map | Global Soil Organic Carbon Map
GSOCseq | Global SOC Sequestration Potential Map
GSP | Global Soil Partnership
HNCV | high nature and culture value
IAS | Invasive Alien Species
IBS-bf | Soil Biodiversity Index of the protocol "Biodiversity Friend"
IDEAM | Instituto de Hidrología, Meteorología y Estudios Ambientales
iDiv | German Centre for integrative Biodiversity Research
IFAD | International Fund for Agricultural Development
IGAC | International Global Atmospheric Chemistry Project
INECOL | Instituto de Ecología A.C.
INEGI | Instituto Nacional de Estadística y Geografía
INIA | National Institute of Agricultural Research
INRA | Institut National de la Recherche Agronomique
INSAI | National Institute of Integral Agricultural Health
IPBES | The Intergovernmental Science-Policy Platform on Biodiversity and Ecosystem Services
IPCC | Intergovernmental Panel on Climate change
IPN | Instituto Politécnico Nacional
ISO | International Organization for Standardization
ISPRA | Istituto Superiore per la Protezione e la Ricerca Ambientale
ISRIC | International Soil Reference and Information Centre
ITPGRFA | International Treaty on Plant Genetic Resources for Food and Agriculture
ITPS | Intergovernmental Technical Panel on Soils
IUCN | International Union for Conservation of nature
IYS | International Year of Soils
JRC | Joint Research Centre
K | potassium
K$^+$ | potassium ion
KJWA | Koronivia Joint Work on Agriculture
LABO | Federal-Länder Working group on Soil Protection
LAC | Latin America and the Caribbean
LANDIS | national soil information system
LAUHA | Steering Group for Evaluation and Monitoring of threatened species
LD | Land degradation
LDN | Land Degradation Neutrality

LOCTI | Ley Orgánica de Ciencia, Tecnología e Innovación
LPI | Living Planet Index
LPJ | Lund-Potsdam-Jena Dynamic Global Vegetation Model
LRI | Le Laboratoire des RadioIsotopes
LUCAS | European Union's Land Use/Cover Area frame statistical Survey Soil
MAEC | Mesures agro-environnementales et climatique
MAOM | mineral-associated organic matter
MBCAs | Microbial biological control agents
MEA | Millennium Ecosystem Assessment
MEP | Ministry of Environmental Protection of China
MICINN | Spanish Science Ministry
MINAGRI | Ministry of Agriculture and Irrigation
Mg | magnesium
Mg^{2+} | magnesium ion
MonViA | Nationales Monitoring der biologischen Vielfalt in Agrarlandschaften
MOOC | Massive Open Online Courses
N | Nitrogen
N_2 | Dinitrogen
N_2O | Nitrous Oxide
Na | Sodium
NBA | National Biodiversity Authority
NBAIM | National Bureau of Agriculturally Important Microorganisms
NBRIP | National Botanical Research Institute's phosphate growth medium
NSB | National Strategy for Biodiversity
NBSAPs | National Biodiversity Strategies and Action Plans
NDCs | nationally determined contributions
NECR | Natural England Commissioned Report
NENA | Near East and North Africa
NEON | The United States National Ecological Observatory Network
NGOs | non-governmental organizations
NH_3 | ammonia
NH_4^+ | ammonium soluble ions
Ni | Nickel
NIAES-NARO | The Institute for Agro-Environmental Sciences
NIFA | National Institute of Food and Agriculture
NNR | National Nature Reserve
NPAB | National Plan for Agricultural Biodiversity
NO_3^- | nitrate soluble ions
NO_x | nitrogen oxides
NRM | Natural Resource Management
NRP | Nominal Rate of Protection
NRSAS | National Research Service Award
NRCS | Natural Resources Conservation Service
NSC ISSAR | National Scientific Center " Institute for Soil Science and Agrochemistry Research"
NSF NEON | National Science Foundation
NSTDA | National Science and Technology Development Agency
NSW | New South Wale
NT | Northern Territory
NUE | Nitrogen Use Efficiency
O_2 | Oxygen
OAB | Observatoire agricole de la biodiversité
OECD | Organisation for Economic Co-operation and Development
OM | organic matter
ONB | Observatoire National de la Biodiversité
OPVT | Observatoire Participatif des Vers de Terre
OTUs | Operational Taxonomic Units
P | Phosphorus
PAHs | polycyclic aromatic hydrocarbons
Pb | lead
PBDEs | polybrominated diphenyl ether
PCBs | polychlorinated biphenyls
PCF | Pan-Canadian Framework on Clean Growth and Climate Change
PET | Potential Evapotranspiration
PFG | Degree Training Program
PFOS | Perfluorooctanesulfonic acid
PFOA | Perfluorooctanoic acid
PGPB | plant growth promoting bacteria
PGPR | plant growth-promoting rhizobacteria
PGSS | Politica para la Gestion Sostenible del Suelo (Sustainable Soil Management Policy)
PIA | Project Implementation Assistance
PLFA | Polimorfismo de Longitud de Fragmentos Amplificados
PNCTI | Technology and Innovation Plan

POM | particulate organic matter
POPs | persistent organic pollutants
PSB | Phosphate-solubilizing bacteria
PSGE | Plan Stratégique Gabon Emergent
QBS | Soil Biological Quality
Qld | Queensland
RECSOIL | Recarbonization of Global Soils
REDD | Reducing Emissions from Deforestation and forest degradation
REVA | Réseau d'Expérimentation et de Veille à l'innovation Agricole
RMQS | réseau de mesure de la qualité des sols
RNA | Ribonucleic Acid
RUBIOS | Red Uruguaya de Biodiversidad de Suelos
RubisCO | ribulose-1,5-bisphosphate carboxylase/oxygenase
SA | Southern Australia
SDGs | Sustainable Development Goals
SE | South East
SNIB | National System of Biodiversity
SinoBON | Chinese Biodiversity Monitoring and Research Network
SOC | Soil Organic Carbon
SOER | State of the Environment Report
SOFR | State of the Forests Report
Ss | Spongospora subterranea
S$_2$- | sulphide
SA | South Australia
SAG Culture Collection | Culture Collection of Algae at the Georg-August-University Göttingen
SALT | Slope Agriculture Land Technology
SEM | Scanning electron microscopy
SIR | substrate induced respiration
SIRD | Programa de Recuperación de Suelos Degradados
SISS | Italian Society of Soil Science
SO$_3$$^{2-}$ | sulfite
SO$_4$$^{2-}$ | sulfate
SOM | soil organic matter
SO$_x$ | sulfur dioxide
SUITMA | soils of urban, industrial, traffic, mining and military areas
Ss | Spongospora subterranea
SSA | Sub-Saharan Africa
SSM | Sustainable Soil Management
SSMP | Sustainable Soil Management Program
TEM | Transmission electron microscopy
TNT | trinitrotoluene
TEEB | The Economics of Ecosystems and Biodiversity
TEV | Total Economic Value
TRF | Thailand Research Fund
TRFLP | Terminal Restriction Fragment Length Polymorphism
UBA | German Environmental Agency
UK | United Kingdom of Great Britain
UKNEA | UK National Ecosystem Assessment
UKSO | UK Soil Observatory
UN | United Nations
UNAM | Universidad Nacional Autónoma de México
UNEA | United Nations Environment Assembly
UNCCD | United Nations Convention to Combat Desertification
UNFCCC | United Nations Framework Convention on Climate Change
UNCITRAL | United Nations Commission on International Trade Law
U.S. | United States
USD | US Dollar
USDA | United States Department of Agriculture
USEPA | United States Environmental Protection Agency
UV | Ultraviolet
UWI | University of the West Indies
VGSSM | Voluntary Guidelines for Sustainable Soil Management
VOCs | Volatile organic compounds
Zn | Zinc
WA | Western Australia
WFCC | World Federation for Culture Collections
WHO | World Health Organization
WRB | World Reference Base for Soil Resources
WSD | World Soil Day
WWF | World Wildlife Fund

CHAPTER 1
INTRODUCTION

A wealth of new scientific, technical and other types of knowledge relevant to soil biodiversity has been released since the establishment of the International Initiative for the Conservation and Sustainable Use of Soil Biodiversity in 2002 and the Global Soil Partnership in 2012, and the publication of the Status of the Worlds Soil Resources and the Global Soil Biodiversity Atlas in 2016.

This new wave of research is a consequence of tremendous growth in the methods available for the study of soil organisms by the scientific community. This research has placed soil biodiversity at the heart of international policy frameworks, including the Sustainable Development Goals (SDGs). Furthermore, soil biodiversity and ecosystem services will be pivotal for the success of the recently declared United Nations Decade on Ecosystem Restoration (2021–2030).

This report is a result of the work of more than 300 soil scientists and experts on soil biodiversity from all regions of the world, and it presents the best available knowledge on soil biota and their ecosystem functions and services The report is a contribution to a decision . of the 14th Conference of the Parties (COP) to the Convention on Biological Diversity (CBD), that invited the Food and Agriculture Organization of the United Nations (FAO) to prepare a report on the state of knowledge on soil biodiversity.

1.1 | A GROWING AWARENESS OF THE IMPORTANCE OF SOIL BIODIVERSITY

FAO had previously cooperated with the CBD on soil biodiversity under the International Initiative for the Conservation and Sustainable Use of Soil Biodiversity. This initiative was established in 2002 by the COP to the CBD at its sixth meeting in Nairobi, Kenya. FAO organized a number of workshops and educational events under this initiative.

Action on soils was given a major boost in 2012 with the creation of the Global Soil Partnership (GSP). The GSP is a partnership between FAO member countries and non-FAO organizations including government organizations, universities, NGOs, and farmer organizations. The work of the GSP is supported by the Intergovernmental Technical Panel on Soil (ITPS), which is composed of 27 experts in soil science drawn from the seven UN regions.

One of the first products of the GSP/ITPS was a revised statement of principles governing the management of soils, the revised World Soil Charter (FAO, 2015). The critical importance of soil biodiversity was fully recognized by the World Soil Charter. The first point of the Charter states that "Careful soil management is one essential element of sustainable agriculture and also provides a valuable level for climate regulation and a pathway for safeguarding ecosystem services and biodiversity". Point 8 of the Charter states the following:

> Soils are a key reservoir of global biodiversity, which ranges from micro-organisms to flora and fauna. This biodiversity has a fundamental role in supporting soil functions and therefore ecosystem goods and services associated with soils. Therefore it is necessary to maintain soil biodiversity to safeguard these functions.

The Status of the World's Soil Resources report (FAO and ITPS, 2015) examined the major threats to soil, including threats to soil biodiversity. The authors of that report judged the evidence available for an assessment to be very limited and the consensus on trend to be low. They emphasized the importance of monitoring programs for soil biodiversity, which can provide the scientific evidence needed for evidence-based management and policy recommendations.

The importance of soil biodiversity was also highlighted in the Voluntary Guidelines for Sustainable Soil Management (FAO, 2017). The report emphasized that soil biodiversity provides a wide range of biological functions which are key attribute of a sustainably managed soil. The report also listed specific measures that could be undertaken to preserve and enhance soil biodiversity.

Major developments on soil biodiversity were also occurring outside of the FAO/GSP umbrella. The Global Soil Biodiversity Initiative (GBSI) was created by five founding organizations: The School of Global Environmental Sustainability (Colorado State

University); Wageningen Centre for Soil Ecology; The University of Manchester; ETH-Zurich; and the European Commission, Joint Research Centre. The GBSI has had a major impact on policy formulation, education, and research in soil biodiversity. One of the most significant products of the GBSI was the Global Soil Biodiversity Atlas, which was produced jointly with the European Commission Joint Research Centre (Orgiazzi et al., 2016).

There has also been increasing recognition by the non-scientific community of the essential role that soil organisms play in supporting life on earth. Soil biodiversity can be clearly identified as a cross-cutting topic; it is at the heart of the alignment of several global agendas such as the United Nations (UN) Sustainable Development Goals (SDGs) and many multi-lateral environmental agreements, in particular those related to biodiversity, desertification and climate change. Furthermore, soil biodiversity and ecosystem services will be pivotal for the success of the recently declared UN Decade on Ecosystem Restoration (2021-2030).

1.2 | STRUCTURE OF THE REPORT

The report is organized into seven chapters. Chapter 2 summarizes current information on the different classes of soil organisms, ranging from the microscopic bacteria through to large soil-dwelling organisms such as ants, termites and earthworms. Each group is examined as to its contributions to ecosystem processes, such as carbon transformations and nutrient cycling, and the state of current knowledge and future challenges for each class are summarized.

Chapter 3 examines the contributions of soil biodiversity to ecosystem services and functions. Ecosystem functions refer to the full range of natural-biological processes carried out by soil organisms; ecosystem services refer to the subset of processes that contribute to human well-being. For example, one of the key ecosystem functions that organisms drive is the cycling of nutrients through various forms and pools in the soil; this ecosystem function is critical for the growth of crops and hence for the provision of food to human population, which is an ecosystem service. Many soil organisms (especially microorganisms) have been developed as commercial products to enhance crop growth and soil remediation, and these products are reviewed in this chapter as well.

The degradation of soil biodiversity can have highly negative consequences for multiple ecosystem functions and services. Currently, soil biodiversity is threatened by global anthropogenic changes, such as land-use intensification, deforestation and extreme climatic events. Recent evidence on the threats to soil biodiversity and the direction of change is examined in Chapter 4 of the report.

The management of soil biodiversity offers many opportunities to address significant societal issues such as environmental remediation of polluted soils and sediments, plant production, and food quality. These opportunities arise in large part from the suite of enhanced techniques that scientist can now deploy to examine soil organisms at the community level. These advances in knowledge, and the opportunities that they create, are examined in Chapter 5 of the report.

Chapter 6 summarizes the results of the survey undertaken by FAO of its Member Nations. The aim of the survey was to collect information at the country level on the status of soil biodiversity, to better understand concerns and threats to soil biodiversity, to compile relevant policies, regulations or frameworks that have been implemented, and to catalogue current soil biodiversity management and use efforts.

Fifty-seven (57) countries submitted their responses. All of the following regions had at least one representation: North America, Latin America and the Caribbean (LAC), Europe and Eurasia, Near East and North Africa (NENA), sub-Saharan Africa (SSA), Asia and South West Pacific.

Chapter 7 presents the conclusions of the report addressing the status, challenges and potentialities of soil biodiversity. It also proposes a way forward with concrete actions to unlock the potential of soil biodiversity. Despite the progress summarized in the report, it is clear that significant knowledge gaps remain. For instance, more comprehensive understanding of the relationship between soil biodiversity and ecosystem functioning is of crucial importance in order to link above- and below-ground compartments in ecosystem modelling and hence to better predict the consequences of biodiversity change and loss.

The development and implementation of effective policy and sustainable soil management strategies that can support biodiversity and ecosystem functions is necessary. Protecting above-ground biodiversity is not always sufficient for soil biodiversity. Above-ground and below-ground (soil biodiversity) are shaped by different environmental drivers and their linkages are not always predictable. The authors hope that the knowledge contained in this report will facilitate the assessment of the state of soil biodiversity as an integral part of national- and regional-level biodiversity reporting.

CHAPTER 2
GLOBAL DIVERSITY AND DISTRIBUTION OF SOIL BIODIVERSITY

2.1 | WHAT IS SOIL BIODIVERSITY?

The Convention on Biological Diversity defines biological diversity as "the variability among living organisms from all sources including, *inter alia*, terrestrial, marine and other aquatic ecosystems and the ecological complexes of which they are part; this includes diversity within species, between species and of ecosystems" (CBD, 1992, Article 2).

We define soil biodiversity as the variety of life belowground, from genes and species to the communities they form, as well as the ecological complexes to which they contribute and to which they belong, from soil micro-habitats to landscapes. The concept is conventionally used in a taxonomic sense and denotes the number of distinct species, but may be extended to encompass genetic, phenotypic (expressed), functional, structural or trophic diversity. It is accepted that the total biomass below ground equals or potentially exceeds that above ground. And while biodiversity in the soil exceeds that of other terrestrial systems by orders of magnitude, particularly at the microbial scale, it remains remarkably undervalued.

Terminology

For the purpose of this report, the terminologies *soil biodiversity, soil biological diversity* and *below-ground biodiversity* have been used interchangeably, and they include soil microbes and soil fauna. Likewise, the terminologies *microbial diversity, soil microbes, soil microbiota, soil microorganisms* and *soil microbiome* are used interchangeably specifically to describe soil microbial diversity.

2.2 | SOIL COMMUNITIES

Soils are one of the main global reservoirs of biodiversity, more than 40% of living organisms in terrestrial ecosystems are associated during their life-cycle directly with soils (Decaëns *et al.*, 2006). In fact, soils are one of the main global reservoirs of biodiversity (Bardgett and van der Putten, 2014; Carey, 2016). This reservoir includes bacteria and *Archaea*, fungi, protists and many more eukaryotes, such as nematodes, oribatid mites, centipedes and millipedes, enchytraeids, tardigrades, springtails, ants, ground beetles and earthworms (Zhang, 2013; Stork, 2018; Coleman and Whittman, 2005). The soil is a complex and heterogeneous system, comprising organo-mineral aggregates of different sizes and organic components, that creates habitats for soil biodiversity across multiple spatial scales; the diversity in habitat composition with pores of different sizes filled with air and/or water allows an incredible number of taxa of different sizes and ecology to inhabit it (Andre, Ducarme and Lebrun, 2002).

Soil communities are hierarchical systems where various types of organisms populate critically different volumes of soil. This includes the micrometre-thick water film around soil particles that house aquatic organisms like bacteria, protists, nematodes and tardigrades (that is, the microfauna), the air-filled pore space for soil animals between 100 µm and 2 mm width (that is, the mesofauna), the hot-spots of nutrients and other resources around plant roots for microorganisms, and the macrofauna and megafauna that perceive soil as a whole in which they make passages that can penetrate all soil horizons across significant soil volumes (Pokarzhevskii *et al.*, 2003). The main driving force of the high diversity of soil animals is this body size fractionation, but also their functional differentiation (Figure 2.2.1, and Figure 2.2.2). A variety of ecological niches in the soil, both in terms of size and in the range of resources provided, leads to a significant functional differentiation of soil organisms.

Box 2.2 | Size-structured soil communities

Soil organisms vary from 20 nm to 20-30 cm body width and are traditionally divided into four size classes (Gilyarov, 1949; Swift *et al.*, 1979).

Microbes including virus, bacteria, Archaea, fungi (20 nm to 10 μm) and **Microfauna** like soil protozoa and nematodes (10 μm to 0.1 mm) mostly live in soil solutions in gravitational, capillary and hygroscopic water, and participate in decomposition of soil organic matter, as well as in the weathering of minerals in the soil. Their diversity depends on the conditions of microhabitats and on the physicochemical properties of soil horizons.

Mesofauna (0.1 mm to 2 mm) are soil microarthropods (e.g., mites, springtails, enchytraeids, apterygota, small larvae of insects). They live in soil cavities filled with air and form coprogenic microaggregates, increase the surface of active biochemical interactions in the soil, and participate in the transformation of soil organic matter.

Macrofauna (2 mm to 20 mm) are large soil invertebrates (e.g., earthworms, woodlice, ants, termites, beetles, arachnids, myriapods, insect larvae). They include litter transformers, predators, some plant herbivores and ecosystem engineers, moving through the soil, thus perturbing the soil and increasing water permeability and soil aeration and creating new habitats for smaller organisms. Their faeces are hotspots for microbial diversity and activity.

Megafauna (greater than 20 mm) are vertebrates (mamalia, reptilian and amphibia). They create spatial heterogeneity on the soil surface and in its profile through movement.

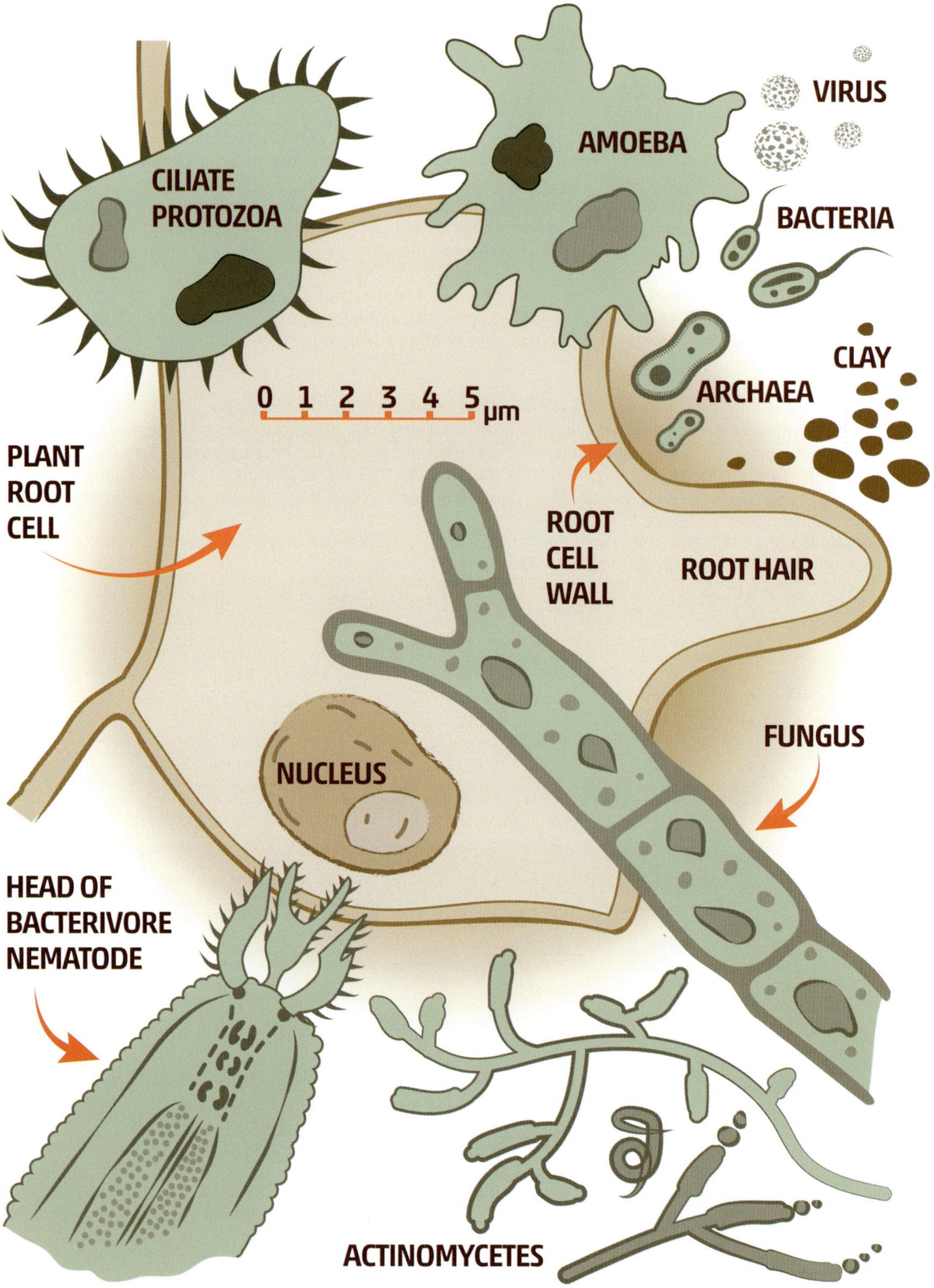
VIRUS
AMOEBA
CILIATE PROTOZOA
BACTERIA
CLAY
0 1 2 3 4 5 µm
ARCHAEA
PLANT ROOT CELL
ROOT CELL WALL
ROOT HAIR
FUNGUS
NUCLEUS
HEAD OF BACTERIVORE NEMATODE
ACTINOMYCETES

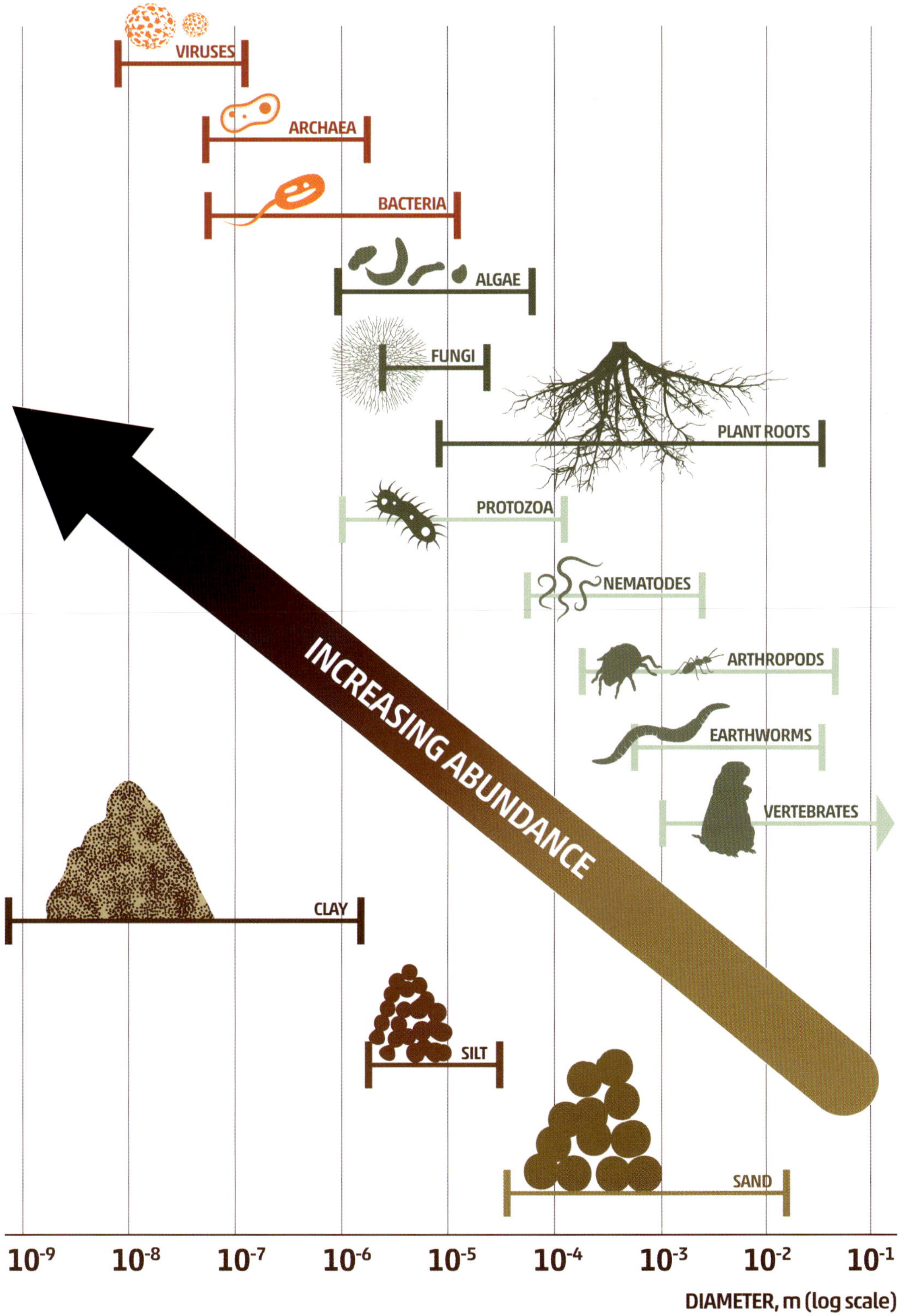

Figure 2.2.1 | From viruses to mammals

Conceptual model illustrating soil biota and its relationship with spatial scales.
Adapted from Weil & Brady, 2017.

Global diversity and distribution of soil biodiversity

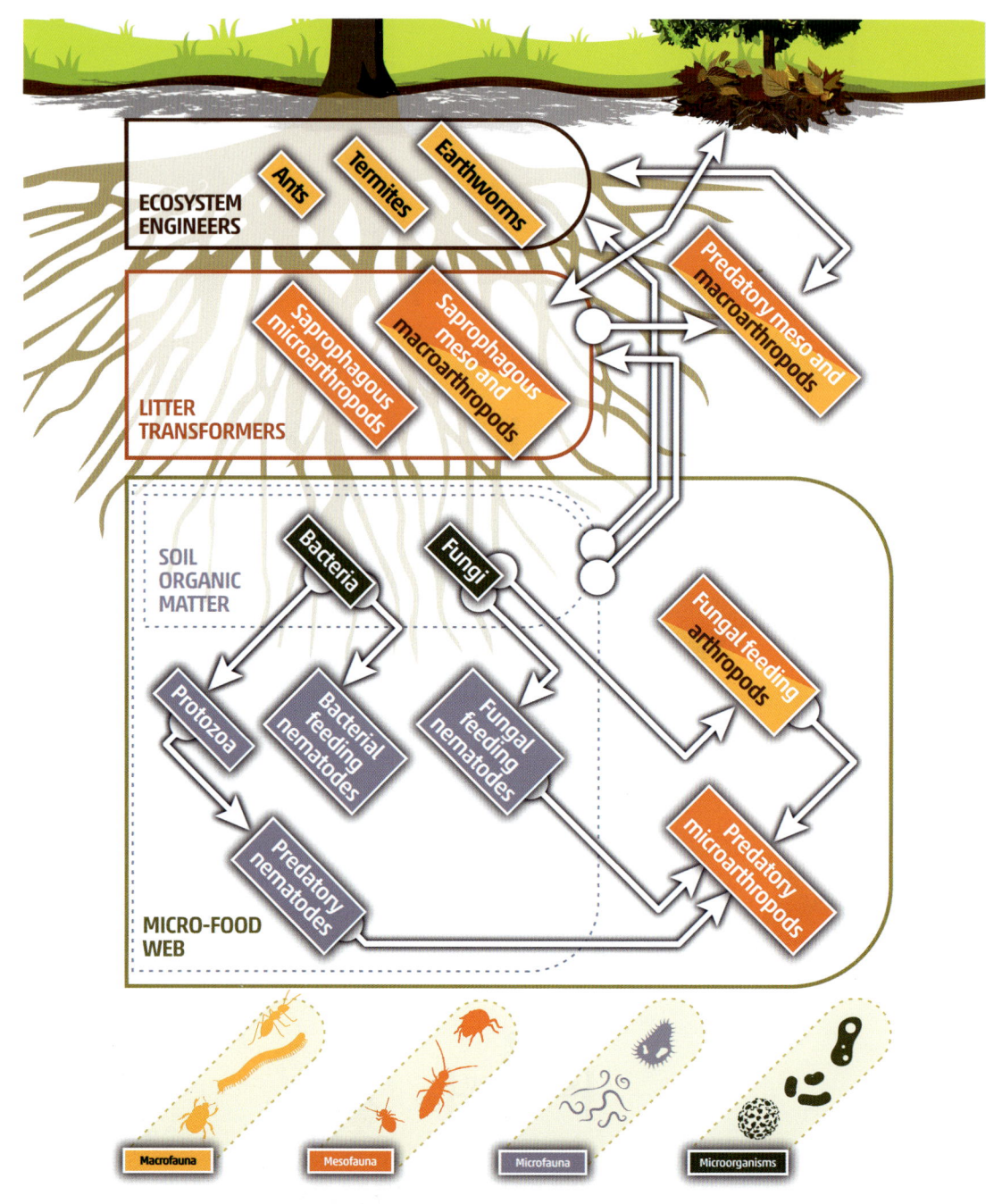

Figure 2.2.2 (previous page) | **Organization of the soil food web**

Simplified model of the different groups of soil organisms: microorganisms, micro, meso and macrofauna grouped into three categories in the food web and its functional differentiation. Firstly, the micro-food web (dotted lines) includes bacteria and fungi, which are at the base of the food web and decompose soil organic matter, which represents the basic resource of the soil ecosystem, and their direct predators, protozoa and nematodes. Secondly, litter transformers include microarthropods that fragment litter, creating new surfaces for microbial attack. Finally, ecosystem engineers, such as termites, earthworms and ants, modify soil structure by improving the circulation of nutrients, energy, gases and water. Adapted from Coleman and Wall, 2015.

2.3 | BIODIVERSITY IN THE SOIL

Soils are considered among the most biologically diverse habitats on Earth. It has been estimated that 1 gram of soil contains up to 1 billion bacteria cells, comprising tens of thousands of taxa, up to 200 metres of fungal hyphae, and a wide range of organisms including nematodes, earthworms and arthropods (Figure 2.3).

Figure 2.3 | Soil Biodiversity

One teaspoon of soil contains more living organisms than there are people in the world.

MICROBES

2.3.1 | BACTERIA AND ARCHAEA

Bacteria are single-celled organisms classified as prokaryotes that existed on the Earth billions of years before plants and animals. They emerge in different shapes such as spherical, rod-shaped or spiral cells, and possess diverse forms of metabolism that make them highly adaptable to different environmental conditions (Madigan *et al.*, 2015). They are found in all environments on the Earth, from hot springs and deep-sea ocean vents to the atmosphere and arctic snow. They also colonize on and in the bodies of other organisms, such as plants, animals and humans, as pathogens, symbionts, or merely making up a commensal community. It is estimated that the Earth hosts approximately 2.5×10^{30} cells. The collective carbon content of all these bacterial cells is comparable to that of all plants on the Earth, and their total nitrogen and phosphorus contents are far greater than that of all vegetation (Madigan *et al.*, 2015; Wang *et al.*, 2017), making these microorganisms the primary source of indispensable nutrients for life.

From a functional perspective, bacteria are classified into three main groups: *photoautotrophs* such as cyanobacteria that use atmospheric CO_2 as their carbon source and fix CO_2 using light energy, producing organic compounds that can be used by other organisms (similar to photosynthesis by plants); *chemoautotrophs* or *chemolithotrophs* that use atmospheric CO_2 as their carbon source (similar to photoautotrophs), but obtain energy from oxidation of inorganic compounds such as ammonia, iron and sulphur, and use this energy for CO_2 fixation and production of organic compounds; and *heterotrophs* or *chemoorganotrophs* that use organic materials as both their energy and carbon sources (similar to animals) (Madigan *et al.*, 2015; Graham *et al.*, 2016).

In terrestrial ecosystems, heterotrophy is regarded as the dominant trophic mode in which bacteria tend to prevail at the basal layers of the food webs (Steffan and Dharampal, 2018). These distinct trophic groups are intricately linked with plants and animals through symbiotic or parasitic associations, or as decomposers (Steffan *et al.*, 2015). For example, at a local scale, some of the dominant controls on bacterial community composition and their activities are plant type and litter quality (Myers *et al.*, 2012; Sun *et al.*, 2014; Jankowski, Schindler and Horner-Devine, 2014). Bacteria as major soil heterotrophs play key roles in carbon transformations and nutrient cycling, improving soil fertility. They can also regulate soil structure and create healthy soil environments to protect plants from pathogenic agents and increase crop yields (Sylvia *et al.*, 2005).

Carbon transformations - A major portion of the plant biomass (fallen on the ground and not consumed by herbivores) and other soil living biomass is disseminated by invertebrate detritivores and microorganisms across the food webs (Moore and de Ruiter, 2012; Paul, 2016). In addition to decomposing vegetal detritus in the soil, releasing essential elements for plant growth, many bacteria can transform different types of saturated and aromatic hydrocarbons such as oil and synthetic chemicals/pesticides (Brzeszcz and Kaszycki, 2018). Soil bacteria together as syntrophic consortia play an important role in breaking down environmental pollutants and circulating such complex recalcitrant compounds through the food web (Madigan *et al.*, 2015). Thus, they have the potential to be used for the bioremediation of the contaminated ecosystems (Dong and Lu, 2012).

Nutrient cycling - The energy and nutrient flow within an ecosystem occurs across a wide range of scales and trophic levels, from above ground to below ground, that define the hierarchy of food webs (Steffan *et al.*, 2015). The activity at one trophic level might influence the process happening at other trophic levels (Wollrab *et al.*, 2012). The correct functioning of the food webs is highly dependent on nutrient cycling, and bacteria are major engines of such substance turnover and biogeochemical cycles on the Earth (Graham *et al.*, 2016). While chemoorganotrophs as decomposers are characterized for their capabilities to mineralize the organic compounds and carbon cycling, chemolithotrophs get their energy from inorganic compounds, and are key drivers of sulphur, nitrogen, iron and other elements cycling transforming the inorganic compounds into the forms usable by plants (Ingham, 2009; Madigan *et al.*, 2015). Bacteria greatly influence soil fertility in different types of ecosystems, and their functional diversity is the key to the maintenance of soil biodiversity, stability of food webs and functioning of the ecosystem (Li *et al.*, 2019; Graham *et al.*, 2016).

Nitrogen is often considered as one of the limiting growth factors in soil for establishing and developing plants (Ågren *et al.*, 2012). Symbiotic and free-living (non-symbiotic) nitrogen fixing bacteria, found in most (undisturbed) soil types, can convert atmospheric nitrogen to ammonia, which nitrifiers convert into nitrate that is readily assimilated by plants (Orr *et al.*, 2011; Mus *et al.*, 2016). The soil environments dominated by bacteria usually have higher pH and nitrogen content that promote plant growth and further stabilize soil cohesion above ground through established vegetation (Madigan *et al.*, 2015).

Soil structure - Soil bacterial communities directly affect soil structure and functionality through various mechanisms. As part of their life cycle, many soil bacteria secrete extracellular polymeric substances (EPS) that bind soil particles together and improve soil aggregation. This enhances soil porosity, water holding capacity, aeration and plant root penetration in the soil (Costa *et al.*, 2018). Bacteria generally have a high growth rate with very short generation time, and compete for the same resources in the soil (Madigan *et al.*, 2015). In a healthy, undisturbed soil environment, non-pathogenic bacteria usually outcompete the pathogenic ones (Lowenfels and Lewis, 2010). Additionally, *Actinomycete* bacteria in soil (for example, *Streptomycetes*) secrete antibiotics that kill or inhibit the growth of plant pathogens, providing a healthy environment for plant growth (Bhatti *et al.*, 2017).

Biological regulation - In the tangled networks of natural food chains, soil bacteria acting as decomposers, chemolithotrophs, pathogens and symbionts can be found at different trophic levels (Steffan and Dharampal, 2018). The trophic diversity of soil microbial communities such as bacteria is likely to regulate the trophic identity of higher-level organisms (Steffan and Dharampal, 2018). Within the food chains, microbial footprints are found in the protein and lipid components of cell structure of higher-level consumers (Larsen *et al.*, 2009; Arthur *et al.*, 2014). As such, bacteria are ubiquitous primary components of all food webs and have a strong impact on the balance of soil food webs and the stability of ecosystem functioning (Moore and de Ruiter, 2012; Crotty, 2011).

Current knowledge and future challenges - Soil bacteria are the pivotal players in food webs (Steffan and Dharampal, 2018). They are rapidly becoming integral components of the model systems investigating the mechanistic links between biodiversity and ecosystem functioning (Birtel *et al.*, 2015). However, the studies of global microbial diversity using culture-independent molecular techniques suggest the existence of a potentially extensive group of unclassified bacterial species (Mora *et al.*, 2011; Yarza *et al.*, 2014; Louca *et al.*, 2019). These undescribed microorganisms may have a wide range of metabolic capabilities influencing ecosystem services, and may even be of particular interest for specific industrial and environmental management purposes (Birtel *et al.*, 2015; Louca *et al.*, 2019). Better understanding of how biodiversity loss by anthropogenic activities might affect the stability of food webs and ecosystem functioning requires a knowledge of current gaps in soil microbial diversity (Louca *et al.*, 2019). It is also a prerequisite in measuring the trophic position of a microbe within its respective community while integrated within the food web. The knowledge of trophic diversity of soil microbiome and macrobiome enables us to more comprehensively evaluate the functional diversity and productivity of ecosystems (Steffan *et al.*, 2015).

Box 2.3.1 | The Rhizobium and nitrogen fixation

The rhizobiome is the microbial community closely interacting with plant roots. Roots provide habitat for microorganisms living inside (endophytes) and in the immediate surroundings of roots (rhizosphere) (Philippot *et al.,* 2013). In fact, roots evolved in a microbial world, and ecosystem functions and services for roots cannot be considered in isolation from these microorganisms.

The rhizosphere is composed of a huge diversity of bacteria and Archaea. While most of the rhizobium is commensal, a small portion fo the community can be pathogenic to plants or mutualistic. Mutualistic associations between microbes and plants are especially critical for the nitrogen cycle. Nitrogen (N) is a limiting nutrient in many soils, and nitrogen from the atmosphere can only be used by plants once it is fixed to ammonium - either by microbes or artificially through the haber-bosch process. While the majority of agricultural systems currently rely on artificial fertilizers, there is a major environmental need to decrease artificial N additions and instead return to farming that can rely more on the microbial community. Therefore, N-fixing microbes remain critical for both managed and natural systems.

Nitrogen fixing microbes: Free living soil microbes are able to convert atmospheric nitrogen to ammonia, which can then be readily assimilated by plants. Of specific value are **rhizobium**, specific nitrogen fixing bacteria that form root nodules where this valuable process occurs. The most familiar rhizobium-plant interaction is with crop legumes.

Ammonia oxidizing bacteria and *Archaea*: Though ammonium is important for plant growth, this is not the end to the nitrogen cycle. Ammonium must then be oxidized to nitrate, a key process in the global nitrogen cycle. This process is largely carried out by ammonia oxidizing bacteria and Archaea. It is only in the last 20 years that the value of *Archaea* to the nitrogen cycle was discovered.

Box 2.3.1.1 | They are the oldest known evidence of life on Earth!

Stromatolites are laminated organo-sedimentary structures -of calcium carbonate- due to the growth of microbial mats on sedimentary substrates, with a great morphological variety and fossilization potential. Cyanobacteria, the most common microbial colonies that form stromatolites, are widely recognized as panchronic organisms, meaning that they have not changed much over the past 3.5 billion years. As a result, stromatolites are among the most important fossil records of early microbiological life. The existence of these panchronic organisms influenced important planetary processes that enabled life on Earth, such as the abrupt change from a reducing atmosphere -with methane and ammonium- to an oxidizing atmosphere -with oxygen, as it is currently the case now- approximately 2.5 billion years ago, which is attributed to these simple organisms.

A mix of three biocrust-forming cyanobacteria species (*Nostoc commune*, *Scytonema hyalinum*, and *olypothrix distorta*) on a wet soil observed under a lopue. Las Amoladeras, Cabo de Gata (Almería, Spain).

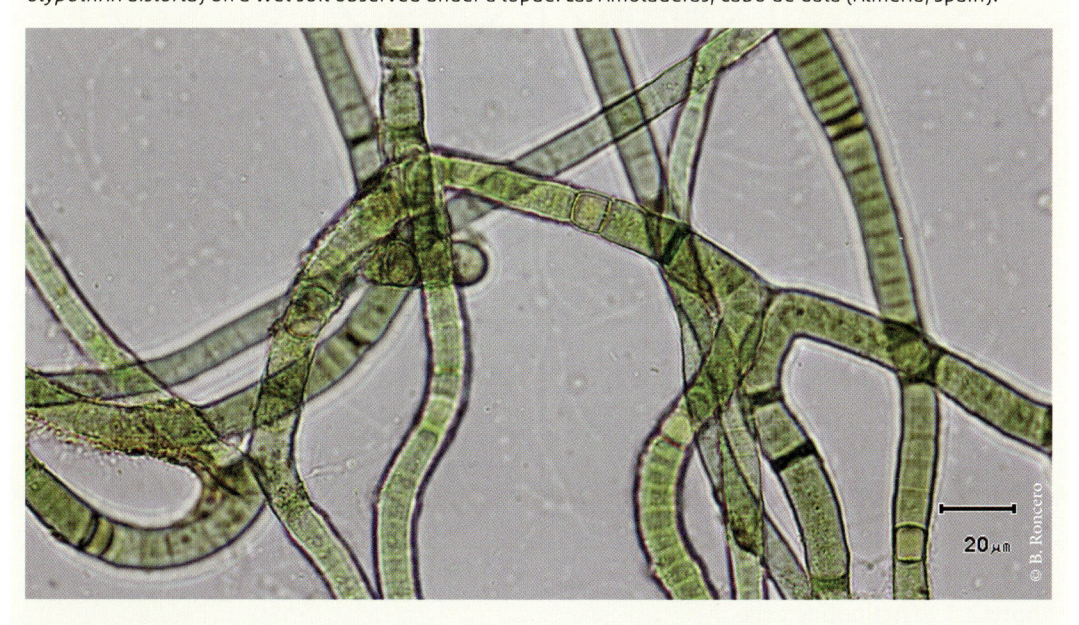

Microphotographs showing cyanobacteria species *Scytonema hyalinum*, from the Tabernas desert (Almeria, Spain).

SEM photographs showing the exopolysaccharides secreted by the biocrust-forming cyanobacteria species *Nostoc commune*, which binds soil particles. Las Amoladeras, Cabo de Gata (Almería, Spain).

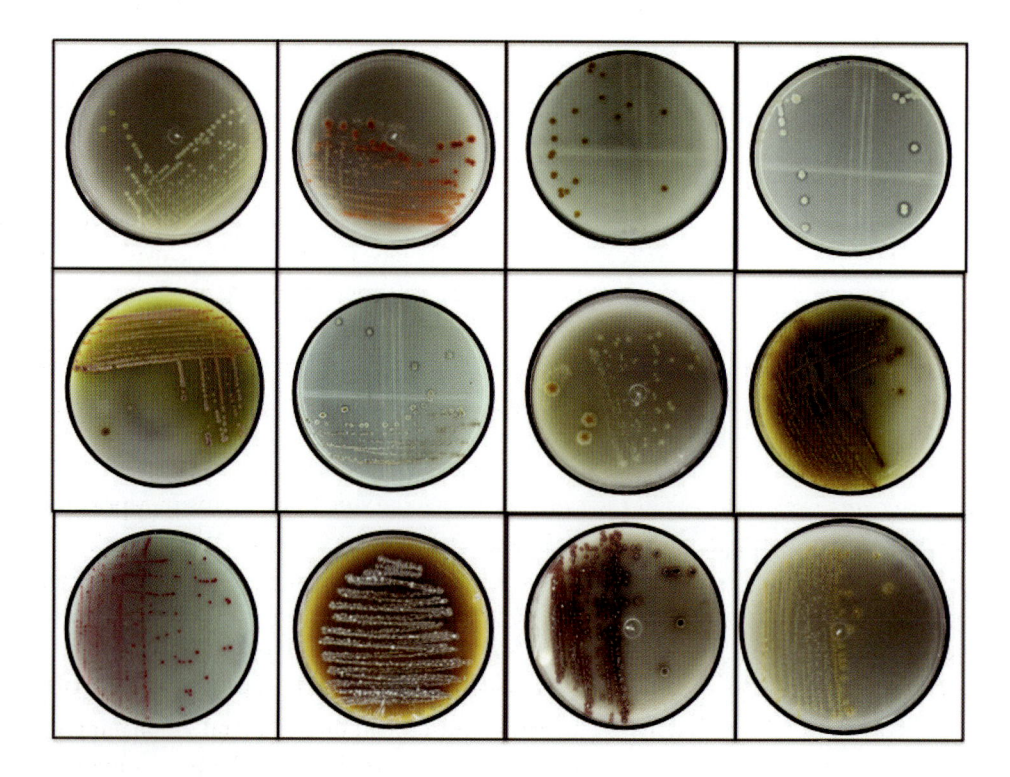

Photo 2.3.1 | Actinobacteria

Actinobacteria isolated in NBRIP agar medium plate from semi-arid soil in Pernambuco, Brazil.

2.3.2 | FUNGI

Fungi are a highly diverse group of organisms, encompassing a wide range of life forms. About 100 000 species have been described so far, with estimates on the total number of species ranging from 0.8 to 3.8 million (Blackwell, 2011; Hawksworth and Lücking, 2017). Fungi can be microscopic, like yeasts, or possess the capability of forming very large fruiting bodies such as the giant puff-balls. Mycelium of fungi have been reported to span astonishingly vast areas. For example, *Armillaria ostoyae* hyphae have been found to inhabit an area of 880 hectares in the state of Oregon in the United States of America (Barnard, 2000), making it arguably the biggest single organism ever recorded. This vast diversity means that fungi are not only numerous, but also fill many important niches in nature (Peay *et al.*, 2016). As a ubiquitous group of organisms, fungi provide a selection of vital functions in organisms, habitats, ecosystems, biomes and the biosphere as a whole (Lange, 2014).

The great majority of fungal species recorded so far are likely to spend at least some portion of their life cycle in soil. Soil fungi have fundamental ecological roles as decomposers, mutualists, or pathogens of plants and animals. Fungi drive soil carbon cycling and mediate mineral nutrition of plants in both natural and anthropogenic ecosystems. As fungi are heterotrophs that rely on photosynthetic carbon as their source of food, both direct and indirect interactions with plants are an important part of fungal ecology. Climatic factors, followed by edaphic and spatial variables, constitute the best predictors of fungal richness and community composition at the global scale (Tedersoo *et al.*, 2014). At local scales, plant diversity can be seen as one of the main drivers of fungal richness (Hiiesalu *et al.*, 2014; Prober *et al.*, 2015).

The functions and services that soils provide are tightly intertwined with soil biodiversity (Wagg *et al.*, 2014). Fungi, as a major constituent of soil biodiversity, provide valuable opportunities to tackle some of the great global challenges of the twenty-first century. With key functions in carbon transformations, nutrient cycling, soil structure formation and biological regulation (Nagy *et al.*, 2017), fungi are also key players in the fight against climate change and land degradation. In addition, fungi offer opportunities for ensuring food security and keeping globally dangerous pathogens in check.

Carbon transformations - Fungi are an integral part of soil food webs as they possess capabilities for decomposing many complex substances (Eastwood *et al.*, 2011). Moreover, mutualists such as mycorrhizal fungi are an important pathway for augmented carbon transformation by plants and the subsequent sequestration into the soil. For instance, it has been demonstrated that up to 70 percent of carbon in boreal forest soils is derived from plant roots and root-associated fungi (Clemmensen *et al.*, 2013). Saprotrophic fungi that decompose litter and dead plant material tend to inhabit the upper portion of soil, while mycorrhizal fungi extend to deeper soil layers, providing a more stable carbon stock. Through plant-fungal interactions and carbon cycle of the soil food web, fungi are major contributors to the carbon stocks of soils around the world (Six *et al.*, 2006). They also interact with calcium carbonate through biomineralizations, thus affecting the inorganic carbon pool (Bindschedler *et al.*, 2016). Whether or not they can provide this carbon cycling and sequestration function depends on land use and anthropogenic influences. Deforestation, land conversion to agriculture, soil degradation and sealing are all processes that hamper the ability of soil fungi to enact the vital role that they have in Earth's ecosystems. Careful consideration of the fungal pathway is a necessity if global aspirations for lower CO_2 emissions and C sequestration are to be met (Wurzburger and Clemmensen, 2018).

Nutrient cycling - Both mutualistic symbionts and saprophytic fungi living at the root-soil interfaces, the rhizosphere, or in the plant-associated soil, are recognized as essential drivers of nutrient cycling, availability and capture (Azcón-Aguilar and Barea, 2015). Saprotrophic fungi are recognized for their abilities to propel nitrogen fixation and phosphorus mobilization, two fundamental processes for sustaining plant productivity. Establishment of mycorrhizal associations between plants and fungi change the biological and physical-chemical properties of the rhizosphere, developing the so-called mycorrhizosphere. It is through the mycorrhizosphere that much of the nutrient cycle takes place, with saprotrophic fungi releasing nutrients from litter and mycorrhizal fungi completing the cycle by assisting with the uptake and transport of nutrients to plants. Fungi therefore inhabit many levels of the soil food web. Understanding the mechanisms and relationships they have with other organism groups is key to developing environmentally sound management solutions for soil biodiversity (Bender *et al.*, 2016; Thirkell *et al.*, 2017).

Soil structure - Soil fungi are an integral part in soil aggregate formation and the resulting structure (Lehmann *et al.*, 2017). There are several pathways through which this occurs, but a key aspect is the physical force of hyphae entangling smaller soil particles to larger aggregates. In the biochemical pathway, substances such as glomalin-related soil protein are left in the soil after hyphae die (Rillig *et al.*, 2014). Glycoprotein glomalin is a key constituent in hyphae of arbuscular mycorrhizal fungi (Wright and Upadhyaya, 1998), the most widespread mycorrhizal association (Öpik *et al.*, 2013). This sticky protein is resistant to decomposition and acts as a glue to form larger aggregates from smaller soil particles. As soil fungi also mediate plant performance in sub-optimal conditions (Smith and Read, 1998), they facilitate plant succession on bare substrates and therefore help to combat erosion and desertification (Yu *et al.*, 2017).

Biological regulation - As mutualists, pathogens, providers of food for microorganisms and as food for others, fungi are a central part of the soil food web. This means that interactions with the fungal realm are important for many other organisms. Whether it is mycorrhizal fungi alleviating competition between plants and widening their niche space (Bever *et al.*, 2010; Klironomos *et al.*, 2011; Gerz *et al.*, 2017), pathogenic fungi of arthropods (Butt *et al.*, 2016) or nematodes (Stirling, 2018) controlling plant pathogen abundance, or mutualistic relationships with nitrogen fixing bacteria (Van der Heijden *et al.*, 2016), fungi have demonstrated to be a vital link in the multifunctionality of the soil ecosystem (Delgado-Baquerizo *et al.*, 2016).

Current knowledge and future challenges - Currently, only a small proportion of the world's estimated fungal diversity has been described (Hinchliff *et al.*, 2015; Blackwell *et al.*, 2018). Moreover, functional traits of many fungal taxa remain elusive even after description. Soil fungal ecology has focused on describing and explaining large-scale macroecological patterns of various functional groups of fungi (Davison *et al.*, 2015; Bahram *et al.*, 2018). Although molecular approaches have played a vital role in this work, and methods have become not only more powerful but also cheaper to use (Lindahl *et al.*, 2013), there is still a lot to discover in the fungal kingdom. This is true not only for ecology but also for applied biotechnology of soil fungi in fields such as agriculture, medicine, restoration and conservation. Through science-based solutions, soil fungi could be the key to steering agricultural towards sustainable production. Ecological intensification through soil ecological engineering (Bender *et al.*, 2016) is seen as a possible solution for enhancing the natural functions of soils in agricultural land use. Soil ecological engineering requires an intimate knowledge of the functions and characteristics not only of fungal functional groups but also of species, strains, and perhaps even more importantly, communities of fungi. Although efforts towards this goal are ongoing and intensive, the high diversity of soil fungi asks for both small-scale and large-scale studies in many ecoregions of the world, if any generalizations are to be made.

Globalization and the great challenges of the twenty-first century also bring about new research needs that require global attention. As land use intensifies, there is a dire need for fast and economically viable methods for determining the health status of soil biodiversity. Molecular approaches have developed to the point that technology is no longer the bottleneck, but data is. More research is needed globally to calibrate our knowledge of soil fungal diversity in different biomes, ecosystems and anthropogenic settings. It is only through extensive background knowledge that we begin to understand, manage and conserve soil fungi, soil biodiversity and the functions it provides for humanity.

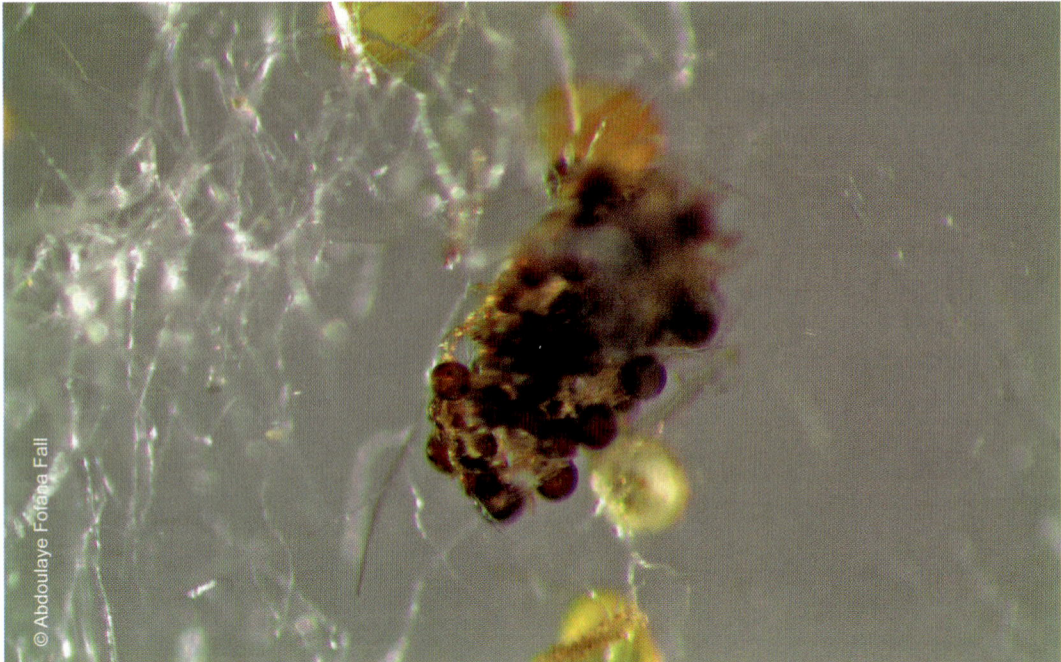

Photo 2.3.2a | Mycorrhizal spore

Spores of an arbuscular mycorrhizal fungus (*Glomus aggregatum*) used as a soil inoculant in agriculture.

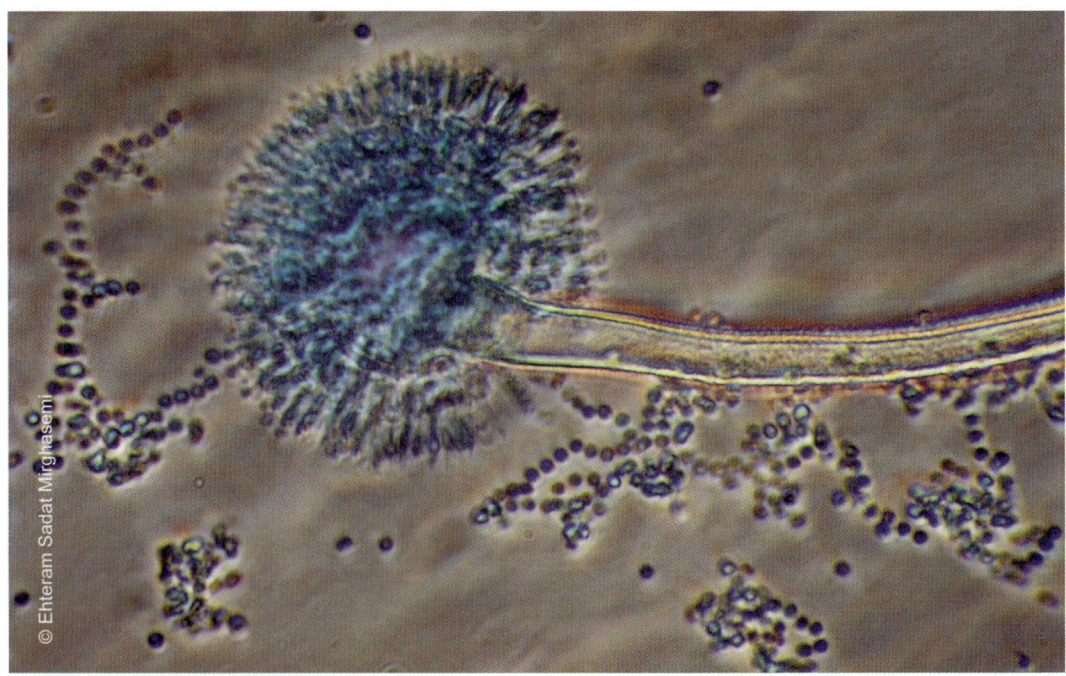

Photo 2.3.2b | Aspergillus spore

Aspergillus is a genus consisting of hundreds of mould species with a wide distribution. The image shows an asexual spore forming structure from Trichocomaceae family.

Photo 2.3.2c | Saprophytic fungus

A saprotrophic fungus (*Marasmius elegans*).

Photo 2.3.2d | Basidiomycete fungus

A basidiomycete fungus (*Marasmius oreades*) in a mossy forest in Poland.

Photo 2.3.2e | Basidiomycete fungus

A basidiomycete fungus (*Chlorophyllum rhacodes*), taken in Slapton, South Devon United Kingdom.

Box 2.3.2 | Arbuscular mycorrhizal fungi

Arbuscular mycorrhizal fungi (AMF) are recognized as an important and widespread component of most terrestrial ecosystems (Treseder and Cross, 2006; McGuire *et al.*, 2008), as they physically colonize roots of host plants and promote plant growth due to improved uptake of nutrients (Farzaneh *et al.*, 2011). The diversity of AMF communities is positively linked to plant diversity and plant productivity (van der Heijden *et al.*, 2008). The diversity and abundance of AMF changes by season and soil depth (Muleta *et al.*, 2007; Birhane *et al.,* 2010), as well as with the above-ground vegetation community (Muleta *et al.,* 2007; Hailemariam *et al.*, 2013). Because AMF are tightly linked to plant roots, the AMF are strongly impacted by anthropogenic disturbances such as tillage, soil compaction and other land-use factors (Tipton *et al.*, 2019). Strategies for AMF conservation include reductions in physical soil disturbance and enhanced agroforestry practices, such as the planting of shade-tolerant trees (Turrini and Giovannetti, 2012; Muleta *et al.*, 2007; Hailemariam *et al.,* 2013; Birhane *et al.*, 2017ab; Welemariam *et al.*, 2018; Birhane *et al.*, 2010).

The AMF play a crucial role in producing fundamental ecosystem services such as soil fertility, soil formation and maintenance, nutrient cycling and plant community dynamics (Wubet *et al.*, 2003). This is facilitated through several mechanisms of the fungal-plant symbiosis. First, the AMF increase the soil volume that can be exploited by the host plant. This leads to increased water and nutrient uptake, which in turn enhance acquisition of other nutrients (Smith and Read, 2008; Birhane *et al.*, 2010). This can be especially important in tropical systems where soils are often phosphorus (P) limited (Soka and Ritchie, 2014; Cardoso and Kuyper, 2006). The AMF are also involved in nutrient transformation through the activities of enzymes. For instance, AMF phosphatases can mineralize organic P sources, then releasing inorganic phosphorus in the cytoplasm (Aggarwal *et al.,* 2011). They also contain enzymes such as nitrogen reductase that break down organic nitrogen in soil (Barea, 1991).

The AMF are involved in the formation and maintenance of soil structure (Rillig, 2004), and increase C input to soils (Zhu and Miller, 2003). The compound glomalin that is produced by AMF binds to soil, producing aggregates that retain nutrients and water, and facilitate root penetration through the soil system.

This aggregation can reduce soil erosion and compaction (Wright and Upadhyaya, 1998), and the stability of the soil aggregates helps maintain soil porosity, which provides aeration and water infiltration rates favourable for plant and microbial growth, further increasing soil stability against wind and water erosion. These aggregates and soil stabilizing processes also help carbon storage by protecting soil organic matter from microbial decomposition (Bird *et al.*, 2002).

Mycorrhizas in general, of which the AMF are one of three broad categories (the others being ectomycorrhizae and ericoid mycorrhizae) play a key role in regulating abiotic and biotic stresses to plants (Soka and Ritchie, 2014) such as drought (Augé *et al.*, 2001; Birhane *et al.,* 2011; Gianinazzi *et al.*, 2010), salinity stress (Pande and Tarafadar, 2002), heavy metal phyto-accumulation (Andrade *et al.*, 2003) and protection against pathogens (Elsen *et al.,* 2008). Plant protection against drought is especially important in arid and semi-arid environments where soil moisture is considered a major limiting factor to plant establishment, distribution and abundance (Smeins *et al.*, 2015).

Box 2.3.2.1 | Natural soil cover

Biological soil crusts or biocrusts are complex communities of autotrophs (bryophytes, lichens, cyanobacteria, microalgae) and heterotrophs (bacteria, archaea, fungi, microarthropods) living in a close relation with soil particles on the soil surface. They cover about 25 percent of drylands soils (about 12 percent of the global land surface) and play an essential role in the functioning of many ecosystems. Biocrusts regulate soil water availability and, by fixing atmospheric carbon and nitrogen, influence soil nutrients cycles, thereby increasing soil fertility and biodiversity. They also secrete compounds that contribute to the formation of soil aggregates, enhancing soil surface stability and reducing soil erosion.

View of a cyanobacteria-dominated biocrust in the Tabernas desert (Almeria, Spain).

Lichens-dominated biocrust covering interplant spaces in the Tabernas desert (Almeria, Spain).

Detail of a lichens-dominated biocrust (dominant large terricolous lichens as Diploschistes diacapsis and Squamarina lentigera in the Tabernas desert (Almeria, Spain).

Box 2.3.2.2 | Laying the groundwork

Lichens are symbiotic associations between fungi and cyanobacteria as well as blue-green algae. The fungi provide the cyanobacteria with water and nutrients, and the cyanobacteria share fixed nitrogen and sugars from photosynthesis with the fungi. Because of their high adaptive capacity, lichens can develop in almost all known ecosystems, from polar areas to deserts. As primary colonizers, they are capable of promoting mineral weathering and releasing nutrients from the bare soil surface for the subsequent arrival of other plant organisms.

Detail of a lichen [Squamarina lentigera (Weber) Poelt.] from el Cautivo (Almería, Spain).

Detail of a lichen [Squamarina lentigera (Weber) Poelt.] from el Cautivo (Almería, Spain).

2.3.3 | VIRUSES

Viruses are packets of DNA or RNA genomic material inside a protein coat, sometimes contained within a lipid envelope. Viruses rely on host cell machinery for replication, such that a free viral particle is biologically inert and subject to biological and environmental degradation. Taxonomically, viruses are grouped according to their genomic composition, and classification according to genome-wide shared predicted protein content is also gaining attention, particularly for bacteriophages (viruses of bacteria) (Bin Jang *et al.*, 2019; Bolduc *et al.*, 2017). Virus taxonomy is not necessarily linked to host taxonomy, as diverse viral taxa can infect the same host species, and members of the same viral taxon can infect diverse hosts. In the context of soil food webs, it may therefore be better to group viruses functionally, according to the host taxa that they infect.

Viruses in soil food webs: Although recent methodological advances have vastly improved our ability to study soil viral communities in the last few years (Emerson, 2019; Emerson *et al.*, 2018; Paez-Espino *et al.*, 2016; Pratama and van Elsas, 2018; Trubl *et al.*, 2018; Trubl *et al.*, 2016; Williamson *et al.*, 2017), soil viral ecology as a field is in its infancy, and our knowledge of soil viral diversity and contributions to food webs is still very limited (Emerson, 2019). In general, viruses are predators (consumers), and because they can infect all soil biota, they have the potential to impact prey across trophic scales (Emerson, 2019; Schoelz and Stewart, 2018). Bacteriophages can burst their host bacteria, releasing cellular contents into the environment as potential resources for other microbes, while viruses of fungi and metazoa (such as nematodes) and plants may slow their hosts' growth and metabolism (Ghabrial *et al.*, 2015; Schoelz and Stewart, 2018). As organic particles in the environment, viruses themselves are also potential resources; grazing on viruses has been demonstrated in marine systems (Deng *et al.*, 2014; González and Suttle, 1993) and presumably occurs in soil as well.

Carbon transformations and nutrient cycling - Early evidence supports direct and indirect viral contributions to soil carbon and nutrient cycling (Emerson, 2019; Emerson *et al.*, 2018; Kimura *et al.*, 2008; Pratama and van Elsas, 2018; Trubl *et al.*, 2018; Williamson *et al.*, 2017). For example, glycosyl hydrolase genes, which are involved in the catabolism of complex carbon polymers into simple sugars, have been recovered from soil bacteriophage genomes, suggesting that phages could express these genes during the infection cycle to assist their hosts in the degradation of complex organic matter (Emerson *et al.*, 2018; Trubl *et al.*, 2018). By culling specific host populations, viruses can also alter carbon and nutrient cycling processes under the metabolic control of their hosts. For example, evidence for active viral infections of methane-generating methanogens and methane-consuming methanotrophs led to the hypothesis that viral predation of these microbes could impact methane cycling in thawing permafrost soils (Emerson *et al.*, 2018). Lysis (bursting) of microbial cells due to viral infection can liberate cellular contents, potentially rendering cellular carbon and nutrients bioavailable to smaller microorganisms but inaccessible to higher trophic levels through a process known in marine systems as the "viral shunt" (Wilhelm and Suttle, 1999). Though we have preliminary evidence for all these processes in soil, their rates and magnitudes are unknown.

Soil structure – Little is known about viral feedbacks to soil structure, but it has been suggested that movement between and compartmentalization within different aggregate size fractions could drive viral community composition, propagation and infectivity (Wilpiszeski *et al.*, 2019). Some viruses have been rendered inactive due to irreversible binding to soil particles, and yet other viruses seem to gain stability from interactions with clay particles (Duboise *et al.*, 1979; Marsh and Wellington, 1994).

Biological regulation - In better-studied marine systems, viruses burst and kill an estimated 20 to 40 percent of global ocean microbial cells daily, impacting food webs, carbon and nutrient cycling, marine snow (aggregates) and climate (Brum *et al.*, 2015; Danovaro *et al.*, 2011; Guidi *et al.*, 2016; Roux *et al.*, 2016; Suttle, 2007; Suttle, 2013; Weitz *et al.*, 2015). At ~10^7 to 10^{10} viruses per gram (Williamson *et al.*, 2017), soil viruses likely play similarly important but as-yet largely unknown roles in terrestrial ecosystems (Emerson, 2019; Fierer, 2017; Pratama and van Elsas, 2018; Schoelz and Stewart, 2018; Williamson *et al.*, 2017). While we assume that viruses of bacteria (bacteriophages) dominate soil viral communities (Emerson *et al.*, 2018; Paez-Espino *et al.*, 2016; Williamson *et al.*, 2017; Williamson, Radosevich and Wommack, 2005), our DNA sequencing-based methods fail to recover viruses with RNA genomes, including most known fungal and plant viruses (Schoelz and Stewart, 2018). We may be missing abundant viruses of fungi, plants, nematodes, protists and other soil fauna simply because we lack the methods to recover or even recognize them (Schoelz and Stewart, 2018).

For viruses of bacteria and Archaea, replication generally proceeds by one of two strategies (lysis or lysogeny), each with different impacts on microbial community composition, function and/or microbial evolution (Howard-Varona *et al.*, 2017). Temperate phages can undergo the lytic or lysogenic cycles, either replicating immediately and bursting the host cell (lysis) or being maintained in the host through lysogeny (for example, integration in the host chromosome as a prophage) unless or until triggered to undergo lysis. It has been hypothesized that these temperate phages, able to switch replication strategies, dominate in soil (Ghosh *et al.*, 2009; Ghosh *et al.*, 2008; Kimura *et al.*, 2008), but this hypothesis has not been thoroughly tested.

Soil microbial diversity loss results in a significant decrease in specialized functional capacity, such as potential nitrification and denitrification activities (Philippot *et al.*, 2013), greenhouse gas fluxes (Trivedi *et al.*, 2019) and pesticide mineralization capacity (Singh *et al.*, 2014). In contrast, reduction in microbial diversity has been shown to have only a small impact on overall soil respiration, which is a broad function carried out by most members of microbiomes (Griffiths *et al.*, 2001; Wertz *et al.*, 2007).

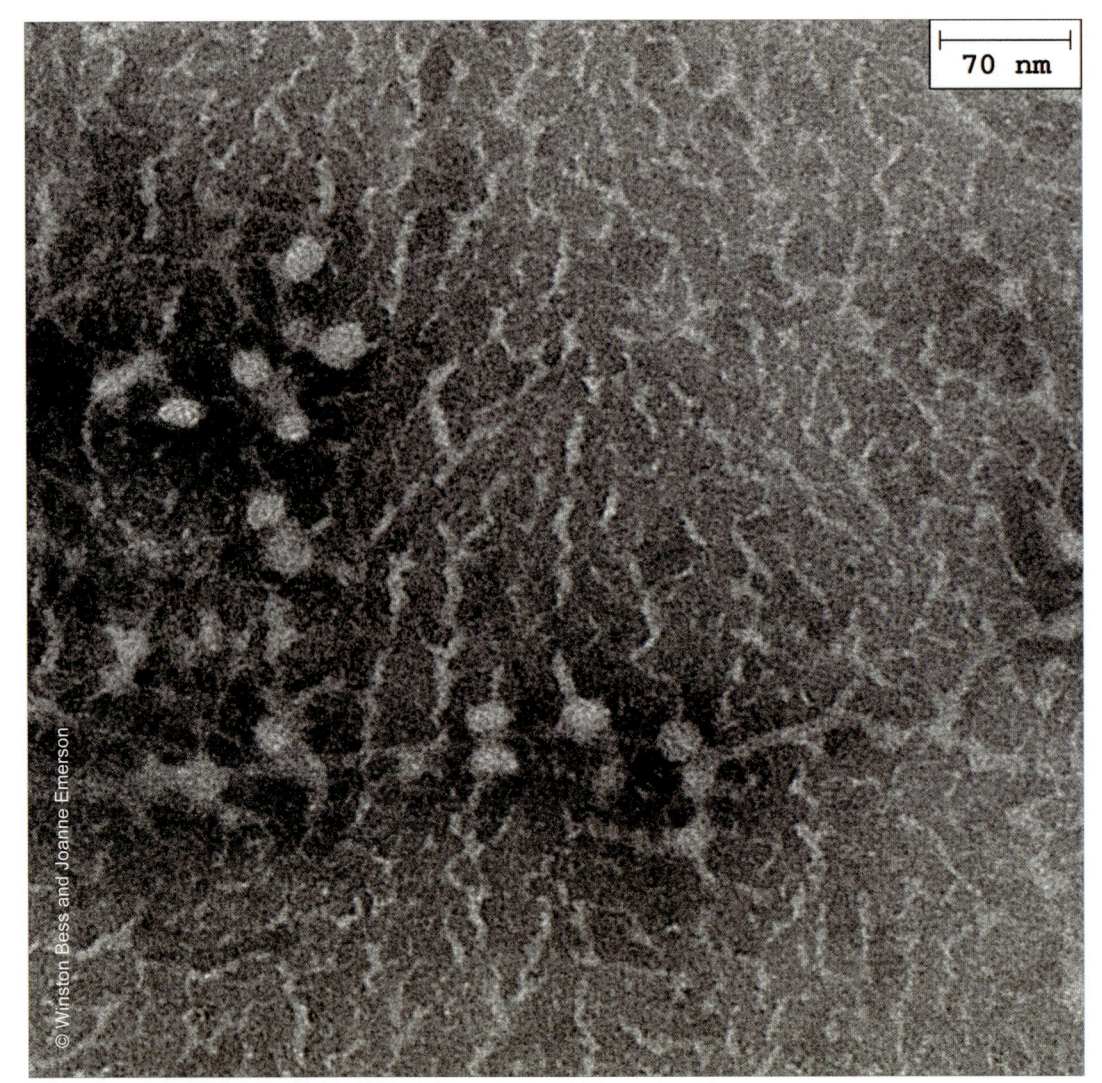

Photo 2.3.3 | Bacteriophage virus

Transmission electron microscopy (TEM) images of viruses (bacteriophages), from woodland soils at Stebbins Cold Canyon, near Winters, CA, USA. The image shows the viruses as a fuzzy, roundish light spots (~30-60 nm diameter).

2.3.4 | ALGAE

Soil algae are phylogenetically diverse, predominantly microscopic free-living, symbiotic organisms capable of oxygenic photosynthesis. Traditionally, soil algae include photoautotrophic prokaryotes (cyanobacteria, blue-green algae) (see also part 2.3.1 Bacteria and Archaea) and photoautotrophic protists (eukaryotic algae) (see also part 2.3.4 Protists). The soils are home to about 170 genera with ca. 1000 species of eukaryotic algae (Ettl and Gärtner, 2014) and about 500 species of cyanobacteria (Pankratova, 2006).

The most abundant groups in the soils are eukaryotic green and stramenopile (xanthophytes, eustigmatophytes, and diatoms) algae as well as cyanobacteria. Microscopic red, cryptophyte, chrysophyte, euglenophyte, and dinophyte algae are also often part of soil communities (Ettl and Gärtner, 2014). In soils in temperate zones, algal biomass can reach up to 500 kg ha^{-1} (Shtina, 1974).

Carbon transformations
Free-living algae and lichen photobionts in the soil, as photoautotrophic primary producers, play an important role in the fixation of CO_2 and the production of organic carbon compounds as well as in the release of O_2 into the soil. The role of stramenopile algae (Xanthophyceae, Eustigmatophyceae, diatoms) in the biogeochemical cycling of soil C has probably been underestimated (Yuan *et al.*, 2012). Cyanobacteria and eukaryotic algae are also involved in the biomineralization of carbon dioxide by calcium carbonate precipitation (Riding, 2002). Although phototrophic processes are restricted to the top few millimeters of the soil profile, microbially-assimilated C can move from the soil surface to the subsoil (Ge *et al.*, 2013). Facultative heterotrophy is also quite common among the various groups of soil algae. Various green, red, xanthophyte algae as well as diatoms and cyanobacteria are known to utilize organic substrates (e.g., glucose, acetate, glycerol, and so on) in the dark (Pribyl and Cepak, 2019).

Nutrient cycling
Soil algae represent a functionally relevant source of soil carbon for soil invertebrates and phagotrophic protists (Schmidt *et al.*, 2016, Seppey *et al.*, 2017, Potapov *et al.*, 2018). Beyond plants, soil microalgae represent a significant but rarely considered input of carbon into soils, that should be taken into account when modelling soil nutrient cycling (Seppey *et al.*, 2017).

The contribution of nitrogen-fixing cyanobacteria to the soil nitrogen pool reaches 30 kg/ha (Pankratova, 2006). N_2-fixing cyanobacteria, in addition to enriching soils with N and organic carbon, can modify a number of chemical and electro-chemical properties of soils, resulting in changes in the availability of micronutrients (Das *et al.*, 1991).

Algae promote the release of nutrients from insoluble compounds and the weathering of silicates by creating a slightly acidic environment (Hoffmann, 1989). They also provide organic matter and are responsible for soil formation, which is especially important in deserts, alpine, and polar regions (Hoffmann, 1989, Borchhardt *et al.*, 2017).

Soil structure
Most of the algae are concentrated on the soil surface and in the upper soil layer where they contribute to the formation of biological crusts (Büdel *et al.*, 2016). The formation of algal crusts on the soil interferes with soil erosion, increases rainwater storage and reduces water loss by evaporation during dry periods, but enhances overland flow by diminishing leaching during wet periods (Lichner *et al.*, 2013, Hoffmann 1989). Algae contribute to aggregate formation and stabilization and thus help protect the soil surface.

The growth of microalgae primarily favors the formation of large macro-aggregates, as small pre-existing aggregates are physically enmeshed by networks of algal filaments (Crouzet *et al.*, 2019).

Biological regulation

Algae and cyanobacteria excrete a variety of substances (including enzymes, growth factors, phytohormones, toxins, and antibiotics) which influence the development of other soil organisms (Abinandan *et al.*, 2019). They are able to metabolize and detoxify pesticides (Caceres *et al.*, 2008, Megharaj *et al.*, 2000). At the same time, herbicides can disturb the growth of soil microalgae and thus alter their functional role in soil aggregate formation (Crouzet *et al.*, 2019).

Plant viruses (e.g. Cauliflower mosaic virus) are naturally present in algae, thus airborne and free-living algae should be considered as an important plant virus shuttle in addition to the dispersion of free viral particles (Petrzik *et al.*, 2015).

Terrestrial algae synthesize and accumulate antioxidants and UV-absorbing sunscreens and UV-screening compounds, and protect other subsurface organisms from strong diurnal and seasonal fluctuations in ultraviolet radiation (Hartmann *et al.*, 2016).

Current knowledge and future challenges

Soil algae are less studied than their marine and freshwater relatives. Macroscopic algal forms and algal blooms in water reservoirs mostly have directly visible environmental and economic impacts, while changes in soil algal biomass and communities indirectly influence other soil organisms, soil fertility and surface runoff.

Although methodologically similar to mycology and bacteriology, soil phycology, as a part of phycology, is a branch of botany. Consequently, soil phycologists are not sufficiently involved in integrated soil studies.

Most of the taxa have been described on the basis of morphological features. Modern molecular-genetic studies show the incongruence of morpho-taxa and phylogenetic groups. Although the morphological identification of soil algae is time-consuming (in most cases isolation of uni-algal culture and observation of the life cycle is inevitable) and inaccurate, it is still widely used for studies of soil algae communities. The application of high throughput molecular methods (e.g., metabarcoding or metagenomics) is limited by the insufficiency of reference databases. Photosynthetic microorganisms are most often neglected in modern metagenomics and metatranscriptomics soil studies. A combination of taxonomic studies based on morphology with state-of-the-art sequencing methods is essential to resolve the issue of soil algae biodiversity in the future.

Photo 2.3.4a | Yellow-green algae

Macroscopic view of the Xanthophyte algae *Botrydium sp.* on the soil surface in Germany.

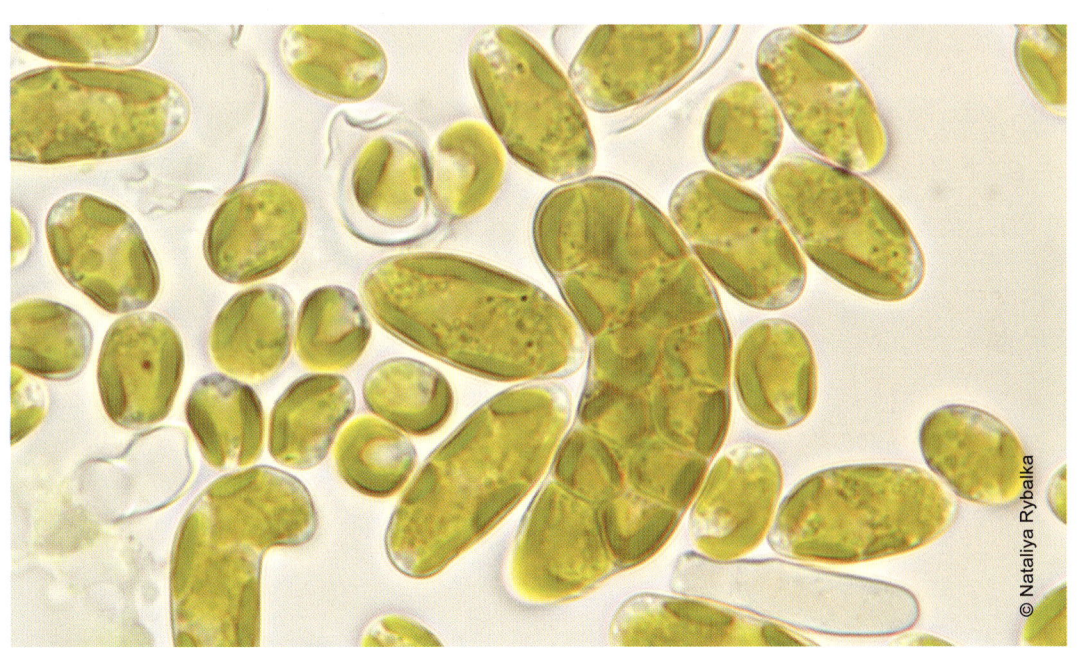

Photo 2.3.4b | Yellow-green algae

Bumilleriopsis sp. (xanthophyte or yellow-green algae).

Photo 2.3.4c | Yellow-green algae

Xanthonema montanum (xanthophyte or yellow-green algae).

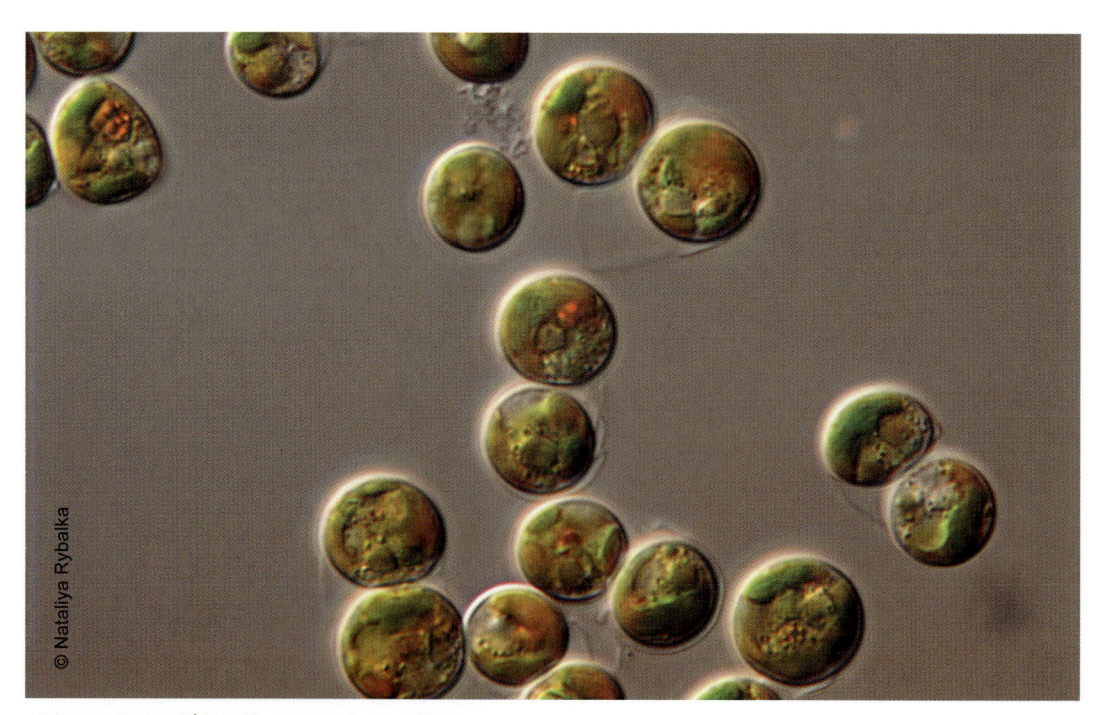

Photo 2.3.4d | Eustigmatophyte algae

Vischeria magna (eustigmatophyte algae).

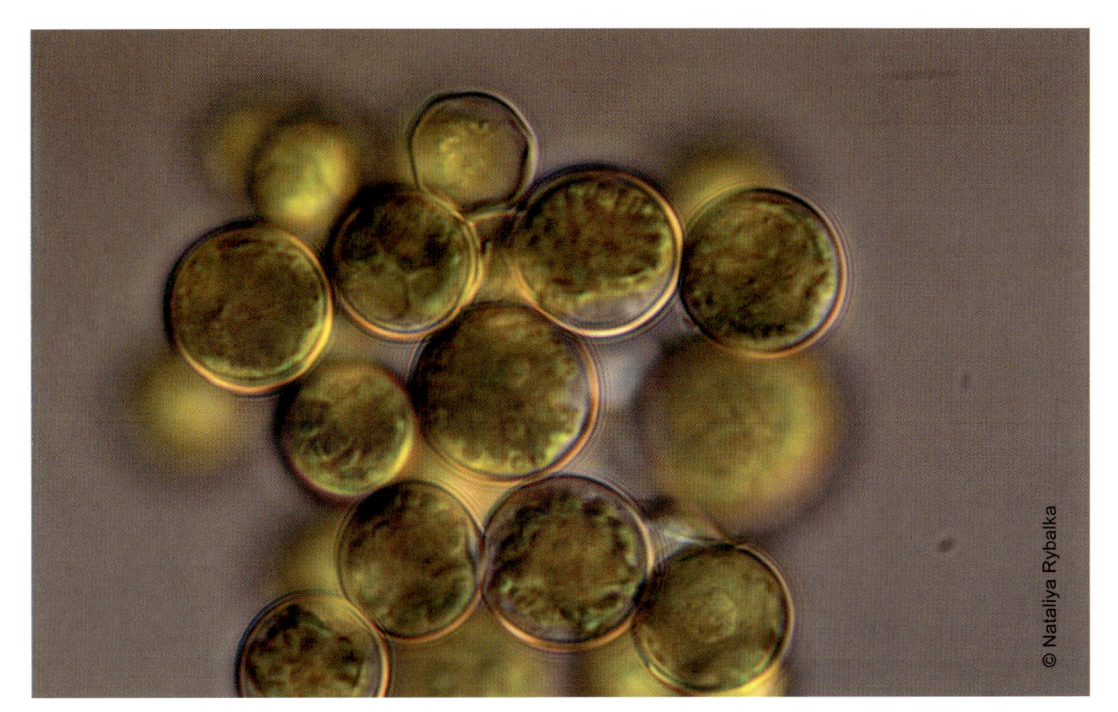

Photo 2.3.4e | Green algae

Trebouxia aggregata (green algae, lichen photobion).

Photo 2.3.4f | Algae cultures

Cultures of various soil algae in test tubes at the SAG Culture Collection, University of Goettingen, Germany.

Box 2.3.4 | Microfossils, a reference to the past

Diatoms and chrysophytes are the remains of single-celled organisms that live in both freshwater and saltwater. As they have a siliceous composition, they can resist in the soil for several millennia after all the organic material that made up the original organisms has weathered. This gives us the possibility to identify and study them, gathering important information on the environment, the climate and the hydraulic regime in which a soil has developed. Soils that have formed in very wet conditions, for example with periodic stagnating water on the surface, will exhibit a diatom chrysophyte assemblage that will reveal temporary shallow and muddy water. These microfossils, nevertheless, can also be transported, for example in floodwaters, and therefore become part of the parent material of a soil, something that is "inherited" from older environments.

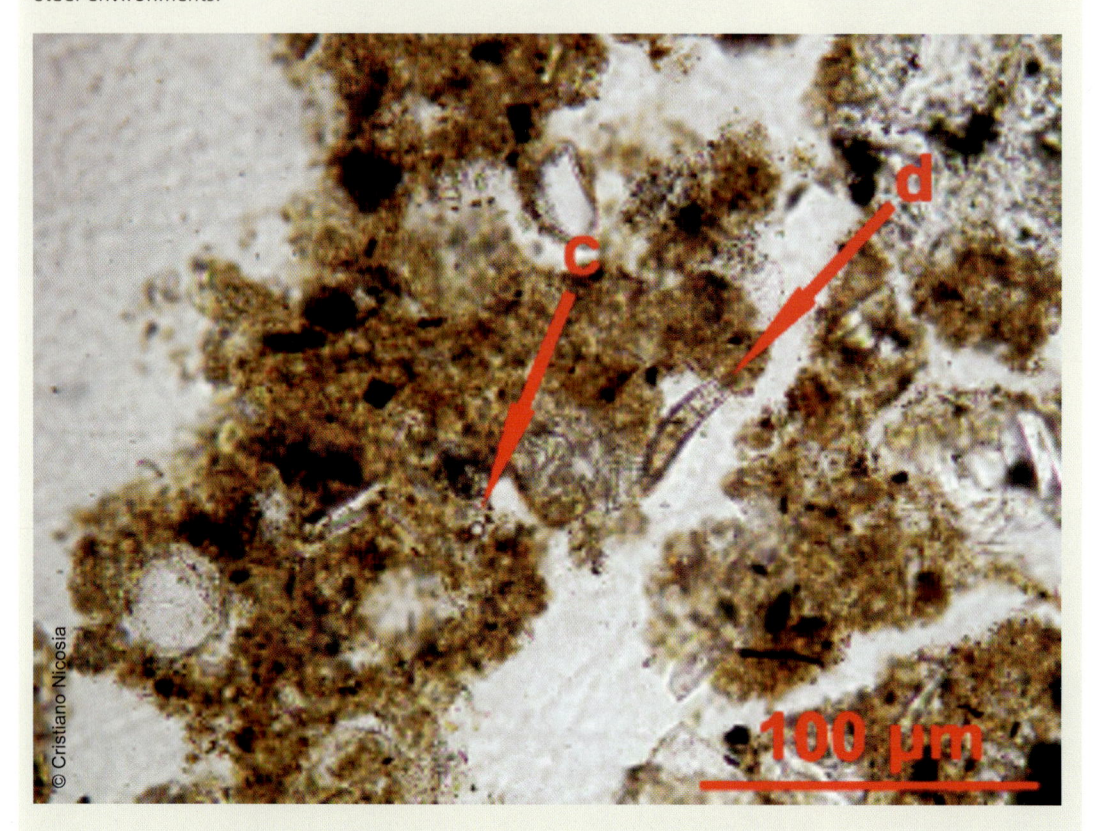

The image shows (d) diatoms and (c) chrysophytes from a volcanic soil in Pompeii. These microfossils derive from periodic inputs of sediments brought by floodwaters at the margins of the alluvial plain.

MICROFAUNA

2.3.5 | PROTISTS

As protists are among the least studied soil biodiversity members (Geisen *et al.*, 2017), we have a limited understanding of their role in soil functions. A protist is any eukaryotic organism that is not an animal, plant, or fungus. Soil protists include amoeba, ciliates and flagellates. Their numbers can exceed hundred of thousands of individuals in one gram of soil.

Carbon transformations - Few protist taxa are direct decomposers in soils causing significant mortality rates in soil microbial populations (Moore and de Ruiter, 2012), but further knowledge on the importance of protists in soils is limited (Geisen *et al.*, 2018). However, there is increasing evidence that especially in habitats with lower plant coverage, phototrophic protists (algae), often in association with fungi in lichens, contribute strongly to CO_2 fixation (Geisen *et al.*, 2018; Seppey *et al.*, 2017). Indirectly, protists might have a much more profound role in carbon transformations as they stimulate the activity of decomposers and change the growth of plants. Through their induction of microbial activity, protists are likely to increase decomposition, whereas by increasing plant performance they stimulate carbon incorporation into plant biomass (Geisen *et al.*, 2018).

Nutrient cycling - Most studies looking at the role of protists in soil functions have focused on nutrient cycling. An interesting notion emerging from these studies is the microbial loop concept, which suggests that protist predation releases nutrients (especially nitrogen) from bacteria, making it available for plant uptake (Clarholm, 1985). This concept is now widely accepted and shows the pivotal importance of microbial predation in the cycling of nitrogen. Far less is known on the role of protists in the cycling of other nutrients, but other elements that are common in consumed prey, particularly phosphorus, is likely of similar importance. Protists are also engaged in the cycling of silica as some incorporate silicon into their biomass to build stable shells (Wilkinson and Mitchell, 2010). Overall, however, most of the work focusing on the role of protists in nutrient cycling has been done in the past. There is a need to reinforce this research field to incorporate the main group of microbial predators into nutrient cycling.

Soil structure - No information exists on the role of protists in the formation of soil structure. However, an indirect role is plausible as protists change bacterial and fungal communities, such as stimulating bacterial biofilm production (Matz and Kjelleberg, 2005). Overall, we propose that trophic interactions between protists and microbes should be included in future studies exploring their indirect effects in soil formation.

Biological regulation - Protist predation is known to shape the composition of bacterial communities in soils (Geisen *et al.*, 2018). Recent work also shows that different protist taxa have a specialized feeding preference and thus likely structure the microbiome in a species-specific manner (Schulz-Böhm *et al.*, 2017). As such, protists have been proposed as a potential biocontrol agent, for instance by reducing the abundance of plant pathogens (Gao *et al.*, 2019). We still have a long way to go to identify those protists that can specifically be used to control different plant pests.

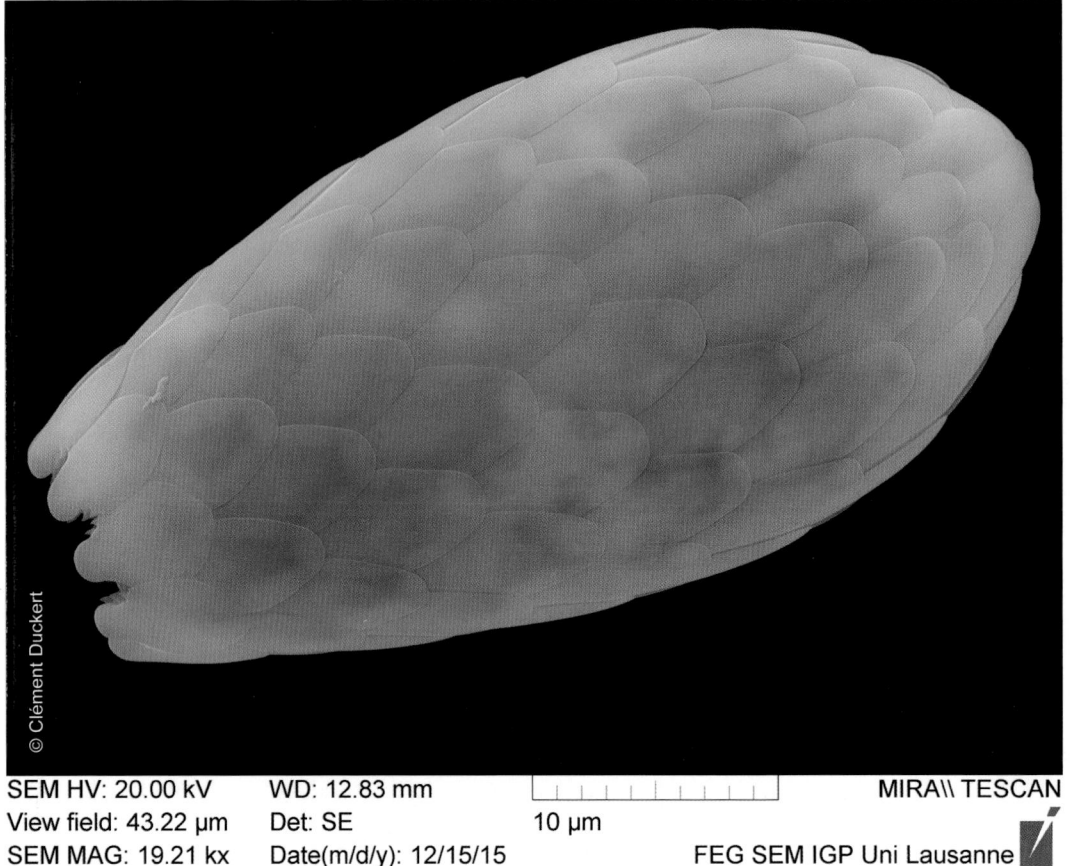

Photo 2.3.5a | Testate amoeba

Scanning electron microscopy (SEM) of a testate amoeba (*Euglypha* cf. *rotunda*).

Photo 2.3.5b | Free-living protist

Scanning electron microscopy (SEM) of a free-living protist (*Raineriophrys fortesca*) with radial symmetry. It feeds on other protists, immobilizing and capturing them with extrusive organelles (extrusomes) located on axopodia. It lives all over the world in soil, fresh and sea waters. The photo shows the silica skeletal elements of a fixed cell.

Photo 2.3.5c | Ciliated protist

Photomicrograph of ciliated protist (*Sterkiella tricirrata*) isolated from an Italian agricultural soils.

Global diversity and distribution of soil biodiversity

Box 2.3.5 | Testate amoebae

Testate amoebae are unicellular organisms enclosed in a shell (test) belonging to three far (or distantly)-related eukaryotic supergroups (i.e. Rhizaria, Amoebozoa and Stramenopiles). Due to their abundance in soils (generally thousands of individuals per gram of soil) combined with high turnover rates, they are significant actors in biogeochemical cycles. Most species feed on bacteria, fungi or small eukaryotes. They are major actors in the soil microbial loop (regulation of bacterial populations and nutrient redistribution), especially in nutrient-poor environments such as peatlands. Furthermore, in forest soils, silica-rich testate amoebae have been shown to participate in the cycling of silica as much as the trees themselves.

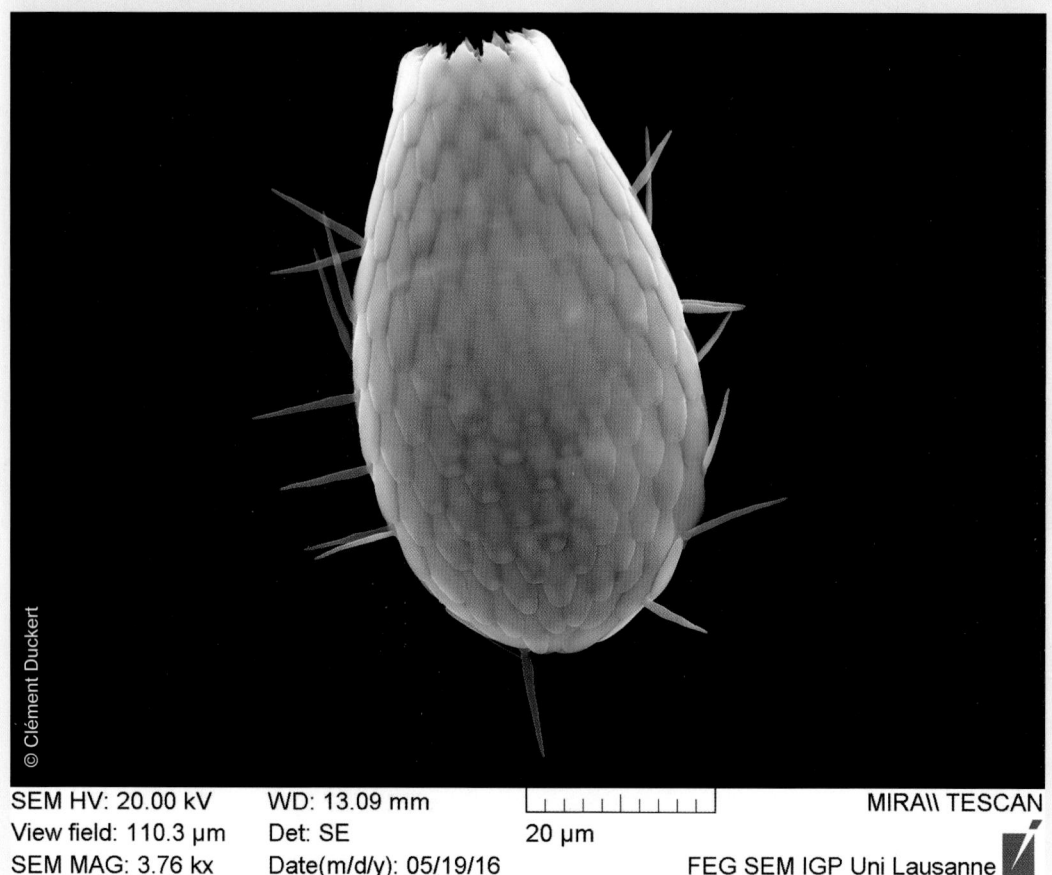

Scanning electron microscopy (SEM) of a testate amoeba (*Euglypha* cf. *compressa*). These microorganisms build their test with self-secreted silica scales and play a major role in the turnover of silica in soils

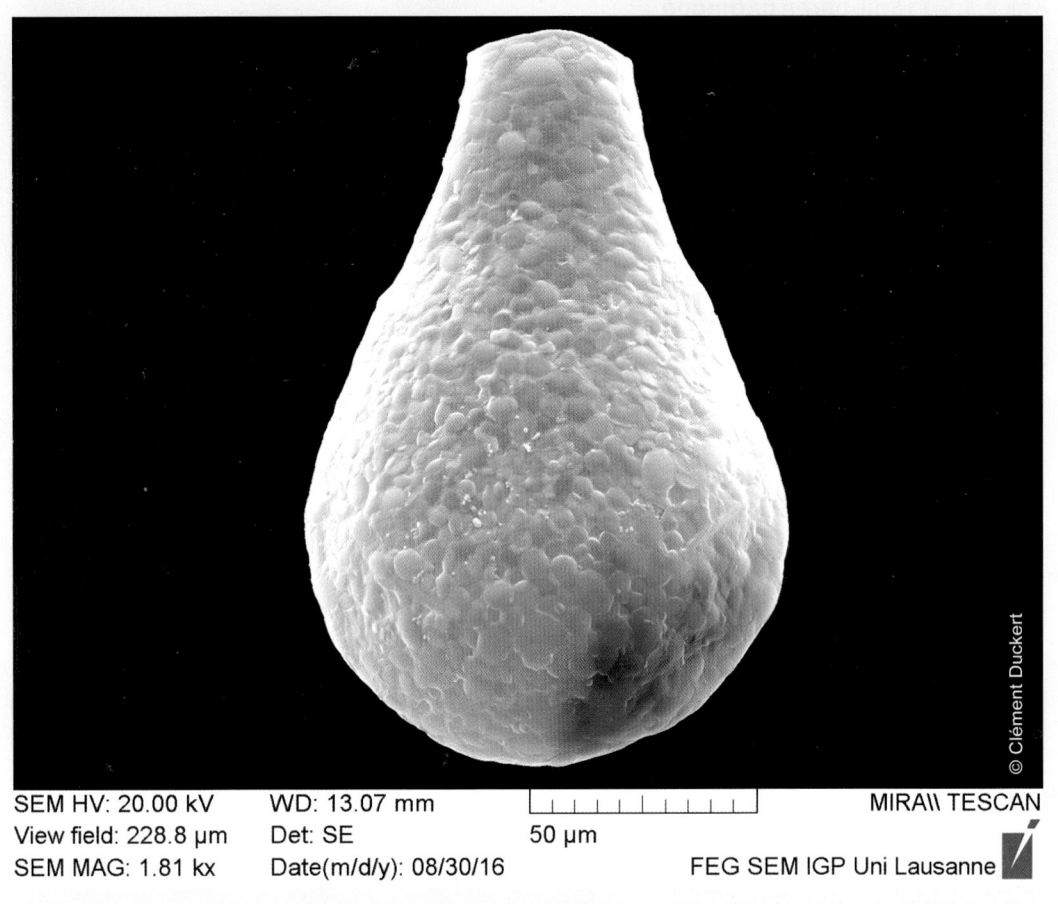

Scanning electron microscopy (SEM) of a testate amoeba (*Gibbocarina* sp.). Testate amoebae feed on bacteria and other small eukaryotes. Some species even prey on other testate amoebae and steal their scales to build their own test.

Box 2.3.5.1 | Soil-borne pathogen

The plasmodiophorid *Spongospora subterranea* (Ss) has become an important soil-borne pathogen in potato-growing areas in the U.S. because it has a broad host range and its dormant resting spores persist in the soil for many years. Ss causes tuber lesions and root galls on potato plants and transmits the potato mop-top virus. When environmental conditions are favorable, primary zoospores are released from resting spores to initiate infections on the root, shoot or stolon to form plasmodia which can further develop into zoosporagia and/or sporosori. Recently, Ss has been detected in a peat-based potting mix, which is used in the initial steps of seed potato production and which is typically spread on seed potato fields after having been used for seed production in greenhouses. As a result, seed potato fields throughout the country are now probably contaminated with this pathogen.

Disease symptoms of (**A**) potato tuber blemishes caused by Spongospora subterranea in a greenhouse trial compared to a healthy tuber and (**B**) root galls of a potato plant.

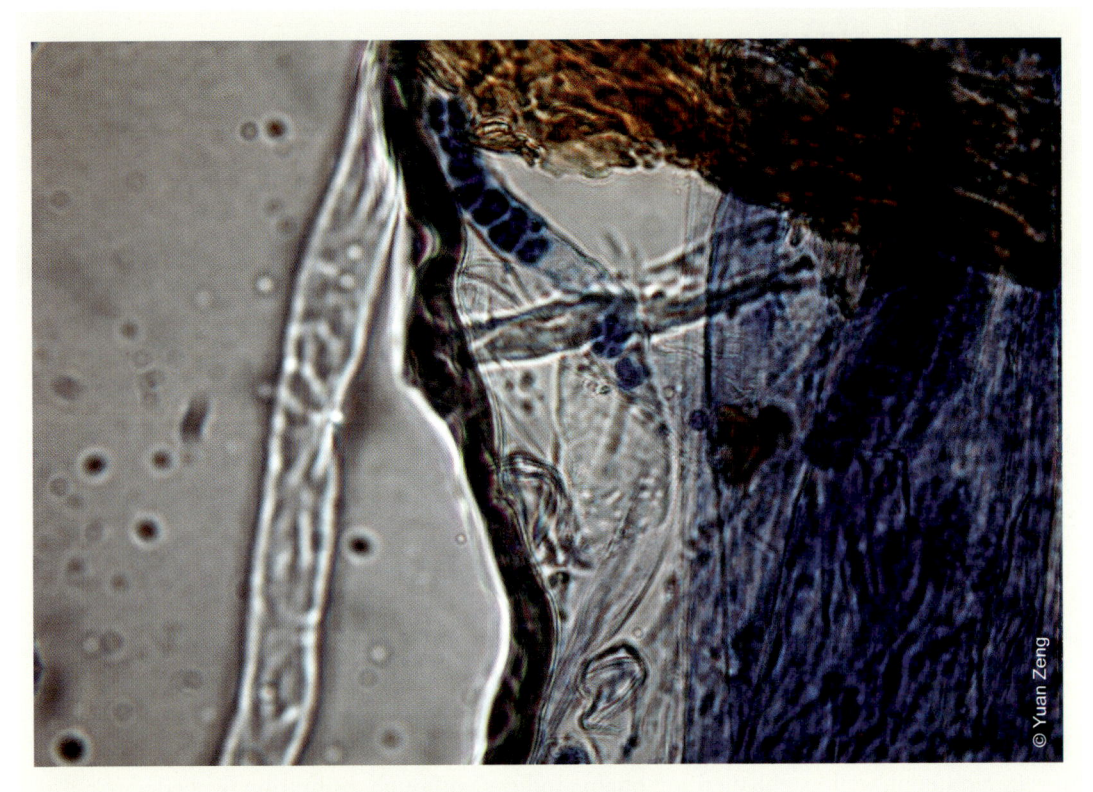

A root hair of a potato plant infected with Spongospora subterranea plasmodium.

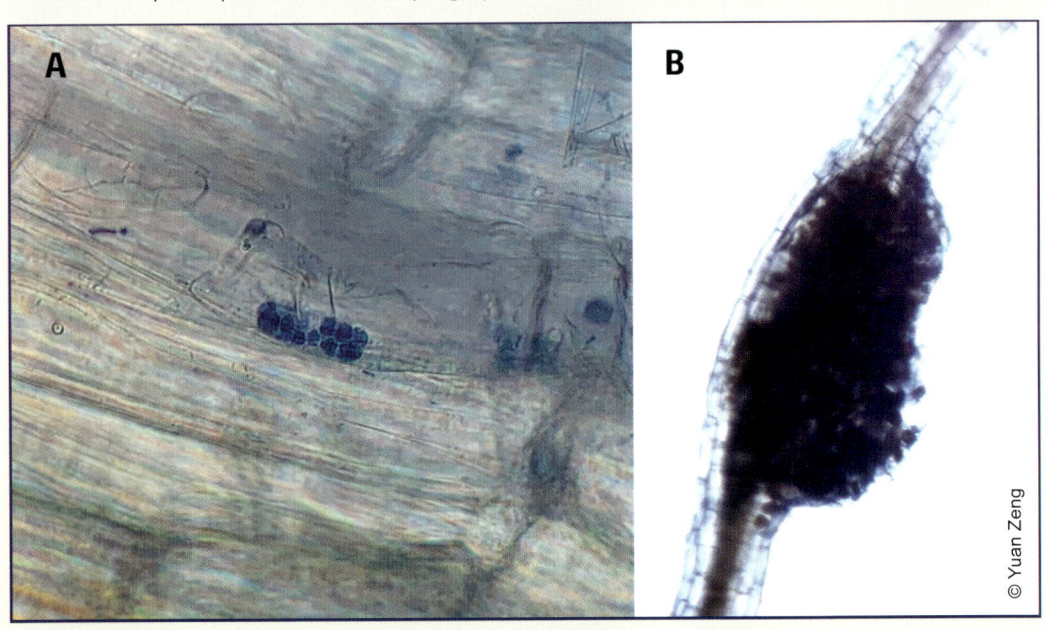

Structures associated with Spongospora subterranea. (**A**) Zoosporangia in tomato root hairs and (**B**) sporosori in a root gall.

2.3.6 | NEMATODES

Nematodes (or roundworms) are the most abundant group of animals in soils and in aquatic systems, representing around 80 percent of all animals on Earth. Nematodes are ubiquitous in aquatic sediments, glaciers and soils worldwide (Kionte and Fitch, 2013), with an estimated global population of 4.40×10^{20} individuals in soils alone (van den Hoogen *et al.*, 2019). The word nematode means "thread-like" (Kionte and Fitch, 2013), which aptly describes their thin, unsegmented body-shape; they are generally very small, often around only 0.3 mm to 3 mm long in non-parasitic species (Whalen and Sampedro, 2010). Nematodes fill a number of trophic roles in soils, including as bacterivorous and/or fungivorous grazers, predators of other nematodes and smaller animals, plant-feeders, omnivores and parasites of both animals and plants (van den Hoogen *et al.*, 2019). Given the wide range of trophic roles held by nematodes, they are important contributors to a number of ecosystem functions.

The long history of work on nematodes has resulted in the development of soil quality indices based on their community composition. Consequently, the community composition of nematodes can directly inform us about the state of the soil (Sieriebriennikov *et al.*, 2004). Nematodes are currently identified microscopically, but a slow shift towards molecular community profiling is occurring. Profiling can be integrated in well-established soil quality indices, although calibration is needed (Griffiths *et al.*, 2018). Recent methodological advances that can build upon profound knowledge about soil nematodes will facilitate the development of user-friendly high-throughput soil quality assessments with these soil worms.

Carbon transformation - Nematodes are important in soil carbon (C) dynamics. Partly this is due to the sheer size of the global nematode population. It is estimated that globally, soil nematodes operate on a monthly C budget of 139 Mt. Of this, nearly 110 Mt C is respired, which is equal to approximately 15 percent of the C released from fossil fuels (van den Hoogen *et al.*, 2019). This respired C may be equivalent to between 40 and 60 percent of C consumed by an individual (Ferris, 2010). Carbon transformation is best exemplified in polar environments, where nematodes are the dominant soil-dwelling animals (Barrett *et al.*, 2008; Simmons *et al.*, 2009). Certain nematode species account for between 2 percent and 7 percent of daily C flux in the McMurdo Dry Valleys of Antarctica (Barrett *et al.*, 2008). However, these populations are also under threat from climate change, with numbers of populations declining with increasing temperatures (Barrett *et al.*, 2008; Simmons *et al.*, 2009). Continued reduction in nematode populations under warming will have far-reaching consequences for Antarctic terrestrial communities. As consumers, nematodes are important links for the flow of C from microbes and decomposing matter to higher-trophic level animals (Ferris, 2010). Free-living nematodes are an important food source for micro-arthropods (Heidemann *et al.*, 2014). Food web analysis has demonstrated greater levels of dissolved organic C leaching with increased biomass of bacterivorous nematodes (de Vries *et al.*, 2013). Nematodes are therefore critically important to soil trophic dynamics.

Nutrient cycling - Nutrient cycling is also well studied in soil nematodes. As with soil C dynamics mentioned previously, nematodes are responsible for the transfer of many nutrients through trophic networks. For example, nematodes retain higher amounts of nitrogen (N) than they need; this is expelled as ammonia, which after nitrification can be readily consumed by plants and bacteria (Ferris *et al.*, 1998). Similarly, N leaching has been shown to increase with nematode biomass (de Vries *et al.*, 2013). Conversely, grazing of denitrifying bacteria by nematodes has been shown to cause a nearly 8 percent decline in denitrification rates in a mesocosm experiment (Djigal *et al.*, 2010). The presence of nematodes significantly increases net N (+25 percent) and net P (+23 percent) availability, as well as plant biomass production (+9 percent) compared to their absence (Gebremikael *et al.*, 2016). Therefore, free-living soil nematodes are an important component in soil nutrient cycles.

Soil structure - Direct interactions between soil structure and nematodes are long established and have shown that size of water-filled pores is the major structural limitation for soil nematodes (Jones and Thomasson, 1976). Nematode communities are often used as a marker of soil quality. There are a number of indices that have been developed to determine the status of soil based on the nematode community composition. These include the Maturity Index (Bongers, 1990) and its derivatives (Vonk, Breure and Mulder, 2013), which assess nematode community composition based on the abundance of various functional groups. These groups are based on functional traits, which can be simplified as a combination of trophic level and reproductive strategy (Vonk, Breure and Mulder, 2013). Although these metrics do not give a direct measure of soil quality, they can be used to infer the impacts of disturbance on a soil community. For example, these metrics are effective at differentiating different management regimes in agriculture (Vonk, Breure and Mulder, 2013) and forestry (Zhao *et al.*, 2014).

Similarly, anthropogenic disturbances such as agricultural intensification are known to degrade soil structure (Hamza and Anderson, 2005) and negatively impact nematode communities (Postma-Blaauw *et al.*, 2010). Such disturbances also result in a loss of larger-bodied organisms (Turnbull, George and Lindo, 2014) and metrics based on changes in body size of the nematode community are becoming a popular method for assessing changes in community structure and its response to a changing environment. For example, various body-size-based assessments of nematode communities have demonstrated the effects of clear-cutting (George and Lindo, 2015a), aridity (Andriuzzi *et al.*, 2020) and fertilizer application (Liu *et al.*, 2015).

Biological regulation - Many nematode species are parasitic and are important disease agents in crops, livestock and humans (Yeates *et al.*, 2009). Certain parasitic nematodes have been identified as biocontrol agents of insect populations, especially those with a soil-dwelling larval stage (Yeates *et al.*, 2009). *Steinernema* and *Heterorhabditis* are two genera that have become popular choices for this purpose. They kill by releasing toxin-producing bacteria inside their hosts (Forst and Clarke, 2002). These taxa are even sold commercially (Abate *et al.*, 2017), for example as a means of controlling invasive beetle outbreaks in Canada (George and Lindo, 2015b). The use of nematodes as a biocontrol agent has proven popular as they offer an alternative to chemical pesticides (Abate *et al.*, 2017). However, there has been little research in understanding the long-term impacts of nematode biocontrols. There are concerns that the global use of nematode biocontrols could lead to unintended killing of non-target insects, especially predators, and may lead to the establishment of invasive nematode populations when species are sold and used outside of their natural range (Abate *et al.*, 2017). Therefore, it is important to consider the long-term risks before the decision to use these biocontrol agents is made.

Photo 2.3.6a | Nematode

A nematode, United Kingdom.

Photo 2.3.6b | Nematode

Nematode, New Zealand.

Photo 2.3.6c | Nematode

A nematode, United Kingdom.

Photo 2.3.6d | Enchytraeid

An enchytraeid, commonly known as potworms (oligochaeta).

Photo 2.3.6e | Springtail carrying nematodes
A collembola (*Dicyrtomina fusca* var) with a load of phoretic nematodes.

MESOFAUNA

2.3.7 | MITES

Mites (Acari) are some of the most abundant microarthropods (0.1 mm to 2 mm) in soil ecosystems. As a subclass of the arachnids, they are exceedingly diverse, with more than 20 000 described species, with undescribed diversity likely exceeding 80 000 species (Coleman, 2008). Soil mites can number 100 000 per square metre, but their overall biomass is low (~2 g wet weight/m², Curry, 1994). The majority of knowledge about soil mites comes from temperate zones (both forest and grassland habitats), but even for this broad region, the diversity and functionality of soil mite communities are poorly known.

Mites are an integral part of soil food webs, and span several trophic levels including fungivores, bacterivores, detritivores and (top) predators. Studies using stable isotopes (^{13}C and ^{15}N) for soil food webs show that bodies of soil mites are enriched in root-derived C and N, suggesting that much of the soil food web is fuelled by nutrients entering the soil from below ground as roots/root exudates, in addition to above-ground leaf litter (Scheunemann *et al.*, 2015; Zieger *et al.*, 2015, 2017). However, relatively few soil mites feed directly on live plant tissues such as roots, bulbs and corms. Instead, the root-derived C and N are taken up by soil microflora (the primary decomposers; bacteria and fungi) and then passed to microbivores and then to predators feeding on microbivores or on root-feeding nematodes. Some species of soil mites do incorporate litter-derived C, which indicates that they feed either directly on litter, or on organisms involved in litter decomposition (Scheunemann *et al.*, 2015; Zieger *et al.*, 2017).

Detritivorous and microbivorous mites

One of the better-studied groups of soil mites are Oribatida (beetle mites, or moss mites) – relatively large, slow-moving mites encased in hard exoskeletons. They are often the most abundant mites in the soil, especially in forest soils rich in organic matter. Oribatida feed on soil fungi, bacteria, yeasts, algae, lichens and decomposing plant detritus; some facultative predation is also possible (Schneider *et al.*, 2004). Food preferences differ between different species, and many Oribatida use a range of food sources opportunistically, so it is often difficult to classify species as true fungivores or detritivores (Schneider *et al.*, 2004; Schneider and Maraun, 2005; Bruckner *et al.*, 2018). Oribatida have unusual life-histories for mites and other small-bodied organisms, such as slow development (life cycles up to several years long), low fecundity and long adult life span (Behan-Pelletier, 1999). These life histories make oribatid mites particularly sensitive to soil disturbance and structural degradation, so they are good indicators of soil quality (Gergocs and Hufnagel, 2009; van Leeuwen *et al.*, 2019). In a recent study of the effects of soil erosion on soil functions (water and nutrient supply, carbon sequestration, organic matter recycling and soil biodiversity) in European vineyards, abundance of Oribatida was the most sensitive indicator of soil degradation among all the soil taxa considered (Costantini *et al.*, 2018).

Carbon transformations - The total biomass of soil fauna is less than 4 percent of the average microbial biomass across biomes (Fierer *et al.*, 2009); however, a global meta-analysis shows that soil fauna consistently enhance litter decomposition, on average by 27 percent across biomes (García-Palacios *et al.*, 2013). Decomposition is enhanced when grazing by soil fungivores (predominantly soil mites and Collembola) stimulates growth and activity of fungi, the primary agents of decomposition and nutrient cycling in terrestrial ecosystems (Scheu *et al.*, 2005; Crowther *et al.*, 2012; A'bear *et al.*, 2014). Oribatid mites enhance the growth of fungi in leaf litter (Kun, 2015) and stimulate the recovery of soil microbial communities after strong disturbances (Maraun *et al.*, 1998). In experiments, the oribatid mite *Scheloribates moestus* stimulated soil enzyme activity and enhanced microbial respiration rates by between 17 percent and 19 percent in different litter types (Wickings and Grandy, 2011). Soil mites also carry fungal propagules on their bodies; as they move about, mites disperse fungi within soil and litter layers (Renker *et al.*, 2005; Kun, 2015). By grazing on fungi and dispersing spores, soil mites influence fungal-mediated decomposition and nitrogen mineralization within the soil (Scheu *et al.*, 2005). The summary of 101 litter decomposition experiments shows that "microarthropods" (combined soil mites and Collembola) accelerate decomposition moderately but significantly (Kampichler and Bruckner, 2010). For example, in alpine forest-tundra in China, soil microarthropods (dominated by Oribatida and Prostigmata mites) contributed from 12 percent to 26 percent of the total C release from litter decomposition (Liu *et al.*, 2019). That said, the direct relative contribution of soil microarthopods to C flux (that is, heterotrophic respiration and methane production) is negligible (Sustr and Simek, 2009).

Nutrient cycling - Through releasing of N and P during litter decomposition, detritivorous and microbivorous microarthropods enhance nutrient cycling, specifically increasing N availability through enhanced mineralization rates, and therefore support ecosystem productivity (Scheu *et al.*, 2005; Carrillo *et al.*, 2011; Soong *et al.*, 2016). In model simulations, lower plant productivity, lower total soil C and lower N mineralization is predicted under reduced abundances of soil microarthropods (Soong *et al.*, 2016). In an experimental food web, the presence of soil fungivores (diverse Oribatida and Collembola community) and their predators increased net N mineralization in the soil (Lenoir *et al.*, 2007), yet in the absence of predators, net N mineralization can be negative (net uptake and immobilization of N) (Staddon *et al.*, 2010), suggesting that the whole soil food web plays a role in nutrient cycling. However, the measurable contribution of soil mites to N mineralization is minor in comparison to the release of N by soil microbes (de Vries *et al.*, 2013). This holds true for the production of nitrous oxide (N_2O) that contribute significantly to global warming; previous experiments showed that soil mites have no significant effect on soil N_2O emissions (Kuiper *et al.*, 2013; Porre *et al.*, 2016), which was attributed to their relatively low biomass.

Soil structure - Unlike "ecosystem engineers" such as earthworms, soil mites are not capable of creating habitable pores in the soil. Mites are limited to utilizing available pore space and do not have a direct effect on the soil porosity (Porre *et al.*, 2016). However, grazing mites such as Oribatida contribute to the fragmentation of plant litter and produce abundant faecal pellets (Hagvar, 2016). These pellets are enriched in nutrients and wrapped in chitin-rich peritrophic membrane; they decompose slowly, persist in the soil profile for a long time, act as slow-release fertilizer and contribute to the formation of stable soil aggregates (Coleman, 2008; Wickings and Grandy, 2011; Maaß *et al.*, 2015).

Biological interactions - Much of the information on biotic interactions refers to oribatid mites. In addition to dispersing fungal spores, Oribatida affect specific composition of fungi when they selectively feed on different fungal species (Schneider and Maraun, 2005). Some data show that oribatid mites can affect growth pattern and morphology of fungal mycelium even when they do not graze the fungi (A'bear *et al.*, 2010). Numerous chemical secretions produced by oribatid mites were identified, with possible anti-fungal properties (A'bear *et al.*, 2010). In tropical America oribatid mites contain a number of alkaloids which they presumably accumulate from their food sources, and these mites provide a major dietary source of alkaloids for poison frogs (Saporito *et al.*, 2007). As fungi accumulate heavy metals from the polluted environment, mites feeding on fungi accumulate metals too. Oribatid mites can accumulate heavy metals (Zn, Cu, Cd, Pb) to very high internal concentrations (Zaitsev and van Straalen, 2001), but accumulation patterns vary for different metals and mite species (Skubala and Zaleski, 2012). Some oribatid mites are of epidemiological and medical importance, as several species are known to serve as intermediate hosts for tapeworms of wild and domestic animals (Vaclav and Kaluz, 2014).

Predatory mites

Predatory mites are found primarily in two orders: Mesostigmata (Parasitiformes) and Prostigmata (Acariformes: Trombidiformes). Mesostigmata include both soil- and plant-living mites and are dominated by predatory taxa, to the extent that "predatory mite" is sometimes used synonymously for Mesostigmata (and especially the suborder Gamasina) (Ruf and Beck, 2005). Prostigmata have a wide variety of feeding habits and life histories; however, the group includes many predatory families, most of which include species that are partly or entirely found in soil or litter (Walter and Proctor, 2013).

Predatory mites are the most important predators of the soil micro- and mesofauna, feeding voraciously on nematodes, enchytraeids and microarthropods (including springtails, mites, and small insects and their larvae). They also serve as prey to larger arthropod species, thus providing an important link between the soil microfauna and macrofauna. Predatory mites are highly mobile, and many are r-selected, allowing them to track resource pulses (Cao *et al.*, 2011).

Carbon transformations - Predatory mites play a role in carbon transformations through their top-down effects on the herbivorous and decomposer fauna on which they feed, as well as through respiration. The contribution of predatory soil mites to CO_2 emissions through respiration is thought to be minimal; mesocosm studies have supported this view, showing only weak and inconsistent increases in CO_2 with the addition of predatory mites (Thakur *et al.*, 2014; Zhu *et al.*, 2016b). Predatory mites are efficient assimilators of carbon (Moore, McCann and de Ruiter, 2005; Osler and Sommerkorn, 2007), and also affect carbon cycles by controlling herbivorous fauna such as root nematodes (resulting in increased plant growth) and by consuming detritivores. However, their contribution to carbon transformations has rarely been quantified.

Nutrient cycling - In comparison to the fungivorous and detritivorous mites, there has been considerably less research on the role of predatory mites in nutrient cycling. In general, the effect of predatory mites on nutrient cycles has been attributed to trophic cascades, whereby their feeding on microbial- or fungal-grazers alters the microbiome of the soil (Bardgett and Wardle, 2010; Schmitz, 2010; Staddon *et al.*, 2019). Since the addition of nitrogen fertilizers can greatly increase the populations of soil mite predators (Minor and Norton 2004; Cole *et al.*, 2005; Salamon *et al.*, 2006; Cao *et al.*, 2011), it is essential to understand the impact this may have on nitrogen cycling, and particularly on emissions of N_2O, a major greenhouse gas. In mesocosm experiments, the results have been inconsistent. For example, adding a mesostigmatan predator to soil mesocosms with enchytraeids and fungivorous mites greatly increased N_2O emissions; however, the same effect was not seen when only one of the fungivorous groups was included (Thakur *et al.*, 2014) and was not replicated by a subsequent study (Porre *et al.*, 2016). Similarly, Zhu *et al.* (2016b) found the effect of predatory mites on N_2O emissions varied depending on the combination of taxa included in the mesocosm. In terms of nitrogen cycling within the soil, it has been hypothesized that an increase in predatory mites would prevent the overgrazing of microbes, thus leading to increased mineral N availability (Schmitz *et al.*, 2010). However, some studies have shown that increased grazing in fact stimulates microbial and fungal growth (Krivstof *et al.*, 2003), and increasing predator numbers may therefore result in the opposite effect, as was observed by Staddon *et al.* (2019).

While most studies have focused on nitrogen, predatory soil mites are also involved in the cycling of other nutrients. For example, cadmium has been found to bioaccumulate in predatory mites, with a resultant decrease in reproduction (Zhu *et al.*, 2016a). Understanding the role of predatory mites in nutrient cycling is complicated by the diversity of trophic interactions with which they are involved. For example, microbial-feeding nematodes and mites and predatory microarthropods are thought to be efficient at mineralizing nitrogen, in contrast to fungal-feeding taxa (fungivorous nematodes, mites and enchytraeids) (Osler and Sommerkorn, 2007). However, all of these groups are consumed by a diverse assemblage of predatory mites (Sánchez-Moreno *et al.*, 2009), suggesting that any top-down effects of these predators on nutrient-cycling could be mediated by bottom-up effects of the microbial and fungal communities and by the specific composition of the herbivore, decomposer and predator communities.

Soil structure - Predatory mites live in the litter layer and in the air-filled pore space of soils (Ruf and Beck, 2005). Consequently, they have little direct impact on soil structure. They may have top-down effects through feeding on taxa such as fungivorous mites and enchytraeids that do modify the soil structure; however, this has not been specifically studied.

Biological regulation - Soil-dwelling predatory mites are best known for their role in regulating herbivorous fauna. In managed ecosystems, they are bred and used as biocontrol agents of herbivorous invertebrates that spend life entirely in the soil (for example, plant-feeding nematodes and root-feeding mites) or that spend part of their life cycle in the soil (for example, spider mites, thrips, fungus gnats, cutworms and root aphids) (Gerson *et al.*, 2003; Hoy, 2011). Predatory mites may also be important control agents of noxious flies in manure and agricultural waste products (Azevedo *et al.*, 2018).

In natural soil systems, predatory mites are equally important for the regulation of both herbivores and decomposers (Walter and Proctor, 2013), though as the species diversity increases, the trophic interactions become necessarily more complex. While most predatory mites are generalists and opportunistic predators, different species do have preferential prey. For example, many Mesostigmata species feed preferentially on nematodes, while other predators (Mesostigmata and predatory Prostigmata) feed on arthropods (Walter and Proctor, 2013). Intraguild predation is also common, further complicating trophic interactions.

Studies have shown a positive effect of increased diversity and habitat complexity above ground on the diversity of soil mite predators (Postma-Blaauw *et al.*, 2012; Tsiafouli *et al.*, 2015; de Groot *et al.*, 2016). The assumption is that a diverse predatory community will have a stabilizing effect on the ecosystem (Finke and Denne, 2005); however, the complexity of soil habitat makes this challenging to assess in natural ecosystems. In a mesocosm experiment, Schneider, Scheu and Brose (2012) show that small predators have a negative effect on decomposer prey, whereas larger predators have a positive effect due to a trophic cascade. However, Klarner *et al.* (2013) demonstrated that for predatory mites, a link could not be determined between body size and trophic level, with closely related species feeding on different prey. Specific taxonomic and biological information on the taxa present is therefore necessary to a fuller understanding of the mechanisms by which predatory mites regulate their prey species and, subsequently, ecosystem processes.

Photo 2.3.7a | Oribatid mite

An oribatid mite (*Liacarus subterraneus*) from South Somerset, United Kingdom.

Photo 2.3.7b | Opilioacarid mite

An opilioacarid mite, from Mexico.

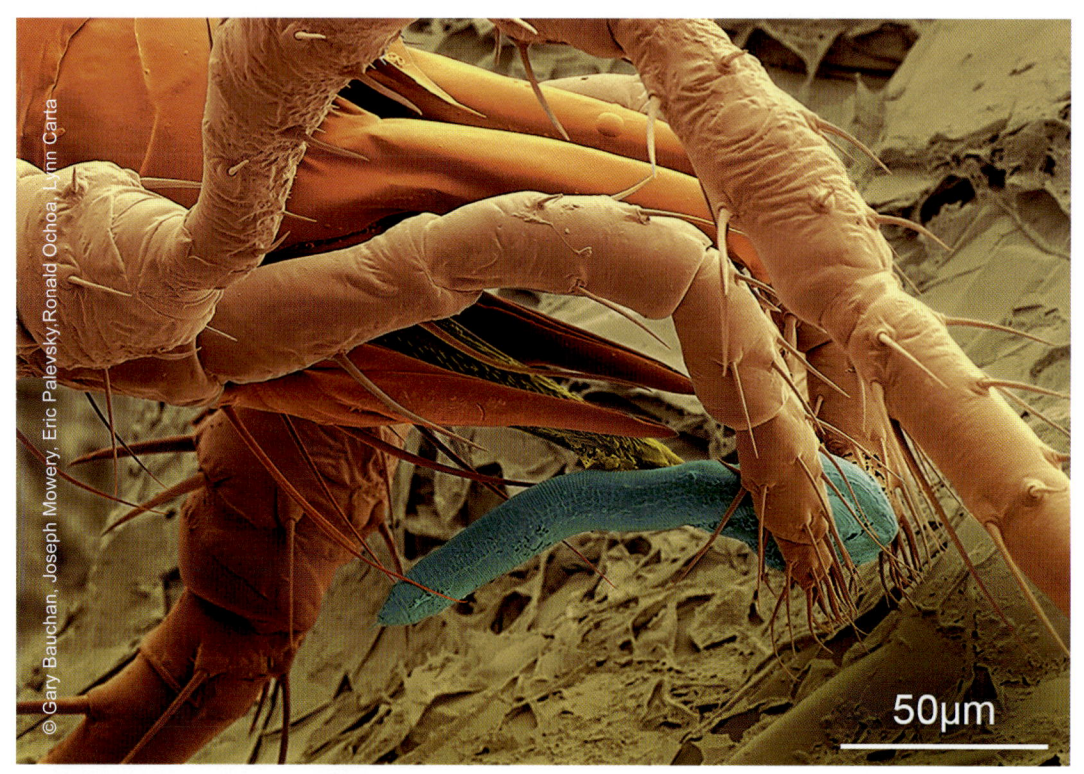

Photo 2.3.7c | Predatory mite

Scanning electron microscopy (SEM) of a mite (*Stratiolaelaps scimitus*) preying on the phytoparasitic nematode (*Meloidogyne incognita*), a major pest in many crops.

Photo 2.3.7d | Mite

An *Erythracarus* species of mite, New Zealand.

Photo 2.3.7e | Predatory mite

A Laelapidae mite, carrying prey- a juvenile *Lepidocyrtus* species of Collembola, Tasmania, Australia.

2.3.8 | SPRINGTAILS

Springtails (Collembola) are microarthropods that are just a few millimetres in size and are commonly named springtails because of an obvious forked jumping organ called a furca that helps them jump to avoid predators. Nowadays, these wingless, entognatha hexapods (that is, having six legs) belong to the class named Collembola within the phylum Arthropoda and are no longer included with the Insect class. They are classified into four orders: Entomobryomorpha, Poduromorpha, Symphypleona and Neelipleona. There are around 8 000 described species world-wide (Deharveng, 2004; Hopkin, 2007).

Most springtails are consumers of fungal hyphae, micro-algae and/or decaying vegetation. A few species can feed directly on plant material or are predaceous. In turn, springtails are prey for predatory mites, other Arachnida, Coleoptera and even Vertebrata such as reptiles and frogs, and can be hosts of parasitic Protozoa or nematodes, bacteria and fungal pathogens (Rusek, 1998). Consequently, Collembolan are a relevant part of the soil food web and contribute to ecosystem functioning, as they directly and indirectly regulate the soil microbial activity and nutrient cycling.

Springtails have four ecomorphological life forms according to their trophic niches (Gisin, 1943; Potapov *et al.*, 2016). The first of these is the atmobiotic life form, which corresponds to species that inhabit plants, trunks and branches of trees, but that can also be found on the litter surface. These are generally large species with round or elongated, pigmented bodies and well developed furca and eyes (full set of eight ocelli). Second is the epedaphic life form that corresponds to species that inhabit the upper litter layers or the surface of fallen logs. These pigmented Collembola are of medium or large size with well-developed furca, but smaller legs and antenna than the previous life form. Third is the hemiedaphic life form that corresponds to medium- or small-size species with less pigmentation, shortened legs and reduced number of ocelli, inhabiting decomposed litter or rotten wood. Fourth is the euedaphic life form, which corresponds to blind and unpigmented species that inhabit the upper mineral layers of the soil, but often also occupy upper horizons. These springtails have an elongated body of medium or small size with a reduced or absent furca. The atmobiotic and epedaphic species are also referred to as the "epigeic" group, while the hemiedaphic and euedaphic are considered the "edaphic" group.

Carbon transformations - The impact of Collembola on energy flow might appear to be quite small because of their very low biomass and small respiration rates that represent a small fraction of total soil CO_2 efflux (Filser, 2002). However, springtails directly accelerate organic matter decomposition by ingestion of organic material (such as litter and animal excrements) and by producing faeces. The quantification of the influence of absence/presence and density of microarthropods on litter decomposition rate has been carried out using chemical analyses and litterbag exclusion studies (Kampichler and Bruckner, 2009). Collembola alter carbon cycling indirectly, since they help microbes to decompose the material by increasing the surface area of ingested dead plant material to become more accessible to microbial attack; they can also directly inoculate microbes on material to decompose. Moreover, springtails are among the most abundant arthropods in the world, and in most terrestrial ecosystem they occur at densities between 10 000 and 100 000 individuals per square metre. Collembolan population density can then affect carbon turnover and soil carbon composition at the molecular level (Chamberlain *et al.*, 2006). The use of ^{13}C labelled material can help identify the amount, location and transformation of organic matter input to the soil, but also the effects of soil microarthropods on the litter decomposition with a microarthropod suppression treatment (Soong *et al.*, 2016). In a review of the role of Collembola in carbon cycling, Filser (2002) explained that effects on C transformation of Collembola are species- and life-stage specific, and that little attention has been paid to the effect of Collembola on dissolved organic carbon leaching, even in laboratory experiments. Finally, abiotic factors (physico-chemical parameters of soil and other environmental characteristics) and biotic factors (relationships with other species) might influence the effects of Collembola on carbon cycling.

Nutrient cycling - Collembola enhance nitrogen mineralization directly through their excreta and indirectly by interacting with microorganisms and thus can increase the plant nutrient supply. Indeed, Collembola excrete nitrogenous waste as ammonia or uric acid and thus increase nitrate availability and plant growth (Kaneda and Kaneko, 2011). Microarthropod assemblages have also proved to significantly affect the release of phosphorus during litter decomposition. In the presence of Collembola in a laboratory experiment, the decomposition of the remaining litter proceeded faster and the amounts of mobilized nutrients (Ca^{2+}, K^+, Mg^{2+}) and nitrate concentrations were higher than in the controls (Pieper and Weigmann, 2008). In this experiment, two Collembola species belonging to different life forms implied contrasting effect on the distribution of calcium within the soil layers at the end of the experiment: the calcium mobilized from the litter was retrieved at the soil surface for an epigeic species and at the bulk soil layers below with an euedaphic species.

Through their population density, their feeding activity and grazing intensity and their role in dead organic matter degradation, springtails can directly alter nutrient cycling and participate actively in humus formation. Collembola influence microbial species composition and biomass and thus indirectly impact mineralization rates that can lead to greater leaching and export of nutrients. For instance, by grazing upon pathogenic fungal hyphae, Collembola can stimulate fungal growth and induce the increase of fungal decomposing activity in soils. Consequently, they can increase the mobilization of nutrients from fungal biomass, but also help in the release of nutrients that were locked in fungal biomass. They also participate in the dispersal of active fungal spores and bacteria cells on their bodies or into their digestive tracts, and they may shift the composition of the rhizosphere microbiome (Crowther *et al.*, 2012; Soong *et al.*, 2016).

Soil structure - Springtails do not actively tunnel or burrow like other organisms such as earthworms, ants or termites, but some euedaphic species can make microtunnels in the soil (Rusek, 1998). Moreover, springtails play a role in soil structure alteration through litter comminution, casting and other disintegration mechanisms. Certain soils can contain millions of Collembolan faecal pellets per square metre, which in addition to adding structure to soils is likely beneficial to plants as they provide a slow nutrient release by microbes.

Maaß, Caruso and Rillig (2015) highlight the fact that the impact of the microarthropods on soil aggregation is nearly unknown. The authors propose potential mechanisms by which springtails could influence soil aggregation. Some effects on soil structure are direct when they process organic matter; there are also potential effects through egg integuments, molts, or necromass that might serve as a source of fresh organic material to microorganisms; this can potentially influence soil aggregation, especially given the high local abundances. Furthermore, since springtails impact fungi and bacteria community activity, they indirectly contribute to soil structure dynamics through mucilage secretion and hyphal network. Finally, a laboratory experiment has shown that Collembola stimulate soil aggregate formation through their interaction with arbuscular mycorrhizal fungi even in the absence of roots (Siddiky *et al.*, 2012).

Biological regulation - In laboratory experiments, some species have been found to be able to control soilborne plant fungal diseases through their direct feeding activity; for instance, they can reduce the foot and root fungal disease complex on cereal. Indeed, they feed preferably on pathogenic rather than on antagonistic or arbuscular mycorrhizal fungal propagules. However, according to field studies and analyses of the gut content of sampled springtails, these fungivores were found to be less selective than in laboratory conditions, and their biocontrol effect against the pathogen might depend on their density and on field conditions (Coleman, Callaham and Crossley, 2018; Innocenti and Sabatini, 2018).

Another type of biological regulation driven by Collembola was studied by Scheu, Theehuas and Jones (1999), who observed that Collembola might indirectly reduce plant-sucking aphid reproduction depending on the plant host. In the laboratory, Scheu, Theehuas and Jones (1999) observed a 45 percent decrease of aphid reproduction in the presence of Collembola with the legume *Trifolium repens*, while an increase in aphid reproduction by a factor of three was measured on the grass *Poa annua*. In the same context, Schütz *et al.* (2006) suggest that the reduction of aphid reproduction is caused by impact of Collembola on resource allocation and growth of wheat (*Triticum aestivum*). Additional studies are needed to improve the understanding of Collembola impact on plant growth and on soil ecosystems in interaction with other soil biota.

Photo 2.3.8a | Springtail

A Collembola, *Dicyrtomina minuta*, standing on snail eggs. East Pennard, United Kingdom.

Photo 2.3.8b | Springtails

A springtail (*Bourletiella arvalis*), Nova Scotia, Canada.

Photo 2.3.8c | Springtail

A *Adephoderia regina*, from an unusual family of Collembola with neck organs. Tasmania, Australia.

Photo 2.3.8d | Springtail

A Collembola, *Lobellinae*, showing a defensive frozen posture to an ant. Northern Queensland, Australia.

Photo 2.3.8e | Springtail

A juvenile *Platanurida* species of Collembola.

Photo 2.3.8f | Springtail

An *Acanthanura* species of giant springtail, a Tasmanian endemic species.

Photo 2.3.8g | Springtail

A large *Pseudachorutes* species of Collembola from tropical Queensland, Australia.

MACROFAUNA

2.3.9 | EARTHWORMS

Earthworms are invertebrate, segmented animals belonging to the phylum Annelida, class Clitellata, order Crassiclitellata. About 6 000 species have been described to date, from around 20 families. Earthworms feed on dead organic matter, either deposited as litter at the soil surface or mixed within the soil. Earthworms can ingest large amounts of litter and/or soil and are classified as saprophagous or detritivorous animals. While ingesting soil and organic matter, they also ingest bacteria and fungi.

Earthworms ingest large amounts of soil for their physiological needs: food for energetic and nutrient requirements, aeration, shelter, breeding and so forth. The annual soil consumption rates may reach important values, for example 1 250 Mg/ha in Ivory Coast (Lavelle *et al.*, 2006), equivalent to 8 cm to 10 cm soil thickness. This intense burrowing and feeding activity can profoundly transform the surrounding environment, resulting in marked changes in soil biodiversity and associated functions. Moreover, earthworms are known to deeply affect soil structure, organic matter dynamics and the activity of other organisms in the soil food web and thereby to contribute to soil health, for example, in man-made systems (Tondoh *et al.*, 2019). For that reason, they have been classified in the functional groups of soil "ecosystem engineers" (Lavelle *et al.*, 2016). The soil modified by earthworms, called the drilosphere, is made up of burrows, casts, and middens that represent important functional domains or hot spots in soil. However, there is not always a direct relationship between earthworms and soil health improvement (Tondoh *et al.*, 2019).

Carbon transformations - Earthworm activities have two main consequences on soil carbon transformations: (i) breakdown and decomposition of ingested organic matter; and (ii) incorporation of organic matter and mixing with the soil. The digestion of organic matter in earthworm guts is based on a mutualistic interaction with microorganisms that inhabit their digestive track (Lavelle *et al.*, 2006). In the gut, earthworms provide an energy-rich mucus and water to microbes that can decompose ingested organic matter and release carbon and nutrients that can be assimilated by earthworms. Earthworms assimilate a small proportion of ingested carbon; the remaining being egested within casts. In casts, organic matter is further decomposed by microorganisms but at a decreasing rate, so that decomposition becomes lower than in the bulk soil after some days (Lavelle and Martin, 1992). The physical protection of organic matter in casts prevents the mineralization of organic carbon. However, there is still a debate as to whether earthworms can simultaneously enhance decomposition (in the short term) and stabilization of organic matter; that is, a reduced decomposition in the long term (Coq *et al.*, 2007; Lubbers, Pulleman and Van Groenigen, 2017). Anecic earthworms feeding on soil-surface organic matter incorporate carbon to depth in the soil. Earthworms also produce calcium carbonate granules in their guts (Canti and Piearce, 2003; Versteegh *et al.*, 2017), thereby interacting with soil inorganic carbon.

Nutrient cycling - The decomposition of organic matter in earthworms' guts and casts, indirectly mediated by an intensification of microbial activity, leads to a mineralization and release of nutrients that become available for plants and microbes. Earthworms consistently increase nitrogen turnover (van Groenigen *et al.*, 2019), and large amounts of mineral nitrogen (ammonium and nitrate) are released by earthworms (Wu *et al.*, 2017). For instance, a five-fold higher ammonium content in the fresh casts of *Pontoscolex corethrurus* has been observed compared to non-ingested soil (Lavelle *et al.*, 1992). Similar results were found for phosphorus, with levels of available P being higher in (fresh) casts than in the bulk soil (Ros *et al.*, 2017), but this effect strongly differs among species and habitat. This release generally improves plant growth and can inhibit the competition between intercropped plants (Coulis *et al.*, 2014). Other soil elements such as silicon can also be affected, with content available for plants generally increased by earthworms (Hu *et al.*, 2018). The recent review by van Groenigen *et al.* (2019) confirms that earthworm casts are much more fertile than bulk soil.

Soil structure - The activities of burrowing, soil ingestion and cast egestion result in a deep modification of soil structure when earthworms are present. The effects differ whether the community is dominated by anecic worms (that is, earthworms living in semi-permanent, mostly vertical burrows), or by endogeic worms (earthworms digging into the soil and filling burrows with their casts). Both soil aggregation and porosity are affected by earthworm activity (Blanchart *et al.*, 2009; Guéi *et al.*, 2012; Schneider and Schröder, 2012) with significant consequences for the flow of water, nutrients and gases, and for soil erodibility. Regarding soil aggregation and consequences for soil erosion, two endogeic earthworm functional groups are recognized: compacting and de-compacting species (Blanchart *et al.*, 2009; Guéi *et al.*, 2012). Indicators of earthworm bioturbation have recently been proposed, considering both burrows and casts in the soil profile (Piron *et al.*, 2017). Similarly, earthworms have been shown to be able to regenerate soil structure after soil compaction (Capowiez *et al.*, 2012).

Biological regulation - Soil engineers such as earthworms have a strong impact on the diversity and activity of other soil organisms. This effect can be direct through consumption, or indirect through a modification of soil structure and transformations of carbon and nutrients. All organisms of the soil food web are affected by earthworms. The recent review by Medina-Sauza *et al.* (2019) revealed that earthworms generally promote the growth of some copiotrophic bacteria such as Flavobacteriaceae, Firmicutes, and g-Proteobacteria due to the mucus they produce in their gut. Chitilinolytic bacteria (*Chitinophagaceae*) are also promoted in earthworm guts, probably due to the degradation of chitin-rich fungal hyphae during gut transit (Bernard *et al.*, 2012). Earthworms also interact with other organisms of the soil food web which affect soil chemistry and soil nutrient availability: arbuscular mycorrhizal fungi with contradictory effects on root colonization (Zaller *et al.*, 2011; Medina-Sauza *et al.*, 2019), nematodes (Demetrio *et al.*, 2019) and mesofauna (Zhu *et al.*, 2018). Earthworms can directly affect soilborne plant pathogens, for example the decomposition of *Sclerotinia sclerotiorum* (Euteneuer *et al.*, 2019). They also affect plant growth and immunity not only through the release of available nutrients but also through the emission of signal molecules leading to the production of phytohormones (Puga-Freitas and Blouin, 2015). Recent studies confirm that earthworms can regulate plant pathogens and disease development through different physical, chemical and biological mechanisms (for instance, Blanchart *et al.*, 2019, for rice blast disease, and Xiao *et al.*, 2019 for thrips attack).

Photo 2.3.9a | Earthworm

A juvenile earthworm, South Devon, United Kingdom-

Photo 2.3.9b | Flatworm

A *Rynchodemus sylvaticus*, a United Kingdom native land flatworm, or Planarian under a stone in South Devon, United Kingdom.

Photo 2.3.9c | Predacious flatworm

A *Bipalium* species of hammerhead flatworm (Planarian) from Japan.

Box 2.3.9 | Micromorphology of biological activity in soils

Soil micromorphology is a technique initiated by Walter L. Kubiëna (1897-1970), which consists of studying thin sections of undisturbed soil (30 µm thick) under a microscope. It allows us to observe the microsites where soil organisms live and how they interact with the mineral matter. The **first picture** shows the activity of soil engineers, as large excrements of earthworms and smaller excrements of enchitraeid worms, filling in channels and chambers (biopores). The **second picture** shows a sclerotium and hyphae of fungi (saprophytic organisms) developing on decaying organic matter. The **third picture** shows the activity of oribatid mites (primary decomposers) as oval excrements in soft plant tissues

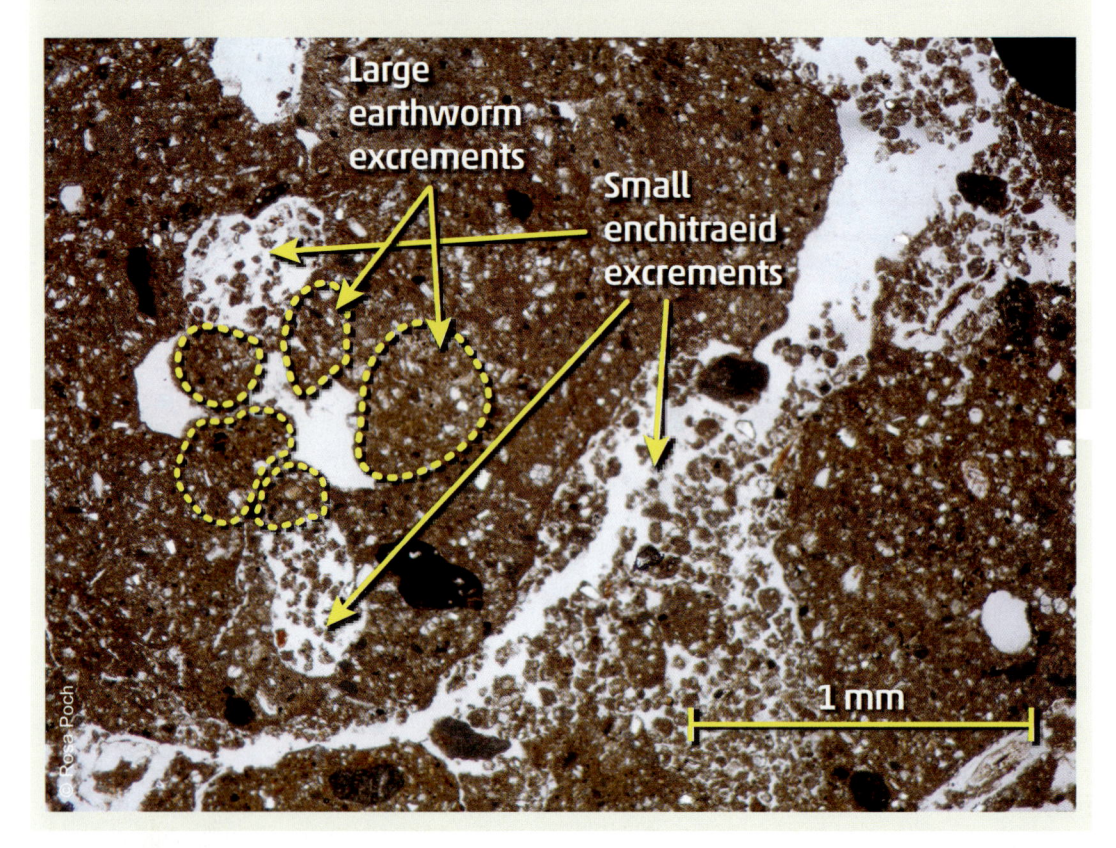

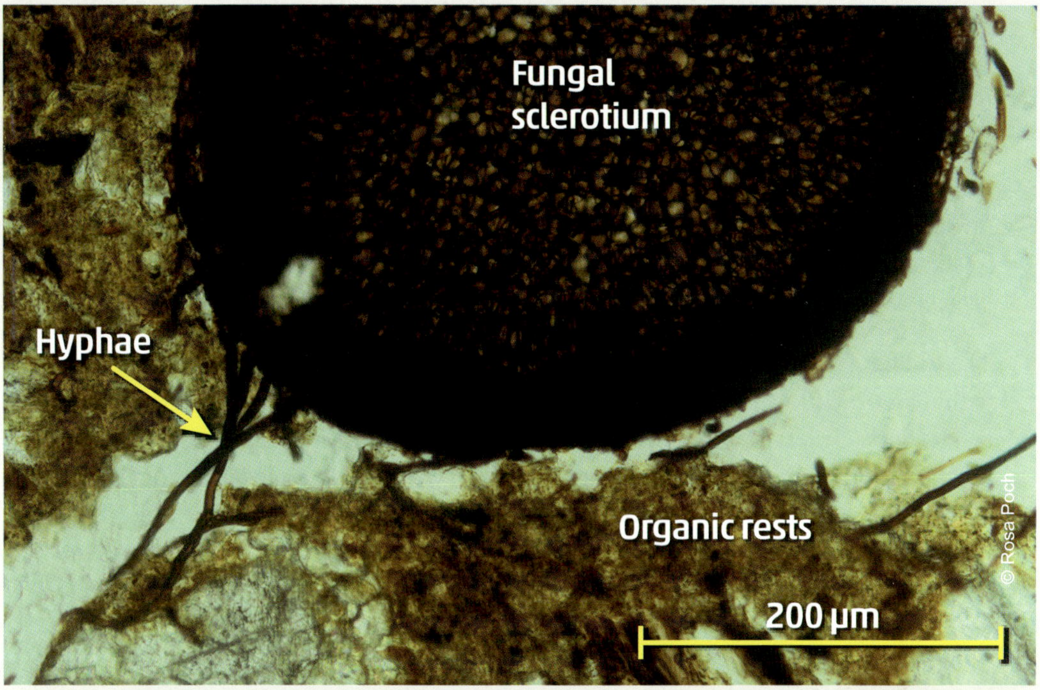
Fungal sclerotium
Hyphae
Organic rests
200 μm
© Rosa Poch

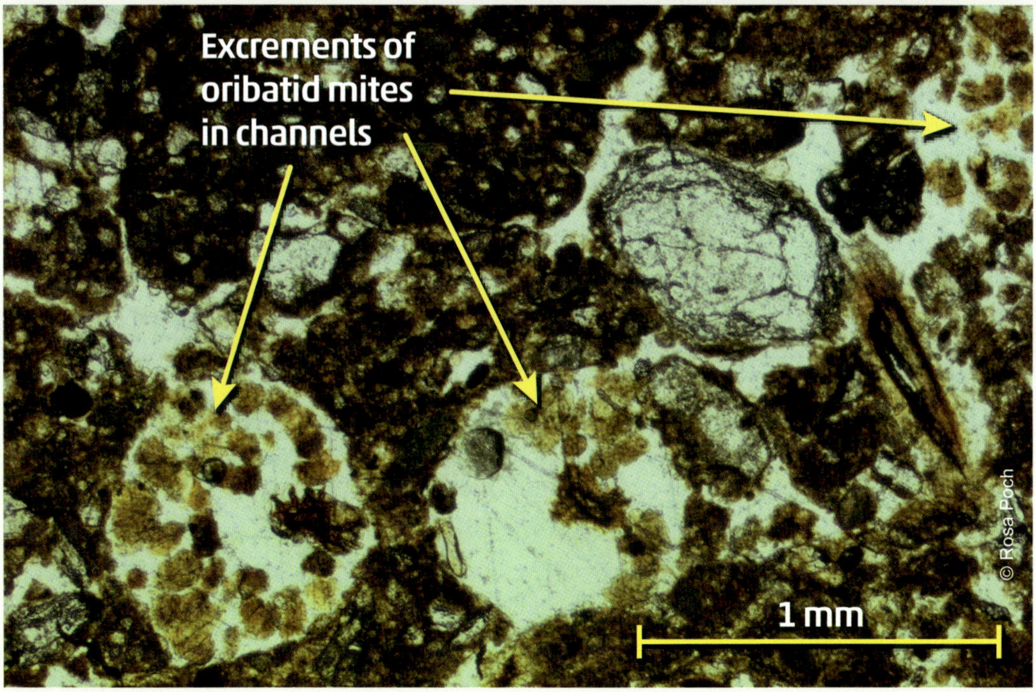
Excrements of oribatid mites in channels
1 mm
© Rosa Poch

2.3.10 | ISOPODS, MILLIPEDES, INSECTS AND SPIDERS

Members of these diverse and understudied taxa have the potential to influence soil functions in a variety of ways, including through their burrowing and tunnelling activities; through consuming detritus, carrion and soil; through the deposition of their own waste; and through the regulation of the composition and activity of lower trophic level organisms (that is, fungi and bacteria).

The soil food web includes a number of other organisms that are not represented within Figure 2.4.1, such as isopods (Order: Isopoda), millipedes (Class: Diplopoda) and spiders (Order: Araneae). Additionally, many species of insects (Class: Insecta) inhabit the soil either permanently or during part of their life cycles. Immature stages of many insects can be particularly abundant in soil and litter. This is a diverse group, including but not limited to Blattodea (cockroaches and termites), Coleoptera (beetles), Dermaptera (earwigs), Diptera (mosquitoes and flies), Hemiptera (true bugs), Homoptera (cicadas, aphids and mealybugs), Hymenoptera (ants, wasps and bees), *Lepidoptera* (moths and butterflies), Orthoptera (grasshoppers, crickets and mole crickets), Neuroptera, Psocoptera, and Thysanoptera.

Organisms within the macrofaunal group feed across all trophic levels and can have meaningful impacts on soil functions. For example, isopods and millipedes are omnivores and detritivores that feed in areas of accumulated detritus and adequate soil moisture, while spiders that inhabit soil and detrital habitats are predators that opportunistically feed upon the relatively larger-bodied organisms within the soil food web (for example, Collembola and others). Soil-dwelling insects are a diverse group whose feeding behaviours span all trophic levels (for example, herbivores, scavengers, decomposers, predators or parasites). Many insect species are also economically important pests due to the feeding activities of their soil-dwelling larvae (for example, beetles, flies and moths).

Carbon transformations - There is a growing body of work that demonstrates how isopods, millipedes, spiders and insects influence soil carbon dynamics, especially decomposition rates. Members of these taxa are likely to have meaningful impacts in regulating soil carbon dynamics through their direct consumption of decaying plant material and animal carrion, through deposition of their own waste, and perhaps most importantly by regulating the activity of their prey and the microbial community via trophic interactions. However, while the impacts of some organisms on processes such as decomposition are fairly well understood (for example, by carrion-associated flies and dung beetles), the number of studies that address the potential impacts of most other organisms of these taxa on soil C is limited. This lack of data highlights the need for more scientific research in order to better understand the functional roles of these diverse members of the soil food web, particularly in terms of their potential cascading impacts on soil C dynamics via the microbial community.

Saprophagous macrofauna such as isopods, millipedes and many insects transform litter and contribute to decomposition by consuming decaying matter and by moving organic matter throughout the soil profile. Macrofaunal presence has repeatedly been found to increase decomposition rates (Hättenschwiler and Bretscher, 2001; Toyota, Kaneko and Ito, 2006; Riutta *et al.*, 2012; Ott *et al.*, 2012; Villisics *et al.*, 2012; David and Handa, 2010). For example, Diptera larvae (Bibionidae) can consume up to 40 percent of the annual litterfall in alder forests (Frouz *et al.*, 2015). Isopods and millipedes are among the most abundant and important decomposers in temperate and tropical ecosystems (David and Handa, 2010). Macrofaunal effects on decomposition are variable and depend on population densities and resource quality. Under some circumstances, their impacts on C dynamics can even rival those of abiotic conditions. For instance, a recent mesocosm experiment show that millipede presence can affect decomposition rates more strongly than elevated CO_2 or temperature (Rouifed *et al.*, 2010). However, these millipede effects varied according to the species composition of the litter they were consuming and did not necessarily lead to higher C mineralization (Rouifed *et al.*, 2010). Species interactions between multiple soil organisms can also change soil respiration rates (for example, between millipedes and earthworms; Snyder *et al.*, 2009). Most of the litter that is ingested by macrofauna is returned to the soil as excrement. Excrement decomposes differently than the original litter (Frouz *et al.*, 1999) and therefore may affect soil C and overall litter decomposition.

Insect larvae and other organisms also affect C transformations in ecosystems by consuming the carrion and waste (that is, faunal coprophagy) of other animals. Among terrestrial invertebrates, insects are the most important scavenging group for facilitating carrion decomposition. Flies are particularly important in this role, as they are typically the first to colonize carrion, and their larvae remove more carrion biomass than any other group (Barton and Bump, 2019). In addition to many other insect species, beetles (Order: Coleoptera) are important members of carrion fauna that contribute to carrion decay, either as consumers of the carrion or as predators of decomposers (Barton and Bump, 2019). Faunal coprophagy can increase litter mineralization by increasing microbial activity and overall C assimilation (Frouz *et al.*, 1999). Our understanding of the extent of coprophagy and its impacts on soil function is still limited, with the exception of the large body of work on coprophagous beetles in the subfamily Scarabaeinae (Coleoptera: Scarabaeidae). These beetles feed on animal dung as adults and larvae, and contribute to C transformation in several ways. Specifically, by increasing the transfer of nutrients from dung to the soil, their activity has been found to stimulate litter decomposition (Tixier *et al.*, 2015; Manning *et al.*, 2016), increase soil C (Menéndez *et al.*, 2016) and reduce greenhouse gas emissions from soils (Slade *et al.*, 2016; Penttilä *et al.*, 2013; Verdú *et al.*, 2019; but see Evans *et al.* 2019 for an example of dung beetles having the opposite effect on soil GHG emissions). The abundance and biomass of dung beetles within a particular habitat is typically correlated with the strength of the beetles' impact on ecosystem functioning (Frank *et al.*, 2017). For example, dung beetle biomass – and therefore dung removal rates – has been found to be higher in forests than in grasslands (Frank *et al.*, 2017). Some evidence even suggests that dung beetle abundance may be more important for ecosystem functioning than species richness (Manning and Cutler, 2018; Alvarado *et al.*, 2019). However, functional group richness, species composition and maintenance of interactions among co-occurring species (such as tunnelling and dwelling dung beetles; Nervo *et al.*, 2017) have been shown to be important for ecosystem function (Larsen *et al.*, 2005; Milotić *et al.*, 2018; Menéndez *et al.*, 2016) and for sustaining multiple ecosystem functions (O'hea *et al.*, 2010; Nervo *et al.*, 2017; Manning *et al.*, 2016; Piccini *et al.*, 2018; Santos-Heredia *et al.*, 2018) in both perturbed (Beynon *et al.*, 2012) and unperturbed systems (Manning *et al.*, 2017; Beynon *et al.*, 2012).

In addition to their direct consumptive and shredding activities, macrofauna can indirectly affect soil C dynamics by altering the abundance and activity of their prey and/or regulating the activity or composition of soil microbial communities. For example, recent studies have shown that grazing isopods modify the composition and activity of fungal communities (Crowther and A'Bear, 2012; Crowther *et al.*, 2013) and thereby alter fungal-mediated wood decomposition rates (Crowther *et al.*, 2015). These impacts vary according to environmental context, whereby isopod control over fungal decomposer activity increases with increasing inorganic soil nitrogen availability (Crowther *et al.*, 2015). Larger invertebrate predators, such as predatory beetles and spiders, can also have cascading effects through the food web that affect carbon transformations. Notably, as with some of the other examples above, the direction and magnitude of predator effects are variable (positive, negative, or neutral) depending on the species present and environmental context. For example, the presence of some predatory beetles (Wu *et al.*, 2011) and web-building spiders (Liu *et al.*, 2014) sometimes results in slower decomposition rates (but see Miyashita and Niwa, 2006 for an example of web-building spiders having no effect on decomposition), while wolf spiders in tropical, temperate and arctic ecosystems can indirectly enhance decomposition (Lawrence and Wise, 2004; Koltz *et al.*, 2018; Liu *et al.*, 2015). However, changes in environmental conditions such as moisture availability, warming, or fertilization can alter or even reverse predator effects on decomposition and soil respiration (Lensing and Wise, 2006; Wu *et al.*, 2011; Koltz *et al.*, 2018; Maran *et al.*, 2016; Melguizo-Ruiz *et al.*, 2019; Kashmeera and Sudhikumar, 2018; Liu *et al.*, 2018; Liu *et al.*, 2015). Studies of predator effects on carbon transformations have been limited to very few taxonomic groups; this is an area that requires further study.

Nutrient cycling - Isopods, millipedes, insects and spiders affect nutrient cycling in a variety of ways by consuming litter and soil, depositing faeces, and dispersing nutrients between habitats, and through interactions with the microbial community. While direct effects of these organisms on C and N mineralization may not be particularly high in comparison to those of the microbial community (Koltz *et al.*, 2018), results from recent studies highlight the potential importance of macrofaunal-microbial interactions for driving nutrient dynamics (Bastow, 2011; Crowther *et al.*, 2011; Crowther *et al.*, 2012; David, 2014). The impacts of certain groups, such as millipedes, isopods and dung beetles, are well studied; much less is known about the roles of many of the other organisms that inhabit the soil on nutrient cycling.

Millipedes play an important role in soil nutrient dynamics by feeding on litter and soil and consequently promoting N release and microbial activity (Makoto *et al.*, 2014; da Silva *et al.*, 2017; Fujimaki *et al.*, 2010). Isopods can also increase microbial respiration and soil nutrient availability (for example, soil C, N, P, K, Mg and Ca; Kautz and Topp, 2000). Isopod digestion of detritus changes detrital C:N as well (Kautz *et al.*, 2002). By promoting the decomposition of mammalian dung, dung beetles enhance nutrient transfer from dung to the soil. Dung beetles reduce the volatilization of ammonia (NH_3) from soil (Kazuhira *et al.*, 1991), thereby increasing the availability of soil N (Maldonado *et al.*, 2019). Dung beetle activity can also result in higher soil phosphorus content (Maldonado *et al.*, 2019) and has been linked to higher emissions of N_2O from dung pats (Penttilä *et al.*, 2013). Organismal effects on nutrient cycling can be significant even in comparison to abiotic drivers (Meyer *et al.*, 2011). For example, in recent studies, millipedes increased the concentration of inorganic N in soil by 40 percent (Makoto *et al.*, 2014) and isopods increased nitrate leaching by 50 percent (Hättenschwiler and Bretscher, 2001). These effects can vary depending on the environmental context, on life stage and on biotic interactions. For example, warming alters millipede effects on soil N dynamics (Makoto *et al.*, 2014). Organisms may also utilize different habitats within the soil profile during different life stages or in response to increased predation risk, and these behavioural shifts can change their effects on soil nutrients. Soil-feeding millipedes, for instance, promote mineralization of soil N, while the activity of adults, who primarily feed on litter, inhibits N mineralization (Toyota and Kaneko, 2012). Likewise, depending on the burrowing abilities of their detritivore prey, predaceous beetles can either increase or reduce levels of soil-soluble N (Wu *et al.*, 2015).

Some insects alter soil nutrient dynamics by dispersing nutrients from one area or ecosystem to another. Mass emergence events of aquatic insects and their dispersal to riparian and terrestrial habitats can transfer large amounts of C and N to nearby soils and thereby alter nutrient cycling (Dreyer *et al.*, 2015). Various insect scavengers also relocate nutrients such as C, N and P away from carrion (Barton and Bump, 2019).

There is evidence that beetle and spider predators can have indirect effects on soil nutrient cycling via their detritivore prey as well. For example, beetle predators in alpine systems have been shown to indirectly reduce soil N and P (Wu *et al.*, 2011), while higher densities of wolf spiders in arctic tundra are associated with higher soil N availability (Koltz *et al.*, 2018). Yet in other cases, spider and beetle predators have not been found to influence soil N (Wu *et al.*, 2014; Maran and Pelini 2016), soil P, or soil organic matter content (Wu *et al.*, 2014). One important mechanism by which predators can affect nutrient cycling is by altering the behaviour of prey. Recent work shows that Collembola reduce their activity when exposed to spider predators; this behavioural change results in lower soil N and lower fluxes of CO_2 from soil (Sitvarin and Rypstra, 2014). Likewise, predatory beetles can modify the position of earthworm activity within the soil profile. By altering the behaviour of their prey, beetle presence indirectly causes lower total N and P content in the upper soil layer and higher N, P and organic matter content in the lower soil profile (Zhao *et al.*, 2013). The community composition of above-ground predators has also been linked to soil nutrient dynamics (Hawlena and Schmitz, 2010) and soil C retention (Schmitz *et al.*, 2017). Overall, however, studies addressing the impacts of organisms from higher trophic levels on nutrient cycling are limited. This is an area that requires further study in order to improve our understanding of the generality of how predators affect nutrient cycling in soil and detrital systems.

Soil structure - Aside from earthworms, termites and ants, the effects of most macroinvertebrates on soil structure are largely unexplored. However, a few key groups have well-documented impacts on soil properties such as soil aggregation, pore structure and hydrology. For example, there is ample evidence that the presence of dung beetles increases soil aeration (Bang *et al.*, 2005) and decreases soil compaction (Manning *et al.*, 2016; see Nichols *et al.* 2008 for review of soil functions provided by dung beetles). Such changes to soil structure as a result of dung beetle burying activities also cause improved soil hydrological properties such as water infiltration (Badenhorst *et al.*, 2018) and soil porosity, which in turn reduce surface runoff and soil losses (Brown *et al.*, 2010; Forgie *et al.*, 2018). The burrowing activities of the larvae of some other insect species can also alter soil porosity. Cranefly larvae create burrows within the soil that cause higher water infiltration rates (Holden and Gell, 2009). Likewise, burrows of ground-dwelling beetle larvae change soil pore structure and infiltration patterns, particularly during the initial stages of soil development (Badorreck *et al.*, 2012). Millipedes also modify soil properties, although effects might vary depending on life stage (Toyota *et al.*, 2006). For example, larvae of the train millipede (*Parafontaria laminate*) increase the development of soil aggregates (Fujimaki *et al.*, 2010), while adults have been shown to increase carbon accumulation within the soil (Toyota *et al.*, 2006). Activity of a different millipede species (*Glyphiulus granulatus*) from Brazil causes increased formation of larger soil aggregates but fewer smaller aggregates (da Silva *et al.*, 2017).

Species interactions among soil invertebrates can also change species-specific impacts on soil structure. Experiments that have investigated earthworm effects in combination with other organisms have found that earthworms produce fewer large soil aggregates when isopods are present (Loranger-Merciris *et al.*, 2008) and that there are fewer aggregates in the 0-2 cm soil layer from earthworms (and less carbon contained in the aggregates) when millipedes are present (Snyder *et al.*, 2009).

For other organisms, such as spiders and predatory beetles, little to nothing is known about how their activities might impact soil structure, but they may indirectly affect soil structure by changing the densities or behaviour of their prey. For example, a recent study showed how predatory beetles strengthen the effects of their earthworm prey on soil properties. In the presence of these predatory beetles, earthworms moved lower within the soil profile, which resulted in reduced soil bulk density and higher soil water content in this lower soil layer (Zhao *et al.*, 2013).

Biological regulation - Research over the last few years has highlighted the potentially important role of isopods, millipedes, spiders and some insect groups in affecting microbial activity and community composition. Such effects can be due to interactions between macrofauna and the microbial community or they may be more indirect. For example, isopod grazing on fungal cord systems alters fungal community structure and reduces the soil fungal:bacterial ratio (Crowther *et al.*, 2013). Through a variety of mechanisms, dung beetle activity also increases microbial biomass and facilitates soil microbial activity (Slade *et al.*, 2016; Menéndez *et al.*, 2016). Indirect effects of bark beetles and wood-boring insects on fungi include modifying woody habitats, which facilitates colonization by fungi and other insect species and causes faster decomposition of woody debris (Strid *et al.*, 2014; Ulyshen *et al.*, 2016). However, interactions among macrofauna and the microbial communities and the mechanisms for the associated changes in soil function are not always clear or can be context dependent. For instance, isopod grazers have top-down effects that often limit fungal biomass, but these effects vary by season and can be negligible if nutrient availability is limiting (A'Bear, Johnson and Jones, 2014; Crowther *et al.*, 2015). In another case, the faecal pellets produced by millipedes can alter microbial community structure in leaf litter (Coulis *et al.*, 2013). Yet these changes are not associated with microbial activity in the pellets and do lead to short-term changes in decomposition rates (Coulis *et al.*, 2013). Spiders can indirectly affect microbially-mediated processes such as decomposition, but there tend not to be significant effects on microbial biomass by web-building or active hunting spiders (Liu *et al.*, 2014; Koltz *et al.*, 2018; Maran and Pelini, 2016). Whether spider activity has cascading effects on microbial composition or community structure has not been thoroughly addressed. Overall, this is a currently active area of research and one that warrants further study.

Photo 2.3.10a | Centipede

A *Lithobius variegatus* or banded centipede from South Devon, United Kingdom.

Photo 2.3.10b | Millipede

A *Polydesmus* species of millipede from South Devon, United Kingdom.

Photo 2.3.10c | Dung beetle

A Coleopteran (*Scybalophagus* sp.) on a sandy soil in the Torotoro National Park in Potosi, Bolivia.

Photo 2.3.10d | Beetle

A Coleopteran (*Scarabaeus auratus*).

Photo 2.3.10e | Millipede

A millipede (*Cylindroiulus* sp.). The organism helps to increase the nutrients in the soil as well as the feces of earthworms and snails.

Photo 2.3.10f | Millipede

A species of millipede (*Anadenobolus monilicornis*) that inhabits leaf litter.

Photo 2.3.10g | Pseudoscorpion

A moss neobisiid (*Neobisium carcinoides*) pseudoscorpion or false scorpion, hunting and trapping a Symphypleona Collembola (*Sminthurus aureus*).

Photo 2.3.10h | Spider

A spider (*Megaphobema sp.*) found about 10 cm below the soil surface (in a tunnel/nest) within a pasture in the Meta Department of Colombia.

Photo 2.3.10i | Wolf spider

Hippasa is a genus of wolf spiders in the family Lycosidae.

Photo 2.3.10l | Woodlouse

Unidentified woodlouse (Isopoda) in a native regenerative bush, Stewart Island, New Zealand.

Photo 2.3.10m | Woodlouse

An isopod (*Armadillidium vulgare*) feeding on leaf litter, collected in a coastal forest near Edinburgh, Scotland.

Photo 2.3.10n | Woodlouse

A common Isopod species (*Porcellio scaber*) found in Western Europe.

Photo 2.3.10o | Woodlouse

Isopods living deeper in the soil tend to be smaller and less pigmented than species living on the soil surface, as for example *Haplophthalmus danicus*.

Photo 2.3.10p | Woodlouse

Myrmecophilous woodlouse *Platyarthus hoffmanseggi* from a nest of Formica ants in northwestern Bohemia.

2.3.11 | TERMITES

Termites are eusocial terrestrial insects organized in colonies and presenting a very high polymorphism. The individuals of a colony are differentiated into three morphologically, behaviourally and functionally different casts: workers, soldiers and alates (Noirot and Alliot, 1947). Termites are widely distributed throughout the world, especially in tropical, sub-tropical and semi-arid regions, while some species are known to be living in high latitudes (Krishna *et al.*, 2013). Africa presents the highest diversity with more than 1 000 species encountered (Langewald, Mitchell and Kooyman, 2003). Termites now belong to the order Blattodea and the sub order Isoptera, which includes 2 934 living species distributed among 282 genera, 16 subfamilies and 9 families (Krishna *et al.*, 2013). Some termite species called "lower termites" are totally dependant on endosymbionts flagellates living in their hindgut to digest cellulose-based products, while the "higher termites" live in exosymbiosis with a mushroom of genus *Termitomyces* (Grassé, 1986).

Termites feed on cellulose contained in organic matter (Grasse, 1986). Some termites consume living plants, while others alter wood, litter or humus. The cellulose that they assimilate serves as the main source of energy. According to the process of assimilation of the food, one distinguishes four trophic groups: wood feeders, soil feeders, interface soil-wood foragers and mushroom growers. Forage termites feed primarily on Poaceae. They consume plants that are not degraded by microorganisms. They allow the fixation of nitrogen (Breznak, 2000). This group is characterized by the genus *Trinervitermes* including five species (*T. geminatus*, *T. occidentalis*, *T. oeconomus*, *T. togoensis* and *T. trinervius*) (Tano *et al.*, 2005). Wood feeding termites consume degraded wood and can digest cellulose (Eggleton, 2000). They digest these compounds thanks to zooflagellates and bacteria contained in their digestive tract (Grasse, 1982). Soil feeding termites ingest decomposed organic matter mixed with mineral particles (Brauman, 2000). These termites consume decomposing organic particles in the humic fraction of soils. Some humivores such as *Cubitermes* consume the products resulting from the degradation of plants by bacteria and flagellates (Brume, 2011). Apicotermitinae sometimes consume decomposing wood (Grasse, 1982), and in a recent study were found in galleries of the cocoa tree.

Mushroom termites include species in exosymbiosis with a superior fungus (Basidiomycete) of the genus *Termitomyces* for the decomposition of cellulose (Rouland, 2000). Their diet consists mainly of litter including dead wood and leaves that are harvested and brought back to the nest for the workers to ingest. These termites use predigested food in the form of a faecal pellet called a "millstone" on which to grow cultures of the fungus *Termitomyces* (Grasse, 1982) as a food source for the colony.

Carbon transformation – Yapi (1991) suggests that the distribution of termites is a function of the size of the species, with small species living in the upper stratum, rich in organic matter, and the larger species in the deeper layers of the soil, poor in organic matter. Within the same horizon, termites seem to select the finer particles (silt and clays), richer in organic matter (Garnier-Sillam and Tessier, 1991). The degree of selection depends on the richness of organic matter of the surrounding soil, with savanna termites showing more selective behaviour than forest termites (Goksoyr and Daae, 1990).

Nutrient cycling - The activity of termites in soil enhances microbial activity and the release of mineral nutrients such as ammonium and nitrate (Petal *et al.*, 2003). Soils handled by these "engineers" are often enriched with fine particles as well as with soil organic matter and exchangeable cations (Ca, Mg, K and Na) compared to the surrounding soil (Lobry de Bruyn and Conacher, 1990). Further, their mounds influence resilience across landscapes: not just agricultural but also forests and African dry grasslands becoming more resilient because the termite mounds lead to islands of fertility and plant production (Ashton *et al.*, 2019; Bonachela *et al.*, 2015).

Soil structure - Many species of termites (for example *Odontotermes*, *Macrotermes*) can transform and improve physicochemical properties and aeration of soil by building complex subterranean galleries and nests (Tano *et al.*, 2005), and thus can enhance soil fertility (Dibog *et al.*, 1999). According to Majer (1989), termites influence the friability and distribution of pores, which are crucial for aeration, soil drainage and root penetration. In fact, by digging galleries and casting the ground, termites regulate the movement of water and air, and thereby enhance the physical properties of the soil (Greenslade, 1985). Soil stratification ensures the availability of nutrients at depths where they are used by plants. By returning to the surface of the lower horizons, these animals increase the nutrients available to the plants and bring back to the cycle the elements that would be lost in deeper soil layers if they had not been there (Humphrey and Mitchell, 1983).

Biodiversity regulation - Many control strategies have been proposed to reduce the effect of termites in agricultural systems. Most of them rely mainly on a broad spectrum of persistent organochlorine insecticides (Logan *et al.*, 1990). However, serious limitations and increasing legal restrictions are associated with the application and efficacy of these chemicals (Djuideu, 2017). Living solely in soil, these insects are well protected both from insecticidal sprays and from predation, making them difficult to control. To avoid the employment of these chemicals, improved ecologically acceptable methods, such as biological control using natural enemies, are needed. Many groups have been cited as parasites of termites in different parts of the world, such as fungi (Blackwell and Rossi, 1986; Rath, 2000), mites (Eickwort, 1990; Zhang *et al.*, 1995), nematodes (Wang *et al.*, 2002) and Diptera Calliphoridae (Sze *et al.*, 2008).

Photo 2.3.11a | Termites

Termite (*Macrotermes* sp.) collected from a mound on an island near the northern edge of Lake Victoria, in the Central Region of Uganda.

Photo 2.3.11b | Termites

The photo shows a termite (*Aparatermes thornatus*) of the Termitidae family, in a primary forest in the Colombian Amazon. These termites feed on leaf-litter and create their nests in decaying wood and standing trees.

Photo 2.3.11c | Worker termite

Close-up of a worker and an asexual fungus nodule of an unknown fungus-growing *Macrotermes* termite species.

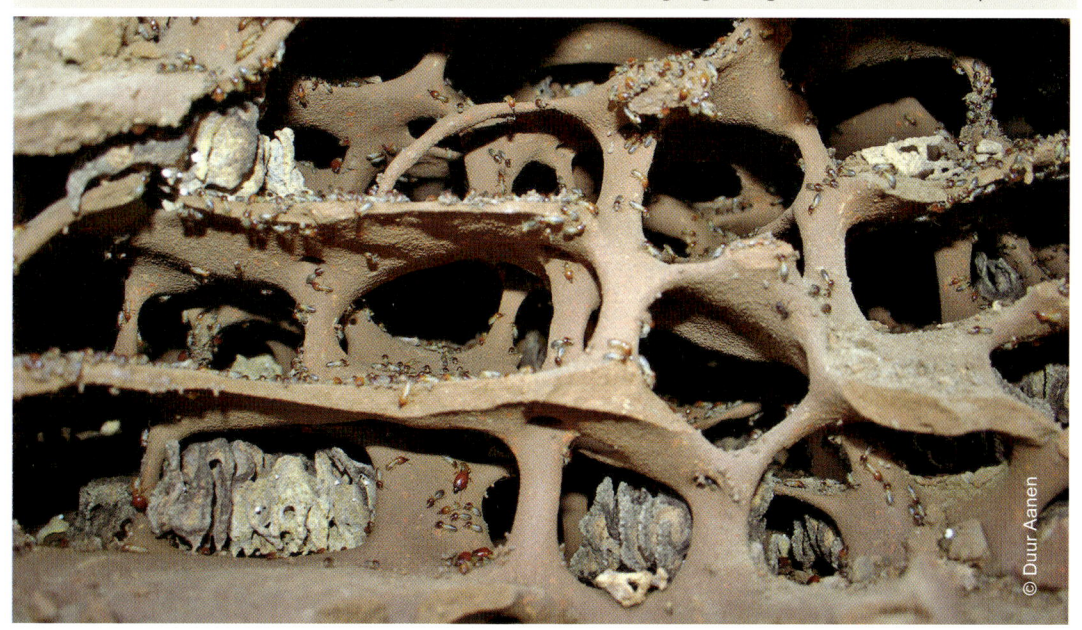

Photo 2.3.11d | Termite nest

Soldiers and a fungus comb inside the nest of the fungus-growing termite *Macrotermes natalensis*.

Photo 2.3.11e | Termite mound

Termite hill from the Northern Ghana in Africa.

Photo 2.3.11f | Termite mound

Termite hill from the Northern Ghana in Africa.

2.3.12 | ANTS

Ants comprise a single insect family (Formicidae), sister to the Apoidea (honeybees), within the insect order Hymenoptera. There are some 20 000 ant species, from which some 15 000 are described. The main characteristics uniting ants as a Family are eusociality, having a petiole and a metapleural gland, and female reproductive castes (queens) that can shed their wings after mating. Ants are a group of thermophilic and predacious insects feeding on other ants, termites, Collembolans, mites and other insects, but there are also ants specialized in honeydew secreted from sap-sucking insects, nectar, seeds and fungi raised from plant leaves. With high species diversity, ants perform myriad ecological interactions in most ecosystems, including plant pollination, seed dispersal and plant protection from herbivores, all with direct or indirect effects on ecosystem processes. As one of the most abundant insect taxa, especially in tropical forests, ants are both primary food for many small vertebrates as well as top predators in soil food webs.

Ants, along with termites and earthworms, have been named "ecosystem engineers" (Del Toro *et al.*, 2012; Prather *et al.*, 2013; Del Toro *et al.*, 2015; King, 2016). Due to their enormous abundance and large species diversity in natural environments (that is, functional diversity including a large spectrum of sizes, ecological traits and ecological strategies), ants heavily modify their surroundings and can influence soil functions in many direct and indirect ways (Prather *et al.*, 2013). For example, Sanabria, Lavelle and Fonte (2014) found that ants can affect (and are effective indicators of) five soil-based ecosystem services in Colombian savanna ecosystems: nutrient provision, water storage and regulation, maintenance of soil structure, climate regulation services, and soil biodiversity and biological activity.

Projected impacts of climate change on ants tend to fall into two categories. First, Gibbs *et al.* (2018) suggest that disturbance (including increases in temperature) has a greater toll on the smallest and largest ants. However, a second study by Kaspari *et al.* (in press) looks at community-level changes of temperate ants after 20 years from a first census, and an average ambient increase of +1 °C in temperature, and suggests that ant communities will first increase in abundance/activity before a general crash. The structure of tropical communities, however, does not appear to have changed in recent times (Donoso, 2017). Being both thermophilic and ectotherms, there is clearly a lack of research both in theoretical and experimental grounds on the response of ants to changes in temperature, and its effects on ecosystem processes (Kaspari, 2019).

Carbon transformations - Through their daily activities, ants can impact the soil portion of the carbon cycle. Del Toro, Ribbons and Ellison (2015) found that, in a warming experiment (+3.5 °C and +5 °C), the activity of the formicine ant species *Formica subsericea* increased soil respiration by a half. Furthermore, it increased decomposition rate of red maple and red oak litter by a third. Similar results were found by Diamond *et al.* (2013) and Stuble *et al.* (2014). Recent studies attest to the interactive effect of nutrient limitation and insect activity on the carbon cycle (Kaspari *et al.*, (2014).

Many ant activities are concentrated around their nests; thus nests (and nest building) can impact surrounding habitats. For example, Clay *et al.* (2014) investigates the decomposition rates and detrital communities below large arboreal nests (refuse piles) of the dolichoderine ant *Azteca trigona*, a common canopy ant in the Neotropics. They show that both artificial and leaf litter substrates decomposed faster below the ant nests than they did 10 m away.

Nutrient cycling - Ants can impact nutrient cycling in at least two ways. First, nest building (in soil, litter, logs, stumps and live vegetation) usually involves comminution of live and dead plant material. Second, ants represent a large percentage of food items for small vertebrates like frogs and lizards. Del Toro *et al.* (2015) found that nitrogen (nitrate) availability decreased moderately (-11 percent to -42 percent) in warming treatments when *Formica subsericea* was present. However, levels of nitrogen availability were decreased during the experiment with respect to natural levels (Del Toro *et al.*, 2012; Prather *et al.*, 2012; Kendrick *et al.*, 2015). Clay *et al.* (2014) also found that relative to leaf litter, refuse piles were enriched in P, K and N, all elements limiting decomposition rates in that forest.

Ants can serve as food for other animals and thus affect the way nutrients transfer from one trophic level to the other. McElroy and Donoso (2019) were the first to compare ants in a frog's (*Rhinella alata*) diet to those available in surrounding soil communities. They found that the frog's diet comprised only the largest ants, compared with those in the surrounding community. These results are troubling since large ants are the most impacted by climate change (Gibbs *et al.*, 2015; Gibbs *et al.*, 2018).

Soil structure - Where they occur, ants are major determinants of soil structure, because many ant species build nests of various sizes and depths. Del Toro *et al.* (2015) show that soil displaced by ants (for example, when digging a nest chamber) double in warming treatments, maybe because ants dig deeper nests to escape from heat. Lavelle *et al.* (2014) show that degraded pastures as opposed to natural savannas along a Colombian Orinoco river basin presents higher soil bulk density, in part due to a reduction in soil engineers such as termites and ants.

Biological regulation - As top predators, ants can exert top-down control on invertebrates lower down in the food web. Tiede *et al.* (2017) studied the responses of ant species richness and incidence across seasons in an altitudinal gradient in south Ecuador. They show ant activity to be responsive to season, but not land degradation.

Photo 2.3.12a | Ant

A micro portrait of an ant (Formicidae) from Southern Ural, Russia.

Photo 2.3.12b | Jack Jumper ant

Jack Jumper ant (*Myrmecia pilosula*) from Tasmania, Australia.

Photo 2.3.12c | Ant

An ant (Formicidae) on the sand dunes of Diamantina National Park, Australia.

Photo 2.3.12d | Leaf-cutter ant

Leaf-cutter ant (*Atta cephalotes*) from the tropical rainforests of South America. On the forest floor, the colony occupies up to 7 metres in diameter.

Photo 2.3.12e | Carpenter ant

A carpenter ant member of the genus *Camponotus*, from Punta Negra, Maldonado department, Uruguay.

Photo 2.3.12f | Ants

An ant species (*Lasius fuliginosus*) commonly associated with tree trunks and dead wood on forest floors, present in a large part of the Palearctic.

Photo 2.3.12g | Carpenter ant

Carpenter ant *Camponotus* sp. from a rocky area of Eastern Morocco.

Photo 2.3.12h | Workers carpenter ants

Workers of the carpenter ant *Camponotus ligniperda* carrying eggs, North-West Bohemia.

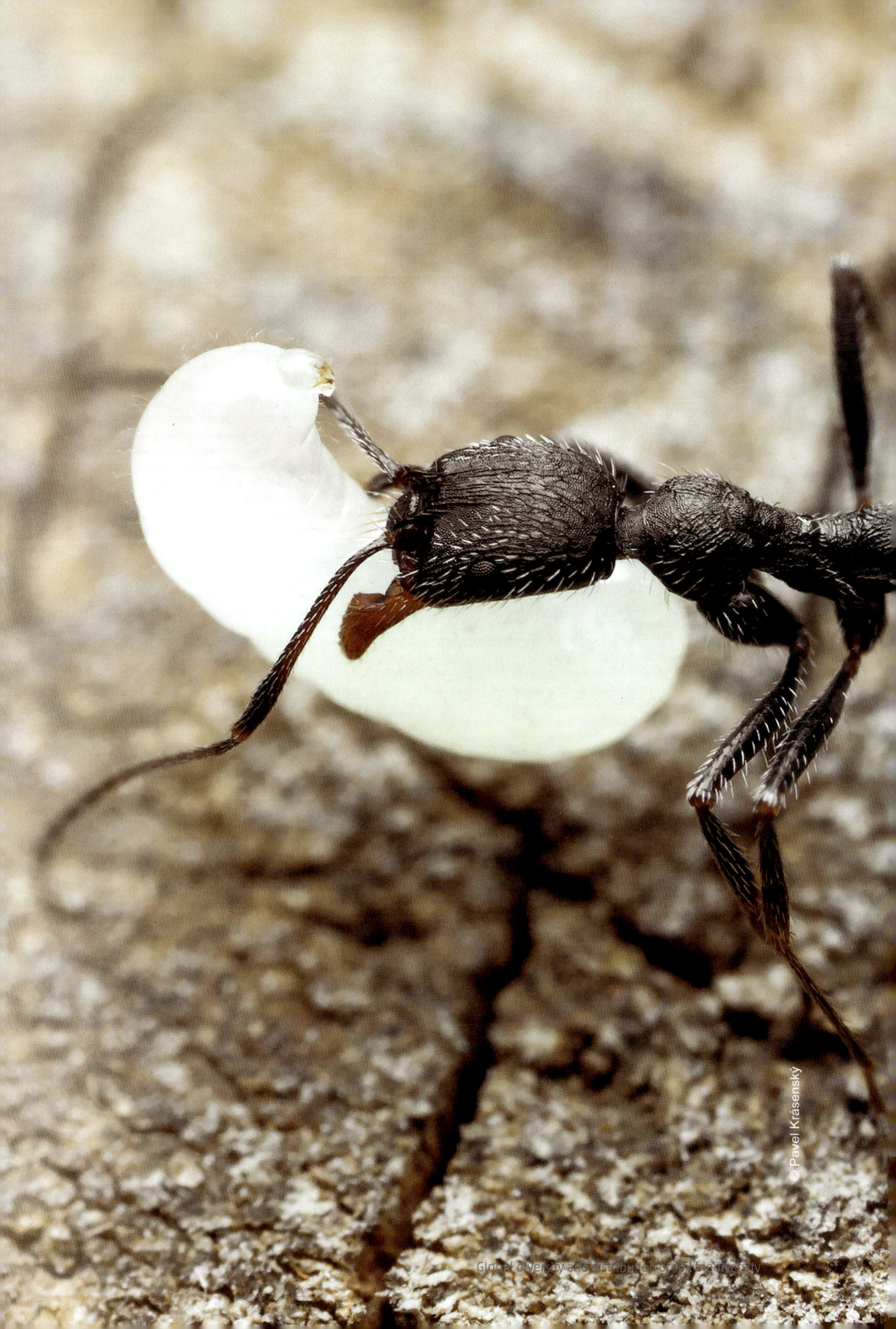

© Pavel Krásenský

MEGAFAUNA

2.3.13 | SOIL VERTEBRATES

Vertebrates are critically important for maintaining a range of soil functions such as enhancement of nutrient cycling, decomposition, carbon sequestration and infiltration. Vertebrates influence soils directly while digging for food or creating shelter, or indirectly, while foraging. A wide range of vertebrates occur at different trophic levels and have different body sizes. European livestock have mixed effects (generally reducing soil function) while fossorial (soil living) vertebrates generally enhance soil function. Vertebrates also influence plant growth and biomass, enhance soil heterogeneity and influence other vertebrates

Vertebrates have substantial impacts on soils. Soil-disturbing vertebrates range in body size from the Common vole, *Microtus agrestis* (< 5 g) to the Superb lyrebird, *Menura superba* (~ 1 kg), to the African Elephant, *Loxodonta africana* (6 000 kg). Soil-disturbing vertebrates span a wide range of trophic groups, from herbivores (Nubian ibex, *Ibex nubiana)* and domestic livestock such as cattle (*Bos taurus*) to primary and secondary consumers (Giant anteater, *Myrmecophaga tridactyla*) to omnivores (Grizzly bear, *Ursus arctos*; Tardiff and Stanford, 1998) and carnivores (American badger, *Taxidea taxus*). The functional consequences of soil disturbance depend on vertebrate body size and on the extent and longevity of the structures or disturbances that they produce (Whitford and Kay, 1999).

Recent reviews of the effects of vertebrates on soils and soil processes have focussed on different vertebrate groups (mammals, rodents, birds), their origins (native versus exotic) disturbance types (foraging pits, hoof marks, resting sites, ejecta mounds, burrow systems) and ecosystem properties (soil chemistry, plant community composition) (Whitford and Kay, 1999; Barrios-Garcia and Ballari, 2008; Root-Bernstein and Ebensperger, 2013; Romero *et al.*, 2015; Platt *et al.*, 2016; Mallen-Cooper *et al.*, 2018). The impacts of vertebrates on soils vary depending on whether they are constructing habitat or foraging (semi-fossorial vertebrates, livestock, large native ungulates). Thus, different animals have different effects on soils and soil processes, as will be discussed below.

Carbon transformation - The capture and retention of organic matter and its decomposition within physical structures are probably the most important processes controlled by soil-disturbing vertebrates. Seeds, organic matter, insect frass, animal faeces and sediment are often trapped within pits and depressions excavated by mammals seeking or caching food or constructing bedding sites. Soil and water also accumulate within these depressions, bringing organic matter into contact with bacteria and fungi.

A combination of greater soil moisture and warmer temperatures results in increased organic matter decomposition (Eldridge *et al.*, 2012). Compared with the surface, where organic matter breaks down through photo-oxidation by ultraviolet light, decomposition in the disturbances created by vertebrates returns carbon (C) and nitrogen (N) to the soil.

Surfaces disturbed by vertebrates support different microbial communities compared with undisturbed surfaces, due to differences in litter type and mass and exposure to the surface. For example, the foraging pits of the Small-beaked echidna (*Tachyglossus aculeatus*) trap large amounts of litter and soil, leading to resource-rich patches that are dominated by bacteria. Copiotrophs, bacteria that are associated with high levels of organic matter, or dung saprobes, fungi that decompose organic matter, thrive in these resource-rich pits (Eldridge *et al.*, 2015). Echidna pits are also associated with lower levels of extracellular enzymes, such as phosphatase and glucosidase, which can affect plant performance. Compared with semi-fossorial animals such as echidnas, disturbance by livestock (sheep and cattle) has variable effects on the soil microbial community composition by altering the soil pH and C levels, either suppressing or enhancing the dominant microbes (Eldridge *et al.*, 2017).

Disturbance initially increases soil respiration by removing subsoil and activating soil microbes. However, as pits and depressions age, the capture of litter and moisture increases different forms of soil C, particularly the more labile forms (James *et al.*, 2009). Soil disturbance and compaction by livestock, however, generally reduces soil C and can have variable effects on respiration depending on the season of grazing (Wang *et al.*, 2017). Analysis of livestock effects on soil C across China demonstrate that once livestock are removed from grasslands, soil C increases markedly, indicating that soil C declines under most livestock management practices (Deng *et al.*, 2017). The activity of kangaroos (*Macropus* spp.) while constructing shelter can lead to increases in soil C (Eldridge and Rath, 1996), and decomposing carcasses can increase soil C and N (Wilson and Read, 2003).

Nutrient cycling - The impacts of vertebrates on soil nutrients depends on whether they are semi-fossorial or grazing and browsing ungulates. For example, the foraging pits of semi-fossorial vertebrates from arid central Australia contain higher levels of total and mineralizable N than surface soils, but those of the European rabbit (*Oryctolagus cuniculus*) and Sand goanna (*Varanus gouldii*) had more nutrients than those of the Bilby (*Macrotis lagotis*) and Bettong (*Bettongia lesueur*; James *et al.*, 2009). Globally, seabirds can also generate large pools of total N (591 Gg N/yr) and phosphorus (99 Gg P/yr), of which up to one-fifth is soluble and available for plants (Otero *et al.*, 2018). Trampling and grazing-induced disturbance by sheep and cattle has been shown to influence soil N availability, directly through the addition of dung, and indirectly by altering plant community composition (Liu *et al.*, 2018).

Soil structure - Digging by vertebrates while foraging or constructing habitat, travelling, or creating dust baths, removes vegetation and surface crusts, and predisposes the surface to erosion by wind and water. Digging also breaks larger pieces of soils (aggregates) into smaller micro-aggregates that are more easily eroded by water or wind. Material removed from the surface often smothers existing vegetation, bringing resource-poor soil, with lower levels of C and N to the surface where it is often lost through erosion (Whitford and Kay, 1999; Platt *et al.*, 2016). Deposition of disturbed soil on existing vegetation initiates a process of decomposition that may increase the C:N ratio of the soil, leading to nutrient immobilization. Digging also homogenizes surface soils, by breaking up the resource-rich upper layer. Compaction by herbivores can also reduce infiltration by destruction soil macropores (Rietkerk *et al.*, 2000).

Biological regulation - Disturbances by vertebrates can influence plants directly by trapping seeds (James *et al.*, 2009) or indirectly by altering soil structure. Digging by pigs (*Sus scrofa*) in seasonally damp meadows, for example, increases soil bulk density and alters species composition (Wang *et al.*, 2018). Complex flow-on effects also affect both the vertebrates and plants. In the Negev Desert for example, Indian crested porcupines (*Hystrix indica*) disturb soils when unearthing tulip (*Tulipa systola*) bulbs. Although they consume some bulbs, the pit that they create captures water and increases the survival and seed set of the remaining bulbs (Gutterman, 1987). Foraging pits may provide refugia for plants requiring higher levels of moisture as rainfall declines with climate change (Whitford and Kay, 1999).

Irrespective of vertebrate type, their disturbances lead to increasing spatial heterogeneity by increasing the diversity of soil microhabitats, resulting in a greater diversity of plants and animals (Stein, Gerstner and Kreft, 2014; Platt *et al.*, 2016). The trapping of seeds, sediment and runoff water in echidna pits leads to the development of resource-rich patches and therefore increased small-scale heterogeneity in semi-woodlands (Eldridge, 2011).

Vertebrate disturbance that leads to altered soil physical and chemical properties can alter habitat for other organisms. For example, surfaces disturbed by prairie dogs (*Cynomys ludovicianus*) influence foraging by ants and Tenebrionid beetles and provide nesting sites for burrowing owls (*Athene cunicularia*; Ray *et al.*, 2016). Disturbances by livestock can have variable effects on the distribution, abundance and diversity of soil-disturbing organisms (Abba *et al.*, 2015).

Knowledge gaps - Several questions of pressing importance remain regarding soil vertebrates in our understanding of soil processes: whether the impacts of vertebrates on soils, particularly the mass of soil moved, scale according to their body size; and how we scale up vertebrate effects on soil function from the patch to the landscape scale (Mallen-Cooper *et al.*, 2008). We also do not know what the long-term effects of reintroduced vertebrate soil animals have on soil and ecosystem functions in human-altered environments (Coggan, 2018), or what the impacts of exotic soil-disturbing vertebrates are on soil functions.

Photo 2.3.13a | Pocket gopher

Botta's pocket gopher (*Thomomys bottae*) in Arizona.

Photo 2.3.13b | Mole

A common European mole (*Talpa europaea*) from Western Europe.

Photo 2.3.13c | Gopher

A gopher (Rodentia) from the Elbrus region, Russia.

Photo 2.3.13d | Prairie dog

A prairie dog.

2.4 | SPATIAL AND TEMPORAL DISTRIBUTION OF SOIL BIODIVERSITY

2.4.1 | SPATIAL PATTERNS OF SOIL ORGANISMS

The majority of soil biodiversity remains undescribed, and information on the functional abilities and ecology of most soil taxa is far from being complete. Yet in the last decade, our knowledge of the structure of soil biodiversity and its contribution to terrestrial ecosystems has started to expand (Tedersoo *et al.*, 2014; Delgado-Baquerizo *et al.*, 2018; van den Hoogen *et al.*, 2019; Phillips *et al.*, 2019). Studies on the relationships between microbial community composition and abundance, and soil biotic and abiotic variables, have highlighted consistent patterns in microbe-environment associations. Similar patterns related to the significance of biodiversity-ecosystem function relationships have also been shown across multiple soil taxa and community assemblages (Soliveres *et al.*, 2016; Schuldt *et al.*, 2018).

Soils are inherently complex systems that include a range of scales, from microsites (for example, in the contiguous space of a single root tip) to continents (Fierer 2017; Thakur *et al.*, 2019). This spatial heterogeneity may vary from microbes (with higher fine-scale spatial variability) to soil fauna, but it creates the substrate to sustain the highest terrestrial biodiversity pool globally (Eisenhauer and Guerra, 2019). At the local scale (that is, plot to landscape), soil pH, organic matter content, nutrient availability, moisture and biotic interactions (vegetation in particular) have been reported to influence the structure of bacterial and archaeal communities (Lauber *et al.*, 2013; Prévost-Bouré *et al.*, 2011; Prescott and SJ, 2013; Norman and Barrett, 2016; Dassen *et al.*, 2017; Stempfhuber *et al.*, 2015). Similarly, soil fungal communities are affected by plant host, soil pH, cation concentrations, soil nitrogen and deposition of nitrogen and potassium (Glassman, Wang and Bruns 2017; Zhang *et al.*, 2018; Patterson *et al.*, 2019; Camenzind *et al.*, 2016; Lladó, López-Mondéjar and Baldrian 2018).

Similar drivers have been observed for the small-scale distribution of soil fauna, but at larger spatial scales climatic conditions and habitat cover may be particularly relevant (van den Hoogen *et al.*, 2019; Phillips *et al.*, 2019). Another consistent pattern emerging from recent literature is the considerable variability in relative abundances of various taxa found in soil biodiversity, even in similar pedogenic conditions (Fierer 2017). Such variability is particularly evident at very fine scales (that is, centimetres to metres; O'Brien *et al.*, 2016). This is chiefly related to the strong spatial variation of soil resources, which can generate hotspots of biological activity and abundance in soil, for example in and on plant debris, or on and around plant roots (Kuzyakov, 2015; Baldrian, 2014; Baldrian *et al.*, 2010). Each of these habitats is characterized by specific properties and consequently supports a unique soil community. Upscaling microscopic scale processes carried out by soil biota to ecosystem and global scales remains challenging, but combined efforts of improved analytical approaches and modelling can improve accuracy.

As a result of such efforts, recent studies show that at the global scale, the variability in soil properties, combined with climatic factors, leads to large heterogeneity in soil fauna communities (van den Hoogen *et al.*, 2019; Phillips *et al.*, 2019). In fact, soil moisture and soil organic carbon are the main drivers of many soil microbial functions (for example, basal respiration) as well as – together with climate – of the distribution pattern of larger soil organisms, such as nematodes (van den Hoogen *et al.*, 2019) and earthworms (Phillips *et al.*, 2019). This results in local richness and abundance patterns being inversely correlated to those observed for plants and above-ground animals – with local soil bacterial, fungal and earthworm richness peaking in temperate to boreal regions, where soils with neutral pH and relatively high carbon to nitrogen ratios are found (Tedersoo *et al.*, 2014; Bahram *et al.*, 2018; Phillips *et al.*, 2019). However, specific functional groups with strong climatic and biogeographic preferences might contrast with these overall patterns. For example, the biogeographic distribution patterns of mycorrhizal fungi, which grow in close symbiosis with vascular plants, are mostly driven by their host plants. Consequently, the richness of ectomycorrhizal fungi peaks in high-latitude forests, where the greatest proportion of ectomycorrhizal symbiotic trees is found, while arbuscular mycorrhizal fungal diversity peaks in the tropics (Steidinger *et al.*, 2019; Tedersoo *et al.*, 2014). Despite the increasing wealth of knowledge of soil biodiversity and the main environmental (edaphic and climatic) variables affecting structure and diversity of soil communities at small scales, identifying which of these biotic or abiotic factors (as well as their interactions) dominate in driving soil biodiversity distribution remains a challenge, particularly so in a rapidly changing world.

Box 2.4.1 | Pedodiversity

Soils are classified according to their features and properties – referred to as soil types, pedologic taxonomies, or pedotaxa. Soil inventories and maps inform us of the properties of soil landscapes and their diversity and variability, and are utilized to plan agricultural activities and land management. The higher the number of edafotaxa, the greater the diversity of a given soil cover and the capacity to respond to environmental factors and impacts. This type of study is called pedodiversity analysis (Ibáñez *et al.*, 1995).

The quantification of pedodiversity is not limited to enumerating the pedotaxa present in a given geographical space, requiring deep statistical analysis. Pedodiversity experts have made use of the same mathematical tools as biodiversity experts. There are different ways of measuring diversity, the following being the most basic and relevant:

Indices of richness: number of taxa (for example, biological species, communities, pedotaxa, soilscapes) known to occur in a defined sampling area.

Indices based on proportional abundance of each taxon. Not only the number but also their relative abundance (for example, the relative area occupied by each pedotaxon) is taken into account.

Indices based on sets of parameters and models describing the distribution of abundance of categories of organisms in a given ecosystem or soilscape.

Indices based on distribution models addressing how biological diversity increases according to increase in the size of the studied area (richness-area interrelationships).

Of the 300 to 400 publications on pedodiversity carried out to date, it appears that the spatial and temporal patterns of the distribution of soil types in the landscape are strikingly similar to those of above-ground biodiversity, creating the following macroecological patterns: (i) species (taxa)-area relationships; (ii) local-regional richness relationships; (iii) latitudinal gradients in species richness; (iv) altitudinal gradients in species richness; (v) species-range size relationships; (vi) nestedness of species occurrence; (vii); abundance-range size distributions; (vii) species–abundance distributions (see Ibáñez et al., 2013 and chapters therein). However, it is still unknown whether macroecological patterns of above-ground and below-ground biodiversity are also comparably similar, as the existing publications are too scarce to reach scientifically sound conclusions (Decäens, 2010; Caruso et al., 2019; van den Hoogen, 2019).

Much less studied has been the relationship between pedodiversity and below-ground biodiversity (Wardle et al., 2004). The same can be said of the relationship between the assemblages of the different soil horizons in which a pedotaxa can be stratified (genetic pedodiversity) and the communities that host each of them (Doblas-Miranda et al., 2010). This is one of the alarming omissions in our understanding of soil biodiversity that should be addressed in the future. However, several studies have shown that different soil types (Garbeva et al., 2004; Fierer and Jackson, 2006; Gagelidze et al., 2018) as well as different soil horizons (Ekelund et al., 2001; Fierer, 2003; Rosling et al., 2003; Blackwood and Buyer, 2004; Hansel et al., 2008; Doblas-Miranda et al., 2010; Eilers et al., 2012) house different assemblages of soil organisms, which demonstrates that soil horizons should be considered in defining the variety in the habitats for soil organisms. It has been demonstrated that the geographical space with a scarce pedodiversity is also poor in above-ground biodiversity (species richness), (see Ibáñez, 2018; Ibáñez and Bockheim, 2013; and chapters therein). Therefore, if we want to preserve biodiversity, we must also preserve pedodiversity (Ibáñez et al., 2012). For this reason, various soil scientists have proposed the creation of networks of soil reserves and/or the inclusion of pedodiversity in other programs related to the preservation of nature (Lobo and Ibáñez, 2003; Gerasimova et al., 2014; Costantini and L'Abate, 2009, 2016).

The huge degradation of the biosphere and geosphere has profoundly altered the pedosphere. Several studies show a loss of pedodiversity in certain countries and geographical areas, and the transformation of some pedotaxa in others (Amundson et al., 2003; Dobrovolsky and Nikitin, 2009; Lo Papa et al., 2011). However, human actions have transformed the pedosphere to such an extent that new anthropogenic taxa are included in soil taxonomies, such as the WRB Technosols (FAO). Therefore, some publications show the loss of certain types of natural soils and their replacement of others of anthropogenic origin (Xuelei et al., 2003, 2007).

2.4.2 | TEMPORAL PATTERNS OF SOIL ORGANISMS

Different communities of soil organisms differ in lifespans and ecologies, as well as in the various ways in which they interact with each other and with their environment. This diversity makes soil an extremely dynamic ecosystem. While globally the high degree of local heterogeneity in soil properties and the immobile nature of many soil organisms result in greater spatial variability than temporal variability, at the local scale the temporal variation is expected to be higher, particularly for microbial communities (Fierer, 2017). Still, climatic variation throughout the year, such as seasonality, drought, freezing and distribution of rainfall, significantly affect the soil properties that structure soil communities (Carini *et al.*, 2018). Due to experimental and observational limitations, the temporal factor in structuring soil communities is relatively poorly understood (Fierer, 2017; Eisenhauer *et al.*, 2018). To explore the temporal variability in soil communities in greater detail, repeated sampling is required across different temporal resolutions and, ideally, at different spatial scales (Eisenhauer *et al.*, 2018; Guerra *et al.*, 2019).

In the absence of disturbances, soil dynamics have been most commonly studied throughout seasons. Across biomes, seasonality is tied to changes in soil moisture, leaf litter input, plant growth and freeze-thaw cycles. Accordingly, seasonal changes have been linked to the temporal distribution patterns of soil organisms, although different aspects of seasonality affect each group of soil biota. For example, the bacteria involved in the nitrogen cycle exhibit significant changes in their community structure across the seasons (Lauber *et al.*, 2013), being able to undergo consistent seasonal changes in abundance, but with a stable level of ecosystem function provisioning and related ecosystem services (Regan *et al.*, 2017). While the diversity and distribution of protists remains understudied, one recent study show that seasonal changes in the abundance of protists from the Cercozoa group depends on their trophic strategy, and suggests that predator-prey dynamics act in concert with the seasonality of bacteria to drive the abundance of bacterivore protists (Fiore-Donno *et al.*, 2019). The seasonal shifts in fungal communities seem to be intricately connected to soil moisture (Zhao *et al.*, 2017). In a coniferous forest, the activity of fungi was highest in the summer and lowest in the winter, and these differences were stronger in soil than in litter (Žifčáková *et al.*, 2016). Seasonal variations have also been found for soil invertebrates. For example, seasonal changes in the abundance and composition of microarthropod communities are reported across a range of montane environments (Wu *et al.*, 2014), while another shows that seasonality partially explains nematode community dynamics (S. Zhang *et al.*, 2019).

The strong ecological relationship between above- and below-ground biota drives the community composition of both compartments, giving way to complex top-down and bottom-up feedbacks (Wardle *et al.*, 2004). In fact, the temporal variability at the microsite level (for example in root exudation profiles) is usually very high (due to root growth) and thus resource quality and quantity of soil microsites as well as species interactions are likely to be highly dynamic (Badri and Vivanco, 2009; York *et al.*, 2016; Eisenhauer *et al.*, 2018). Conversely, the temporal variability of resources and the strength of species interactions at the local and landscape scales, although still variable, will occur over longer time scales (Thakur *et al.*, 2019). These relations have also been shown for interactions between below-ground and above-ground species, with increases in plant diversity resulting in higher fungal and bacterial biomass (Eisenhauer *et al.*, 2017). At the same time, soil biota as a whole and mycorrhizae in particular have been shown to influence the diversity of Mediterranean-type shrublands and temperate forests respectively (Bennett *et al.*, 2017; Teste *et al.*, 2017). The effects of these interactions persist in time, causing long-term shifts in composition for both above- and below-ground biota. In one case, sowing a grassland with a different soil biota and seeds resulted in a shift in both the nematode and plant communities, which was still detectable 20 years after the treatment (Wubs *et al.*, 2019).

Disturbance experiments in soil have employed a wide range of perturbations, including drought, flood, freeze-thaw cycles, temperature shifts, fumigation and contamination (Griffiths and Philippot 2012), and importantly, sampling the soil biota at time scales which are often in line with the temporal dynamics of the communities of interest (hours to a year). For example, a study of bacterial and fungal responses to wetting after a prolonged drought shows that the active microbial community shifted drastically within 2 hours after wetting consistent with the resulting CO_2 pulse (Barnard, Osborne and Firestone 2013). Drought has been shown to increase the ratio of fungi to bacteria rapidly (Mariotte *et al.*, 2015). Another branch of disturbance experiments has examined the effect of disturbances on the soil biota across much larger time scales, more in line with macro-ecological succession (that is, from 5 years to 100 years). While these studies often find shifts in soil communities with each successional stage, these are often non-linear over the entirety of the experiment, and tend to reflect the adaptation to changing environmental parameters, such as soil organic matter or soil pH (Roy-Bolduc *et al.*, 2016; Zhang *et al.*, 2018).

Human activities can impose selective pressure(s) and thus change the long-term stability of soil community composition and functioning, by causing a range of press and pulse disturbances. The current distributions of soil bacteria have stronger correlations with climate from ~50 years ago than with current climate (Ladau *et al.*, 2018). This lag is likely associated with the time it takes for soil properties to adjust to changes in climate. Further, experimental warming is shown to lead to increasingly divergent succession of the soil microbial communities, with possibly higher impacts on fungi than on bacteria (Guo *et al.*, 2018). Long-term continuous monitoring programs are needed, with explicit consideration of ecosystem types, climatic zones and management practices to distinguish temporal variations from the actual impact of environmental changes and a better understanding for long-term community adaptation (Guerra *et al.*, 2019).

2.4.3 | BELOW-GROUND DISTRIBUTION DOES NOT FOLLOW ABOVE-GROUND PATTERNS

Recent work shows that the global distribution of soil biodiversity does not follow that of commonly observed above-ground biodiversity patterns (Tedersoo *et al.*, 2014; Delgado-Baquerizo *et al.*, 2018; Cameron *et al.*, 2019; Crowther *et al.*, 2019; Phillips *et al.*, 2019). In fact, major mismatches have been found between above-ground and soil biodiversity at the global scale for over 27 percent of Earth's terrestrial surface (Cameron *et al.*, 2019) and during ecosystem succession (Delgado-Baquerizo *et al.*, 2019). Temperate broadleaf and mixed forests (Cameron *et al.*, 2019) as well as tropical forests (Delgado-Baquerizo and Eldridge, 2019) have shown to harbour high above-ground biodiversity but low soil biodiversity, whereas the boreal and tundra biomes have shown intermediate to high soil biodiversity but low above-ground biodiversity. While more data on soil biodiversity are needed, both to cover geographic gaps and to include additional taxa, results by Cameron *et al.* (2019) and others (Ciobanu *et al.*, 2019) suggest that protecting above-ground biodiversity may not sufficiently reduce threats to soil biodiversity. Moreover, there is growing evidence that different soil organisms have different environmental preferences (Tedersoo *et al.*, 2014; Fierer *et al.*, 2017; George *et al.*, 2019; Delgado-Baquerizo *et al.*, 2018) suggesting that further investigations are needed in order to generate adequate policies and management practices to protect soil biodiversity. Taken together, these recent insights suggest that soil biodiversity should be included in policy agendas and conservation actions, by adapting management practices to sustain soil biodiversity and considering soil biodiversity when designing protected areas (Nielsen *et al.*, 2015; Cameron *et al.*, 2019). Assuming that protecting above-ground biodiversity will be sufficient to conserve below-ground biodiversity is misleading and poses additional threats to soil life.

Three decades of above-ground biodiversity-ecosystem functioning research have shown that we need to understand the environmental and biotic context-dependency of biodiversity-ecosystem functioning relationships (Isbell *et al.*, 2017b). For instance, recent theories and concepts predict that biodiversity-ecosystem functioning relationships do vary across gradients of environmental stress (Baert *et al.*, 2018) and with spatial and temporal scale (Isbell *et al.*, 2017b). Moreover, meta-analyses of experimental results reveal context sensitivities of biodiversity-ecosystem functioning relationships to soil conditions and across time (Guerrero-Ramírez *et al.*, 2017) and across environmental stress gradients (Baert *et al.*, 2018). For soil biodiversity-ecosystem functioning relationships, potential context-dependencies are likely, but have rarely been investigated (Tiunov and Scheu, 2005; Eisenhauer *et al.*, 2012b; Nielsen, Wall and Six, 2015; Delgado-Baquerizo *et al.*, 2016; Trivedi *et al.*, 2016). Moreover, experimental evidence linking soil microbial biodiversity to ecosystem functioning is still weak (Philippot *et al.*, 2013; Trivedi *et al.*, 2019), as microorganisms are traditionally projected to be highly functionally redundant in soils. On the contrary, most recent scientific evidence suggests that soil biodiversity contributes directly to significant ecosystem functions and services (Bardgett and van der Putten 2014; Wagg *et al.*, 2014; Nielsen *et al.*, 2015; Orgiazzi *et al.*, 2016; Delgado-Baquerizo *et al.*, 2016; 2017). Even so, the strength of soil biodiversity effects may depend on specific traits that are lost from the system through environmental change, and further investigations to support this are needed. For instance, one of the most important facets of soil biodiversity has been shown to be the functional dissimilarity among soil organisms, spanning large gradients from microorganisms to macrofauna (Bradford *et al.*, 2002; Heemsbergen *et al.*, 2004; Eisenhauer *et al.*, 2010). Thus, environmental changes that reduce this functional dissimilarity, for example by detrimentally affecting soil organisms (tillage, land degradation), are likely to compromise a plethora of different soil-mediated ecosystem functions.

2.4.4 | INTERACTIONS AMONG SOIL ORGANISMS

The earthworm drilosphere (that is, the part of the soil influenced by earhworm secretions, burrowing and castings) is generally acknowledged as being a soil hotspot for microbial activity (Figure 2.4.4.1), and a recent review (Medina-Sauza et al., 2019) shows that specific bacterial groups consistently increase in soils where earthworms are present, regardless of the earthworm functional group (Yakushev et al., 2009; Chang et al., 2016). In contrast, the effect of earthworms on richness and diversity of the microbial community is not consistent and can be neutral (for endogeic see de Menezes et al., 2018), negative (Furlong et al., 2002) or positive (epi-anecic burrows, Hoeffner et al., 2018). These microbial responses depend on the substrate, the earthworm functional group and the microbial taxa: a positive effect on bacterial diversity is not always associated with an increase of fungal diversity (Hoeffner et al., 2018). The microbial response also depends on the microhabitat, with contrasting results among burrow, casts, middens and guts (Medina-Sauza et al., 2019). Furthermore, earthworms modify the abundance of specific taxa within the microbial community (Lavelle et al., 2016). In turn, all these modifications of microbial community (abundance, diversity, activity) affect nutrient cycling through an increase of nutrient mineralization in the soil and the modification of the microbial functional genes involved in nutrient cycling (Lavelle et al., 2016; Medina-Sauza et al., 2019).

The effect of earthworms on soil fauna is very complex and is due to several processes, including habitat modifications. The effect of epigeic species depends on the earthworm density going from positive to negative with increasing density due to substrate competition. Endogeic earthworms mainly negatively affect microarthropods, also due to competition for food resources (litter material rich in nitrogen), with Oribatida being more detrimentally affected than Collembola. Impacts of anecic earthworm species do not differ depending on density but rather on scale: they are neutral on the habitat scale and positive on the microhabitat scale, the latter mainly attributed to the formation of stable microhabitats (burrows/middens) which are rich in nutrients and microorganisms (Salmon and Ponge, 2001) and providing more pathways for Collembola to explore deeper layers and avoid predation (Salmon et al., 2005). All these impacts of earthworm on microarthropods have consequences for soil functioning because microarthropods impact nutrient cycling and decomposition processes (Figure 2.4.4.1), and are important in linking components of the soil food web because they feed on soil microorganisms and soil fauna (protozoa, nematodes, Collembola and other soil microarthropods) and are also predated by macroarthropods (Coleman et al., 2004). If earthworms impact on Collembola and Mites, the reverse is also true, and the presence of Collembola could lead to a loss of earthworm body mass (Scheu et al., 1999).

Effects of ants and termites on soil microorganisms and fauna are less well documented, but it is reported that within the biogenic structures of termites and ants (nest, mounds) there is increased microbial activity and release of mineral nutrients such as ammonium and nitrate (Folgarait, 1998; Jouquet et al., 2006), depending on the ant species (Dauber and Wolters, 2000). Ants modify the habitat structure and change the nutrient content of the soil (organic matter, P, N and K) because of food storage, aphid cultivation and accumulation of faeces and ant remains (Lavelle *et al.* 1997). This leads to positive effects for soil biota, such as bacteria, nematodes, mites and *Collembola*ns, which reach higher densities in harvester ant nests (Lavelle and Spain, 2001; Boulton and Amberman 2006). This positive effect is also observed at different trophic levels of the above-ground macro-arthropod community (Sanders and van Veen, 2011).

The continuous modification of the top layer of the soil, including its physical and chemical characteristics as well as the modification and destruction of microhabitats (Figure 2.4.4.1), leads to a one-sided selection of soil biota. If thresholds (tipping points) of resilience are exceeded due to high-impact disturbance, a regime shift occurs towards an impoverished status, which is at least in the short term irreversible (Vogel *et al.*, 2018). Not only soil disturbance, but also plant species richness loss, leads to below-ground extinction cascades, which, in turn, lead to pronounced deterioration of fundamental below-ground functions (Weisser *et al.*, 2017).

Figure 2.4.4.1 (next pages) | **Interactions between organisms through the soil food web**

Interactions between organisms through the soil food web in terrestrial ecosystems.

Through photosynthesis, plants transform light energy into chemical energy, allowing them to produce their own organic compounds -carbohydrates- to live. Plant and microbial activity account for 90 % of the energy transfer in terrestrial ecosystems.

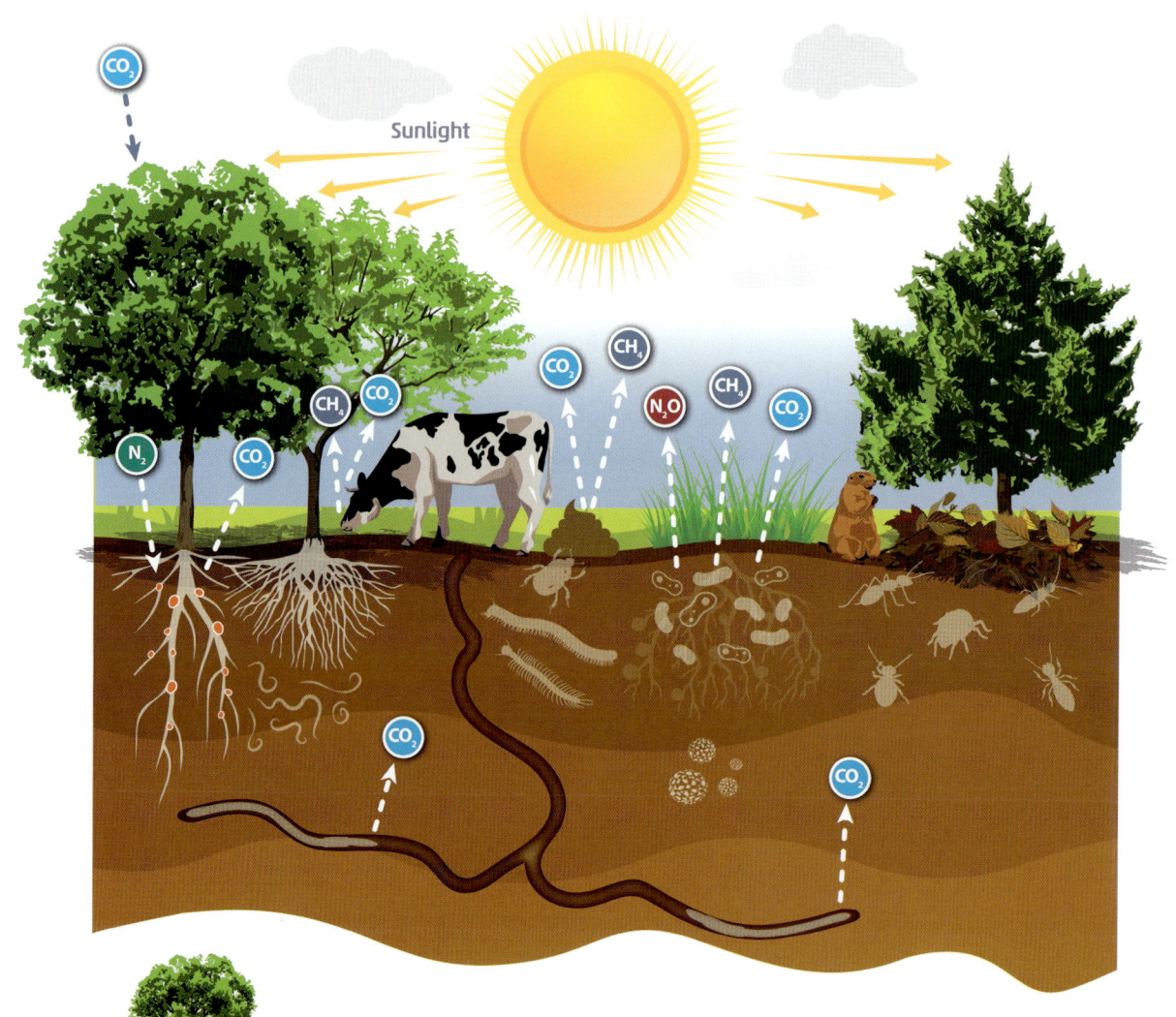

Most carbon and nutrients are transferred to the soil via litterfall, root exudation and decaying organisms. Bacteria, fungi and archaea that inhabit the root microbiome or are root symbionts benefit from this input. Plants return about half of this carbon to the atmosphere through respiration in the form of carbon dioxide (CO_2).

Leguminous plants receive nitrogen in assimilable forms, since symbiont N-fixing bacteria such as Rhizobium form nodules in their roots transforming dinitrogen (N_2) into ammonium (NH_4^+). Nitrifying microorganisms such as Archaea, transform ammonium into nitrate that is readily available for plants.

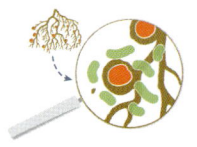

Different microorganisms return carbon and nitrogen to the atmosphere, making the continuity of life possible. Most carbon entering the soil is lost through microbial respiration in the form of CO_2. A broad array of soil denitrifying bacteria produce progressively reduced products ending in N_2, which returns to the atmosphere. In the absence of oxygen and on organic substrates, some soil bacteria produce methane (CH_4).

Viral infections produce lysis (bursting) of microbial cells releasing carbon and nutrients into the soil solution as potential resources for other microbes, fungi and nematodes.

Nematodes fill a number of trophic roles in soils, including bacterivorous and/or fungivorous grazers, predators of other nematodes and smaller animals, plant-feeders, omnivores, and parasites of both animals, and plants.

Most collembola are consumers of fungal hyphae, micro-algae and litter, but can also be the prey of mites, pseudoscorpions and coleoptera. They accelerate litter decomposition and enhance nutrient cycling, by excreting nitrogenous waste as ammonia or uric acid, thus increasing nitrate availability for plant growth.

Mites (Acari) are an integral part of the soil food web, and span several trophic levels including fungivores, bacterivores, detritivores and top predators. Mites accelerate litter decomposition and enhance nutrient cycling and mineralization rates, especially in the case of nitrogen.

As ecosystem engineers, earthworms are a major driver of soil ecosystem services. The activities of burrowing, soil ingestion and cast egestion result in a deep modification of soil structure and organic matter dynamics, affecting soil aggregation and porosity with significant positive impacts in the flow of water, nutrients and gases. Additionally, earthworms can regulate plant pathogens.

Saprophagous macrofauna such as isopods, millipedes and many insects transform litter and contribute to decomposition by consuming decaying matter and by transporting it throughout the soil profile. Predator organisms like arachnids, carabid beetles and centipedes can also regulate the activity of lower trophic levels, and often aid in biological control of pests.

Many termites improve physicochemical properties of soil by casting the ground into mounds, by concentrating chemical fertility and by building complex subterranean galleries and nests, thus enhancing soil fertility, aeration, drainage and root penetration.

Ants impact nutrient cycling by building nests in soil, litter, logs, or stumps that usually involve the crushing of living and dead plant material. They also represent a food source for small vertebrates like frogs and lizards, affecting the nutrient transfer from one trophic level to the other.

Digging activities of some vertebrates can be harmful since they increase the vulnerability of topsoil to wind and water erosion and could produce a breakdown of aggregates, a disturbance that makes them more easily erodible. Excavated material often smothers existing vegetation, bringing resource-poor soil, with lower levels of carbon and nitrogen to the surface where it may be lost through erosion.

Animal wastes and other organic residues can be an important source of organic carbon and nutrients in some ecosystems. Soil bacteria and fungi decompose manure, releasing up to 75% of this carbon to the atmosphere, and can release up to 85% of the carbon by decomposing other residues.

Symbiotic relations

Rhizobium

Mycorrhizae

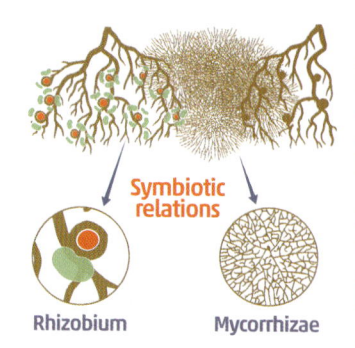

The rhizosphere is the narrow region of soil directly influenced by root secretions/exudates and associated soil microorganisms. This is also known as the root microbiome. This dynamic zone is composed of different communities of bacteria, archaea and fungi, which are the primary components of the soil food web and play a key role in the functioning, balance and stability of the soil ecosystem.

Leguminous plants receive nitrogen –a limiting nutrient in many soils-, in the form of ammonia, thanks to rhizobium, an N_2-fixing bacteria that forms nodules in the plants' roots. In return, rhizobium receives nutrients and habitat from the roots.

Mycorrhizae (symbiont fungi in roots) play a key role in providing ecosystem services such as soil fertility, soil formation and maintenance, nutrient cycling and improving plant root exploration of the soil.

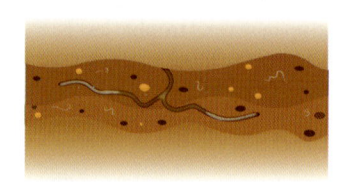

The drilosphere is the soil volume modified by earthworms, made up of burrows, casts, and middens and the earthworms themselves, that represents an important functional domain or hot spot in soil.

CO_2 — Carbon dioxide is assimilated from the atmosphere by autotrophic organisms such as plants and cyanobacteria via photosynthesis. It is then returned into the atmosphere via the respiration of living aerobic organisms.

CH_4 — Methane is produced by bacteria and archaea in the absence of oxygen (O_2). These methanogenic microorganisms live in freshwater sediments in the interfaces of organic-rich anaerobic sites, such as wetlands. Another portion live in the rumen of cattle and the digestive tract of termites, allowing them to degrade and assimilate cellulose. Those microorganisms generate methane as a metabolic product which is then released into the atmosphere. Metanotrophic bacteria are another type of soil microorganisms living in well-aerated forest soils that feed on methane, therefore these soils can act as methane sinks.

N_2 — Dinitrogen or molecular nitrogen accounts for 79% of the earth's atmosphere. However, before nitrogen can be used by plants it must be first transformed into a soluble form. Natural nitrogen fixation can occur atmospherically, by lightning, or be achieved biologically, by certain microorganisms, capable of converting gaseous nitrogen (N_2) into ammonia (NH_3). These include some algae, bacteria and archaea which may or may not be associated with plant roots.

N_2O — Gaseous forms of nitrogen oxides are produced by a wide array of denitrifier microorganisms, most of them facultative and heterotrophic anaerobes. Although denitrification is a natural process of the nitrogen cycle, overuse of fertilizers in agricultural soils enhance the production of N_2O, a long-lived greenhouse gas, which significantly contributes to global warming.

2.5 | NOVEL TECHNOLOGIES – FROM INDICATORS TO MONITORING

Studies have identified taxonomic groups that may serve as potential indicators to assess the sustainability of agricultural soil management and to monitor trends in soil condition and functions over time (Paula *et al.*, 2014; Kaiser *et al.*, 2016; Trivedi *et al.*, 2016). For example, the relative abundance of *Acidobacteria* has been shown to be higher in natural systems, while the abundance of Chloroflexi was higher in agricultural systems. The abundance of these bacterial phyla is easily measured, since well-established molecular protocols for quantification are available (Fierer *et al.*, 2012); they are sensitive to soil management actions and are integrative – that is, they provide adequate coverage across a relatively wide range of ecological variables such as soil types, climate and crop sequence. In the future, development of appropriate tests for simple monitoring of proposed candidate biological indicators that can be integrated into a minimum data set for soil microorganisms will facilitate measuring the short- and long-term impacts of agriculture on soil health. However, significant background work including identifying the context of monitoring (environmental sustainability versus productivity), selection of parameters for biological indicators (positive or negative) needs to be tested and validated before an efficient indicator of primary productivity can be developed for monitoring purpose. For soil fauna, such well-established and informative bioindicators already exist, with a strong focus on earthworms, Collembolans and nematodes.

Although there is an increased awareness that soil biodiversity is an important component of overall soil health and productivity, the integration of this knowledge into ecosystem management is lacking (Lemanceau *et al.,* 2015). Meaningful metrics by which soil biodiversity can be measured and assessed must be decided upon. Unlike standard soil testing that can assess a nutrient deficiency, the role of soil biodiversity in plant production is less straightforward. A venue where experts could come together from different regions of the world and agree upon a set of bioindicators would be a necessary first step. A 2010 report from the European Commission includes criteria for selecting valuable indicators (Turbe *et al.,* 2010):

- ***Meaningful:*** *indicators must relate to important ecological functions and use good surrogates (for example, recognized high-value organisms as indicator groups). This ensures the indicators will serve their purpose accurately, that is, monitor trends in soil biodiversity.*

- ***Standardized:*** *the selected parameters should be readily available and (almost) standardized. This ensures the comparability of data among sites.*

- ***Measurable and cost-efficient:*** *the selected parameters must be easy to investigate in the field and to sample and affordable, and must not be restricted only to experts or scientists but should also be assessable by all interested people. This ensures the indicators will be used in practice and can be routinely collected.*

Given the importance of molecular techniques to the field of soil biodiversity, bioindicators that target DNA or RNA are likely to be included. Incorporation of molecular techniques will require retooling soil testing labs to include the related equipment and expertise to extract nucleic acids from soils and to sequence or quantify targeted genes of interest. Protocols for characterizing microbial communities are now commonplace, and the use of metabarcoding for soil fauna continues to increase (Orgiazzi *et al.*, 2015). These approaches have already been applied in large scale efforts that include:

- *The European Union program on Ecological Function and Biodiversity Indicators in European Soils (EcoFINDERS)*
- *The United States National Ecological Observatory Network (NEON)*

And could be incorporated into efforts such as:

- *China's Soil Ten Plan that is designed to protect soil and limit contaminates*
- *Australia's Biome of Australian Soil Environments (BASE)*
- *The African Soil Health Consortium*

A standard set of bioindicators linked to soil health will inevitably result in large data sets that will need to be curated and translatable to producers. Similar to the drastic decrease in sequencing costs, computing power and storage makes this feasible. For example, the European Union's Land Use/Cover Area frame statistical Survey Soil (LUCAS) (Orgiazzi *et al.*, 2018) has recently included measures of soil biodiversity and includes an open access database. This project is a model for other regions. Data accessibility will need to be then paired to tools that translate biodiversity into specific recommendations for farming practices and other types of land management. Although a number of tools exist to assess ecosystem services in the context of land management, few fully integrate soil biodiversity and most are applicable only to developed countries (Grêt-Regamey *et al.*, 2017).

Novel technologies at farm and landscape scales may become powerful tools that can promote the sustainable management of soils. Knowledge and technological advances at the microscale or macroscale may provide new perspectives on soil functions that may ultimately be transferred to novel technologies. Sustainable land management requires sound resource management at the watershed and landscape levels and beyond, which in turn requires models based on big data generated from soil-water-plant-atmosphere information. Until now, information on soil biodiversity has not yet been included, but once it is aggregated into the model it may increase management strength, providing that sufficient knowledge is available regarding the diversity and functions of the soil microbiome. Artificial intelligence has great potential in the assembly of data and the aggregation of information from multiple databases.

Research devoted to the definition of biological indicators is making great progress, but it remains a challenge to develop robust, reliable and resilient biological indicators. The abundance and/or diversity of earthworms or nematodes have been well delineated. Some recent studies have even produced global maps (Philips *et al.*, 2019; van den Hoogen *et al.*, 2019). However, the soil microbiome (or functional groups/keystone species) as an indicator is complex. Some specific functional groups have the potential to be indicative of relevant soil processes, such as the N cycling-related microbiome and arbuscular mycorrhizal fungi in relation to phosphorus nutrition. The emerging novel technologies such as metagenomic, metabolomic and volatilomic approaches are expensive, but they may provide useful information on microbial functions in addition to the taxonomic diversity of the soil microbiome.

Given the high heterogeneity at temporal and spatial scales in complex soil systems (Ettema and Wardle, 2002), the challenge lies in how to connect the diversity of the soil microbiome with its potential functions, and in understanding how the diversity and functional traits of below-ground soil microbes link with their above-ground biodiversity and productivity. In practice, it may be more plausible to develop a holistic biological indicator based on plants and other above-ground properties that may indicate the taxonomic and functional composition of the local microbiome and of the invertebrates in the soil community.

Effective and efficient monitoring tools are important in recording changes in soil biodiversity and establishing databases to link diversity with soil functions. Monitoring programs may help to identify key physical and biological soil constraints to plant productivity, in addition to routinely measured soil nutrients. This is useful for the development of management strategies to address constraints in the short term and the long term. Visualization of biological information in combination with digital soil mapping tools may be effective in providing management-related soil biological information, and this may have the potential to increase the transfer and adoption rates of knowledge by different stakeholders.

The soil contains arguably the most diverse terrestrial communities on the planet and is home to more than 25 percent of global biological diversity; moreover, it supports most life above ground by means of increasingly well understood above- and below-ground linkages. Evidence supports these vital connections between soil biodiversity and the Sustainable Development Goals (SDGs), and emerging knowledge is shaping actions we can take for a more sustainable future. Soil organisms play many important roles in managing invasive species. Soil biodiversity can help avoid, reduce and reverse land degradation, sustaining and improving habitat for people and other life on Earth. Long taken for granted, soil biodiversity should be embraced as part of the urgent need to develop a more sustainable future for all.

Box 2.5 | The bait-lamina test

Developed in 1990 in Germany, the bait lamina test is used to measure the biological activity of the soil. It enables the monitoring of zootic or microbial processes in the soil. The test system is based on the visual assessment of feeding on small portions of thin laminated bait substrate exposed to edaphic processes. Therefore, special bait portions are fixed in holes pierced in strips, which are then exposed to the decomposition activity in the soil. The disappearance of the bait material is directly associated with feeding activity. Compared to others, the test has the advantage of providing comparable, easy, quick and inexpensive screening.

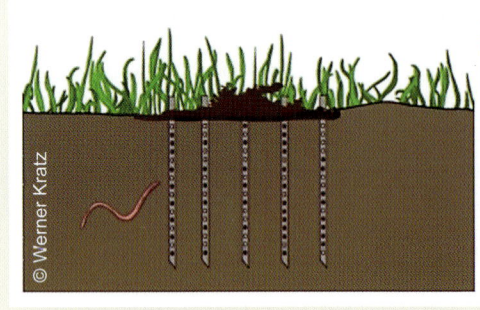

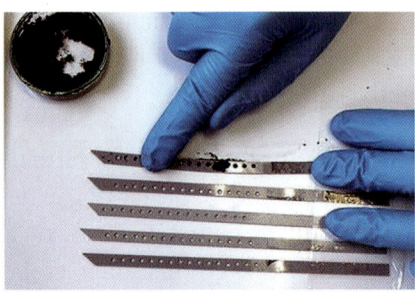

The bait-lamina test

© Andy Murray

© Andy Murray

CHAPTER 3
CONTRIBUTIONS OF SOIL BIODIVERSITY TO ECOSYSTEM FUNCTIONS AND SERVICES

3.1 | DEFINING SOIL FUNCTIONS

The definition of soil function and the determination of its boundaries is not an easy task. Various disciplines provide different perspectives; some focus only on the outcome, ignoring the real producers of these functions (see below). Moreover, the terms "functions" and "services" are also often confounded. Functions refer to natural-biological processes, while services refer to those functions that contribute to human well-being. Banwart *et al.* (2019) thoroughly described soil functions as "flows and transformations of mass, energy, and genetic information that connect soil to the wider critical zone, transmitting the impacts of human activity at the land surface and providing a control point for beneficial human intervention." In simple words, the soil functional outcome is a result of interactions among physical, chemical, biological and human factors.

Soil functioning is principally defined by the parent material, the climate, and the topography, as these specify the physical and chemical environment and determine the conditions for living organisms (Dominati *et al.*, 2010; van Leeuwen *et al.*, 2019; O'Connor *et al.*, 2009). While the limits are set by external factors, the ability of soil to function and to support life is provided by soil organisms interacting in complex food-webs (Figure 3.1.1.1). One essential, complex ecosystem process that encompasses several soil functions and is supported by the whole soil food web is decomposition. Soil organisms are directly or indirectly (that is, through one or more trophic levels) involved in the decomposition of dead organic matter for covering their needs in energy, carbon and nutrients (de Ruiter *et al.*, 1994), but there are also other sources of carbon and nutrients deriving from roots and rhizodeposits/root exudates (Haichar *et al.*, 2014) that feed soil organisms involved in the soil food web.

Through their biological activity, soil organisms change their micro-environment. They transform complex chemical forms into simple molecules or compounds that can be absorbed by plants, providing feedback to plant productivity and re-growth. They structure soil by changing conditions of aeration and drainage and they control populations of other organisms, thus regulating above-ground biodiversity. Although their biological activities operate at small scales, their activities aggregate at larger scales, resulting in major functional outputs (Kibblewhite *et al.*, 2008) such as: a) carbon transformation, b) nutrient cycling, c) formation of soil structure and d) biodiversity regulation (biocontrol). These biodiversity-based soil functions define soil health, sustain soil ecosystem processes, influence above-ground diversity and contribute to climate regulation, and at the same time underpin regulating, supporting and provisioning ecosystem services that are essential for human well-being (Hedlund *et al.*, 2004; Lavelle *et al.*, 2006; Jeffery *et al.*, 2010; Bardgett and van der Putten, 2014; Orgiazzi *et al.*, 2016).

As a basis for understanding soil biodiversity and its role in ecosystem functioning, soil organisms have been approached from a functional diversity perspective in the context of the soil food web. Thus, soil organisms interacting in the soil food web are assigned to functional groups, based on characteristics such as their feeding source, their form, size, taxon and trophic level.

Each of these functional groups may contribute to one or more soil functions, but there is also high functional redundancy, (that is, a variety of soil organisms having the same functional ability). Decomposer organisms such as bacteria and fungi facilitate chemical breakdown of substrates and play a significant role in the functions of carbon transformation, soil structure formation and nutrient transformation. In the case of the latter, nitrogen-fixing bacteria and mycorrhizal fungi are significant players. Organisms that feed on microbes (for example, bacterivorous and fungivorous nematodes) regulate the populations of bacteria and fungi, and affect key ecosystem functions, such as nutrient transformation, indirectly. Together with organisms at higher trophic levels (for example, predatory nematodes, oribatid mites), members of the soil food web play a significant role as biological regulators. Organisms that feed directly on dead organic matter (detritivores) as well as other macro-faunal organisms (such as isopoda, earthworms) also affect several soil functions, and specifically soil structure formation, and are considered ecosystem engineers.

That said, human activities and interventions play a critical role on the outcome of soil functioning. Human-induced changes, such as the intensification of land use, lead to the modification of soil structure and abiotic properties as well as to tremendous changes in the structure, composition and diversity of the soil food web. Soil health (Kibblewhite *et al.*, 2008), as well as related Sustainable Development Goals (SDGs), depend on the maintenance of the four major biodiversity-based soil functions (carbon transformation, nutrient cycling, formation of soil structure, and biodiversity regulation). Unfortunately, these functions are recognized as being under threat (CEC, 2006; Gardi *et al.*, 2013). To overcome these obstacles and to sustain soil functions at specific levels, knowledge on how soil food webs respond to specific management and restoration regimes under the perspective of global climate change is essential. To this end, it is crucial to focus research on better understanding the links among biodiversity attributes and soil functions and ecosystem services (see for example de Vries *et al.*, 2013), and among abiotic properties, soil organisms and climate (Bhusal *et al.*, 2015; Orgiazzi and Panagos, 2018), and to develop efficient monitoring tools and maps by up scaling the bio-indication potential to the scales that are important for management decisions (Stone *et al.*, 2016; van Leeuwen *et al.,* 2017; van den Hoogen *et al.,* 2019).

© Andy Murray

Table 3.1 | Examples of organisms, their main soil functions and ecosystem services, and gaps and opportunities

Organisms	Main Soil Ecosystem Functions	Ecosystem Services	Gaps and Opportunities
Megafauna (e.g. moles, beavers, armadillos)	**Bioturbators** Soil, organic matter and sediment redistribution to greater depths		
Macrofauna (e.g. earthworms, ants, termites, millipedes, insects)	**Ecosystem engineers** Fragment, rip, and tear organic matter, providing smaller pieces for decay by organisms; mineralization of organic matter; **Bioturbators** Moving and mixing soil, increasing water permeability and soil aeration	**Supporting services:** *Soil formation* *Nutrient cycling* **Regulating services:** *Climate regulation* *Disease and pest regulation* *Water regulation (water availability, including regulating extremes – drought and flood)* *Remediation* *Pollination* **Provisioning services:** *Food* *Freshwater* *Fuel* *Fibre* *Biochemicals* *Genetic resources* **Cultural services:** *Spiritual, recreational, symbolic values of landscapes*	**Lack of data and limited knowledge on:** Organisms and their functional roles Interaction between organisms and their communities, as well as with terrestrial biodiversity (including crops) Distribution of soil organisms globally Better understanding of how biodiversity loss by anthropogenic activities might affect stability of food webs and ecosystem functioning Better understanding of the impact of climate change on soil communities Improvement of biological control practices for pests and pathogens **Opportunities:** Undescribed microorganisms may have a wide range of metabolic capabilities influencing ecosystem services Application of soil microorganisms for specific industrial, agricultural or environmental purposes
	Bioremediation *Eisenia fetida* earthworms can accumulate cadmium and some other pollutants like polycyclic aromatic compounds (PAHs)		
Mesofauna (e.g. mites, springtails)	**Decomposers** All mesofauna modify the fine-scaled structure of the soil through their feeding activities. **Foodweb stabilizers** Serving as predators, fungivores and preys to different communities		
Microfauna (e.g. tardigrades, nematodes,)	**Decomposers** Recycle nutrients and increase nutrient availability for primary production		
	Nitrogen fixers arbuscular mycorrhizal fungi (AMF) biologically fix atmospheric N_2		
Microbes (e.g. virus, bacteria and archaea, fungi)	**Bioremediation:** Break down, removing, altering, immobilizing, or detoxifying various chemicals and physical wastes from the environment like PAHs (see chapter 5) **Decomposers:** Weathering minerals; Carbon transformation by decomposition of organic matter and storage, nutrient cycling by transforming inorganic compounds into forms usable by plants, regulate soil structure and pathogens		
	Gas producers: Methanogenic archaea transfer of C, N_2, N_2O, CH_4 denitrification		
	Nitrogen fixers: Rhizobia bacteria found on legume roots helping to increase nitrogen uptake.		

3.1.1 | THE SOIL FOOD WEB APPROACH

The soil food web approach provides a way to describe and quantify the soil biodiversity and its role in soil ecosystem functioning. The food web approach occupies a central position in community ecology. The strengths of food web interactions (that is, those between resource species and consumer species) affect the distribution and abundance of organisms in fundamental ways, since the success of populations is largely a function of benefits from the acquisition of food and losses from predation. In this way, food webs provide a way to analyze the dynamics and persistence of soil biodiversity. The food web approach also enables the analysis of the relationship between the structure of the soil biological community and soil ecosystem processing, as the food web interactions represent flows of matter, energy and information. For this reason, the soil food web approach contributes to two central aspects of soil biodiversity: its preservation; and its role in local, regional and global cycling of materials, energy and nutrients.

The *Soil Food Web* described soil biodiversity in the form of an ecological network. Figure 3.1.1.1 gives a schematic representation of a food web (de Ruiter *et al.*, 1994). Boxes denote the presence of a functional group. Functional groups aggregate species that have similar diets and life-history characteristics, such as growth and death rates and energetic efficiencies. Arrows denote feeding rates pointing to the consumers. The first trophic level in soil food webs consists of microorganisms; that is, bacteria and fungi, decomposing soil organic matter, and herbivorous nematodes feeding on roots. At the higher trophic levels, we see a large variety of faunal organisms: protozoa (amoebae, flagellates, ciliates), nematodes (bacterivores, fungivores, omnivores, herbivores and predators), mites (bacterivores, fungivores, predators), collembola (fungivores and predators), enchytraeids and earthworms.

The soil food web diagram also reflects the large variety in soil organisms in connection to the large heterogeneity in soils (Moore and Hunt, 1988). The red arrows in the soil food web diagram form the so-called bacterial channel. In this channel, the fluxes of material all originate from bacterial productivity. The pink arrows form the fungal channel and originate from fungal productivity. The green arrows form the root channel, originating from root productivity. The black arrows form linkages among the different channels. The red channel are mostly aquatic organisms such as bacteria, protozoa, and nematodes, living in the water film around soil particles. The fungal channel are mostly terrestrial organisms – fungi, mites, and insects. Exceptions are the bacterivorous mites in the bacterial channel and the fungivorous nematodes in the fungal channel. These energy channels show that the soil organisms form one biological community, while different components of the community can be organized in relatively separated and independent ways; that is, in soil food web compartments.

Soil food webs provide an excellent model for studying the biological mechanisms underlying soil ecosystem functioning. Soil food webs are directly responsible for soil processes that originate from the decomposition of soil organic matter and the food consumed and processed in the other trophic interaction in the soil food web (de Ruiter *et al.*, 1994). Together, the organisms in the soil food web are directly responsible for soil ecosystem processes, such as carbon sequestration and nutrient cycling. Given the sizeable amounts of materials they process, soil food webs are responsible for major components in the global cycling of materials, energy and nutrients (Wolters *et al.*, 2000, Griffiths *et al.*, 2000). Furthermore, soil food webs play a key role in terrestrial systems in their ability to provide services in the form of food productivity (Brussaard *et al.*, 2007) and the conservation of above-ground biological diversity (Hooper *et al.*, 2000; de Deyn *et al.*, 2003).

Figure 3.1.1.1 (next page) | **Diagram of a soil food web**

Boxes denote the presence of a functional group. Functional groups are aggregations of species on the basis of similar diets and life-history traits. Arrows denote feeding rates pointing to the consumers. The colour of the arrows refer to 'energy channels', which are defined of the basis of the primary resource of food chains within the food web. **Red** arrows: root channel, **orange** arrows: detritus channel, **blue** arrows: bacterial channel, **green**: fungal channel, **black**: energy channels linked by omnivorous and polyphagous organisms

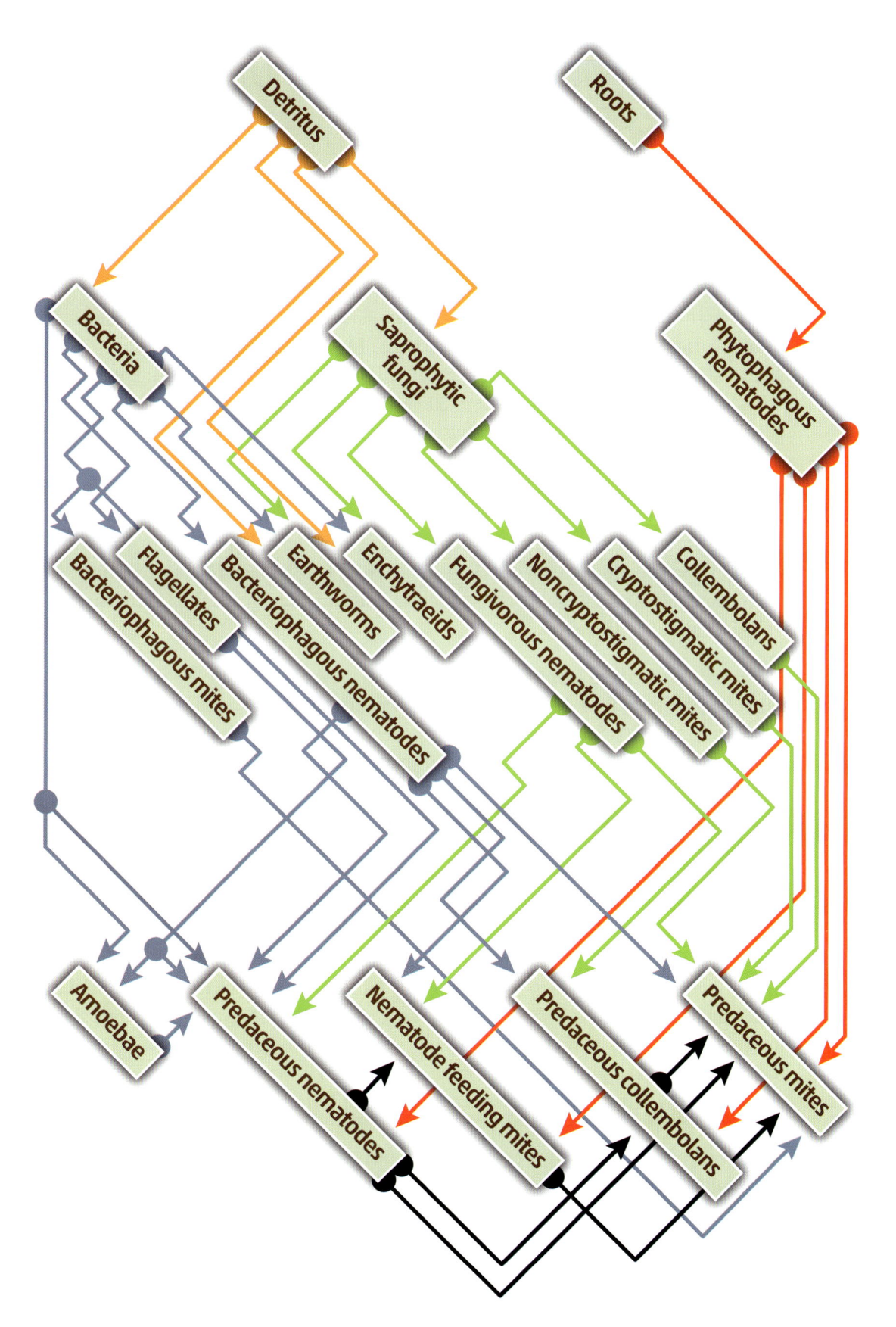

Contributions of soil biodiversity to ecosystem functions and services

3.1.2 | THE MULTIFUNCTIONALITY OF SOIL BIODIVERSITY

It has been long debated whether soil biodiversity is superfluous insofar as many species are redundant. Currently evidence is increasing that there is less functional redundancy than previously supposed. For example, mycorrhizal fungi form symbiotic associations with the majority of plant roots and help plants to acquire limiting nutrients such as phosphorus and nitrogen (van der Heijden *et al.*, 2015). Nitrogen-fixing bacteria associate with legume roots and fix large amounts of nitrogen that are of pivotal importance for plant productivity in a wide range of ecosystems across the world. Other soil organisms decompose organic matter (Heemsbergen *et al.*, 2004; Lindahl *et al.*, 2006), contribute to soil carbon stabilization, soil aggregate formation and soil structure (Rillig and Mummy 2006), cause disease (van der Putten *et al.,* 1993) which can enhance plant species diversity in nature (van der Putten *et al.,* 2013), or suppress disease (Mendes *et al.,* 2013). Moreover, soil biota break down contaminants such as pesticides (Fenner *et al.,* 2013), produce antibiotics (Nesme *et al.,* 2014), clean water that percolates through the soil profile, and prevent leaching of nutrients into ground and drinking water (Bender and van der Heijden, 2016). Thus, it is important to consider that soil organisms play multiple roles in ecosystems and influence multiple ecosystem functions (multifunctionality).

Experiments carried out in microcosms (van der Heijden *et al.,* 1998; Philippot *et al.,* 2013; Wagg *et al.,* 2014; Rillig Semchenko *et al.,* 2018) and at global observational scales (Jing *et al.,* 2015; Delgado-Baquerizo *et al.,* 2016) revealed that soil biodiversity and soil microbial diversity promotes or is directly linked to ecosystem multifunctionality and functioning. These observations mirror results obtained in plant communities that revealed that plant diversity enhances ecosystem sustainability (Tilman *et al.*, 1996) and multifunctioniality (Hector and Bagchi 2008). A recent study showed that grassland microcosms with poorly developed soil microbial networks and reduced soil microbial richness had the lowest multifunctionality (Wagg *et al.,* 2019). This was due to fewer taxa present that support the same function (redundancy) and lower diversity in the abundance of taxa that support different functions (reduced functional uniqueness; Wagg *et al.,* 2019). This indicates that high soil biodiversity provides an insurance that is required to sustain multiple functions (sensu Yachi and Loreau, 1999). Interestingly, an increasing number of studies also show that cross-kingdom interactions between various groups of soil biota (for example between bacteria and fungi or between different guilds of soil arthropods) and soil food web structure and composition are of pivotal importance for plant health (Van der Putten *et al.,* 2006; Duran *et al.,* 2018), decomposition (Heemsbergen *et al.,* 2004), or overall ecosystem functioning (Wagg *et al.,* 2019). Microbial decomposition is traditionally supposed to be the result of non-specialized microbes. However, recent work shows considerable specialization among decomposers (Veen *et al.,* 2017), making decomposers less functionally redundant than was previously supposed.

Soil biota are involved in multiple functions simultaneously, and thus functional redundancy is likely to fade as more functions or environmental conditions are considered. For example, in a mesocosm study, extreme drought or extreme rainfall conditions revealed that part of the microbial community is more drought-tolerant, while the other part is more tolerant to extreme moisture (Meisner *et al.*, 2018). Most likely, when a larger variety of environmental conditions are eventually examined, there will be less functional redundancy in the soil community than originally assumed. Also, the loss of soil bacteria has been shown to negatively affect the nitrogen cycle (Philippot *et al.*, 2013), of which some aspects are only carried out by few species (Swift *et al.*, 1998).

To understand how changes in soil biodiversity affect ecosystem functioning, it is therefore important to consider not only whether the total number of taxa present relates to a function, but how the reduction in the number of species that support a single function relates to the loss of multiple functions simultaneously. In addition, the influence of an individual species on an ecosystem function is not independent of other species present and is a result of a myriad of positive and negative, direct and indirect associations among the different species that as a whole drive ecosystem functioning. In conclusion, there is increasing evidence that soil biodiversity plays a pivotal role in the functioning of the Earth's ecosystems. However, precise mechanisms and general patterns are still unclear and there is poor knowledge about how to promote and use soil biodiversity for the development of sustainable agricultural ecosystems.

3.2 | DEFINING SOIL ECOSYSTEM SERVICES

Ecosystems are living systems, which interact with one another and their surrounding environment. They provide benefits, or services, to the world. Ecosystem services, then, are the multitude of benefits that nature provides to society. Ecosystem services are classified into supporting, provisioning, regulating, and cultural ecosystem services. Despite an estimated value of USD 125 trillion (Costanza *et al.,* 2014), these assets are not adequately accounted for in either political or economic policies, which means there is insufficient investment in their protection and management.

Biodiversity losses can negatively affect the supply of ecosystem services. Even if soil organism biomass is very low compared with the mineral fraction, the activity is crucial for a functioning soil because the soil organisms support many fundamental processes and enhance key ecosystem provision services, such as food and fibres, water quality, biodiversity conservation, and ecosystem supporting services such as nutrient cycling and soil structure formation (Kibblewhite *et al.,* 2008; Pulleman *et al.,* 2012; El Mujtar *et al.*, 2019). Biodiversity, as with ecosystem services, must be protected and sustainably managed (FAO, 2019). The links from soil and soil biodiversity to ecosystem services are not well described despite the underlying significance for soil organisms, soil functions and soil biodiversity. A conceptual scheme of the relationship between soil biodiversity and ecosystem services is provided in Figure 3.2.1.

Soil biodiversity is generally under pressure due to some threats such as intensification of agriculture, which has a negative impact on several provisioning and supporting ecosystem services (Tsiafouli *et al.,* 2017). Bender *et al.* (2016) highlighted that agriculture has the potential to restore the sustainability of agricultural ecosystems by stimulating soil life and regulating ecosystem processes. Despite this collective understanding, however, soil biodiversity still receives little recognition in agricultural management strategies and in support of ecosystem services. Forest ecosystems also benefit from below-ground soil biodiversity for the vast majority of interactions studied among provisioning, regulating, supporting and cultural ecosystem services (Bakker *et al.*, 2019).

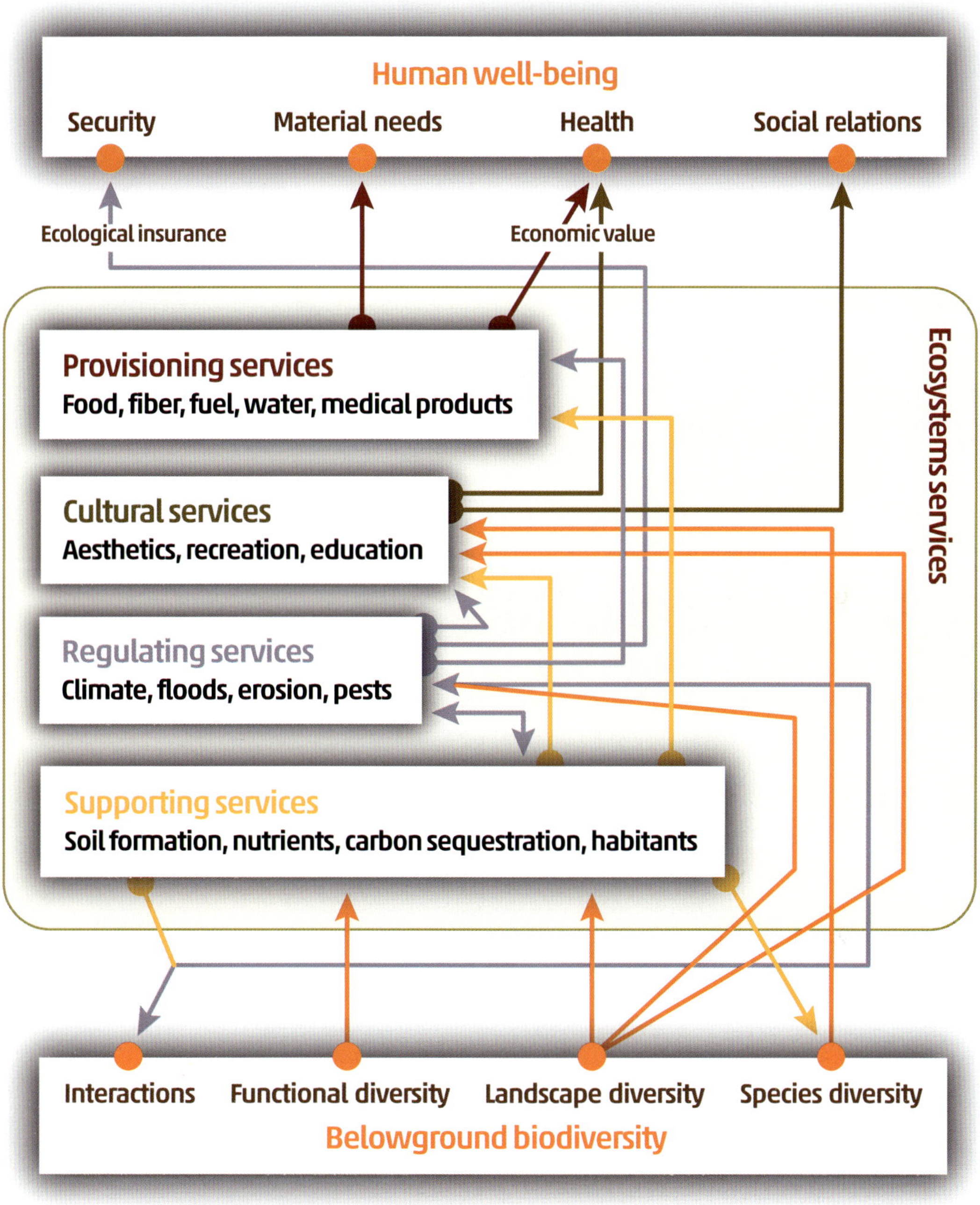

Figure 3.2.1 | Relationship between soil biodiversity and ecosystem services

A conceptual scheme of the relationship between soil biodiversity, ecosystem services and human wellbeing (Bakker et al., 2019, modified from Scholes et al., 2010).

©FAO / Matteo Sala

3.3 | SOIL BIODIVERSITY SUPPORTS THE *SUSTAINABLE DEVELOPMENT GOALS*

Soil biodiversity is integral to sustaining all life on Earth, especially humanity. With the adoption of the 2030 Agenda for Sustainable Development, United Nations Member States have agreed to work toward accomplishing 17 ambitious Sustainable Development Goals (SDGs) globally to improve and sustain life for people and the natural world on which we depend. In this chapter, we explore how soil biodiversity plays a key role in meeting many of these SDGs (Figure 3.3.1).

Although the SDGs do not refer directly to soil, the strong connection between soil biodiversity and the SDGs is clear. Many SDGs such as food security, water scarcity, climate change, biodiversity loss and health threats are closely linked to or dependent on soil biodiversity (Table 4.0.1, Keesstra *et al.,* 2016). Deterioration of relationships between humans and soil has resulted in unsustainable management of agricultural, forest and urban ecosystems, leading to environmental degradation and major societal consequences. Soil cannot be untangled from its biodiversity, though these intimate interconnections are not always recognized

Soil biodiversity supports human health and well-being (SDG 3) through regulation of many disease-causing organisms as well as playing a central role in agricultural production, supporting SDG 2 (no hunger) as well as agricultural livelihoods contributing to SDG 1 (no poverty), SDG 5 (gender equality), SDG 8 (decent work and economic growth) and SDG 12 (responsible consumption and production). Life in the soil mediates water flow and uses excess nutrients and pollution, advancing SDG 6 (clean water and sanitation) and SDG 14 (life below water). Carbon cycling from soil biological activity, both in terms of C sequestration and greenhouse gas emissions, are central to regulating climate (SDG 13). The soil contains arguably the most diverse terrestrial communities on the planet, and is home to more than 25 percent of global biological diversity. Moreover, it supports most life above ground via increasingly well understood above- and below-ground linkages (SDG 15). Evidence supports these vital connections between soil biodiversity and the SDGs, and emerging knowledge is shaping actions we can take for a more sustainable future. Soil organisms play many important roles in managing invasive species. Soil biodiversity can help avoid, reduce, and reverse land degradation, sustaining and improving habitat for people and other life on Earth. Long taken for granted, soil biodiversity should be embraced as part of the urgent need to develop a more sustainable future for all.

Flood regulation

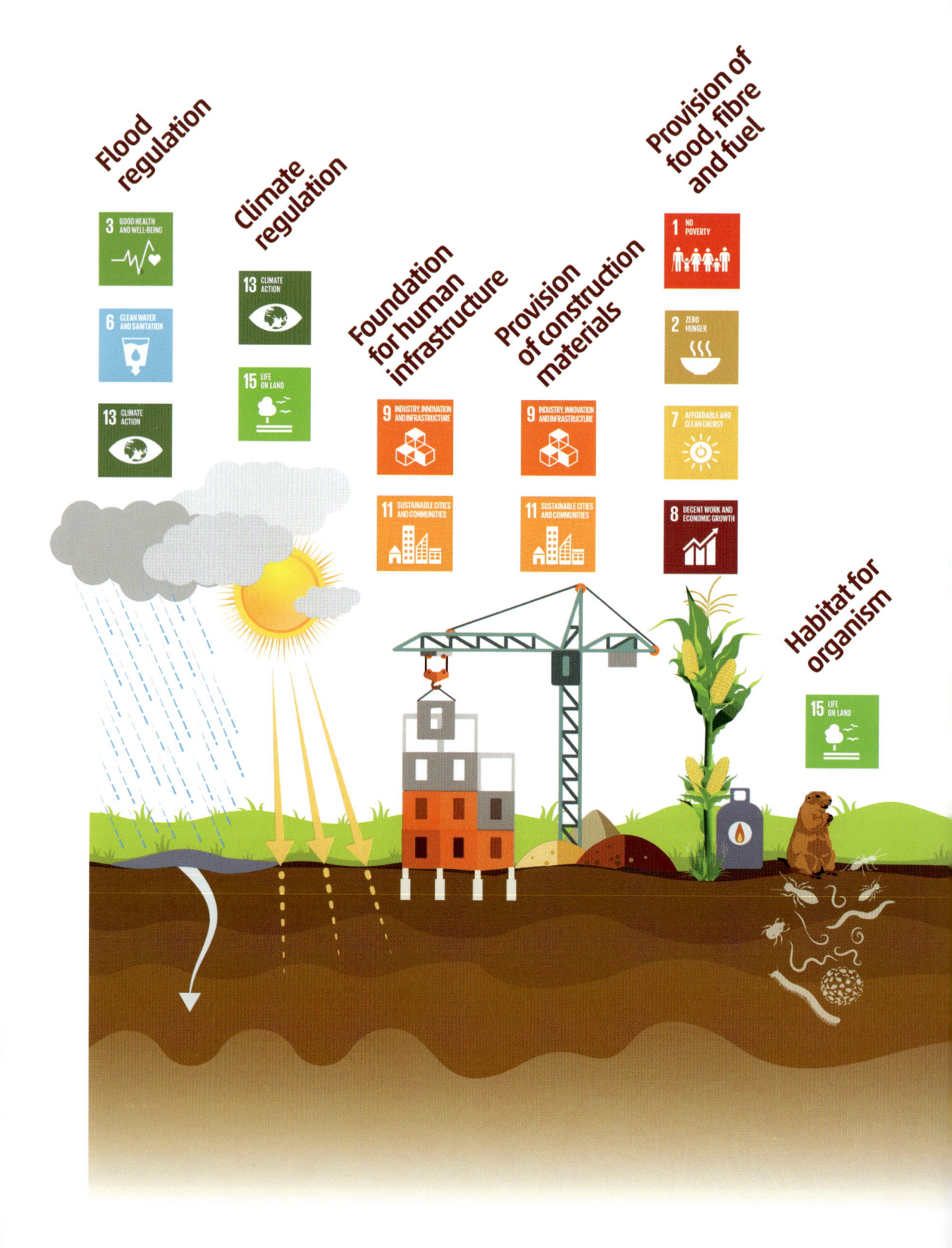

- 3 GOOD HEALTH AND WELL-BEING
- 6 CLEAN WATER AND SANITATION
- 13 CLIMATE ACTION

Climate regulation

- 13 CLIMATE ACTION
- 15 LIFE ON LAND

Foundation for human infrastructure

- 9 INDUSTRY, INNOVATION AND INFRASTRUCTURE
- 11 SUSTAINABLE CITIES AND COMMUNITIES

Provision of construction materials

- 9 INDUSTRY, INNOVATION AND INFRASTRUCTURE
- 11 SUSTAINABLE CITIES AND COMMUNITIES

Provision of food, fibre and fuel

- 1 NO POVERTY
- 2 ZERO HUNGER
- 7 AFFORDABLE AND CLEAN ENERGY
- 8 DECENT WORK AND ECONOMIC GROWTH

Habitat for organism

- 15 LIFE ON LAND

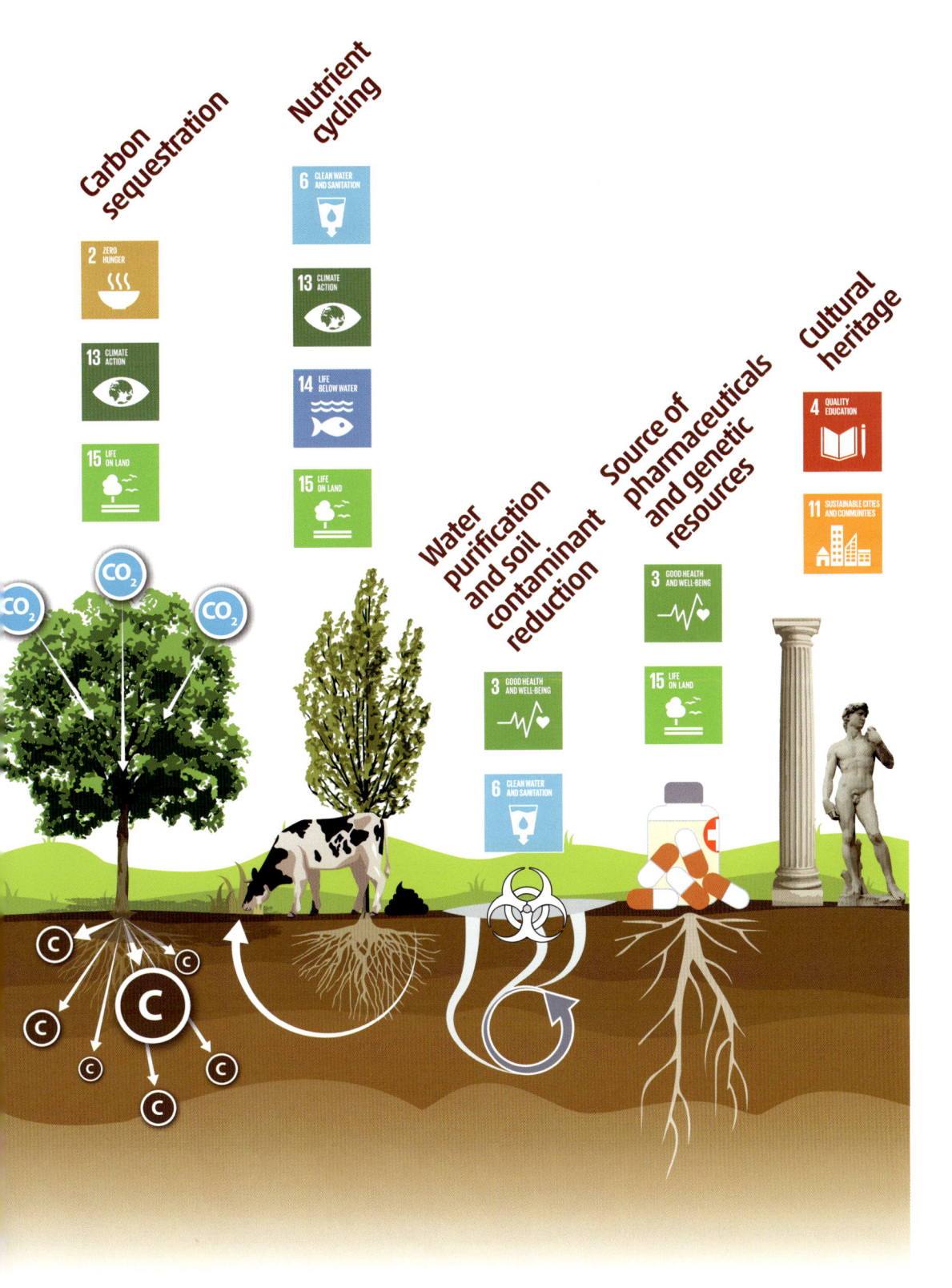

Figure 3.3.1 | Healthy soils, a prerequisite to achieve the SDGs

A healthy soil is capable of providing most ecosystem services and therefore achieving compliance with SDGs and human well-being.

Contributions of soil biodiversity to ecosystem functions and services

Box 3.3.1| Soil Biodiversity and the SDGs

Food Production (SDG 2): Achieving global food security is one of the greatest challenges of our time (Burns et al., 2010; Lobell et al., 2008). The mismatch between the rate of human population growth and global food production has long been recognized (Malthus 1798). We need to double global food production in the coming decades, on less land and using less inputs. The majority of the world's human population eats a soil-grown, plant-based diet. Even where diets are rich in animal products, livestock rely on soil-grown plants (fish being a notable exception). The quantity and nutritional quality of crops is very much a product of the soils in which they grow. The link between crop production and soil quality is well established. For example, that 30 percent of the world's human population is affected by Zn deficiency (Alloway 2008) is not surprising, given that the soils of almost half of the world's cereal growing regions are low in Zn (Cakmak, 2002; Graham and Welch, 1997). Taken together, it is clear that efforts focused on achieving food security cannot ignore the link between soil and food.

Water quality (SDGs 6, 14): Nutrient exports from agricultural systems are a large contributor to water quality issues in many parts of the world. For example, reports of up to 160 kg of nitrogen (N) and up to 30 kg of phosphorus (P) per hectare can be lost via leaching and surface runoff (Herzog et al., 2008; Sims et al., 1998). When nutrients reach waterways, they can contaminate groundwater and can also cause eutrophication which can lead to algal blooms and the loss of terrestrial and aquatic biodiversity (Carstensen et al., 2014). Soil biota play an important role in regulating the movement of water into and through soil as well as cycling of nutrients in the soil and water. This can impact their risk of nutrients being lost via leaching. Similarly, some soil microbes (such as mycorrhizas) play an important role in helping plants to interact with such nutrients, thereby reducing the risk of nutrient leaching (Cavagnaro et al., 2015). Wider use of agricultural practices that leverage soil biota to manage nutrient availability and loss could contribute significant progress toward water quality goals

Air quality (SDGs 3, 13): Soil plays an important role in air quality and soil greenhouse gas emissions. When agricultural soils are tilled, the increased oxygen can spur biological activity and respiration of carbon dioxide that can contribute to global warming. Moreover, certain soil microbes under anaerobic conditions (such as flooded or very wet soils) can transform nitrate into the potent greenhouse gas nitrous oxide. Similarly, other soil microbes can release methane from soil, which also contributes to global warming. Soil microbes have also been reported to help purify air (Khaksar et al., 2016). It is also worth noting that soil microbes and soil fauna can also help to bind soil particles together and improve soil structure in some situations. In doing so they can reduce the risk of wind erosion, thereby helping to reduce levels of dust in the air we breathe.

Soil health (SDGs 2 and 3): The prevention of soil pollution could reduce soil degradation, increase food security, contribute substantially to the adaptation and mitigation of climate change, and contribute to the avoidance of conflict and migration. By taking immediate actions against soil pollution we can thereby contribute to the achievement of almost all the SDGs, with a significant impact on goals 2 and 3.

Climate mitigation (SDG 13) and adaptation: Soil, and the biota it supports, have a central role to play in climate mitigation, and agricultural soils play a key role because they cover such a large portion of the earth. In addition to reducing greenhouse gas emissions from soil (see air quality above), the soil has the potential to sequester large amounts of carbon. Agricultural practices such as reducing tillage and maximising plant cover can improve soil biological activity and C sequestration. Not only should we mitigate climate change, we must also adapt to it. Soil-provided ecosystem services can help to buffer systems against (resistance) and rebound from (resilience) external perturbations (Jackson et al., 2003).

Affordable and clean energy (SDG 7): Biofuel can be produced using plant-based feedstocks, using plants grown expressly for this purpose or grown for other purposes and waste materials used in energy production systems. For example, waste materials, such as those from forestry, food processing, wine production, and many others can be processed in various ways (such as pyrolysis, gasification, liquefaction) to extract the energy they contain. The production of such materials for the most part has their genesis in the soil (for example, crop residues). Moreover, many of these processes produce bioproducts that can potentially be used as a soil amendment (often rich in carbon) and (potentially) nutrients (Dreake et al., 2016). Furthermore, such materials can provide habitat for soil organisms.

Above-ground biodiversity/biodiversity loss (SDG 15): It is increasingly well understood that above- and below-ground communities are closely linked, and that a change in one can affect the other. For example, a reduction in below-ground diversity can reduce above-ground plant diversity (van der Hejiden et al., 1998). Similarly, changes in above-ground vegetation can alter below-ground communities. For example, in a recent meta-analysis it was found that by reducing soil tillage and planting a cover-crop, the formation of beneficial mycorrhizal associations (symbiosis between plant roots and soil fungi that improve plant nutrient acquisition) was improved (Bowles *et al.*, 2017). Soil fauna including nematodes, collembola, and mites have been shown to increase grassland succession and plant diversity (De Deyn *et al.*, 2003).

Gender Equality (SDG 5) and Decent Work and Economic Growth (SDG 8): Women are important members of farming communities around the world. Men and women relate to land differently, and their unique perspectives are driven by varying roles, responsibilities, access to resources and control. Understanding these roles, along with power relations in land management, is a primary requirement to achieving effective outcomes when addressing sustainable agricultural development (Chotte *et al.*, 2019). Efforts to mainstream gender as proposed by the United Nations Convention to Combat Desertification Gender Action Plan (GAP) as well as recommendations from the United Nations Framework Convention on Climate Change (UNFCCC), the United Nations Convention on Biological Diversity (UNCBD), UN Women, the International Union for Conservation of Nature (IUCN), and the Convention on the Elimination of All Forms of Discrimination against Women (CEDAW) among others convey the importance of gender equality and gender inclusive action in supporting agricultural production and protecting land from degradation. The numerous recommendations promoting gender equality and human rights and empowerment of women and girls in environmental governance have been proposed by the United Nations Environment Assembly (UNEA, 2019). Women form a major part of agricultural development (UNCCD, 2017b) with traditional knowledge and skills in farming being closely tied to the maintenance and improvement of land productivity (UNCCD – Global Land Outlook: Gender-responsive land Degradation Neutrality, 2017). Women's contributions can include knowledge and respect for soil organisms and their role in supporting farming practices. These vital roles of women need to be understood and addressed, on the one hand to enable communities to support women as farmers and as leaders, and on the other hand to ensure that men and women benefit equally, and that inequality is not perpetuated. It is important to enable more equal access to natural resources and to facilitate women in becoming active users and managers of natural resources (Okpara *et al.*, 2019), goals that are interlinked with SDG 5 and SDG 8.

3.4 | PROVISIONING ECOSYSTEM SERVICES

Provisioning ecosystem services are produced within agriculture, forestry and fisheries and include food, fodder, fibres and wood, fresh water, raw materials, and genetic medicinal and ornamental resources (Adhikari and Hartemink, 2016). The significance and support of soil biodiversity of these provisioning ecosystem services will be the focus in this section.

3.4.1 | NUTRIENT CYCLING

Nutrients are chemical elements that are used by plants and other organisms for their growth. Soils are the major terrestrial reservoir of organic carbon and nutrients such as nitrogen and phosphorus. The soil nutrient pool includes macronutrients that plants need in a large quantity (such as nitrogen, phosphorus, and potassium) and micronutrients, which are needed in smaller quantities (iron, manganese, boron, molybdenum, copper, zinc, chlorine, nickel, sulfur, and cobalt). Nitrogen is a critical nutrient for plant growth, since it is part of the chlorophyll molecule, and plants require nitrogen for amino acid and protein production. Phosphorus is present in DNA and is important in photosynthesis, respiration, healthy root systems and seed development. Potassium is used in photosynthesis, protein synthesis, regulation of water usage, and disease resistance. Micronutrients are required as cofactors for enzyme activity. Many scientific studies have focused on the ratio C:N:P for quantifying the fluxes of elements between compartments and then highlighting the functioning of a given ecosystem (for example, Zechmeister-Boltenstern *et al.*, 2015). A deficiency of one of the nutrients may result in decreased plant quality and/or productivity. As a consequence, nutrient deficiency can induce the reduction of overall biodiversity since plants underpin above-ground and below-ground food webs.

Nutrient cycling is the transformation of nutrients contained in minerals from the original bedrock and in dead biomass into simple molecules or compounds that are assimilable by plants and other organisms. For instance, nitrogen inputs are naturally made through soil organic matter (SOM) transformation by decomposer organisms and biological fixation from the atmosphere through symbiosis between bacteria and plants, especially legumes. Organic matter also provides natural chelates that maintain micronutrients in forms that plants can uptake. All soil organisms contribute to nutrient cycling through a number of physical (fragmentation of litter, bioturbation and transport of soil particles) and chemical (organic matter mineralization, transformation of molecules by a large number of enzymes) processes. Plant uptake of nutrients stimulates plant growth, in turn increasing the amount of plant-based inputs back into the soil. In total, 80 to 90 percent of primary production enters the below-ground system via detritus (Bardgett, 2005; Coleman *et al.*, 2017). Dead plant material (litter, woody debris, roots), animal excrements and carrion make up the majority of potential sources of nutrients for plants, soils and their biodiversity. Soils also receive inputs of nutrients from human activities. For instance, agricultural soils can receive large inputs of fertilizer (mineral and/or organic) K, N or P; N and S can be indirectly added at the soil surface of acid rain while NOx and SOx can enter soils through from the burning of fossil fuels.

Turnover processes and nutrient movement (mineralization, immobilization) depend on the climate, physical properties (texture, structure) and chemical parameters (pH, SOM, cation exchange capacity) of soils, plant cover and plant chemical composition, and taxonomical and functional diversities of soil animals and microorganisms (Colman and Schimel, 2013; Coleman *et al.*, 2017). There are still many gaps in our understanding of the mechanisms of the process of decomposition and its regulation. Nutrient cycling can be altered by soil moisture content and temperature, since these factors directly impact microbial biological activity. In this context, global warming can modify nutrient cycling such as N pools and fluxes, leading to higher N losses from soil because of the increase of net N mineralization (Bai *et al.*, 2013). Furthermore, soil degradation is one of the four major global challenges currently facing humanity and has been shown to impact the interrelated biogeochemical cycles of C, N, and P (Quinton *et al.*, 2010; Berhe *et al.*, 2018). Urbanization can also have an impact on nutrient cycling through indirect effects (such as increase of temperature, increase of N deposition because of traffic) and direct effects through management practices that alter bio-physico-chemical soil parameters

(Lorenz and Lal, 2009; Bittman *et al.*, 2019). Finally, non-native plant invasions can also alter nutrient cycling (Jo *et al.*, 2017). Long-term field experiments might help to follow potential acclimation of biodiversity and thus evolution in nutrient cycling functioning in the face of human disturbance. (Figure 3.4.1.1).

According to the studied spatial scale, excess of nutrients can induce toxicity for plants, but also alteration of global nutrient cycling (Lavelle *et al.*, 2005). For instance, in N cycling, plants and microbes take up nitrate produced by transformation of SOM, but a significant amount may also be lost from soil by leaching or runoff, especially in the case of excess fertilizer. This leads to eutrophication of soils, ground waters and streams. Systems of crop-livestock production are the largest cause of human alteration of the global N and P cycles (Bouwman *et al.*, 2013), since it is common for farmers to apply nutrients in excess to make it more available to crop plants. (Figure 3.4.1.1). Global studies show that P deficits covered 29 percent of the global cropland area, and 71 percent had P surpluses. This nutrient overloading causes not only environmental damage but also a financial loss. Supporting soil users in any interventions to reduce nutrient overuse is a first necessary step.

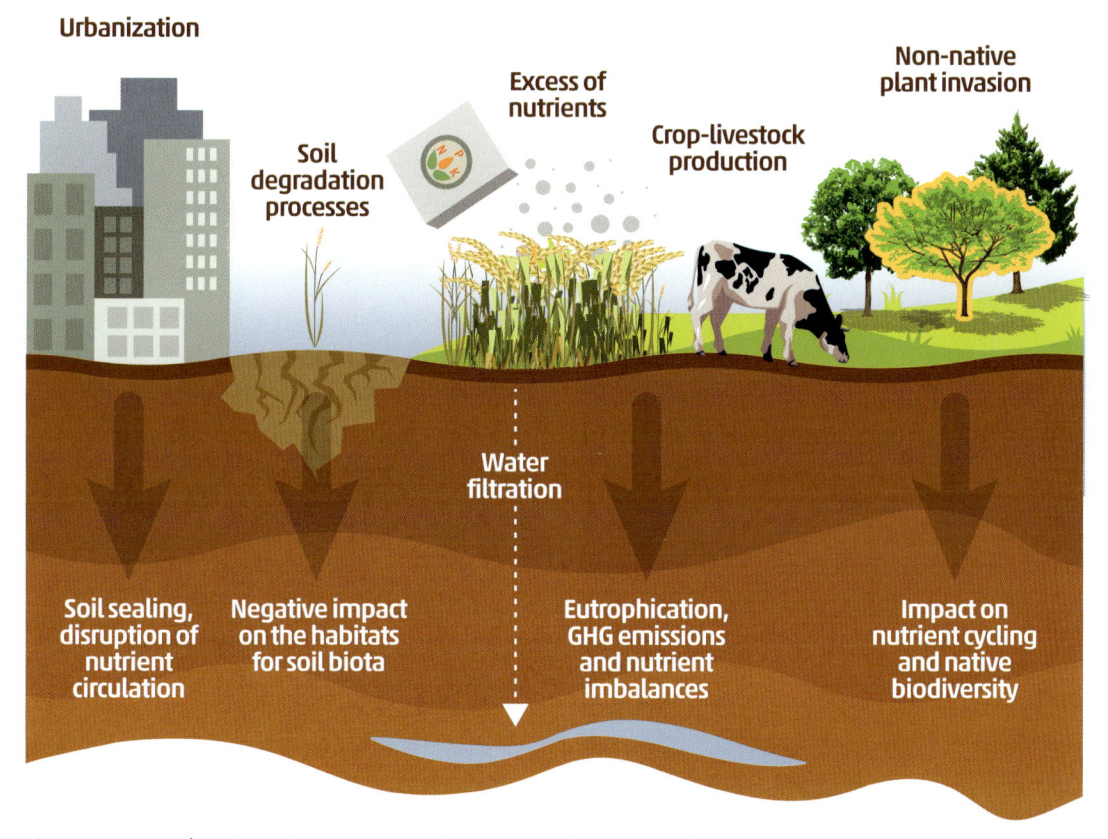

Figure 3.4.1.1 | Main drivers for the alteration of global cycles

Main drivers for the alteration of global cycles and their side effects on the environment.

3.4.2 | FOOD PRODUCTION

Achieving global food security is one of the greatest challenges of our time (Burns *et al.*, 2010, Lobell *et al.*, 2008). The mismatch between the rate of human population growth and global food production has long been recognized (Malthus, 1798). We need to ensure food production to meet future demands, on less land, using fewer inputs and safeguarding ecosystems. Achieving this goal in a time of significant environmental change makes the challenge all the more significant. Food security by its very definition demands that enough food of sufficient quality be produced. We need enough food to meet the energy and nutritional requirements of the global human population (SDG 2) while ensuring environmental sustainability.

The majority of the world's human population eats a soil-grown, plant-based diet. Even where diets are rich in animal products, livestock rely on soil-grown plants (fish being a notable exception). The quantity and nutritional quality of crops is very much a product of the soils in which they grow. The link between crop production and soil quality is well established. For example, that 30 percent of the world's human population is affected by Zn deficiency (Alloway, 2008) is not surprising, given that the soils of almost half of the world's cereal growing regions are low in Zn (Cakmak, 2002; Graham and Welch, 1997). Taken together, it is clear that efforts focused on achieving food security cannot ignore the link between soil and food.

Agriculture comes in many shapes and forms, ranging from high-input industrial-scale agriculture to subsistence farming. In some cultures, soil invertebrates are an important direct source of protein (Paoletti *et al.*, 2000). Most of the world's agriculturists are subsistence farmers, where the food farmers eat is the food they produce. Whereas industrialised large-scale agriculture depends heavily on complex resources (such as inputs of water, nutrients, and pesticides) and distribution and supply chain and networks, subsistence farmers may not have ready or reliable access to supply chains and markets. They lack the access to resources that can be used to boost production (such as mineral fertilizer) or to protect crops (such as pesticides), or to participate in markets.

The lack of access to industrial inputs (such as synthetic fertilizers and pesticides) means subsistence farmers are organic farmers by default (Cardoso and Kuyper, 2006). As such they rely heavily on soil biota and the ecosystem services they provide to support production. Similarly, soil biota play an important role in high-input agricultural systems. For example, soil organisms play a key role in nutrient cycling, including the transformation of nutrients into forms that are more or less available to plants (ammonium versus nitrate), more readily leached into waterways (nitrate), or converted into greenhouse gasses (nitrous oxide). Soil biota also play a key role in the cycling of C in soils, including increasing soil C which can help mitigate climate change, while improving soil structure and hence water retention and thereby reduce risk of soil erosion. Further, soil biota can form beneficial associations with plants that can symbiotically fix N (rhizobia and legumes) and take up and deliver nutrients including P, Zn and N to plants (mycorrhizas). Soil biota can also be significant pests and pathogens that cause significant crop losses. Taken together, soil biota provide important ecosystem services that every form of agriculture relies on, or is affected by, in differing ways and to different extents.

The development of high-yielding crop varieties and access to pesticides and synthetic fertilizers has resulted in substantial increases in global food production following the Green Revolution (Khush, 1999). However, if we are to achieve global food security in a sustainable manner, much more needs to be done (Lynch, 2009). Aside from the fact that many of the world's farmers do not have ready access to synthetic fertilizers, there are also increasing concerns around access to easily mined sources of some mineral fertilizers (such as P) and the embedded energy in others (such as N). This has significant implications for large-scale industrial agricultural systems that rely on such inputs. Consequently, attention is turning to alternative sources of nutrients, such as compost, manure and crop residues (Cavagnaro, 2015). In addition to providing nutrients, organic materials can also increase soil C, thereby providing other benefits. While such sources of nutrients may be more readily available to subsistence farmers (and indeed are currently in use), they rely on soil biota to make the nutrients in them available to plants (via mineralization; Ng *et al.*, 2014). Moreover, this reliance on soil biological processes to drive and regulate nutrient supply to plants can make organic amendments less predictable compared to synthetic fertilizers.

Plant growth-promoting bacteria (PGPB), phytobacteria found in the soil that stimulate plant growth and improve soil and plant health, were described over one hundred years ago (Lugtenberg and Kamilova, 2009; Backer *et al.*, 2018; Olanrewaju *et al.*, 2017). A subset of the PGPB are the plant growth-promoting rhizobacteria (PGPR) which colonize the plant's roots. PGPR have increased nutrient uptake, provided biocontrol, and improved plant health. Likewise, drought- and saline-tolerant plants have been linked to microbes that provide the plants with ability to tolerate these abiotic stresses. When these microbes were added to susceptible cultivars and then exposed to these stresses, the plant biomass and yield increased compared to the non-inoculated stressed plants (Naylor and Coleman-Derr, 2018; Kearl *et al.*, 2019).

3.4.3 | RAW MATERIALS

Production of raw materials from biomass, especially wood, fibres and biofuels, are the major provisioning services of soil following food production. Fibres are used for such products as textiles and rope, while forestry has been a major provider of raw material for thousands of years for heating, building materials and even decoration. Within the last decades, wood, agricultural crops and agricultural waste have increasingly been used for biofuel production. It is anticipated that the use of raw materials for biofuel production will increase in the near future to meet the demands of reduced GHG emissions

(Hoekman and Broch 2018; Qin *et al.*, 2018). Hence, these raw materials are very important to produce carbon neutral energy.

The production of the majority of raw materials has the same or even more severe environmental issues as does agriculture for food production with use of agrochemicals, such as increasing risk of erosion and reducing soil carbon. This is due to the increase in monoculture and removal of all organic matter from the agricultural field (Hoekman and Broch, 2018).

Various climate mitigation initiatives of green energy (supporting SDG 13) by using raw plant and wood material for biofuel potentially change the land and soil management. This was described by Mishra *et al.* (2019), who evaluated the ecosystem services in alternative landscape scenarios. Land use systems that integrate woody vegetation with livestocks and/or crops represent high nature and culture value (HNCV) agroforestry, have higher above-ground biodiversity (Moreno *et al.*, 2018), reduce erosion risk and nutrient leaching, and have high production of raw materials. Crops (including trees) that are grown for bioenergy production often have a longer growth time, which generally results in increased soil carbon content. Thus, HNCV agroforestry will probably benefit soil biodiversity.

3.4.4 | CLEAN WATER

Essential for agriculture and critical for the survival of people, water quality and availability are crucial for numerous SDGs, from SDG 6 (clean water and sanitation) and SDG 14 (life below water) to SDG 2 (zero hunger) and SDG 3 good health and well-being) for its essential role in agriculture and food security. However, the pollution of water and the increasing frequency and severity of droughts and floods resulting from climate change, unsustainable management of natural resources and inadequate sanitation constitute major threats to our human societies and to biodiversity in general.

The influence of soil biodiversity on water dynamics and quality is often complex and varies with the environment. The influence of microorganisms is usually indirect and results from their impact on soil organic matter dynamics, which in turn affects soil aggregation and soil porosity dynamics as well as the composition of the soil solution (such as the amount of dissolved organic carbon and minerals). An exception to the rule is, however, biological crusts produced by cyanobacteria and mosses that control water dynamics in arid environments (Ram and Aaron, 2007). Conversely, larger soil organisms have a direct impact on soil structure through their ability to move inside the soil, then impacting a large number of ecological functions, including those that control water infiltration, diffusion and retention, as well as the susceptibility of soil to resist to wind and water erosion. Among soil organisms, animals from the soil macrofauna group (that is, soil invertebrates larger than 2 mm) such as earthworms, termites and ants are considered to play important roles in controlling soil structure dynamics (Bottinelli *et al.*, 2015). They are commonly named soil "bioturbators" and "engineers" because of their large population and activities in temperate and tropical ecosystems (Lavelle *et al.*, 1997; Jouquet *et al.*, 2006). Their foraging and burrowing activities, as well as their ability, in the case of social insects, to create nest structures with specific soil properties in soil, has an important impact on soil structure dynamics and hence on water dynamics and quality in soil. Although less widespread, other soil macro-invertebrates can also significantly impact soil structure and water dynamics because of their role in alteration of porosity, soil surface microrelief, soil aggregation and bulk density – for example, beetles' larvae (Brown *et al.*, 2010; Chcik *et al.*, 2019) and millipedes (Toyota *et al.*, 2006; Fujimaki *et al.*, 2010).

The sensitivity of ants to environmental disturbance, combined with their great functional importance as ecosystem engineers, make them not only powerful monitoring and assessment tools in restoration programs but also key actors in ecological engineering (Bulot *et al.*, 2014). The size of the entrance of ant nests as well as their architecture also have been shown as impacting bulk density (Cammeraat *et al.*, 2008) and water infiltration (Eldrigde *et al.*, 1993); indeed, large nests decrease bulk densities and increase water infiltration. At a smaller spatial scale, soil mesofauna (that is, soil invertebrates smaller than 2 mm) can actively make microtunnels in the soil, and they play a role in soil structure alteration through litter comminution, casting and other mechanisms of micdisintegration. Their interaction with microorganisms can also indirectly influence soil structure. However, in comparison with earthworms, termites and ants, the effects of these smaller organisms remains poorly known.

Soil macrofauna can influence soil hydrological properties at different scales of observation and through antagonistic processes. At a small scale, any changes in clay and soil organic matter contents, as well as in soil porosity, are likely to influence water-holding capacity. At a medium scale, the production of a dense network of foraging galleries connected to the soil surface usually improves water infiltration. This network can occupy a significant volume of soil in some situations (Capowiez *et al.*, 1998; Mando *et al.*, 1999; Buhl *et al.*, 2004; Perna *et al.*, 2008), and obviously these large macropores are of primary importance in the regulation of water infiltration, the diffusion of solutes and therefore water quality (Ehlers, 1975; Nkem *et al.*, 2000; Léonard and Rajot, 2001; Cammeraat *et al.*, 2002; Dominguez *et al.*, 2004; Zehe *et al.*, 2010). However, the translocation of soil on the ground in the form of earthworm casts, termite and ant sheeting or mounds can either increase soil roughness and favour water infiltration, or rapidly generate structural crusts that foster water runoff and soil detachment (Jouquet *et al.*, 2012; Bargués Tobella *et al.*, 2014). The degradation of mound nests by rain or organisms (for example, bears and pangolins breaking termite nests for feeding purposes) can also form seals that locally reduce water infiltration (Traoré *et al.*, 2019). Therefore, the final impact of soil fauna on soil hydrological properties results from the balance between processes favouring soil infiltrability (such as the formation of open galleries on the soil surface) and those fostering the formation of impermeable erosion crusts and the detachment of soil aggregates. Finally, at a larger scale, the concentration of nutrients and the presence of specific vegetation on mounds produced and maintained by soil fauna activity can also have an impact on the hydrological characteristics of watersheds in certain situations. They can contribute to better water infiltration, preferential flow and groundwater recharge (Ackerman *et al.*, 2007; Bargués Tobella *et al.*, 2014). Recent studies also suggest that termite mounds may increase the robustness or resilience of African dryland ecosystems against water shortage and desertification (Bonachela *et al.*, 2015).

Compared to the abundant literature focusing on the impact of soil macrofauna on soil structure, the effect of soil organisms on water quality remains poorly studied. Soil bioturbation and the creation of interconnected soil macropores by soil macrofauna can increase nutrient leaching below the root zone. This has been observed mainly with earthworms and using short-term experiments and lysimeters, where the addition of earthworms significantly increased the amount of soil leachate and the leaching of NH_4^+, NO_3^- and dissolved organic nitrogen in cultivated environments (Subler *et al.*, 1997; Dominguez, 2004; Jouquet *et al.*, 2010). Bioturbation by earthworms can also impact the mobility and leaching of heavy metals from casts (Udovic *et al.*, 2007), but in an experiment carried out in controlled conditions, Amossé *et al.* (2013) did not find any significant impact on the amount of bacteria and viruses in water. These examples are, however, to be considered with caution because the impact of soil fauna is often site- and species-specific, because interactions between organisms have never been taken into account, and because these studies were carried out in controlled or semi-controlled conditions, making it difficult to generalize from them.

Two examples of the utilization of soil biodiversity for improving ecosystem functioning or for the rehabilitation of degraded lands can be found in (i) the intensification and externalization of earthworm activity and (ii) the stimulation of termite activity in the field. The intensification and externalization of earthworm activity using epigeic earthworms is at the core of the production of vermicompost and the treatment of liquid manure. In the last decades, vermitechnology has been applied to the management of various types of wastes and sludge, to convert them into vermicompost for increasing land fertility. Although this practice is gaining in popularity, surprisingly few studies have questioned their impact on water quality and availability. Studies carried out in northern Viet Nam showed that the leaching of mineral nutrients and then the pollution of water can be significantly reduced if mineral nutrients or compost are replaced by vermicompost (Jouquet *et al.*, 2011; Doan *et al.*, 2013). Similarly, vermifiltration is an efficient and low-cost technology for processing organically polluted water and reducing the transfer of minerals and pathogens to the aquatic system (Li *et al.*, 2008; Morand *et al.*, 2011).

In arid or semi-arid environments, bioturbation is mainly carried out by termites. The ability of termites to promote water infiltration in crusted soils as part of soil rehabilitation and vegetation cover regeneration has been strikingly demonstrated in Africa (Kaiser *et al.*, 2017), Asia (Pardeshi and Prusty, 2010) and Australia (Dawes, 2010). In these studies, the application of mulch or organic matter on or into the soil, as in the case of the agricultural and forestry "zaï" systems (see Roose *et al.*, 1999 for a description of the Zaï agricultural practice), triggered termite activity which then created burrows through the crusted soil surface. Increase in termite activity can therefore result in an enhancement in water infiltration by a factor of 1.5 to 25, an increase in water retention and a reduction in the bulk soil density of the upper soil layer (Cheik, 2019).

Since the capacity of soil biodiversity to enhance the water quality and quantity is indirect, research studies are scarce and the awareness of this capacity is low. Better understanding of the functional roles of soil biodiversity could help stakeholders to remediate soils by considering soil organisms in their sustainable ecosystem restoration plan. Moreover, it is obvious that the interactions between microbial and faunal communities and between below-ground and above-ground compartment should be considered in ecological restoration context. In this vein, long-term study cases should be carried out *in situ* to help scientists and stakeholders to go forward together and find indicators of restoration success. There is also the need to create a global database of management projects to connect the initiatives around the world. In parallel, transdisciplinary approaches, such as putting together ecologists and economists, might increase awareness. Finally, one hot topic to focus on could be emerging pollutants such as plastics and microplastics and their interaction with soil biota, in order to find a solution to fight this issue by considering both soil and water.

3.4.5 | SOIL BIODIVERSITY AND HUMAN HEALTH

Ever since the nineteenth-century microbiologist Louis Pasteur discovered that soil can harbour *Bacillus anthracis*, the bacterium that causes Anthrax, scientists have recognized that life in the soil is a source of human disease (Thorkildson *et al.*, 2016). From the perspective of biomedicine, *risk* and *disease* continue to be the primary focus of studies of the link between soil and human health (Jeffrey *et al.*, 2011). But recently, a countervailing narrative has emerged: far from making us sick, most soil creatures help protect and maintain our health (Wall *et al.*, 2015).

As discussed elsewhere in this report, soil biodiversity helps prevent erosion, filter and conserve water, and break down environmental pollutants. It also plays a role in capturing and storing atmospheric carbon—which might help fight climate change. Even if this were all soil organisms did, they would clearly be central to our well-being and survival on this planet. But emerging research suggests that soil biodiversity has a more direct impact on our health by boosting the nutrient content of our food, protecting us from foodborne illness, and modulating our immune response. This section focuses on these three associations and discusses how greater collaboration between health science and agriculture is needed to better understand these links, and to protect soil and human health.

Soil biodiversity, nutrient cycling, and nutrition

The phytobiome – a region surrounding the roots of plants comprised of non-living structures, plants, and micro- and macro-fauna – influences plant yield and nutrition and, by extension, human health and nutrition (Leach *et al.*, 2017). Plants secrete compounds that feed nearby organisms, and those organisms have a dynamic exchange with the plant. This relationship enables plants to capture essential minerals and to manufacture various chemicals, such as antioxidants, that protect them from pests and other stressors (Garcia *et al.*, 2018). When we consume these plants, these antioxidants benefit us by stimulating our immune system, regulating our hormones, and slowing the growth of human cancer cells (Manach *et al.*, 2004).

The abundance and profile of microorganisms can vary across plant habitats and plant genotypes, but one consistent finding is that biodiversity within the phytobiome hastens plant growth, increases plant yield, and increases plant nutrient density. Complementing these findings, several studies show that organic farming methods (which are known to promote soil biodiversity) produce plants with a higher concentration and variety of antioxidants (Reganold *et al.*, 2010; Barański *et al.*, 2014; Ren *et al.*, 2017). While there is much to discover in the phytobiome, this complex zone of plant-organism-soil interaction has the potential to enhance food security and improve human nutrition.

Soil biodiversity and food safety

Soil biodiversity helps mitigate the risk of foodborne illness by boosting plant defenses against opportunistic infections. For example, *Listeria monocytogenes* is found in low concentration in many agricultural soils, but its pathogenicity depends on the richness and diversity of soil microbial communities (Vivant *et al.*, 2013; it also depends on soil type, pH and other soil factors). In one study, bacteria spread rapidly in autoclaved soil but was suppressed in soil containing an abundance of other species. Other research shows that small groups of rare species might have a disproportionately large effect on counteracting pathogenic soil organisms (Cernava *et al.*, 2019). In general, plants surrounded by greater diversity of microbes in their root structure have greater fitness and resilience against pests and pathogens (Johnson *et al.*, 2013). Understanding and promoting these soil dynamics could help protect humans from foodborne illness.

Soil biodiversity and immune modulation

In the early 1900s, scientists began identifying antibiotic substances in soil that could fight specific microbial infections or more generally modulate human immune response. Since then, a wide range of therapeutic drugs and vaccines have been derived from (or inspired by) substances produced by soil organisms. For example, amphotericin, the principal systemic medication used to treat invasive fungal infections, was first recovered from a strain of bacteria growing in the soil of a riverbed in Venezuela. Bleomycin, also originally identified in soil, is used as chemotherapy for a variety of common cancers (Vitorino *et al.*, 2017).

While most biopharmaceutical research is focused on identifying unique microbes (or their by-products) that can be developed into biotherapeutics, new technologies that make it possible to study the metagenome – or collective genome – in an environmental sample have sparked an interest in exploring how complex microbial communities in soil and other indoor and outdoor environments influence human immune and nervous response via the skin, gut, and lungs. The route of exposure and the profile of microbes vary from study to study, but species diversity consistently emerges as an important health mediator. This may be due to the fact that we co-evolved with a constellation of environmental microbes (often referred to as "old friends") and they continue to communicate with our internal physiology. (Langgartner *et al.*, 2019).

A series of studies focused on farm children in Europe and the United States of America illustrates this phenomenon. Researchers observed that children raised on traditional dairy farms (farms that use little fossil fuel or chemical inputs) have lower rates of allergic disease than children raised on conventional farms. Findings suggest that messages from organisms found in soil and on farm animals "program" the immune system and determine how farm children respond to allergens later in life. This programming likely starts *in utero* and continues through the first few years of life (von Mutius, 2016; Stein *et al.*, 2016; von Mutius, 2018).

Separate evidence suggests that environmental microbial inputs from soil and other niches can bolster resilience in humans and modify the risk for neurodevelopmental and psychiatric pathology. This line of inquiry is in its infancy but suggests that early exposure to a diverse collection of soil microorganisms might help prevent chronic inflammatory diseases including allergy, asthma, autoimmune diseases, inflammatory bowel disease, and depression (Lowry *et al.*, 2016).

Transdisciplinary collaboration is needed to protect human health

Soil biodiversity plays a central role in preserving human health through a range of pathways including water purification, climate stabilization, nutrient and food security, and immune regulation. Additionally, the dramatic global rise in food-related chronic diseases – including obesity, diabetes, stroke and heart disease – can largely be traced back to crops produced using soil-degrading agricultural practices (Franck *et al.*, 2013; Mattei *et al.*, 2015; Smith *et al.*, 2015). Such practices include encroaching on wild space, deforestation, contaminating and diverting water supplies, monocropping, excessive tilling and the overapplication of pesticides, herbicides and fertilizers. Ongoing efforts to do true cost accounting have increased awareness about the direct and indirect health costs generated by these agricultural practices and the cost savings that would accrue from regenerating soil and promoting human health from the soil up (http://www.fao.org/nr/sustainability/full-cost-accounting/en/). By most estimates, these savings (measured in years of disability and health care expenditures) could be enormous (IPES Food report, 2017).

Despite this evidence of a strong link between soil health and human health, it is rare for health institutions – international, national, or regional – to take an active and visible role in protecting soil. The World Health Organization, for example, has issued dozens of white papers on the dangers of soilborne illness (with a particular focus on helminthiases) but only one statement encouraging Good Agricultural Practices (GAP) and other activities that build soil biodiversity (WHO guidelines, 2003). In the United States of America, the National Institute of Food and Agriculture (NIFA) awards nutrition research grants through their Agriculture and Food Research Initiative (AFRI), but these rarely attract human health researchers. A few non-profits and academic centres, including Health Care Without Harm, the Planetary Health Alliance and One Health Initiative, attempt to bridge the artificial divide that exists between soil health and human health, but these groups have limited resources and influence. We need more collaboration between the health and agricultural sectors to protect soil and to address mounting health crises related to climate change, soil loss, and industrial farming.

Health institutions urgently need to educate their members about the role soil plays in human health, and about activities (including those outlined elsewhere in this report) that preserve soil or contribute to its destruction (Dwivedi *et al.*, 2017). They need to participate in transdisciplinary research and innovation that builds and preserves soil structure and biodiversity. And most importantly, the health sector needs to leverage its immense influence and financial resources to shape policies, public health education programs, land use planning, and procurement practices that create healthy soil and that protect this rapidly disappearing health resource.

Environmental exposures are responsible for 70 percent of morbidity and mortality world-wide. Allergies, chronic lung diseases and other conditions caused or triggered by environmental factors are attributable to a subject's genetic responsiveness to distinct environmental stressors. Thus, it is obvious that environmental quality is a key factor driving human health, which has been acknowledged in SDG 3 (ensure healthy lives and promote well-being for all at all ages). Well-known examples of such complex interactions include emerging infectious diseases, food safety, the selection of antimicrobial resistant pathogens and the large number of pollutants from the environment, which strongly impact our health (Xi *et al.*, 2017). This is in fact not a new finding but has been already proposed almost 2 500 years ago by Hippocrates. He claimed as one of the first that our health is triggered by the kind of land where we are living (Brevik and Sauer, 2015).

Nowadays it is well accepted, both by scientists and decision makers and by some members of the general public, that our health will be more and more difficult to maintain on a polluted planet with diminishing resources (Destoumieux-Garzón *et al.*, 2018): a document provided in 2017 as a result of a joint meeting of the European Union ministers for environment and health, as well as the World Health Organization stated that "environmental factors that could be avoided and/or eliminated cause 1.4 million deaths per year in the EU" (Landrigan *et al.*, 2018). The authors postulated that "public authority shares the common responsibility for safeguarding the global environment and for promoting and protecting human health for all environmental hazards across generations and in all policies." This has been summarized as the "One Health Concept", which is an important contribution to the discussion of how to improve and sustain our health on a planet that will soon host 10 billion people. As one of the first consequences, "One Health" has been implemented in the framework of ecosystem services provided by soils (Keith *et al.*, 2016). In addition, the new focus on interactions between humans and their environment as a trigger for health and disease also stimulated new research efforts in medicine to unravel fundamental processes of immunity, repair and regeneration, tolerance and disease development. It is obvious that a consequent implementation of the "One Health Concept" into medicine requires a very strong personalized component, taking individual exposure levels with different environmental factors into account as well as individual genetic disposition. Based on such data, strategies for a general improvement of environments should be developed in combination with new personalized strategies for therapy of diseases caused by losses in environmental quality. Some important examples of how the abiotic and biotic soil factors can influence human health are given in the following.

Soil microbes for human, plant and animal health

A handful of soil may contain between 10 and 100 million microorganisms, belonging to more than 5 000 different taxa (Ramirez *et al.*, 2015) and providing a wealth of metabolites with many applications. Due to the immense potential that soil microorganisms hold for pharmaceutical industries, bioremediation, and agriculture, characterizing and preserving soil biodiversity are essential in order to facilitate the discovery of new drugs and therapeutic measures for treating and controlling diseases. The object of the present subsection is to present examples of scientific observations illustrating the impact of soil microbes on human health and on disease prevention. Once transferred to the public discourse, these observations and other similar findings could coalesce into a solid argument in favour of soil biodiversity preservation.

An important contribution of soil microorganisms to human and animal health is the provision of antibiotics. An antibiotic is a low molecular weight product from a microorganism, with the capacity to kill or inhibit the growth of other, susceptible, microorganisms. As antimicrobial agents, antibiotics are used widely to treat infectious diseases (Aminov, 2010; Crofts *et al.*, 2017). Another potential use of soil microorganisms in health care stems from viruses known as bacteriophages, which are parasitic non-cellular agents that can kill their bacterial host, and therefore are being investigated as a cure for bacterial infections (Burrowes *et al.*, 2011; Frampton *et al.*, 2012; Jones *et al.*, 2007). Harbouring an immense diversity of viruses, soils can be considered as a promising reservoir of bacteriophages useful to phage therapy. While this subsection mainly deals with antibiotic-producing microorganisms and bacteriophage with potential for phase therapy, it should be mentioned here that soils also serve as reservoirs of fermentative bacteria and fungi useful to food processing, and of health-sustaining components of the gut microbiota (on this last point, see Blum *et al.*, 2019).

Soil microorganisms for disease control

Soil microorganisms have been contributing to protect humans and animals against infectious diseases for thousands of years, particularly through antibiotic production. Although industrial-scale production of antibiotics started only after the 1940s, human exposure to antibiotics has a very long history. Ancient human populations could have been exposed to antibiotics through a number of mechanisms: (i) application of soil on skin; (ii) traditional remedies (composed of soil and herbal products); (iii) ingestion of material contaminated with soil or antibiotic-producing microorganisms through diet; (iv) direct ingestion of soil for medicinal or spiritual purposes (referred to as '*geophagy*').

For instance, the presence of tetracycline has been detected in human skeletal remains from ancient Sudanese Nubia that date back to 350-550 CE (Nelson *et al.*, 2010) and in skeletal remains from the Dakhel Oasis, Egypt, that date back to the late Roman period (36 BCE to 400 CE) (Cook *et al.*, 1989). Scientists suggest that lack of traces of bone infections in skeletal remains from the ancient Sudanese Nubian population and the Dakhel Oasis indicate that exposure to antibiotics may have provided some protection against infectious diseases in ancient civilizations (Cook *et al.*, 1989; Nelson *et al.*, 2010).

In another example, the presence of antibiotic-producing actinomycetes was reported in Red Soil from Jordan, which was used for thousands of years for treating skin infections (Falkinham *et al.*, 2009; Aminov, 2010). In a study conducted with Jordan Red Soil, *Staphylococcus aureus* and *Micrococcus luteus*, bacteria that cause skin infections, were applied to sterilized and unsterilized soil and a bactericidal effect on these pathogens was observed with unsterilized soil only (Falkinham *et al.*, 2009). In the same study, a number of antibiotic-producing *Lysobacter*, actinomycetes and *Bacillus* species effective against *M. luteus*, *S. aureus*, *Escherichia coli*, *Mycobacterium smegmatis*, *Saccharomyces cerevisiae* and *Aspergillus niger* were isolated from the soil. Guo *et al.* (2015) reported that acidic Red Soil of China hosts various groups of bacteria producing secondary metabolites with antimicrobial properties.

In addition to antibiotics, other biotic factors such as protozoan or myxobacterial predation, lytic microorganisms and bacteriophages, as well as antimicrobial abiotic components such as clay minerals and ions of zinc, iron, copper, silver and other metals, may also provide disease control traits to soils used in traditional medicine (Otto *et al.*, 2013).

The modern antibiotic era

Prior to the discovery of antibiotics, infectious disease control depended on inorganic or synthetic antimicrobial agents such as inorganic mercury salts, copper salts, organo-arsenic derivatives of Atoxyl (Salvarsan and Neosalvarsan) and sulfonamidochrysoidine (Prontosil), with many side effects and poor efficacy (Aminov, 2010). The production of antibiotic compounds by *Pseudomonas aeruginosa*, a bacterium frequently reported in soils, was known by the late 1890s, when Emmerich and Low obtained an antibiotic called pyocyanase from extracts of this bacterium. It was used in hospital for treatment of bacterial diseases, but was eventually abandoned owing to inconsistent results and to side effects. In 1929, Alexander Fleming discovered the antibiotic activity of *Penicillium*, a fungus commonly found in soils. It was not until 1940 that the technique for mass production of penicillin was developed by Howard Florey and Ernst Chain. Penicillin is earmarked as the first antibiotic successfully trialed and mass produced. Its therapeutic use launched the modern antibiotic era. As the techniques developed by Fleming to identify antibiotic-producing microorganisms came into general use, a wide variety of antibiotics was discovered in the following years. Currently, nearly 80 percent of all antimicrobials in use are of microbial origin, and soils are the primary source of bacteria and fungi producing new molecules potentially useful to human, animal and plant disease control (Grossbard, 1952; van Elsas *et. al.*, 2008; Falkinham *et al.*, 2009; Aminov 2010; Guo *et al.*, 2015; Rogozhin *et al.*, 2018).

Antibiotic discovery was at its peak from the 1950s to the 1970s, and all the classes of antibiotics we know of today were discovered during this period. Since then, the rate of new discoveries has declined (Aminov, 2010; Crofts *et al.*, 2017). Antibiotics are currently used in prevention and control of infectious diseases in human healthcare, animal husbandry, veterinary care, cropping (mainly fruit, vegetable and flower production), and postharvest management of perishables (McManus *et al.*, 2002). Lack of policies and of proper regulations have resulted in mismanagement and misuse of antibiotics in the above-mentioned sectors (McManus *et al.*, 2002; O'Neill 2016), with antibiotic residues being released in soil and water environments.

The consequences of widespread antibiotic use

In 1945, Alexander Fleming cautioned about potential resistance development to penicillin (Aminov, 2010). It is now commonly accepted that all microorganisms will eventually develop resistance to any antibiotic. The widespread use and misuse of antibiotics maintains high selection pressure for antibiotic resistance in a variety of environments (including human and animal gut, soil, waterbodies and others). The spread of antibiotic resistance has become one of the major challenges facing human and animal health care, with acquisition of multidrug resistant traits by aggressive human pathogens being of concern (Crofts *et al.*, 2010). According to the Food and Agriculture Organization of the United Nations (FAO), antimicrobial resistance by disease-causing microorganisms is responsible for 7 million human deaths annually on a global scale (O'Neill 2016). O'Neill (2016) emphasizes "that by 2050, 10 million lives a year and a cumulative USD 100 trillion of economic output are at risk due to the rise of drug-resistant infections if we do not find proactive solutions now to slow down the rise of drug resistance."

The soil is the recipient of a spectrum of antimicrobial resistance determinants (namely residues and genes) that we release in the environment. Some soil-dwelling heterotrophic bacteria belonging to orders *Burkholderiales*, *Pseudomonadales* and *Actinomycetales* use antibiotics as a carbon source (Dantas *et al.*, 2008), implying that their activity might decrease the antibiotic residue concentrations in soil environments.

Phylogenetic evolution, rather than horizontal gene transfer, is thought to determine the genetic structure of the "soil resistome" (van Goethem *et al.*, 2018). Hence, the genetic flow of antibiotic resistance genes would be largely vertical in the soil, and as such would be uncoupled from the mostly horizontal flow occurring in clinical settings (Forsberg *et al.*, 2014). This observation, associated with the ubiquity of antibiotic resistance in diverse environments, with the genetic relatedness of the soil and of the human gut resistomes (Nesme and Simonet, 2015) and with the capacity of gut bacteria such as *Escherichia coli* to survive for extensive periods in soils (Joergersen and Wichern, 2018), suggests that biodiversity and antibiotic resistance are two independent characteristics of the soil microbiome, influencing one another. The effect of antibiotic resistance determinants on soil biodiversity has been investigated (Martinez 2009; van Goethem *et al.*, 2018), and several studies have emphasized the importance of antibiotic resistance genes for bacterial ecophysiology at the ecosystem level (see the review by Nesme and Simonet, 2015). Reciprocally, diverse populations of soil organisms have been postulated to limit pathogen numbers (Brevik *et al.*, 2018; Wall *et al.*, 2015), and the same may be true for survival of antibiotic-resistant bacteria passing from the gut to the soil.

Opportunities and knowledge gaps

Until recently, discovery of antibiotic-producing microorganisms depended on culture-based technologies. However, the high-resolution power of metagenomics led researchers to take full measure of the depth and breadth of soil biodiversity and to realize that over 90 percent of soil microorganisms, including many of ecological importance, escaped detection by culture-based techniques (van Elsas *et. al.*, 2008; Xin *et al.*, 2015). A comparison of culture-based and metagenomic characterization of soil fungi from a dry, mixed evergreen forest indicated that the fungi commonly occurring in culture, such as *Aspergillus*, *Penicillium* and *Trichoderma*, formed a lesser percentage of the detected fungal community when the diversity was analyzed using metagenomics, as compared to a culture-based approach (Dandeniya and Attanayake, 2016). Through culture-based methods, at most ten different fungal types were distinguished based on colony morphology, whereas the metagenomic analysis indicated a hundred-times greater diversity. *Pestalotiopsis*, *Cladosporium*, and *Chaetomium*, genera known to include antibiotic producers, were among the fungi reported in these forest soils (Dandeniya and Attanayake, 2016). Many other studies yielded similar observations, showcasing the soil as a virtually untapped source of antibiotic-producing microorganisms.

Soils from extreme environments appear particularly promising in this respect. Extreme conditions require unique metabolic adaptations. Rogozhin *et al.* (2018) isolated a new antibacterial and antitumour substance from a strain of the fungus *Emericellopsis alkalina* obtained from an alkaline soil. Working with soils with pH values ranging from 2.6 to 6.6, Guo *et al.* (2015) identified a diverse group of acidophilic actinobacteria producing a wide range of novel antimicrobial compounds.

In addition to promising discoveries of new antibiotics from highly diverse soil microbial communities, soil biodiversity holds other potentialities for medical practice. While it remains largely unexplored, the biodiversity of soil viruses offers the promise of bacteriophage therapy for alternative treatment of bacterial infections in humans and plants (Burrowes *et al.*, 2011; Frampton *et al.*, 2012; Jones *et al.*, 2007). Reported viral abundance in soils ranges from 2.2×10^3 gdw^{-1} (grams per dry weight) in desert sands from Saudi Arabia to 5.8×10^9 gdw^{-1} in forest soil from eastern Virginia in the United States of America, with the majority of soil viruses being bacteriophages (defined as viruses infecting bacterial hosts). Distinct viral communities exist in different soils, with most communities being dominated by tailed bacterial viruses (Williamson *et al.*, 2017). For the control and prevention of infections caused by *Pseudomonas aeruginosa* (such as pneumonia, bloodstream infections, urinary tract infections and surgical site infections) such tailed bacteriophages isolated from environmental water, hospital wastewater and sewage wastewater have been used (Pires *et al.*, 2015). Soil is a largely underexplored, yet promising, environment for isolating bacteriophages for phage therapy.

Soil microorganisms contribute to the healthy living of humans, animals and plants. Several soil bacteria and fungi are being used traditionally in the production of soy sauce, cheese, wine and other fermented food and beverages. Lactic acid bacteria that could potentially be used to produce heavy metal (cadmium and lead) -absorbing probiotic products have been discovered from mud and sludge samples (Bhakta *et al.*, 2012). Soils provide habitats for a variety of lactic acid bacteria belonging to *Lactobacillus*, *Lactococcus* and other genera (Chen *et al.*, 2005), opening the possibility that probiotic bacteria useful in food fermentation or other processes be isolated from soils.

Preserving the richness of soil biodiversity will keep the gates open for healthy living.

3.5 | REGULATING ECOSYSTEM SERVICES

Regulating ecosystem services are services by which ecosystems act as regulators of the natural processes that make life possible (such as clean air and water or waste management). They are often invisible and hence underappreciated. When they are damaged, the resulting losses can be substantial, and the services can be very difficult to restore. Soil biodiversity is essential for the continuous function of the regulating services.

3.5.1 | CLIMATE REGULATION

The regulating service of soil on climate is mainly focused on GHG emission and uptake (that is, emission of carbon dioxide, methane and nitrous oxide) and uptake or sequestration of carbon, and is important for SDG 13 (climate action). For climate regulating services, soil biodiversity has an important impact. With the ambitions of reducing GHG emissions to combat climate change, this ecosystem service has come into focus as a key function to achieve the Paris Agreement goals. Agriculture, forestry and natural environments all have significant impacts in this regard.

Globally, agroecosystems contribute 10 to 12 percent of all direct anthropogenic GHG emissions each year, with an estimated 38 percent resulting from soil N_2O emissions and 11 percent from CH_4 in rice cultivation (Smith *et al.*, 2008, 2014). These emissions come largely as a result of intensification that relies heavily on agrochemicals (fertilizers, herbicides and pesticides) and has consequently reduced functional group diversity in soils, particularly for larger soil biota (Postma-Blaauw *et al.*, 2010). In addition, the practice of recirculating organic waste and manure to the field also increases emissions of GHG. This has reduced the biological capacity for soils to self-regulate, leaving them vulnerable to changing environmental conditions (Brussaard *et al.*, 2007; Tsiafouli *et al.*, 2015). While these issues have been long recognized (Black and Okwakol, 1997; Giller *et al.*, 1997), specific attempts to design agricultural practices around sustainably managing soil biodiversity are relatively recent (Bender *et al.*, 2016). However, early indications suggest that this soil ecological engineering can create more resilient (Ciancio and Gamboni, 2017; Dubey *et al.*, 2019), adaptive (Birgé *et al.*, 2016) and resource-efficient (Leff *et al.*, 2016; Bowles *et al.*, 2017) agroecosystems that in turn reduce associated GHG emissions (Pagano *et al.*, 2017). Indeed, organic farming approaches that limit/exclude agrochemical use can reduce soil N_2O emissions by up to 40 percent (Skinner *et al.*, 2019). Similarly, new approaches in rice cultivation indicate that targeting cultivars to increase root porosity can increase methanotroph abundance (Jiang *et al.*, 2017), and application of purple non-sulphur bacteria can increase competition with methanogens for substrate availability (Kantachote *et al.*, 2016); each can significantly reduce CH_4 emissions relative to conventional practices. Current initiatives in agricultural practices are investigated with the aim of reducing climate negative impact of agriculture, especially focusing on reducing GHG emissions (Olesen *et al.*, 2018). These initiatives include adding nitrification inhibitors to manure and chemical fertilizers to reduce nitrous oxide emissions, control storage of manure and modifying barn conditions to decrease methane emission. (3.5.1.2) These initiatives are all dependent on well-functioning soil biota.

In addition to direct effects of climate change on soil organisms and biodiversity, indirect effects through above-ground vegetation changes could also threaten soil biodiversity. For example, there is accumulating evidence that warmer and drier conditions will benefit vascular plants at the expense of the peat-forming vegetation such as *Sphagnum* mosses (see for example Dieleman *et al.*, 2015). A recent study (Juan-Ovejero *et al.*, 2019) has shown that, under this scenario, mites in detriment of the more hydrophilic species (for example, enchytraeids dipteral larvae and many collembolans) will increasingly dominate soil communities, due to their different physiological adaptations to water stress. Similar shifts have also been observed in fungal communities, from ericoid mycorrhiza and arbuscular mycorrhiza under shrubs and grasses to ECM in forests (Hagedorn *et al.*, 2019). These functional shifts in soil biota communities are likely to exacerbate C losses from these systems through mining for nutrients.

The report produced by the Convention on Biological Diversity (CBD Technical Series no. 10) concluded that functionally diverse communities with high biodiversity would be able to better adapt to climate change than impoverished ones. However, in these C rich systems, where a low number of species performs soil functions and hence where ecosystem stability is low, designing activities that aim to conserve these key functional groups should be a high priority. This urgent activity, along with preventing habitat loss, fragmentation and degradation, will increase resilience of these vulnerable ecosystems and in turn will result in a greater likelihood of adapting and mitigating to climate change. Indeed, the first proposed target to achieve SDG 13 on climate action is to strengthen resilience and adaptive capacity to climate-related hazards and natural disasters in all countries.

Box 3.5.1.1 | Climate Change and Soil Biodiversity

Climate change and the conservation of biodiversity are recognized globally as being a common concern of humankind (UNFCCC, CBD). The ways we manage our land and soil, and its biodiversity, have the potential to have a profound influence at a local, regional and global scale. The relationship between climate change and soil biodiversity is complex. We are only beginning to scratch the surface of understanding the connections, feedbacks, threats and opportunities posed by the interactions. Nevertheless, science points to certain key messages that have fundamental implications for policymakers: climate change must be addressed as a matter of urgency; above ground biodiversity is declining at a staggering rate; and land management practices, and climate change itself, can disrupt crucial plant-soil biodiversity-carbon feedbacks. The science is clear that whether natural or working lands, undisturbed or restored systems with perennial vegetation and zero tillage management typically have greater ecosystem-level carbon and soil biodiversity, and resilience to climate change, than their more disturbed and degraded counterparts.

There is also increasing acknowledgment that addressing climate change is inextricably linked with critical issues of food security, secure water supplies, poverty eradication and sustainable livelihoods, and requires transformational change (Campbell *et al.*, 2018; Ripple *et al.*, 2019). The IPCC *Special Report on Climate Change and Land* points to the dire consequences of continuing with a business-as-usual approach (IPCC, 2019). However, by acknowledging the interconnections, and reconnecting land and food systems with environmental, social and cultural dimensions that take into account soil biodiversity, we can identify ways to address climate change, with synergistic impacts across a range of SDGs, especially SDG 1 (no poverty), SDG 2 (zero hunger), SDG 3 (good health and well-being), SDG 6 (clean water and sanitation), SDG 11 (sustainable cities and communities), SDG 12 (responsible consumption and production patterns), SDG 15 (life on land, including Land Degradation Neutrality (LDN)) and with other global agendas supporting these objectives (IPCC 2019; Lal *et al.*, 2018).

SDG 13, the UNFCCC and the Paris Agreement

The urgency of the need to take climate action is recognized in SDG 13, which calls on states "to take urgent action to combat climate change and its impacts" and includes the target to integrate climate change measures into national policies, strategies and planning. It acknowledges the United Nations Framework Convention on Climate Change (UNFCCC) as the primary international, intergovernmental forum for negotiating the global response to climate change. The UNFCCC, and the Paris Agreement concluded under it, require Parties to provide national inventory reports of anthropogenic GHG emissions by sources and removals by sinks; to communicate their nationally determined contributions (NDCs) to the global temperature reduction goal, as well as national adaptation plans (NAPs); and to report on progress with them. NDCs and NAPs can include commitments that aim to harness and increase the contribution of soil biodiversity to climate mitigation and adaptation, and may also contribute to fulfilling other international commitments, for example, under the Convention on Biological Diversity (CBD), Convention to Combat Desertification (UNCDD) and Ramsar Convention on Wetlands (Ramsar Convention).

Link between soil health, soil biodiversity and mitigating and adapting to climate change

The importance of adaptation in many areas cannot be overstated, and the Paris Agreement set out a new global adaptation goal (Paris Agreement, Article 7). Fostering diverse communities of soil organisms to maintain soil health and support vegetation is crucial to climate adaptation and resilience. Increasing soil organic matter (SOM) is a triple win for climate change adaptation, mitigation and food security (Sommer, 2014). Soil health is enhanced by SOC storage because of the many co-benefits of soil organic matter with ecosystem properties such as soil drainage, nutrient retention, compaction mitigation. Whether natural or productive, virtually all ecosystems are managed (whether through protection/restoration or intensive cultivation). Native ecosystems, grazing lands, agroforests,

and croplands may all be protected or managed in ways that maintain or improve soil health. In productive systems, adaptive soil health management systems are based on four principles: 1) continuous living roots, 2) minimize disturbance, 3) maximise soil cover, and 4) maximise diversity, which are supported by an array of management practices to help build soil organic matter (Manter *et al.*, 2018). The multiple benefits of increased SOM underscore the objectives of the 4 per 1000 Soils for Food Security and Climate initiative launched by France in 2015 at the UNFCCC COP21 with an aspiration to increase global soil organic matter stocks by 4 per 1000 (or 0.4 percent) per year to compensate for the global emissions of greenhouse gases by anthropogenic sources. Regardless of whether that target is realistic, increases in SOM as a result of the initiative will benefit resilience and adaptation to climate change and food security. Healthy soils unify productive and natural systems by focusing on maintaining an intact, diverse below-ground ecosystem that ensures the flow of carbon, nutrients, and energy necessary for resilient landscapes (Kibblewhite *et al.*, 2008). As a terrestrial system accrues soil organic matter, both soil health and soil organic carbon increase and a multitude of co-benefits emerge.

Future plans, challenges and gaps

While the potential for synergies from harnessing soil biodiversity for climate action with other global agendas is enormous, transformative policies and action and enabling conditions are necessary for success. Existing nature protection policies need to be fully implemented, and there is a pressing need for support and security of tenure for those at the front line of protecting, restoring and caring for land and soil and conserving its biodiversity, including indigenous peoples and local communities. There is also a need for education and awareness raising, capacity strengthening and full engagement across the public and private sector towards overcoming financial barriers and implementing natural capital accounting and payment for ecosystem services. Fulfilment of, and increases in, existing international climate finance commitments is essential, as is an increase in the allocation of funding for the implementation of natural climate solutions, which can provide cost effective, low risk, mitigation with multiple co-benefits including water filtration, flood buffering, soil health, above- and below-ground biodiversity habitat as well as adaptation and resilience (Griscom *et al.*, 2017; IPCC 2019). Ongoing work in the context of the Koronivia Joint Work on Agriculture (KJWA or Koronivia process) that was initiated at UNFCCC COP 23 in 2017, and which operates as a platform for agriculture related discussions under the UNFCCC, requires broad participation and interdisciplinarity to achieve transformative outcomes. A promising option in the framework of KJWA for mitigate GHGs emission could be achieved by supporting the implementation of actions on the ground through RECSOIL (Recarbonization of Global Soils), which constitutes an implementation tool for scaling up SOC-centred Sustainable Soil Management (SSM), based on collaborative efforts under the Global Soil Partnership (GSP). The main priorities of RECSOIL and associated multiple benefits are: a) to prevent further SOC losses from carbon-rich soils (peatlands, black soils and permafrost) and, where feasible (in agricultural and degraded soils), to increase SOC stocks; b) to enhance farmer income by increasing soil productivity; c) to contribute to improved food security and nutrition; and d) to mitigate climate change through NDCs of Parties within the framework of the UNFCCC.

The complexity of GHG accounting in the land sector, and lack of harmonization in use of terminology between different initiatives and across different disciplines in the land, climate change, and biodiversity arenas is a challenge.

Our knowledge of how SOM is formed and stabilized is changing rapidly. New technologies and studies highlight key interactions among plant carbon allocation, soil biota, and microaggregates and soil structure. A deeper understanding of the ecological controls on soil carbon is needed if we are to create models of carbon balance under changing patterns of climate, land use, and other factors (with a view to more accurately identifying geographic areas with high potential for increased and long term sink capacity, as well as those at high risk of becoming net sources of GHG emissions) and to calculate carbon budgets, and assess the impacts and potentialities of land use practices and bioengineering techniques and innovations to maintain and increase levels of SOC. In particular, more knowledge is required about SOC equilibriums and saturation points in mineral soils, and the influence of vegetation, and relevance of depth.

Understanding the critical role that soil biodiversity plays in carbon sequestration and stabilization of SOM, and the implications of a changing climate, is an important aspect of this work. Lack of comparable definitions and of global data regarding tillage or crop rotations is a challenge (Jackson *et al.*, 2017). Soil scientists and data technologists are working to address gaps in the science and data and to map soil types, biogeographical and land use areas, and SOC stocks and baselines, and to provide relevant guidelines (Vargas-Rojas *et al.*, 2019). Accessible methods for measuring changes in SOC levels are also important (Smith *et al.*, 2019). A good example of this is the development of the Global Soil Organic Carbon Map (GSOCmap) (FAO and ITPS, 2018). This map allows the estimation of SOC stock from 0 to 30 cm, and it is the first global SOC assessment which was produced through a bottom-up approach where countries developed their capacities and stepped up efforts to compile or collect all available soil information at the national level. Using the GSOCmap as input, the GSP is currently developing the GSOCseq, another global data generation initiative that is intended to provide updated information on the potential for carbon sequestration in the world's soils. The general procedures to develop this map will be through a bottom-up approach (based on best available data), collaboration with country experts and implementation of widely used and validated soil organic matter simulation models.

Seeking to take into account fuller (CO_2, CH_4 and N_2O) GHG balances in research, and further work on soil biodiversity and climate change related aspects of N and P cycles, and on the role of black soils, could add significant value. Limiting factors and tradeoffs must also be identified and addressed (Zomer *et al.*, 2017; Campbell *et al.*, 2018; IPCC 2019). Accelerating ambition and action on climate change is critical. The sink capacity of land must not be seen as an alternative to an ambitious emissions reduction strategy (Fee, 2019), and soils can be expected to have reduced capacity to act as sinks for carbon sequestration at higher temperatures. Adopting value chain strategies, such as dietary change and sustainable sourcing, alongside land-based strategies as outlined by the IPCC (IPCC, 2019) could also have positive impacts for soil biodiversity. Recognizing the value, opportunities and risks posed by terrestrial and coastal soils and mainstreaming these critical GHG sinks and reservoirs, and their vital soil biodiversity communities, into decision making in all sectors is fundamental; we need to protect, restore and fund.

Box 3.5.1.2 | Nitrification inhibitors

An adequate supply of nitrogen (N) is essential to maintain/increase crop yields. Plant roots take up N from the soil solution as nitrate (NO_3^-) and ammonium (NH_4^+) soluble ions. The production of NO_3^- occurs with the oxidation of NH_4^+ through the nitrification process carried out predominantly by autotrophic soil bacteria (i.e. Nitrosomonas and Nitrobacter). An excessive supply of N through fertilizers increases the rate of NO_3^- formation, which is highly prone to leaching causing water contamination and nitrous oxide emissions (N_2O), a potent greenhouse gas.

Nitrification inhibitors are compounds that hinder the activity of the Nitrosomonas bacteria, in the nitrification process. This action boosts the efficiency of the use of N fertilizers by slowing down the transformation from NH_4^+ to NO_3^-. This helps to keep N in the form of NH_4^+ reducing potential contamination and greenhouse gas emissions, by a gradual release of N without affecting the supply of N to crops. However, N fertilizer additions in alkaline pH conditions, low soil moisture content, and high temperatures promote the volatilization of N in the form of NH_3.

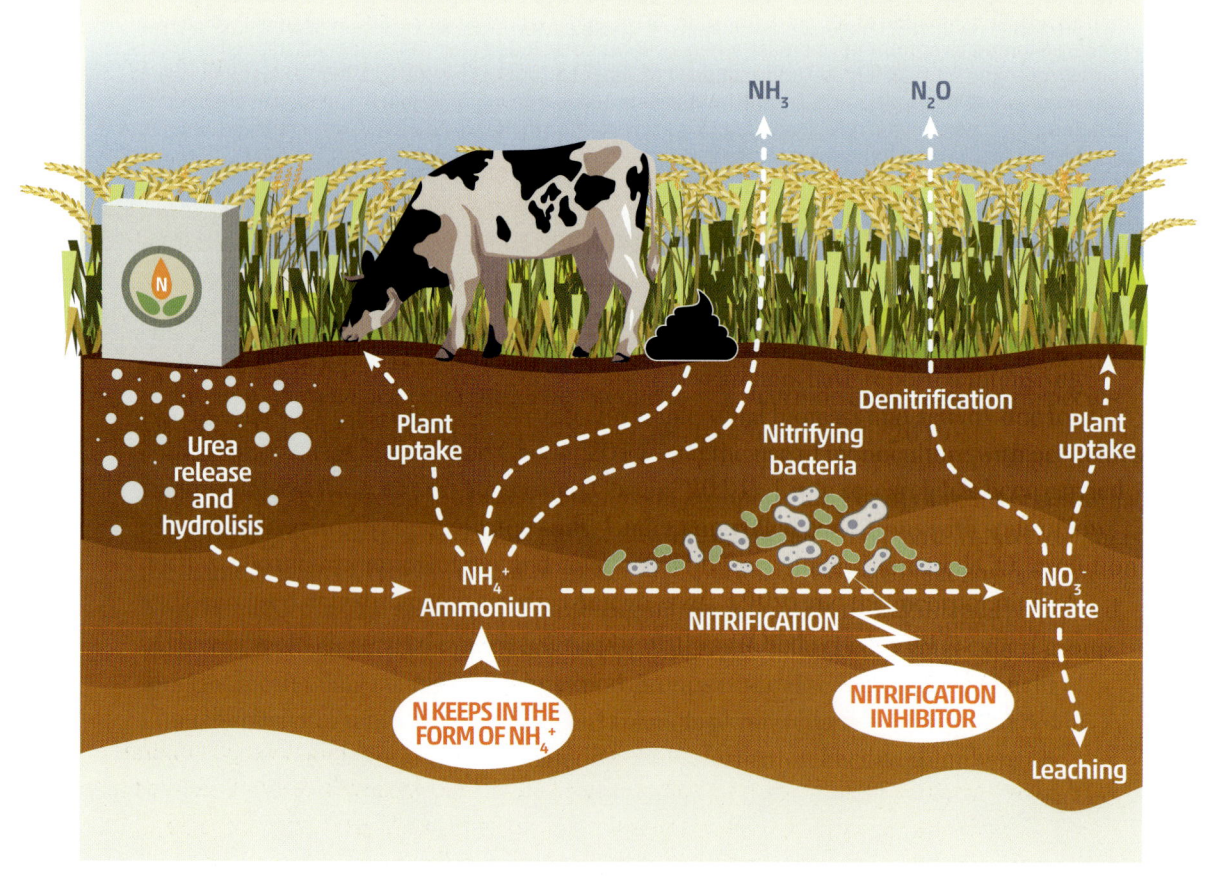

3.5.2 | SOIL CARBON CYCLES

Soils comprise the largest carbon stocks on earth, with an estimated total of at least 1 500 Gt carbon (C) (Crowther *et al.*, 2019). Contrary to above-ground biomass patterns, the largest below-ground carbon stocks are found in cold or waterlogged conditions (Crowther *et al.*, 2019). The living component of soils is dominated by fungi, totalling a global biomass of approximately 12 Gt C, followed by bacteria (7 Gt C), soil animals (2 Gt C), and archaea (0.5 Gt C) (Bar-On *et al.*, 2018). Whereas above-ground metazoan biomass distribution is shaped primarily by climatic conditions, the distribution of below-ground microbial biomass is governed primarily by edaphic characteristics (Crowther *et al.*, 2019). Globally, fungal and bacterial biomass are generally higher in regions with high soil organic matter (SOM) contents and lower pH (Bahram *et al.*, 2018). These SOM stocks are among the primary drivers of the accumulation of microbial biomass (Xu *et al.*, 2013) and high abundances of nematodes (van den Hoogen *et al.*, 2019) in the high latitudes.

Soils across the globe contain highly variable soil organic matter (SOM) levels, ranging from very low percentages in desert soils to extremely high percentages in boreal forest soils. SOM is mainly derived from above-ground plant and animal inputs (mainly at soil surface layers), and below-ground plant, animal and microbial inputs; a small fraction forms from abiotic sources. In SOM, soil organic carbon (SOC) is a major constituent, followed by organic nitrogen, phosphorus and sulphur. However, much of SOM is biologically inaccessible as a basal resource for soil organisms, and while other SOM components are considered accessible, they vary in degrees of biodegradability (often termed labile and recalcitrant) that largely depends on the relative elemental concentrations (C:N) and the nature of the carbon compounds. (Figure 3.5.2.1). The organic carbon in the soil occurs in the form of cellular (macro) molecules such as polysaccharides, phenolics, organic and amino acids.

The transformations of carbon in the soil all boil down to cycling between abiotic (mineral) and biotic (organic) soil carbon pools. The key forms of mineral carbon in the soil are carbon dioxide (CO_2 - the final product of carbon oxidation processes) and methane (CH_4 - the final product of reductive processes). The aforementioned mineral compounds both constitute greenhouse gases (GHGs) and are produced and released to the atmosphere under particular conditions. For instance, CO_2 can be removed (fixed back into organic carbon), whereas CH_4 can be back-oxidized in the soil microbiome. Thus, depending on the conditionsl soils can either function as a net CO_2 source or as a net sink. Similarly, local oxygen conditions will determine the fate of CH_4 in the system. Proper assessments of these processes are key to our understanding of the global impact of these soil functions. Below, we discuss the role of the soil microbiome in the key soil carbon cycling processes.

The role of the soil microbiome in the soil carbon cycle

In a broad sense, the carbon in soil is recycled within a microbe-driven soil food web. A key process is the respiration of carbonaceous substances with molecular oxygen serving as the terminal electron acceptor; this major process returns C to the atmosphere as CO_2. Microorganisms are thus central players in the transformation of plant and animal residues and are also key reservoirs of organic C in soil. As described above, there is another large reservoir of SOM, which can – to some extent – be degraded, and for another part may condense over time to form humus. Much of the latter is typically protected from microbial degradation, but it can be made more available by tillage (ploughing). In undisturbed soils, though, the organic C may accumulate into humus.

CO_2 fixation - Some soil bacteria (and archaea) fix CO_2 autotrophically, and so up to 5 percent of respired CO_2 may be re-fixed. The Calvin-Benson-Bassham (CBB) cycle, responsible for photosynthesis in green plants as well as microorganisms, is the predominant pathway in this process, with RubisCO (ribulose-1,5-bisphosphate carboxylase/oxygenase) being the key enzyme involved. In addition, phototrophs of the *Proteobacteria, Chlorobi, Chloroflexi, Firmicutes, Acidobacteria* and *Gemmatimonadetes* phyla also contribute fixed C to the soil carbon reservoir. These bacteria carry the RubisCO large-subunit gene *cbbL* and fix CO_2 during autotrophic growth. Another, often overlooked, process is carbon monoxide (CO) fixation. CO is present in soil in relatively low concentrations compared to CO_2. Up to 20 percent of it can be oxidized by *Proteobacteria* and Actinomycetes that possess the CO dehydrogenase gene *coxL* (Hirsch, 2019). Thus both CO_2 fixation and CO oxidation play important roles in the soil C cycle.

Methane production and oxidation - Methane is generated in soil in a process called methanogenesis which occurs in wet, carbon-rich soils. Anaerobic conditions in such systems spur processes like fermentation by bacteria and anaerobic respiration by specific methanogenic archaea (Hirsch, 2019). A common mechanism uses CO_2 as a substrate, whereas other mechanisms use methylated compounds or organic acids such as like acetate. Wetlands, including rice paddies, have been estimated to generate around 160 Tg CH_4/y (Hirsch, 2019), which outweighs the production by ruminant animals (100 Tg CH_4/y). Conversely, methanotrophic bacteria in soil are estimated to be able to re-oxidize 30 Tg of CH_4/y (Hirsch, 2019). These methanotrophs (members of the Alpha- and Gammaproteobacteria) produce the key enzyme methane monooxygenase. Methanogenesis and methane oxidation are mechanisms of key relevance to global warming. In fact, CH_4 emission rates are found to rise globally due to global warming and thawing permafrost.

Other anaerobic processes - Both denitrification and sulphate reduction can be involved as electron capture processes in organic-rich soils in anaerobic (wet) conditions. Denitrification is a main cause of GHG (N_2O) emissions. Denitrification is beneficial for facultative anaerobes, as they can switch to this process in waterlogged, anoxic soil conditions when organic matter is abundant. Fungi that contain a bacterial-type nitrite reductase gene in their mitochondria can also denitrify. The *nosZ* gene is important for the conversion of N_2O to N_2. Thus, next to managing drainage, organic matter and fertilizer inputs into soil, maintaining conditions that promote microorganisms with a functional *nosZ* gene is thought to be important for reducing GHG emissions (Hirsch, 2019).

With respect to sulphate reduction, soil bacteria and archaea from several phyla (including *Deltaproteobacteria*, *Firmicutes* and Nitrospirae) can oxidize carbon compounds via this pathway, including the reduction of sulfate (SO_4^{2-}) to sulphide (S^{2-}) via sulfite (SO_3^{2-}) in a dissimilatory reaction. Typically, this process occurs in waterlogged soils (including rice paddies) and sediments and results in the familiar "rotten egg" odour when such soils are disturbed.

Outlook - Carbon is either fixed or released from soils, depending the activity of the soil microbiomes and driven by abiotic conditions such as water content, temperature, oxygen level and pH. Moreover, soil type and soil management regime have strong effects. For instance, in agricultural practice, the large-scale removal of plant material (grazing by herbivores, agricultural practices) may reduce soil carbon. The grazing, in conjunction with the activity of the soil biota, may also affect carbon release at the roots. With respect to the diverse forms of permafrost, so-called pleistocene-aged permafrost appears to be a major contributor, as it has high OM content, yet low abundance of methanogens. Thus, thawing is predicted to result in high CO_2 release yet low CH_4. Over the long run, the respective fluxes will strongly depend on the soil conditions, with respect to water (soils becoming wetter or dryer), plant cover and disturbance regime (Waldrop and Creamer, 2019). Extreme factors (such as temperature) may affect soil function (Hirsch, 2019). A better understanding of the organisms involved is important as global climate change coupled with the needs of an ever-increasing population place pressure on soil functions and agricultural sustainability. Carbon cycling in soil is clearly carried by suites of bacteria, archaea, fungi, protists, micro- and meso-fauna. Their relative involvement depends on soil type and conditions. Thus, from the large diversity of microorganisms in soil there is a capacity for selection of appropriately adapted microbiomes as the environment changes.

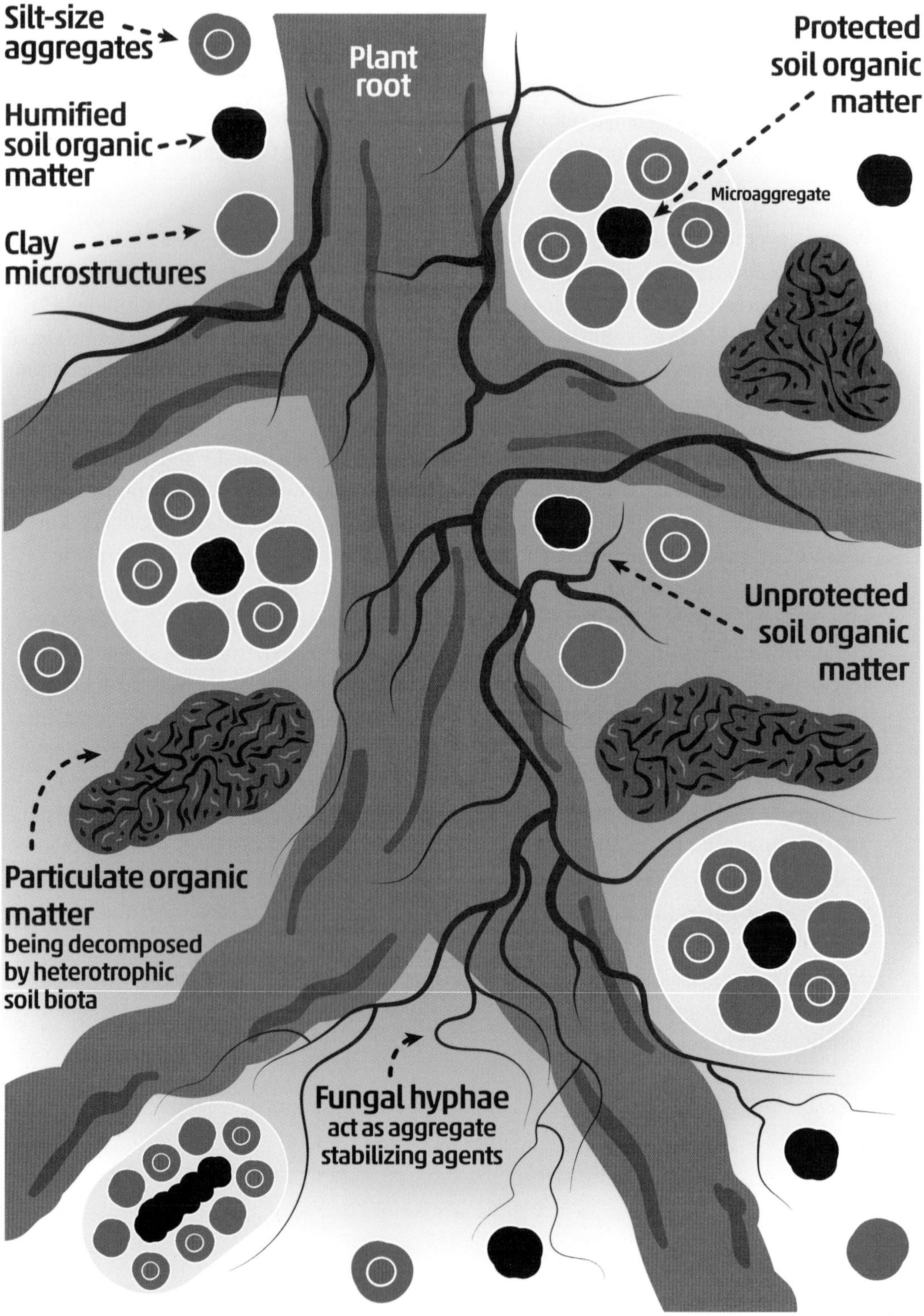

Figure 3.5.2.1 | Role of soil biodiversity in aggregate stability and carbon storage

Most soils contain 1 to 6 percent of organic matter, however not all this amount can be used by heterotrophic soil organisms. SOM accessibility to biota depends upon the physical protection that microaggregates provide from microbial attack, the biochemical composition and consequent resistance to microbial decomposition, and the chemical bonds to organo-mineral particles

State of knowledge of soil biodiversity

3.5.3 | CARBON SEQUESTRATION AND STORAGE

Soil has a tremendous potential for regulating the atmospheric carbon content by sequestering carbon and thus mitigating climate change for the benefit of SDG 13 (climate action). Effective biodiversity conservation and management can lead to higher levels of carbon sequestration and hence climate change mitigation. Very few studies have focused on the role of specific soil organisms and biodiversity on C sequestration and GHG mitigation, with the exception of some studies reporting that specific microorganisms have specific control of many atmospheric trace gasses (Conrad, 1996).

The global changes in temperature and rainfall patterns have a great impact on soil biodiversity abundance, distribution patterns and activities, with strong implications on soil organic matter decomposition and the carbon balance (see for example Crowther, 2017). According to the Status of the World's Soil Resources report (FAO and ITPS, 2015), SOC loss is the second largest global threat to soil functions. Global estimates indicate that soil organic carbon is being lost at a rate equivalent to 10 to 20 percent of total global carbon dioxide emissions (Olivier *et al.*, 2015).

Since organic matter decays faster with increasing temperatures, the areas surrounding the pan-Arctic region (including tundra, boreal forests and peatlands ecosystems) are of special concern because they store vast amounts of carbon that can be released into the atmosphere (~1700 Pg total C; Parish *et al.*, 2008; Tarnocai *et al.*, 2009; Joosten, 2015). Furthermore, carbon accumulation in these areas occurs relatively slowly (~21 g C m^{-2} yr^{-1}; Chaudhary *et al.*, 2017) due to harsh environmental conditions limiting soil biological activities (such as waterlogging conditions, acidity and cold temperatures).

In these systems, soil invertebrate communities are dominated by small sized organisms, collectively known as "mesofauna" (that is, with body diameter < 2 mm and including enchytraeid worms, mites and collembolans). In particular, enchytraeid populations are very sensitive to increasing temperatures and frequency of droughts, leading to important losses of non-resilient species and changes in the vertical distribution of more adaptable species (Briones *et al.*, 2007a). Such changes have important implications for C storage, as a result of an increase in the decomposition rates at deeper soil layers, and the potential release of "previously locked carbon" (Briones *et al.*, 2007b, 2010). Consequently, peatlands drainage and fires are responsible for almost 10 percent of greenhouse gas emissions from the Agriculture, Forestry and Other Land Use sector (AFOLU) according to FAO (2015). Therefore, they are likely to continue to be a hotspot of GHG emissions of global importance (Tuniello *et al.*, 2016).

In the case of C sequestration, it is widely accepted that high fungal to bacteria ratios are usually associated with higher carbon sequestration (Briones *et al.*, 2014; Malik *et al.*, 2016), a typical condition of tundra and boreal regions. In particular, it has been shown that AFM fungi promote soil carbon storage through a positive effect on soil aggregation (Wilson *et al.*, 2009), but also via increased fungal grazing by collembolans (Duhamel *et al.*, 2013). However, using global data sets, Averill *et al.* (2014) showed that soil in ecosystems dominated by ectomycorrhizal and ericoid mycorrhizal-associated plants contains 70 percent more carbon per unit nitrogen than soil in ecosystems dominated by AM-associated plants. In tropical and temperate areas, ecosystem engineers (earthworms, ants and termites) are the main promoters of C stabilization by creating biogenic structures (casts, galleries, nests and mounds) that form organo-mineral associations (Vidal *et al.*, 2016). Similarly, other soil animals such as mites producing slowly decomposing excrements may contribute to stable, carbon-storing humus (Hågvar, 2016).

Due to the variation in spatial scales at which different carbon substrates in soils can be processed and protected either physically, biologically or chemically (Jastrow *et al.*, 2007), soil fractionation has long been recommended to separate particulate organic matter (POM) from mineral-associated organic matter (MAOM) when evaluating soil carbon sequestration (Cambardella and Elliott, 1992). However, a growing body of empirical research has led to a shift in understanding of the mechanisms for SOM formation and persistence (Cotrufo *et al.*, 2013). This new understanding highlights two discrete biologically mediated pathways to each of the POM and MAOM fractions (Cotrufo *et al.*, 2015). A physical pathway to POM formation is dominated by litter fragmentation and bioturbation as a result of larger soil fauna and microarthropods (ecosystem engineers), whereas the microbially-mediated pathway to MAOM formation is dominated by bacteria, archaea and microbial communities that biochemically alter carbon substrates through decomposition and condensation/polymerization.

This new understanding has brought several additional insights about microbial community dynamics that can increase microbial carbon utilization efficiency (CUE), increase soil carbon sequestration in the relatively stable MAOM fraction, and even help to identify the microbial taxa associated with healthy soils that are more resilient to the impacts of climate change (Dubey *et al.*, 2019; Sokol *et al.*, 2019). While this level of mechanistic understanding is not typical in biogeochemical soil models, recent attempts to include representations of microbial function and POM/MAOM pools in mathematical models show promise (Wieder *et al.*, 2014; Robertson *et al.*, 2019). Being able to use ecosystem models to represent soil biodiversity and leverage the wealth of data that metagenomic techniques provide is a crucial step in being able to predict any potential feedbacks with climate change (Dubey *et al.*, 2019).

Box 3.5.3.1 | Keeping the carbon in the soil

Keeping the carbon already contained in the soil is more effective than any other practice aiming at capturing atmospheric carbon. Peatlands where peat-forming vegetation is intact or has been restored are likely to be more resilient to climate change impacts than degraded ones (Parish et al., 2008). Therefore, the conservation of C-rich soils like peatlands, forest soils and permanent grasslands should become a priority.

Recent estimates indicate that agricultural land uses have resulted in the loss of 133 Pg C from the soil (Sanderman et al., 2019). On 1 December 2015 at the COP21, the international initiative "4 per 1000" was launched, with the aim of demonstrating that agriculture, and in particular, agricultural soils can play a crucial role where food security and climate change are concerned. Accordingly, good agricultural practices that revolve around increasing C retention and decrease nutrient losses should be adopted (such as no-till and conservation agriculture). However, many limitations to achieving SOC increase at a rate of 4‰ yr-1 over large areas (de Vries et al., 2018; Poulton et al., 2018) and likely to require large investments (Paustian et al., 2016). Therefore, soil management practices should aim to ensure all soil functions rather than specifically maximise carbon sequestration. General guidelines to make soils more resilient to climate change should include the following:

(1) Conservation of C-rich soils like peatlands, forest soils and permanent grasslands should become a priority.

(2) Improve SOC pool by adopting less intensive agricultural practices that limit/exclude agrochemical use and maintain a live plant cover (such as conservation agriculture) or use organic amendments such as mulch, compost.

(3) Adopting 'paludiculture' practices to cultivate organic soils without drainage, whereas land use practices that require drainage should, if possible, be relocated to areas with mineral soils.

(4) Increase soil biodiversity by protecting natural areas, keeping habitat heterogeneity and diversifying land use options.

The great recent advances in collating and mapping soil biodiversity data at global scales (see for example Tedersoo et al., 2014; van den Hoogen et al., 2019; Phillips et al., 2019) means that it will now be possible to link specific soil biota groups to terrestrial carbon stocks. From this, new management strategies that enhance C retention across different ecosystems can be developed and finally incorporated in environmental policies.

Box 3.5.3.2 | Biochar as C-sequestration

Biochar is carbon-rich solids produced during pyrolysis of biomass at oxygen-limited conditions and may be beneficial soil amendments and increase soil quality as biochar sequester carbon, increase water infiltration, mediate soil pH, elevate CEC, increase nutrient uptake, and modify GHG emissions (Li et al., 2019; Zhang et al., 2018). The feedstock is important for the chemical and physical properties of the produced biochar. Examples of feedstock are agricultural waste as straw and coconut shells, wood, animal manure and wastewater sludge. The pyrolysis conditions such as temperature in combination with oxygen availability are significant for the biochar properties and have to be carefully controlled. When controlling the pyrolysis conditions, problematic compounds such as heavy metals can be removed while production of problematic polycyclic aromatic compounds can be kept to a minimum. Biochar is reported to reduce availability of polluting compounds such as heavy metals and organic contaminating compounds by absorption, contributing to soil remediation (Kavitha et al., 2018).

Biochar has a priming effect on soil organic matter and microbial activity, which can be both negative and positive, probably depending on the physical structure of the soil. This is suggested to be due to a higher stimulation of microbial activity in sandy soil (Wang et al., 2016).

After introduction into the soil, about 3 percent of the biochar is relatively easily degraded, with a lifetime of 108 days, while the remaining 97 percent has an estimated lifetime in soil of >500 years (Wang et al., 2016). Hence, it will remove carbon from the current carbon cycle, and in this way reduce atmospheric CO_2 and mitigate climate change.

3.5.4 | SOIL FORMATION AND EROSION PREVENTION

While parent rock determines soil type as well as soil texture through the size of primary particles, biological activity directly affects the aggregation of mineral grains into secondary units. Farmers have observed the role of soil organisms in soil structure modifications for a long time. The crucial role of organic binding agents often produced by bacteria and fungi are affected by size, quality and stability of soil aggregates (Tisdall and Oades, 1982). The activity and diversity of soil biota are essential to soil structuring, thus contributing to soil formation and regeneration. When grassland is converted into cropland or when peatland is converted into agriculture, loss of soil structure can be observed due to mechanical disturbances and increased oxidation. Soil stability is strongly depending on soil texture, as finest soil texture (clay, loam) leads to highest stability, and soil organic matter content (Le Bissonais, 1995). Nevertheless, soil organisms, including microorganisms and fauna, have a large share of responsibility for soil stabilization (Pérès et al., 2013). A good and stable soil structure is one of the main objectives of farmers and is also the goal of SDG 12 (responsible consumption and production), SDG 13 (climate action) and SDG 15 (life on land).

Stabilization processes are required in organic carbon degradation, along with complex chemical, physical, and biological interactions within the soil matrix (see for example Kögel-Knabner et al., 2008). Soil macrofauna play an important role in soil structure modification through bioturbation and the production of biogenic structures (Brussaard et al., 1997; Lavelle and Spain, 2001). Earthworms, ants and termites, but also small mammals, modify their surrounding environment and thus affect soil water and nutrient dynamics through their interactions with other soil organisms. In terms of biomass and function, earthworms play a key role in agricultural soils with strong effects on soil structure and processes through borrowing, feeding and casting (Blouin et al., 2013, Pulleman et al., 2012). The burrow network allows increased water infiltration, especially when they are provided by anecic earthworm species (see the following section on water flux) and by consequence could take part of decreasing soil erosion by 50 percent (Shuster et al., 2002). Deposited cast of anecic species, by increasing the soil roughness, decrease the speed of runoff and therefore lead to a fourfold decrease in soil erosion (Le Bayon and Binet, 2001). However, other authors suggest that over longer time-scales (thousands of years or more), the erosion of cast material could lead to vast amounts of sediment accumulation in alluvial soil or floodplains (Feller et al., 2003). No matter what, the effect of earthworm biostructure on soil erosion has to be reconsidered with respect to the rain event intensity, the slope, the soil surface cover, and at larger scale, the landscape organization (presence of hedges). Enchytraids also form aggregates, but at a smaller spatial scale (Marinissen and Didden, 1997).

Soil crusting is considered to be an important form of soil degradation that limits land productivity because it considerably reduces soil water infiltration and root growth. Soil-crusting risk depends on the texture, as loamy soils present the highest risk (Le

Bissonnais, 1996). The application of organic mulch has allowed the recuperation of surface crusted soils in Burkina Faso (Mando *et al.*, 1996). Many other examples have been described, highlighting the importance of appropriate residue management on farmland to foster biological activity benefiting soil structure.

Despite the important role of soil organisms in soil formation and erosion, little attention has been given to the essential organisms in intensively used agricultural soils, where natural and biologically mediated processes have been largely replaced by human activity. In particular intensive tillage practices affecting soil structure with mechanical soil loosening and winter furrow leaving a coarse surface to be broken up by the frost have significant negative impacts. Stabilization of aggregate depends on many parameters such as biotic (microorganism, fauna) and abiotic (texture, organic matter) and formation and stabilization of macroaggregates result from combinations of functional groups (Frazão *et al.*, 2019). Macroaggregate breakdown increases with age because the action of organic binding agents is progressively disrupted. Microaggregates are far more stable, and as long as soil organisms are active, in particular bacteria producing organic binding agents in combination with fungal hyphae, they can become building blocks during the formation of new soil macroaggregates (Barrios, 2007). If this is no longer the case, erosion processes will accelerate with soil particles from disrupted soil aggregates or even soil horizons being lost or transported vertically (for example by gravity or water) and horizontally (by water, ice or wind). Aggregate stability is linked to plant diversity. Plant mixtures with high frequency of grasses increase soil aggregate stability by increasing root biomass, soil organic carbon concentrations, and soil microbes, while legumes may be less favourable for soil aggregate stability (Pérès *et al.*, 2013). A healthy soil ecosystem with high biodiversity will help reduce erosion and secure soil formation. If the soil quality is deteriorating and erosion processes are increasing, it might take a very long time for the regulating service to be restored again. Consequently, in order to maintain long-term sustainability of soil fertility and soil structure, soil management plans need to more carefully address soil as a habitat and not only as a substrate for cropping.

3.5.5 | REGULATION OF WATER FLOW

The ability of soils to store and release water is a widely acknowledged regulating ecosystem service and is part of SDG 6 (clean water and sanitation) and SDG 15 (life on land). Furthermore, soil moisture is the driver of many chemical and biological processes and is therefore essential for soil development and functioning (Dominati, 2013). As we are experiencing dramatic climatic changes, we must enhance water storage to increase resilience to weather extremes. Only soils with diverse and abundant soil life and high biodiversity can act as efficient filters to produce clean drinking water. Soil organisms, both fauna and flora, facilitate water infiltration and drainage thanks to the creation of macro- and micropores. In addition, a greater accumulation of organic matter in the soil by biota increases the water holding capacity of the soil. Soil organisms thereby improve soil aeration and water infiltration (Brussaard *et al.*, 2007, Pulleman *et al.*, 2012, El Mujtar *et al.,* 2019). By regulating the water flow, they contribute to flood mitigation and climate control (Vogel *et al.*, 2019). These multiple water services are closely connected to continuously undergoing soil formation processes. A favourable soil structure facilitates the germination and the establishment of crops, helps to prevent water logging, reduces the risks of water shortage and maximises resistance against physical degradation (El Mujtar *et al.*, 2019).

Regulation of water flow is threatened by compaction of agricultural soil, which can be caused by traffic and loss of soil carbon due to agricultural practices. Also xenobiotic compounds as pesticides and organic contaminants can reduce the soil structure and reduce the regulation of the water flow.

The water flow regulation is increased by increasing the soil carbon content by return of organic matter including plant residues to soil, by reduced and low tillage, which both will increase biodiversity. Interestingly, sequestering carbon by incorporation of biochar in soil has a very positive effect on the water filtration and water holding capacity of the soil (Mao *et al.*, 2019).

3.5.6 | WASTEWATER TREATMENT

Ecosystems such as wetlands filter effluents, decompose waste through the biological activity of microorganisms, and eliminate harmful pathogens. Despite this direct regulating service of soil biodiversity, the most efficient treatment of wastewater is in man-made installations. Wastewater treatment processes are engineered installations that receive wastewater (agricultural, domestic, and industrial) and use biological and chemical methods to produce aqueous effluents that can be discharged safely to the environment; largely oceans, rivers and lakes (Olsson and Newell, 1999). A "solids" component comprised of biological cells/cellular debris and/or chemical precipitates is also generated and handled by different methods depending upon composition and volume. The activated sludge process (Nielsen and McMahon, 2014) and wetlands (Wu, *et al.*, 2015) are common treatment options, with the choice of process dependent upon many variables including wastewater volume and pollutant composition (concentration and types) (Prasse, *et al.*, 2015).

Wastewater from agricultural and domestic sources will likely contain animal and/or human pathogens and the treatment process chosen will need to mitigate pathogens. Biological processes dominate wastewater treatment methods, particularly of agricultural and domestic wastewaters, which have substantial organic components that can be bio-decomposed largely by microorganisms but aquatic plants contribute in wetland processes. The source of microbes in biological treatment processes come partially from the wastewater influent but are also opportunistically sourced from soils in the vicinity of the installations. Thus, soil biodiversity substantially contributes to this major biotechnological industry. Nutrients (such as nitrogen and phosphorus) are common pollutants in wastewater. Nitrifier/denitrifier and phosphorus accumulating microbes largely sourced from soils ensure that most nitrogen in wastewater is converted to N_2 gas and most orthophosphate is bio-accumulated inside specific bacteria as polyphosphate. For example, polyphosphate accumulators were identified as Betaproteobacteria, *Candidatus* Accumulibacter phosphatis (Crocetti *et al.*, 2000; Hesselmann *et al.*, 1999), and found responsible for mitigating discharge of phosphorus, which is a substantial environmental eutrophying agent (Oehmen *et al.*, 2007). These bacteria were likely sourced from soils rather than the wastewater itself. Likewise, nitrogen removal is mediated by a community of soil-sourced bacteria and archaea (Limpiyakorn *et al.*, 2013), with wide-ranging nitrogen metabolic capacities (Schmidt *et al.*, 2003, van Loosdrecht and Jetten, 1998). Soil microbes dominate wetland ecosystems, underscoring their high value (Rajan *et al.*, 2019).

The discharge of wastewater effluents to lake and river environments impact soil microbial communities. Particularly of concern is the rise in antimicrobial resistant genes from wastewater treatment systems, which could be a source of these genes into other ecosystems (Subirats *et al.*, 2019, Yin *et al.*, 2019), especially if the treated effluent is used for agricultural irrigation.

3.5.7 | INVASIVE SPECIES

The majority of our knowledge of invasive soil species concerns agricultural pests, of which many contribute to huge economic losses globally (Coyle *et al.*, 2017; SDG 1 (no poverty), SDG 2 (zero hunger)). In addition, when certain species are introduced to pristine areas as biocontrol agents, this can lead to massive biodiversity loss, affecting terrestrial ecosystem functioning (SDG15). Invasive soil species are also often found in urban or disturbed environments (for example, after fire), which have habitats that encourage the spread and existence of invasive species. However, due to undersampling and a general lack of taxonomic knowledge, the general distribution of invasive species in the soil are largely unknown, while the invasion status of many species is uncertain, being listed as having a 'cosmopolitan' distribution.

Soil taxa listed under the 100 of the World's worst invasive alien species (Lowe *et al.*, 2004) include five species of ants, including the Argentine ant (*Linepithema humile*, crazy ant (*Anoplolepis gracilipes*), big-headed ant (*Pheidole megacephala*), little fire ant (*Wasmannia auropunctata*) and red imported fire ant (*Solenopsis invicta*). Others on the list include the Rosy wolfsnail, (*Euglandina rosea*), soilborne fungal pathogens such as

Phytophthora cinnamomimi, a termite (*Formosan subterranean* termite, *Coptotermes formosanus shiraki*) and a soil-dwelling flatworm species (New Guinea flatworm, *Platydemus manokwari*).

Impacts

Although gaps in our understanding of invasive species in soil exist, there are numerous examples of soil microorganisms and invertebrates that have substantial ecological and socioeconomic impacts when they invade ecosystems (reviewed in Cameron *et al.*, 2016; Thakur *et al.*, 2019). The effects of invasive species in soil vary depending on the invader's trophic position and functional role, including whether they are pathogens, herbivores, detritivores, omnivores, and predators. As a result, soil invaders are likely to affect our ability to meet a number of the SDGs. First, through their impacts on species, communities, and ecosystems, invasive organisms in the soil are affecting both life on land (SDG 15) and life below water (SDG 14). For example, the Argentine ant (*Linepithema humile*) is listed as one of the world's worst invaders and can cause substantial declines in diversity of native ant species (Suarez *et al.*, 1998). The effects of organisms invading soil can further extend to impact aquatic systems. For instance, *Phytophthora lateralis* is a soilborne plant pathogen that causes Port Orford cedar root disease. It has a range of impacts including reducing native diversity and reducing shading along streams, which increases water temperature and leads to invertebrate and salmon killing, as well as increasing soil erosion (Robin *et al.*, 2011). Secondly, effects of invasive soil organisms can directly impact food resources and thus their control is important for progress on addressing SDG 2 (zero hunger) and SDG 3 (good health and well-being). For example, Phytophthora cinnamomi disperses through soil and plants and causes root rot. It has more than 500 host species and consequently can cause changes in vegetation that may cascade to affect entire food webs (Hardham and Blackman, 2018). In particular, it affects food crops and cash plants such as avocado, pineapple, and eucalyptus. Often, invasions of soil organisms may have impacts that affect multiple SDGs. For example, non-native earthworms are ecosystem engineers and their invasions can cause cascading effects that impact plant communities (SDG 15), forest productivity (SDG 15), carbon sequestration (SDG 13), wildlife and human disease (SDG 3), and soil and water quality (SDG 2, SDG tab6) (Frelich *et al.*, 2019).

Pathways of introduction

With the broadening of international trade since the early twentieth century, an increased trade and importation of plants globally caused a progressive increase in the number of alien terrestrial invertebrates introduced (Faulkner *et al.*, 2016). The majority of soil invertebrates' introductions appear to have been accidental, as contaminants or stowaways (Faulkner *et al.*, 2016). A wide range of human activities can act as vectors for transport of soil, and consequently of organisms living within the soil. For instance, soil can be transported as a contaminant of shipping containers (Godfrey and Marshall 2002), shoes (McNeill *et al.*, 2011), and plants (McNeill *et al.*, 2006). Soil animals easily can be transported through the horticultural trade surviving as eggs, while many small soil arthropods are desiccation resistant, surviving months in an anhydrobiotic state. The majority of non-indigenous taxa found in soil associated with commodities, cargo, vehicles, or footwear, included fungi, bacteria, moss, invertebrates, nematodes, and rotifers (reviewed in McNeill *et al.*, 2011). There are few border and quarantine interception records for soil organisms but, in general, the majority of soil invertebrates' introductions appear to have been accidental, as contaminants or stowaways (McNeill *et al.*, 2011). This is partly due to their small size, which has resulted in many species having been imported undetected, but it is also because taxonomic identification of some introduced taxa remains problematic and it is often difficult to determine whether a soil organism is non-indigenous in a particular region (McNeill *et al.*, 2011).

Control and biosecurity

Soil is recognized as a biosecurity risk; for example, the International Plant Protection Convention states that soil is a high-risk pathway for the introduction of invasive species/pests and consequently that guidelines are needed to minimize risk of introduction of pests with soil movement (IPPC, 2007). Given the impacts of invasive soil organisms on other organisms and ecosystem services on land and in aquatic systems, appropriate biosecurity methods are critical for maintaining life on land (SDG 15), food production (SDG 2), and health and well-being (SDG 3), as well as clean water and life below water (SDGs 6 and 14). However, the relative risk levels of different pathways and vectors are poorly understood for invasive soil organisms, which limits the ability of officials and regulators to target the pathways that are most likely to lead to introductions of invaders (McNeill *et al.*, 2017). A further issue is the lack of taxonomists trained in identification of soil organisms, and consequently training of taxonomists is a key priority to facilitate the detection of newly introduced species and to aid in their eradication and control (Convention on Biological Diversity, 2014). Taxonomic skills can also be supplemented, although not replaced, by identification of species through molecular approaches, such as DNA barcoding (www.boldsystems.org), which has been successfully used globally as an early detection and management tool for invasive species (Armstrong and Ball 2005; Bergstrom *et al.*, 2018). Ongoing survey work is needed to increase detection of new invasive species and to understand risk levels of different pathways, in order to improve management and control of the impacts of soil invaders on terrestrial and aquatic biodiversity, and ecosystem services.

Climate change and invasive species

It has long been suggested that climate change (SDG 13) will exacerbate the impact of invasive species (Cannon, 1998), facilitating the spread and establishment of invasive species. Understanding the physiological traits of invasive species may shed some light on how to better manage or prevent the introduction of invasive species (Karsten *et al.*, 2016), especially pest species which are predicted to change in distribution with climate change (Bebber *et al.*, 2013; Pecl *et al.*, 2017). Several physiological studies using Collembola (springtails) as model organisms have indicated that invasive species are generally more tolerant of warmer, drier conditions than are indigenous species (Chown *et al.*, 2007; Slabber *et al.*, 2007; Janion *et al.*, 2010; Janion-Scheepers *et al.*, 2018). A predicted increase in frequency of extreme temperature events will likely have an impact on soil community composition and functioning (IPCC, 2007). Indeed, some recent experimental evidence suggests that variation in thermal tolerance traits could lead to trophic mismatches, although some interspecific variation could lessen the severity of this (Franken *et al.*, 2017).

3.5.8 | BIODIVERSITY REGULATION AND BIOCONTROL

The interactions of soil organisms, directly and indirectly through competitive, facilitative, mutualistic, pathogenic or predatory effects, affect the overall structure of the soil food web; these interactions are self-reinforcing and self-regulating processes that lead to emergent community properties such as community stability and biological control or biocontrol. Biocontrol of plant disease is the reduction in the numbers and activity of a plant pathogen or pest, using one or more organism. This process often occurs below ground within plant roots, at the rhizosphere, or more generally within bulk soil.

The interactions that lead to biodiversity regulation (that is, the reinforcement of stable populations within the soil system, including pest control) can be positive or negative. For example, between soil microorganisms, facultative (positive or beneficial) associations between different bacterial strains (including *Bacillus*, *Paenibacillus*, *Pseudomonas* and *Rhizobia)*, and different arbuscular mycorrhizal (AM) fungal species stimulate growth of AM fungi and the germination of their spores, leading to increased root colonization of the host plant by AM fungi, increased solubilization of phosphate, and the suppression of pathogens around the rooting zone. Biological regulation does not require direct contact between organisms, and rather often arises through indirect antagonistic interactions.

Microbial biological control agents (MBCAs) employed for commercial plant production act via multiple modes of pest-host interference mechanisms (see Kohl *et al.*, 2019 for current review). Some MBCAs act via nutrient or space competition (that is, exploitative competition) that can modulate the growth conditions for the pathogen or pest (such as insects or weeds). For instance, strains of *Fusarium oxysporum* that are non-pathogenic can be superior competitors for carbon and root colonization sites. Another mechanism of biocontrol is growth interference of a pathogen through antibiosis, where volatile or non-volatile substances produced by one organism, such as enzymes or other metabolites, kill or inhibit growth of another organism.

Such interactions are highly regulated by compounds such as signalling compounds, enzymes and other interfering metabolites. Antibiotics are both volatile and non-volatile substances produced by one species of organism that in low concentrations kill or inhibit growth of another organism.

Volatile organic compounds (VOCs), are typically small, odorous compounds (<C15) with low molecular mass (<300 Da). Microbial volatile organic compounds belong to different chemical classes including alkenes, alcohols, ketones, benzenoids, pyrazines, sulfides and terpenes. Because many of these compounds are secondary metabolites or simple products of metabolism such as alcohols and aldehydes, they can diffuse easily through gas-filled pore space, and can operate at larger spatial scales as they do not need direct contact to play an important role in long-distance microbial interactions. *Trichoderma* (Fungi) species are particularly active producers of several antifungal volatiles, and are an effective biocontrol agent against several phytopathic fungi (Saravanakumar *et al.*, 2017).

Soluble antibiotics are secondary metabolites with high polarity, which makes them soluble in water, a characteristic that can lead to strong biocontrol efficacy but at smaller spatial scales than VOCs. Examples of organisms that produce soluble antibiotics and have been used as MBCAs are several soil fungi (such as *Aspergillus and Penicillum*, *Trichoderma*), which produce soluble antibiotics such as citrinin, patulin, gliotoxin and penicillic acid, peptide antibiotics and trichodermin. Fluorescent pseudomonads also have been shown to suppress a variety of plant pathogens by secretions of soluble antibiotics.

Siderophores – specialized iron-binding molecules – can also play a role in biocontrol. Siderophores are low molecular weight secondary metabolites produced by microbes under iron deficiency, to help bind and supply iron to the organism. Siderophores produced by plant growth promoting bacteria (PGPB) can play a role in the prevention or reduction of the effects of pathogenic microbes in plants by depriving such pathogens of iron if they produce higher rates of siderophores than the pathogen. (Figure 3.5.8.1). Both *Pseudomonas* and *Bacillus* spp. are important biocontrol agents because they produce siderophores that are very competitive for Fe binding.

There are different classes of siderophores such as hydroxamate, catecholate and mixed ligand siderophores. Siderophores produced by PGPB can play a role in the prevention or reduction of the effects of pathogenic microbes in plants by depriving such pathogens of Fe. In this case, PGPBs must produce higher rates of siderophores, and they must be specific to the producing organism and very competitive for Fe binding, relative to the pathogen. This is especially true for *Pseudomonas* and *Bacillus* spp, which are important biocontrol agents.

Direct antagonisms

Several microbial groups are effective biocontrol agents because they exhibit direct contact phenomena that lead to the destruction of organism and the elimination of competitors. For example, aggressive litter-decomposing and wood decay basidiomycetes (fungi) can destroy the mycelia of other fungi by growing over them, thus capturing their resources by replacement. This can be physical or combined with chemical antibiosis. For instance, *Collybia peronata* will overgrow and out-compete *Cladodporium cladosporoides* in laboratory cultures *in vitro*, by first producing hyphal runners which traverse the mycelia of *Cladodporium cladosporioides* for the purpose of capturing resources and spreading.

In addition to strong competitive interactions, antagonistic microbial interactions also include acting through parasites, which invade and kill microfauna, mycelium, spores and resting structures of fungal pathogens and cells of bacterial pathogens, or direct predation. The most common example of predation in soil is protozoa and nematodes preying on bacteria and fungi, particularly in the rhizosphere, where numbers of bacteria and fungi are high. Additionally, many soilborne fungi have been demonstrated to be antagonists of nematodes, including predacious fungi, endoparasitic fungi, parasites of nematode eggs and cysts, and fungi that produce metabolites toxic to nematodes. More than 150 fungal species have been isolated from the females or cysts of *Heterodera glycines (*the soybean cyst nematode), including *Exophiala tusarium* and species of *Glicocladium, Neocosmospora, Paraphoma, Phoma, Stagonospora, Verticillium, Dictyochaeta* and *Pyrenochaeta*. *Fusarium oxysporum* was able to colonize sclerotia of *Sclerotinia sclerotiorum* surfaces under soil substrate conditions, and antagonist re-isolation from sclerotia and viability reduction were low. Sclerotia were planted in the *Fusarium oxysporum* colony, and a significant reduction in sclerotia viability was detected over time, resulting in a reduction of the inoculum source.

Biological control is a way to relieve the pressure on soil biodiversity and restore the ecological balance (Ruiu, 2018). The United States Environmental Protection Agency (US EPA) defines biological pesticides as naturally occurring substances that control pests (biochemical pesticides), microorganisms as bacteria, fungi, viruses, protozoa and nematodes that control pests (microbial pesticides), and pesticidal substances produced by plants containing added genetic material (plant-incorporated protectants) (US EPA, 2019). The basic concept of biological control is to facilitate the natural ecosystem to counteract the potential of pests (Ruiu, 2018) and generally increase biodiversity and ecosystem functioning. Biological control agents can be very target-specific and can limit non-target effects while being benign to the ecosystem. Ruiu (2018) reviewed entomopathogenic microorganisms used for biological control;these organisms include bacteria, fungi, baculoviruses and nematodes.

Worldwide, the largest commercial success of a biological control agent is without doubt *Bacillus thuringiensis* (Bt), a common bacteria isolated from soil (Jouzani *et al.,* 2017). Bt is a biological control agent with insecticidal activity against a range of different insects, and different strains and marketed products have different specificity towards different pest insects which increases the specificity against the target organisms (reviewed by Bravo *et al.,* 2011). Bt produces an intracellular toxin thatm upon ingestion by an insect, is released in the insect gut where the deadly activity happens. (Figure 3.5.8.2). This specificity means that the Bt toxin does not affect warm-blooded animals such as birds and humans, and Bt products have even been sprayed from aircrafts over cities and other populated areas as late as May 2019 (Williams 2007, CityNews 2019). The Bt toxin producing genes have been inserted into agricultural crops, especially maize, and the GMO Bt-maize is presently grown in large areas of the world (Romeis *et al.*, 2019). Lately, the soil bacterium Bt has been found to have other plant growth-promoting properties and may in the near future be involved in other environmentally friendly plant growth-stimulating products and agricultural practices (Azizoglu 2019; Jouzani *et al.*, 2019). This is not surprising, as many other *Bacillus* species are known to have plant-beneficial traits (Saxena *et al.*, 2019).

Box 3.5.8.1 | Anaerobic soil disinfection (ASD) - an alternative fumigation technique

An alternative method to fumigants for managing soil pathogens is a technique that manipulates the soil microbiome. Anaerobic soil disinfestation (ASD) is a biologically based chemical fumigant alternative. ASD consists of incorporating organic soil amendments into the soil, covering the soil with a plastic mulch, and saturating the soil to field capacity with water (Rosskopf *et al.*, 2015). The soil undergoes a shift from aerobic to an anaerobic environment, during which facultative anaerobes and anaerobes increase in population. These microbes produce anti-bacterial and fungal metabolites such as short chain organic acids, methyl sulphide compounds, hydrocarbons, and dimethyl disulphide (Hewavitharana *et al.*, 2019). Plants grown in ASD treated soil had yields similar to or in excess of those for plants planted in chemically fumigated soil, and ASD yields were 30 times greater than the untreated controls (Shrestha *et al.,* 2016). Multiple studies have demonstrated that the microbial community changes with the use of ASD. The technique has been proven to work in multiple crops worldwide, yet the mechanics behind ASD are not fully understood and there remains a significant knowledge gap in pest control research. ASD offers broad-spectrum and non-specific pest control, like methyl bromide, and has been shown to manage soilborne bacterial and fungal plant pathogens and root-knot nematodes, and to provide weed control. However, just like chemical fumigants ASD impacts most soil life and biodiversity, so with the ambition of preserving biodiversity ASD should have limited use

Box 3.5.8.2 | Biological control and crops

Conventional agriculture with low crop diversity and extensive use of chemical pesticides tends to reduce soil biodiversity, unbalance the ecosystem with an oversimplification of the species present, and pave the way for pathogenic organisms to prevail.

Soilborne pathogens are one of the major biological causes of yield loss, decreases in fruit quality, and plant mortality. Plant pathogens are challenging to manage, as the same microbe can infect multiple hosts and it can remain in the soil for years. Growers will lose entire crops and grow less profitable crops or even abandon their fields because of the severity of soilborne diseases.

The abundance of pathogen suppressive microorganisms is significantly greater in a diverse crop rotation compared to monocultures (Peralta *et al.*, 2018). Biological control of pathogens cannot be attributed to just one aspect but rather by multiple factors of both biochemical, physical and biological origin. Hence, many strategies to fight the plant pathogens should be considered.

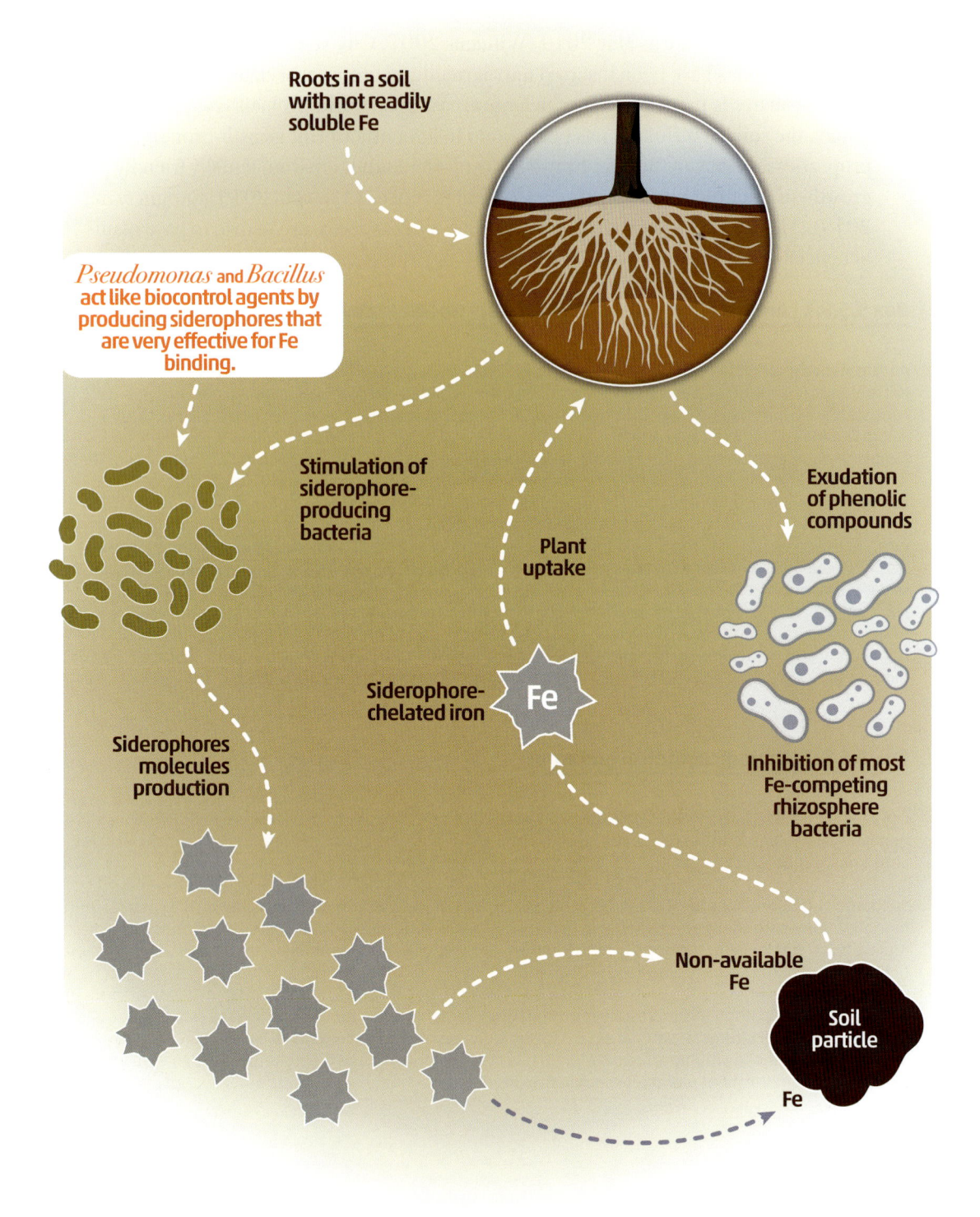

Figure 3.5.8.1 | Synergistic and antagonistic effects of soil microbes and plants

Chelating agents play important roles in soil ecology. Some plants, when growing in soils with non-readily soluble iron (Fe) (as calcareous soils), can either produce it by themselves or stimulate a specific group of soil bacteria to synthesize Fe-chelating agents named siderophores. They are organic molecules able to capture and carry iron into the plant roots. (Adapted from Wail and Brady, 2017).

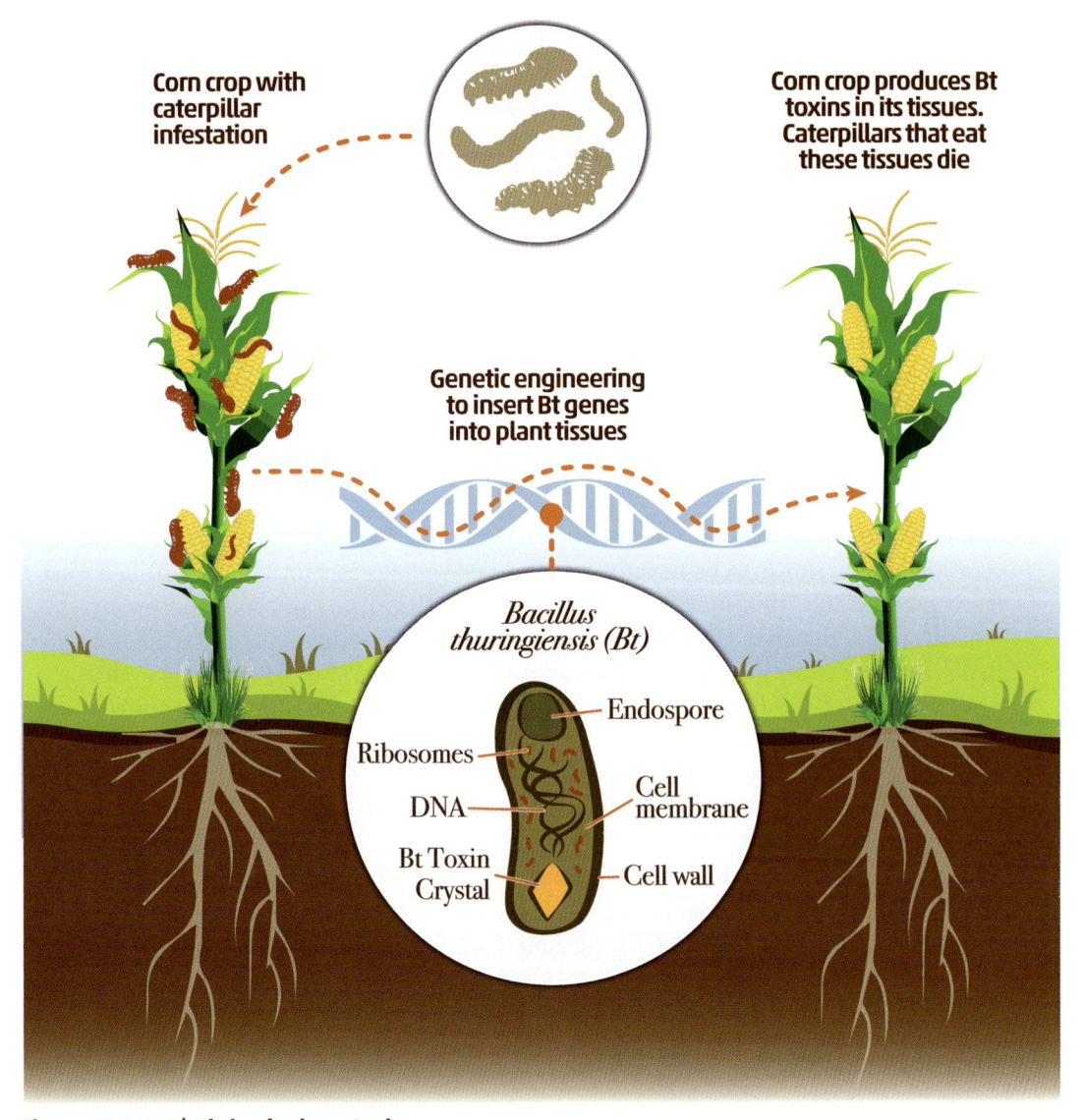

Figure 3.5.8.2 | Biological control

Bacillus thuringiensis (Bt), a bacterium species isolated from the soil, has been successfully used as a biological control agent against insects. Bt produces an intracellular toxin that, when ingested by an insect, is released in the insect's gut, killing it. The genes that produce the Bt toxin are inserted into agricultural crops, particularly maize, giving the plant the ability to avoid attack by certain pathogenic organisms.

3.5.9 | GENETIC DIVERSITY

Provisioning and regulating ecosystem services arise from a biodiverse ecosystem. Maintenance of genetic diversity, a supporting ecosystem service, underpins provisioning and regulating services, providing the raw material that enables humans to modify our environment in beneficial ways in response to change. Soil biodiversity is essential to the maintenance of genetic diversity as a supporting ecosystem service. Soil microbial and faunal biomass is equivalent to or exceeds above-ground biomass in most ecosystems (van der Heijden *et al.*, 2008; Fierer *et al.*, 2009). These microbes and animals have been referred to as the "unseen majority" (Whitman *et al.*, 1998; van der Heijden *et al.*, 2008) and likely encompass at least as much genetic diversity as has been documented in above-ground ecosystems. Wagg *et al.* (2014) demonstrated that as soil biodiversity is lost, ecosystem functions are reduced. With at least half of terrestrial genetic diversity housed below ground, conservation of soil biodiversity is a key component of maintaining genetic diversity.

Soil biodiversity can play a protective role in the emergence of disease. Biodiversity has repeatedly been shown to be negatively correlated with disease prevalence through what has been called the 'dilution effect' (Civitello *et al.*, 2015). Greater soil biodiversity is associated with enhanced control of pest and pathogen populations and reductions in disease incidence (Ferris and Tuomisto 2015; Wall *et al.*, 2015). There is some evidence that soil biodiversity may even directly mitigate human illness. In a study of teenagers in a small town in Finland, Hanski *et al.* (2012) found that exposure to greater soil biodiversity was associated with a reduced risk of allergies. Fungal plant pathogen diversity increases with plant diversity, yet fungal disease incidence and severity decline (Rottstock *et al.*, 2014). One possible mechanism for this observed decline is interference between pathogens (Ostfeld and Keesing, 2017), which is more likely in more diverse systems. The same relationship is expected for soil biodiversity and disease incidence and severity, although data are currently lacking.

Because most species are rare, taxonomic diversity tends to be negatively correlated with species abundances. Consequently, if higher soil biodiversity reduces pathogen abundance, then disease risk should be lowest in the most diverse communities (Ostfeld and Keesing, 2017). Extending to soil ecosystems, the finding that more diverse microbiomes within hosts suppress strains resistant to antimicrobial compounds (Keesing *et al.*, 2010) suggests that conservation of soil biodiversity is of critical importance to slowing the emergence of new diseases and antimicrobial resistance of disease agents for crops, domesticated animals and humans. Recent work by Kinkel and others (Schlatter *et al.*, 2017) advances the hypothesis that soils that are generally suppressive of plant disease result from competitive interactions among pathogen populations and diverse networks of saprophytic populations. Such general suppression would depend on the diversity of potential competitors for pathogens. If soil organisms at higher trophic levels are more susceptible to extinction than those at lower trophic levels (such as plant or animal pathogens), as has been observed in animal systems (Duffy, 2003), then conservation of soil biodiversity is also essential for the maintenance of organisms capable of natural regulation of populations of disease agents.

3.6 | CULTURAL SERVICES

Decades ago, the influential pedologist, Hans Jenny, noted the growing cultural gap between soils and humans, further heightened by the decline in the ability of soil scientists to feel connected to and communicate about soils: "we don't mention our emotional involvements. In fact, our soil language is lifeless, and the soil descriptions in our publications are utterly boring." He goes on to say that our "intellectual isolation and ... invisibility have to do with the lack of formulating exciting ideas about soils themselves and their relations to people" (Jenny, 1999).

Cultural ecosystem services in the Millenium Ecosystem Assessment refer to the "the nonmaterial benefits people obtain from ecosystems through spiritual enrichment, cognitive development, reflection, recreation, and aesthetic experiences" and include not only services, but benefits and values (Chan *et al.*, 2012, Milcu *et al.*, 2013). The FAO Ecosystem Services and Biodiversity (ESB) framework includes the following among cultural services: aesthetic inspiration, cultural identity, sense of home, and spiritual experience related to the natural environment. Cultural services are deeply interrelated and also connected to provisioning and regulating services (Chan *et al.*, 2012). Good health and well-being (SDG 3) is also strongly connected to many cultural services. The knowledge that soil is alive expands the possibilities for human-soil relationships. Discovery of the commonalities shared by the diverse microbiomes that thrive in the bodies of both humans and soils makes these connections even more pertinent and compelling. The concept of "soil health" reflects society's increasing recognition of soil's aliveness which stems, of course, from soil biodiversity (Kirschenmann, 2005). With that has come concern for and increasing desire to take better care of soils (Krzywoszynska, 2016, 2019).

3.6.1 | SPIRITUAL EXPERIENCE AND SENSE OF PLACE

A sense of place is tied, in part, to its landscapes and soils. The sensory information that defines a place - its sights, sounds, smells - stems in part from biodiversity of its soils. This includes the local insects that spend part of their life cycle below ground and the scents of post-rain geosmin from actinomycetes or hydrogen sulfide from sulfate reducers in swampy soils. Among rural communities connected to land, consideration of soil is less as an asset and more as a partner in a relationship (Puig de la Bellacasa, 2019). Values and associated principles emerging from their specific landscapes and ways of life have guided and sustained the traditions and practices of farmers (Fitter *et al.*, 2010; McNeill and Winiwarter, 2004). Globally, farmers value the importance of soil biodiversity, either for its own sake or as translated through its impact on soil properties and functions. In many cases, the traditional knowledge of farmers has been disconnected, or lost, with the introduction of "improved" seeds, agrichemicals, equipment and cultural practices brought about by development and engagement in global markets (Barrera-Bassols and Zinck, 2003). In reaction to widespread degradation and loss of soils via erosion and nutrient depletion over the past decades, however, the concept of soil health has emerged, with many farmers recultivating human-soil relations (Krzywoszynska, 2019).

Soil has long been part of philosophical systems and religions throughout the world (Minami, 2009). The origin or creation stories of many cultures include involvement of clay, soils or earth. Throughout the world, soil is the home or origin of numerous spirits who protect the land, crops and harvest. The "energy" or "vitality" that some cultures believe give rise to soil's generative power (Minami, 2009) would appear to be associated with soil's life; with its biodiversity.

Specific soil properties and functions combine to define a sense of place in crop-growing regions via the foods and wines grown there. *Terroir*, or the interaction of climate, soil, and the vine in a given place (van Leeuwen *et al.*, 2004, 2018) determines the character and quality of grapes grown in a particular region. For wine, we know that the soil microbiome plays a central role in *terroir*, both in viticulture (for example, in soil fertility and crop health) and enology (for example, in fermentation and flavour; Belda *et al.*, 2017). The concept of *terroir* has also been described for other foods including vegetables, fruits and cheeses (Trubek, 2008).

Historically and continuing today, many human conflicts and wars are fought over land and often arise in regions with seriously degraded soils (due to such factors as erosion or drought; Lal, 2015). Though not called out explicitly, declines in the below-ground biodiversity of these regions undoubtedly contributed to soil degradation and provided additional reasons for conflict. Collapses of civilizations in some regions of the world have followed periods of long-term conflicts over land and soil (Hillel, 1992; Minami, 2009).

3.6.2 | AESTHETIC APPRECIATION AND INSPIRATION FOR CULTURE, ART AND DESIGN

For centuries, a diverse variety of art forms and traditions have engaged with soils, using the soil as a medium itself (for example in pottery and as pigment) or as the subject of artwork (such as in landscape paintings). A number of publications have compiled and catalogued many of these creations and activities (Landa and Descola, 2010; Toland *et al.*, 2018; Feller *et al.*, 2015). Soil is featured or embodied in paintings, ceramics, sculpture, literature, philosophy, cinema, architecture, performance art and multi-media productions, among other forms of art (Minami, 2009). Direct engagement with soil is part of many recent art exhibitions and public engagements (Puig de la Bellacasa, 2019). Soil biodiversity contributes to many art works, either responsible for some of the shapes, reliefs, and colours depicted in landscapes, or as the origin of earth-derived colours that wind up on palettes or cave walls. (Figure 3.6.2.1). Examples are pigments originating from iron and manganese-oxidizing or reducing bacteria (Tuli *et al.*, 2015).

© Christopher Marley

Figure 3.6.2.1 (next pages) | Soil biodiversity and inspiration for art

The beauty of organisms, their wide variety of shapes, colors and patterns has inspired different forms of art and soil biodiversity is not the exception. Author: Christopher Marley.

© Christopher Marley

Contributions of soil biodiversity to ecosystem functions and services

© Christopher Marley

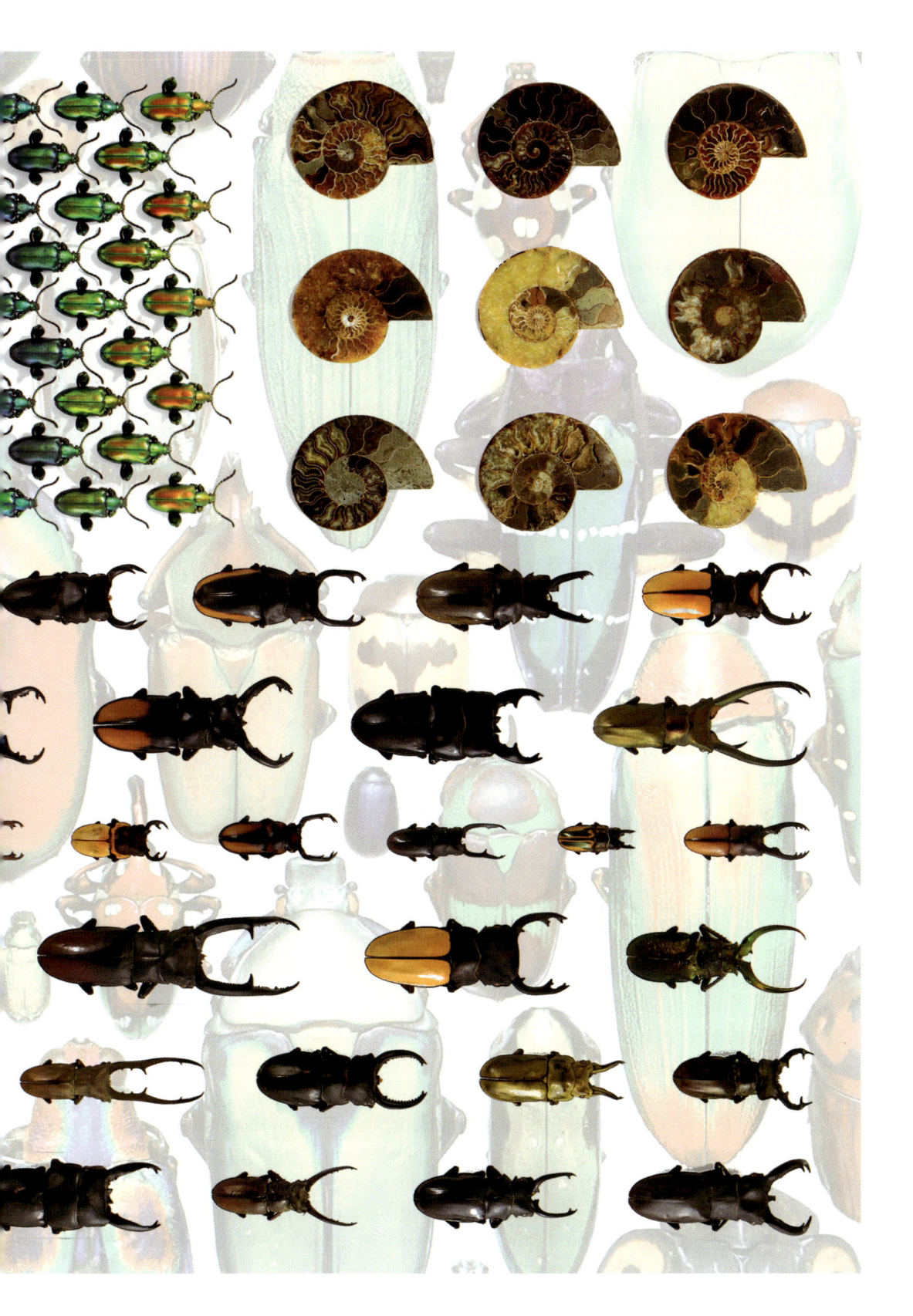

Contributions of soil biodiversity to ecosystem functions and services

3.6.3 | CULTURAL HERITAGE, KNOWLEDGE AND EDUCATION

Soils are major repositories where physical artefacts of human cultural heritage (such as structures, objects and organisms) become archived. These sites are important for understanding culture and history, and are the subject of archaeology and paleoecology (Adhikari and Hartemink, 2016). Soil biota play important roles in both preservation and deterioration of these artefacts. The microbial processes involved are both direct – via decomposing organic material or corrosion – and indirect – through governing soil environmental conditions such as redox, pH, and nutrient availability. Understanding these soil biological processes could lead to tools and practices to better preserve these historical sites and archives.

Ethnopedology is the study of local knowledge of soils, and recognizes the value of the cultural context in local sustainable land management. This includes "beliefs, myths, rituals and other symbolic meanings, values and practices related to land management, and soil quality evaluation" (Barrera-Bassols and Zinck, 2003). Ethnopedology underscores how local technical knowledge cannot be abstracted from its cultural context, as well as how essential both the natural and social sciences are in understanding how humans "engage symbolically, cognitively and practically with soil and land resources" (Barrera-Bassols and Zinck, 2003).

3.6.4 | RECREATION AND MENTAL AND PHYSICAL HEALTH

Natural spaces provide social, recreational, spiritual, and sometimes therapeutic benefits for society. There has long been considerable public support for green spaces, national parks, and wilderness areas. More recently, the many human health benefits associated with contact to Nature - including soil - have been documented, and these encounters are being prescribed, for example with Nature Rx and creation of gardens at clinics, prisons, schools and places of employment (Frumkin *et al.*, 2017). Many positive attributes of these spaces (such as tranquility, wildness and visual and olfactory stimulation) depend on soil biodiversity, particularly in less managed areas. Soil can be an essential part of what makes a tourist destination desirable, for example in agrotourism and wildlands. Impoverishing biodiversity not only jeopardizes these spaces but negatively impacts human enjoyment and the intrinsic value of such experiences (Fitter *et al.*, 2010). Soil is vulnerable and can be severely degraded through compaction by feet and vehicles, use of chemicals, careless waste disposal, and destruction of vegetation (McCool and Moisey, 2001). Sustaining heavily impacted locations year after year depends on the activities of soil organisms to regenerate them through re-building organic matter and soil structure, cleaning up pollution, and supporting reestablishment of plant communities.

Though not the primary goal, the economic value of these ecosystems for tourism and recreation is often greater than their ecological value for provisioning services (Fitter *et al.*, 2010).

Human-soil relationships

Kirschenmann (2005) quotes the philosopher Aldo Leopold, who reminds us that "we are not "conquerors" of the biotic community, we are simply "plain members and citizens" of it. Leopold notes that we can learn from soil: "it is the soil that helps us to understand the self-limitations of life, its cycles of death and rebirth, the interdependence of all species. And our task as humans is not to 'save' the environment, nor preserve things as they are, but to engage the environment in ways that revitalize the biotic community" (Kirschenmann, 1997).

The scholars Anna Krzywoszynska and Maria Puig de la Bellacasa, bringing new perspectives from anthropology and geography, describe the valuing of soils and soil biodiversity through the idea of "care networks" that connect humans and non-humans, such as soil biota (Krzywoszynska, 2016). Soils are viewed not simply as providers of services for human exploitation and consumption but as "living worlds with an intrinsic value for themselves beyond human use" (Puig de la Bellacasa, 2019). Krzywoszynska (2019) writes about the need for more ethical relations between humans and non-humans, especially with soil organisms who "pervade and create liveable environments" and where "care for non-human lives (e.g. soil biota) becomes part and parcel of caring for human well-being."

Krzywoszynska and Puig de la Bellacasa's ideas have been inspired, in part, by scientific advances in our understanding of soil food webs, and through application of non-invasive technologies for visualizing soil (Puig de la Bellacasa, 2015). Building on the knowledge that soil biota is organized into food webs provides a platform to integrate thinking about the role of all organisms and flows of material into ethical systems that are not solely human-centred and materialist.

Importance for future

Inclusion of social-cultural perspectives in considerations of soil biodiversity broadens public engagement by increasing social acceptance and legitimacy of management decisions, finds common ground to engage actors with different values and goals, and bridges gaps among different disciplines and cultures. Public support and policy change for soil and its biodiversity will not be driven by science alone. For society at large, the sense of discovery, connection and inspiration rooted in soil biodiversity - knowing that soil is alive – is the best means to expand enthusiasm and advocacy for soils.

3.7 | ECONOMIC VALUE OF SOIL BIODIVERSITY

Soil biodiversity provide a plethora of values to human society by contributing significantly to ecosystem services from soil biota. Values in this context are purely anthropocentric, as opposed to ethical or bio-centric reasoning on biodiversity. It follows the framework of the Millennium Ecosystem Assessment and TEEB.

Soil biodiversity can be considered a natural capital asset from which a flow of soil ecosystem services is produced (Turner and Daily, 2008). When the soil biodiversity asset is reduced, for example due to degradation of soils, this will lead to costs to society (such as requiring mitigation of environmental impacts such as GHG emissions) or to landowners (through increases of costly inputs to alleviate the decline in soil ecosystem services).

If society omitted to account for the value of soil biodiversity and the costs of running down the natural capital from soil biota, policies would be misguided and society would misallocate scarce resources (Pascual *et al.*, 2015). Continuing along the economic concept of soil biodiversity as a natural capital asset, soil biodiversity represents a portfolio of resources that build up soil natural capital; the flow of ecosystem services provides a return or interest received from the asset (Perrings *et al.*, 2006) cited by (Pascual *et al.*, 2015).

Society and private landowners directly or implicitly make trade-offs between enhancing soil biodiversity or investing in other economic activities on the land, depending on how they perceive the expected net returns. In other words, landowners and society contribute directly and indirectly through regulation and consumption choices to maintaining or reducing the natural capital asset of (high) soil biodiversity. Shedding light on the values at stake across space and time is crucial for making optimal decisions. These optimal decisions may differ depending on whether we consider public or private goods flowing from soil biodiversity to be assets.

An example of a private good from soil biodiversity is crop yield, which is dependent on the support of nutrient delivery and water regulation provided by soil biodiversity.

Examples of a public good from well-functioning soil biodiversity are carbon sequestration and nutrient retention, which contribute directly to two essential planetary boundaries: global nitrogen cycle and climate (Rockström *et al.*, 2009; Steffen *et al.*, 2011).

Different approaches exist to attribute monetary values to changes in soil biodiversity, depending on the type of value. Pascual *et al.* (2015) offer a framework based on the Total Economic Value of soil biodiversity, which is the cumulative value of total output value. This consists of use and non-use values and the natural insurance value, which relates to the capacity of soil biodiversity to maintain the production of ecosystem services over time under risk and uncertainty. It is important to recognize that the economic values of ecosystem services do not exist without the human values, and that human inputs into land utilizations also contribute to the generation of value of the ecosystem service provision. Ecosystem services are not generated purely from natural processes (Figure 3.7.1).

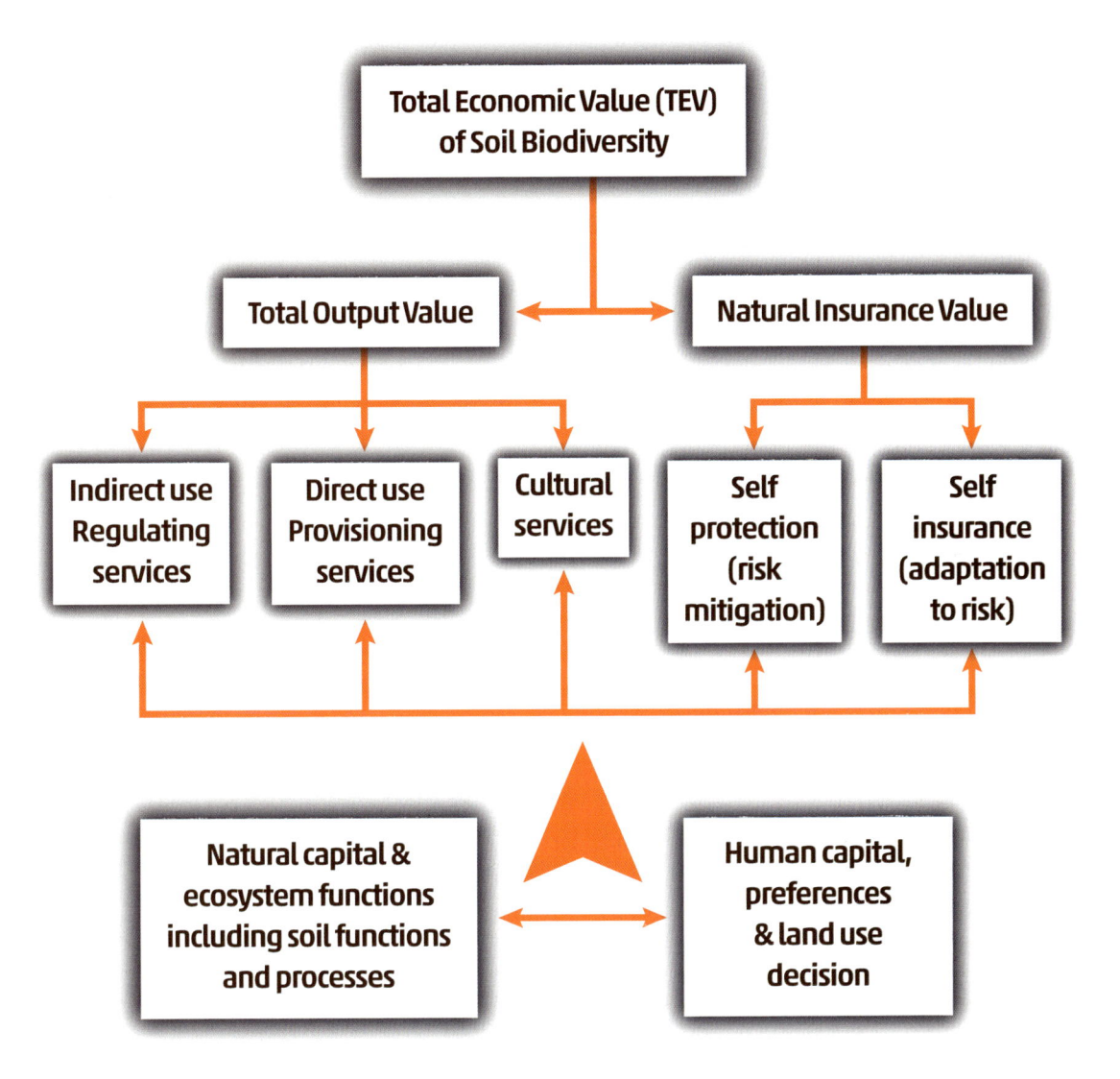

Figure 3.7.1 | Economic value of ecosystem services

Components of Total Economic Value in relation to ecosystem services, ecosystem functions and land use decisions making. Adapted from Pascual *et al.* (2015).

While the Total Economic Value framework is accepted and widely used in environmental assessments (such as UKNEA), the emphasis has been on quantifying the Total Output Value (Figure 3.7.1) and less attention has been given to evaluation of the value of accounting for risks and uncertainty in the provision of ecosystem services, that is, the natural insurance value.

Furthermore, the literature on economic values of ecosystem services has focused on measuring different *components* of value rather than on quantifying the role of the supporting functions underpinning those values. One of the main reasons for this has been the concern that quantifying supporting services, intermediate services and final services independently would lead to double (or triple) counting and result in estimates that would not be usable for economic prioritizations in decision-making processes (Fisher and Turner, 2008). This remains a valid concern (Yang *et al.* 2018) as the value of the final services such as food production already include the value of the biological processes underpinning the production. However, this does not imply that the interdependences among the multiple services supported by soil biodiversity can be ignored or that they are not economically relevant. It is essential that valuations of alternative land utilization plans take account of how investments in the provision of some services may have synergistic effects on other services, whereas investment in other services may result in trade-offs.

A fairly recent development in natural resource economics has been the concept of natural insurance value, originally proposed in the context of biodiversity economics by Baumgärtner (2007). Baumgärtner shows that biodiversity can be interpreted as a natural insurance value to the risk-averse natural resource manager and that this value component is an additional component to use and non-use values in the total economic value framework. This implies that including natural insurance values from biodiversity would not lead to double counting and that accounting for natural insurance would lead to higher investments in biodiversity.

In more recent work, the concepts have been developed further and have distinguished insurance value into two types – a component relating to mitigation of risk (reducing the probability of adverse outcomes) and adaptation to risks (lowering the magnitude of impacts in the case of an adverse event; Baumgärtner, 2007). The empirical quantifications of natural insurance values remain scarce, but research is emerging using integrated ecological-economic models on the value of investment in regulating soil functions (Sidibé *et al.*, 2017). A real policy concern in this field is the argument that provision of market insurance to adapt to increases in risk due for example to climate change may potentially reduce farmers' incentives to invest in sustainable soil practices. There is some empirical evidence that this should be a concern (Wu, 1999). In response to this it has been suggested that development of market insurance products should consider implementation of sustainability conditions to ensure that long-term provision of private and public goods from soils is protected (Jørgensen *et al.*, 2020).

Public and private goods from soil biodiversity

The optimal decisions about preserving or enhancing soil biodiversity may differ depending whether we consider public (collective) or private goods flowing from soil biodiversity assets through bundles of soil ecosystem services.

Private goods are characterised by being exclusive and rivalrous, such as crop production that is appropriated by private landowners, while collective goods are the opposite (non-exclusive and non-rivalrous), such as soil biodiversity's contribution to reducing soil erosion or eutrophication, benefitting people outside the farm gate. Optimization of private goods (for example, maximizing yields through intensive farming practices) typically leads to an undersupply of public goods, causing negative externalities such as nutrient pollution and GHG emissions. The notion of public and private goods is therefore central to the management of soil biodiversity. The present soil biodiversity crisis is aggravated by the property rights regime (Bartkowski *et al.*, 2018), although there is a growing recognition of the social obligations of property owners to manage land more oriented towards the common good.

The natural insurance value of soil biodiversity exhibits both private and public goods characteristics. A private landowner can benefit from the natural insurance value of mitigating and/or adapting to risks on his or her own land. Also, future owners benefit from current sustainable management practices, giving incentives to adopt a long-term perspective. This is however complicated by the extended practice in developing countries of rented agricultural land, where the tenant, compared to the landlord, is found to be disincentivised to invest in soil biodiversity assets generating long-term private benefits and instead overexploit soil biota services (Foudi, 2012). In the face of risk and uncertainty, the natural insurance value of soil biodiversity for collective goods include for instance the capacity over time to maintain carbon sequestration services or reduce eutrophication and soil erosion, which is appropriable by society at large, but controlled by private landowners (Pascual *et al.*, 2015).

3.7.1 | THE CHALLENGE OF VALUING SOIL ORGANISMS

Many would say that it is not possible to place a value on ecosystems, nor on their natural capital and on the services they provide to people, or that it is meaningless to estimate the value of natural capital stocks at large scales because this value is essentially infinite (Costanza *et al.*, 1997; Robinson *et al.*, 2013). Valuing total ecosystem services at a national level invites a similar criticism (Robinson *et al.*, 2013). It is the effect of change in stocks or flows, or margins, rather than estimating a total value, that is important. Nevertheless, Costanza *et al.* (1997) did estimate the annual value of services provided by ecosystems of the world, and this was reckoned to be somewhere between USD 16 and USD 54 trillion, in comparison to an annual global gross national product of around USD 18 trillion. These ecosystem services included all the benefits that humans derive from processes acting upon natural capital (that is, ecosystem assets and natural resources). Since this landmark paper, there has been a steady rise in the number of publications on ecosystem services, but only a small proportion include studies that directly link soil properties to the services (Adhikari and Hartemink, 2016).

Making a robust assessment of the economic value of nature and biodiversity is a huge challenge because there are many different ways to derive value and because value is not the same as price (*Nature* Editorial, 2019). The concept of economic valuation of ecosystem services leads to a hierarchy of categories of value with their sum equalling what is termed the total output value (see Baveye *et al.*, 2016; Jónsson and Davíðsdóttir, 2016). Total output value can be divided into instrumental value (that value directly benefiting humans) and intrinsic value (the non-anthropocentric value, for example conferred by genetic diversity and ecological processes). Instrumental value itself can be split into use value and non-use value. Use values are those related to direct consumption or use of services, or regulation services. Non-use values are those detached from the actual use of the service (for example option, bequest and existence values).

The soil ecosystem service valuation framework in Pascual *et al.* (2015) included a further category of "natural insurance value." This is value that one places on the ability to reduce risk of reductions in ecosystem service flows in uncertain conditions. Hence, in addition to using stocks to value ecosystem services, the management of soil natural capital and link to services may also be used as an approach to increasing system resilience and mitigating production risks in agriculture (Cong *et al.*, 2014). We can also therefore think about the massive diversity and likely functional redundancy found in soil communities from the perspective of greater stocks of soil biodiversity delivering resilient ecosystem service flows under disturbance or environmental stress.

The variety of values is accompanied by a variety of valuation methods. Market-based valuation can be made using market prices, production functions or replacement costs; valuations based on market prices require a well-functioning market (POST, 2011) and, for example, regulating services are not typically marketed and their valuation is more challenging. Non-market valuation can be made using a revealed preference approach (such as hedonic pricing or travel costs) or a stated preference approach (such as contingent valuation or price modelling; Baveye *et al.*, 2016). Ecosystem service valuation is also dependent on market values; these differ geographically and culturally, with services having different values (Robinson, 2013). It is also argued that assets managed for monetary values only, as opposed to more intangible values, may risk being mismanaged, with increased chances of degradation (POST, 2011).

The actions of soil organisms are strongly linked to important tangible and intangible economic values. Focusing on the stocks of soil natural capital is valuable because change in soil natural capital stocks can be used to estimate the value of potential gains or losses of ecosystem service flows. We cannot, however, estimate natural capital stocks by examining changes in ecosystem service flows. Baseline data on the stock of organisms provides a powerful gauge against which future changes in soil properties and threats to their function, including climate change, may be quantified. Regular measurements over time can provide evidence as to whether the soil organism stock is in decline, and can provide inference on changes in the value of ecosystem services.

Examples of valuations

At the local scale, Dominati *et al.* (2014) used a range of valuation methods (market price, replacement costs, provision cost, defensive expenditure) to value soil services in a pasture system in New Zealand. For example, flood mitigation was valued using the costs associated with building an on-farm water-retention dam, to substitute for the water retention capacity of the soil. They presented six guiding principles for such valuation: differentiating soil services from supporting processes; identifying key soil properties and processes behind each service; distinguishing natural capital from added/built capital; identifying how external drivers affect natural capital stocks; analyzing the impact of degradation processes on soil properties; and basing the economic valuation on measured proxies (Dominati *et al.*, 2014). The recent paper by Pascual *et al.* (2015) demonstrated two examples of valuation of soil biodiversity and ecosystem services, with the specific cases of the value of earthworms to water infiltration, and the value of earthworms to crop productivity and greenhouse gas regulation. This study was carried out in the context of soil biodiversity as an insurance against climatic variability and the implications for agricultural outputs.

Agricultural activities have been depleting or degrading soil natural capital for centuries. Improving or increasing soil natural capital is therefore a major step towards more sustainable agricultural systems (Pretty, 2008). Our land management practices determine whether the stocks of natural capital are being degraded (in the case of non-sustainable practices), maintained or improved. Degradative changes in soil ecosystems impinge on their condition or quality and, therefore, on their functional capacity. In Europe, there has been much focus on the impacts of soil erosion and organic matter depletion.

Eight major threats that were identified and based on the Impact Assessment (European Commission, 2006) were estimated, on an annual basis, to cost the following to the European Union: erosion: EUR 0.7 to 14.0 billion; organic matter decline: EUR 3.4 – 5.6 billion; salinization: EUR 158 – 321 million; landslides: up to EUR 1.2 billion per event; contamination: EUR 2.4 – 17.3 billion. No estimate was possible for compaction, sealing and biodiversity decline. Degradation of soils in England and Wales was estimated to cost between GBP 0.9 billion and GBP 1.4 billion per year (Graves *et al.*, 2015). Having a thorough knowledge of the stock of soil organisms means that the impacts of degradation threats on ecosystem service provision and valuation can be better understood. Indeed, Adhikari and Hartemink (2016) suggest that future research on ecosystem services should focus on exploring functional diversity of soil biota.

Opportunities and knowledge gaps

Understanding the value of ecosystem services linked to soil organisms is vital for decision-makers when considering soil use and land management changes. We need to sift through the multitude and complexity of approaches when thinking about economic valuation of services mediated by soil organisms. It is vital that clear and influential messages be presented to different groups of stakeholders. The solution is not to generate another framework to value the benefits derived from soil organisms, but to make better use of the data and examples we have and to communicate these in a coherent and effective manner.

Two general (an not mutually exclusive) routes towards this goal are suggested, with contrasting coverage and depth. For the first option, a key group of soil organisms should be selected and their diversity and abundance considered as holistic indicators of a capacity for provision of sustainable soil services. The economic value of these indicator organisms should then be derived at a global scale. A good example would be to use earthworms. Economic values for the ecosystem services that earthworms provide were shown to be substantial. Bullock *et al.* (2008) suggest that earthworms add EUR 723 million per year to livestock production in Ireland; adding the equivalent value for food crops could raise the total value of earthworms to over EUR 1 billion. Bailey *et al.* (1999) estimated the value of earthworms for soil structuring service at GBP 0.48 per kilogram of earthworms under reduced tillage. If such valuations of earthworm-mediated ecosystem services could be generalized across the globe, we could begin to develop an appreciation for their value. For the second option, efforts must be made to generate a global map of case studies based on a pluralistic valuation approach (Pascual *et al.*, 2015b), where location-based valuations for soil biodiversity are made for areas representing different biomes and different land management practices. This approach should take into account the fact that soil ecosystem service assessment is dependent on market values, which can change over time, as well as between countries or regions where services have different values (Robinson, 2013). We must invest in soil biodiversity.

©FAO/Ronald Vargas

CHAPTER 4
THREATS TO SOIL BIODIVERSITY - GLOBAL AND REGIONAL TRENDS

4.1 | INTRODUCTION

The vast diversity and the important role of soil biodiversity in ecosystem functioning and ecosystem service delivery can be deeply affected by human activities as well as by natural disasters, though the latter may also be influenced by human-induced changes (for example, deforestation or road building causing landslides). Most threats to soil biodiversity and function are directly related to human activities and associated with land use cover, management and change. These include deforestation, urbanization, agricultural intensification, loss of soil organic matter/carbon, soil compaction, surface sealing, soil acidification, nutrient imbalance, contamination, salinization, sodification, land degradation, fire, erosion and landslides (Figure 4.1.1).

Land-clearing is a major global threat to soil ecosystem services. This threat has many disguises depending on the specific characteristics of the world's ecoregions. For example, deforestation to make way for food and fibre production systems can lead to massive erosion and nutrient depletion in ecoregions characterized by mountainous terrain with steep hillsides and major precipitation events such as in Latin America and New Zealand.

Land clearing for agricultural intensification in large, low relief ecoregions in Australia (for example tropical and subtropical grasslands) can lead to wind erosion or to acidification. Besides losing tree cover, some agricultural practices result in rapid loss of soil C and microbial biomass, particularly of soil fungi. Simplified and often highly dynamic communities suffer from increasing rates of erosion and leaching due to reduced capacities to absorb water and mineral nutrients therein. Monocultures commonly result in proliferation of above-ground and below-ground pests and pathogens, which require introduction of pesticides in intensively managed fields. These have variable and largely unpredictable effects on natural soil biota. Agricultural practices also reduce soil nutrient concentrations, which requires fertilization. Excess fertilizer applications typically reduce the abundance of mutualistic soil biota, which enables increase in pathogenic microbiota

(Wall *et al.*, 2015). Modern intensive agriculture demands a continuous and constant trade-off between provisioning and regulating/supporting services. Productivity aims to increase the rate of provisioning services to the detriment of regulating services; however, when regulating and supporting ecosystem services are disrupted, food production is seriously affected, the result being a vicious downward spiral (FAO, 2011).

Global trends in urbanization and infrastructure-building bury much of the land surface under concrete, strongly reducing biodiversity and ecosystem services. The same applies to mining and waste dump areas. Two other threats – climate change and invasive species – are usually indirectly associated with human activities, but have become increasingly worrisome in the last couple of decades. The effects of changes in climate and invasive species are poorly understood for much of the soil biota. Accumulating evidence suggests that global change effects may largely differ by taxonomic and functional groups, specific factors and their combinations.

The last decade has shown that extreme climatic events, such as drought and floods, are the aspect of climate change that may be most relevant, overriding gradual shifts in temperature and precipitation. Orgiazzi *et al.* (2016) also discuss variations among groups of soil organisms in responding to these environmental change drivers. For example, soil acidification may affect soil microorganisms more strongly than meso- and macrofauna (Orgiazzi *et al.*, 2016), while for other threats, such as land-use intensification, larger soil fauna may be more strongly affected than microorganisms (Gossner *et al.*, 2016; Phillips *et al.*, 2019a). This points to the fact that the effects of land use intensification and climate change on soil biodiversity are organism-dependent (George *et al.*, 2019), and therefore that detailed information on multiple soil species and traits is urgently needed to better understand and predict threats to the different facets of soil biodiversity (Pey *et al.*, 2014; Salmon *et al.*, 2014).

The level of impact to soil biodiversity and function is not the same for all types of threats and for all regions of the world, and the effects of global change on soil biodiversity may be direct or indirect – via altered vegetation and nutrient availability. Importantly, climate change and land use intensification drivers are not completely independent of one another and thus co-occur. For instance, soil erosion is a process that is particularly relevant in disturbed ecosystems like agricultural lands, with approximately 80 percent of the Earth's agricultural lands experiencing significant levels of soil erosion (Orgiazzi *et al.*, 2016; Borelli *et al.*, 2017; Sartori *et al.*, 2019).

Furthermore, intensively used agricultural lands that are treated with soil tillage often also receive high levels of mineral fertilizers and pesticides. Land degradation is often related to other drivers like overgrazing by livestock and/or intensive agricultural use. This means that greater efforts are needed to understand the multiple direct (such as intensive land use) and indirect (such as climate change) anthropogenic impacts (Veresoglou *et al.*, 2015; Orgiazzi *et al.*, 2016) on soil biodiversity. Another important implication is that threats to soil biodiversity do not only co-occur but can have additive, interactive or synergistic effects (Thakur *et al.*, 2018), reducing soil biodiversity to even lower levels than what we would expect to find based on single driver studies. For example, effects of increasing temperature were shown to be minor under ambient water conditions, but detrimental for soil biological activity under drought (Thakur *et al.*, 2018).

Important interactions among several of the individual threats listed above and the combination of factors may synergistically affect soil biota and its functioning. For example, plants under drought stress may be more vulnerable to invasive pathogens and pests. At the same time, altered climatic conditions may promote invasion of co-introduced microbial species. Fragmentation of natural communities may reduce migration of both macro- and microorganisms, with increasing risks of extinction. Many of these combined effects may be unpredictable, because of our poor knowledge of ecophysiology and functioning of key soil organisms. Taken together, it is likely that the combined global change factors reduce biodiversity of native species, which is partly compensated for by increasing spread of cosmopolitan species. The combined global change effects are predicted to be context-dependent (that is, they differ by biome, organism group and relative effect on dominant vegetation or its shift).

Unfortunately, the level of knowledge of the impacts of these threats on soil biodiversity and function are highly variable, depending on the threat and the region, as well as the target biota (macro-, meso- or microfauna, microbes). Notably, despite slowly accumulating evidence for the ubiquity of significant interactive effects of environmental change drivers (Eisenhauer *et al.*, 2012a; Thakur *et al.*, 2018, 2019), there are currently almost no mechanistic understanding nor well-informed predictions of interactive impacts of multiple drivers (Borelli *et al.*, 2018; Thakur *et al.*, 2019) on soil diversity and consequences for ecosystem functions.

Despite the mounting scientific evidence warning about major threats to soil biodiversity and function in response to climate change and land use intensification, soil biodiversity has been omitted from many global biodiversity assessments and conservation actions (Cameron *et al.*, 2019; Eisenhauer and Guerra 2019), and understanding of global patterns of soil biodiversity remains limited (Delgado-Baquerizo *et al.*, 2018; Cameron *et al.*, 2019; Crowther *et al.*, 2019; Phillips *et al.*, 2019b; van den Hoogen *et al.*, 2019).

In the following sections, specialists from around the world tackle each of these threats and their potential impacts on soil biodiversity and ecosystem functions, highlighting knowledge gaps to address in future research.

Figure 4.1.1 (next pages) | **Major anthropogenic threats to soil biodiversity**

The major threats to soil biodiversity are caused by human-induced changes and the negative impacts can be amplified by the synergistic and additive effects that might occur among such threats.

Deforestation
drivers and effects on soils

- Land use change
- Loss of SOM and nutrients.
- Changes in soil physical properties.
- Disruption of suitable habitat.
- Changes in pH.

Impacts on soil biodiversity

- Loss of specialist species and increase in generalist taxa.
- Decrease in predator species.
- Reduced soil and functional diversity.
- Recovery could take decades.

Agricultural intensification
drivers and effects on soils

- Greater use of external inputs (pesticides, fertilizers) and more soil disturbance.
- Greater risk of soil erosion, contamination, land degradation, compaction and salinization.
- Alteration of hydrological and biogeochemical cycles.
- Disturbance of soil structure.
- Loss of SOM.

Impacts on soil biodiversity

- Decrease in soil biodiversity.
- Smaller and less complex belowground food webs.
- Recovery of soil communities may take years or decades.
- Less efficient and functional soil food webs.
- Loss of soil carbon and nutrients through leaching.

Nutrient imbalances
drivers and effects on soils

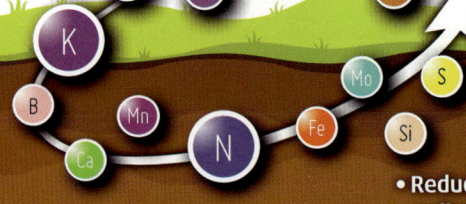

- Change in the availability of essential nutrients.
- Excessive use of mineral fertilizers.

Impacts on soil biodiversity

- Reduces the growth capacity of soil microorganisms.
- Reduces nutrient flow through the soil food web.
- Alteration of the nutritional content of primary producers and litter inputs.

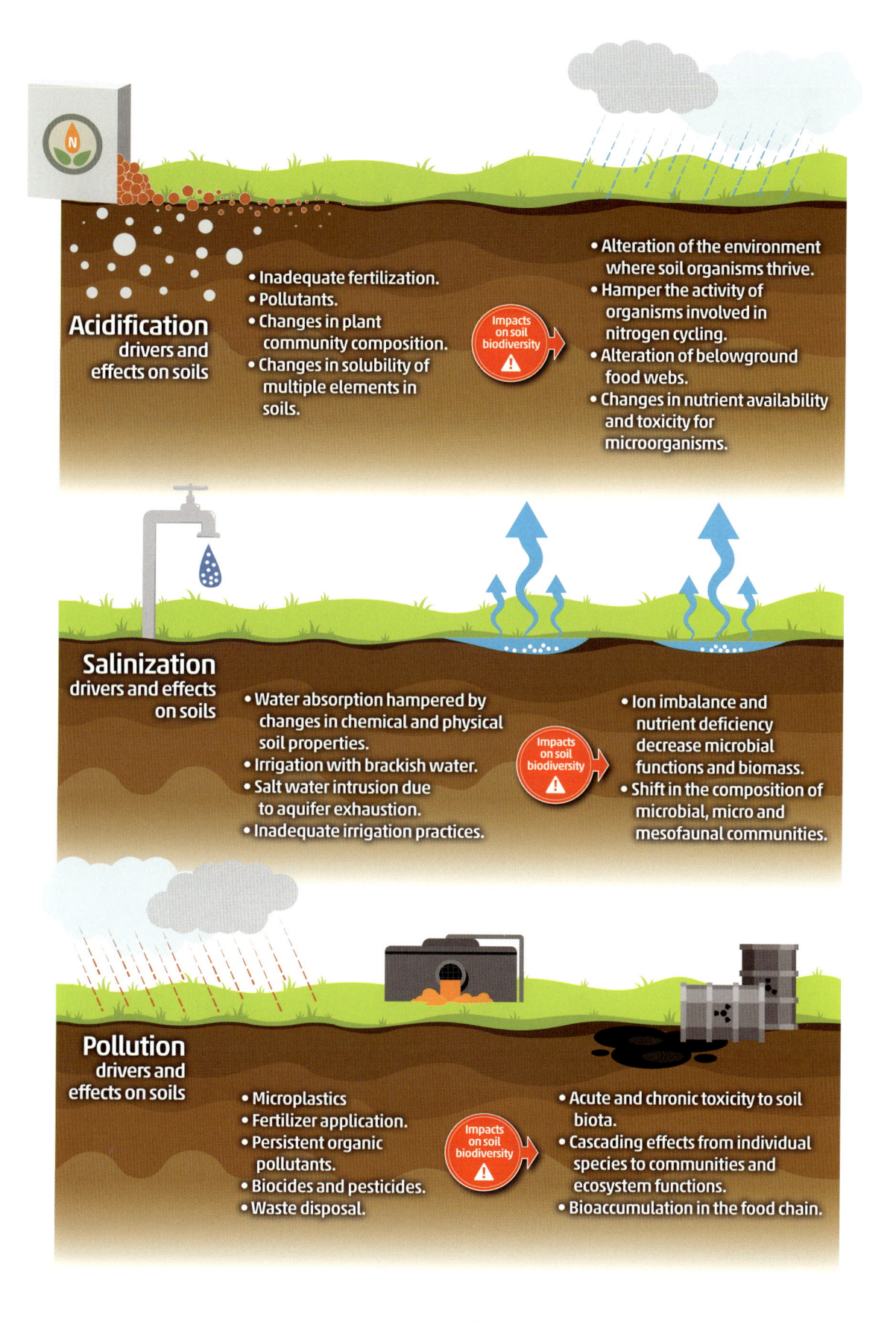

Acidification
drivers and effects on soils

- Inadequate fertilization.
- Pollutants.
- Changes in plant community composition.
- Changes in solubility of multiple elements in soils.

Impacts on soil biodiversity

- Alteration of the environment where soil organisms thrive.
- Hamper the activity of organisms involved in nitrogen cycling.
- Alteration of belowground food webs.
- Changes in nutrient availability and toxicity for microorganisms.

Salinization
drivers and effects on soils

- Water absorption hampered by changes in chemical and physical soil properties.
- Irrigation with brackish water.
- Salt water intrusion due to aquifer exhaustion.
- Inadequate irrigation practices.

Impacts on soil biodiversity

- Ion imbalance and nutrient deficiency decrease microbial functions and biomass.
- Shift in the composition of microbial, micro and mesofaunal communities.

Pollution
drivers and effects on soils

- Microplastics
- Fertilizer application.
- Persistent organic pollutants.
- Biocides and pesticides.
- Waste disposal.

Impacts on soil biodiversity

- Acute and chronic toxicity to soil biota.
- Cascading effects from individual species to communities and ecosystem functions.
- Bioaccumulation in the food chain.

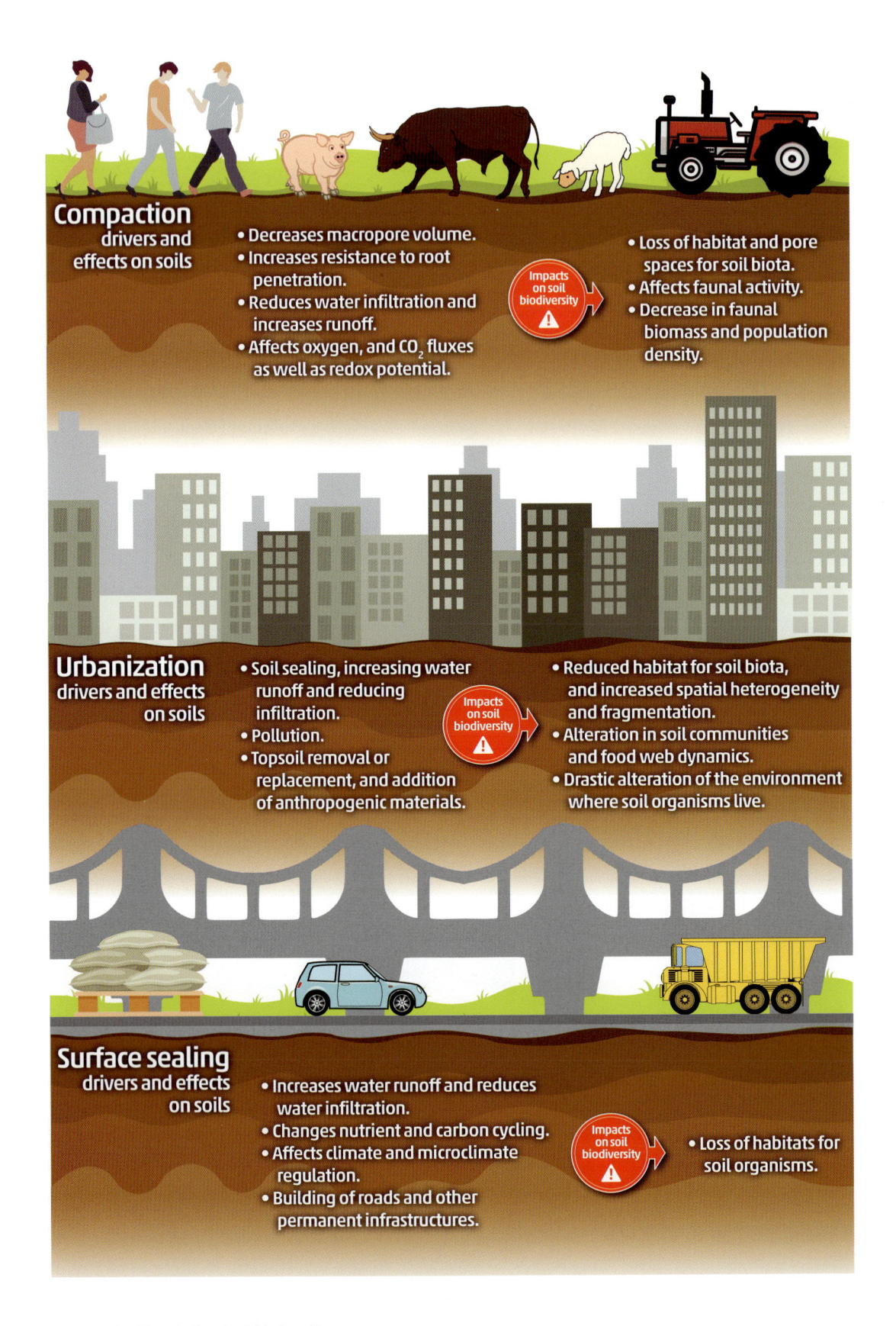

Compaction
drivers and effects on soils

- Decreases macropore volume.
- Increases resistance to root penetration.
- Reduces water infiltration and increases runoff.
- Affects oxygen, and CO_2 fluxes as well as redox potential.

Impacts on soil biodiversity

- Loss of habitat and pore spaces for soil biota.
- Affects faunal activity.
- Decrease in faunal biomass and population density.

Urbanization
drivers and effects on soils

- Soil sealing, increasing water runoff and reducing infiltration.
- Pollution.
- Topsoil removal or replacement, and addition of anthropogenic materials.

Impacts on soil biodiversity

- Reduced habitat for soil biota, and increased spatial heterogeneity and fragmentation.
- Alteration in soil communities and food web dynamics.
- Drastic alteration of the environment where soil organisms live.

Surface sealing
drivers and effects on soils

- Increases water runoff and reduces water infiltration.
- Changes nutrient and carbon cycling.
- Affects climate and microclimate regulation.
- Building of roads and other permanent infrastructures.

Impacts on soil biodiversity

- Loss of habitats for soil organisms.

Fire
drivers and effects on soils

- Wildfires.
- Anthropogen burning for land clearing.
- Removal of topsoil organic matter.

Impacts on soil biodiversity ⚠️

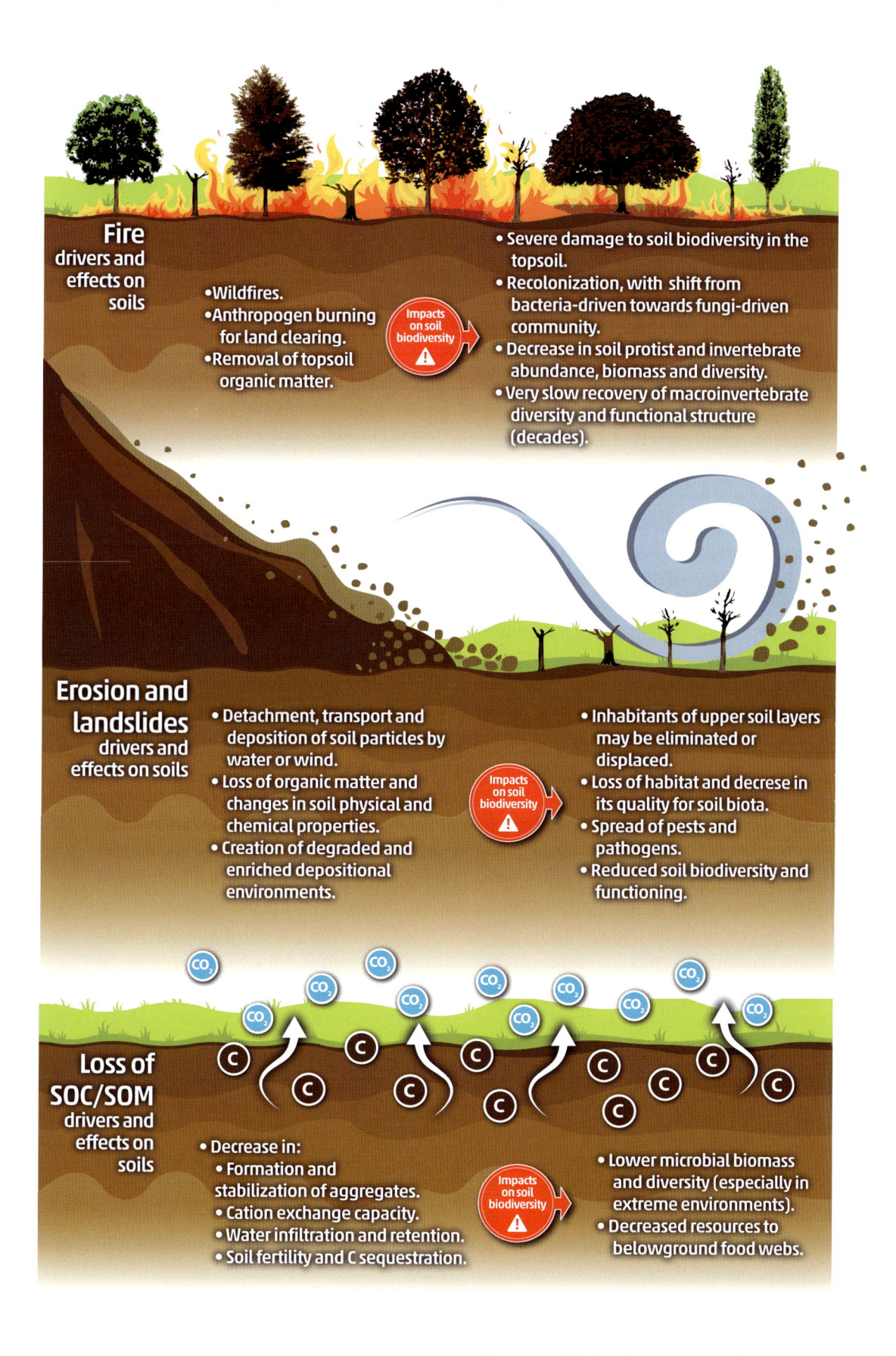

- Severe damage to soil biodiversity in the topsoil.
- Recolonization, with shift from bacteria-driven towards fungi-driven community.
- Decrease in soil protist and invertebrate abundance, biomass and diversity.
- Very slow recovery of macroinvertebrate diversity and functional structure (decades).

Erosion and landslides
drivers and effects on soils

- Detachment, transport and deposition of soil particles by water or wind.
- Loss of organic matter and changes in soil physical and chemical properties.
- Creation of degraded and enriched depositional environments.

Impacts on soil biodiversity ⚠️

- Inhabitants of upper soil layers may be eliminated or displaced.
- Loss of habitat and decrese in its quality for soil biota.
- Spread of pests and pathogens.
- Reduced soil biodiversity and functioning.

Loss of SOC/SOM
drivers and effects on soils

- Decrease in:
  - Formation and stabilization of aggregates.
  - Cation exchange capacity.
  - Water infiltration and retention.
  - Soil fertility and C sequestration.

Impacts on soil biodiversity ⚠️

- Lower microbial biomass and diversity (especially in extreme environments).
- Decreased resources to belowground food webs.

4.2 | THREATS TO SOIL BIODIVERSITY

4.2.1 | DEFORESTATION

Forest ecosystems cover roughly 30 percent of the Earth's land surface and contain highly diverse and poorly studied soil communities. These systems are increasingly under threat, with over 1.3 million square kilometres lost in the last three decades (World Bank, 2016). The negative environmental impacts of deforestation are most evident in the tropics, where the majority of future deforestation is anticipated (Laurance *et al.*, 2014). In Amazonia, the largest intact tropical forest in the world (Lapola *et al.*, 2014), about 17 percent of the rainforest has been destroyed over the past 50 years, with recent losses again on the rise (INPE, 2017). Deforestation often involves the removal of plant biomass through logging of high-value wood trees and slashing and burning of low-value trees prior to consolidation into cattle ranching operations or mechanized agriculture with highly disturbed soils. This results in loss of soil organic matter and nutrients and changes to soil physical properties that disrupt resource supply and habitat suitability to a variety of soil organisms (Neill *et al.*, 1997; Garcia-Montiel *et al.*, 2000; Cerri *et al.*, 2004; Smith *et al.*, 2016). Consequently, deforestation can dramatically alter the structure of soil communities (Crowther *et al.*, 2014), commonly through the loss of specialist species (Mueller *et al.*, 2016), which in turn leads to decreased functional diversity and functional homogenization (Clavel *et al.*, 2011; Nordén *et al.*, 2013).

The increased prevalence of generalist taxa is a consistent response to deforestation across broad taxonomic groups. However, recent studies in tropical rainforests have shown that responses of soil biodiversity to deforestation can be remarkably different from those of above-ground plants and animals. For example, the rapid invasion of a single peregrine earthworm species following deforestation and pasture establishment in Amazonia can enhance earthworm abundance and biomass while decreasing species richness (Barros *et al.*, 2002, 2004). In Central Amazonia, deforestation and the establishment of pastures leads to a dramatic fall in the diversity of ecosystem engineering taxa, with approximately 70 percent of the original taxa disappearing and being replaced by large populations of invaders such as the earthworm *Pontoscolex corethrurus*, a species that can cause profound changes to soil structure and functioning (Barros *et al.*, 2004; Chauvel *et al.*, 1999). Logging of old growth Bornean forest has been shown to reduce termite abundance and diversity, with studies indicating a reduction of 65 percent in termite species richness following forest disturbance (Donovan *et al.*, 2007) and broad effects over all termite functional groups (Luke *et al.*, 2014). These impacts, together with climate change, can have important implications for ecosystem function and resistance to drought, since termites are key regulators of decomposition, nutrient heterogeneity and moisture retention (Ashton *et al.*, 2019).

Most tropical rainforest soils are naturally acidic, and often receive large quantities of lime following deforestation to neutralize pH, especially with the establishment of more intensive cropping systems. However, large shifts in pH impose stress to native microorganisms, affecting their growth (Fierer and Jackson 2006; de Carvalho *et al.*, 2016). This process results in the loss of endemic species of soil microbial decomposers and homogenization of soil communities after conversion of tropical rainforests to pastures and croplands, altering C sequestration and element cycling, and reducing ecosystem resilience to disturbance (Rodrigues *et al.*, 2013).

The abundance and biomass of soil predators such as spiders and predatory insects consistently decreases following deforestation (Franco *et al.*, 2019), indicating that the conversion of forests to arable land affects key organisms involved in population regulation and may favour a few groups that can tolerate disturbance (Franco *et al.*, 2016; Rousseau *et al.*, 2013). These benefited organisms are often plant pests that can harm crops or existing forest. For example, Silva *et al.*, 2008, showed increased populations of plant-parasitic nematodes following forest conversion to pasture in the Brazilian Amazonia.

A recent meta-analysis focused on Amazonian deforestation reported that the abundance, biomass, richness and diversity of soil fauna and microbes are all reduced following deforestation, with greater losses in wetter Amazonian regions and sites with acidic soils (Franco *et al.*, 2019). No evidence of soil biodiversity recovery was found in converted areas over time; biodiversity losses were still evident up to 30 years after forest conversion to arable land. However, limited geographic coverage, omission of micro and mesofauna, and low taxonomic resolution reported in most studies impede our ability to make more specific predictions of deforestation responses and associated management recommendations (Franco *et al.*, 2019).

A cross-biome study in North America showed that the conversion from forest to pasture has consistent directional effects on microbial community composition and catabolic profiles relevant to ecosystem function. Both bacterial and fungal biomass decreased in response to land-use conversion, and although the diversity of both groups increased, the effect size was moderated by soil texture with lesser effects observed on fine-textured soils (Crowther *et al.*, 2014).

Finally, not only deforestation, but also most forms of within-forest degradation (such as wildfires and selective logging) can have pronounced impacts on biodiversity (Gibson *et al.*, 2011). Recent research shows that soil biodiversity and related ecosystem processes may be lost after even very-low, reduced-impact logging intensities (de Carvalho *et al.*, 2016; França *et al.*, 2017). With logging operations rapidly expanding across public lands and more frequent severe dry seasons increasing the prevalence of wildfires in tropical forests, the question of how these within-forest disturbances in intact primary forests affect soil species and their functions emerges as an important research priority for conserving soil biodiversity.

4.2.2 | URBANIZATION

Around the world, urbanized environments – those dominated by residential, commercial and industrial land uses, including cities, towns, villages and suburban and exurban landscapes – continue to expand in conjunction with growing urban human populations (UN, 2019). The initial process of urbanization significantly alters soils and their biodiversity in many ways, especially through removal and replacement of topsoil, compaction, sealing (paving) and addition of anthropogenic materials (Marcotullio *et al.*, 2008; Pickett and Cadenasso, 2009). Within urbanized environments, pollution, landscape management, invasive species and the urban heat island effect, among other variables, further directly and indirectly affect soil properties, including those in remnant native habitat patches that have become surrounded by urban land uses. The multiple interacting and long-term outcomes of urbanization can be perceived as threats to soil biodiversity because urban environmental conditions may degrade soil communities through reduction and loss of populations and shifting communities in ways that affect food web dynamics and ecosystem processes. In turn, soil-derived urban ecosystem services are often negatively affected (Pavao-Zuckerman and Pouyat, 2017). Because such services are critical to supporting the well-being of urban residents, a focus on urban soil biodiversity must become an integral part of global and local efforts to support Sustainable Development Goal (SDG) 11: creating a more sustainable future for cities and other urban communities.

Unfortunately, knowledge about urban soil biodiversity needed to guide sustainable planning and management of urbanized environments is woefully underdeveloped. In a recent review, Guilland *et al.* (2018) identified approximately 100 scientific articles (since 1990) that focused on urban soil organisms and their functional aspects, about half of which focused on arthropods. Even if this review underestimates the amount of relevant research, it does suggest an overall scarcity of basic research about urban soil biodiversity. In particular, there are few, if any, studies that have examined patterns in one place before and after urbanization or how diverse urban variables interact to shape soil communities. Many studies about urban soils examine physicochemical conditions and biogeochemical processes without also investigating the biota. In this context, it is not currently possible to provide robust, generalized conclusions and predictions about how urbanization impacts soil biodiversity patterns, especially at a global scale because of the geographical imbalance of research: 88 percent of studies have been in Europe and North America, with 7 percent, 4 percent, 1 percent and 0 percent from Asia, Australia, South America and Africa, respectively (Guilland *et al.*, 2018). Thus, in addition to increasing the total amount of research about urban soil biodiversity, a major challenge is to increase the breadth of examined biomes and regions. This is a critical need given that many of the fastest growing urban areas and human populations, and thus most pressing concerns about urban sustainable development, are in regions for which nearly nothing is known about urban soil biodiversity (UN, 2019).

Though it remains limited, research advances about urban soil biodiversity over the past three decades have led to a few emerging foundational principles. A key insight is that many urban soils, despite their potentially degraded quality, are inhabited by abundant and diverse organisms from across all taxonomic and functional groups (see Chapter 2), sometimes at levels similar to or greater than other land uses including agriculture (Ramirez *et al.*, 2014; Joimel *et al.*, 2017) (Figure 4.2.2.1). Important drivers of population and community patterns are the environmental conditions created by human management of above-ground habitat structure (for example, plants, detritus layers, impervious surfaces), which influences organic matter inputs and physicochemical conditions such as soil temperature, moisture and pH (Byrne, 2007). Diverse combinations of management goals and activities by many managers across urbanized environments help generate high levels of spatial habitat heterogeneity (alongside background environmental conditions including the underlying native soil template) which likely influences biodiversity patterns (Ossola and Livesley, 2016). This heterogeneity is associated with high habitat fragmentation due to many small, isolated soil patches created by impervious surfaces (roads, buildings); such landscape structural patterns interact with other factors (including pollutants) to create unique conditions that determine which organism can colonize and persist in which patches (Reese *et al.*, 2015). The overall nature and strength of this "urban filtering process" (*sensu* Aronson *et al.*, 2016) for determining the structure and dynamics of soil communities across diverse urban land covers is not well characterized which prevents robust assessment of the degree to which urbanization threatens soil biodiversity from local through global scales. This is also hindered by the lack of studies examining community structure with lower levels of taxonomic resolution (genera, species), especially for protozoans and animals. It is, however, safe to assume that not all soil species are able to pass through the filter such that urban soil biodiversity is degraded to some degree as compared to native communities. Regarding this, which specific soil organisms may need targeted conservation attention is unknown for many places around the world. Future research should aim to investigate how ecological filtering of key functional groups (for example, soil structure formers or population regulators) in different urban conditions affects food web dynamics, ecosystem processes and the associated ecosystem services desired for a specific location.

Given the already large percentage of people that live in urbanized environments and predicted continuation of growth in urban human populations worldwide (UN, 2019), our overall ignorance about urban soil biodiversity may be a bigger threat than urbanization itself. Without more knowledge about how soil organisms are "filtered" by diverse urban variables, we cannot know how to more sustainably plan and manage current or future urbanized environments in ways that conserve and restore crucial soil-based ecosystem services. Indeed, urban soil restoration represents a major opportunity for providing solutions to help urbanized communities reach SDG 11 (Byrne, 2020). To support this, major investments in basic urban soil biodiversity research, including how urban biota contribute to ecosystem services and human health (Li *et al.*, 2018), are urgently needed, especially in tropical biomes and developing countries. Policies and urban planning that integrate the sustainable management and restoration of soils are rare but also needed for reducing urban threats to soil biodiversity (da Silva *et al.*, 2018). On a rapidly urbanizing planet, the well-being of humanity depends in large part on how well we can quickly improve our knowledge, appreciation and management of urban soil biodiversity.

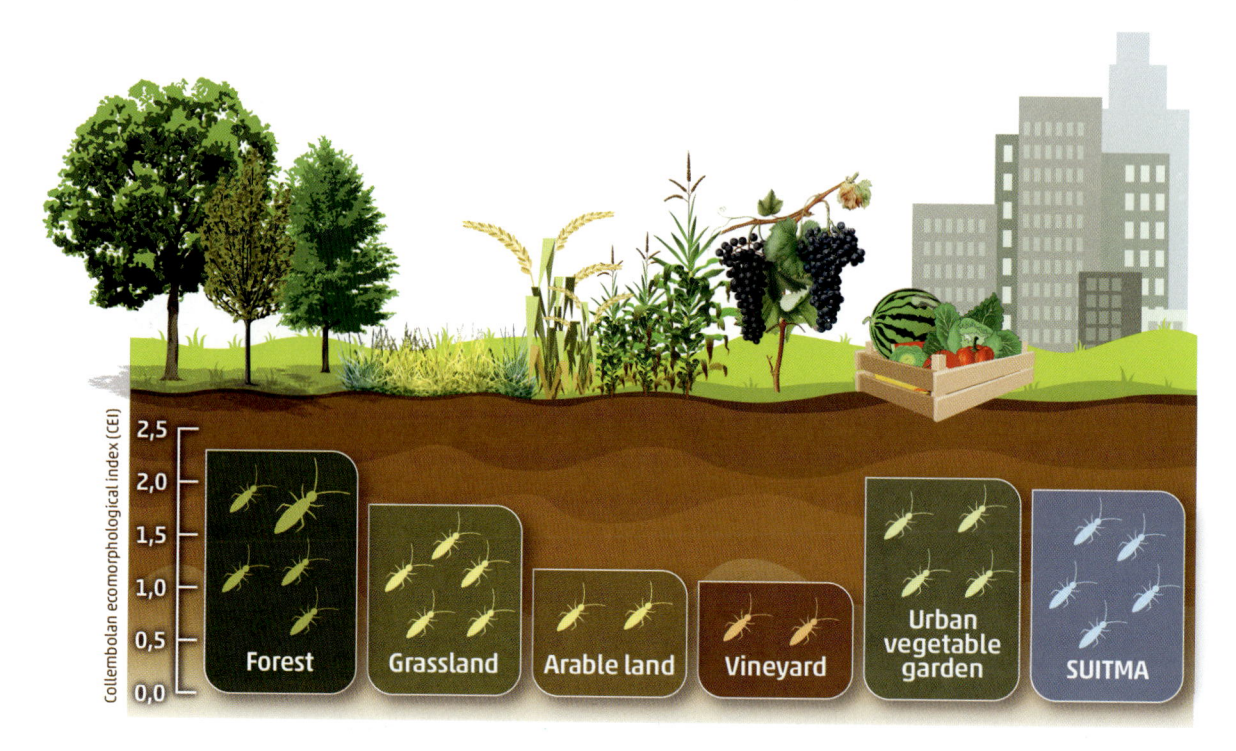

Figure 4.2.2.1 | Collembolan Ecomorphological Index

Soil invertebrates such as microarthropods, including Collembola and Acari, are not just considered as biological indicators of soil quality but also as bioindicators of anthropisation including urbanization and contamination. Indices are useful tools to compare soil biological quality. The higher the Collembolan Ecomorphological Index (CEI), the greater is the abundance of microarthropods adapted to their habitat in the soil. The CEI shows that microarthropod communities are more constrained in agricultural ecosystems compared to urban and forest ecosystems. SUITMA: soils of urban, industrial, traffic, mining and military areas. Adapted from Joimel *et al.* 2017.

4.2.3 | AGRICULTURAL INTENSIFICATION

Agricultural intensification is defined by the Food and Agriculture Organization of the United Nations (FAO) as the "increase in agricultural production per unit of inputs." Related land use management includes among other practices simplified cropping systems (monocultures and few varieties), use of heavy machinery, high input of chemicals such as fertilizers and pesticides, soil tillage and slash and burning. All these practices are driving forces that pose a range of threats to soil organisms and soil functions. In fact, arable lands, which cover extensive terrestrial areas, have been identified as ecosystems where soil organisms and functions are most threatened (Orgiazzi *et al.*, 2016a).

Agricultural intensification is placing tremendous pressure on ecosystems, leading to large-scale ecosystem degradation and loss of productivity in the long term (Tilman *et al.*, 2001; Vitousek *et al.*, 2009). For example, conversion of natural ecosystems to agricultural lands has resulted in substantial environmental costs, including land degradation, increased emissions of greenhouse gases, decreased organic matter in soils, loss of biodiversity and alterations of biogeochemical and hydrological cycles (Balmford *et al.*, 2005). Modern agriculture thus faces great challenges not only in terms of meeting the food, fibre and fuel demands of an ever-increasing human population, but also in mitigating environmental costs, particularly in the context of inappropriate management practices, a changing environment and growing competition for land, water and energy (Chen *et al.*, 2014). Understanding the mechanisms that control the extent to which soil properties and biological communities change following the conversion of natural to agricultural systems and management practices is of paramount importance to comprehend the consequences of land use changes for soil functions and agricultural productivity (Sala *et al.*, 2000).

Agricultural management practices act and interact with each other in different ways, and affect the soil ecosystem to different degrees and to different extents. In general, they alter soil environmental properties and disturb the soil structure, leading to loss of Soil Organic Matter (SOM; see the following section), degrading micro-habitats that are important to many soil organisms. As the application of agricultural management practices is frequent, biological processes are constantly disrupted and the soil ecosystem is not allowed to recover. The magnitude of effects of specific agricultural treatments on the soil ecosystem depends on their level of intensity, application frequency, timing and extent (Snapp *et al.*, 2010; Roger-Estrade *et al.*, 2010). Agricultural intensification may impact soil organism abundance, biomass, community structure, species richness, species diversity, functional diversity and distribution, and effects of the same disturbance are not equal for all organisms. Relatively larger soil animals and those at higher trophic levels such as earthworms, mites, Collembolans and predatory nematodes are usually more affected (Postma-Blaauw *et al.*, 2010; de Vries *et al.*, 2013; Tsiafouli *et al.*, 2015).

Negative impacts of agricultural intensification have consequences on the specific functions that soil animals perform, including soil structure formation and ecosystem engineering, population regulation by predation, and feeding on fungal hyphae. Considering the entire soil food web, intensive agriculture reduces the biomass and number of functional groups, thus decreasing the links (interactions) between them. Moreover, within the functional groups intensive agriculture reduces species richness, Shannon diversity and taxonomic distinctness (Tsiafouli *et al.*, 2015). Smaller and less complex food webs may negatively impact on ecosystem functioning, with important implications for the services ecosystems provide. For example, a shift from "slow" fungal-based to "fast" bacterial-based soil food webs (Thiele-Bruhn *et al.*, 2012; de Vries *et al.*,

2013), leads to losses of C and N from soil in the form of gases. Furthermore, there may be a decline in the resistance and resilience of food webs to environmental stressors, such as drought (De Vries *et al.*, 2012). Adopting sustainable agricultural practices might lead to recovery of biological communities, but recovery might take years or even decades depending on organisms (de Groot *et al.*, 2016).

The threats posed by agricultural intensification are often multiplied due to the interactive effect of other threats. For example, losses of carbon, soil structure and soil biodiversity can reduce an ecosystem's ability to sequester carbon (Wiesmeier *et al.*, 2019). Changes in these soil properties also decrease water infiltration capacity, root penetration and access to nutrients for plants. In concert, all of these changes increase the risk of soil erosion, land degradation, compaction and salinization, thus reducing agricultural productivity which threatens the achievements of Sustainable Development Goals, particularly SDG 2 (zero hunger). The excessive use of fertilizers and pesticides affects the quality of water (Foster and Custodio, 2019) posing several threats to other ecosystems, and also poses direct threats to animal and human health. Ecosystem services provided by beneficial crop-associated organisms, such as regulation of pest and diseases (Tamburini *et al.*, 2016) and pollination, are also reduced (Bretagnolle and Gaba, 2015). Monocultures and the use of few varieties reduces local variety traits and this above-ground loss is likely coupled to loss of soil biodiversity, though the magnitude of these impacts is still uncertain.

Several syntheses and meta analyses have been conducted to evaluate how agricultural intensification affects soil organisms. Examples include analyses of nitrogen (N) additions on soil microbial biomass (Treseder, 2008), of nutrient inputs on mycorrhizal abundance (Treseder, 2004), and of agricultural intensification on soil biodiversity (de Graaff *et al.*, 2019). Results from these analyses indicate that agricultural intensification can significantly alter soil biodiversity, with negative impacts of synthetic N fertilization on microbial biomass, arbuscular mycorrhizal fungal (AMF) and faunal diversity, and a reduction in soil faunal and bacterial diversity with tillage (Treseder, 2008; de Graaff *et al.*, 2019). Results also indicate that soil biodiversity may be enhanced by agricultural practices if agricultural management practices promote soil organic matter (SOM) accumulation and retention (de Graaff *et al.*, 2019), highlighting the importance of implementing sustainable agricultural management practices to promote soil health.

Agricultural intensification can negatively impact ecosystem functioning through its effect on soil microbial properties. For example, a meta-analysis showed that an N fertilization-induced reduction in microbial biomass also affected ecosystem carbon (C) fluxes by reducing carbon dioxide emissions (Treseder, 2008). Reductions in AMF abundance (Treseder, 2004) and diversity (de Graaff *et al.*, 2019) following agricultural intensification are likely to significantly impact ecosystem functioning, because AMF are crucial to plant nutrient acquisition, plant production, and C transfer from the atmosphere to soil (Smith and Read, 1997). While many studies have quantified the effects of agricultural management practices on ecosystem functioning, fewer have linked changes in soil organism diversity directly to changes in ecosystem functioning.

However, agricultural intensification impacts on functional microbial diversity have been evaluated by community-level physiological profiling of heterotrophic bacterial or fungal assemblages (Zak *et al.*, 1994; Lupwayi *et al.*, 2017). A recent meta-analysis of these studies found that microbial functional diversity significantly increased if N fertilizer inputs promoted soil carbon retention (de Graaff *et al.*, 2019). We caution that methods employed to study functional diversity use standardized incubation conditions that are likely not be optimal for all soil communities and may bias results when extracted communities rather than whole soils are measured (Chapman *et al.*, 2007). Future research exploring a more direct link between soil biodiversity and ecosystem functioning will improve our understanding of agricultural intensification impacts on biodiversity and ecosystem functioning.

Given the dearth of studies that directly link changes in soil biodiversity and ecosystem functioning following agricultural intensification, some have synthesized data from studies that experimentally manipulated changes in soil biodiversity and measured consequences for ecosystem functioning (de Graaff *et al.*, 2015; Kardol *et al.*, 2016; Nielsen *et al.*, 2011). These biodiversity manipulation studies indicate that changes in soil biodiversity affect ecosystem process rates (de Graaff *et al.*, 2015; Nielsen *et al.*, 2011), but the manipulations used in these studies tend to exaggerate biodiversity losses and possibly overestimate consequences for ecosystem functioning relative to measured biodiversity losses from agricultural intensification. Finally, while many studies focus on the impact of biodiversity loss within a trophic group on ecosystem functioning, others have shown that loss of interactions among species can supersede these effects (Valiente-Banuet, 2014), thus highlighting the importance of understanding the soil food web for the sustainable provisioning of ecosystem services. Advances in analytical techniques (for example, meta-genomics) to identify soil organisms and link their structure to their function, coupled with an increase in soil biodiversity manipulation experiments that manipulate diversity within and across energy channels, trophic groups, functional groups, taxa and genetic differences should help solidify links among agricultural intensification, soil biodiversity and ecosystem functioning.

4.2.4 | LOSS OF SOIL ORGANIC MATTER AND SOIL ORGANIC CARBON

Soil organic carbon (SOC), as a main resource for soil organisms, affects several soil functions, including the support of biodiversity (Wiesmeier *et al.*, 2019). There is evidence from global analyses that soils with higher SOC harbour larger microbial biomass (Maestre *et al.*, 2015; Crowther *et al.*, 2019; Wiesmeier *et al.*, 2019), and SOC also appears as one of the main drivers of soil microbial diversity at the global scale (Delgado-Baquerizo *et al.*, 2016), with a generally positive effect of SOC content on microbial diversity (Fierer and Jackson 2006; Maestre *et al.*, 2015), particularly in extreme environments with low plant productivity such as polar (Siciliano *et al.*, 2014) and dryland regions (Maestre *et al.*, 2015). The pattern, however, differs between taxonomic groups (Tecon and Or, 2017). Global patterns for the distribution of diversity in soils are poorly understood, so that the factors driving them are difficult to understand (Decaëns, 2010). For some groups, patterns of diversity respond to a latitudinal gradient that can be partly explained by organic matter variation (Decaëns, 2010; Caruso *et al.*, 2019 for Oribatid mites). Global-scale patterns of diversity for other groups are not explained by SOC content (Nielsen *et al.*, 2014 for nematodes) or productivity gradient (Decaëns *et al.*, 2010). Earthworm diversity, for instance, was not related to SOM/SOC content across Europe (Rutgers *et al.*, 2016) and the world (Phillips *et al.*, 2019), although at the local scale it may be important (Hendrix *et al.*, 1992).

Nonetheless, the generally positive relationship between soil C stock and soil biodiversity suggests that soil carbon loss is a threat to soil biodiversity. In this regard, Orgiazzi *et al.* (2016a) identified SOC decline as a major threat to both soil microbial and fauna biodiversity. But the underlying causes may be different, as the main drivers of SOC loss, land use change and climate change (see below) also directly impact soil biodiversity. For instance, soil biodiversity was higher in agricultural soils than in carbon-rich northern forests (Griffiths *et al.*, 2016), but the main factor explaining biodiversity was pH, and low pH soils tend to have higher carbon content. Several authors also highlight the importance of soil carbon quality in addition to quantity for below-ground diversity on a global scale (Crowther *et al.*, 2019). For instance, Szoboszlay *et al.* (2017) found evidence of associations between particular SOC fractions (especially particulate organic matter) and specific bacterial taxa across a large range of European soil with various land uses. In their study, SOC content explained 5 percent of the variation in bacterial diversity, while SOC quality explained 22 percent (Box 4.2.4.1). SOC loss results in a decline of several soil functions, including soil fertility and C sequestration (maintaining and increasing SOC storage in soil is crucial in climate change mitigation), and SOC loss is an important indicator of soil degradation (Lal 2015; Lorenz *et al.*, 2019).

There are global maps of soil carbon available (FAO and ITPS, 2017), but there are currently no global maps of carbon loss directly, and data available on SOC dynamics is unbalanced (Jandl *et al.*, 2014), even though carbon loss is highly related to land use change, particularly conversion of natural environments into agricultural or pastoral use. Mapping carbon loss can therefore be done quite reliably by mapping land use change. Climate change also threatens soil carbon, and has been mapped; this is discussed elsewhere in this report. A map of threats to soil carbon could be made by overlaying these maps. This would require a way of calculating the effect of these factors. There are many ways of doing this, from simple statistical models to IPPC methods and dynamic simulation models. The most commonly used soil carbon models, CENTURY (Parton *et al.*, 1993) and RothC (Jenkinson *et al.*, 1990) as well as the IPCC method (IPCC, 2006) have been set up in a dynamic simulation tool linked to GIS maps (Easter *et al.*, 2007). There are also earth system models such as LPJ (Sitch *et al.*, 2003) and CLM (Oleson *et al.*, 2010) that can simulate the land surface on a global grid. The large uncertainties caused both by model structure and input data means that results from models must be interpreted and used cautiously. However, models do provide a consistent way of simulating SOC as a function of soil texture, climate and land use. This can give indications about where the risk of loss is high, and what management decisions can minimize the risk.

Overall, the effects of SOC loss on soil biodiversity are globally poorly understood, because (i) data on soil biodiversity and the patterns of distribution at large scale are insufficient (especially for soil fauna), (ii) the biological mechanisms involved in how SOC affects soil biodiversity are poorly known, and (iii) many of the other threats to soil biodiversity also lead to changes in SOC content, especially land use and climate change. Furthermore, although we address the role of SOC/SOM content for soil biodiversity, several studies suggest that shifts in SOC/SOM quality and heterogeneity is more important and should be considered for a comprehensive understanding of SOC/SOM impacts on soil biodiversity. This would also require a better understanding of which aspects of SOM/SOC quality are important for soil organisms, and how to assess this. Finally, long-term experiments designed to investigate these effects are needed, to disentangle the effect of soil organic matter directly from those of other factors causing SOC loss in addition to affecting soil biodiversity directly. Especially more data from non-agricultural systems are needed.

Box 4.2.4.1 | How do SOC fractions influence soil biodiversity?

Soil organic carbon (SOC) and its different fractions affect soil microbial diversity across different land-use types. The figure shows associations among different SOC fractions - such as particulate organic matter (POM) - and specific bacterial taxa, highlighting the importance of soil carbon quality – set of indicators that allow to establishing how easily SOM can be mineralized- over the total carbon quantity for belowground microbial diversity. POM represents a substrate and a microhabitat for soil microbial communities strongly influencing bacterial community structure. Isolated SOC fractions with different functional traits and turnover rates (POM included) explained 22% of the variation in the soil bacterial community in contrast with total SOC that explained 5%.

Circles represent Operational Taxonomic Units (OTUs), and hexagons represent SOC fractions. The size of the circles is indicative of microbial abundance, and different colors show their taxonomic classification. Green edges indicate positive and red edges mean negative associations. SC-rSOC: particle size between 0.45 and 63 µm, oxidizable; DOC: dissolved organic carbon, particle size <0.45 µm; POM: particulate organic matter, particle size >63 µm, low density, SA: sand and stable aggregates, particle size >63 µm, high density). Szoboszlay *et al.*, 2017

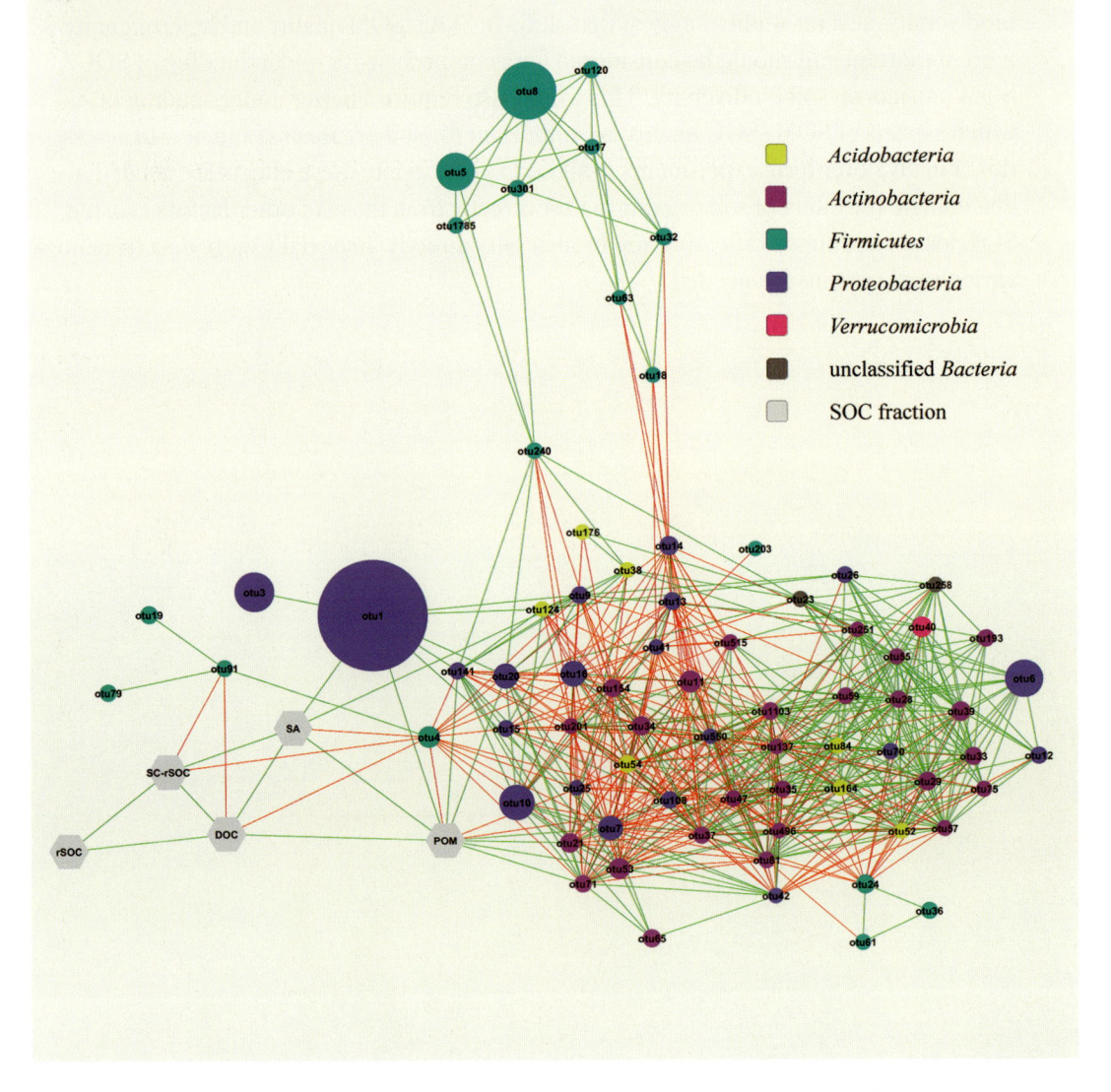

©FAO/Matteo Sala

4.2.5 | SOIL COMPACTION AND SEALING

Compaction is a soil physical degradation process affecting soils in agricultural and urban areas. Soil compaction is associated with loss of crop productivity but can also affect grasslands and tree plantations. This degradation process can occur at the soil surface or at the subsurface affecting root elongation and water and air exchanges. According to the *Status of the World's Soil Resources* (FAO and ITPS, 2015) the status of soil compaction around the world is highly variable according to each world region and varies from fair to poor, whilst the trend in most regions is classified as deteriorating.

Soil compaction decreases the volume of macropores and, consequently, alters soil structure, penetration resistance, soil pore distribution and bulk density. As a result of this decrease, the proportion of water and air volumes are modified, affecting oxygen and carbon dioxide concentrations as well as redox potentials (Figure 4.2.5.1). These changes affect faunal activity and cause a decrease in biomass and population density (abundance; Beylich *et al.*, 2010). Increase in penetration resistance and bulk density affect the burrowing action of macrofauna, especially of burrowing species such as earthworms. They also impair the action of ecosystem engineers by reducing available habitats and access to water and oxygen. The reduction in macropore volume and its consequences on other soil physical attributes also affects the habitable space for mesofauna. Soil compaction and soil biodiversity are interdependent, compaction affects soil biodiversity, but soil organisms can counteract compaction. Ecosystem engineers can counteract the effects of soil compaction and contribute to the regeneration of compacted zones with time (Turbé *et al.*, 2011).

Soil microbial activity and biomass are also affected by soil compaction. The effects of soil compaction on soil microorganisms and microbial processes are complex and depend on many factors (Nawaz *et al.*, 2013). Changes in water and air volumes, waterlogging and redox potential affect microbial processes which, for example, cause changes in carbon gas effluxes (CO_2 and CH_4) and net nitrogen mineralization processes..

Soil biodiversity and soil fauna may be more affected by soil compaction than plant growth and plant yield. Nevertheless, many threshold values of soil compaction can be found for soil physical processes affecting plant growth but are non-existent for soil biota and biological processes. The increasing need of soil protection and protection of soil functions demands the need for the identification and development of threshold values related to soil organisms and biological processes (Beylich *et al.*, 2010).

The migration of rural populations to urban environments during the last two centuries has led to the growth and expansion of cities worldwide, impacting the landscape and soil resources. Urbanization has caused an increase in soil sealing, which is defined as the permanent covering of the soil surface by an impermeable material impeding changes between above-ground and below-ground environments (Turbé *et al.*, 2011). Soil sealing can be considered as total soil loss, permanently affecting many soil functions related to water production and regulation, food production, biodiversity and climate regulation.

With the exception of the South West Pacific and Southern African regions, the status of soil sealing in the other regions varies from fair to very poor and the main trend is classified as deteriorating (FAO and ITPS, 2015). The South West Pacific and Southern African regions are classified as good but with a deteriorating trend.

Natural soil sealing occurs by soil crusting, which impedes soil infiltration, but most sealing occurs due to anthropogenic activities related to urbanization. Sealed soils are considered non-functional, reducing infiltration and increasing runoff, decreasing organic matter input, and isolating the soil from the above-ground environment. Consequently, water and gaseous exchanges are affected, as well as nutrient cycling related to organic matter dynamics. Soil biota can survive with residual water and organic matter after recent sealing, but when exhausted, bacteria can enter an inactive state and soil fauna may disperse or die off (Turbé *et al.*, 2011). Reduction in soil carbon and nitrogen contents, soil respiration, changes in soil physico-chemical attributes and enzymatic activity negatively influence microbial activity. Most sealed soils have their topsoil removed, causing a reduction in soil organic matter, increasing moisture stress, and creating an alkaline environment and poor ventilation affecting soil biota and their activity (Piotrowska *et al.*, 2015).

Fragmentation of native ecosystems and the implementation of green areas with non-native species cause impacts on soil organisms and their activity (Scalengle *et al.*, 2009). Sealed ecosystems under cities are replaced by pavement and concrete infrastructures, isolating small to medium areas with native ecosystems and/or green areas with exotic species, such as parks, affecting above and below-ground biodiversity connectivity (see the earlier section on urbanization). The use of concrete and asphalt pavements to seal the soil elevates the soil temperature, which exerts pressure on soil biota and biological processes. Overall, soil sealing affects the hydrological cycle, nutrient and carbon cycling, climate, and microclimate regulation, resulting in the loss of habitats for soil organisms, soil biodiversity and all services and functions with the exception of the capacity to support infrastructure (FAO and ITPS, 2015).

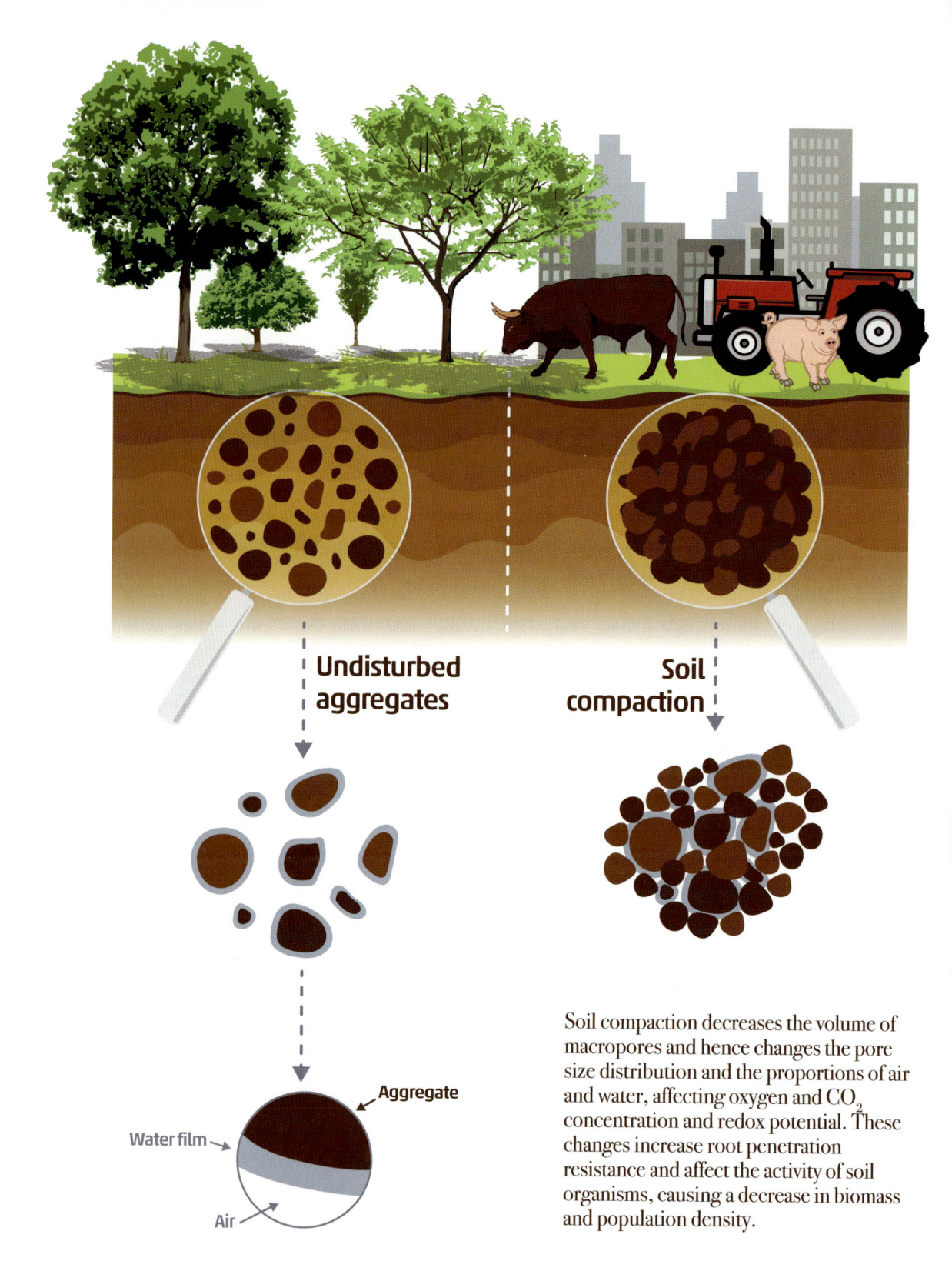

Soil compaction decreases the volume of macropores and hence changes the pore size distribution and the proportions of air and water, affecting oxygen and CO_2 concentration and redox potential. These changes increase root penetration resistance and affect the activity of soil organisms, causing a decrease in biomass and population density.

Figure 4.2.5.1 | Soil compaction

Compaction and sealing adverse impacts on soil.

4.2.6 | SOIL ACIDIFICATION AND NUTRIENT IMBALANCES

Soil acidification

The acidification of ecosystems is a natural process that is driven by the metabolic activity of soil organisms and plants through the ecosystem succession and that is also linked to the build-up of soil organic matter (Delgado-Baquerizo *et al.*, 2019). This natural acidification drives changes in the biodiversity and abundance of soil communities across decades, centuries, millennia or even millions of years of ecosystem development (Delgado-Baquerizo *et al.*, 2019). The resulting acidophylic or acid-tolerant species present in these soils are adapted to these conditions but may still be threatened by pH changes in the soil, be they positive (for example, liming to raise pH) or negative (such as acidification due to acid rain). This process is, however, very different from the acidification of ecosystems that has been historically linked with the emission and further deposition of pollutants since the beginning of the industrial revolution and, later on, the green revolution (Bobbink *et al.*, 2010; Greaver *et al.*, 2012; Tian and Niu, 2015).

The industrial revolution started a period in which massive amounts of sulphur- and nitrogen-rich compounds derived from the burning of fossil fuels were emitted to the atmosphere, and were deposited back to the soil as sulphuric and nitric acids dissolved in rainwater, or acid rain (Fowler *et al.*, 2013). This impacted the vitality and structure of forests across very wide areas (Menz and Seip, 2004), with unknown but potentially negative consequences for the abundance and biodiversity of soil organisms that thrive in these soils (Lv *et al.*, 2014). Due to abatement policies, the amount of oxidized nitrogen, and particularly sulphur, that is emitted to the atmosphere has now been reduced in many world regions, particularly in Europe and in the United States of America, although it remains as a problem in areas of China and India, which are also amongst the less studied areas in terms of soil biodiversity (Lv *et al.*, 2014; Menz and Seip, 2004). The emission and further deposition of reduced nitrogen compounds derived from agricultural practices and three-way catalytic converters is still a threat to ecosystems in many regions of the world and is also a main agent of acidification, given the ability of NH_4^+ to release protons in the soil solution (Forest *et al.*, 2013). Other human-induced causes of soil acidification include acid mining for the extraction of minerals.

The effects of soil acidification on soil organisms can be direct, via alterations of the physicochemical environment in which the soil organisms thrive. For example, many bacterial taxa are known to be highly selective for the soil pH range in which they can grow, which is typically associated with the importance of pH for the regulation of their metabolic activity (Fierer *et al.*, 2009; Lauber *et al.*, 2009). The greatest abundance and diversity of active bacteria is typically found in soils with pH around 7 (Lauber *et al.*, 2009; Ochoa-Hueso *et al.*, 2018). Moreover, acidification is known to hamper the activity of soil microorganisms involved in N transformations, such as mineralization of organic N and biological N_2 fixation, while low soil pH promotes the production of N_2O, a potent greenhouse gas, during nitrification and denitrification (Granli and Bøckman, 1994), with potential consequences for the global climate. Indirect effects of soil acidification on soil organisms can operate by different mechanisms: the first are changes in plant community composition, which may result in the complete alteration of soil communities which very typically depend on plant litter inputs and rhizodeposits as the main carbon source supporting brown food webs; thus changes in plant community structure can cascade through the whole soil food web (that is, across trophic levels) by altering the abundance, composition and activity of those soil organisms that are at the base of the food web. A second set of mechanisms involves changes in soil pH, which determine changes in the solubility of multiple elements in soil, including trace elements, many of which are typically needed in low concentrations but that are toxic at high concentrations (Stevens *et al.*, 2011). Soil acidification results in the leaching of base cations, particularly in poorly buffered soils, making them unavailable for soil organisms, including microbes (Velthof *et al.*, 2011). Acidification may thus lead to deficiencies of nutrients such as phosphorus, calcium, magnesium and molybdenum, and a release of toxic compounds, including aluminum, iron, manganese and heavy metals that are immobile at higher pH values.

A global meta-analysis showed that the decrease in soil pH in response to the addition of mineral N was more evident in grasslands, whereas boreal forests were more resilient to the N-induced soil acidification (Tian and Niu, 2015). This suggests that the consequences of acidification for soil organisms may also be more important in acidic and poorly buffered soils, such as those from many natural and semi-natural grasslands in the United Kingdom of Great Britain and Northern Ireland and in central Europe, which have seen a dramatic loss of their plant biodiversity in the last 150 years (Stevens *et al.*, 2004), with likely mirroring consequences for the abundance and biodiversity of below-ground communities. In agroecosystems, acidification considerably reduces soil fertility, affecting microbial transformations, which may ultimately cause depression of crop growth, and yields (Marschner, 1995; Bolan, Bolan *et al.*, 2003).

The effects of acidification on soil organisms are, however, often very difficult to separate from those of the direct effects of the agents of acidification. For example, a study carried out in a calcareous semi-arid shrubland in central Spain (pH ~7.5) showed that mineral nitrogen additions up to 20 kg N/ha annually increased the abundance of soil microarthropods, a response attributed to an incipient soil acidification (Ochoa-Hueso *et al.*, 2014). Beyond that load, the addition of nitrogen resulted in a decrease in soil faunal abundance, attributed to the negative effects of excessive nitrogen, particularly to high concentrations of ammonium, which is known to be toxic for many soil organisms. This response was driven by collembolans, the most abundant group in those soils.

Nutrient imbalances

Ecosystem productivity is co-limited by the availability of N and P at the global scale (Elser *et al.*, 2007). Thus, changes in the absolute and relative availability of key essential nutrients such as N and P can greatly affect soil biodiversity and their functioning (Elser *et al.*, 2018). Nutrient imbalances occur when one or more essential nutrients are in short supply in relation to other essential nutrients. This situation is now widespread in soils worldwide due to the above mentioned increase in the availability of mineral N and P from polluting sources (atmospheric N deposition, runoff water) and in agroecosystems due to excessive use of mineral fertilizers. This increase in the availability of essential nutrients has consequences for plant growth and microbial decomposition that can cascade to more complex effects on soil food webs. This is because energy and nutrient imbalances (typically C:N, C:P or N:P, but also N:K and so on) between consumers and their resources strongly constrain nutrient cycling and limit consumer reproduction and growth (Andersen *et al.*, 2004; Frost *et al.*, 2005; Person *et al.*, 2010). The effects of nutrient imbalance on soil organisms and food webs can be direct (in the case of soil organisms such as bacteria and fungi that take up their nutrients directly from the soil solution) or indirect, via alterations of the nutritional content of primary producers and their litter inputs, with cascading consequences for both green and brown food webs.

The ecological stoichiometry theory has been applied to study the balance of energy and chemical elements such as C, N and P in ecological interactions (Sterner and Elser, 2002). This theory can help us to better understand trophic interactions by analyzing the imbalances in the relative supplies of key elements between organisms and their resources, yet the mechanisms that control elemental stoichiometry in different taxonomic groups and the effects of nutrient supply imbalances are not yet clear. It has been observed that the concentration of major elements in the soil can be high or low, without this altering the natural ecosystem functioning. However, what is really important is the degree of C:N:P nutritional imbalance that affects biodiversity, causing a cascade of unknown ecosystem effects. The main mechanism by which nutritional imbalance affects soil biodiversity is associated with the growth capacity of organisms. Elser *et al.* (1996) proposed the growth rate hypothesis that postulates that cellular stoichiometry varies according to growth rate due to increased allocation to P-rich ribosomal RNA to support rapid growth. To date, this hypothesis has been a powerful tool for understanding variation in biomass C:N:P ratios in microbes and small consumers, important components of soil biodiversity (Elser *et al.*, 2018) and the base of the food webs.

Additionally, C:N:P ratios for different organisms have been proposed (Table 4.2.6.1) as a reference of nutritional needs and immobilization capacities of organisms in different ecosystems (Redfield, 1958; Elser *et al.*, 2000; Cleveland and Lipzin, 2007; Zhang and Elser, 2017). Deviations from this elemental stoichiometry reflect a nutritional imbalance and therefore a greater energy investment to acquire the limiting nutrient; this energy investment may not necessarily be feasible for all microorganisms and small consumers, thereby causing a decrease in soil biodiversity.

Table 4.2.6.1

C:N:P ratios for different organisms

Organism	C:N:P	Reference
Ocean phytoplankton	106:16:1	Redfield (1958)
Terrestrial plants	968:28:1	Elser *et al.* (2000)
Aquatic plants	307:30:1	Elser *et al.* (2000)
Earthworm	127:26.5:1	Marichal *et al.* (2011)
Soil bacterial biomass	60:7:1	Cleveland and Lipzin (2007)
Fungal biomass	250:16:1	Zhang and Elser (2017)

In soil, organic matter represents the main energy and nutrient input via decomposition (solubilization, depolymerization and mineralization). Downing and McCauley (1992) established that P limitations occur when N:P exceeds ~30. In a study carried out in a calcareous semiarid grassland in the Chihuahuan desert in northern Mexico, Hernández Becerra *et al.* (2016) found that land use change (grassland to alfalfa crop) modified soil microbial N:P stoichiometry from 5.3 in grassland to 33.2 in alfalfa crop, increasing soil acidification (pH from 9 to 7) and reducing bacterial diversity (12 and 9 phyla respectively). Interestingly, they found that there were no OTUs shared between the agricultural plot and the native grassland, which may indicate a change not only in taxonomic diversity but also in functional diversity, associated with nutrient imbalance. However, further research is needed to better understand how to reduce nutrient imbalance in human managed ecosystems.

4.2.7 | POLLUTION

A recent study identified environmental pollution as the largest cause of premature death in the world, killing more people than AIDS, malaria, and tuberculosis combined, and accounting for one in four deaths in the poorest countries (Landrigan *et al.*, 2018). Chemical pollutants are also further known to affect wildlife species and ecological communities including those in the soil. This can lead to local- or regional-scale losses in biodiversity that can be explicitly linked through evidence to impacts on ecosystem functions and associated services (Cardinale *et al.*, 2012; Hayes *et al.*, 2018). These real impacts shape the public debate on the use and safety of chemicals, fueling concerns even in those cases where limited impacts may actually exist.

Recognition of the effects of chemical contaminants on ecosystems underpins a desire to improve the chemical condition of our environment, encouraging the mitigation of some but not all of these effects. Landmark policies on abatement of acid rain, nutrient management strategies, control on the use of certain persistent organic pollutants (POPs), biocides and pesticides and improvement in wastewater treatment, as well as economic and cultural shifts affecting industrial sectors, such as energy production, transportation and metal processing, have changed the type and amount of chemicals entering the soil environment, often reducing loads and mitigating impacts. The aim of these risk-based policies is to restrict inputs to levels below those expected to cause biological effects on soil species populations and humans.

By preserving structure, the assumption is that function will be protected, especially for those ecosystem processes for which there is recognized functional redundancy.

Despite management effort, legacies of past activities or unmanaged chemical use and release, poor governance and gaps in knowledge lead to chemicals still being released into the soil environment. Progressive advances in chemical analysis methods for soils allow increasingly accurate measurements of soil contaminants supporting monitoring applications. Monitoring programs and the development of concepts such as 'pesticidovigilance' – the practice of monitoring the effects of pesticides after approval for use – are developing, placing greater emphasis towards post-approval assessment to allow regulatory decisions to be refined, and to make the trade-offs between environmental costs and intended food security more explicit (Milner and Boyd, 2017).

To date the majority of large-scale regulatory contaminant monitoring programs are designed to assess the chemical status of water bodies. The initiatives are directed to support major policies such as the Water Framework Directive in Europe. Nutrients, pesticides and trace elements are the primary focus. With few national frameworks for soil protection published and implemented, large-scale contaminant monitoring of soil status at the national scale is limited to programs in only a few countries (Gardi *et al.*, 2013; Hassanin *et al.*, 2005; Spurgeon *et al.*, 2008). Most research instead remains the domain of academic groups conducting small pilot programs. There are literally thousands of such studies describing concentration of macronutrients, trace elements, pesticides, plant protection products and biocides, other persistent organic pollutants (POPs), industrial chemicals, pharmaceuticals, veterinary medicines, nanomaterials and recently plastics in soils collected from individual sites, transects and local and regional surveys. If compiled together with any associated georeferenced and other meta-data, these studies would present a fantastic resource by which to study the characteristic of soil contaminant loads across biomes, continents, countries, landscapes and land-uses across the world. However, lack of consensus on terminology and analytical methodologies make such a task a non-feasible activity. 4.2.7.1 shows that chemicals that reach soils can come from a number of sources.

Box 4.2.7.1 | Main Sources of contaminants that impact soil biodiversity

Direct inputs to agricultural land occur as a result of the widespread use of pesticides and fertilizers. The complexity of such inputs is rising. For example, in the United Kingdom of Great Britain and Northern Ireland there has been approximately a 50 percent rise in the average number of pesticide active ingredients applied to arable crops (from 11 in 2000 to 17 in 2015; FERA). Fertilizers are also well known to have resulted in the eutrophication of soils leading to changes in plant, invertebrate and microbial community structure (Rowe *et al.*, 2014; Stevens *et al.*, 2010). The use of alternative sources of fertilizers, such as manure or sewage sludge can also be a source of soil contaminants, such as trace elements and emerging contaminants like pharmaceuticals, veterinary drugs and antibiotics and plastics, if their quality is not controlled.

Diffuse inputs of contaminants include polycyclic aromatic hydrocarbons (PAHs), trace elements (such as zinc) and also emerging contaminants, such as rare earth elements to soils, especially in urban areas. Fossil fuel burning from domestic heating, cooking, vehicle emissions, and tire and brake wear are also important sources of these types of contaminants.

Raw material extraction can result in the wide-scale pollution of soils surrounding mining sites. Mining is responsible for soil pollution in many countries. Coal, oil, metal and ore extraction all lead to the production of waste materials (slags and waste and drilling fluids) that transfer contaminants including hydrocarbons and trace elements to soil. Even though the impacts of mining have been known since the first pollution investigations, prevention practices are still poorly managed in many countries, resulting in soil contamination and pollution.

Industrial and transport activities result in the direct release of contaminants to soil. For example, the United Kingdom of Great Britain and Northern Ireland harbours a significant legacy of contamination resulting from past industrial activities. An audit report estimated that there are 325 000 contaminated land sites in the United Kingdom of Great Britain and Northern Ireland (House of Commons Environmental Audit Report). This level would be typical for a post-industrial country of similar size and population density.

There is good evidence to suggest that deposition of some major contaminants, such as trace elements and organic contaminants associated with combustion products (for example, polycyclic aromatic hydrocarbons, PAHs) have significantly declined in developed areas (such as Europe) (Cape *et al.*, 2003; Dore *et al.*, 2014). However, these are increasing in rapidly developing economies in Asia, South America and Africa (Yu *et al.*, 2017).

Chemicals present in waste streams can also reach land. More than 10 million tonnes of biosolids, composts and digestates are applied to the United Kingdom of Great Britain and Northern Ireland land each year. These wastes can contain established contaminants, such as PAHs and polychlorinated biphenyls (PCBs). However increasingly, the presence of newer POPs, such as polybrominated diphenyl ether (PBDEs), fluorinated organics (PFOS, PFOA), chlorinated naphthalenes, veterinary medicines, antibiotics, human pharmaceuticals and nanomaterials in these wastes is being recognized.

With large areas of the world involved in armed conflicts in the present and recent past, chemicals associated with military applications, such as energetic compounds (explosives and propellants), trace elements and oil products occur as contaminants in soil, where they can have impacts on soil species and processes (Kuperman *et al.*, 2017, 2018).

Building on recent observations of the widespread occurrence of plastics in the marine environment, similar studies in soils have also found the widespread presence of large- and small-scale (microplastic and nanoplastics) plastic debris in terrestrial ecosystems (Horton *et al.*, 2017).

To improve our knowledge of the contamination status of soils and the impacts on species and processes, progress is needed in the characterization of contaminant presence and concentration and in understanding the resulting effects in realistic exposure scenarios (for example, long-term exposure to mixtures for interacting species). Progress in analytical methods means that the detection of ultra-low concentrations of inorganic and organic contaminants in soil is now possible. What is less well understood is, first, how available these contaminants are for uptake by organisms, and secondly whether exposure to these concentrations over extended time as mixtures has any impacts. To assess bioavailable exposure, significant progress has been made in understanding real bioavailability. Concepts such as the biotic ligand model as applied for trace elements (Thakali *et al.*, 2006a; Thakali *et al.*, 2006b) and chemical activity as applied for organic contaminants (Mayer and Holmstrup, 2008; Schmidt *et al.*, 2013) have advanced the ability to understand pollution exposure in specific soil chemistry contexts (such as pH and organic carbon content).

Contaminants represent only one aspect of the anthropogenic pressure to which ecosystem may be exposed. As many of these pressures can lead to changes in ecosystem structure and function, it can be difficult to attribute those effects of contaminant perturbations to those caused by other stressors. The effect of some contaminants is known based on the reasons for their application; examples include nematocides, molluscicides and carbamate insecticides for earthworm control on golf courses that target specific invertebrate groups. Other established examples of pressures include the widely reported impacts of copper fungicides on soil fungal and invertebrate communities in vineyards (Hayes *et al.*, 2018) and the effect of energetic materials on soil communities. Thus, a study by Kuperman *et al.* (2014) showed that, although overall soil microarthropod or nematode communities abundance was not affected by exposure to polynitramine EM CL-20 (China Lake compound 20), greater sensitivities were found for predatory mesostigmatid mites and predatory nematodes. Similar greater sensitivity of predatory nematodes to chemical exposures compared with other trophic groups of the nematode community was observed in studies with copper and p-Nitrophenol by Parmelee *et al.* (1993). In the same microcosm assay, total microarthropod numbers were reduced by 50 percent in the 30 mg/kg 2,4,6-trinitrotoluene (TNT) treatment compared with numbers in control oak-beech forest silt loam soil (Parmelee *et al.*, 1993). Adverse effects of exposure to EM were also determined in single-species toxicity tests with earthworms (Simini *et al.*, 2003), enchytraeid worms (Kuperman *et al.*, 2006, 2013) and springtails (Phillips *et al.*, 2015). These cases provide an illustration of the cascade of possible effects from individual species to communities and functions. Measurements of functional endpoint have shown that the activities of soil microorganisms, which are critical to terrestrial biogeochemical cycles and are key to sustaining the functioning of the terrestrial ecosystems, were inhibited in EM-polluted soils. Basal respiration was the more sensitive endpoint for assessing the effects of nitroaromatic compounds on microbial activity in a sandy loam soil, whereas substrate-induced respiration and microbial biomass were more sensitive endpoints for assessing the effects of nitroglycerin (Kuperman *et al.*, 2017). Litter decomposition was inhibited in soil polluted with dinitrotoluene or nitroglycerin (Kuperman *et al.*, 2018).

Understanding of the effects of long-term exposure on species is still hindered by the lack of methods to assess soil community change over extended time-scales. In ecotoxicology, short-term single species testing still dominates. Further, even with our increasing knowledge, the lessons learned from the cases where contaminants have been found to impact species have taught us that unexpected effects can and do occur. The development of the ecosystem services approach provides an opportunity to establish protection goals that are more explicit in their aims, such as species or local population conservation for certain species, or a broader protection of ecosystem function. Targeting of assessment to vulnerable species and landscapes can prioritize areas and critical receptors for protection, allowing efforts to be better tailored to meet specific goals.

Soil contaminants are of high concern to the public; as a result, their sustainable management is a high-profile and demanding task. What challenges regulators is that contaminants are found ubiquitously in soils in all biomes and regions. These contaminants may, however, have no effect if they do not reach a threshold for toxicity. Defining these thresholds and making them robust for variation in soil types, mixtures of soil contaminants, extended exposure time and for multiple interacting soil species remains a challenge. Risk assessment tools allow having a balance between different exposure scenarios and the ecotoxicological costs. This balance needs to be continuously scrutinized as evidence changes and tools develop.

It is also important to note that the same approaches to the management of soil pollution are not applied worldwide. Recent years have seen the large-scale transfer of heavy industry and an associated outsourcing of pollution to the newly industrialising economies. In these countries, the issues being faced with poor air quality and pollution of soil and water now mirror those faced by the industrialized world over the twentieth century. To meet growing demands for access to cleaner air, water and land, these nations can draw on the policies and practices applied in countries that have already gone through the process of improving pollution management and control.

The implementation of chemical management policies is recognized within the United Nations SDG 12: ensuring sustainable consumption and production patterns. However, rather than follow an incremental path, there is an opportunity to short-circuit the process of policy development by taking note of the lessons of past failures. By working to promote the idea that new opportunities for industrial innovation take place within an environment where both the costs and benefits are assessed, the negative impacts of pollution can be limited. This can help to mitigate the negative environmental impacts of chemicals, while gaining benefits from their use and creating an environment in which there is better understanding and advocacy of chemicals, based on an acceptance that occurrence is not risk, but in which real and recognized impacts can be identified and mitigated.

4.2.8 | SALINIZATION AND SODIFICATION

Soil salinization is a term used for the accumulation of salts in soils at a level that negatively impacts agricultural productivity, environmental health and economic welfare (Rengasamy, 2006). Generally, a soil is described as saline if the electrical conductivity measured in a saturated soil paste (ECe) is higher than 4 dS/m at 25°C (US Laboratory Staff, 1954), while 8-16 dS/m and levels >32 dS/m correspond to medium and hypersaline environments, respectively (Brouwer *et al.*, 1985). The USDA (1954) classification of salt-affected soils is shown in Table 4.2.8.1

Table 4.2.8.1

The USDA classification (USDA, 1954) of salt affected soils

Characteristics	Saline soils	Saline-sodic* soils	Sodic* soils
Electrical conductivity (EC$_e$) (dSm^{-1} at 25°C)	> 4	>4	<4
Exchangeable Sodium Percent (ESP)	<15	>15	>15

* Sodic soils are also alkaline if their pH is over 8.5

The Food and Agriculture Organization of the United Nations (FAO) estimates that globally over 830 M hectares of arable land are affected by salinization (Szabolcs, 1989; Martinez-Beltran and Manzur, 2005), corresponding to about 10 percent of the globe's arable land (Szabolcs, 1989). Salinization affects up to 3 M hectares of land in Europe, the 17 western states of the United States of America, more than 5 percent of the land in Africa, about a fifth of the arable land of western Asia, and 30 percent of the Australian land area (Chhabra, 1996; Rengasamy, 2006; Ladeiro, 2012), making it a world-wide environmental challenge. Of the global threats that collectively compromise about 10 hectares of arable land per minute (Griggs *et al.*, 2013), salinization contributes about 30 percent (Buringh, 1978). The distribution of salt-affected soils in the world is shown in Table 4.2.8.2.

Table 4.2.8.2

Worldwide distribution of salt-affected areas (Million ha) (Metternicht and Zinck, 2003)

Area	Saline Soils	Sodic Soils	Total	Percentage
Australia	17.6	340.0	357.6	38.4
Asia	194.7	121.9	316.5	33.9
America	77.6	69.3	146.9	15.8
Africa	53.5	26.9	80.4	8.6
Europe	7.8	22.9	30.8	3.3
World	351.2	581.0	932.2	100

Salt-affected soils are an important ecological entity in the landscape of any arid and semi-arid region in the world, and these naturally-occurring saline soils have an ecological value, as a habitat for hallophytic plant, animal and microbial communities. For instance, microbial communities that inhabit ecosystems of naturally saline conditions are structured to function well at high salt concentrations, thus maintaining both high growth rates and other ecosystem processes (Rath *et al.*, 2019). Composition of microbial communities and their abundance are significantly different in saline and normal soils. However, bacterial diversity systematically decreases with an increase in soil salinity (Rath *et al.*, 2019), suggesting that ecosystem function provisioning grows less resilient at high salinities.

Salt accumulation in the surface soil is often found in agricultural areas in arid and semi-arid regions, where it is caused by irrigation with brackish or saline water in poorly drained soils (Allison, 1964) (Figure 4.2.8.1). In areas with shallow groundwater, evaporation can also lead to higher salt concentration in the soil surface layer (Rengasamy, 2006). In addition, soil salinization can be the result of changes in vegetation cover that alter ecosystem water balances. Saltwater intrusion from marine environments is also an important cause of soil salinization (Chandrajith *et al.*, 2014) that has resulted in the salinization of 53 percent of coastal regions in Bangladesh (Haque, 2006).

Figure 4.2.8.1 | Salinization

Salt efflorescences in an irrigated wheat field in Chaplanay in Kandahar Province, Afghanistan. The main effect of excessive salt content in the soil solution is the increase of osmotic pressure that impedes the absorption of water by roots and other organisms and eventually causes plasmolysis of cells.

Soil salinization has direct impacts on plants and has subsequently been a research priority for crops for decades (Ayers and Westcot, 1976; Chhabra, 1996; Stevens and Partington, 2013). For instance, salt exposure is known to reduce crop yield under both greenhouse and field conditions in barley (Pal *et al.*, 1984; Richards *et al.*, 1987), wheat (Richards, 1983; Bajwa *et al.*, 1986), cotton (Meloni *et al.*, 2001; Soomro *et al.*, 2001), sugar cane (Choudhary *et al.*, 2004), rice (Bajwa *et al.*, 1986), maize (Bajwa *et al.*, 1986) and sugar beet (Ghoulam *et al.*, 2002). Crops and cultivars differ in their tolerance to salinity, and this is also modulated by environmental and soil factors. Furthermore, indirect consequences of salinization are ion imbalance and nutrient deficiency (Marschner, 1995), further aggravating the negative effects on plant productivity. Although crop resistance to salt exposure is a promising development (see for example, Bennett *et al.*, 2013), overall plant productivity will be impeded by salinization. Salinity not only adversely affects agricultural production but also influences the naturally occurring plant community assembly. These salt-affected lands are either devoid of any vegetation or have very meagre cover. High salt concentration in soil results in high osmotic potential, which affects the metabolic processes of vegetation. In these areas biosaline agriculture, which involves salt-tolerant conventional and non-conventional tree, shrub and herbaceous crops. has emerged as an alternative option. A restricted number of highly salt-tolerant plant species such as *Prosopis julifora* (Sw.) DC, *Salvadora persica* L., *S. oleoides*, *Acacia nilotica* (L.) Willd., *Capparis decidua* (Forssk.) Edgew., *C. sepiaria* L., *Ziziphus nummularia* Aubrev., *Clerodendrum phlomidis* L. and *Maytenus emerginatus* were reported in alkali soils. Some herbaceous species *Desmostachya bipinnata* (L.) Stapf, *Sporobolus marginatus* Hochst., *Cynodon dactylon* (L.) Pers., *Chloris virgata* Sw., *Trianthema triquetra* Willd., *Suaeda fruticosa* Forssk. and *Kochia indica* Wight are prominent, particularly during the rainy season (Dagar *et al.*, 2001).

While the influence of salinity on plants has received much attention, less is known about the effects on soil microorganisms. Soil microorganisms are negatively affected by high salt concentrations, which are reflected in decreased microbial functions such as respiration and growth after salt exposure (Setia *et al.*, 2011; Rath *et al.*, 2015). However, microorganisms can counteract some of the negative effects of salinity through physiological adaptations. Organisms can adapt their physiology through the synthesis of osmolytes (Kakumanu *et al.*, 2014; Turk *et al.*, 2007) and changes in the composition of cell membranes (Turk *et al.*, 2007; Zhang *et al.*, 2008). In addition to physiological responses of the resident community, selection for more salt-tolerant species can lead to a shift in the taxonomic composition (Rath *et al.*, 2018). The changes in both community composition together with physiological adaptations manifest as an increased community salt tolerance in response to salt exposure (Rath and Rousk, 2015). As community salt tolerance increases, microbial process rates that were inhibited in response to acute salt exposure could, at least partially, recover. Fungi and bacteria are reported to be differently affected by salt exposure (Rath *et al.*, 2016; Kamble *et al.*, 2014). Generally, fungi are considered to be more resistant to short-term exposure to salinity (Rath *et al.*, 2016). However, it is unclear whether an increased resistance to short-term exposure would indeed translate to a shift towards a more fungal-dominated system, as both increasing fungal (Kamble *et al.*, 2014; Wichern *et al.*, 2006) and increasing bacterial dominance (Sardinha *et al.*, 2003; Pankhurst *et al.*, 2001) in response to high soil salinity have been reported. While bacteria and fungi fulfill similar roles as decomposers of organic matter, they differ in the range of substrates they can decompose. Fungal and bacterial biomass also differs in their chemical composition (Six *et al.*, 2006) and nutrient content (Mouginot *et al.*, 2014; Stickland and Rousk 2010). Thus, shifts in the relative contribution of fungi and bacteria to decomposition in response to salinity could have implications for C and nutrient dynamics in soil (Strickland and Rousk, 2010; Schmidt *et al.*, 2011). Given these physiological constraints, microbial communities and subsequently ecosystem function are affected by salt in saline soil ecosystems.

Most general microbial metrics systematically decrease in more saline soils, including respiration (Rath and Rousk, 2015), soil enzyme activities (Batra and Manna, 1997) and microbial biomass (Batra and Manna, 1997; Wichern *et al.*, 2006; but also see Rath and Rousk, 2015). It is still debated whether soil bacteria and fungi are differently affected by soil salinity (Rath and Rousk, 2015), with reports of both higher fungal sensitivity (see for example Wichern *et al.*, 2006), and higher fungal resistance to salinity (see for example Rath *et al.*, 2016). However, when combined with plant litter, it was reported that fungal growth was maximum when bacterial growth was inhibited by the highest salinity, and fungal growth was lowest when the bacterial growth rate peaked at intermediate salt levels, which shows a competitive interaction between bacteria and fungi (Rath *et al.*, 2019a). Additionally, incorporation of an easily available and decomposable source of energy will improve the ability of microbes to withstand salinity (Mavi and Marschner, 2013) and reduce the negative effect of salinity on soil microbes. Microbes belonging to phylum Proteobacteria, Spirochaetes, Tenericutes, WS3 Plantomycetes, Bacteroidetes, Halobacteria, Nitriliruptoria, Gammaproteobacteria and Alphaproteobacteria were found more in saline soils (Canfora *et al.*, 2014; Rath *et al.*, 2019; Zhang *et al.*, 2019).

4.2.9 | FIRE

Wildfires are catastrophic events that occur in most biomes of the world. Burning results in dramatic changes in both functional structure and species composition of terrestrial ecosystems both above and below ground; this is unavoidably linked to shifts in their functioning and provision of key ecosystem services (Niklasson and Granström, 2000). The United Nations clearly recognized fires among the key threats to global biosphere sustainability (United Nations Forum on Forests, 2007). Although the total area burned each year has been decreasing due to the ongoing campaign to prevent burning for agricultural purposes and forest protection, around 340 million hectares are still damaged by fire every year (Willis, 2017). Moreover, fire frequency has remained stable or even increased, and up to one-sixth of the entire area of certain biomes like tropical savannas and grasslands are annually burned, while boreal forests are burned every 15-25 years. This means that soils in large areas are subject to drastic thermal and toxic effects, which may result in severe damage to below-ground biota (Zaitsev *et al.*, 2016).

However, there are contrasting opinions about the real impact of fires on below-ground ecosystems and their functionality (Pressler *et al.*, 2019). On one side, numerous studies demonstrated that fires reduce soil biodiversity and biomass of below-ground organisms (see reviews by Zaitsev *et al.*, 2016 and Pressler *et al.*, 2019). At the same time, certain soils demonstrate remarkable resilience against burning due to various mechanisms like presence of microrefugia and patchiness (Gongalskym and Zaitsev, 2016). However, there is a general consensus that burning has negative implications on soil functions and ecosystem services such as organic matter mobilization and immobilization safeguarded by soil organisms (de Vries *et al.*, 2013).

With the exception of very few organisms, such as the so-called pyrophilous animals, most invertebrates reduce their abundance and biomass immediately after a fire event. Gongalsky *et al.* (2012) showed that organisms living deeper in the soil have higher survival rates after fire than surface-dwelling species. Sterilization effects of the fire on the topsoil is strongly modulated by fire intensity and may result in the almost complete extinction of soil bacteria and fungi in the topsoil immediately after burning. During recolonization, which is rather quick, due to large volumes of unburnt dead wood after the fire (Bastias *et al.*, 2006), there is an overall shift from bacteria-driven towards fungi-driven community with the associated distribution of organic matter flow in detrital food webs. Soil protist and microinvertebrate abundance and biomass are significantly decreased (by 25 percent on average) after single fires and in repeatedly burnt ecosystems rarely achieve 50 percent of the values typical for control (unburnt) sites. Similar reduction levels affect the taxonomic diversity of these groups. Due to their higher mobility macroinvertebrate abundances recover after a relatively short time (a few years) and are mainly limited by microhabitat availability in the burned sites. However, recent assessments clearly demonstrated that in boreal forest ecosystems, recovery of macroinvertebrate diversity and functional structure may require up to 75 years depending on the fire intensity and ecoregion, with longer times associated with higher latitudes (Zaitsev *et al.*, 2016). Overall, current research suggests that in the first years after a fire event soil biodiversity may decrease two-fold and may never recover to initial levels if fires are repeated (Zaitsev *et al.*, 2016).

Therefore, under such conditions, overall functioning of detrital food webs may be considerably reduced. However, there are certain compensatory mechanisms ensuring surprising functional resilience of below-ground communities after burning. Detrital food web modelling showed that soil protists, enchytraeids and associated micropredators form the most sensitive channels of element and energy flow below ground (Zaitsev *et al.*, 2017). Recent studies also showed remarkable stability of microbial community functioning and associated greenhouse gas emission levels due to compensatory mechanisms of overall microbial activity driven by changing physical and chemical soil properties and mobilization of additional carbon and nitrogen sources (Goncharov *et al.*, 2020).

The current level of knowledge about fire effects on soil biota brings us to a conclusion that in the short run, burning strongly reduces below-ground biomass and functioning. Both, however, may recover within a few years after burning. There are multiple and complex mechanisms behind the functional resilience of soil ecosystems and consequent sustainability of ecosystem services provision in fire-prone ecosystems. However, soil biodiversity requires more time for recovery and may never return to an initial level in case of repeated fires. Thus, multiple fires of predominantly anthropogenic nature, especially in agricultural lands, grasslands and some forest ecoregions, represent one of the greatest threats to below-ground biodiversity and stability of soil ecosystems.

4.2.10 | EROSION AND LANDSLIDES

Erosion involves the detachment, transport and deposition of soil particles through water or wind. Erosion, the natural process that has shaped the earth's landscape, is now one of the main drivers of degradation of the upper layers of the soil due to its acceleration by anthropogenic activities (such as agriculture, deforestation and soil sealing) (Figure 4.2.10.1).

Landslides are gravitational movements of a mass of rock, earth or debris down a slope. Landslides displace great volumes of soil, and can be triggered by natural processes (for example, heavy or prolonged rainfall, earthquakes, volcanic eruptions, rapid snowmelt and permafrost thawing), but the likelihood of their occurrence is magnified by human actions (such as slope excavation and deforestation).

At global scale, approximately 36 Pg (1×10^{15} g) of soil are estimated to be eroded each year by water (Borrelli *et al.*, 2017). Soil loss due to wind erosion in arable land has been estimated at about 2 Gt (1×10^{12} g) (FAO and ITPS, 2015). There are no global estimates of yearly soil losses due to landslides as they are very variable in volume (from a few cubic metres to several cubic kilometres), depending on the area and depth of the layers involved. Over the last years, climate changes have significantly affected the frequency of erosive and landslide occurrences. As the possibility of extreme climate events has increased, so has the concern about their negative impact on soil biodiversity (Orgiazzi and Panagos, 2018).

Erosion creates both degraded-eroded and enriched depositional environments. Following landslides, the upper parts of the affected slopes are usually stripped of soil, while fallen soil and mineral masses are accumulated and mixed at the foot of the slopes. In both cases, the impact on soil organisms can be direct or indirect. As a direct effect, the inhabitants of the upper soil layers may be eliminated or displaced far away from their original environment.

Water erosion comprises several processes (such as water splash and sheet, rill and gully formation) that selectively affect different soil species. Landslide effects on soil organisms may be considered similar to those caused by water erosion. Splash detaches soil particles and degrades soil microstructure (due to microaggregate disruption and soil pore clogging), worsening the quantity and quality of habitable soil microhabitats. Habitat destruction affects microorganisms and micro-invertebrates living near the soil surface, while endogeous meso- and macro-invertebrates (such as earthworms) are vulnerable to clogging of the burrows they excavate (Baxter *et al.*, 2013). Sheet, rill and gully erosion result in mid- and long-distance transport of great quantities of soil and associated soil biota from degraded-eroded to depositional sites. In many cases, the depositional sites are favoured by the arrival of nutrients and biological propagules, but soilborne plant pests, such as plant-pathogenic nematodes (Chabrier and Quénéhervé, 2008), can also be spread.

The predisposition of various taxa to be transported by wind depends on their size, abundance and location in the vertical soil profile. Billions of tonnes of desert dust, made of very fine soil particles and astonishing quantities of soil organisms, are transported yearly across continents (Griffin, 2007). In the case of the microscopic and very abundant soil prokaryotes (that is, bacteria and archaea), every year up to 4.3×10^{11} per square metre of prokaryotic cells may be wind-blown from the upper layer of forest, 1.6×10^{12} square metres per year from pasture, and 1.9×10^{12} square metres per year from arable soils (adapted from Torsvik *et al.*, 2002). A rich variety of soil animals has been found in the aeroplankton, including nematodes, rotifers, collembolans, tardigrades and mites (Nkem *et al.*, 2006; Ptatscheck, 2018). Effects of wind erosion on transport and dispersal of soil organisms has been studied in particular for soil nematodes, whose eggs and larvae can be transported as far as 40 kilometres from their origin (Carroll and Viglierchio, 1981). Among soil micro-arthropods, collembolans and mites are the most abundant. Collembolans have been found in air samples taken at a height of 3 000 m (Glick, 1939) and wind dispersal is proposed as a significant dispersal route for both epigeic and endogeic springtails and oribatid mites (Querner *et al.*, 2013; Lehmitz *et al.*, 2011).

Soil and soil organism mobilization by wind has net negative effect on the biodiversity and functioning of the eroded soils worldwide. That vulnerability is particularly relevant in some ecosystems, such as drylands, where plant cover is very scarce and soil life is carbon-limited. Drylands are protected against erosion by soil biological crusts formed by bacteria (cyanobacteria and heterotrophic), algae, mosses, liverworts, fungi and lichens (Maestre *et al.*, 2011). These biological structures are crucial not only for stabilization of the underlying soil, but also the regulation of water cycles and provision of nutrients to the rest of the below-ground communities. Biological crusts are extremely vulnerable to physical disturbance by human activities (such as trampling, cattle raising and off-road vehicles), to the point that their disruption is associated to growing desertification (Pointing and Belnap, 2012)

Overall, soil microorganisms and microscopic eukaryotes show a very high vulnerability to wind and water transportation (Finlay, 2002). This phenomenon, along with the ability of several soil organisms to survive many weeks in seawater and freshwater (Coulson *et al.*, 2002), suggest that wind and water erosion may be a key vector in long-distance colonization of soils and are at the basis of the ubiquitous geographic distribution of many soil organisms (Figure 4.2.10.2).

Erosion and landslide events can lead to changes in soil physico-chemical properties, which indirectly affect living communities (Baxter *et al.*, 2013). In fertile soils, most of the organic matter is close to the soil surface in the form of decaying plant litter, or associated to fine (silt and clay) soil particles (Plante *et al.*, 2006). Both wind/water erosion and landslide events remove the fine organic particles in the soil, leaving behind large particles and stones (Lal, 2001). Erosion of the topsoil layer (0-10 cm) significantly decreases soil organic matter to the point that soil removed by either wind or water erosion is 1.3 to 5.0 times richer in organic matter than the soil left behind (Pimentel and Kounang, 1998). Since across biomes as well as at local scales, soil microbial and faunal abundance positively correlate with soil organic matter content (Schnürer *et al.*, 1985; Fierer *et al.*, 2009), erosion is bound to lessen soil biota biomass.

Effects of landslides and erosion on organisms living in deeper soil layers are less clear. There is some evidence that soil impoverishment (reduction of soil organic carbon inputs) results in reduced below-ground food webs as they are predominantly regulated by bottom-up forces (Moore and de Ruiter, 2000). However, there is no conclusive proof of the correlation between carbon content and soil faunal and/or microbial composition, which might be more dependent on other soil properties (such as soil pH and texture) (Cole *et al.*, 2005; Fierer and Jackson, 2006; Johnson *et al.*, 2003).

Both the richness and the abundance of soil organisms may have different impacts on soil susceptibility to erosion. For instance, the dense network of mycorrhizal fungi present in a grassland may reduce the amount of soil loss by rain or windstorms (Burri *et al.*, 2013). On the other hand, at the opposite size extreme, the extensive excavation activity of some mammals (such as moles and pocket gophers) may weaken soil structure and thus accelerate erosive processes (Reichman and Seabloom, 2002). Other groups of organisms have less definitive effects on soil loss. For instance, the burrowing activity of earthworms can reduce erosion by favouring water infiltration (Shuster *et al.*, 2002). Simultaneously, surface cast production by some earthworm species may accelerate erosion, as this material can be easily transported by water or wind (Shipitalo and Protz, 1987). Therefore, the quantity of material loss in a living soil is different from that in an inert one. Only a few applied studies, mainly targeting plants (Allen *et al.*, 2016; Berendse *et al.*, 2015), have analyzed the effects that soil organisms have on erosion, and have estimated the potential amount of material eroded/preserved due to the presence of soil organisms.

Large-scale studies of soil erosion go through the application of models that permit estimation of the amount of soil loss by combining different factors (such as soil physico-chemical properties, rainfall erosivity and land management). So far, none of these models has included a biological factor accounting for the diversity of organisms living in the soils. Nevertheless, the possibility to include an "earthworm factor", taking into account both the abundance and richness of this group of soil organisms, has been recently proposed (Orgiazzi and Panagos, 2018). Therefore, the incorporation of a biological component is feasible, upon the availability of distribution data of soil biodiversity at large scales.

In recent years, the importance of soil organisms in shaping large-scale processes, such as climate regulation through impact on the carbon cycle, has been recognized (Luo *et al.*, 2016). The derived models benefited from the integration of biological factors, as confirmed by successive validations through ground data collection (Wieder *et al.*, 2015). A similar path is desirable for soil erosion and biodiversity, in order to ensure more accurate estimates of soil loss.

© Christian Thine

Figure 4.2.10.1 | Satellite images of natural and anthropogenic-enhanced water erosion

A. True-colour image captured three days after heavy rainfall in Rome and the surrounding area of Lazio, Italy. The image shows sediment gushing into the Tyrrhenian Sea, part of the Mediterranean Sea. Image captured on February 2019. **B.** Satellite image of northern Brazil showing the sediment-laden water that appears brown as it flows from the lower left to the open ocean in the upper right. Image captured on August 2017. **C.** Image of Yukon Delta in the US state of Alaska showing how the river branches off into numerous channels flowing to the sea. The sandy colour of these channels and of the coastal water illustrates how much sediment the river carries to the sea at this time of year. Image recorded on August, 2017.

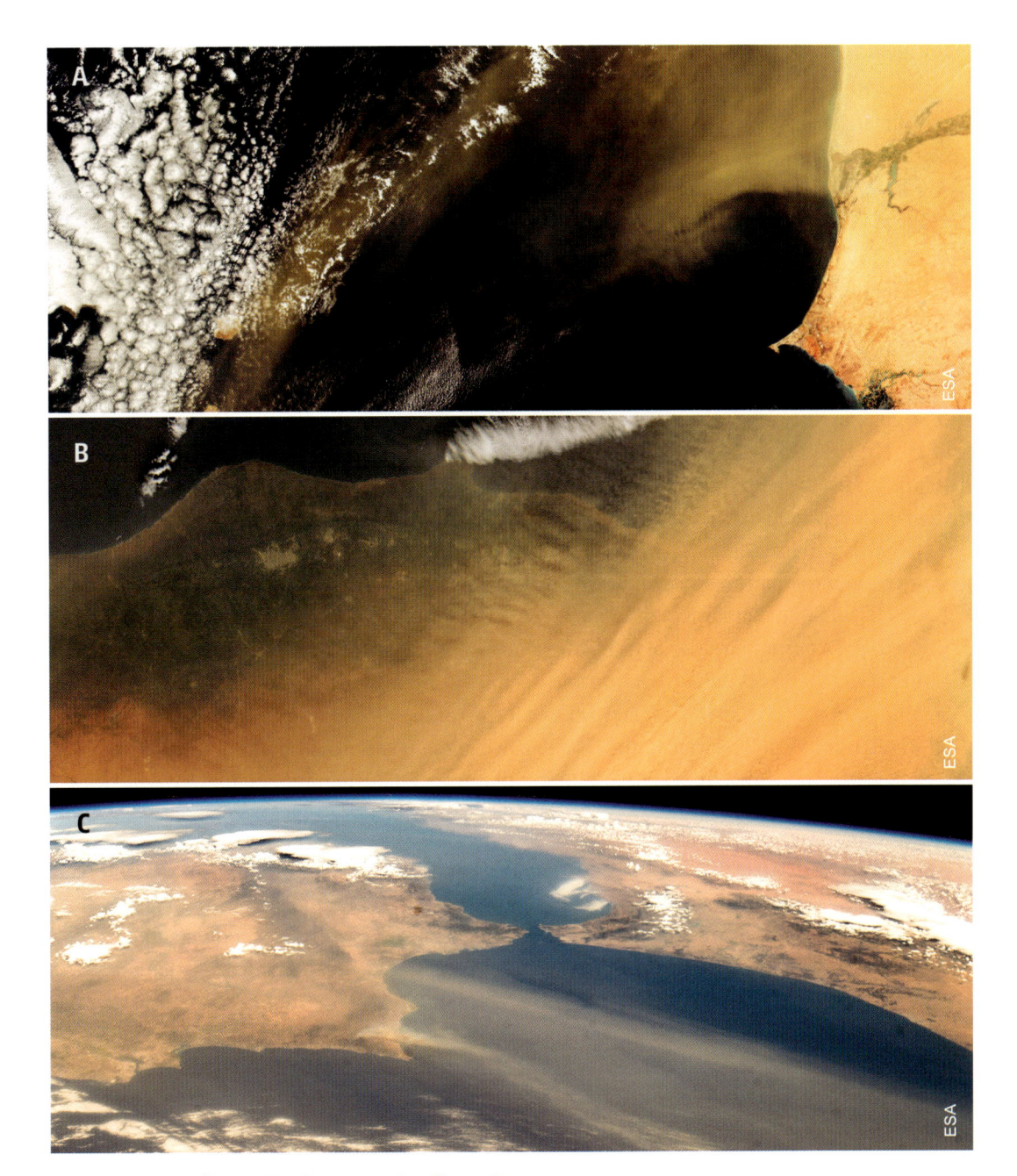

Figure 4.2.10.2 | Satellite images of soil erosion

A. Satellite images of the Atlantic Ocean and the Cape Verde archipelago peeking out from under the clouds, 570 km off the west coast of Senegal and Mauritania, seen on the right of the image. The dust and sand coming mainly from the Sahara and Sahel region are being carried by the wind towards Cabo Verde from Africa. Image captured on May, 2018. **B.** Dust, carried by the wind from Desert storms in North Africa was blown northwards across the Mediterranean Sea causing snow in eastern Europe to turn orange. Dust reached as far afield as Greece, Romania, Bulgaria and Russia. Image captured on March, 2018. **C.** This image of Portugal and Spain was captured by ESA astronaut Alexander Gerst who commented "it looks like a mixture of dust, sand and smoke" International Space Station on August 2018.

4.2.11 | CLIMATE CHANGE

Climate change is associated with uniformly rising CO_2 levels and, in most ecosystems, increased temperature and water limitation. Generally, elevated CO_2 initially enhances photosynthesis, which aggravates limitation of macro- or micronutrients in soil and intensifies plant-microbe competition for soil resources. Elevated CO_2 may thus alter the balance of mutualistic and free-living groups or favour certain mutualists over others (Terrer *et al.*, 2016). Elevated temperature *per se* may promote growth and biodiversity of most soil communities, especially in cold ecosystems. However, increased fluctuations of temperature seasonally and annually are likely to enhance stress from water limitation, especially when coupled with reduced rainfall or longer dry periods. Reduced water availability may directly reduce overall soil biomass and biodiversity (Maestre *et al.*, 2015; Bahram *et al.*, 2018). Long drought periods render the native ecosystems more vulnerable to natural or human-induced burning, which may have devastating effects on ecosystems not adapted to wildfire. Combined, altered temperature and precipitation patterns and fire regimes shift biomes, potentially with enormous changes in plant growth forms and vegetation types. These changes in vegetation may result in cascading effects on all soil biota that largely determine the soil functional potential, including nutrient cycling. In particular, loss of tree cover due to drought stress, pest outbreak or intensive fire results in excessive soil drying and decline of soil organic material and fungal biomass. Moderate nitrogen pollution may act as a fertilizer to counteract soil nutrient limitation, whereas heavy pollution may alter the balance among taxonomic and functional groups and ecosystem nutrient cycling. Pollution of nitrogen and sulphur acidify soil, which favours saprotrophic groups.

Climate change has different implications for different areas of the globe, and therefore the potential impacts of climate change will take a number of different forms; each of these, in turn, will have different potential effects on soil biodiversity. For example, global models of climate variously predict, for different geographic regions, that future conditions will be hotter, drier, wetter, or have more frequent droughts and/or extreme temperature, wind, precipitation events, and so on (IPCC, 2014). Since these climatic changes will drive changes in vegetation type and community composition (mediated by the associated disturbances such as drought, flooding, wind or fire), there is clear potential for major alterations to ecosystem-defining processes like organic matter accumulation, decomposition and cycling, which are known to influence the diversity and composition of soil biota (Coyle, *et al.*, 2017; Coleman *et al.*, 2018). In any case, there will almost certainly be large geographic expanses consisting of novel combinations of climate, vegetation and soil (Hobbs *et al.*, 2009), and there is general consensus that climate-driven changes will result in major losses of biodiversity for above-ground biota (Bellard *et al.*, 2012).

Soil is a remarkably rich reservoir for biodiversity, and the mechanisms by which this diversity arises are still not fully understood. Nevertheless, soil microbes (bacteria, fungi, archaea) are responsible for many critical ecosystem processes which humans (and indeed the entire terrestrial biosphere) rely upon. Recent advances in understanding the diversity of soil bacteria suggest that a number of co-varying factors can explain global patterns of diversity (Bickel *et al.*, 2019), and chief among these are temperature and climatic water content (a composite of soil water-holding capacity, number of consecutive dry days and potential evapotranspiration or PET). Climatic water content thus relates strongly to the heterogeneity of niche space produced by differences in water-filled pore space and the connectivity of such habitats, which in turn can influence soil pH, also known to correlate with bacterial diversity (Lauber *et al.*, 2009). Thus, changes in factors influencing climatic water content (consecutive dry days, PET), may be reasonably expected to also change soil bacterial diversity.

Similar to bacterial diversity, soil fungal biodiversity is still in the process of being documented and is not particularly well known at the global scale. However, there is evidence that many fungal communities are dominated by a few taxa world-wide, and that these taxa possess genes that are related to stress resistance (Edigi *et al.*, 2019). This is consistent with findings from a mesocosm study performed in peat soils from Canada that became dominated by a few taxa from the Ascomycota and Basidiomycota after 18 months of warming treatments (Asemaninejad *et al.*, 2018). Climate change responses of fungal communities and biodiversity are likely to be biome-specific, however. Other workers in Mediterranean-type ecosystems found that fungal diversity increased in certain parts of the landscape with warming treatments (Birnbaum *et al.*, 2019), so it is difficult to generalize about how fungal diversity may respond to climate change. The implications of these findings are that fungi, from a functional (and perhaps even biodiversity) standpoint, may be particularly well suited to adapt to changing climate conditions, but this may be dependent upon ecosystem type. Indeed, microbial diversity was unchanged when soils from dryland ecosystems around the globe were subjected to warming treatments, although microbial community composition shifted in response to the treatments (Delgado-Baquerizo *et al.*, 2017). However, the capacity of soil microbial communities to buffer ecosystem responses to climate change has not been evaluated for other global biomes.

Faunal components of the soil biota are also globally diverse, and most groups are still not well characterized in terms of their biodiversity, including the nematodes and microarthropods, although molecular approaches are producing better estimates of genetic diversity on the global scale (for nematodes, for example, see Nielsen *et al.* 2014). These advances notwithstanding, most studies that deal with the effects of climate change on the soil mesofauna address faunal responses in terms of communities, assemblages, functional groups or trophic groups. As such, studies addressing species-level biodiversity responses to climate change treatments are relatively rare, but a few trends are noteworthy. For example, Caruso *et al.* (2019) found that oribatid mite diversity generally increased along a south to north gradient in the United Kingdom of Great Britain and Northern Ireland, and that precipitation and soil organic matter were positively (but not strongly) associated with this increased diversity. Considering that both precipitation and soil organic matter content are expected to be influenced by changing climate, it is possible to infer that mite diversity could reasonably be expected to respond to changes in these conditions.

Macroinvertebrates, many of which are ecosystem engineers (capable of modifying habitat to create niche space for other taxa), are also likely to be responsive to climate changes, as their distribution and abundance are frequently well predicted by general patterns of temperature, moisture and vegetative cover. Local earthworm species richness, abundance and biomass, for instance, were positively related to precipitation at the global level, so that any climate change involving changes in precipitation level of frequency may have important impacts on earthworm communities and their potential for ecosystem service delivery (Philips *et al.*, 2019). Termites, for example, are most diverse in tropical rainforest systems, and their diversity decreases as ecosystems become more arid (Bourguibignon *et al.*, 2017). Thus, if climate change causes drier conditions to predominate, for example, in tropical dry forests where termite diversity is highest at elevations where rainfall is more abundant (Casalla and Korb, 2019), then negative effects on termite diversity can be expected. On the other hand, there is some evidence that the presence of ecosystem engineering macroinvertebrates may provide some buffering of ecosystems to climate changes. For example, Ashton *et al.*, (2019) found that termites were associated with greater soil nutrient heterogeneity (and soil moisture and decomposition rates) during a drought, relative to soils where termites were experimentally excluded, suggesting that when termites were present, these soils experienced less relative change when conditions were drier, with likely positive results for the biodiversity of other soil organisms present at the sites. Similar patterns have been observed for another group of ecosystem engineers, where the presence of earthworms in simulated agricultural soils reduced the negative effects of warming on total below-ground meso- and macrofauna taxon richness (Siebert *et al.*, 2019).

In light of other anthropogenic threats to soil biological diversity (such as land cover change, agricultural intensification and atmospheric deposition of pollutants), which have been relatively well-documented (Coyle *et al.*, 2017), the expected responses of soil biodiversity to climate change are barely known and difficult to predict. It is clear that climate-change-driven impacts on soil biodiversity will be context dependent, and will be strongly influenced by the starting condition (that is, by what vegetation and/or ecosystem type is present), and by the degree and direction of climate change. Focusing on conservation of the diversity of particular groups of soil invertebrates, specifically the ecosystem engineers, may have cascading positive effects on pools of soil biodiversity among other soil flora and fauna.

4.2.12 | INVASIVE SPECIES

Terrestrial invasive species can arise from any level of biological organization ranging from viruses and microbes (bacteria and fungi) to plants, invertebrates and mammals, and each type of invasive species has the capacity to alter soil biological diversity either directly (for example, through competitive displacement), or indirectly (for example, through changes in vegetation composition and/or habitat modification). Introductions of non-native species have been ongoing for centuries, if not millennia, around the globe, and the frequency at which new introductions are made is also increasing (Simberloff *et al.*, 2013). Worse, once established, there appears to be a synergistic relationship between climate change, specifically warming, and the ability of introduced species to increase their ranges in the invaded area (Walther *et al.*, 2009).

Globalization, especially global trade, increases intentional and unintentional introduction of new plant, animal and microbial species into new environments. Import of potted seedlings or saplings is of particular concern, because of undeliberate co-introduction of thousands of microbial species. Although only 10 percent of introduced plant species become naturalized and 1 percent become invasive (Gallagher *et al.*, 2014), these figures may be much higher for microbial species that usually have larger distribution range and ecological amplitude.

It is well recognized that invasive trees, such as species of pines, eucalypts and wattles, may transform entire ecosystems, partly via their recalcitrant litter, stimulation of burning and activities of root-associated microorganisms. Symbiotic biota may further facilitate the invasion process and switch from their intimate plant partners to local potential host trees and invading native soil communities. Introduced pathogens and perhaps endophytes may find new naïve hosts and become serious pathogens of native plants in the new environment. Furthermore, these antagonistic microorganisms may occasionally hybridize with local pathogens and evolve pathogenicity in new hosts. Introduction of animal-associated pathogens pose similar threats to those of plants.

Microbial introductions

Exotic microbes can have profound, ecosystem-changing consequences when pathogens from one continent are introduced to another continent where potential hosts have no inherent resistance. For example, in Australia, more than half of all the putative species from the genus *Phytophthora* sampled in a recent continental-scale sampling were thought to have been introduced, and in fact the two most widespread species were non-native (Burgess *et al.*, 2019). These findings have wide-ranging implications, as *Phytophthera* is a relatively well-known group of fungal pathogens which can have important economic (agricultural and forestry) as well as ecological consequences. Another example of microbial introduction is the case of the parasitic fungus *Cryphonectria parasitica*, known as chestnut blight, which resulted in the total removal a dominant canopy tree (*Castanea dentata*, American chestnut) in North American forests. The loss of chestnut from forests of eastern North America resulted in major changes in forest composition, with clear changes in the abundance of species with different chemical composition of leaf litter (for example, *Rhododendron maxima*). Considering that studies of soil fauna have shown that "single-tree influences" can be observed for the distribution of earthworm species in eastern North America (Boettcher and Kalisz, 1991), and also for eastern Europe (Hobbie *et al.*, 2006), as well as for broader measures of soil and litter biodiversity including for example mites, nematodes and spiders (Mueller *et al.*, 2016), it is clear that any microbial pathogen that significantly impacts the occurrence of a canopy dominant tree species could have cascading effects on soil biodiversity.

Plant introductions

Plant invasions have strong potential to influence soil biological diversity through various mechanisms. Plant invasions can alter the overall plant community, often resulting in a monoculture stand consisting of only the invasive species (both in overstory and understory plant communities), which changes the diversity of below-ground plant roots and root exudates, which in turn may influence the diversity of organisms that rely on roots and exudates as food or habitat resources. Thus, invasive plants can influence soil biological diversity at nearly all levels, ranging from microbial to vertebrate animal diversity. One example of soil microbial diversity impacts is from Canada where Bugiel *et al.* (2019) found a negative impact of dog-strangling vine (*Vincetoxicum rossicum*) invasion on soil bacterial composition and diversity as measured by variation in terminal restriction fragment length polymorphism data relative to uninvaded sites. Interestingly, not all plant invasions have negative effects on soil microbial species richness, and in some cases (albeit rarely encountered in scientific reports) plant invasions can actually increase species richness of mycorrhizal fungi as demonstrated for Hawaiian subtropical forest ecosystems (Gomes *et al.*, 2018).

Invasive plants can also exert impacts at the meso- and microfauna level, as observed in central Europe (Slovakia), where invasion of the herbaceous knotweed, *Reynoutria japonica* (syn. *Fallopia japonica*) negatively affected soil nematode species richness compared to uninvaded soils (Čerevková *et al.*, 2019). Similarly, in southern Europe (in the Tuscan Appenines), invasion of native oak-dominated stands by black locust trees (*Robinia pseudoacacia*) was associated with decreases in richness in nematodes and microarthropods, as well as decreased plant species richness (Lazzaro *et al.*, 2018). It is notable that mesofauna responses to invasive plants, as with microbes mentioned above, are not always negative. In the Guangdong Province of China, invasion of abandoned farmland by *Artemisia artemisiifolia* was accompanied by an increase in abundance of soil fauna, but this may be partly related to the relatively degraded condition of these populations following agricultural use of the soils (Qin *et al.*, 2018).

Above ground, monocultures of invasive plants can change the timing, chemistry and decomposability of leaf litter inputs into soil ecosystems, and this too has potential to influence the below-ground biotic community. For example, Lobe *et al.* (2014) found that invasion of riparian forests in the state of Georgia in the United States of America by Chinese privet (*Ligustrum sinense*) altered surface soil pH, and that this apparently favoured populations of non-native European earthworms, but that when the invasive plant was removed, soil pH trended back toward those observed in uninvaded forests, and the abundance of native earthworms rebounded. Recovery of wetland soil invertebrates negatively impacted by non-native plants was also shown in ecosystems in Kwa-Zulu National Park in South Africa when aggressive efforts to remove the invasive plants were undertaken (Eckert *et al.*, 2019). These case studies suggest that although plant invasions have capacity to negatively affect soil biodiversity, active management of invasive plant populations can result in cascading benefits to soil biota.

Soil invertebrate invasions

In large areas of the globe, non-native soil invertebrates have been introduced. Depending on the ecosystem invaded, these organisms can have dramatic negative impacts on native plants, microbial communities and other soil animals. Perhaps the best-studied group of invasive soil animals are the earthworms, and this is owing to their relatively large size and their ability to act as ecosystem engineers; thus their large impacts on soil ecosystems where they invade (Hendrix *et al.*, 2008). Invasive European lumbricid earthworms were reported to reduce species richness and diversity of litter- and soil-dwelling microarthropods (specifically oribatid mites) by 50 to 75 percent in forests of the Allegheny Plateau in the eastern United States of America (Burke *et al.*, 2011). In a study in the southeastern United States of America, Snyder *et al.* (2011) found that millipede species richness was negatively impacted by an invasive Asian earthworm species (*Amynthas agrestis*), and they attributed this to the rapid consumption of forest floor leaf litter by the earthworm. Overall, managing earthworm invasions is quite difficult, and the best approach is thought to be prevention of introductions, as removing invasive earthworms after their populations become well established may be very cost and labour intensive (Callaham *et al.*, 2006).

4.3 | REGIONAL STATUS OF THREATS TO SOIL BIODIVERSITY

There are important regional differences in the importance and role of threats to soil biodiversity and functioning, depending on various abiotic and human factors such as climate, extent of industrialization, area in different types of native vegetation and anthropogenic land uses (especially urbanization, agriculture and forestry), and level of protection of soil resources, among others. These differences are explored in the following sections, in six of the main world regions (using the classification of FAO, 2015): Asia, South West Pacific, Latin America and Caribbean, North America (excluding Mexico), Europe and sub-Saharan Africa. Eurasia, North Africa and the Near East received no input from specialists, so they were excluded from this analysis.

In each region, the impact of thirteen threats (deforestation, urbanization, agricultural intensification, loss of SOM and SOC, soil compaction and sealing, soil acidification and nutrient imbalance, contamination, salinization and sodification, land degradation, fire, erosion and landslides, climate change and invasive species) were evaluated per ecoregion, following the ecoregions used by Orgiazzi *et al.*, (2016b) in the *Global Soil Biodiversity Atlas*, modified from the original maps of the World Wildlife Fund ecoregions of the world (Olson *et al.*, 2001). Hence, ten ecoregions, separated mainly by their relationship to vegetation types but also generally associated with specific types of climates and often soils, were evaluated: tropical and subtropical forest; tropical and subtropical grassland savanna and shrubland; temperate grassland, savanna and shrubland; montane grassland and shrubland; Mediterranean forest, woodland and shrubland; temperate broadleaf and mixed forest; temperate and boreal coniferous forest; tundra; desert and dry shrubland (Figure 4.3.1; Orgiazzi *et al.*, 2016).

For each region of the world, and for each ecoregion therein, the current status and future trends of each threat, as well as the potential impact of the threats on soil biodiversity and function and the present knowledge level (such as the extent of available literature and number of studies) were obtained from expert opinion and consensus among various experts from each world region. For Europe, the exercise was further performed using GIS-based quantitative data, following methods based on Orgiazzi *et al.* (2016a), but for the rest of the world, only expert opinion was used. From the list of thirteen threats, the most important ones were singled out: those that presented high or fair level of geographic spread within the ecoregion, an increasing trend over time, and a high or fair level of impact on soil biodiversity.

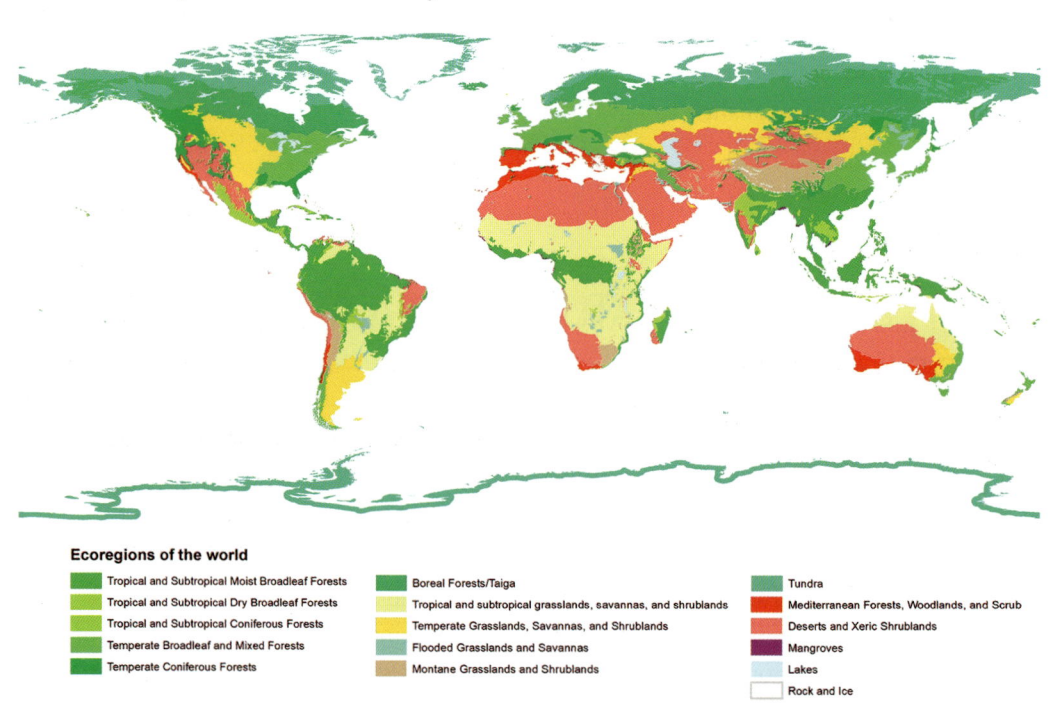

Figure 4.3.1 | Terrestrial Ecoregions of the World used for the assessment of regional threats

Source: World Wildlife Fund - US

4.3.1 | SUB-SAHARAN AFRICA

Sub-Saharan Africa (SSA) is diverse in terms of relief, climate, lithology, soils and agricultural systems. A combination of some of these has been used to stratify the region into agro-ecological zones (Fischer *et al.*, 2002; Otte and Chilonda, 2002; Global Harvest Choice, 2010), including the sub-humid zone, the humid zone, the highland zone and the arid and semi-arid zones. These represent the same basic zones as the ecoregions listed below.

Tropical and subtropical grassland

The sub-humid zone occupies 22 percent of SSA, mainly in southern and central Africa. The zone receives 1 000 mm to 1 500 mm of rain annually. This zone is very diverse in terms of climate, soils and land use. It is mainly covered by Luvisols, Cambisols Ferralsols, and Acrisols which are developed from parent material that is strongly weathered, and the levels of plant nutrients as well as the clay fraction are low. Their natural vegetation cover consists of medium height or low woodland with understory shrubs and a ground cover of medium to tall, perennial, grasses. Agriculture is the mainstay of the communities that cultivate both food and cash crops, including cassava, yams, maize, fruits, vegetables, rice, millet, groundnut, cowpeas and cotton. From these crops, products such as cottonseed cakes and the residues of the crops are available as feed for livestock. In some areas of this zone farmers grow soybean and leguminous forage crops. The use of fire and deforestation, and the associated loss of soil organic carbon and compaction with changes in land use and urbanization, as well as erosion and landslides and invasive species, were selected as the most important threats to soil biodiversity and function in this ecoregion (Table 4.3.1.1).

Tropical and subtropical forest

The humid zone occupies 19 percent of SSA, mostly in central and west Africa at low latitudes north and south of the equator and receives more than 1 500 mm of rainfall annually. Soils in this zone include Ferralsols, Acrisols and Luvisols, the last of which are commonly encountered at the forest-savannah boundary. Vegetation consists of rainforest and derived savannas with natural vegetation dominated by tall, closed forest which may be evergreen or semi-deciduous and which is often floristically rich. The herbaceous vegetation often contains large amounts of the major nutrients. The soils are strongly weathered and hence have high levels of iron and aluminum oxides and low levels of phosphorus. The organic matter content is therefore generally low and the soils are fragile and easily degraded when the vegetative cover is lost. Deforestation and agricultural intensification may lead to significant decreases in soil biodiversity and are the main threats to soil biodiversity and function in this ecoregion (Table 4.3.1.1).

Montane grassland and shrubland

The highland zone represents 5 percent of SSA's land area, most of which is in eastern Africa and half in Ethiopia. This zone includes areas above 1 500 m altitude that have a mean daily temperature of less than 20°C. The main highland areas are in Ethiopia, Kenya, Uganda, Rwanda, Burundi, the eastern Congo, Tanzania, Angola and Lesotho. The highland areas vary in climate, topography, soils and land use with topography varying from gently rolling hills to deeply incised valleys and steep slopes. Soils are sometimes deep and fertile Andisols and Nitosols, but shallow soils of inherently low fertility are widespread. In many mountain grassland areas, soils only have a very shallow surface horizon that is fertile. Cultivating these so-called shallow low-fertility soils forms a surface crust which reduces water infiltration, resulting in high runoff causing soil erosion, and unless soil conservation measures are taken and soils are sufficiently covered with vegetation, overland flow removes large amounts of topsoil, carrying with it soil organic carbon and its associated biodiversity. The zone receives bimodal rainfall (more than 1 000 mm annually) and there are two growing seasons. Livestock rearing is widespread: farmers grow fodder, and animal traction is of increasing importance. Population pressure is encouraging crop–livestock integration, for which the cool highlands have high potential. Threats to biodiversity in highlands and humid and sub-humid zones include deforestation due to rising population, overgrazing and burning of above-ground cover leading to soil erosion, and loss of plant species with potential negative effect on rhizosphere biodiversity (Table 4.3.1.1). These areas are also particularly threatened by climate change.

Desert and dry shrubland

The arid and semi-arid zones occupy 54 percent of the land area of SSA, most of which is in West and East Africa. Rainfall is low and extremely variable ranging from 500 mm to 1 000 mm annually. Due to high temperatures and evapotranspiration rates, these are mostly associated with Arenosols and Cambisols, sandy and loamy sandy soils poor in plant nutrients and with low water-holding capacity. Vegetation cover consists of short annual grasses, legumes, scattered shrubs and trees. The main livelihood activity of the communities living in the drier zone is keeping livestock including sheep, goats, cattle and camels that browse the herbage and shrubs and move from place to place in search of fodder. Where rainfall is higher and more reliable in the semi-arid zone, there is better vegetation cover of open low-tree grassland and a relatively healthy environment for humans and livestock. Cropping and crop–livestock systems dominate these areas and farmers commonly grow millet, sorghum, groundnut, maize and cowpeas. Threats to soil biodiversity in this ecoregion include wind and water erosion, loss of soil organic matter and soil nutrients, salinization and sodification and waterlogging in low areas (Table 4.3.1.1).

Table 4.3.1.1 | Threats to soil biodiversity in Sub-Saharan African Ecoregions

The main threats to soil biodiversity and function, and the level of scientific knowledge of the impacts of these threats on soil biodiversity in each of the ecoregions of sub-Saharan Africa (SSA)

Ecoregion	Main threats	Knowledge level
Tropical and subtropical forest	Deforestation	High
	Agricultural intensification	High
	Loss of SOM and SOC	Fair
	Erosion and landslides	Fair
Tropical and subtropical grassland	Deforestation	High
	Loss of SOM and SOC	Fair
	Soil compaction and sealing	Low
	Fire	Low
	Erosion and landslides	Low
Montane grassland and shrubland	Deforestation	Low
	Agricultural intensification	Low
	Loss of SOM and SOC	Low
	Erosion and landslides	Low
	Climate change	Low
Desert and dry shrubland	Loss of SOM and SOC	Low
	Salinization and sodification	Low
	Land degradation	Low
	Fire	Low
	Erosion and landslides	Low
	Climate change	Low

4.3.2 | ASIA

The current status of soil biodiversity varies immensely within Asia. Deforestation for traditional agriculture has caused huge losses of soil biodiversity in the distant past and conventional high-input agriculture, urbanization and contamination in the recent past (Table 4.3.2.1). Climate change is a new threat, particularly affecting highlands and coastal areas. Coexisting with the threats is the opportunity of conserving and restoring biodiversity by organic farming. Knowledge on impacts of the threats suffers from several knowledge gaps: (i) coverage of selected socio-ecological scenarios, taxa and functional groups; (ii) lack of long-term monitoring following a common protocol such as long-term soil fertility experiments and permanent forest plots revealing soil biodiversity/ecosystem function relationships; and (iii) lack of analysis of interaction among different threats.

Tropical and subtropical forests

Deforestation is a threat more to taxonomic diversity of earthworms, termites and mycorrhiza than to bacteria, saprophytic fungi and nematodes, and it reduces functional diversity of all groups of soil organisms (Wong *et al.*, 2016; Kerfahi *et al.*, 2016; Kimber and Eggleton, 2018). Further, detrimental impacts are more pronounced in primary forests than in secondary forests, and when natural forests are replaced with shifting agriculture with short cycles (4 to 8 years) in the humid tropics and settled annual crop systems with high agrochemical inputs/low organic inputs in the dry tropics (Bagyaraj *et al.*, 2015; Yimyam *et al.*, 2016; Bhadauria, 2016). Though the deforestation rate in Asia is twofold higher than the global average (0.1 percent), countries differ in the magnitude of impact of the threat of deforestation. The threat is not so high in Bhutan, China, India, Laos and Vietnam, where reforestation/afforestation rates have exceeded deforestation rates for the past 10 to 15 years. In Indonesia and Malaysia natural forests are, by and large, replaced by oil palm/rubber plantations, a change that reduces the diversity and abundance of macro-invertebrates (Mumme *et al.*, 2015) but not necessarily of bacteria, fungi and nematodes (Kerfahi *et al.*, 2016). Much of tropical Asia has islands of primary forests conserved for cultural/religious purposes (Lyngwi and Joshi, 2015) and equally structurally complex and species-rich home gardens (Mohan Kumar, 2016). Intensive use of agrochemicals invariably correlate with loss of soil organic carbon, and salinization in many situations is a persistent threat in alluvial plains, but its impact is declining with increasing adoption of organic farming, rotation of crops with positive below-ground interactions and integrated nutrient/pest management (Venkateswarlu, 2016). Rapid urbanization coupled with industrialization and contamination in developing countries and changing precipitation patterns are potential current threats that have not been properly analyzed in the available studies.

Temperate broadleaf and mixed forests

Deforestation, agricultural intensification and loss of soil organic matter are not as extensive in temperate forests as in tropical and subtropical regions. Deforestation reduces invertebrate abundance and diversity (Ma and Yin, 2019) with decreases in termite (Thakur, 2016) and earthworm populations (Bhadauria *et al.*, 2016). Crustaceans (for example, Talitridae and *Ligidium japonicum*) are less likely to persist in the absence of *Cryptomeria japonica* forests in Japan (Ohta *et al.*, 2015). Loss of inland forests results in a greater magnitude of decline in generic richness of nematodes than coastal forests (Kitagami *et al.*, 2018). The region is witnessing agricultural intensification in accessible areas and abandonment in remote areas. Bacterial and fungal communities (Zhang *et al.*, 2018) appear to be more resilient to intensification, abandonment and deforestation/forest degradation than are macrofauna (Bhadauria *et al.*, 2012) and mesofauna (Miura *et al.*, 2008). Urbanization has caused soil biodiversity loss over large areas, but urban parks can conserve substantial biodiversity (Song *et al.*, 2015; Wang *et al.*, 2018). Increase in salinity in coastal areas due to sea level rise/climate change will reduce diversity of both macro- and microfauna (Wu *et al.*, 2015).

Desert and dry shrublands

This ecoregion suffered extensive degradation of natural vegetation in the past and is inherently poor in soil biodiversity. At present loss of natural vegetation is balanced by afforestation and, with the development of irrigation facilities, degraded lands are being increasingly restored (artificial oases). Agrochemical inputs are quite low and intensification of organic production is leading to recovery in soil biodiversity along with vegetation cover and enhancement of soil carbon stocks. Studies tracking changes in biodiversity following deforestation are lacking but some insights can be seen from tree planting. Planting of the shrub *Haloxylon ammodendron* increased the diversity of predators, collembola and fungi, decreased the diversity of insect herbivores and oribatida and did not change bacterial diversity, while *Populus gunsuensis* planting increased the diversity of virtually all groups in north-west China (Li *et al.*, 2018). A switch over to wheat-maize intercropping from wheat and maize monocultures is leading to more efficient biocontrol (Liu *et al.*, 2018). In India, mixed planting of leguminous and non-leguminous trees resulted in an 8 to 65-fold increase depending on season and taxon (Tripathi *et al.*, 2009) and organic cowpea/maize/Lucerne mixed cropping a 7-26 percent increase in collembola, oribatid mites, nematodes, actinomycetes and fungi populations (Roy *et al.*, 2012). The ecoregion is also facing increasing urbanization and climate change but studies evaluating them are lacking.

Table 4.3.2.1 | Threats to soil biodiversity in Asian Ecoregions

The main threats to soil biodiversity and function, and the level of scientific knowledge of the impacts of these threats on soil biodiversity in each of the ecoregions of Asia

Ecoregion	Main threats	Knowledge level
Tropical and subtropical forest	Deforestation	High
	Agricultural intensification	High
	Loss of SOM and SOC	Fair
Tropical and subtropical grassland	Deforestation	High
	Loss of SOM and SOC	Fair
	Soil compaction and sealing	Low
	Fire	Low
	Erosion and landslides	Low
Montane grassland and shrubland	Deforestation	Low
	Agricultural intensification	Low
	Loss of SOM and SOC	Low
Desert and dry shrubland	Loss of SOM and SOC	Low
	Salinization and Sodification	Low
	Land Degradation	Low
	Fire	Low
	Erosion and landslides	Low
	Climate change	Low
Temperate broadleaf and mixed forest	Urbanization	Low
	Agricultural intensification	Fair
	Loss of SOM and SOC	Fair
Temperate grassland	Climate change	Low

Montane grasslands and shrublands

This ecoregion suffered extensive degradation because of overgrazing in the past, and while currently degradation is outweighed by restoration, the region is facing perhaps the highest rate of global warming and expansion of woody cover. Studies in this region are few and confined largely to China and Mongolia. Degradation has impacted beta diversity more than alpha diversity and composition more than richness of bacterial communities. Prevalence of many populations associated with human diseases is a major threat to both livestock and humans (Zhou *et al.*, 2019). Tree (*Pinus tabuliformis*) planting reduced diversity of fungi but had insignificant impact on bacterial diversity and favoured ECM fungi at the expense of decline in biotrophic fungi (Wang *et al.*, 2019). Tree planting alone thus may not be an effective way of restoring soil biodiversity. Climate change may have dramatic impacts on soil biodiversity and its functions evident from earthworm (*Pheretima aspergillum*) invasion and soil organic carbon depletion around Zoige peatlands in China (Wu *et al.*, 2017). Threats of urbanization and agrochemical based intensification are considered quite low.

4.3.3 | EUROPE

Within the European Union (EU), pressures on soils that affect their function have been highlighted by the EU Soil Thematic Strategy (EC, 2006). The Strategy, adopted by all Institutions and Member States explained why further action is needed to ensure a high level of soil protection. To this end, the European Commission highlighted soil erosion, organic matter decline, compaction, salinization and landslides as issues that should be reduced, together with preventing further contamination and limiting or mitigating the effects of sealing, for instance by rehabilitating brownfield sites. Reviews of threats to soil for Europe are provided every five years by the European Environment Agency as part of the State of the Environment Report (SOER), the JRC's State of Soil in Europe (JRC, 2010) and the Status of the World's Soil Resources (FAO and ITPS 2015).

There are several studies assessing soil threats (such as erosion, compaction, pollution, land degradation) at the European level, but indicators related to soil biodiversity are rarely measured at an appropriate scale or resolution. The main sources available are The European Atlas of Soil Biodiversity (Jeffery *et al.*, 2010) and the Global Soil Biodiversity Atlas (Orgiazzi *et al.*, 2016b) that contain maps with rather coarse information on soil biodiversity. Rutgers *et al.* (2019) predicted soil biodiversity at the scale of Europe, using data for soil biological (earthworms and bacteria) and chemical (pH, soil organic matter and nutrient content) attributes in a soil biodiversity model. Aksoy *et al.* (2017) also made an assessment of soil biodiversity potential in Europe, showing that the main threats to soil biodiversity are soil degradation, land use management and human practices, climate change, chemical pollution as well as genetically modified organisms (GMOs) and invasive species. Gardi *et al.* (2013) and Orgiazzi *et al.* (2016a) further specified habitat fragmentation, intensive human exploitation, soil organic matter decline, soil compaction, soil erosion, soil sealing and soil salinization as important threats. European soils are a widely used resource, submitted to a number of relatively well identified threats (ENVASSO, 2008), and therefore soil biodiversity can be threatened by all previously mentioned processes in Europe. The potential importance of several of these threats to soil biodiversity as defined by a group of experts are presented in Figure 4.3.3.1.

Agricultural land management is one of the most significant anthropogenic activities that greatly alters soil characteristics, including physical, chemical and biological properties (Jangid *et al.*, 2008; Garcia-Orenes *et al.*, 2010). This is particularly relevant in Mediterranean environments, that take up a great part of southern Europe, where unsuitable land management together with climatic constraints (scarce and irregular rainfall and frequent drought periods) can contribute to increased rates of erosion and other soil degradation processes of agricultural land (Caravaca *et al.*, 2002). These conditions can lead to a loss in soil fertility and a reduction in the abundance and diversity of soil organisms. More than 45 percent of Europe's land is used for agricultural production (EUROSTAT, 2019) and 12.7 percent of European arable lands have soil loss >5 t/ha annually requiring protection. Panagos *et al.* (2015) estimated the mean soil loss rate in European Union as 2.46 t/ha annually and the monetary loss for agriculture in Europe due to soil erosion is about 1.25 billion Euros per year (Panagos *et al.*, 2017). Hence, agricultural management in Europe is one of the most important threats to soil biodiversity (Figure 4.3.3.1 and Table 4.3.3.1).

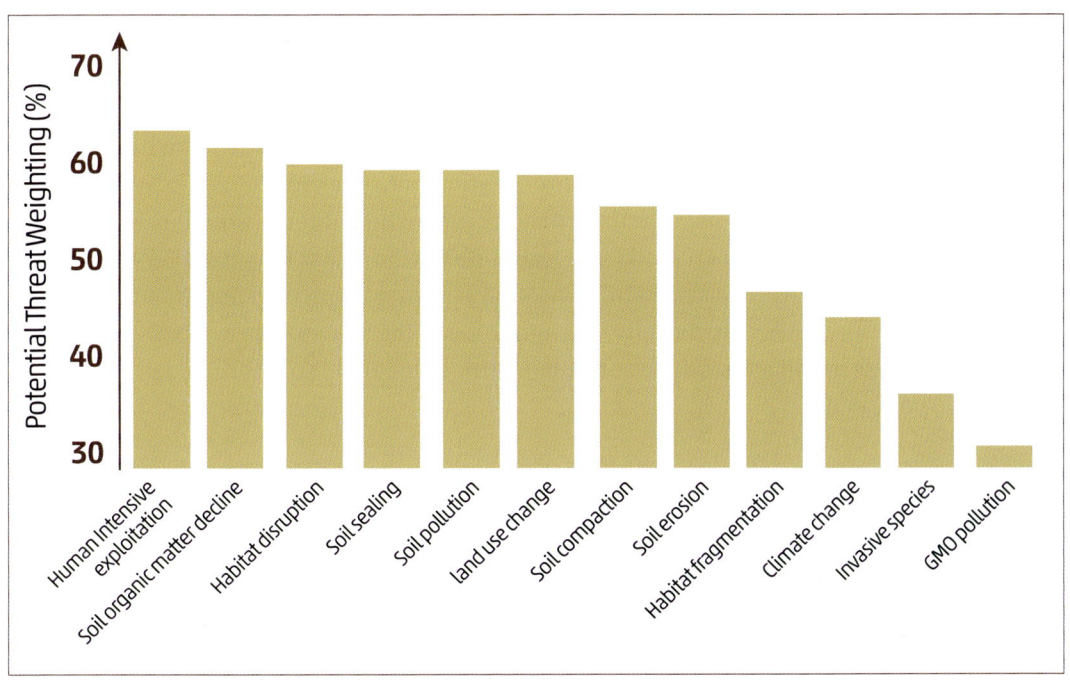

Figure 4.3.3.1 | Importance of threats to soil biodiversity in Europe

The potential threat weighting given by specialists to a selection of soil threats to soil biodiversity in Europe (after Jefferey *et al.*, 2010).

Salinization in Europe affects an estimated area of several millions of hectares (4 dS m^{-1} is the threshold to define saline soils), and has consequences not only for crop productivity, but also for soil organisms (Jeffery *et al.*, 2010).

The loss of soil organic matter, is especially relevant in Mediterranean soils with semi-arid climate (Novara *et al.*, 2011; Laudicina *et al.*, 2015). According to Turbe *et al.*, (2010) the largest emissions of CO_2 from soils are resulting from land use change (for example, from grassland to agricultural fields) and the related drainage of organic soils in Europe. However, precise future estimations are difficult to extract from the literature, given the number of uncertainties, including the dynamic trends in land-use change in Europe. Given the political importance of the management of soils for carbon storage, some recent works have estimated the potential for agricultural soils to sequester more carbon through changes in management, and this has been recently considered in the context of different biological strategies for C sequestration (Woodward, 2009).

As a consequence of human practices many land areas of Europe have suffered forest fires. After the fires, the post-fire management of the burned soil can be a key to promote the recovery of soil biodiversity recovery. Different studies have shown that there are several post-fire management actions, such as salvage logging, that is a common practice in most fire-affected areas in Europe, that can retard soil biodiversity recovery compared with other types of management (Garcia-Orenes *et al.*, 2017; Pereg *et al.*, 2018).

Climate change is probably one of the main environmental problems facing the world, causing major known and unknown effects on all ecosystems in our planet. In this sense, there is an important knowledge gap about the impact of the storage and release of greenhouse emissions on soil biodiversity (Vries and Griffiths, 2018). Several studies carried out in the 1990s found that plant growth and below-ground allocation of C, particularly of rhizodeposits, increased under elevated CO_2 levels, which had consequences for microbial biomass and respiration rates (Zak *et al.*, 1993; Newton *et al.*, 1995). These authors hypothesized that the proportion of fungi would increase under elevated CO_2 because of increased plant litter production. It has long been recognized that soil moisture, that it has been influenced by climate change is an important driver of the composition and activity of soil communities, and the first studies to assess the effect of fluctuations in soil moisture on soil communities did not do so from a global climate change perspective.

Several studies reported since the 1990s have shown context dependent effects, highlighting the need to understand the role of how different soil and vegetation types drive soil biodiversity response to climate change. While it might be hard to see consistent patterns, some generalities are starting to appear. In particular, microbial groups and bacteria taxa that are associated with oligotrophic or K-strategist life-history strategies seem to be consistently increasing in abundance under drought and increasing temperatures, while they decrease in response to elevated CO_2. In contrast, under pulse disturbances such as drought followed by rewetting, the more copiotrophic or r-strategist groups, with high maximum growth rates, are able to rapidly regain their abundance (De Vries and Shade, 2013).

Biological invasions and introduced exotic species are becoming a problem in the invaded areas because they develop excessive abundance over native species. An overview of invasive species in Europe can be found on the DAISIE European Invasive Alien Species Gateway (http://www.europealiens. org), where the current estimate is that approximately 11 000 species are invasive in Europe. For the EU, knowledge on the distribution of regulated invasive alien species is collected through the European Alien Species Information Network. The potential impact of exotic invasive species can be particularly worrying in rare ecosystems and there is a critical knowledge gap concerning impacts of invasive species on soil biodiversity. Considering the immense biodiversity of organisms present in one gram of soil, it is irrelevant to simply describe how invasive species influence the total numbers of soil organism species. It is more insightful to consider what sort of species exotic invaders influence and what the functions of those species are. In this sense The European Atlas of Soil Biodiversity discusses different concrete examples of the effect of exotic species on soil biodiversity.

Table 4.3.3.1 | Threats to soil biodiversity in European Ecoregions

The main threats to soil biodiversity and function in each of the ecoregions of Europe

Ecoregion	Main threats
Temperate and boreal coniferous forest	Urbanization Contamination Land Degradation Invasive species
Temperate broadleaf and mixed forest	Deforestation Urbanization Contamination Land degradation Invasive species
Temperate grassland	Contamination Land Degradation
Mediterranean forest, woodland and shrubland	Urbanization Contamination Land Degradation
Desert and dry shrubland	Deforestation
Tundra	Contamination
Boreal Forests/Taiga	Deforestation

4.3.4 | LATIN AMERICA AND THE CARIBBEAN

Temperate and boreal coniferous forest

In Latin America and the Caribbean (LAC) these ecoregions are located in Argentina and Chile in a narrow region along the Andes, ranging between the 37th and the 52nd parallels south, and in Mexico they are known as the "bosques de Oyamel y de Pinus". In the former countries, the weather is temperate to cold humid, with rainy winters.

Coniferous forest are dominated *Araucaria araucana*, *Austrocedrus chilensis* and *Fitzroya cupressoides*. In general, these species constitute mixed forests with species from the family Nothofagaceae. These trees have small, needle- or scale-like, acidic leaves. The main threat to soil biodiversity in this region are deforestation and fires, which cause great changes in vegetation composition in big areas (Table 4.3.4.1). Furthermore, agricultural intensification and urbanization has increased in the past years and are therefore important threats to soil biodiversity in coniferous forests.

Temperate broadleaf and mixed forest

These forests are located in LAC in part of Argentina, Chile and Mexico. In Argentina and Chile, temperate broadleaf forests occur in the Chaco and Pampean provinces (Chacoan dominion) and in the Maule and Valdivian Forest provinces (Andean Region) (Arana, 2017). In Mexico, they occur in the mountain regions of the Sierra Madre Oriental and Occidental, and consist mainly of pine-oak forests. The weather is temperate with distinct warm and cool seasons and variable precipitation. The input of C to the soil from the forest (as leaf litter or dead wood) allows high abundances of decomposers. The main threats to soil biodiversity in these forests in LAC are related to very high rates of (historical and current) deforestation, in part due to wood extraction and the expansion of agricultural frontiers that are leading now to the agricultural intensification. These activities usually co-occur with fire, which is another of the main threats to soil biodiversity. The concomitant loss of SOC and soil compaction are also important threats to soil biodiversity. As a consequence, a great reduction in surface covered with broadleaf forests has occurred.

Temperate grassland

The main temperate grasslands in LAC are located in the Pampas of Argentina, a wide plain with more than 52 M ha in extension. Rains range from 1200 mm in the northeast to 400 mm in the southwest and are concentrated mainly from late spring to early autumn, with dry winters. Annual mean temperature is about 16 ºC with warm summers and severe winters. Temperate grasslands support high levels of soil microbial and faunal diversity but the impact of current threats is reaching worrisome levels. Many of the grasslands have been replaced with annual crops, strongly reducing the original grassland cover. High rates of deforestation have occurred to enable the expansion of cropland frontier. This phenomenon, together with agricultural intensification associated with soybean monocropping without rotations and high rates of pesticide use, have caused important losses of SOM, soil compaction and sealing and soil contamination. All these processes are threatening soil biodiversity and jeopardizing soil ecosystem processes relying on soil biology.

Montane grassland and shrubland

High altitude grassland and shrubland biomes are mainly associated to the Andes Mountains in South America, which include the ecosystems known as Páramo (Ecuador, Columbia, Peru, Venezuela), Puna (Argentina, Chile, Bolivia and Peru), Estepa (in southern Argentina and Chile). In North and Central America (Mexico, Guatemala), they are known as Zacatonales. Unique climatic conditions occurring at high altitude, including elevated radiation and extreme temperature fluctuation, among others, provide unique niches resulting in a high degree of endemism. The main threats of these ecosystems are climate change, erosion and landslides, land degradation, loss of SOC and contamination, all of which affect not only soil biodiversity but also their ecosystem functioning and services. The latter is of particular interest for these ecosystems, as soil microbiota adapted to harsh environmental conditions important for biotechnology, human health and agricultural adaptation to climate change, among others.

Tropical and subtropical grasslands, savannas and shrublands

Tropical and subtropical grasslands, savannas and shrublands tend to receive from 900 to 1 500 mm yearly rainfall, have prolonged dry seasons conditions, and often frequent burning, that do not allow development of extensive tree cover. They include the large area of Cerrado in Brazil and Bolivia, as well as the Paraguayan and Argentinean Chaco, grasslands in the Pampa and Llanos of Venezuela and Colombia, and seasonally flooded regions along the Pacific and Gulf coasts of Mexico. They are well known for their complexity of habitats and unusually high levels of endemism and beta diversity. Many of these regions have been extensively occupied for agricultural and pastoral uses, as well as for mining, greatly reducing original cover of this ecoregion, as well as impacting above and below-ground biodiversity. The greatest threats to soil biodiversity and function in this ecoregion are agricultural intensification and associated erosion, invasive species and climate change.

Mediterranean forest, woodland and shrubland

Mediterranean forest, woodland and shrubland occur in part of Mexico (Baja California Peninsula) and central Chile, in areas under climatic conditions characterized by rainfall and droughts concentrated during winter and summer, respectively. Due to geographical isolation, these regions have a remarkable level of biodiversity and endemism, and as such are considered hotspots for global biodiversity. Mediterranean ecosystems in LAC are highly susceptible to human interventions such as urbanization and agricultural intensification. Moreover, the effects of climate change, land burning and erosion processes also represent important pressures to soil biodiversity in these ecosystems, at all trophic levels.

Tropical and subtropical moist and dry forests

Tropical and subtropical moist and dry forests are characterized by low variability in annual temperature and can be found from Argentina to Mexico, but with the vast majority of remaining forest in the Amazonian basin. Both forests tend to receive 1500 mm or more rain per year but moist forests have mostly semi-evergreen and evergreen deciduous tree species, while dry forests tend to have a prolonged dry season and mostly deciduous trees that lose their leaves in the dry season. Both forests are well known for their high alpha and beta biodiversity, and may be home to half of all species on the planet. They are highly sensitive to burning and deforestation, mainly for agriculture and pastoral land use, which have already greatly reduced original forest cover in LAC, impacting local and regional climate patterns, and having major impacts on both above and below-ground biodiversity both locally and regionally.

Table 4.3.4.1

The main threats to soil biodiversity and function, and the level of scientific knowledge of the impacts of these threats on soil biodiversity in each of the ecoregions of Latin America and the Caribbean

Ecoregion	Main threats	Knowledge level
Tropical and subtropical forest	Deforestation	Fair
	Agricultural intensification	Fair
	Contamination	High
	Fire	Fair
	Erosion and landslides	Fair
	Climate change	Fair
Tropical and subtropical grassland	Deforestation	Fair
	Agricultural intensification	Fair
	Loss of SOM and SOC	Fair
	Soil compaction and sealing	Low
	Contamination	Fair
	Fire	Fair
	Erosion and landslides	Low
	Climate change	Low
	Invasive species	Low
Mediterranean forest, woodland and shrubland	Deforestation	Low
	Urbanization	Low
	Agricultural intensification	Low
	Loss of SOM and SOC	Low
	Land Degradation	Low
	Fire	Low
	Erosion and landslides	Low
	Climate change	Low
	Invasive species	Low
Montane grassland and shrubland	Loss of SOM and SOC	Low
	Land Degradation	Low
	Erosion and landslides	Low
	Climate change	Low
Desert and dry shrubland	Contamination	Low
	Land Degradation	Low
	Erosion and landslides	Low
	Climate change	Low
Temperate and boreal coniferous forest	Deforestation	Low
	Agricultural intensification	Low
	Loss of SOM and SOC	Low
	Fire	Low

Temperate broadleaf and mixed forest	Deforestation	Low
	Agricultural intensification	Low
	Loss of SOM and SOC	Low
	Soil compaction and sealing	Low
	Fire	Low
	Erosion and landslides	Low
Temperate grassland	Deforestation	Fair
	Agricultural intensification	Fair
	Loss of SOM and SOC	Fair
	Soil compaction and sealing	Fair

Desert and dry shrubland

These ecoregions are present in part of Mexico, northern Venezuela, northeast Brazil, central Argentina and along the Pacific coast in Peru and northern Chile. Due to natural climatic conditions of these habitats, they have rather low population densities as compared to other ecosystems; however, environmental pressures due to human intensification is significantly growing. The main threats to soil biodiversity to these habitats in LAC are climate change, land degradation, pollution, salinization and sodification and erosion and landslides. Although low or lack of rainfall is a distinctive feature of these ecoregions, counterintuitively, water soil erosion could be of special significance in particular zones of the hyper-arid Peruvian and the Atacama Desert in Chile, where intense rains from the "invierno altiplánico" occurring at high altitude or derived from the El Niño events originate considerable floods that dramatically affect bare and dry soils. Thus, soil microorganisms adapted to these environments such as cyanobacteria, lichens and a range of extremophiles, among others, are highly affected.

4.3.5 | NORTH AMERICA

Currently, threats to soil biodiversity associated with the direct and indirect effects of climate change represent the largest threat to North American soils (Table 4.3.5.1).

Boreal Forests

Boreal forests are the most widespread ecoregion in North America and because the rates of deforestation have remained relatively unchanged over the past several decades (Alig *et al.*, 2003), threats to soils are largely due to climate change, although land conversion remains an important factor (Dyk *et al.*, 2015). Changes in fire intensity and frequency have resulted in rapid and recent net carbon loss from soils (Walker *et al.*, 2019). Such losses compound the effects of climate change by increasing the susceptibility of young forests to fire and further soil carbon loss. Additional risks include gas and oil expansion, which includes urbanization, land conversion and contamination (Yeung *et al.*, 2019).

Broadleaf forests

Change in vegetation cover through insect defoliation is a major concern, particularly for bark beetles in both eastern deciduous and western coniferous forests. The combined effect of these agents is expected to grow due to changes in historical climate no longer limiting establishment of invasive species (Potter *et al.*, 2018). Fire in these systems is increasing, particularly for continental forests, which have had steadily increasing fire incidence since 2000 (Potter, 2018).

Grasslands

Because most of the land in the temperate grassland regions is privately owned, conservation of these systems is challenging. More than 80 percent of North American temperate grasslands have been converted to agriculture since European settlement (Glasser *et al.*, 2012). Current threats to grassland soils include agricultural intensification, including a resurgence of monocropping (Wang *et al.*, 2019) and increased biocide use (FAOSTAT, http://www.fao.org/faostat/). Grassland soils outside of agriculture are threatened by increasing land conversion chiefly due to urbanization and oil and gas development (Schaeffer and DeLong, 2019).

Mediterranean shrublands

North American chaparral, although one of the smallest ecoregions in North America, is a biodiversity hotspot. Chaparal soils are increasingly threatened by shorter fire intervals associated with climate change (Sypard *et al.*, 2019). Short fire cycles lead to conversion of shrublands by invasive herbaceous plants (Park *et al.*, 2018). Increasing land conversion due to urbanization is also a growing concern as the region overlaps with one of the most densely populated areas of the continent.

Deserts

The deserts of North America comprise both warm and cool deserts and both are affected by altered fire regimes and subsequent encroachment of invasive species. More frequent and severe fires, accompanied with altered moisture regimes means that native plants are being replaced both by woody vegetation (Juniper in the Great Basin; Davies and Bates, 2017), or invasive grasses (cheat grass).

Tundra

The biggest threat to soil biodiversity in the tundra is the loss of soil organic carbon due to climate change (Plaza *et al.*, 2019). This region has experienced the most rapid warming, leading to earlier phenology, warmer soil temperatures and changes in vegetation (Myers-Smith *et al.*, 2019). Even with conservative increases in global air temperature, increases in soil temperature will lead to accelerated losses of SOM (Biksaboom *et al.*, 2019).

Table 4.3.5.1

The main threats to soil biodiversity and function, and the level of scientific knowledge of the impacts of these threats on soil biodiversity in each of the ecoregions of Canada and the United States of America

Ecoregion	Main threats	Knowledge level
Temperate and boreal coniferous forest	Fire	High
	Climate change	High
	Invasive species	Fair
Temperate broadleaf and mixed forest	Climate change	Low
	Invasive species	Fair
Temperate grassland	Urbanization	High
	Agricultural intensification	High
	Salinization and sodification	High
	Erosion and landslides	Fair
	Climate change	Low
	Invasive species	Fair
Mediterranean forest, woodland and shrubland	Urbanization	High
	Fire	High
	Climate change	High
	Invasive species	High
Desert and dry shrubland	Deforestation	Low
	Urbanization	Low
	Agricultural intensification	Low
	Salinization and sodification	Low
	Land Degradation	Low
	Fire	Low
	Erosion and landslides	Low
	Climate change	Low
	Invasive species	High
Tundra	Loss of SOM and SOC	High
	Climate change	Fair
	Invasive species	None

4.3.6 | SOUTH WEST PACIFIC

The South West Pacific region includes the 22 island nations of the Pacific, New Zealand and Australia (*Status of the World's Resources, Main Report*; Chapter 15: Regional assessment of soil change in the Southwest Pacific). The soils of this region are diverse and cover a wide breadth of latitudes and altitudes. These soils include highly weathered soils in humid tropical areas and continental Australia and relatively young volcanic soils of Indonesia, Papua New Guinea, New Caledonia, Norfolk Island and New Zealand. The main ecoregions and associated threats are listed in Table 4.3.6.1 Some of these threats are interactive as noted.

Agricultural intensification was mentioned in all the ecoregions as a major threat to soil biodiversity. Invasive species was also considered important in all but one ecoregion, and deforestation was considered a major threat in all the forested ecoregions (Table 4.3.6.1). The reduction in threats associated with land clearing can be promoted by the protection of remaining forests (old growth), the replanting of new forests and the ongoing assessment of growth stage (SOFR, 2018). Old growth forests have significant habitat, nature conservation and aesthetic value and contribute to C storage and water production. According to SOFR (2018), there has been no national survey of old growth forests since 1995-2000 when it was estimated that they made up 5M ha of a total survey area of 19M ha (excluding Northern Australia; Qld and NT). Australia claims that it has met target 11 of the Aichi Biodiversity targets about preserving Australia's native forest (SOFR, 2018). Partial information is available for 60 percent of Australia's forest dwelling vertebrate fauna and vascular plants, but the report (SOFR, 2018) indicates that there are 'no comprehensive lists of invertebrate fauna, non-vascular flora (including algae, liverworts and mosses, fungi and lichens) or microorganisms that occur in forests, even though these species play key roles in ecological processes'.

Table 4.3.6.1

The main threats to soil biodiversity and function, and the level of scientific knowledge of the impacts of these threats on soil biodiversity in each of the ecoregions of the South Western Pacific

Ecoregion	Main threats	Knowledge level
Tropical and subtropical forests	Deforestation	Low
	Agricultural intensification	Low
	Fire	None
	Erosion	None
	Climate change	Low
	Invasive species	Low
Tropical and subtropical grasslands	Deforestation	None
	Urbanization	None
	Agricultural intensification	None
	Fire	None
	Erosion and landslides	None
	Climate change	None
	Invasive species	Low
Temperate broadleaf and mixed forest	Deforestation	Low
	Urbanization	None
	Agricultural intensification	None
	Loss of SOM and SOC	Low
	Land Degradation	None
	Invasive species	Low
Temperate grasslands	Agricultural intensification	None
	Loss of SOM and SOC	Low
	Land Degradation	Low
	Invasive species	Low

Mediterranean Forest, Woodland and Shrubland	Deforestation	Low
	Urbanization	Low
	Agricultural intensification	Low
	Loss of SOM and SOC	Low
	Soil acidification	Low
	Land Degradation	None
	Fire	None
	Invasive species	Low
Montane grassland and shrubland	Agricultural intensification	Low
	Loss of SOM and SOC	Low
	Invasive species	None
Desert and dry shrubland	Agricultural intensification	None
	Loss of SOM and SOC	Low
	Fire	Low

Tropical and subtropical forests

These forests occur in large, discontinuous patches on the equatorial belt and between the Tropics of Cancer and Capricorn made up of small fragmented coastal areas in Queensland and patches on Lord Howe, Norfolk Islands, the North and South Cook Islands, New Caledonia, Vanuatu, Fiji, Tonga, Samoa and associated small island chains (Somerville *et al.*, 2017). Characterized by low variability in annual temperature and high levels of rainfall (>200 centimetres annually) these forests are dominated by semi-evergreen and evergreen deciduous trees and predominantly eucalyptus and acacia species (SOFR, 2018). Land clearing (deforestation) for agriculture is considered the most important threat. More than 80 percent of the 1.2 million ha cleared in Australia between 1991 and 1995 was on the coastal periphery in Queensland (Bradshaw, 2012) and while National Parks protect the high diversity of plants and animals in this small ecoregion, the quality of the biodiversity has been significantly impacted by land-clearing for agriculture and mining, introduced pests and diseases and other anthropogenic sources (Sommerville *et al.*, 2018). Significant modification of the surrounding tropical and subtropical grassland ecoregion associated with land clearing of the Burdekin River catchment for cereal and sugar cane production pose significant threats to the quality of this ecoregion and its soil biodiversity. Related to these threats are the interactive threats of erosion and landslides and fire, exacerbated by climate change.

Tropical and subtropical grasslands

This region is described by rainfall levels between 90-150 centimetres per year. In Australia, this is one of the four dominant ecoregions, covering more than 30 percent of the continent and occurring mainly in Queensland (approximately 70 percent), NT (approximately 50 percent) and WA (approximately 20 percent). Although northern NSW coastal regions and an extensive area in SE Qld was cleared between 1950-1980 (Mackenzie *et al.*, 2017), land clearing related to agricultural intensification mainly for grazing and pasture land-uses in Queensland with some cropping and urbanization proceeds at an accelerated rate post 1980 (Bradshaw, 2012; Mackenzie *et al.*, 2017).

One year of increased land-clearing in Queensland has already removed many more trees than will be planted during the entire AUD 50 million Australian Government 20-million trees program. Further, under 'Caring for our Country' and Biodiversity Fund grants, tree planting to restore habitat across Australia since 2013 was just over 42, 000 ha while 296 000 ha was cleared in Queensland alone in 2013-2014 (Australian Government, 2017b). Land clearing for agricultural intensification has been associated with a significant risk of hillslope erosion particularly impacting marine water quality (MacKenzie *et al.*, 2017). A closer examination of this ecoregion by NRM regions within it identifies the Burdekin NRM region having the highest hillslope erosion rates in Australia (Teng *et al.*, 2016), Cape York with emerging erosion with land development (Olley *et al.*, 2013), and Qld Mackay-Whitsunday and Qld SW both having moderate to high rates of erosion due to large areas of bare soil attributed to cropping for sugarcane, overgrazing and the Millennial drought (MacKenzie *et al.*, 2017). Adding to this, vegetation clearance for new banana enterprises in the southern Cape York NRM region in the last decade was identified as a significant factor in water erosion of soil (SoE 2016, 2018). The impact of sediment movement primarily as the result of erosion of 31.1 million hectares of land and over 100 000 kilometres of streambank on the poor quality of the Great Barrier Reef catchments has received considerable attention (Schaffelke *et al.*, 2017). This has been attributed to erosion and resultant loss of ground cover and the adoption of best management practices that exclude cattle from gullies and maintaining ground cover to promote 'healthy soil are needed' (https://www.rccfplan.qld.gov.au/science-and-research/the-scientific-consensus-statement). Native vegetation clearance and increased grazing intensity across large parts of Northern Australia related to the arrival of drought tolerant African cattle breeds in the 1980s, and increased density of watering points were identified as causal factors of gully erosion and a major source of excess sediment in streams and estuaries (SOE, 2016; Teng *et al.*, 2016; Bartley *et al.*, 2014). Despite the availability of management recommendations to adopt grazing and cropping best practices to improve soil conditions (erosion acidity and carbon) and the quality of ecosystem services (Cork *et al.*, 2012) there is little evidence of wider adoption in susceptible areas of this ecoregion.

Climate change has also been associated with higher frequency of intense bushfires in the ecoregion (SOFR, 2018). It has been claimed that climate change is increasing the intensity of extreme weather events in Queensland – drought, bushfires, heatwaves, floods and cyclones. Currently 65 percent of Queensland is drought declared and parts of the state's west and south have been drought-affected for more than six years (https://www.climatecouncil.org.au/wp-content/uploads/2019/08/qld-report-climate-council.pdf).

Temperate broadleaf and mixed forest

This ecoregion has a moderate climate and high rainfall that gives rise to unique eucalyptus forests and open woodlands and extends across Tasmania (approximately 100 percent), Victoria (approximately 60 percent), NSW (approximately 30 percent) southern Queensland (approximately 10 percent) and eastern SA (approximately

<1 percent). It has served as a refuge for numerous plant and animal species when drier conditions prevail over most of the continent, resulting in a remarkably diverse spectrum of organisms with high levels of regional and local endemism. It includes major urban centres or capital cities of Melbourne, Sydney, Brisbane and Hobart and is therefore significantly impacted by human activities such as urbanization and land-use intensification. Land clearing of forests is historical and most of the biodiversity is preserved in National Parks (e.g.; South-west, Boodera, Wollomi and Blue Mountains being the largest) with 40 percent of Tasmania protected by national parks or state reserves. This ecoregion has significant coastal frontage with the highest level of soil loss by water erosion (MacKenzie *et al.*, 2017). In western and central Tasmania and the steep forested areas along the Great Dividing Range (mainland) greater than 25 tonnes of soil/ha/year are lost (Teng *et al.*, 2016). This erosion has been discounted due to the steepness of slopes. From this erosion data however, the greatest percentage nutrient loss for total N, Total P and SOC occurs in these coastal regions (MacKenzie *et al.*, 2017). The soil acidification risk is high particularly where agricultural activities occur in higher rainfall areas and this threatens long-term agricultural viability especially in the SW and NE regions of Victoria and SE regions of NSW where the gross value of production is highest (MacKenzie *et al.*, 2017). The pH of soils especially on coastal fringes are highly acid with pH <4.8 and further inland between 4.8 and 5.5 in others. Soil carbon stocks are declining under current land-use in Tasmania and the Northern reaches of NSW and into Southern Queensland and in other regions it remains steady (MacKenzie *et al.*, 2017).

Temperate grasslands

This ecoregion has cooler and wider annual temperatures and extends from northern Victoria (approximately 20 percent) into NSW (approximately 55 percent) and southern Queensland (approximately 25 percent). These regions are devoid of trees, except for riparian or gallery forests associated with streams and rivers. Positioned between temperate forests and the arid interior of Australia, the southeast Australian temperate savannas span a broad north-south swatch across Queensland, New South Wales and Victoria. Australia's most significant river system, the Murray Darling river system catchment occurs mainly in this region. A combination of drought (attributed to climate change) and water mismanagement has been reported as the cause of mass fish kills reported in December 2019 (Australian Academy of Science, 2019). Most of this ecoregion supports agricultural enterprises such as sheep breeding and grazing and wheat cropping and only small fragments of the original eucalypt vegetation remains. Soil pH data is between 4.8 and >5.5 and soil acidification risk is considered low to medium despite there being insufficient data for modelling and soil Carbon stocks are declining under current management (MacKenzie *et al.*, 2017).

Mediterranean Forest, Woodland and Shrubland

Mediterranean ecoregions are characterized by hot and dry summers, while winters tend to be cool and moist. In Australia, this ecoregion is significant in the southern states of Victoria (approximately 30 percent), central and western NSW (approximately 20 percent), southern SA (approximately 25 percent) and south eastern WA (approximately 25 percent) and represents the most significant cereal production regions in Australia. In the south eastern corner of WA, there is one of the world's 34 internationally recognized biodiversity hotspots with more than 1500 endemic species of plants and only 30 percent of its original habitat (Williams *et al.*, 2011). In SA, the significant wine growing regions (such as The Barossa Valley, McLarenVale). Alongside this hotspot there is significant and expanding intensive cereal cropping region with widespread soil acidification, particularly in the subsoil which threatens long term business viability in these regions if left untreated. In regions of SA, soil acidification risk is also high with pH_{Ca} values of <4.8 in much of the region. Land clearing is largely historical (between 1920-1950; SOFR, 2018) in the SE WA region however since the 1980s this region is extending on the northern and southern margins. All areas in this region have the capacity to store carbon through best crop management strategies except in central Victoria (MacKenzie *et al.*, 2017).

Montane grassland and shrubland (mostly New Zealand)

This ecoregion includes high elevation (montane and alpine) grasslands and shrublands. In Australia montane grassland and shrublands are restricted to the montane regions of south-eastern Australia above 1300 metres. This region occupies less than 3 percent of the Australian landmass and straddles the borders of the Australian Capital Territory, Victoria and New South Wales on the Australian mainland, as well as a significant element in Tasmania. There is very little soils data available for this region however it falls within the zone of highly acid soils where the pH_{Ca} is < 4.8 (MacKenzie *et al.*, 2017). In contrast, this represents a major ecoregion in New Zealand, especially the central region of the south island.

Desert and dry shrubland

These ecoregions vary greatly in the amount of annual rainfall they receive; generally, however, evaporation exceeds rainfall in these ecoregions. In Australia, they represents the largest region covering more than 70 percent of SA, 65 percent of WA, 50 percent of the NT, 25 percent and 10 percent in western central Qld and north western NSW respectively. Land clearing due to mining and exotic weed species incursions are the greatest threats (NT Landcare, 2019) together with climate change and drought and associated wind erosion and fire. According to MacKenzie *et al.* (2017), the bare soil index over a 16-year period is high and the region is prone to significant wind erosion events. The bare soil index developed in Australia to identify the risk of erosion by wind, calculates the proportion of each year when bare ground is equal to or greater than 50 percent. For this ecoregion, a significant area shows fractional cover (or <50 percent) for the entire year. The influence of drought on vegetation cover coupled with the effects of management such as the reduction in stock as dry weather persists, are key in reducing this threat of soil loss. As this ecoregion is not considered agricultural, there is little data on soil conditions.

4.4 | GLOBAL SYNTHESIS

Using the information provided on the main threats in the ten ecoregions present in the six world regions (Table of threats for Sub-Saharan Africa, South West Pacific, Latin America and Caribbean, North Africa and Near East, North America and Europe; FAO and ITPS, 2015), a summary table of the main threats common to these ecoregions was produced (Table 4.4.1). This exercise showed that the most widespread threat to soil biodiversity in the world was the loss of SOM and SOC, and that this could be associated with other threats such as deforestation and agricultural intensification (both linked with land use change) and with climate change (particularly in tundra). This clearly shows the importance of sustainable management and conservation practices, to maintain this resource in soils, which represents one of the bases for the soil's food webs. Deforestation and agricultural intensification were also major threats worldwide, being important in tropical and temperate broadleaf and mixed forests and temperate and montane grasslands and boreal forests/taiga, although the level of available information on the topic was highly variable, depending on the particular world regions where these ecoregions occur. The ecoregions with the highest number of threats were the deserts and dry shrublands, the tropical and subtropical grasslands, and the temperate broadleaf and mixed forests. Invasive species also represented an important threat, particularly in Mediterranean and temperate forests and tundra.

Table 4.4.1 | Threats to soil biodiversity in global Ecoregions

The main threats to soil biodiversity and function in the world's ecoregions

Ecoregion	Main threats
Tropical and subtropical forest	Deforestation Agricultural intensification
Tropical and subtropical grassland	Deforestation Loss of SOM and SOC Soil compaction and sealing Fire Erosion and landslides
Mediterranean forest, woodland and shrubland	Urbanization Land degradation Fire Invasive species
Montane grassland and shrubland	Agricultural intensification Loss of SOM and SOC
Desert and dry shrubland	Loss of SOM and SOC Salinization and sodification Land degradation Fire Erosion and landslides Climate change

Temperate broadleaf and mixed forest	Deforestation Urbanization Agricultural intensification Loss of SOM and SOC Invasive species
Temperate grassland	Agricultural intensification
Temperate and boreal coniferous forest	Fire Invasive species
Tundra	Loss of SOM and SOC Climate change Invasive species
Boreal Forests/Taiga	Deforestation

The only other available diagnosis of the extent of various threats to soil biodiversity was published in the Global Atlas on Soil Biodiversity (Orgiazzi *et al.*, 2016) and also used expert opinion, but this did not include all of the threats listed here. The resulting consensus map of global threats (Figure 4.4.1) was produced using data on: 1) loss of above-ground biodiversity (plant species loss) as a proxy of land use change (such as deforestation); 2) nitrogen fertilizer application (as a proxy for pollution and nutrient overloading); 3) cropland cover and cattle density (as a proxy for agricultural intensification and associated soil compaction); 4) fire density (as a proxy for risk of fires); 5) water and wind erosion vulnerability indices, to assess soil erosion risks; 6) desertification vulnerability index; and 7) global aridity index (as a proxy for climate change).

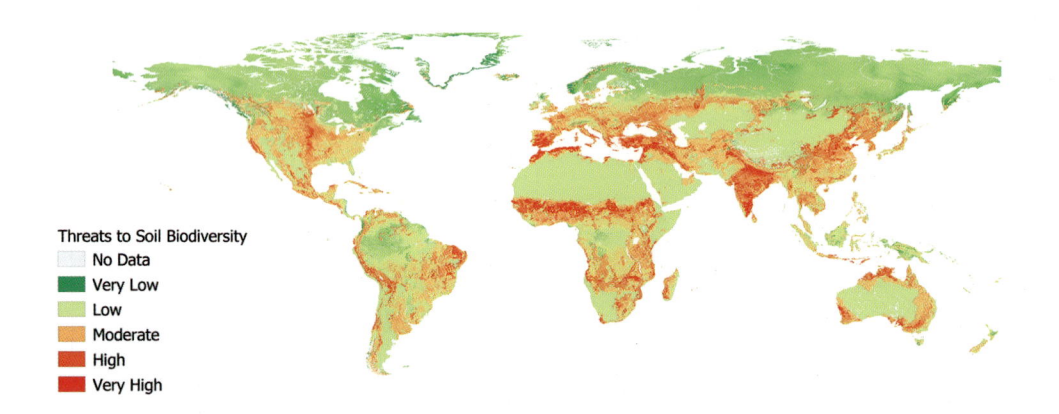

Figure 4.4.1 | Estimated levels of current potential threats to soil biodiversity worldwide

Source: "Global Soil Biodiversity Maps" associated to the Global Soil Biodiversity Atlas, European Soil Data Centre, Joint Research Centre of the European Commission. June 2016.

This map (Figure 4.4.1) should be considered as a first attempt to locate sites with important potential threats to soil biodiversity at a global scale. However, its interpretation should be made with caution, given that the actual extent of the threats to soil biodiversity can be assessed only if we know what is present in these soils; and the expert assessment done for the world's regions (above) showed that there are many places for which very little data or information is available. Clearly, further efforts are needed, taking into consideration various other sources of data that may have become available since then, and using data on other important threats identified in Table 4.4.1. Means of overcoming, or better considering, potential interactions among various threats is also an important issue to take into account in future attempts to better address threats to soil biodiversity worldwide.

Many environmental variables (such as temperature and land cover) can now be mapped and monitored for change relatively easily, using data collected by remote sensing (satellites). However, these still do not provide direct information on the state of the organisms present (diversity, populations). These must be derived from case studies performed throughout the world in the different ecoregions and include a range of taxa, with distinct functions in soils, so that the risk to soil biodiversity and function can be better assessed.

Finally, even if threats can be mapped with their extent of impacts on soil physical integrity and chemical quality as done by the FAO and ITPS (2015) in the *State of the World's Soil Resources* for various drivers (erosion, loss of SOM, soil nutrient depletion, contamination and pollution, soil acidification, salinization and sodification, waterlogging, compaction, crusting and sealing), syntheses of available data on impacts of these to soil biota (as many potential representative groups/taxa as possible) and support to obtain missing data are needed in order to produce accurate maps that reflect the true potential impacts of these threats on soil life worldwide. So far, these were generally found to be absent for most of the world regions, although efforts to map some more well-known taxa such as fungi, earthworms and nematodes worldwide have been made (Tedersoo *et al.*, 2014; Phillips *et al.*, 2019; van den Hoogen *et al.*, 2019), and these could be used as surrogates for the whole soil biota, to help produce more realistic models of impacts of threats on overall soil biodiversity. Nonetheless, even these still show important limitations in available data, particularly for tropical regions, which may limit application of the models where an important part of the world's soil biodiversity may be residing.

As globalization connects markets worldwide, source (producing countries) and sink (consuming countries) relationships may have important consequences to soil biodiversity that need further consideration. Countries with large areas devoted to intensive agriculture, or in the process of major intensification efforts to feed global markets, mostly dominated by commodities (such as sugar and soybean) may be experiencing especially important negative effects on soil biodiversity that need better attention. In addition, invasive species may become increasingly widespread as transport between countries and continents is stimulated by growing markets and improved by countries with emerging economies. These are just a few of a number of important issues that need to be addressed, particularly in international fora and agreements on climate change (IPCC) sustainable development, conservation of biodiversity (CBD) and ecosystem services (IPBES).

© Guillermo Peralta

CHAPTER 5
RESPONSES AND OPPORTUNITIES

5.1 | INTRODUCTION TO THE MANAGEMENT OF SOIL BIODIVERSITY

Given the prominent contribution of soil organisms to key terrestrial processes (Bardgett and van der Putten, 2014; Geisen *et al.*, 2019), information about the relationships between soil biodiversity and environmental variables at both local and global scales is of primary importance for the development of ecosystem-level conservation and management efforts.

While above-ground biodiversity is familiar to most people, and its protection is managed under national and global laws and regulations, there are very few comparable activities that directly focus on the protection of soil biodiversity. Besides "red lists" of mushrooms, which are the fruiting bodies of soil fungi, there is little information on endangered soil biota, let alone "red lists" of species of soil microorganisms or soil invertebrates. In contrast, there is wider knowledge about "unwanted" soil biota, including introduced invasive species such as invasive earthworms in North America, the New Zealand Flatworm in the United Kingdom of Great Britain and Northern Ireland, soilborne pathogens such as *Phytophthora cinnamomi* in Australia, and below-ground weevils and below-ground pests such as cyst, root knot and root lesion nematodes in many agricultural regions. Therefore, management of soil biodiversity will include both the maintenance, protection and restoration of wanted species, and the prevention, suppression or control of unwanted species. (Figure 5.1.1).

Protecting above-ground biodiversity also requires protection of the soil and its soil biodiversity, yet soil biodiversity is less targeted by conservationists, despite the tight linkages between below-ground and above-ground biodiversity. For example, the Millennium Ecosystem Assessment (MEA, 2005) contains very little information on soils and soil biodiversity, and the European Habitat Directives and Natura 2000 do not even mention soil biodiversity (EASAC, 2018). Despite the promotion of sustainable soil management by the Global Soil Partnership since 2012, in many cases soil management is still focused on managing soil fertility rather than on protecting soil biodiversity as a key determinant of soil health and soil-mediated ecosystem services.

There is increasing global recognition of the need to enhance soil-based multiple ecosystem services other than merely production-driven service provision. For example, besides food production, soils also need to produce clean ground water, to store organic carbon and nitrogen thus reducing greenhouse gas emissions, to store and regulate soil water availability to plants and soil organisms, and they should control or prevent the outbreak of plant, animal and human diseases (EASAC, 2018). On the one hand, these services emerge through the self-organizing nature of the soil system in which soil biological attributes (that is, soil organisms and materials derived from their activity) interact intensively with physical and chemical properties at different spatial scales (Ettema and Wardle, 2002). On the other hand, above-ground organisms can have major impacts on the spatial distribution and functional attributes of soil organisms through vegetation cover, litter inputs and plant functional traits (Wardle *et al.*, 2004; Barrios, 2007; Bardgett *et al.*, 2014).

It is increasingly acknowledged that feedback interactions between plants and soil biota drive changes in the composition and functioning in natural and managed ecosystems (Van der Putten *et al.*, 2013). Steering these feedbacks towards enhancing synergies and minimizing trade-offs in agricultural landscapes (Barrios *et al.*, 2012; Veen *et al.*, 2019) could promote nature-based innovations as part of integrative approaches supporting transitions towards sustainable agriculture that avoid, reduce and reverse land degradation (IPBES, 2018). For instance, intensive agriculture is still pursued in places where soil degradation is severe, and soil is still improperly managed in pursuit of high agricultural production. Knowledge on how the physical, chemical and biological components interact is needed in order to restore and manage the integrity and functioning of soils, and how soil biodiversity enhances soil multi-functionality.

The FAO and ITPS (2015) *Status of the World Soil Resources Report* states that 33 percent of land is "moderately to highly degraded." This calls for immediate action to restore degraded lands through the implementation of sustainable soil management, particularly in regions where fragile and degraded soils impose great constraints on soil functions and ecosystem-wide provided services. Increasing or maintaining soil biodiversity is one effective solution that can assist in soil restoration, provided that the causes of soil degradation have been solved. Strategies to enhance soil biodiversity for agriculture, land management, crop growth or ecosystem services may involve soil tillage, irrigation, organic matter addition, crop and tree species, and crop rotational and intercropping systems. These management approaches are environment-specific and need to be adjusted to the local soil, climate and land use history and to the socio-economic system. Measures may include highly specific soil inoculation with beneficial microbes such as symbiotic mycorrhizal fungi and nitrogen-fixing bacteria. This has been successfully used to increase phosphorus (P) and nitrogen (N) availability to crops, for example in Africa, while at the same time reducing the potential harmful effects of excess mineral fertilizer inputs for (amongst others) ground water pollution and greenhouse gas production (Smith and Read, 2008; Giller, 2001). Diversification of agricultural systems through increased tree cover can also contribute to enhancing below- and above-ground biodiversity and the ecosystem services they provide (Barrios *et al.*, 2018).

Significant knowledge gaps and technological barriers exist in the field of management practices for promoting soil biodiversity. The effectiveness of inocula applied to soils is variable, whereas the long-term potential risk of applying engineered microbiomes on soil biodiversity and ecosystem functions is still a matter of debate. Novel technologies such as metagenomic, metabolomic and volatilomic approaches that can help measure the diversity status and functionality of the soil community are expensive, and relatively few references are available to integrate and intepret the obtained values with respect to the state of soil biodiversity and functioning. In addition, applications need to be developed for the recording of farming data, linking the information to remotely sensed databases and storage of data, and analyzing big data in order to provide management advice.

Steering and harnessing these interactions is helpful in the development of nature-based technologies to tackle environmental problems.

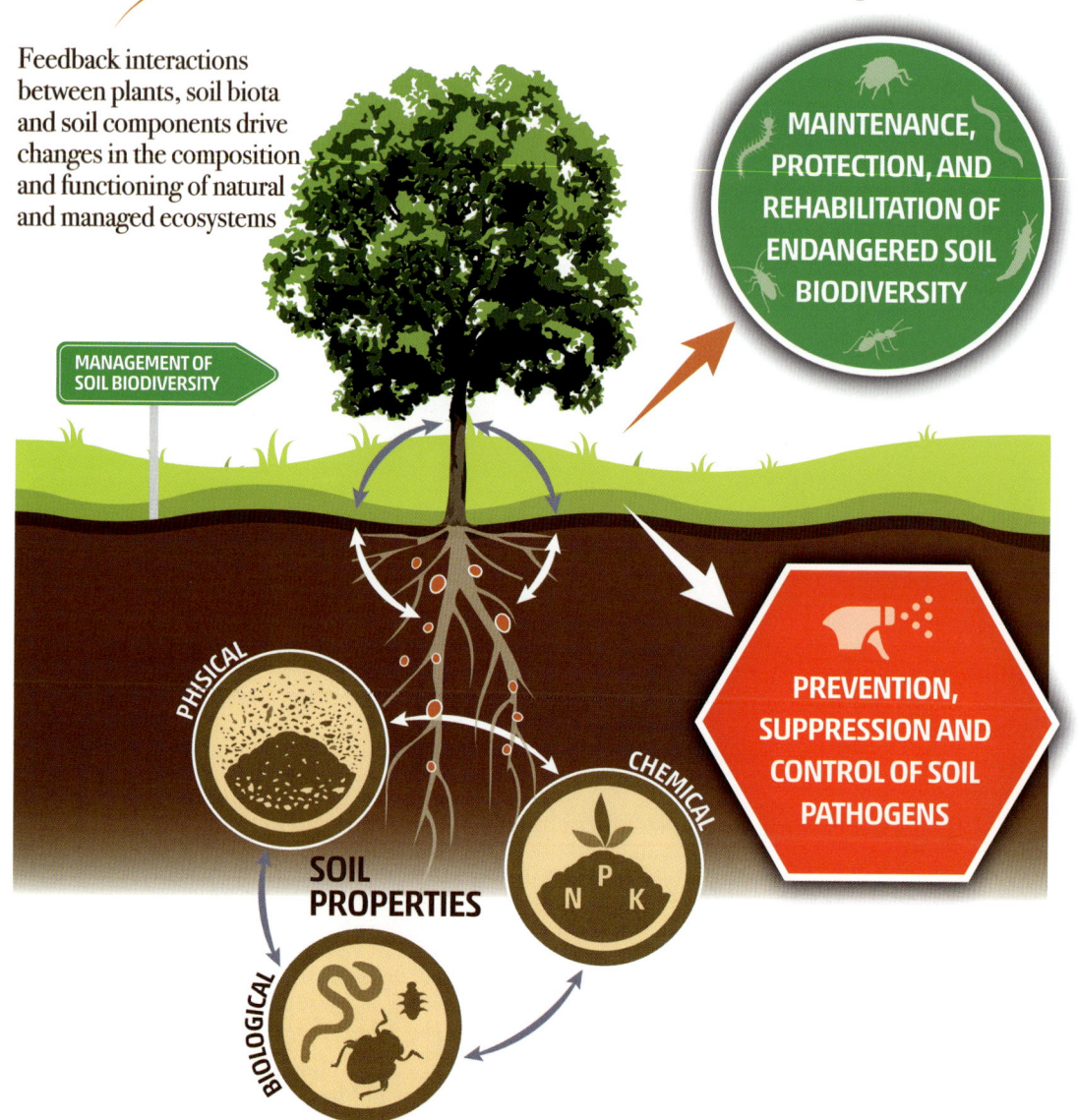

Feedback interactions between plants, soil biota and soil components drive changes in the composition and functioning of natural and managed ecosystems

MANAGEMENT OF SOIL BIODIVERSITY

MAINTENANCE, PROTECTION, AND REHABILITATION OF ENDANGERED SOIL BIODIVERSITY

PHISICAL

CHEMICAL

BIOLOGICAL

SOIL PROPERTIES

N P K

PREVENTION, SUPPRESSION AND CONTROL OF SOIL PATHOGENS

KNOWLEDGE GAPS AND TECHNOLOGICAL BARRIERS

- Long-term potential risk of applying engineered microbiomes on soil biodiversity and ecosystem functions.
- Novel technologies are expensive.
- Applications need to be developed for:
  ○ The recording of farming data.
  ○ Linking the information to remotely sensed databases and storage of data.
  ○ Analyzing big data in order to provide management advice.

Figure 5.1.1 (previous page) | **Management of soil biodiversity**

Along with other strategies, some environmental problems and crop needs can be tackled with nature-based solutions. The development of these types of solutions should be based on comprehensive approaches that consider the interactions of plants with soil biodiversity as well as the management of beneficial and harmful organisms.

5.2 | ECOSYSTEM RESTORATION: STARTING FROM THE GROUND AND LEVERAGING SOIL BIODIVERSITY FOR SUSTAINABILITY

The restoration of degraded land, soil and ecosystems is one of the most promising actions for meeting the Sustainable Development Goals (SDGs) centred on protecting life on land, water quality, global climate change and human well-being. Soil biodiversity plays a central role in avoiding, reducing and reversing land degradation by stabilizing soils, tightening nutrient cycling, increasing soil organic matter content, influencing water infiltration and quality, and supporting biodiversity above and below ground. Knowledge of effective restoration using nature-based solutions (including soil biodiversity) and sustainable land management techniques and approaches lags in many parts of the world, requiring increased research and knowledge sharing to support local actions in order to produce the desired restoration outcomes. The role of soil biodiversity in restoration activities is gaining recognition; continued work is required for restoration goals to meet their potential.

5.2.1 | LAND USE AND LAND DEGRADATION

Ecosystem restoration is recognized as a highly effective tool to reverse land degradation and reach global sustainability goals, reflected in the recent declaration of the United Nations Decade for Ecosystem Restoration (2021-2030). Restoration and soil recovery advances SDG 2 (zero hunger), SDG 3 (good health and well-being), SDG 6 (clean water and sanitation), SDG 13 (climate action), SDG 14 (life below water) and SDG 15 (life on land). Many international organizations focus on reducing and reversing land degradation. Land degradation (LD) is defined by the United Nations Convention to Combat Desertification (UNCCD) as "the reduction or loss of the biological or economic productivity and complexity of rainfed cropland, irrigated cropland, grazing land, forest and woodlands resulting from a combination of pressures, including land use and management practices". It has been recognized in Sustainable Development Goal 15.3, and UNCCD is the custodian agency of indicator 15.3.1 (proportion of land that is degraded over total land area). The UNCCD collaborates closely with the other two Rio Conventions: the Convention on Biological Diversity (CBD) and the United Nations Framework Convention on Climate Change (UNFCCC) to address land degradation, especially in dry lands. Soil biodiversity is a central part of these ecosystem restoration efforts.

Diverse soil communities are foundational to terrestrial ecosystem restoration; they provide multiple ecosystem benefits including pollutant degradation, climate mitigation through soil carbon accrual, and soil erosion prevention. In former agricultural fields, soil biological community recovery under restoration to grassland in Africa, Europe and North America contributes to recovery of soil carbon (C) and N stocks and fluxes (Baer *et al.*, 2015; Morriën *et al.*, 2017). Studies of soil fauna have found shifts in community composition with time since restoration in North American tallgrass prairie (Barber *et al.*, 2017; Wodika and Baer, 2015; Wodika *et al.*, 2014), Costa Rican forests (Cole *et al.*, 2016) and Australian mines (Cristescu *et al.*, 2012). In most cases, though, restored communities did not resemble communities in native reference ecosystems during the course of the study; more research is needed. The societal and ecological benefits of ecosystem restoration outweigh costs nearly ten to one (IPBES 2018), due in large part to the recovery and activity of soil biodiversity.

Broad approaches to combatting land degradation centre around preventing degradation, using best practices to limit degradation of managed lands, and actively restoring ecosystems that have been degraded. In the new Strategic framework 2018-2030, the UNCCD is committed to achieving Land Degradation Neutrality (LDN) in order to restore the degraded land, including soil habitat and biodiversity. Land Degradation Neutrality relies on three entry points in the response hierarchy: avoid, reduce, reverse. (Figure 5.2.1.1). As these tenets extend to ecological restoration beyond the UNCCD strategic framework, we use them to further explore how restoring soil biodiversity supports the Sustainable Development Goals as well as other, integrated, global agendas.

Reverse land degradation

Reversing land degradation is possible through restoration and rehabilitation. Measures can be different, but an important step is political commitment of the country to support those new techniques at the local level with an appropriate policy framework, as mentioned above. The impact of sustainable land management interventions varies across biophysical and ecological contexts, and it becomes more diffuse and challenging to track, particularly with large-scale institutional or collective actions (for example, grazing agreements or community land use planning). Moreover, existing land use policies at the country level are very different, with sometimes significantly different success rates and efficiency. There is an urgent need for more effective land use policies. But this will require a policy environment that engages with people – local communities, indigenous peoples, men, women, youth – and is responsive to their needs on issues such as land rights, urban planning and land management decisions.

Soil biodiversity plays key roles in restoration of degraded lands, including recovery of life on land (SDG 15). Reversal of land degradation can range from stabilizing highly eroded landscapes devoid of vegetation to combatting undesired weeds and animals in order to encourage productivity of desired communities and ecosystems. In former industrial sites, bacteria and fungi within soil can actively degrade chemical pollutants in soil such as diesel (Bell *et al.*, 2013) and tolerate and chelate heavy metals (Mergeay *et al.*, 2003; Rajkumar *et al.*, 2012) to reverse degradation and improve habitat (SDG 15). Plants and soil ecosystem engineers such as earthworms and termites can play a key role in reducing and preventing soil erosion (Jouquet *et al.*, 2012), although many local environmental factors influence the realized effects of earthworms on erosion (Blanchart *et al.*, 2004). In soils with altered hydrology, burrowing animals such as earthworms, ants and termites, and small mammals such as mice, voles and marmots, can recreate macro- and micropore networks that influence drainage (SDG 6, SDG 14). Restored soil biological activity can increase soil carbon pools and storage through decomposition and protection within soil aggregates, functioning as one mitigating activity for global climate change (SDG 13). Soil carbon is often used as a proxy measure for soil quality or soil health; in fact, it is one of the six measures being used to track progress toward the UNCCD land degradation strategy.

Maximize the conservation of natural capital →

REVERSE
Where feasible productive potential and ecological services of degraded land can be restored or rehabilitated

REDUCE
Land degradation can be reduced through application of sustainable management practices

AVOID
Prevent degradation of non-degraded land and confer resilience

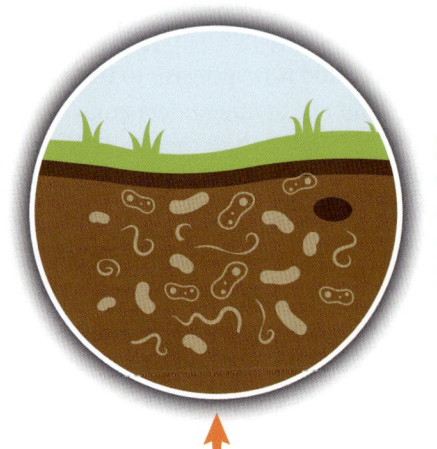

Bacteria and fungi within soil can actively degrade chemical pollutants in soils and tolerate heavy metals

Soil biota activity can contribute to avoid, reverse and reduce land degradation

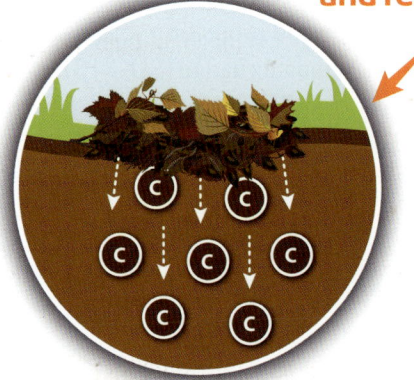

Soil biological activity can increase soil carbon storage through decomposition and protection within soil aggregates helping to reduce land degradation

Activities of soil ecosystem engineers (earthworms, termites) prevent soil erosion

Figure 5.2.1.1 (previous page) | **Soil biodiversity as a tool for nature-based solutions**

The land degradation neutrality response hierarchy encourages broad adoption of measures to avoid and reduce land degradation, combined with localized action to reverse it, across each land type. Soil biodiversity can contribute to avoid, reduce and reverse land degradation due to its close interaction with abiotic factors in soil ecosystems. (Figure adapted from Cowie *et al.*, 2018).

5.2.2 | NOVEL WHOLE-ECOSYSTEM APPROACHES FOR SOIL RESTORATION AND SOIL MANAGEMENT

Human management of agricultural land and other soils is known to reduce soil biodiversity. Examples include negative effects of tillage (Tsiafouli *et al.*, 2015), mineral fertilizers (Ramirez *et al.*, 2012) and pesticides (Thiele-Bruhn *et al.*, 2012) on field level (functional) soil diversity. Insufficient return of organic matter to agricultural soils has led to severe degradation of the productive capacity of naturally poor soils (Vanlauwe *et al.*, 2015). Furthermore, the use of large-scale monocultures also reduces soil biodiversity due to host specificity of many of the soil bacteria, fungi and the higher trophic level organisms (micro- and mesofauna) they attract, facilitating the spread and expression of soilborne diseases (Boudreau, 2013; Brooker *et al.*, 2015).

Most soil processes are regulated not only by soil microbial communities but by the whole soil food web (De Ruiter *et al.*, 1995; Moore *et al.*, 2004). Of key importance are the interactions among below-ground microbes, micro- (that is, protists, nematodes) and mesofauna (that is, springtails, detritivore and predatory mites, proturans, symphylans) as well as invertebrate soil engineers (for example, earthworms, millipedes) (Coleman *et al.*, 2018). The survival of newly introduced bacteria or fungi will depend importantly on the regulation of their abundance by their consumers (De Vries *et al.*, 2013).

Whole-ecosystem approach to soil restoration and management

While basic research is now demonstrating that soil functions are strongly dependent on the full complement of soil biodiversity (Müller *et al.*, 2016; Bender *et al.*, 2016; Raaijmakers and Mazzola, 2016), most applied research and commercial activity into soil restoration is focused on the isolation, characterization, cultivation and application of single microbial isolates (for example, arbuscular mycorrhizae, diazotrophic bacteria such as *Rhizobia*, *Bacillus* spp., *Trichoderma* spp.). In many cases, this classical microbiological approach has led to promising results in laboratory trials (Seo *et al.*, 2009; Vitti *et al.*, 2016; Rendina *et al.*, 2019). However, in the real world many applications have not been successful in improving soil functioning (Chaparro *et al.*, 2012; Raaijmakers and Mazzola 2016). Microbes are fast-evolving and prone to laboratory domestication (adaptation to benign laboratory conditions; Palková 2004; Eydallin *et al.*, 2014; Sterken *et al.*, 2015), which often impairs their ability to survive under competitive field conditions (Corkidi *et al.*, 2004; Berruti *et al.*, 2017). Consequently, the introduced microbes often persist only for a limited time (weeks, months) in the field. Basic research shows the soil food web to consist of an extensive web of biotic interactions that jointly determine many soil processes and, in turn, ecosystem functions (Van der Heijden *et al.*, 2008; Bardgett and Van der Putten 2014). Novel studies are now showing that the integrated management of whole soil communities can be the key to successful restoration of soil biodiversity and functioning. Currently three novel methods are leading to improved ability to manipulate and control whole soil communities for ecosystem functioning: (i) whole soil microbiome plus micro-fauna inoculation, (ii) microbiome engineering and (iii) *ex-situ* cultivation of soil microbiomes.

Whole-soil microbiome inoculation

Until recently, soil biodiversity as a driving factor of nature restoration played an underappreciated role. However, it is now clear that above-ground and below-ground diversity have important connections, and that soil biodiversity is an important contributor to local levels of plant diversity (Petermann *et al.*, 2008; Mangan *et al.*, 2010; Schnitzer *et al.*, 2011; Teste *et al.*, 2017). As in other fields, inoculation studies for nature restoration, or restoring vegetation on former mining sites, were mostly limited to the introduction of single or a few strains of beneficial bacteria or fungi (for example, mycorrhizal fungi) (Neuenkamp *et al.*, 2018). However, it is becoming clear that introduction of whole soil communities leads to more effective restoration (Rowe *et al.*, 2007; Middleton *et al.*, 2015; Emam 2016) and faster responses over time (Neuenkamp *et al.*, 2018).

Recently, field trials conducted at the scale relevant for nature restoration practice (that is, hectares) have demonstrated that whole-soil community inocula, representing the entire soil biodiversity, are a powerful tool in the restoration of terrestrial ecosystems (Wubs *et al.*, 2016) and that the introduced soil biodiversity in fact co-determines which plant species thrive in the restored grassland and heathland communities. In this whole-soil inoculation method, a relatively small amount of solid donor soil (for example, 1 L/m to 2.5 L/m) is introduced on the recipient soil to be restored (for example, an ex-arable soil) using a manure spreader. Consequently, there is no active strain isolation or selection involved (except for the inadvertent disturbance effects that are always induced on soil communities when a soil is disturbed). Complementary field studies have now shown that such a single whole-soil inoculation treatment can affect soil and plant diversity composition for over two decades when they are combined with the sowing of a complementary plant community (Wubs *et al.*, 2019). The effects of this inoculation may be revealed at different rates: when introduced into an impoverished soil, effects can show up within a couple of years (Wubs *et al.*, 2016), whereas inoculation into an existing soil may take more than five years to reveal a maximal effect (Wubs *et al.*, 2019).

Positive results have also been observed in other field experiments (for example, aiming to restore heathlands (van der Bij *et al.*, 2018), biocrusts in arid soils (Chamizo *et al.*, 2018) and species-rich meadows (Vécrin and Muller, 2003)). However, not all field trials have realized the anticipated results (Kardol *et al.*, 2009). An important question now is when and where whole-soil biodiversity inocula are effective, and how their consistent effectiveness may be achieved. Many micro- and mesofauna are sensitive to mechanical disturbance; transplanting entire blocks (turfs) of undegraded soil containing an intact ecological assemblage can be an effective method to effectively reintroduce these groups (Moradi *et al.*, 2018; van der Bij *et al.*, 2018). The long-term efficiency of this method at large scale will depend on the conditions of the surrounding degraded soil that can favour or hinder the establishment of the soil biota dispersed from the transplanted blocks. For example, when the conditions in the degraded soil are too harsh, this may negatively affect survival and, therefore, effectiveness of the inocula (Kardol *et al.*, 2009).

Soil microbiome engineering

An important novel tool that may enable the creation of more consistently effective whole-soil microbiome inocula is artificial microbiome engineering (Mueller and Sachs, 2015; Gopal and Gupta, 2016). Hitherto, artificial selection has been applied to the germlines of crops, but the microbiome engineering approach proposes that selection can also be applied, perhaps more effectively, to the plant-associated microbiota. Each plant species and genotype is, to some extent, associated with its own set of rhizosphere microbes (Bezemer *et al.*, 2010; Bulgarelli *et al.*, 2012; Lundberg *et al.*, 2012), while this is also influenced by management (Zhu *et al.*, 2016; Sofo *et al.*, 2019). Typically, in the order of approximately 10 percent of the total diversity found in the rhizosphere is linked to a particular plant species, and even shoot endophytes can derive from local soil via the plants' xylem (Fausto *et al.*, 2018).

Microbiome engineering proposes that it is the plant species-specific part of the microbiome that may be selected to fit with optimal performance under the targeted environmental conditions (Mueller and Sachs, 2015). In order to do so, a set of plants is jointly cultivated with the soil microbiome. The plants are subsequently phenotyped and those plants are selected that express the desired trait most strongly (for example, yield, disease resistance, phytoremediation capacity). Next, rather than selecting the seeds of those plants for propagation, soil samples are taken from the root zone of those plants and again co-cultivated with new naive individuals of the host plant species, and this is repeated for a number of cycles (Mueller and Sachs, 2015). In this way it has been possible to select for shifts in flowering time and biomass production in *Arabidopsis thaliana*, and these effects can be induced solely by inoculation of the engineered microbiome (Swenson *et al.*, 2000; Panke-Buisse *et al.*, 2015). The effects could be generalized across several *A. thaliana* genotypes and a related crop. New studies are now diving into the potential for this new technique to mediate, amongst others, resistance to insect herbivores (Pineda *et al.*, 2017), human health in urban areas (Mills *et al.*, 2017) and plant salt tolerance (Qin *et al.*, 2016).

Most studies so far have used a single source of soil microbiomes for microbiome engineering; however, it has been argued recently that combining microbiomes from disparate sources, an approach termed community coalescence, may provide a diversity template from which novel synergistic functions may arise (Rillig *et al.*, 2015, 2016). This novel approach has now been tested in the context of nature restoration (Wubs *et al.*, 2018) and horticulture (Ma *et al.*, 2018). The results show that synergistic effects may indeed arise from community coalescence, but this is highly dependent on the plant species and inoculum sources as well as the fitness metric (growth or disease resistance) used.

Ex situ cultivation of soil microbiomes

In various cases, the sources of soil biodiversity in nature may be sensitive to disturbance and may be in conflict with policy aimed at biodiversity conservation to use these areas as sources for beneficial microbes. However, it has been shown repeatedly in greenhouse studies (reviewed in Van der Putten *et al.*, 2013) as well as in controlled field trials (Bezemer *et al.*, 2010) that plant species can select their particular subset of the soil biodiversity from a common soil community. This suggests that from starter inocula sourced in nature it may be possible to start *ex situ* soil biodiversity conservation programmes via targeted co-cultivation with the host plant species or indeed whole plant communities. The extent to which these controlled cultivations mirror the full plant-associated soil biodiversity found in natural conditions needs to be tested before such programmes can be started.

Integration of restoration in the landscape

It is well established in the ecological literature that above-ground and below-ground biota are in constant interaction (Wardle *et al.*, 2004; Bardgett and Wardle, 2010; Fausto *et al.*, 2018), typically mediated via changes in plant resource allocation, defence investments and regulatory hormone expression (Bezemer and van Dam, 2005; Vitti *et al.*, 2016). As a result, the ecosystems services provided by above-ground compartments (for example, pollination, pest control, carbon storage in (tree) biomass) and below-ground compartments (for example, nutrient cycling and soil carbon storage) are the net result of the interactions across these compartments. It is therefore important that any soil biodiversity restoration programme be effectively embedded within its landscape and the expected interactions therein (Kardol and Wardle, 2010; A'Bear *et al.*, 2014). A key aspect is to integrate these interactions at the right spatial and temporal scales, mainly by focusing on the local but long-term legacy effects in the below-ground compartment and the large-scale and temporally dynamic above-ground component (Veen *et al.*, 2019). Utilization and conservation of soil biodiversity will depend on the appropriate integration within landscape-scale processes.

When soils have been extremely degraded (by, for example, mining or public works), biological restoration of the below-ground community will have to be preceded by restoration (or absolute recreation) of the physical and chemical properties of the substrate. In severe cases, this will require construction of man-made "technosols" (Macías and Camps Arbestain, 2010). Technosols are defined as unconsolidated residues derived from anthropogenic activities, with characteristics similar to the geological and biogenic components of soils, which under the influence of soil-forming factors may develop into new soils (Dudal *et al.*, 2002). In soil restoration after opencast mining and quarrying, a particular type of technosol is frequently obtained by mixture of mineral debris with organic materials recycled from urban, agricultural and forestry activities (for example, green compost, biochar, or sewage sludge) (Luna *et al.*, 2016). The recreation of appropriate soil physical and chemical conditions through technosols may serve as a key starting point for restoration of soil biodiversity through colonization from the surrounding undisturbed areas (Andrés 1999). (Figure 5.2.2.1).

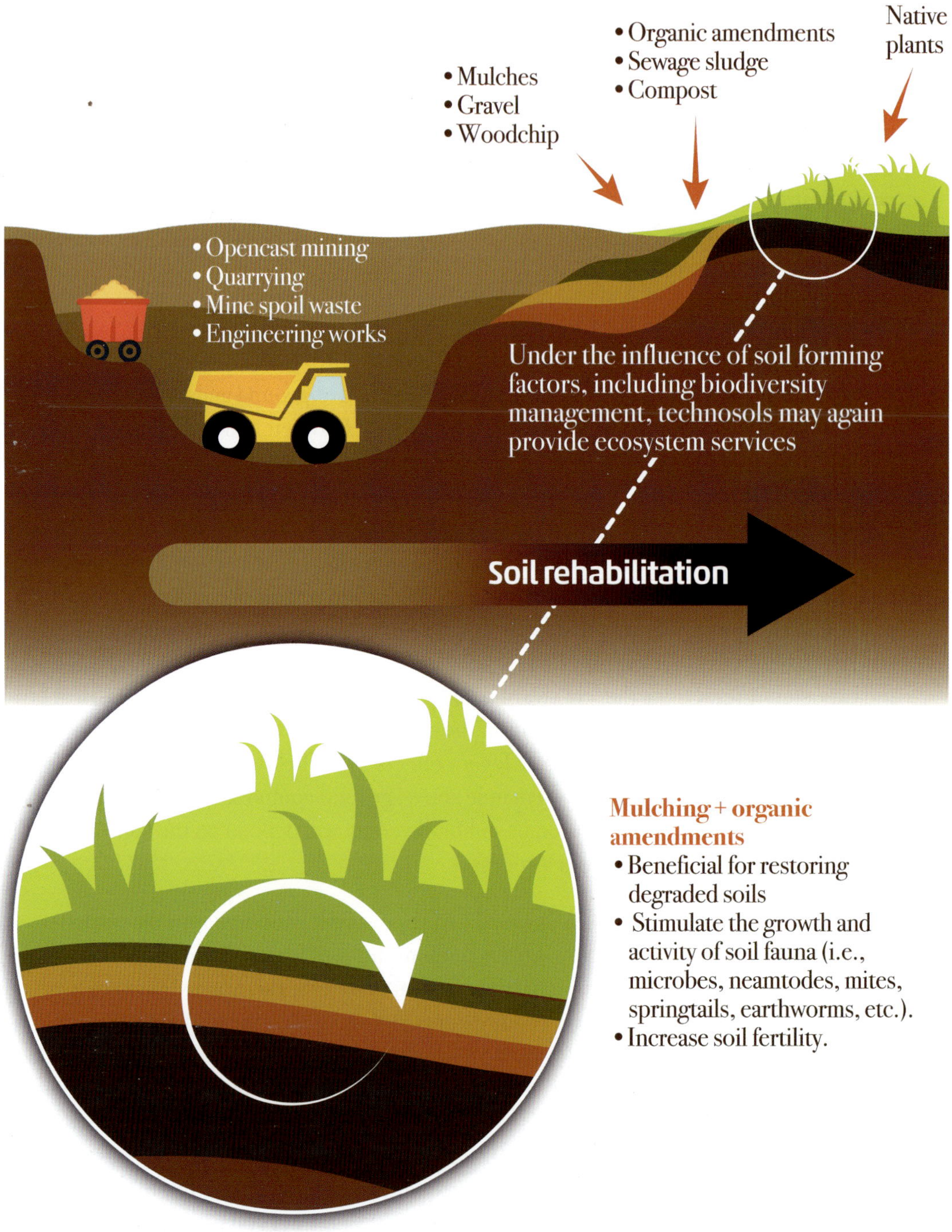

- Mulches
- Gravel
- Woodchip

- Organic amendments
- Sewage sludge
- Compost

Native plants

- Opencast mining
- Quarrying
- Mine spoil waste
- Engineering works

Under the influence of soil forming factors, including biodiversity management, technosols may again provide ecosystem services

Soil rehabilitation

Mulching + organic amendments
- Beneficial for restoring degraded soils
- Stimulate the growth and activity of soil fauna (i.e., microbes, neamtodes, mites, springtails, earthworms, etc.).
- Increase soil fertility.

Figure 5.2.2.1 | Combination of soil rehabilitation strategies

Mining activities have drastic negative effects on soils, especially in arid areas. An alternative to restore the biological communities of the soils is the establishment of technosols that can perform again several ecosystem services. Essential actions in the recovery of soil functionality include the addition of organic matter, which together with the action of pioneer plants favor the growth and activity of soil microbial populations, eventually influencing the improvement of the ability to produce biomass. Study case from Luna *et al.* 2016.

5.3 | POTENTIAL OF SOIL BIODIVERSITY IN THE FIGHT AGAINST SOIL POLLUTION

One important value of soil biodiversity is its use in the design of strategies for detoxifying polluted environments. Biological breakdown of contaminants is considered as both cost-effective and ecologically sound, as long as contaminants are turned into harmless substances, made less mobile or accumulated in biotic tissues that can be safely removed and processed (Gillespie and Philp, 2013).

While the general public might not be aware that microbes are at work at soil decontamination sites, promotion, improvement and dissemination of microbial strategies for environmental management must go hand in hand. Scientific demonstration of the need for preservation of soil biodiversity must be pursued, together with the exposition of this biodiversity as an essential component of environmental maintenance strategies.

5.3.1 | USE OF SOIL BIODIVERSITY AS A SOIL POLLUTION REMEDIATION TOOL

Soil pollution is a serious environmental problem with negative consequences for soil biodiversity and thus for soil ecosystem functioning. Due to the importance of the soil system as a most valuable resource, and in particular of soil biodiversity as the "biological engine" that drives soil functioning, the recovery of polluted soils is a matter of great urgency and relevance.

The term "soil remediation" refers to a group of physicochemical and biological technologies designed to clean up polluted soils. Many physicochemical methods of soil remediation, such as excavation and containment, soil washing or incineration, are expensive. Even more relevant is the fact that such methods may result in a strong adverse impact on soil functioning, and in particular on soil biodiversity. In fact, some of them can even be more damaging to the integrity of the soil system than the contaminants themselves. In contrast, biological methods of soil remediation, such as bioremediation (with microorganisms) and phytoremediation (with plants) are, in general, cost effective and more environmentally friendly.

The objective of all soil remediation methods should focus not only on removing the contaminants from the soil but also on improving soil health. Many remediation technologies, especially those based on physicochemical techniques, eliminate the soil contaminants at the expense of undesirably affecting the integrity of the soil ecosystem. As soil health depends largely on the activities of the soil biota, soil biodiversity should be restored or conserved when remediating a polluted soil. Moreover, some polluted sites, particularly metalliferous mining sites, can harbour a unique biodiversity of, for instance, metallophytes and metal-tolerant microorganisms that must be preserved for its intrinsic and utilitarian value. Furthermore, soil biodiversity and specific soil organisms can act as most suitable indicators of the effectiveness of soil remediation methods and the restoration of soil health.

Bioremediation is defined as the process whereby contaminants are biologically degraded under controlled conditions that enhance plants' or microorganisms' growth and enzymatic activities. Bioremediation technologies can lead to the degradation of a target contaminant to an innocuous state or to levels below concentration limits established by regulatory authorities. Bioremediation is involved in degrading, removing, altering, immobilizing or detoxifying various contaminants from the environment through the action of bacteria, fungi and plants. Bioremediation of pollutants had an estimated (in 1997) global economic value of USD 120 billion per year (Pimentel *et al.,* 1997).

The bioremediation of polluted soils can be carried out according to two main strategies: (i) biostimulation, or the intended alteration of the soil environment (by means of adjusting such factors as soil pH, moisture, nutrient or oxygen contents) to stimulate the degradation capacity of native microbial populations; and (ii) bioaugmentation, via the inoculation of microbial strains (singly or in combination, that is, consortia of microbial strains) with the metabolic abilities to degrade the target contaminants (Adams *et al.*, 2015). (Figure 5.3.1.1). Traditionally, bioaugmentation initiatives have often failed to achieve the desired objective due to the poor survival and/or growth of the inoculated strains in the recipient soil, owing to their lack of competitive fitness under the specific environmental conditions of the polluted soil. Relatively little is known about the required ecological fitness traits of the strains used for bioaugmentation, as well as the design of microbial consortia (sometimes, combining bacteria and fungi).

Soil pollution remediation through soil microorganisms

Microorganisms are widely distributed in the biosphere because their metabolic ability is very varied, and they can easily grow in a wide range of environmental conditions. The nutritional versatility of microorganisms can also be exploited for biodegradation of contaminants (Tang *et al.*, 2007). Microorganisms act as significant contaminant removal tools in soil, water and sediments; they present several advantages over other remediation procedural protocols, such as the capacity to act *in situ* without the removal of polluted soil. Microorganisms can restore the original natural surroundings and prevent further pollution (Demnerova *et al.*, 2005). Soil microorganisms, and the knowledge gained about them, can potentially be applied for soil remediation. The United States Environmental Protection Agency has collected information about the types of remediation technologies used across the United States of America (https://www.epa.gov/remedytech/). In Europe, the EUGRIS portal provides a similar overview (http://www.eugris.info/), as does the Ministry of Environment and Forests of India (MoEF, 2011).

Tolerant microbial strains commonly exist in highly polluted soils, and their presence can be postulated to depend on pre-existing soil microbial diversity, allowing for genetic adaptation and selection of tolerant microbes. Key issues for sustainable use of bioremediation microbiomes are thus (i) preservation of soil biodiversity together with characterization of putative effective microorganisms, and (ii) provision of access to these microorganisms. As stated earlier, a careful description of the source ecosystem of microorganisms used for bioremediation should be an essential element of strain characterization.

Microorganisms are ubiquitous, but this does not imply that any contaminant will be degraded everywhere. Important limitations include absence or scarcity of efficient microbes, or unfavourable environmental conditions. On the other hand, microbes with a desirable catabolic potential can be introduced in contaminated sites, either alone for bioremediation or in combination with plants for phytoremediation.

Microorganisms possess enzymes that allow them to degrade and immobilize different contaminants, or to use them as a source of carbon in the case of organic contaminants, which are used environmental contaminants as a food (Kumar *et al.*, 2011). Certain bacteria and fungi are capable of targeted and efficient breakdown of polycyclic aromatic hydrocarbons (PAHs) (Husain *et al.*, 2009). Polycyclic aromatic hydrocarbons are compounds that originate from petroleum or petroleum-derived products and from incomplete combustion of fossil fuels or biomass. In soils, PAHs are stable, especially those containing many aromatic rings, with half-lives of one year or more (Roslund *et al.*, 2018). Collectively, PAH-degrading organisms have the potential to rapidly degrade a wide spectrum of PAH types, specifically up to 100 percent of the low molecular weight PAHs and up to 50 percent of the high molecular weight PAHs within 70 to 80 days. In Canada, an estimated 38 000 m^3 of soil was contaminated with an oil by-product containing polycyclic aromatic hydrocarbons, cyanide, xylene, toluene and trace elements released by a gasification plant. After application of bacteria and a nutrient mixture (such a dual application corresponding to a combination of biostimulation and bioaugmentation techniques), the various constituent contaminants of the oil tar were reduced by 40 to 90 percent in just 70 to 90 days (Warith *et al.*, 1992). *Biostimulation* refers to the changes in the environmental conditions to favour the growth of local microbial populations, while *bioaugmentation* means the addition to the contaminated soil of living cells able to degrade the target contaminant.

Some microbes can tolerate trace elements and prove useful for remediation of contaminated sites (Gaur *et al.*, 2014; Dixit *et al.*, 2015). Reduced toxicity and removal of trace elements can be achieved via two mechanisms. First, trace elements can be bound by adsorption to microbe-soil colloids; this process is influenced by soil type and organic matter content and enhanced by accumulation inside microbial cells. Second, microorganisms can stimulate the activity of plants used for a process called phytoremediation, which is the use of plants for removing trace elements.

With respect to distribution and generalization of effective microorganisms, the Nagoya Protocol on Access and Benefit Sharing framed in 2014 (CBD, 2019) should be followed, as it outlines best practices for ensuring access to genetic resources. Many universities maintain *ex-situ* collections of microorganisms for access by scientists and enterprises. The World Federation for Culture Collections (WFCC) maintains a Global Catalogue of Microorganisms (http://gcm.wfcc.info/) listing strains and cultures from around the world (to date, the catalogue comprises 447 695 strains from 48 countries). Source information and publications on the strains are given, which makes the catalogue a valuable tool for the scientific and industrial communities. Importantly, one quarter of the organisms in the above-mentioned collections are soil organisms (Wu *et al.*, 2013), suggesting that soil biodiversity is a significant contributor to environmental management potential. Maintaining and documenting this potential is necessary, since new challenges may present themselves in the future with respect to contamination control and bioremediation.

Soil pollution remediation through plants

Microorganisms have limitations for the remediation of soils polluted with trace elements. Trace elements cannot be degraded. Microorganisms can transform them from one oxidation state or organic complex to another, but they cannot remove them from the soil. Therefore, the phytoremediation field has paid much attention to the remediation of trace element-polluted soils through two major approaches:

- phytoextraction, or the utilization of hyperaccumulators (that is, plants that have the capacity to accumulate high amounts of trace elements in their above-ground tissues); and

- phytostabilization, or the use of plants to immobilize trace elements in the rhizosphere through absorption and accumulation by roots, adsorption onto roots, or precipitation within the root zone, resulting in decreased metal bioavailability and ecotoxicity.

Different plant species growing in the same soil differ by their macro- and trace elements' uptake ability (Shtangeeva *et al.*, 2019). Beside the impact on plant growth (Kalingan *et al.*, 2016), trace elements might enter into the food chain (Chrzan A., 2016). Additionally, plants can serve as trace element phytostabilizators (restricting their transport), shield animals from toxic species ingestion, and consequently prevent transmission across the food chain (Usman *et al.*, 2019).

Regrettably, hyperaccumulators are usually small in size and have slow rates of biomass production. Consequently, an effective phytoextraction frequently takes many years and often decades, hindering its practical application. In this respect, chelate-induced phytoextraction is a strategy that deals with the application of chelating agents to the soil in order to increase trace element uptake and translocation by high-biomass plants. Nonetheless, the application of chelating agents to the soil often leads to harmful consequences, such as metal leaching and direct or indirect toxic effects on soil biota, thus also hindering the deployment of this strategy. In turn, phytostabilization does not remove the toxic trace elements from the soil. Moreover, it requires a follow-up programme to monitor its long-term effectiveness (that is, to check for possible changes in bioavailability over time). Furthermore, most legal regulations are based on total trace element concentrations, not on bioavailable concentrations.

Different alternatives are being proposed to overcome the above-mentioned limitations of phytoremediation. For instance, the combination of both strategies (phytoextraction and phytostabilization) can help minimize the disadvantages of each of them. Interestingly, a strong emphasis is currently being paid to the possibility of obtaining economic value during the long periods of time commonly required for effective phytoremediation. In particular, phytomanagement is a gentle remediation option which aims to achieve economic benefits (from products such as timber, resin, or essential oils,) and environmental benefits (ecosystem services such as carbon sequestration, erosion control and regulation of the water cycle) from a proper management of polluted sites through the establishment of a suitable plant cover, while minimizing the linkages between contaminants and biological receptors.

In addition, other nature-based practices are being implemented to improve the performance of phytoremediation. Rhizoremediation involves the application of trace element tolerant rhizospheric microorganisms together with tolerant plants, in order to improve overall remediation capacity. In this association, the microorganisms assist the plants in trace element uptake and in survival. A major advantage of phytoremediation and rhizoremediation is that above-ground tissues can be removed from the contaminated site, and the hyperaccumulated elements can then be recovered in usable quantities, according to the principles of a bio-based economy (Wang *et al.,* 2016). Plants typically used are willow (*Salix sp.*), poplar (*Populus sp.*), *Jatropha curcas*, maize (*Zea mays*), castor bean (Ricinus communis), black nightshade (*Solanum nigrum*), *Arabidopsis halleri* and other *Brassicaceae*. Whereas some of these plants are annuals and others are perennials, all share the capacity for fast growth and adaptation to a variety of climates (Dixit *et al.,* 2015). The rhizosphere microbiome plays an important role in this process, as well as in soil energy transfer and nutrient cycling (Guo *et al.*, 2019). At the same time, plant-promoting bacteria that have a promoting activity in phytoremediation of soil pollution by trace elements might be used in specific remediation applications (Ren *et al.*, 2019).

In recent years, the most innovative approaches within the field of soil remediation are resulting from the combination of physicochemical and/or biological techniques. An example, for the remediation of soils polluted with chlorinated hydrocarbons, is the combination of (i) nanoremediation via the application of nanoscale zero valent iron (a promising technology for the remediation of soils polluted with organic contaminants that is still under development), and (ii) bioremediation via biostimulation and/or bioaugmentation. Similarly, the combination of phytoremediation with bioaugmentation via the inoculation of plant growth-promoting bacteria (rhizobacteria or endophytes) or contaminant degrading strains is showing promising results.

Soil pollution remediation through macroorganisms

While current research mostly focuses on the use of microbes, other soil organisms could also be of interest (Hirano and Tamae, 2011; Velki and Ečimović, 2016). Even if studies on soil fauna and bioremediation are rare, given the importance of soil invertebrates in soil structuration and nutrient recycling and their link to microbial community, their presence can have important indirect effects on the success of bioremediation. For example, earthworms can be good biosensors and accumulate polycyclic aromatic compounds (PAHs) and other contaminants (Matscheko *et al.*, 2002; Slizovskiy and Kelsey, 2010). As long as the concentration of soil contaminants are not lethal, earthworms can be used in bioremediation. *Eisenia fetida* earthworms can accumulate cadmium if soil concentration is below 0.08 mg/g (Aseman-Bashiz *et al.*, 2014). Soil fauna can also be dispersal agents for both microorganisms that degrade organic contaminants and the contaminants themselves through the soil profile. For instance, soil invertebrates such as earthworms have been shown to improve decontamination of organic (for example, pesticides) and inorganic contaminants (metals) by plants and microorganisms (Hickman and Reid, 2008; Jusselme *et al.*, 2012; Morillo and Villaverde, 2017).

Finally, certain organisms are able to bioaccumulate contaminants in specific organs of their body, such as in hepatopancreas of isopods or snails (Hopkin and Marten, 1982; Fritsch *et al.*, 2011), or the chloragogenous tissue of earthworms (Sizmur and Hodson, 2009). Gut microbiota analysis of earthworm *Lumbricus terrestris* described "indicator" bacterial genera of *Paenibacillus* and *Flavobacterium* as biomarkers of "exposure in earthworms inhabiting Cd-stressed soils that might have implications for environmental monitoring and protection of soil resources" (Šrut *et al.*, 2019). This might open the concept of zooremediation with the same way of thinking about phytoremediation.

Controlling and optimizing bioremediation processes is a complex system due to many factors, including the existence of a microbial population capable of degrading the contaminants, the availability of contaminants to the microbial population, and environmental factors (such as type of soil, temperature, pH and the presence of oxygen or other electron acceptors and nutrients) (Abatenh *et al.*, 2017).

The usefulness and potential of soil organisms for environmental management, and particularly for breakdown, detoxification and immobilization of contaminants, is clearly demonstrated and represents a powerful argument in favour of soil biodiversity preservation. Studies on bioremediation and the microorganisms involved have mainly focused on two types of contaminants, PAHs and trace elements. To some extent at least, insights gained on these compounds can be generalized to other types of contaminants, even though the biochemical breakdown mechanism is unique to each compound or class of compounds. To succeed in remediation, one needs to consider not only the organism to be used for contaminant degradation, but also microbial interactions with the physical, biological and chemical components of the soil and with introduced or native plants (Uqab *et al.*, 2016; Xu and Zhou, 2017). Here we highlight metabolic properties and sources of microorganisms useful to bioremediation.

Bioaugmentation: Addition to the contaminated soil of living cells able to degrade the target contaminant.

Biostimulation: Changes in the environmental conditions to favor the growth of local microbial populations.

Trifolium pratense

- Macronutrients
- Micronutrients

Bioremediation

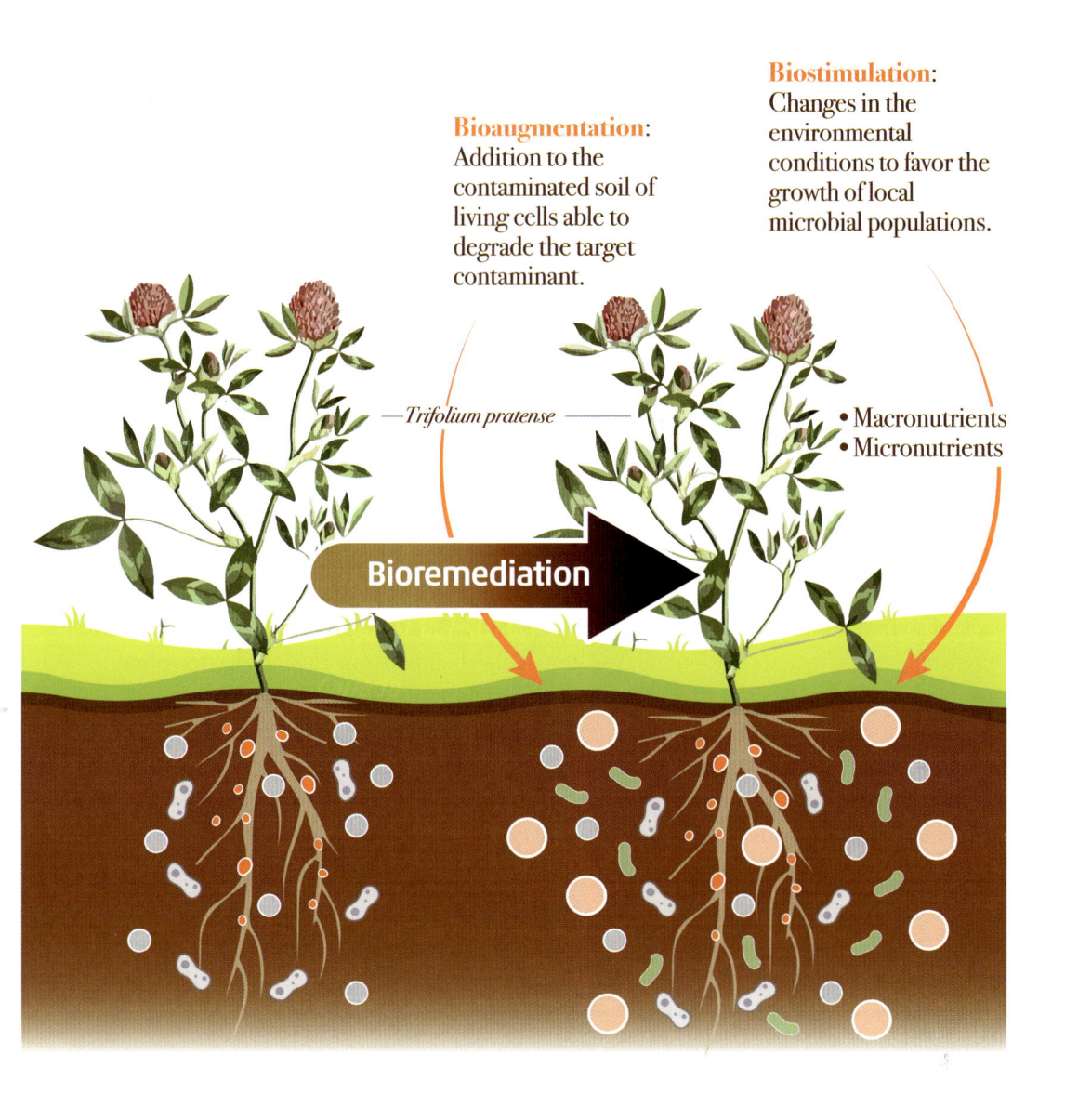

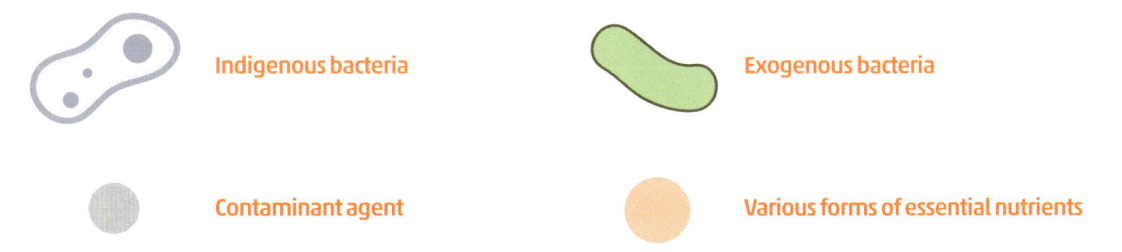

- Indigenous bacteria
- Exogenous bacteria
- Contaminant agent
- Various forms of essential nutrients

Figure 5.3.1.1 | Biorremediation

Soil microorganisms represent a powerful tool in the management of contaminated soils. Biostimulation and bioaugmentation are environmentally-friendly strategies that contribute to the degradation of target contaminants.

5.3.2 | KNOWLEDGE GAPS AND ENVIRONMENTAL RISK ASSESSMENT

In spite of the demonstrated usefulness of soil microbes for environmental cleanup, key knowledge gaps remain about the real potential of soil biodiversity in tackling soil pollution problems, and about how soil pollution actually affects soil biodiversity.

Knowledge gaps on soil biodiversity potentiality

First, while more than 20 bacterial genera and 69 strains, representing a wide taxonomic spectrum of bacteria, have been listed as capable of PAH breakdown (Seo *et al.*, 2009), it is likely that many more bacteria and fungi spanning a large fraction of the microbial tree of life are capable of PAH breakdown but have not yet been discovered (Table 5.3.2.1). A second knowledge gap is that most publications reporting isolation of PAH-degrading microorganisms or their use for bioremediation fail to include detailed information on the conditions prevalent at the site of origin of those organisms. This lack of source information diminishes our capacity for matching microbial properties to environmental characteristics of target sites. It also hampers our capacity to decipher mechanisms for emergence of PAH-degrading microorganisms through natural selection occurring at contaminated sites. Since microbial consortia often prove more effective at bioremediation than pure cultures, a knowledge of the site of origin of the degrading organisms would facilitate the constitution of ecologically compatible and complementary microbial associations. Primary information on isolation sites is critical to the preservation of habitats.

Table 5.3.2.1

Examples of soil PAH-degrading bacteria with their isolation source

Bacterium or bacterial consortium	Isolation source	Key findings	Reference
Thalassospira sp. strain TSL5-1	Petroleum-polluted soil at Shengli Oil Field, China.	Possesses two biochemical pathways, the salicylic acid and phthalate routes, for the breakdown of pyrene. Breakdown occurs at salinity ranging from 0.5 percent to 19.5 percent, with optimal value between 3.5 percent and 5 percent. Degradation activity affected by pH.	Zhou *et al.*, 2016
Acinetobacter, Bacillus, Microbacterium, Ochrobactrum	Soil from industrial effluents and petrol-distribution service dumping sites at four locations in Orissa, India.	Isolates belonging to all four genera were able to break down pyrene. *Bacillus megaterium* YB3 showed the best results. Under laboratory conditions, this strain required 7 days to break down 72 percent of the pyrene (500 mg/L) added to a growth medium.	Meena *et al.*, 2016

Bacterium or bacterial consortium	Isolation source	Key findings	Reference
Amycolatopsis sp. Poz14, *Gordonia* sp. Poz20, and *Rhodococcus* sp. Poz54	Soil at a petroleum-contaminated area in Veracruz, Mexico.	A laboratory incubation experiment showed that three actinobacterial isolates utilized both low molecular weight PAHs (anthracene and naphthalene) and high molecular weight PAHs (fluoranthene and pyrene). *Amycolatopsis* sp. degraded 100 percent of the naphthalene, 38 percent of the anthracene, 25 percent of the pyrene and 18 percent of the fluoranthene in 45 days. These bacteria can potentially serve as a broad-spectrum clean-up agent for PAHs.	Ortega-Gonzalez *et al.*, 2015
Pseudomonas sp. N3, *Pseudomonas monteilii* P26, *Rhodococcus* sp. F27, *Gordonia* sp. H19, *Rhodococcus* sp. P18	*Pseudomonas* strains were isolated from oil-polluted marine sediments. Actinobacteria were isolated from an oil-contaminated soil in Patagonia.	Tests with consortium of two *Pseudomonas* bacteria and three actinobacteria showed that the *Pseudomonas* strains efficiently degraded low molecular weight PAHs (anthracene and naphthalene). Actinobacteria were more efficient at pyrene breakdown. A combination of all five strains degraded 100 percent of the phenanthrene and naphthalene and 42 percent of the pyrene.	Isaac *et al.*, 2015
Achromobacter insolitus, Bacillus licheniformis, Bacillus cereus, Microbacterium sp., *Sphingobacterium* sp., *Pseudomonas aeruginosa*	Soil (0cm to 30 cm) a few metres away from a petrochemical plant in Maharashtra, India.	Under controlled conditions, breakdown of naphthalene, anthracene and phenanthrene (up to 750 mg/L) was possible. The most effective strain was *P. aeruginosa* 16S, but a consortium of bacteria proved more efficient at breakdown than individual isolates.	Fulekar, 2017

In contrast to PAH-degrading microorganisms, microorganisms used in rhizoremediation of trace-element-containing soils are often well characterized with respect to their soil of origin, with information being available regarding soil texture, soil pH and original host plant species. Table 5.3.2.2 provides examples of bacterial strains used in rhizoremediation of trace elements. However, the data remain too scant to yield generalizations about characteristics of the original habitat of microorganisms used in rhizoremediation. In addition to bacteria listed in Table 5.3.2.2., arbuscular mycorrhizal fungi (AMF) are valuable components of rhizoremediation processes (Gaur and Adholeya, 2004; Meier *et al.*, 2012). The symbiotic fungi stimulate host plant growth in soils with high trace element concentrations (Goehre and Paszkowski, 2006).

At comparable trace elements concentrations, sorption to AMF biomass is higher than sorption to a range of other soil microorganisms (Joner *et al.* 2000).

Table 5.3.2.2

Examples of microorganisms for rhizoremediation of trace-elements-contaminated soils with their isolation source

Organism or consortium	Isolation source	Key findings	Reference
Bacillus cereus strains BDBC01, AVP12 and NC7401	Roots of *Tagetes minuta* in a soil adjoining automobile workshops in Kashmir, Pakistan.	Three rhizobacterial *Bacillus* strains were found promising in rhizoremediation. These bacteria harboured high adsorption capacity for chromium (Cr), nickel (Ni) and cadmium (Cd).	Akhter *et al.*, 2017
Bacillus subtilis and *Pseudomonas putida*	Isolates supplied by China Center for Type Culture Collection. The original isolation sources were not provided.	*B. subtilis* was more efficient than *P. putida* at cadmium absorption, while *P. putida* was better at copper absorption. Interaction between bacteria and soil particles provides a bridge for metal ions between bacterial cells and clays.	Du *et al.*, 2017
Agromyces, *Flavobacterium*, *Serratia*, *Pseudomonas* and *Streptomyces*	Soil from a lead mining area in Arnoldstein, Austria. Soil pH, texture and organic matter content were characterized.	Willows (*Salix caprea*) were grown in zinc-, cadmium- and lead-contaminated soil and inoculated with the bacteria. *Streptomyces* AR17 enhanced Zn and Cd uptake. *Agromyces* AR33 promoted plant growth and thereby increased the total amount of Zn and Cd extracted from soil.	Kuffner *et al.*, 2008
Pseudomonas putida biovar B (four strains)	Soil from a nickel-contaminated site in Port Colborne, Ontario, Canada. Soil texture and organic matter content were characterized.	*Pseudomonas putida* biovar B strain HS-2 was tolerant to 13.2 mM Ni in culture medium. Pot experiments showed that canola (*Brassica napus*) inoculated with this strain doubled biomass weight and nickel uptake by shoots and roots as compared to uninoculated control.	Rodriguez *et al.*, 2008

Organism or consortium	Isolation source	Key findings	Reference
Pseudomonas fluorescens G10 and *Microbacterium* sp. G16	Roots of rapeseed (*Brassica napus*) grown on heavy metal-rich site in Nanjing, China. Soil pH and organic matter content were characterized.	Pot experiments showed that rape seedlings inoculated with *Pseudomonas fluorescens* G10 and *Microbacterium* sp. G16 were tolerant to high lead content in soil. Increased biomass production and total Pb uptake in the bacteria-inoculated plants were obtained as compared to control.	Sheng *et al.*, 2008
Ralstonia sp. (J1-22-2), *Pantoea agglomerans* (Jp3-3) and *Pseudomonas thivervalensis* (Y1-3-9)	Roots of canola in a mine wasteland site in China. No soil data provided.	Plants grown in copper-rich soil and inoculated with these bacteria grew well and took up more Cu than uninoculated plants.	Zhang *et al.*, 2011
Arbuscular mycorrhizal fungi (AMF; various *Glomus* species and strains)	AMF isolated from non-polluted soil, heavy metal-polluted soil and soil treated with sludge rich in heavy metals. Source data provided.	Upon inoculation on Trifolium subterraneum, cadmium adsorption capacity was higher in Glomus spp. isolated from the polluted soil as compared to Glomus from non-polluted soil. Intermediate values were found upon inoculation with isolates from the sludge-treated soil. Cd concentration was higher in roots than in shoots but highest in hyphae.	Joner *et al.*, 2000
Arbuscular mycorrhizal fungi (AMF; various species and strains of *Glomus* and *Scutellospora*)	Roots of *Plantago major* grown on uncontaminated and heavy metal-contaminated sites in Montreal, Québec, Canada. Geo-referenced sites, soil chemical data provided.	*Scutellospora aurigloba*, *S. calospora* and some of the *Glomus* species were most abundant in unpolluted soil. *G. etunicatum*, *G. irregular*, *G. intraradices* and *G. viscosum* were found in both polluted and unpolluted soils, while *G. mosseae* was dominant in polluted soils.	Hassan *et al.*, 2011

A significant body of knowledge exists on bioremediation technology applicable to many organic compounds, but further research is needed for trace elements, radionuclides and complex polycyclic hydrocarbons (Atlas and Philp, 2005). Furthermore, as new contaminant molecules will be invented, these will require new degrading enzymes from yet-to-be isolated soil microbes. The role of soil biodiversity is likely to be pivotal in this context. Soil bacteria and fungi contribute 4 000 to 5 000 species to the biodiversity of a typical terrestrial ecosystem (Heywood, 1995). Many microbial strains remain to be isolated in order to expand the capabilities of remediation technology. In addition, available evidence suggests that whole microbiomes (or interacting sets of microbes) may outperform single strains for bioremediation.

Whereas a large and diverse set of bacteria and fungi with potential for environmental management have been isolated and cultured, it remains of paramount importance that native sources of this soil biodiversity be protected and conserved. Microbiome investigations for environmental management are still novel and highly experimental, yet they underline the need for conservation of entire soil biotic communities. Currently, proper means for conservation of soil biodiversity, *ex situ* or *in situ*, remain a matter for research.

Knowledge gaps on environmental risk assessment of soil pollution

Ecotoxicological studies are carried out within a constraining and limiting framework, as they mainly use standardized procedures recommended in international test guidelines adopted by the International Organization for Standardization (ISO) and the Organisation for Economic Co-operation and Development (OECD). These procedures and studies involve easy-to-culture model species, age-synchronized test organisms and artificial soils. The test concentrations are often unrealistically high, and the studies use pure active substances, while commercial formulations are present in the field. The duration of the tests is often short, ranging from some hours to weeks. Finally, very few studies have assessed the impact of pesticides under natural conditions after their market authorization. For instance, fewer than 1 percent of the studies on pesticides and freshwater invertebrates were performed in the field at the community level (Beketov and Liess, 2012).

To bring more realism in laboratory tests, ecologically relevant test conditions that are representative of field conditions are needed, including natural soils and realistic temperature and moisture conditions. These parameters can influence the bioavailability of contaminants for the test organisms (Nélieu *et al.*, 2016; Harmsen, 2007). Harris *et al.* (2014) emphasized the importance of accurately defining exposure in soil ecotoxicological studies. Increasing attention is given to mixtures of contaminants (Werner and Hitzfeld, 2014). However, more work is needed especially on soil organisms in the agricultural landscape where different pesticides are applied simultaneously or in sequence during the growing season, while at the same time soils can be exposed to microplastics, nanoparticles, trace elements and fertilizers. (5.3.2.1). It is challenging to assess the effects on soil organisms of the exposure to dynamic mixtures of fluctuating chemical composition (for example, by using a whole season approach; Van Hoesel *et al.*, 2017).

In laboratory tests, the chosen species should be sensitive and as representative as possible of the ecosystems to be assessed (Romeis *et al.*, 2013; Pelosi *et al.*, 2013). Ecotoxicologists are encouraged to use test methods that encompass the whole life cycle of the organisms. That implies using other measurement endpoints in addition to traditional ones (survival, reproduction) such as behaviour (avoidance, movements). Such tests also have to account for different development stages, since juveniles can be more sensitive than adults (Bart *et al.*, 2018). Entire life cycle and long-term, multi-generation tests would enable a better effects assessment of persistent contaminants.

In the field, the spatio-temporal variability of environmental conditions and a high number of biotic and abiotic factors operating at the same time hinder impact assessment of contaminants. This is one of the main reasons that many studies report contrasted results regarding the effects of one or several pesticide applications on non-target organisms. To limit the influence of confounding factors and to properly address the consequences of contaminant release under natural conditions, different strategies are possible. Experimental trials can use information on the soil, climate, agricultural practices and history of land use. It is also possible to limit the influence of confounding factors by choosing plots with homogeneous agricultural practices (for example, conventional ploughing, the same cover crop, the same amount and type of organic amendments) or land use, soil type and climate. Although very few data are available on the real exposure of non-target organisms in their natural habitat, these data may help with understanding the effects of chronic environmental pollution.

Understanding biological levels of organization to link laboratory/field works

Most of the procedures for the risk assessment of soil contaminants, such as pesticides, compare the "supposed" exposure (predicted environmental concentrations) to ecotoxicological endpoints for some single species. Laboratory data on animal life cycles using realistic conditions and relevant species can be used to parameterize population dynamic models to be used in risk assessment (Bart *et al.*, in press).

The Dynamic Energy Budget (DEB) theory (Kooijman, 2010) offers a realistic and mechanistic description of the acquisition and the use of energy (provided by food) by an organism to ensure different vital biological functions (growth, reproduction, maintenance). As a next step, the effects of contaminants also need to be based on factors such as population structure and (community) interactions in the ecosystem, timing of release into the environment and landscape structure (Boivin and Poulsen, 2017). Characterizing the exposure or bioavailability (for example, by measuring internal contaminant concentrations in organisms) at the landscape level is a way to link laboratory and field approaches.

Box 5.3.2.1 | Microplastics

Microplastics are emerging persistent contaminants extensively documented in aquatic ecosystems. Recently, their occurrence in soils has started to being reported. Sources of microplastics in soils include recycled organic waste, mulching film, sludge, wastewater irrigation, and atmospheric deposition. (Ee-Ling *et al.* 2018). These exogenous materials can influence soil biota at different trophic levels, and affect human health through food chains (He *et al.* 2018). Earthworms can be significant transport agents of microplastics, incorporating this material into the soil via casts, burrows, egestion, and adherence to the earthworm exterior (Rillig *et al.* 2017). This transport could expose other soil organisms to microplastics and move them to deeper parts of the soil profile potentially reaching groundwater (Rillig *et al.* 2017). Further research is needed to find out the fate and effects of microplastics in soils and take action to avoid this input of exogenous materials.

©FAO / Vinisa Saynes

5.4 | SOLUTIONS TO SPECIFIC SOIL BIODIVERSITY-RELATED PROBLEMS

5.4.1 | CURRENT AND FUTURE APPLICATIONS OF SOIL BIODIVERSITY RESEARCH ON SUSTAINABLE FOOD PRODUCTION

Although modern agriculture can be highly productive, it is not sustainable, primarily because of its reliance on chemical inputs to the soil systems and the amount of fossil fuel needed for soil tillage and other cultivation practices. A likely way to move beyond this and to achieve a degree of sustainability is to incorporate techniques that maintain or enhance soil nutrition and health, without the need for additional chemical additives. Soil organisms play a key role in soil health and nutrient turnover, and thus any attempts to improve soil health and ecosystem services in order to enhance agricultural sustainability need to include strategies to enhance and maintain the biodiversity of soils and to use the services provided by the soil biota in a sustainable way.

New molecular techniques using next-generation sequencing allow for the relatively easy and thorough assessment of soil biota, indicating levels of biodiversity, and the subsequent placement of organisms into functional groups. This, coupled with improvements in more traditional approaches such as culture-based techniques, allow for improved knowledge of what organisms are in the soil, and what impacts those organisms are likely to make on associated cropping systems. Impacts can include positive effects such as nutrient mobilization and transfer, improvements to soil structure and water dynamics, and buffering against soil pathogens. New analytical approaches employing advanced computing power allow for complex network analysis of soil communities and how these communities interact with one another, with crops and with the surrounding soil environment (Li and Wu, 2018; Ramirez *et al.*, 2018). This provides a degree of predictive power to our understanding of how the soil systems will respond to changes in variables such as climatic factors, new cropping systems and soil management. Another application for these tools is the determination of which symbiotic organisms (mycorrhizal fungi and nitrogen-fixing bacteria) are present in the soil, and assistance to the field practitioner in assessing the efficacy of these organisms in benefitting their associated crop hosts (Morgan, Bending and White 2005).

These monitoring tools allow for direct interventions. For example, soil systems that are shown to be rich in certain fungal pathogens or root-feeding nematodes specific to a certain group of plants allow for a farmer to avoid the planting of crops susceptible to these pests, or to use tolerant crop varieties. When the native symbiotic fungi and/

or bacteria in the soil are not optimally suited for the desired crop, specific species or strains may be inoculated, albeit such inocula are not usually very persistent following inoculation to the soil. Such inocula can be presented in the form of smart seed coats or powdered inoculum added to the soil at the time of planting (Rocha *et al.*, 2018). Interventions at the field level can have downstream impacts on crop production, for example on post-harvest processes. Soil microbiota have been found to positively (through beneficial microbial interactions) or negatively (through plant pathogens) influence the quality and longevity of harvested crops (Zarraonaindia *et al.*, 2015; Rillig *et al.*, 2018). Thus, the application of screening methods for associated biota, for example by next generation sequencing, and the subsequent necessary interventions would prove valuable in the post-harvest process. This may enhance sustainability of the full agricultural value-chain.

Future scenarios of agricultural systems might see current and novel technologies individually and in combination as common techniques employed by farmers. New technologies might be next-generation sequencing and cost-effective *in situ* assessments of agricultural soil systems. Farmers would perform fine scale monitoring of their crops and soils, and equipment could auto-inoculate the soils with beneficial organisms, treat pests as they are detected, and provide real-time feedback to science-based databases.

5.4.2 | MICROBIOME-BASED APPROACHES TO IMPROVE PLANT PRODUCTION

In the last decade, soil and plant microbiota and microbiomes have received great attention. It has become evident that complex microbial communities are associated with plants, providing important functions for their host such as providing nutrients, outcompeting and antagonizing pathogens, stimulating plant growth or development and protecting plants against abiotic stresses (Bulgarelli *et al.*, 2013; Hardoim *et al.*, 2015). Furthermore, the term *holobiont* has been coined, recognizing that plants and associated microorganisms act together in a concerted manner and that the plant microbiome is an integral component of plant performance (Aleklett and Hart, 2013; Sanchez-Cañizares *et al.*, 2017). The fact that microorganisms provide beneficial activities, together with an increasing demand for sustainable agricultural production, has led to great commercial interest world-wide. Many small companies have been established to support this growth, and large companies have started to invest in the development of microbiome-based solutions for crop protection or nutrition (Sessitsch *et al.*, 2018).

Nevertheless, currently there are still rather few products on the market, and generally many products show great effects when tested under highly controlled conditions but fail to show reproducible results under field conditions (Compant *et al.*, 2019). The reasons are manifold. On the one hand, applied strains are frequently selected for their plant growth-promoting capacities under laboratory conditions, but do not take into account that any applied strain has to survive in the target environment and establish in a highly competitive environment. Furthermore, microbial biofertilizers or biopesticides do not behave and perform in the same way as their chemical alternatives; they require suitable formulations and application approaches, which still need to be developed. Last, but not least, understanding the role of soil and plant microbiomes has just started, and novel approaches can be expected to benefit from microbial diversity and activities.

Knowledge-based selection of inoculant strains and improved microbial delivery

Currently, inoculant strains are mostly selected based on their activities; for example, biocontrol activity against a certain plant pathogen, determined under laboratory and then greenhouse conditions. Such tests are valid as a starting line; however, their establishment in the target environment is crucial. A biocontrol strain, for instance, in most cases (depending on the mechanism employed) must colonize the same niche and at the same time as the pathogen. Furthermore, a certain dose of microbial cells is required to show the desired effect. Efficient colonization is a major bottleneck in the application of microbial inoculants. First, high quality inoculants are required, delivering a high number of viable and active cells. For this, new formulations need to be developed, particularly for highly sensitive microorganisms such as Gram-negative bacteria (Berninger *et al.*, 2018). Alternative application approaches (for example, by packaging inoculant strains within seeds) have also been shown to improve colonization and activity of the applied strains (Mitter *et al.*, 2017). However, characteristics of applied strains need also to be determined with respect to their capacity to colonize the target niches.

An additional and important aspect is that microbial inocula need to interact with different plant genotypes, whereas often beneficial plant-microbe interactions are plant genotype-specific. A major criterion is the capacity of the inoculant strain(s) to compete with the huge diversity encountered in the soil and plant (root) environment, to establish in the consortium and to synergistically interact with other microorganisms. Here, a better understanding of microbiome networks and core microbiota may help in the future to select candidates with a high potential to establish in the plant environment (Poudel *et al.*, 2016; Tohu *et al.*, 2018). Also required is a better understanding of how beneficial activities are regulated under various conditions, such as those occurring in the field.

Microbiome-based precision farming

With increasing knowledge about plant microbiomes, it will be possible to establish predictive models on microbiome activities and functions (Schlaeppi and Bulgarelli, 2015; Stegen *et al.*, 2018). It may be possible to predict which functions can be mediated by a certain microbiome and which functions must be provided externally by (for example) certain amendments such as organic compounds, plant extracts or microorganisms. Predictive models may be needed to assess the performance of an inoculant strain in a given environment. In the future, such models may help to forecast which crop and/or agronomic practice will be best suited for a given soil or soil microbiome.

5.4.3 | SOIL BIODIVERSITY AS A FARMER'S TOOL

The adoption of soil conservation practices such as reduced tillage, improved residue management, reduced bare fallow and conservation reserve plantings have stabilized, and partially reversed, soil organic carbon (SOC) loss in North American agricultural soils (Paustian *et al.*, 2016). Improved grazing management, fertilization, and sowing legumes and improved grass species are additional ways to increase soil C by as much as 1 Mg C/ha/yr in best cases (Conant *et al.*, 2016). Restoration of late-successional grassland plant diversity leads to accelerating annual carbon storage rates on degraded and abandoned agricultural lands that, by the second period (years 13 to 22), are 200 percent greater in the highest diversity treatment than during succession at this site, and 70 percent greater than in monocultures. This was associated with greater above-ground production and root biomass, and with the presence of multiple species, especially C_4 grasses and legumes (Yang *et al.*, 2019). Similarly, in a recent meta-analysis Bowles *et al.* (2017) found that by reducing soil disturbance (tillage) and maintaining plant cover (cover crops), the formation of beneficial mycorrhizal associations could be enhanced in cash crops. With mycorrhizas playing an important role in the growth and nutrition of plants, this is yet another example of the importance of soil biota in conservation agriculture.

Farm incomes are strongly linked to food prices, production costs and access to markets. World prices of food and agricultural inputs are highly volatile and depend on international markets, where farmers have no influence. For example, farmers worldwide experienced vulnerability between 2001 and 2008, when "world prices for nitrogen, phosphorus and potassium fertilizers increased more than world prices for rice, wheat and maize" (FAO, 2011). Farmers seek to reduce this vulnerability by reducing production costs while increasing yield.

While farmers are attempting to reduce vulnerability, environmental policies worldwide are constantly revised to reduce admissible pesticide levels in food, to limit the use of chemical fertilizers and pesticides (FAO Pesticide Registration Toolkit) and to increase organic farming practices and biofertilizers. For example, organic farming acreage increased by 533 percent worldwide in the period between 1999 and 2017 (IFOAM, 2017), and the demand for organic farm inputs is expanding. In India, for example, production of biofertilizers increased from 22 666 tons in 2004-2005 to 38 933 tons in 2007-2008 (Charyulu and Biswas, 2010). The global market of organic inputs is expected to pursue its growth in the next years. With respect specifically to biofertilizers, the global market was valued at USD 1 254.2 million in 2016 and has been predicted to increase at a compound annual growth rate of 12.9 percent during the period from 2017 to 2025 (TMR, 2017).

Collectively, current environmental policies encourage farmers to think in terms of resource efficiency, low input techniques, biological control, carbon sequestration and ecosystem services, all factors strongly dependent on a diverse soil life. New generations of farmers will need to manage soil biota and farm biotic processes in order to remain profitable (Barrios, 2007) and to meet new market demands. Soil biology can potentially be used to address these policies simultaneously, especially when indigenous knowledge is considered.

Current farm uses of soil biology

The market offers a large array of farm inputs based on microbes themselves or on their metabolites, commercialized as biofertilizers, biostimulants or biopesticides. Based on rhizobia, the first biofertilizers were introduced in 1896 by Nobbe and Hiltner (Fred *et al.*, 1932). Since then, the biofertilizer industry has focused on developing single strain-based microbial products. Today, organisms commonly used for stimulation of nutrient cycling include mycorrhizal fungi, which assist host plants in transporting bioavailable phosphate, and symbiotic nitrogen-fixing bacteria, such as rhizobia for legumes and diazotrophic endophytes such as *Acetobacter* or *Azospirillum* for grasses. For biocontrol, *Trichoderma* is now used among farmers worldwide to mitigate a wide range of soil diseases. *Bacillus subtilis* is used to enhance the plant immune defence against *Rhizoctonia solani*.

Recently, new products have become available that propose microbial mixes and consortia of plant growth-promoting rhizobacteria, which outperform single strain products (Thomloudi, 2019; Bradáčová, 2019). The biological toolat the disposal of farmers includes micro- as well as macroorganisms such as nematodes to control insect and slug populations (Askary, 2017). As farmers become familiar with the use of beneficial soil organisms, they extend applications of these organisms to different farm processes (see 5.4.3.1).

Box 5.4.3.1 | Purposes of on-farm use of soil microorganisms

For soils

- Facilitate organic matter decomposition;
- Increase soil aggregation, enhance humus- forming processes;
- Encourage formation of soil colloids, improve soil nutrient retention;
- Suppress root diseases through substrate competition and niche exclusion.

For plants

- Seed and seedling inoculants improve germination, stimulate growth, encourage synergistic plant-microbe interactions;
- Biofertilizers and biostimulants fix nitrogen, increase availability of phosphorus, potassium and other nutrients;
- Biofilm-producing biopesticides mitigate plant pests and diseases.

For animal production

- Probiotics and prebiotics help increasing diversity of the gut flora in animals, enhance feed conversion and stimulate the immune system;
- Silage inoculants improve feed preservation;
- Bed management in stables reduces nitrogen volatility and regulates air quality in stables.

For waste valorization

- Degrade crop residue;
- Stabilize manure, accelerate composting.

While microbial products have been emphatically proposed to farmers as a tool for reducing dependency on chemical inputs (Shah, 2014; FNCA, 2011), their introduction in the market faces some challenges (Parnell, 2016). Farmers often perceive commercial microbial products as less effective than traditional agrochemicals, with good reason since their *in vitro* performance is difficult to reproduce under field conditions. Moreover, biological inputs have reduced shelf life and field persistence as compared to chemical fertilizers and pesticides. In addition to their transient and environmentally dependent effect, the high cost of biological products also restrains their adoption by farmers (Parnell, 2016), and especially by smallholders with little purchasing power and poor access to credit. In response to these limitations, some farmers with proper training attempt to reproduce native consortia of soil microorganisms to assemble biofertilizer, biocontrol and biostimulant farm inputs. To this end, farmers rely on relatively simple, rapid and affordable techniques.

The functionality of major microbiome components is soil- and climate-specific (Steidinger *et al.*, 2019). Therefore, the ability to source native consortia of soil microorganisms gives farmers access to locally adapted and biodiverse inoculants. The use of native microbial species – as opposed to foreign species – as a farm input may be a valid strategy for increasing biotic resistance to invading alien pathogenic microorganisms (Thakur *et al.* 2019).

5.4.4 | INDIGENOUS KNOWLEDGE RELATED TO SOIL ORGANISMS

Traditional oriental farming resorts to diverse techniques to source, reproduce and use soil microbes. As observed in central Java, local farming communities have a wide diversity of ancestral methods involving different inoculants based on soil microorganisms sourced from a range of environments and using various substrates and brewing conditions. Scientific literature and local practices provide examples of such low-technology strategies for beneficial microbe production (Kumar and Gopal, 2015).

The "rice trap" method (Abu Bakar, 2013; Restrepo, 2007) is commonly used to source and reproduce diverse species of soil microbes. Cooked rice is placed in a covered with gauze, often with a spoon of sugar. The is buried in the ground at a depth of 10 cm and covered with the original litter. After a week, rice is colonized by fungi that are later multiplied in water.

In Japan (Higa, 1996) and in South Korea (Cho, 2016) other methods to reproduce soil microorganisms have been described. The Korean methodology involves the juxtaposition of boiled potatoes and forest leaf mold, defined as the forest soil placed immediately beneath leaf litter (Cho, 2016). Potatoes and leaf mold are placed in two separate nets suspended in water at room temperature. For one to two days, those nets are regularly kneaded while immersed in the water, to extract their content in the liquid and allow microbial multiplication. The resulting extract is diluted in water and sprayed on crops.

Latin American traditional farmers source indigenous microorganisms by fermenting a mix of forest leaf mold, bran and molasses for 30 days. The product of this solid-state fermentation is amplified in water through forced oxygenation and sprayed on soil and crops (Restrepo, 2007; Mancini, 2019).

Microbial sourcing has been traditional practice among indigenous communities in Asia and South America. It has proved effective in enhancing plant nutrition and protection (Kumar and Gopal, 2015; Alori and Babaloa, 2018). Inspired by these long-standing practices, smallholder farmers from Asia, Africa (Kumar and Gopal, 2015) and South America (Restrepo, 2007) are adopting sourcing of native microbial consortia, defined as the recovery of microbes from their natural environment, accompanied by their multiplication and followed by their use in surrounding farms belonging to the agroecosystem from which the recovered and multiplied microbes originate.

Opportunities and knowledge gaps

While commercial and large-scale farmers rely on commercial biological products to stimulate nutrient cycling and to enhance crop protection, smallholder farmers from Asia, Africa and Latin America have a long-standing tradition of valuing and exploiting soil biodiversity to produce farm biological inputs. An extensive monitoring of these traditions and a rigorous evaluation of their effect on yield and plant health represent fruitful avenues for popularizing and illustrating the importance of soil biodiversity and the necessity of its preservation.

5.5 | REGIONAL EXAMPLES OF NOVEL APPROACHES AND APPLICATIONS

5.5.1 | AGRICULTURAL GREEN DEVELOPMENT IN CHINA

China has made great efforts to eliminate hunger and poverty, in line with social and economic achievements since the introduction of the reform and opening-up policy. However, these benefits are mostly derived from intensified agriculture, characterized by high levels of external inputs at the expense of environmental cost, that need to be remediated with great efforts. Clearly, there is an urgent need for a transformation from high-input and high-environmental-footprint agriculture to sustainable intensification. In order to facilitate this transition, a new programme has been started, named Agriculture Green Development (AGD). Meeting the projected demand of the country's fast-growing population on the limited arable lands while sustaining food security, food quality and environment quality is a great challenge that demands novel management strategies. These are being developed under AGD. In this context, the promotion of soil biodiversity to enhance soil functions and to deliver multi-functions is becoming increasingly appealing.

Intercropping is a traditional practice in Chinese agriculture and that of other regions that makes more use of available resources (light, nutrients, water) and space. Intercropping is the simultaneous cultivation of two or more crop species in the same field area, producing more biomass than the average of the corresponding monocultures, particularly in low-input systems. Furthermore, intercropping also provides a wide range of ecosystem services such enhanced water efficiency and decreased disease incidence. More than 28 million ha of land in China are devoted annually to intercropping systems. These benefits are mostly due to the trait-based functional design of the cropping systems. For example, cereals and legumes represent a typical crop combination, showing interspecific facilitation in terms of both yield and nutrient usage. The deployment of mixtures of resistant and susceptible rice genotypes are shown to successfully reduce rice productivity lost due to pathogens. Likewise, there is increasing evidence that soil biodiversity is important for soil functions. For instance, increased arbuscular mycorrhizal fungi abundance in intercropping is correlated to the amount of soil macroaggregates. Disease was reduced with greater inter- and/or intraspecific diversity. Intercropping increases soil organic carbon (SOC) content due to an increase in root biomass and modification of decomposition rates.

Continuous monoculture severely inhibits plant growth, and farmers often suffer with yield decline and economic losses. Monoculture-induced soilborne diseases can be controlled by crop rotation, selection of resistant cultivars or the intensified use of pesticides and other chemicals to protect the crop. However, often these measures are practically or economically unfeasible and/or environmentally unsound. The application of biofertilizers is shown to be effective in improving soil health by direct suppression of pathogens or antagonistic interactions via modification of the indigenous microbial community. Biofertilizers are often applied for cash crops including vegetables and fruits due to the high profits in the cropping systems. For example, the Fusarium wilt disease is one of the most serious soilborne diseases constraining the yield of many cropping systems. Biofertilizers enriched with microbial isolates such as Trichoderma spp. and Bacillus spp. are shown to be effective in fusarium wilt disease control, and have been applied to a variety of crops, including Solanaceae, Cucurbitaceae and Malvaceae. Especially in low-diseased soils, the new biofertilizers exhibit satisfactory performance, and application of these biofertilizers is becoming more popular in China as more and more farmers are recognizing the fact that they can both promote plant growth and protect plant roots from soilborne pathogens (Fu *et al.*, 2017). Biofertilizers amended bulk and rhizosphere soils increased the abundance of bacteria while decreasing fungal abundance (Shen *et al.*, 2015). Bacterial richness and diversity were also motivated. Bacterial and fungal community composition was significantly different between treatment and control. In additon to the biological inoculants, other potential taxa in native soil motivated by the amendments were also observed to be involved in disease suppression (Xiong *et al.*, 2017). In practice, the application form of biofertilizer (*Trichoderma* in the form of biomanure or mixtures of solid andliquid fungal agents), the climate condition and soil types should be considered.

In highly soilborne-diseased soil, soil fumigation using ammonium bicarbonate coupled with biofertilizer (BOF) application was conducted in the field to suppress Fusarium wilt disease. Fumigation strongly suppressed soil pathogen abundance, while both lime and BOF addition enhanced the suppression effect. Alterations in the bacterial and fungal community composition were primarily driven by fumigation followed by fertilization. The abundance of the total microbial community also exhibited a positive influence on the survival of certain of microbial populations after fumigation (Shen *et al.*, 2018; Shen *et al.*, 2019a; Shen *et al.*, 2019b). In practice, the climatic condition and soil type should be considered. Moreover, crop rotation disrupting year-to-year characteristics of the monoculture soil has also been used in China to suppress the soilborne disease. Different crop rotation systems have been observed with varying effects in various crops with different diseases, through mechanisms such as interrupting the cycling of pathogens, allelochemicals and antagonistic microbes, and increasing soil microbial biomass and activity through an increase of soil carbon by root exudates and residues (Wang *et al.*, 2015).

Soil amendments also suppressed soilborne disease though altering protist communities. The addition of organic material and beneficial microbes leads to profound changes of protist community composition, and eventually to functional changes. Continuous application of biofertilizer revealed significant differences in bulk and rhizosphere soil protist community structures, and the structures from biofertilizer treatment were obviously distinct from those resulting from other treatments in both bulk and rhizosphere soils (Xiong *et al.*, 2018; Guo *et al.*, 2018).

5.5.2 | AUSTRALIA

Australia is the sixth largest country by area and is one of the driest continents on Earth, with some of the oldest soils having very low fertility. Although much of the continent is semi-arid or desert, owing to its extensive geographical expanse (9° and 44°S; 112° and 154°E) Australia hosts a diverse range of habitats, from alpine heath to desert to tropical forest, leading to high levels of biodiversity at the continental scale. However, human activity and alien species invasion threatens many of Australia's ecoregions and the species therein.

Although our understanding of soil biodiversity is far less than that of above-ground diversity, Australia has been at the forefront of mapping soil biodiversity at the continental scale with the implementation of the Biomes of Australian Soil Environment (BASE) project. This project sampled soils from over 1 500 sites across Australia, encompassing biomes including deserts, agricultural land, tropical and alpine regions and coastal areas, and provides a globally unique resource for environmental research and management. This reference data base of bacterial (16S rRNA), fungal (internal transcribed spacer, ITS) and eukaryote (18S rRNA) sequences provides information on soil biodiversity at multi-tropic levels and is openly available. Such continental-scale studies require a harmonized approach to sample, process and analyze data in order to ensure base-line data are useful for future monitoring. This then allows data to be combined with other environmental data, such as climate and geochemical and vegetation data, to build a better understanding of soil biodiversity. A recent study used this unique data base alongside plant productivity data, derived from satellite spectral imaging data, to build a systems-level understanding of drivers of biodiversity, soil fertility and production. This study provided the first empirical evidence showing that tripartite positive relationships and feedbacks between microbial and faunal diversity, soil fertility and plant productivity were consistent in topsoils across all soil types and climates across the Australian continent. This dataset also provided evidence of strong shifts in the relative abundance and occurrence of soil taxa (bacteria, archaea and eukaryotes) with increasing distance from the equator. Together these studies provide new insights into the mechanisms driving soil biodiversity and demonstrate its functional significance in terms of soil fertility and primary productivity.

Box 5.5.2.1

As Australian soils are old and highly weathered, soil carbon levels are inherently low and prone to erosion. Loss of ground cover and poor agricultural practices in some cases have resulted in accelerated rates of soil C-loss across the continent, which presumably has had negative consequences for soil biodiversity. Recognition of this has led to widespread adoption of no-till farming practices within the last 30 years such that over 84 percent of grain farming in Western Australia operates under no-till farming, with adoption in other states growing rapidly. This brings Australia to one of the top five countries globally with the largest area under no-till. No-till systems minimize aggregate breakdown and compaction, reducing risks of erosion, and thus serve to retain soil carbon and soil structural heterogeneity, and to promote biological diversity and stability. However, there has been some evidence of recent declines in the number of hectares under no-till in Australia in some areas as a result of reduced effectiveness of herbicides with increasing herbicide resistance. Other concerns over no-till include increased incidence of soil- and stubble-borne diseases, and stratification of nutrients and organic carbon near the soil surface. This has led to the development of innovative, non-chemical methods to control weeds in non-till systems, for example the use of microwaves to kill weeds. The evolution of no-till systems that maintain permanent crop cover, use cover-crops and adopt advanced technologies such as precision farming will minimize chemical inputs and reduce disturbance and will serve to promote soil health and biological diversity within agricultural systems within Australia. This reduced disturbance that promotes natural cycles and focusses on building organic matter and resilient microbial communities is the basis for Regenerative Agriculture, a growing movement within Australia and globally.

In 2012, Australia appointed a National Soils Advocate, Major-General the Hon. Michael Jeffery, in recognition of the importance of Australia's soils for national interests and security and in acknowledgement of the current threats imposed by declining soil carbon levels, increased acidification and loss of soil nutrients across the continent. Following extensive consultation with farmers, scientists, Indigenous interest groups, policy makers, consultants, students and community groups, the Hon. Michael Jeffery presented a report to the Australian government whereby ten recommendations were made to protect and improve the health of Australia's soil, water and vegetation. One of the most pressing recommendations was to establish a National Soils Policy to maintain and restore soil health and thus soil biodiversity. It is recognized that development and implementation of such a policy would require a co-ordinated and integrated approach involving numerous portfolios including agriculture, environment, health, education, defence, Australian Aid, Indigenous affairs, regional development and industry.

Government support (AUD 2M, over 4 years) for the Soils for Life initiative, which promotes adoption of regenerative landscape management, is further evidence that soil health and biodiversity has gained recognition on the political forum in Australia. Established and chaired by the Hon. Michael Jeffery, the Soils for Life initiative has published 27 case studies that prove the concept of regenerative landscape management across many sectors of the farming Industry. With growing interest and uptake of regenerative agricultural practices across the farming sector within Australia, the Soils for Life initiative aims to publish 100 case studies that demonstrate best practice and innovation within farming systems within Australia. This will serve to build resilience within Australian farming systems by promoting biological activity and diversity. It will bring these to the forefront of farming, political and public discussion and could lead to wider-scale adoption of best practice management approaches to build soil carbon and improve health and biodiversity.

5.5.3 | LATIN AMERICA

The most significant example of successful application of a biological process in Brazil and other countries in Latin America, mainly in Argentina and Uruguay, is the inoculation of selected *Bradyrhizobium* bacterial strains in soybean (Franco, 2009). Soybean was cultivated in 2018 in an area of about 35 million ha in Brazil, yielding 120 million tons of grain (Conab, 2019). Inoculation of selected *Bradyrhizobium* strains in soybean cropped in Brazil totally replace mineral N fertilizers, saving billions of dollars a year. Besides its huge economic advantage, biological N_2 fixation from the air by *Bradyrhizobium* is a clean biotechnology, avoiding the lixiviation and volatilization of N-compounds due to the low N-fertilizer use-efficiency by plants. A decade ago, use of N_2-fixing bacterial strains in agriculture was almost exclusively restricted to soybeans, because for a long time it was believed that important food crops such as common beans and cowpeas did not respond to inoculation due to their promiscuity with native inefficient strains belonging to many species and genera. However, in recent years, use of inoculants containing N_2-fixing efficient bacterial strains is increasing in these species, although having a high potential application they are still limited by constraints.

Another biotechnology that is gaining credibility with farmers and hence increasing their application in the field, is the co-inoculation of rhizobia with other plant-growth promoting bacteria such as *Azospirillum* and inoculation of *Azospirillum* in cereals. This genus was isolated and described decades ago by Johanna Dobereiner, an outstanding Brazilian microbiologist intrigued by the large areas of ever green pastures without any N-fertilizer application.

A new promising research area is the multifunctionality of microorganisms related to plant growth-promoting traits other than biological N_2-fixation such as phosphate solubilization, biological control of pests and diseases and hormone production, among others (Martins *et al.*, 2018; Costa *et al.*, 2017; Estrada *et al.*, 2013; Jung *et al.*, 2003; Jeasen *et al.*, 2002).

5.5.4 | THE UNITED STATES OF AMERICA

In the United States of America, agricultural productivity has increased dramatically over the last century, largely from innovations associated with the Green Revolution. Higher yielding crop varieties were developed to take advantage of the advent of inexpensive fertilizer. Continued advances in mechanization and harvesting technology, as well as herbicides and pesticides, favoured uniform monocultures, which in turn reduced the reliance on human labour in industrial agriculture. While combined advances in breeding and chemistry supported low food prices and a secure food supply, they coincided with severe soil erosion and degradation, pollinator decline, increased greenhouse gas (GHG) emissions and nutrient runoff leading to eutrophication. Over the course of the Green Revolution, agricultural systems in the United States of America became less reliant on soil biological processes as chemical nutrition supplemented biological nutrient cycling. However, it is increasingly clear that reliance on agrochemistry incurs costly trade-offs and emerging challenges. These concerns contributed to the development of organic standards for farming techniques that utilize biological approaches and minimize non-natural inputs. Consumer demand for organic foods has steadily increased over time, along with other labels and certifications affiliated with sustainable farming claims. In some respects, these approaches mimic traditional farming practices that relied on natural soil functioning and biodiversity. Today, the perceived dichotomy between organic and conventional practices may be dissolving, as hybrid practices that incorporate aspects of both models gain favour as best practices that support long-term sustainability and profitability.

The concept of soil health exemplifies the common ground among different farming philosophies. There is broad support for practices such as no-till and cover cropping that can help regenerate soil organic matter and enhance soil biodiversity and functioning. In parallel, there is now broad recognition of the importance of biological processes for production agriculture. Industry has invested in the development and deployment of biological solutions including biostimulants and biocontrol.

Biostimulants – products that deliver live microbes or natural products to stimulate microbial activity – have a long history of utilization in agriculture. For example, soybeans have successfully been inoculated with N-fixing bacteria to enhance their productivity. Other products deliver undefined and often inconsistent microbial mixtures, with mixed or undocumented efficacy. More recently, biological products with targeted functionality have taken advantage of high-throughput screening, low-cost sequencing and gene annotations. Bacteria or fungi with specific functions such as phosphorus solubilization or production of ACC deaminase have long been isolated using traditional selective media plating techniques. New high-throughput approaches enable screening of synergistic consortia through dilution of soil extracts, which may be more robust and effective across different environments. Informed by genomic information and metabolic mapping, functional consortia could also be synthetically assembled. This new generation of technologies may be more consistently effective in increasing plant yield and vigour, reducing stress responses and inhibiting disease.

5.6 | FUTURE RISKS

It is well established that soils are immensely rich in species, but a high proportion are still to be described, and their ecology is rarely well defined. It is therefore difficult to hypothesize what is being lost and what the consequences may be. There is a great need for further understanding of below-ground organisms to fill this gap before species are lost. Still, although there is yet little evidence, it is most likely that ongoing pressures from global changes such as climate change, land use including urban growth, and increased dispersal of pathogens and pests may result in local extinctions of soil organisms. It is more uncertain whether extinctions will occur at a global scale, given the broad distribution and large number of individuals of most soil organisms. However, it is obvious that land degradation, soil pollution and even agriculture will result in extinction of local soil biodiversity. Most notably, larger organisms are more likely to be at risk than smaller ones (Tsiafouli *et al.*, 2015), given their inherently smaller population sizes and often more restricted ranges. Specifically, fragmentation can result in reduced gene flow among populations, resulting in increased risk of biodiversity loss. The resulting reductions in soil biodiversity at local scales will have unknown knock-on effects on above-ground biodiversity, whole ecosystem functioning and, consequently, human well-being. For example, shifts in microbial assemblages could result in the potential loss of beneficial microbes that aid plant growth and resistance to pathogens and stress, resulting in negative impacts in natural and managed landscapes alike. There is also the potential that biodiversity loss and novel environmental conditions could increase the risk of disease outbreaks and reduced ability of local communities to resist invasive species (Bardgett and Van der Putten, 2014), both above ground and below ground. The use of chemicals may contribute to the development of pesticide resistance, making control of pathogens increasingly difficult. Moreover, the loss of diversity may hinder the development of new technologies, pharmaceuticals and agricultural products through the loss of yet-to-be-described functional capabilities of particularly microbes (for example, enzymes and antimicrobial compounds).

5.7 | EDUCATION, MAINSTREAMING AND POLICY

The increased awareness of the importance of soil for maintaining biodiversity has created a range of actions ranging from awareness-raising and education to policy actions.

In December 2013, the Sixty-eighth Session of the United Nations General Assembly declared 2015 as the International Year of Soils (IYS) and World Soil Day (5 December) as official UN observances (A/RES/68/232). Through its Global Soil Partnership (GSP), FAO received the official mandate to coordinate the implementation of those two global communication campaigns.

The IYS 2015 generated a momentum that represented a turning point in the history of soil. The advocacy activities, over 900 events in 90 countries organized within the year of the campaign, provided an opportunity to position sustainable soil management at the centre of the policy debate. The topic of soil biodiversity was brought to light with communication products such as infographics, flyers and videos launched specifically during the IYS.

Every year on 5 December, World Soil Day campaigns call on governments, education and academic sectors, farmers, scientists, youth, business and civil society to take action and raise awareness on a specific soil threat. Since 2014, WSD has gained great momentum and has become one of the most celebrated days of the UN calendar. Hundreds of millions of persons are reached every year through celebrations across the world and articles in major newswires in several languages.

For the WSD 2020 celebration, the theme focuses on soil biodiversity under the slogan "Keep soil alive, protect soil biodiversity." A wide range of communications products such as flyers, brochures, animations, posters, videos and much more will be prepared in all FAO languages and spread through WSD organizers and the GSP audience. The WSD 2020 campaign will increase awareness about the need to support soil biodiversity, and to communicate what it is, and why and how it is important. It will also contribute to the adoption of a broader approach by linking it to food security, food safety, human health (including the discovery of well-known antibiotics) and climate change. In addition, it will make a connection between soil biodiversity and people's daily lives, highlighting why human life cannot exist without healthy soils. Calls to action through the engagement of key influencers will make the messages more tangible for the public. Content-sharing on social media and digital platforms will allow us to reach millions of people and make the theme of soil biodiversity familiar and of relevance to a larger audience.

Knowledge on soil biodiversity is being more and more integrated into higher education, and many resources are made freely available, including web-based teaching resources such as "It's Alive!" (https://biology.soilweb.ca/), which provides students with information about the effects of several types of forest harvesting on soil microbial groups and their functions. Moreover, several Massive Open Online Courses (MOOCs) on soil science have become available (see for example https://www.classcentral.com/tag/soil-science). The scientific community is also being encouraged to collaborate across fields using multidisciplinary approaches in order to close gaps on global soil species distributions and functions, to assess their vulnerabilities to global change drivers, and to understand the causal links among soil biodiversity, ecosystem functions and their associated services (Guerra *et al.*, 2019). An example stems from the recent global survey of 16 soil chronosequences spanning a wide range of ecosystem types (Delgado-Baquerizo *et al.*, 2019), or the recently published study by a diverse team of scientists on the global abundance of nematodes in the soil (van den Hoogen *et al.*, 2019).

Communicating research and functions of living soil with farmers is an important step in the development of management options for soil biodiversity, which in turn can increase yields and reduce costs (Orgiazzi *et al.*, 2016). Initiatives like the tea bag index (Keuskamp *et al.*, 2013) or the Soil Conservation Council of Canada's "Soil Your Undies" initiative (see https://soilcc.ca/soilyourundies/2017/soil-your-undies.php) are ongoing collaborative projects with citizen scientists designed to assess the health of soils and to raise awareness of the need to protect soils (Keuskamp *et al.*, 2013).

Citizen science is suggested to have the potential to contribute immensely to regional and global assessments of biodiversity by providing large amounts of data for monitoring biodiversity (Chandler *et al.*, 2017). For example, in an effort to find microorganisms useful for therapeutics, citizens sent in soil samples from around the globe (www.whatsinyourbackyard.org). Such initiatives could be applied in the future to help close the gaps in global soil biodiversity data by using global consensus on sampling strategies and methodological approaches, which is currently a major challenge (Cameron *et al.*, 2018). Citizen scientists are also helping to measure the impacts that management practices have on environmental health (Ryan *et al.*, 2018). For example, the use of simple technologies, such as soil kits, is empowering farmers to acquire and practise site-specific nutrient management, resulting in increased yields with reduced fertilizer inputs (Attanandana, Yost and Verapattananirund, 2007).

Education and outreach

In recent times, plenty of resources have become available to make the knowledge of soil biodiversity accessible to society at large, for audiences from young children to schoolteachers, farmers and politicians. These include books, such as the *Global Soil Biodiversity Atlas* (Orgiazzi *et al.*, 2016), a colouring book for children created by the Soil Science Society of America (https://www.soils.org/files/iys/iys-colorbook-for-web.pdf), an educational booklet from Brazil entitled "Curumim and Cunhantã helping the soil biodiversity" (http://repositorio.ufla.br/jspui/handle/1/1476) and playing cards, such as the French card game "The Hidden Life of Soil" (https://rnest.fr/le-programme-gessol/), as well as applications such as ISRIC's SoilInfo app (https://www.isric.org/explore/soilinfo), an approach to generating open soil data.

Outreach projects are being successfully used to inform and educate the public about soil biodiversity. Campaigns by volunteers like the Toronto Wildlife Centre's Backyard Biodiversity project (https://www.torontowildlifecentre.com/backyard-biodiversity/) help people learn more about the flora and fauna in their backyards, including a Wildlife Hotline advising the community about a diversity of wildlife situations. The recently-started campaign "Will Run for Soil" (https://www.willrunforsoil.com/) aims to change the conversation around soils by making a film about soils and the scientists who study them. Science communication initiatives like the University of Waterloo's (Canada) "Let's Talk Science" (https://outreach.letstalkscience.ca/uwaterloo/local-programs/classroom-community-visits.html) teaches soil biodiversity workshops by visiting classrooms and homeschool groups. Moreover, transdisciplinary approaches through the combination of arts and sciences help to raise awareness and educate the public on soil biodiversity-related issues, such as an ongoing project on creating ecological consciousness to climate change through the use of sound (https://www.soundandspaceresearch.com/latestnews/2019/2/7/aural-soilscapes-ckjtc). Initiatives like these will likely inspire future projects to increase the awareness to the importance of soil biodiversity and how to preserve it.

To combine and make better use of existing information and knowledge from relevant disciplines (including biology, ecology, soil science, and agronomy) as a means to guide practical action for conserving and sustaining the functions and value of soil biodiversity in agricultural systems, case studies were compiled in the recent years and made available through FAO and CBD web sites. These can be downloaded and disseminated by partners for use at local and national level. A standard format for the presentation of case studies was prepared as a basis for replication and adaptation including: the type of problem addressed; proposed solutions; specific techniques and management practices; tools and approaches for improved management and assessment; analysis of the principles and lessons learned from such experiences (http://www.fao.org/3/y4810e/y4810e0c.htm#TopOfPage).

As part of policy actions, several networking activities have been initiated in order to mobilize interested stakeholders and to facilitate regional and thematic coordination and cooperation among partners. One of these is the Global Symposium on Soil Biodiversity, (http://sdg.iisd.org/events/global-symposium-on-soil-biodiversity-gsobi20/) a science-policy meeting that aims at filling critical knowledge gaps and at promoting discussion among policy makers, food producers, scientists, practitioners and other stakeholders on solutions to living in harmony with nature. Global reports to highlight the importance of soil biodiversity and the need for its conservation through policy actions are increasingly being released by governmental and non-governmental organizations, including the European Commission's report on *Soil biodiversity: functions, threats and tools for policy makers* (https://ec.europa.eu/environment/soil/biodiversity.htm) as well as the recent warning by scientists to consider (soil) microorganisms in the development of policy and management decisions due to their global importance in climate change regulation (Cavicchioli *et al.*, 2019). Future reports like these as well as updates of databases to include soil biodiversity data, such as the International Union for Conservation of Nature's "Red List of Threatened Species" (https://www.iucnredlist.org/) will be essential steps towards implementing policy actions to preserve soil biodiversity.

Table 5.7.1

Elements of an outreach strategy for soil biodiversity, organized by impact on human well-being

Impact of soil biodiversity on:	Element of strategy
Natural capital	Incorporate soil biodiversity information into natural capital assessments and monitoring.
	Characterize the trajectory of soil natural capital under both agricultural intensification and extensification and under more sustainable soil management practices.
	Analyze how changes in soil natural capital impact ecosystem services provided to other biomes and habitats.
Agriculture	Promote on-farm use of beneficial soil microorganisms for biological control of pests and diseases and enhancement of plant nutrition.
	Monitor traditional and artisanal methods for sourcing of indigenous microorganisms by smallholder farmers and evaluate efficiency of their use.
Environmental management	Preserve soil biodiversity and characterize microorganisms useful to bio- and phytoremediation, to face current and future challenges related to soil contamination.
	Ensure access to microorganisms useful to soil bioremediation.
	Monitor source ecosystem of microorganisms used for bioremediation.
Health	Use a combination of metagenomic soil analysis and advanced culture techniques to isolate microorganisms producing novel antimicrobials.
	Pay attention to extreme soils as a source of new antibiotic-producing microorganisms.
	Explore the biodiversity of soil viruses to devise bacteriophage therapy strategies.
Economy	Sift through the multitude and complexity of approaches when thinking about economic valuation of services mediated by soil organisms. Present clear and influential messages to different groups of stakeholders.
	A key group of soil organisms should be selected and their diversity and abundance considered as holistic indicators of a capacity for provision of sustainable soil services and their. The economic value of these indicator organisms should then be derived at a global scale.
	Efforts must be made to generate a global map of case studies based on a pluralistic valuation approach, where location-based valuations for soil biodiversity are made for areas representing different biomes and different land management practices.

Soil biodiversity is part of an integrated living system driven by mutualisms and food webs, in which humans also participate. A healthy soil is a dynamic system with a diverse and complex assemblage of soil organisms whose interactions determine functional capacity. The integrity of soil biodiversity in all of its many facets, and not only some components of it, must be preserved.

© Guillermo Peralta

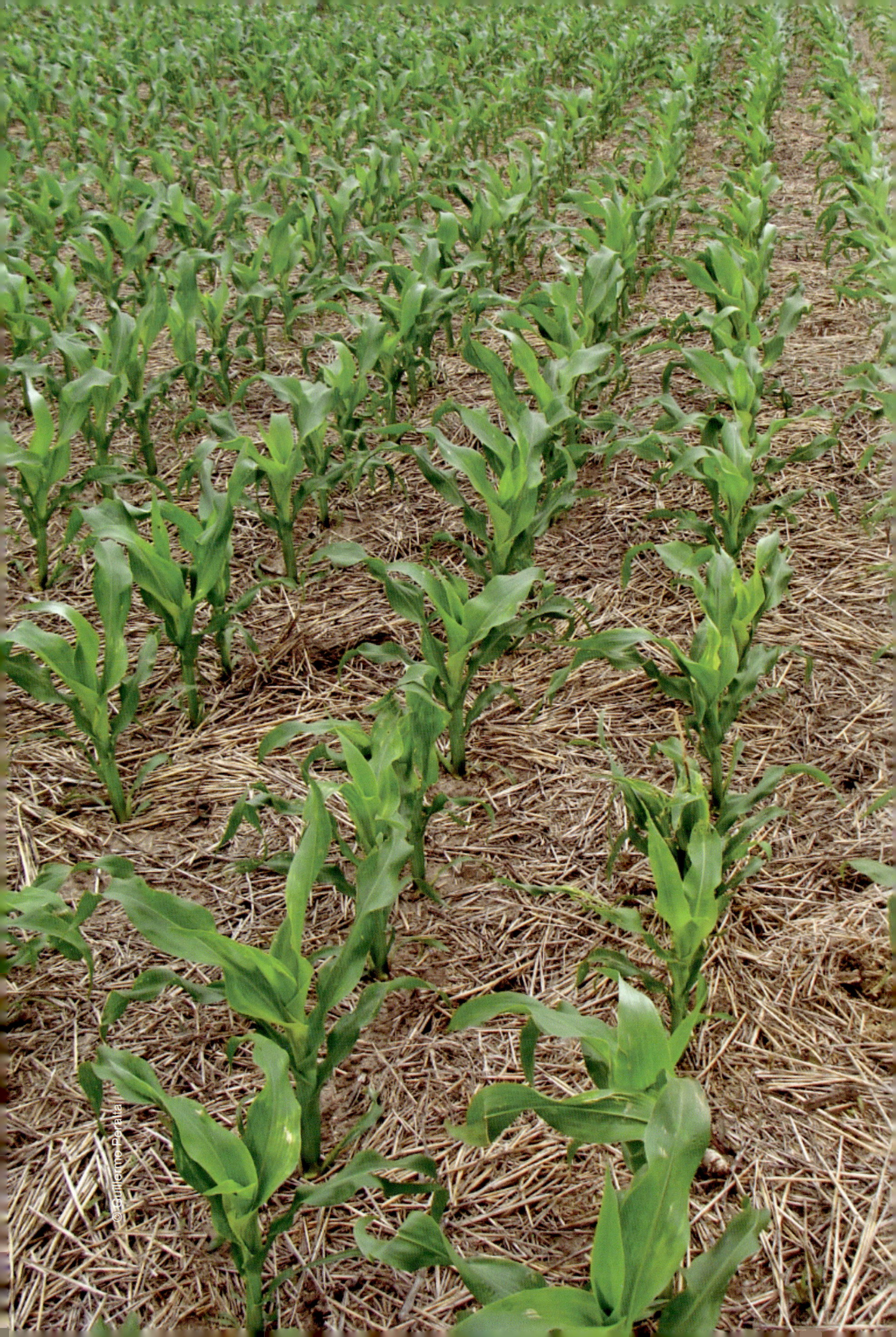
© Guillermo Peralta

CHAPTER 6
STATE OF SOIL BIODIVERSITY AT THE NATIONAL LEVEL

At the Fourteenth Conference of the Parties (COP) to the Convention on Biological Diversity (CBD) (held in Egypt in 2018) the Food and Agriculture Organization of the United Nations (FAO) was invited to prepare a report on the state of knowledge on soil biodiversity covering the current status, challenges and potentialities. Additionally, the COP requested the Secretariat of the CBD, in consultation with FAO under the aegis of the Global Soil Partnership (GSP) as well as other interested partners, to review the implementation of the International Initiative for the Conservation and Sustainable Use of Soil Biodiversity.

From August to October 2019, FAO invited its members countries to participate in the "National Survey on the Status of Soil Biodiversity: Knowledge, Challenges and Opportunities." To enhance further collaboration, the Parties to the CBD were also invited to submit information through the same platform and were encouraged to coordinate with the appropriate line Ministries and relevant institutions at the national level.

The survey was a first step in this process. The aim was to collect information at the country level on the status of soil biodiversity, to better understand the concerns and threats to soil biodiversity, to compile relevant policies, regulations or frameworks that have been implemented and to catalogue current soil biodiversity management and use efforts.

The survey consisted of 16 questions divided into five sections: (i) General information; (ii) Assessment; (iii) Research, capacity building and awareness raising; (iv) Mainstreaming (policies, regulations and governmental frameworks); and (v) Gap analysis and opportunities. The full questionnaire is available in Annex II; the countries that submitted their responses are listed in Annex III.

Fifty-seven (57) countries submitted their responses and all of the following regions had at least one representation: North America; Latin America and the Caribbean (LAC); Europe and Eurasia; the Near East and North Africa(NENA); Sub-Saharan Africa (SSA); Asia; and South West Pacific.

6.1 | ASSESSMENT OF SOIL BIODIVERSITY

The objectives of this section are to present what countries reported on the state of assessments of the level of current knowledge on soil biodiversity, the identification of the main drivers of pressure that have had a negative impact on below-ground biodiversity over the last ten years and the understanding of how soil biodiversity has been monitored.

A detailed report by region is provided in the Annex 1. Some countries provided more details than others, but for the purposes of this report, the information is presented as it was received.

HIGHLIGHTS

- Globally, above-ground biodiversity has been thoroughly studied and well documented, and policies to promote the conservation and sustainable use of terrestrial, marine and other aquatic ecosystems have been developed. However, for some countries, soil biodiversity remains unknown and yet to be assessed.

- There are few national assessments specifically addressing underground biodiversity, and some countries have reported assessments with indirect links to soil biota. There are also few countries that maintain a national soil information system that includes soil biodiversity. Figure 6.1.1 shows some national assessments developed by countries with links to soil biodiversity.

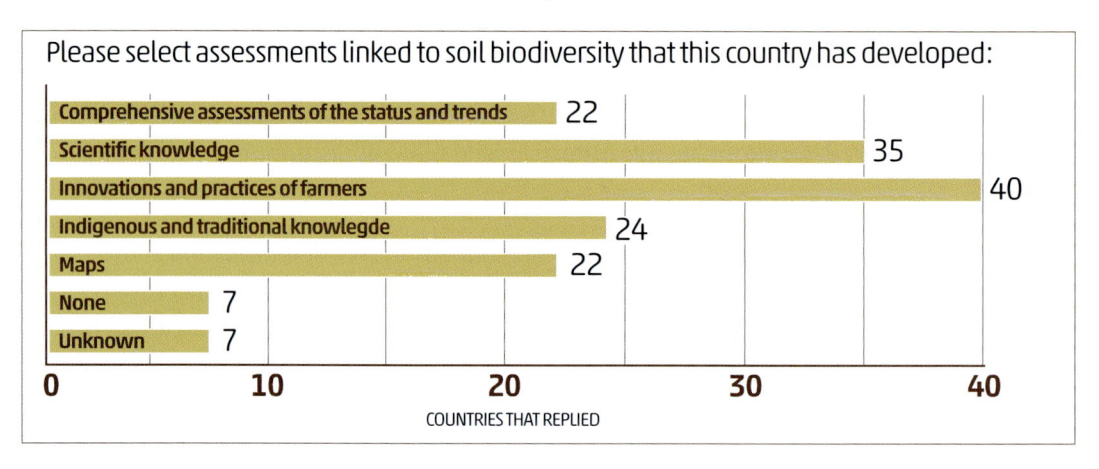

Please select assessments linked to soil biodiversity that this country has developed:

Category	Countries that replied
Comprehensive assessments of the status and trends	22
Scientific knowledge	35
Innovations and practices of farmers	40
Indigenous and traditional knowlegde	24
Maps	22
None	7
Unknown	7

Figure 6.1.1 | National assessments that include soil biodiversity
(Survey Question 2.4, Annex II)

The majority of countries have developed assessments or maps that are directly or indirectly linked to soil biodiversity.
- There is an overall recognition of the potential of soil biodiversity to contribute to ecological restoration, pest control, improvement of plant nutritional quality and human health. Initiatives such as regenerative agriculture, conservation agriculture and organic agriculture are starting to include soil life as a priority.

- High soil diversity is expected to accelerate soil remediation.

- Countries reported on the main practical applications of soil biodiversity and on any available evidence that the enhanced use of soil biodiversity has contributed positively to food production and nutrition.

- The main practical application of soil biodiversity was agreed to be the use of soil ecosystem services that humans have taken for granted. Some examples are soil fertility, water purification, clean air, ecosystems resilience, temperature/precipitation regulation, and climate change mitigation and adaptation. Figure 6.1.2 shows how certain ecosystem services provided by soil biodiversity are perceived by experts in the reporting countries.

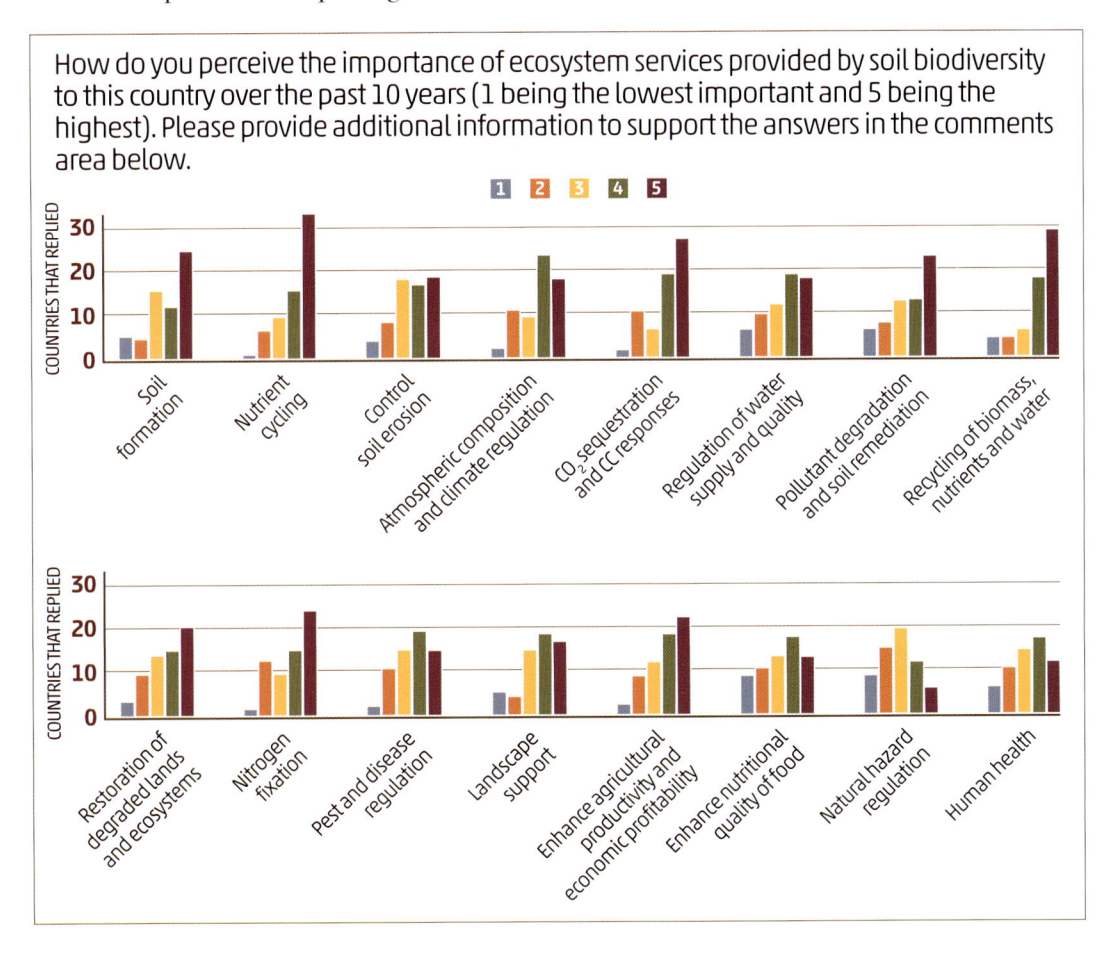

Figure 6.1.2 | Soil biodiversity and ecosystem services perception
(Survey Question 2.1, Annex II)

There is a strong perception that soil biodiversity has high importance for most ecosystem services.

- The current practices that are driving the loss of soil biodiversity are presented in Figure 6.1.3 which shows the perception of national experts on how those practices have had a negative impact on soil biodiversity over the last ten years.

State of soil biodiversity at the national level

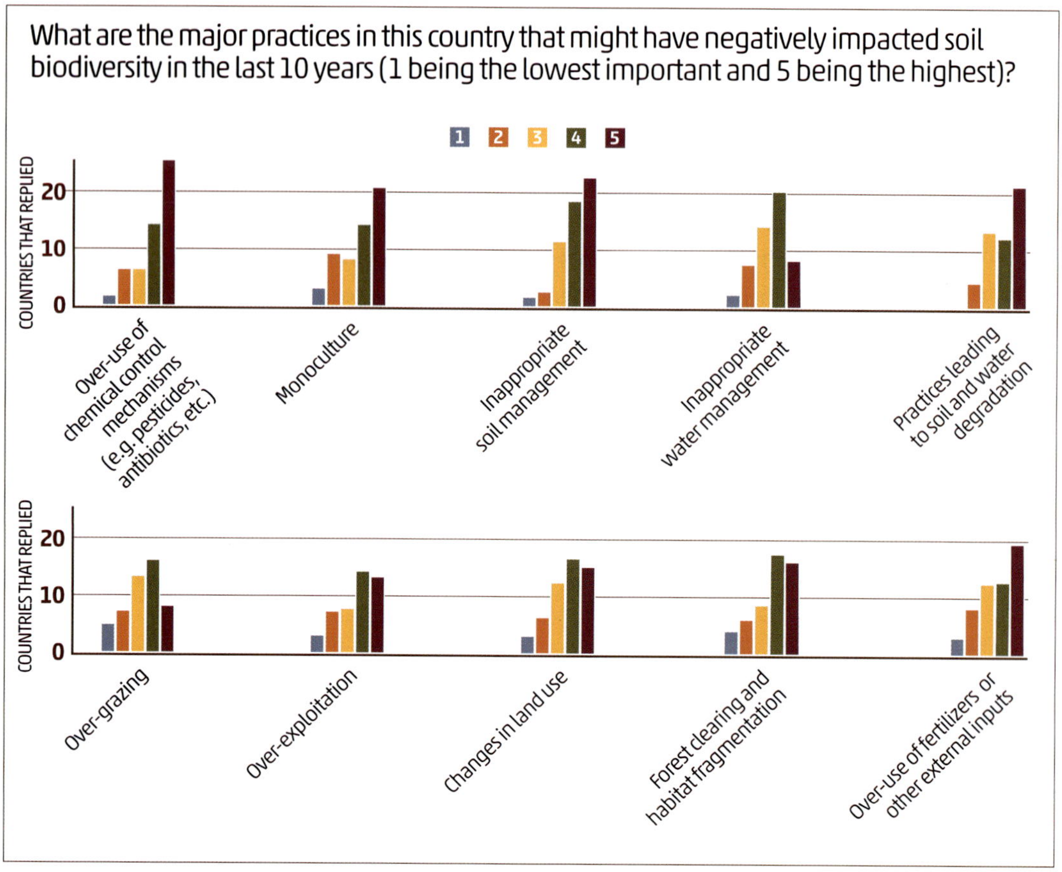

Figure 6.1.3 | Major practices that have a negative impact on soil biodiversity
(Survey Question: 2.8, Annex II)

There is a strong perception among the answers that practices related to land use change and intensive agriculture have a negative impact on soil biodiversity.

- The application of soil-biodiversity friendly practices (for example, no-tillage, biofertilizers, biopesticides, cover cropping and crop rotation) increases and improves the conditions for the soil organisms, their communities and soil ecosystem functioning. It is widely accepted that these practices contribute to the increase of organic matter in agricultural soils, which has led to an increase in microbial biomass and, subsequently, to improvements in soil health, crops productivity and quality.

- The increasing adoption of nature-based solutions can encourage plant and soil biota diversity. This diversity contributes to increasing resilience and improving the control, prevention or suppression of pests and pathogens.

- The use of inoculants for symbiosis contributes to productivity and stress-resistance in agriculture.

- Soil mutualistic microbiota (arbuscular mycorrhizal fungi, nitrogen fixing bacteria) are key components of soil biodiversity as their diversity and abundance can reduce

- the cost and dependence on chemical nitrogen fertilizers in agriculture, enhance soil fertility and environmental sustainability (air, soil, water) notably by reducing greenhouse gas emissions from the energy-intensive manufacture of nitrogen fertilizers and their excessive application in soils .

- There is a rising trend in the market for biological control agents for soilborne diseases over the last decade, in response to the changing environment and the social recognition of the need for increased sustainability.

- Some countries reported that the introduction of invasive alien species has had a significant effect on above-ground biodiversity and has also had negative impacts on the native soil fauna.

- Urbanization has been shown either to decrease macro-invertebrate diversity or to alter it through a greater contribution of non-native species and a homogenization of the fauna with the loss of specialist species.

- Excessive reduction of soil biodiversity, especially the loss of key species and/or species with unique functions, can have cascading ecological effects, leading to the long-term deterioration of soil fertility and the loss of agricultural productive capacity.

- With regard to the inappropriate use of chemical control mechanisms (for example, disease control agents, pesticides, herbicides, and veterinary drugs), further studies are needed. However, there is evidence that some pesticides can have an impact on mesofauna and have sublethal effects on other organisms.

- Fertilizers also can affect soil fauna.

- Although the relevance of soil biodiversity is well recognized, the majority of countries lack the knowledge, capacity and resources to implement soil health principles and adopt best practices, as well as to invest in research, assessments, indicator development, and monitoring.

- Generally, there is a lack of data on soil biodiversity at local, national, regional and global levels. For instance, very few available soil information systems and soil surveys include soil biodiversity as one of the soil properties to be considered.

- Therefore, to better plan successful soil biodiversity monitoring, there is an urgent need to invest in soil biodiversity assessments in most countries worldwide.

- Countries reported that some indicators have been used to evaluate soil biodiversity, such as the indicators presented in Figure 6.1.4.

Are there any implemented indicators for evaluating soil biodiversity related to:

Indicator	Countries that replied
Ecotoxicology	11
Genetic diversity	14
Soil carbon sequestration	26
Soil erosion	22
Soil fertility management	31
Water management	17
Soil pollution	18
Biological control	19
AMR	6
None	13
Unknown	5

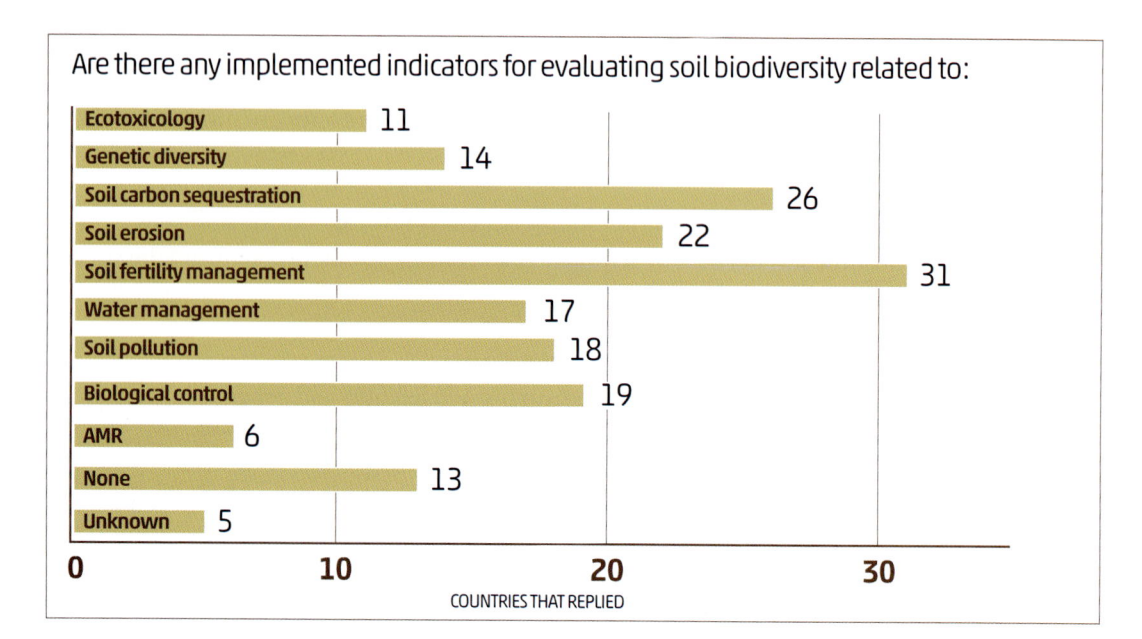

Figure 6.1.4 | Indicators commonly used for soil biodiversity evaluation
(Survey Question 2.7)

Some indicators currently in place can be used to evaluate soil biodiversity, such as indicators to measure soil fertility, soil carbon sequestration, biological control, soil erosion and soil pollution.

6.2 | RESEARCH, CAPACITY DEVELOPMENT AND AWARENESS RAISING

This section identifies and synthesizes some of the soil biodiversity research programmes and initiatives implemented by countries to support sustainable soil management practices. Detailed information can be found in Annex 1. Countries also reported on capacity development, training, extension or interdisciplinary educational programmes and awareness-raising activities that aim at the conservation and sustainable use of soil biodiversity. Some countries have adopted sustainable soil management to prevent land users from impairing essential soil functions, in line with the World Soil Charter [1] and the Voluntary Guidelines for Sustainable Soil Management (VGSSM).[2]

---

[1] http://www.fao.org/3/a-i4965e.pdf
[2] http://www.fao.org/3/a-bl813e.pdf

HIGHLIGHTS

- Taxonomy is essential not only for the future of soil biodiversity research programmes, but also to raise awareness of the collapse of biodiversity.

- The lack of taxonomists in many fields is a real concern in terms of knowledge and assessment capacity for the implementation of soil biodiversity conservation and sustainable use.

- Numerous national institutions, research centres, networks, universities and schools are including soil biodiversity in their programmes. Some of them are also leading research on technological innovations as well as on traditional and agroecological approaches related to soil biodiversity (for example, research, practical application, assessment, indicators and monitoring).

- Tropical conditions are different from those in temperate zones, which require different metrics for assessing vulnerability and sustainability, as well as soil health and quality.

- There are activities to implement the Voluntary Guidelines for Sustainable Soil Management (VGSSM).

- Best agricultural management practices (for example, biological nitrogen fixation; no-tillage; integrated crop-livestock-forestry systems; crops diversification; restoration of degraded lands) have been adopted.

- Awareness raising was achieved through citizen-based scientific initiatives, knowledge transfer, capacity building, symposia and outreach materials (for example, atlases, magazines, maps, games, campaigns and publications).

- Farming systems have been evolving towards environmental and economic sustainability.

- There are initiatives to extract DNA from soils to measure soil macrofauna, analyse and define soil biodiversity indicators and standardize protocols on soil biological quality.

- Special events have been celebrated, including World Soil Day (5 December).

6.3 | MAINSTREAMING: POLICIES, PROGRAMMES, REGULATIONS AND GOVERNMENTAL FRAMEWORKS

This section provides a summary of what the countries reported in terms of their policies, programmes, regulations, enabling frameworks or any legal instruments that refer to soil biodiversity. Countries were also invited to present how they integrate soil biodiversity into sectoral and/or cross-sectoral policies (Annex 1) (such as those related to health, food security, environment, forestry, agriculture, protected areas, land management, climate change, family farmers, indigenous peoples and local communities).

HIGHLIGHTS

- There is an overall interest in promoting sustainable soil management as well as sustainable agriculture and/or sustainable intensification practices through policies, laws and other activities.

- There is a growing interest in examining and understanding the connections between soil biodiversity and other topics such as human health and ecosystem restoration.

- There are some direct and indirect references to soil biodiversity in countries' National Biodiversity Strategies and Action Plans (NBSAPs) and other global commitments.

- There is some direct and indirect integration of soil biodiversity in sectoral and/or cross-sectoral policies.

- Some countries are starting to integrate soil biodiversity in different areas such as food security, environment, forestry, agriculture, protected areas, land management, climate change, indigenous peoples and local communities and family farmers.

- There is a general lack of specific policies, programmes and actions that take into account the particularities of soil biodiversity, as well as measures to promote its conservation and sustainable use.

6.4 | ANALYSIS OF THE MAIN GAPS, BARRIERS AND OPPORTUNITIES IN THE CONSERVATION AND SUSTAINABLE USE OF SOIL BIODIVERSITY

In this section, a summary of what the countries reported in terms of their experiences regarding the barriers to the implementation of better soil biodiversity management strategies is presented. This section also includes the plans of the countries to ensure that soil biodiversity conservation and sustainable use are (directly or indirectly) taken into account in national planning and sectoral policy development (for example, national biodiversity strategies and action plan, national agricultural planning and national health planning). Figure 6.4.1 shows the perception of the main barriers to the implementation of better soil biodiversity management strategies.

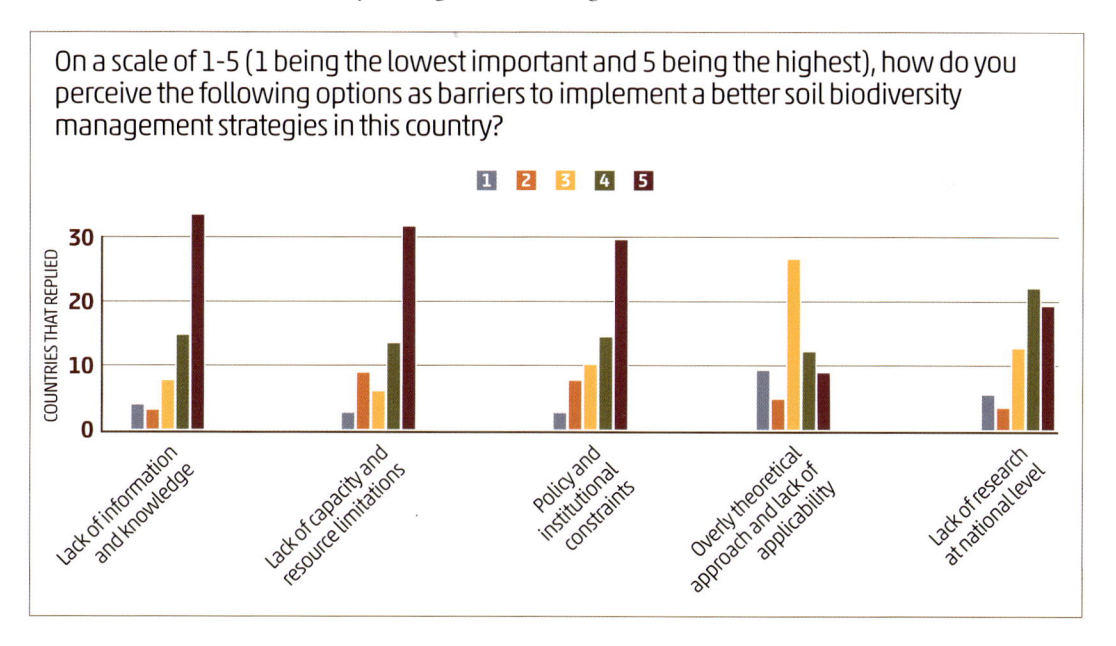

Figure 6.4.1 | Barriers to the implementation of better soil biodiversity strategies
(Survey Question: 5.1, Annex II)

For the vast majority of countries, the most important barriers are related to access to information and resources, and the lack of inter-sectoral articulation to implement policies and strategies.

HIGHLIGHTS

- The partial understanding of the status, phenomena and dynamics of soil ecosystems, as well as their "directions" and stability, under different management systems and in the face of changing climate conditions represent barriers to the implementation of better soil biodiversity management.

- Although few countries reported direct and indirect links to soil biodiversity in their NBSAPs, there is an overall need to reinforce the direct links in their future biodiversity plan.

- Existing public policies have indirect benefits, but there is a need to include soil biodiversity considerations in future policies and programmes.

- Several countries reported that inter-sectoral and inter-institutional articulation was still needed to explore synergies and avoid duplication or fragmentation, as in some countries soil polices are the responsibility of different ministries.

- There are some successful cases of inter-sectoral collaboration, for example in agricultural planning and environmental studies to conserve and sustainably use biodiversity.

- There is still a general lack of understanding on how soil biodiversity influences ecosystem functions and services towards general biodiversity.

- Taxonomists urgently need to characterize soil biodiversity before its loss.

- There is a need to combine the work of ecologists and taxonomists to better correlate the composition of species community with ecosystem processes and patterns.

- There is a need to assemble comprehensive soil censuses that include organisms, their features (that is, numbers, qualities, metabolic roles), their interactions and functioning to better understand the impacts of cropping systems, management systems, climate change conditions and other ecosystems. This will enable the development of relevant management approaches that support the optimization of soil biodiversity for sustainable and even regenerative development of ecosystems and to increase ecosystem functions and services for our world.

- There is a need to raise awareness of soil biodiversity and its loss drivers in order to include it in planning instruments and strategies for environmental and productive sector management.

- There is a need to promote the necessary shift from the use of conventional physical and chemical indicators to the use of biological indicators.

- The adoption of conservation practices by farmers, as a basic premise for soil conservation, remains low.

- There is a need to inform the general public about the importance of soil biodiversity and its conservation.

- There is a need for indicators to measure and monitor soil health and soil biodiversity.

- There is a need for an integral vision of agriculture and forestry sectors, and a collaboration between the government and educational institutions.

- There may be a need for a global convention that would oblige signatories to take action to fully understand each country's soil biodiversity and to identify and address threats to its survival and function, to avoid both the loss of unexplored biodiversity and the loss of the ecosystem functions it provides, which sustain human life.

© Andy Murray

CHAPTER 7
CONCLUSIONS AND WAY FORWARD

7.1 | WHAT IS SOIL BIODIVERSITY AND HOW IS IT ORGANIZED?

We define soil biodiversity as the variety of life belowground, from genes and species to the communities they form, as well as the ecological complexes to which they contribute and to which they belong, from soil micro-habitats to landscapes. Soil biodiversity is essential for most of the ecosystem services provided by soils, which benefit soil species and its multiple interactions (biotic and abiotic) in the environment. Soil biodiversity also supports most surface life forms through the increasingly well understood links between above and belowground. For humans, the services provided by soil biodiversity have strong social, economical, health and environmental implications.

Soils are one of the main global reservoirs of biodiversity, more than 40% of living organisms in terrestrial ecosystems are associated during their life-cycle directly with soils (Decaëns *et al.*, 2006). Soils are home to the most diverse terrestrial communities on the planet and host more than 25 percent of the world's biological diversity; in addition, they support most life forms aboveground through increasingly well-understood links between above and belowground communities.

Soil biodiversity can be divided into different groups: microbes, micro, meso, macro, and megafauna. They include a wide range of organisms, from unicellular and microscopic forms to invertebrates such as nematodes, earthworms, arthropods and their larval stages, as well as mammals, reptiles, and amphibians that spend a large part of their life belowground. In addition, there is a great diversity of algae and fungi, as well as a wide variety of symbiotic associations between soil microbes and algae, fungi, mosses, lichens, plant roots, and invertebrates. These organisms are part of a vast food web that ensures the cycling of energy and nutrients from microscopic forms through the soil's megafauna to organisms that live on top of the soil.

7.2 | STATUS OF KNOWLEDGE ON SOIL BIODIVERSITY

The "Planetary boundaries" framework defines a safe operating space for humanity, based on the processes that regulate the stability of the planet. According to this framework, the loss of biodiversity has already exceeded thresholds that could have disastrous consequences for humanity and the environment. Currently, the extinction rate of species is estimated to be between 100 to 1,000 times higher than what could be considered natural due to rapid climate change, and anthropogenic activities (mainly land use change), which indicates that the Earth cannot sustain the current rate of biodiversity loss without a significant depletion of ecosystem resilience. On the other hand, according to the World Wildlife Fund (WWF) Living Planet Report 2020, the Living Planet Index (LPI) now tracks the abundance of almost 21,000 populations of mammals, birds, fish, reptiles and amphibians worlwide. The 2020 Global LPI shows an average decline of 68% in the monitored populations of the mentioned groups between 1970 and 2016. At present, the LPI contains data only for vertebrate species as, historically, these have been better monitored; however, efforts to incorporate data on invertebrates are underway to broaden our understanding of changes in wildlife populations.

The above highlights that globally the mentioned populations has been thoroughly studied and well documented, and that consequently policies to promote the conservation and sustainable use of terrestrial, marine, and other aquatic ecosystems have been developed. However, there are few national assessments specifically addressing soil biodiversity, while some countries have reported assessments with indirect links to soil biota. There are also very few countries that maintain a national soil information system that includes soil biodiversity. At the regional level, the European Union has a regional soil monitoring system in which soil biodiversity is a key component. However, the lack of data on soil biodiversity for many taxa, and many countries, has led to coarse global estimates of soil biodiversity. Figure 7.1, which presents a soil biodiversity map based on indicators associated with microbes (microbial soil carbon) and the main macrofauna groups, is an example. Although this index provides a preliminary idea of the distribution of soil biodiversity worldwide, it must be complemented with measured and harmonized data from all groups (i.e., microbes and, micro, meso, macro and mega soil fauna) and from all regions of the world, by an inclusive multidisciplinary group of experts from multiple organizations and institutions focusing on key indicators of overall soil biodiversity.

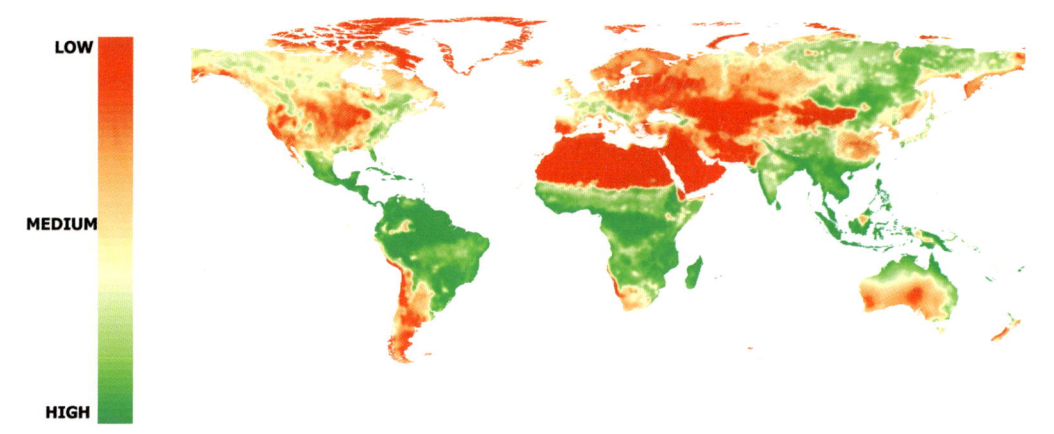

Figure 7.1 | Soil Biodiversity map

The map shows a simple index describing the potential level of diversity living in the planet's soils. In order to carry out this preliminary assessment, two sets of data were used. The distribution of microbial soil carbon developed by Serna-Chavez and colleagues (2013) was used as a proxy for soil microbial diversity, and the distribution of the main groups of soil macrofauna developed by Mathieu (unpublished data) was used as a proxy for soil fauna diversity (adapted from the Global Soil Biodiversity Atlas, JRC, 2019).

Data and information on soil biodiversity, from the national to the global level, are necessary in order to efficiently plan management strategies on a subject that is still poorly known. Soil biodiversity should be an important part of soil surveys and any soil mapping efforts. Guidelines and protocols should therefore be developed and included in soil survey description manuals.

7.3 | SOIL BIODIVERSITY POTENTIAL

7.3.1 | PROVISION OF ECOSYSTEM SERVICES

Soil biodiversity is essential for most of the ecosystem services and functions that soils provide and perform. Soil **microbes** (i.e., bacteria, fungi) and **microfauna** (i.e., protozoa and nematodes) transform organic and inorganic compounds into available forms. These transformations are critical for nutrient cycling and availability, for plants, and other species growth, for cycling of soil organic matter and carbon sequestration, and for the filtration, degradation, and immobilization of contaminants in water and soil. An important part of the food web is represented by **mesofauna**, such as springtails and mites, which accelerate litter decomposition and enhance nutrient cycling and availability (especially nitrogen), and predators of smaller soil organisms. Soil **macro**, and **megafauna** such as earthworms, ants, termites, and some mammals act as ecosystem engineers that modify soil porosity, water and gas transport, and bind soil particles together into stable aggregates that hold the soil in place and thus reduce erosion.

The important role of soil biodiversity in ecosystem functioning and ecosystem service delivery can be threatened by human activities, climate change as well as natural disasters, although the latter can also be influenced by human-induced changes. These include deforestation, urbanization, agricultural intensification, loss of soil organic matter/carbon, soil compaction, surface sealing, soil acidification, nutrient imbalances, pollution, salinization, sodification, wildfires, erosion, and landslides. These co-occurring drivers of environmental change can have synergistic effects and may thus pose a particular threat to soil organisms and ecosystem functions. Soil biodiversity can mitigate threats to ecosystem services, for instance by acting as a powerful tool in bioremediation of contaminated soils. Biostimulation and bioaugmentation are environmentally sound strategies that contribute to the filtration, degradation, and immobilization of target contaminants. Furthermore, the integral use of organisms such as microbes (bioaugmentation), plants (phytoremediation) and earthworms (vermiremediation) as a bioremediation strategy in hydrocarbon-contaminated soils has proven to be a viable alternative for increasing hydrocarbon removal. On the other hand, soil macrofauna, such as earthworms, termites, and ants, play an important role in improving soil structure and aggregation, which can improve resistance to soil erosion caused by wind and water.

7.3.2 | FOOD SECURITY, NUTRITION AND HUMAN HEALTH

For humans, the services provided by soil biodiversity have strong social, economic, health, and environmental implications.

Nutrition and human health

Recently, there is a growing interest in the relation between soils (soil biodiversity) and human health through the microbiome concept. Soil biodiversity supports human health, both directly and indirectly, through disease regulation and food production. Several soil bacteria and fungi are traditionally used in the production of soy sauce, cheese, wine, and other fermented food and beverages. The relationship between plant roots and soil biodiversity enables plants to produce chemicals such as antioxidants that protect them from pests and other stressors. When we consume these plants, these antioxidants benefit us by stimulating our immune system and contributing to hormone regulation. A series of studies and evidences suggests that early exposure to a diverse collection of soil microorganisms can help prevent chronic inflammatory diseases, including allergy, asthma, autoimmune diseases, inflammatory bowel disease, and depression. Furthermore, since the early 1900s, many drugs and vaccines have been derived from soil organisms, from well-known antibiotics such as penicillin to bleomycin used to treat cancer and amphotericin for fungal infections. In a context of increasing diseases caused by resistant microorganisms, soil biodiversity has enormous potential to provide new drugs to combat them.

Agriculture and food security

The role of soil biodiversity through the ecosystem functioning and services they provide are critical for agriculture and food security. The supporting services (carbon and nutrient cycling; soil formation including soil structure) are key for food production and are related to the Availability and Utilization (includes nutrition and food safety) pillars of food security. The regulating services (water and climate cycles regulation, mitigation of greenhouse gases emissions, control of soil pests and diseases, resistance and resilience against disturbance and stress, and decontamination) contribute directly to the Utilization (includes nutrition and food safety) and Stability pillars of food security. Finally, the provisioning services (provision of nutritious food, fiber and fuel, filtering and storage of water and source of medicines and pharmaceuticals) have a positive impact on the Availability, Access and Utilization pillars of food security.

7.3.3 | ENVIRONMENTAL REMEDIATION

Bioremediation technologies can lead to the degradation of a target contaminant to an innocuous state or to levels below the concentration limits established by regulatory authorities. Soil organisms are also used directly to transform toxic compounds into benign forms through bioremediation. Many soil bacteria can transform different contaminants such as saturated and aromatic hydrocarbons (e.g. oil, synthetic chemicals and pesticides). Soil bacteria and fungi can reduce petroleum hydrocarbons by up to 85 percent after a spill.

7.3.4 | CLIMATE CHANGE

Soil microbial taxa play a key and unquestionable role in biogeochemical cycling, plant nutrition, carbon sequestration as well as GHG emission and mitigation. Part of the anthropogenic CO_2 emissions can be absorbed by plants and stored in soils through microbial decomposition, which can allow the retention of soil carbon for long periods of time. This invaluable service provided by soil microbes is essential for climate change mitigation and adaptation. Soil microorganisms also play an important role in the mitigation of non-CO_2 gases such as N_2O. Farming activities are the most important source of this gas emitted by soils, which derives from the overuse or misuse of nitrogen-containing fertilizers. Nature-based solutions involving soil microorganisms have a significant potential to mitigate climate change. An example is the N_2-fixing efficient bacterial strains that could replace mineral fertilizers by increasing the nitrogen available in the soil, reducing N_2O emissions as well as other forms of N (NO_3^-) that contaminate groundwater and coastal ecosystems. Soil microorganisms are involved in every step of the carbon and nitrogen transformations that produce these greenhouse gases and the study and preservation of soil microbial groups is essential for finding innovative solutions to climate change.

7.4 | CHALLENGES AND GAPS

Recognition of Soil biodiversity in the 2030 and Post-2020 agendas

- Soil biodiversity and overall sustainable soil management are not yet fully taken into account when planning interventions for sustainable development. Despite the multiple benefits that soil biodiversity provides, it is still not included as a nature-based solution.

- The increasing adoption of nature-based solutions can encourage plant and soil biota diversity. This diversity contributes to increasing resilience and improving the control, prevention, or suppression of pests and pathogens.

- Soil biodiversity and sustainable soil management is a prerequisite for the achievement of many of the Sustainable Development Goals, particularly SDG 1, 2, 3, 6, 12, 13 and 15. However, this is not reflected in concrete actions, particularly at the global level.

- Although biodiversity loss is at the forefront of global concerns, the biodiversity that is below-ground (soil biodiversity) is not being given the importance it deserves. It is particularly important that the ongoing preparations for the Post-2020 Biodiversity framework clearly recognise the role of soil biodiversity in the global biodiversity and ecosystems, so that this is translated into concrete actions.

- The conservation of threatened soil biota, and the assessment of conservation imperatives including red-listing of threatened species is needed, in order to prevent extinction, particularly of species that have important functional roles in the ecosystem.

Soil Biodiversity data and information: from assessment to monitoring

- Generally, there is a lack of detailed data on soil biodiversity, particularly at the species level of all main groups of soil biota (microbes, micro, meso, macro and megafauna), at local, national, regional, and global levels. For instance, very few available soil information systems and soil surveys include soil biodiversity among the soil properties to be considered. Therefore, in order to better plan effective soil biodiversity monitoring, there is an urgent need to invest in harmonized soil biodiversity assessments in most countries of the world.

- There is a need to standardize sampling and analysis protocols worldwide to enable the collection of large comparable datasets.

- Although a number of tools exist to assess ecosystem services in the context of land management, few of them fully integrate soil biodiversity and most of them are only applicable to developed countries. Research on the definition of biological indicators is making great progress, but the development of robust, reliable, and biological indicators remains a challenge.

- Efforts to advocate for the establishment of national soil information systems that fully includes soil biodiversity data and information should be pursued.

- Local knowledge on soils and soil biodiversity management must also be taken into account when planning data collection.

- There is a need to assess and monitor soil biodiversity and define its status in order to better understand the concerns and threats to soil biodiversity, to compile relevant policies, regulations, or frameworks that have been implemented and to catalogue current efforts in soil biodiversity management and use.

- Effective and efficient monitoring tools are important to record changes in soil biodiversity and to establish databases to link diversity to soil functions. Visualization of biological information in combination with digital soil mapping tools can be effective in providing management-related soil biological information, and this may have the potential to increase the transfer and adoption rates of knowledge by different stakeholders.

- Long-term continuous monitoring programmes are needed, with explicit consideration of ecosystem types, climatic zones, and management practices to distinguish temporal variations from the actual impact of environmental changes and to gain a better understanding of these changes for long-term community adaptation.

- There is a need to promote the necessary shift to include biological indicators of soil health along with physical and chemical.

Policies pro Soil Biodiversity

- There is a general lack of specific policies, programmes, and actions that take into account the particularities of soil biodiversity, as well as measures to promote its conservation and sustainable use.

- Soil biodiversity needs to be reflected in National Reports and National Biodiversity Strategies and Action Plans (NBSAPs). Particularly important ecoregions for soil biodiversity both at global and regional levels, and those lacking in information have been highlighted in this report. These could be used to prioritize sampling and research efforts in order to fill in these gaps.

- Inter-sectoral and inter-institutional collaboration should be strengthened to explore synergies and avoid duplication or fragmentation, as soil polices can be under the responsibility of different ministries.

- Policies and urban planning need to integrate soil biodiversity into sustainable soil management and ecosystems restoration plans to ensure healthy soils for people by reducing urban threats to soil biodiversity.

Agriculture

- The overuse and misuse of agrochemicals constitutes one of the main drivers to soil biodiversity loss, thus reducing the potential of soil biodiversity for a sustainable agriculture and food security.

- Sustainable soil management practices are not widely adopted, thus soil degradation continues to be a global threat.

- Soil biodiversity is not fully used as an alternative solution when managing soils for increasing soil productivity nor for controlling soil-borne pests and diseases.

- Many microbial biofertilizers, biopesticides, and other related products show great effects when tested under laboratory and greenhouse conditions, but fail to provide reproducible results under field conditions. One of the reasons for this is the difficulty for certain organisms to survive in a highly competitive environment.

- The adoption of good and sustainable practices by farmers, as a basic premise for sustainable soil management (including soil biodiversity), remains low due to the lack of technical support, provisions of incentives and enabling environments.

- The market for biological control agents for soil-borne diseases has been on an upward trend over the last decade in response to environmental change and social recognition of the need for greater sustainability. However, this is still limited and its use should be scaled up backed up by policies.

Ecosystem restoration

- The role of soil biodiversity in restoration activities is increasingly recognised; continued work is needed to ensure that restoration goals meet their potential.

- Field studies conducted at scales relevant to ecosystem restoration (i.e., hectares) have demonstrated that a whole-soil biota inoculation method representing all soil biodiversity is a powerful tool for the restoration of terrestrial ecosystems. However, the effectiveness of any soil biodiversity restoration programme depends on appropriate integration into its landscape and the expected interactions within it.

- Ecosystem restoration efforts must include soil biodiversity and sustainable soil management practices; otherwise they will lack a central part of any ecosystem. This is particularly important for the United Nations Decade on Ecosystem Restoration 2021-2030.

Human wellbeing

- The loss of soil biodiversity could limit our capacity to develop new antibiotics and tackle infectious diseases.

- Soil pollution and antimicrobial resistance constitute a serious threat to below-ground biodiversity. Its loss and modification negatively impacts above-ground biodiversity and human wellbeing as it can enter the food-web.

Novel technologies

- While a large and diverse set of bacteria and fungi with potential for environmental management have been isolated and cultured, it remains of paramount importance to protect and conserve the native sources of this soil biodiversity. Microbiome investigations for environmental management are still new and highly experimental, but they underline the need to conserve entire soil biotic communities. Currently, proper means for the conservation of soil biodiversity, *ex situ* or *in situ*, remain a matter of research.

- New emerging technologies such as metagenomic, metabolomic, and volatilomic approaches are expensive, but they can provide useful information on microbial functions in addition to the taxonomic diversity of the soil microbiome. The challenge lies in linking the diversity of the soil microbiome to its potential functions, and in understanding how the diversity and functional traits of below-ground soil microbes are related to their above-ground biodiversity and productivity.

- New molecular techniques using next-generation molecular sequencing allow a better understanding of soil organisms and the effects these organisms may have on associated cropping systems. This knowledge allows us to better understand how soil systems will respond to changes in climatic factors, new cropping systems, and soil management. Further implementation is required.

- Strengthen education and capacity building in the adoption of molecular tools to contribute to human, plant, and soil health.

- Many small companies have been established to support the biological control of soil-borne diseases, and large companies have started to invest in the development of microbiome-based solutions for crop protection or nutrition.

© Andy Murray

7.5 | THE WAY FORWARD

Despite the clear importance of soil biodiversity in the provision of essential ecosystem services (provision of food, fiber and fuel, filtering of water, source of pharmaceuticals, carbon and nutrient cycling, soil formation, GHG mitigation, pest and disease control, decontamination and remediation), its proper use and management is not up to scale. It is only just over a decade ago that initiatives and research networks were established to contribute to the knowhow, conservation, use, and sustainable management of soil biodiversity. These include the establishment of the International Initiative for the Conservation and Sustainable Use of Soil Biodiversity in 2002, the establishment of the Global Soil Biodiversity Initiative in 2011 and the Global Soil Partnership in 2012, and the publication of the Global Soil Biodiversity Atlas by the European Commission in 2016.

Since then, soil biodiversity has started to emerge as an alternative solution to global challenges and not only as an academic field emerged. Some countries are starting to use soil biodiversity in different areas such as agriculture, food safety, bioremediation, climate change, pest and disease control and human health. Some regions, like the European Union, have set up action plans for sustainable production, consumption, and growth to become the first climate-neutral continent in the world by 2050; soils and soil biodiversity are important components of the European Green Deal. In addition, some national institutions, research centres, networks, universities, and schools are starting to include soil biodiversity in their programmes. Some of them are also conducting research on technological innovations as well as on traditional and agroecological approaches related to soil biodiversity (e.g. research, practical application, assessment, indicators, and monitoring).

We must take advantage of this momentum to:

1. Advocate for mainstreaming soil biodiversity into the sustainable development agenda, the Post-2020 biodiversity framework, the UN decade on ecosystem restoration, and all areas where soil biodiversity can contribute;

2. Develop standard protocols and procedures for assessing soil biodiversity at different scales;

3. Promote the establishment of soil information and monitoring systems that include soil biodiversity as a key indicator of soil health;

4. Improve knowledge (including local or traditional) of the soil microbiome;

5. Strengthen the knowledge on the different soil groups forming soil biodiversity (i.e., microbes, micro, meso, macro and megafauna);

6. Establish a global capacity building programme for the use and management of soil biodiversity and the Global Soil Biodiversity Observatory.

Table 7

Summary of the status, potentialities, challenges and the way forward

Theme	Challenges/Gaps	Specific actions	*Cross-cutting actions	*Cross-cutting scopes
Understanding soil biodiversity, from cells to vertebrates	● Better understanding of microbiome (or functional groups/keystone species) networks. ● Better understanding of micro, meso and macrofauna roles in soil functions and nutrient cycling. ● More research is needed to corroborate SB data in different ecosystems and agroecosystems. ● Small and large-scale SB studies in many ecoregions of the world, especially in the southern hemisphere. ● Targeted research about the long-term impacts/risks of methods of biocontrol in the environment. ● Long-term continuous monitoring programs are needed in different ecosystems, climate types and management practices to address the temporal variability of environmental changes. ● It is necessary to develop robust and reliable biological indicators and measurement methods.	● Monitoring tools that include: new analytical approaches; advanced computing power; next-generation sequencing for the assessment of microbial SB coupled with traditional techniques; increase predictive power to changes in climatic factors, new cropping systems, and SSM; digital soil mapping tools in combination with biological information. ● Implement large-scale (watershed and landscape) SB studies. ● Include soil biodiversity in the Guidelines of Soil Survey including standard methods for measurement. ● Implement SB models based on big data generated from soil-water-plant-atmosphere information. ● Obtaining or increasing financial support to implement novel technologies -metagenomic, metabolomic, volatilomic- in developing countries. ● Establishment of a Global Soil Biodiversity Observatory. ● Support the development of community-based monitoring and information systems (CBMIS). ● Simplify methodologies and tools for soil biodiversity assessment that are directly accessible in all regions of the world. ● Mobilize targeted participatory research and development, ensuring gender equality, women's empowerment, youth, gender-responsive approaches and the participation of indigenous people and local communities. ● Increase taxonomic capacity and address taxonomic assessment needs in different regions. ● Support training in the identification and description of SB at all levels, and particularly for lesser-known taxa.	● Advocate for the implementation of SSM under the VGSSM at national level. ● Implement the use/ management and conservation of soil biodiversity as nature-based solutions. ● Promote ecosystem-based approaches that conserve, restore and avoid soil degradation and biodiversity loss. ● Develop partnerships that support multi-disciplinary approaches, foster synergies and ensure a multi-stakeholder perspective regarding SSM and SB. ● Implement the combined use of traditional knowledge, novel technologies and innovation and ensure that all relevant stakeholders have access to these tools and associated policies. ● Develop robust and reliable biological indicators, and monitoring/assessment protocols for SB. ● Raise social awareness on SB loss and recovery; threats to SB including agricultural intensification and best practices for SB assessment; and management and monitoring for all land management activities.	● Guarantee soil health for all ecosystem vitality & human well-being. ● Support agriculture for sustainability, productivity, and resource use efficiency. ● Support farmers to reduce vulnerability by diminishing production costs, increasing yields and strengthening their capacity to design and implement SSM practices. ● SB can significantly contribute to tackle environmental problems. ● The knowledge that soil is alive expands the possibilities for human-soil relationships. ● SB must be considered a natural capital asset from which ecosystem services are produced.

Theme	Challenges/Gaps	Specific actions	*Cross-cutting actions	*Cross-cutting scopes
Contributions of soil biodiversity to ecosystem services and functions	● Economic valuations of SB functions and ecosystem services provided are scarce. ● More attention must be paid to the regulation services -such as carbon storage- that rely on SB. ● It is highly necessary to develop methods to measure SB contribution to all ecosystem services affected, and at different spatial and temporal scales.	● Support projects focused on the economic valuation on SB functions and services. ● Measure SB contribution to different soil functions and services at different scales, and under different conditions. ● Develop baseline data on SB and make regular small and large-scale measurements over time. ● Better analyze the relationship between the structure of SB communities and their role in the ecosystems and agroecosystems functioning. ● Promote the adoption and feasibility of Payment for Environmental Services based on SB, with appropriate policies at various governmental levels.	● Advocate for the implementation of SSM under the VGSSM at national level. ● Implement the use/management and conservation of soil biodiversity as nature-based solutions. ● Promote ecosystem-based approaches that conserve, restore and avoid soil degradation and biodiversity loss. ● Develop partnerships that support multi-disciplinary approaches, foster synergies and ensure a multi-stakeholder perspective regarding SSM and SB. ● Implement the combined use of traditional knowledge, novel technologies and innovation and ensure that all relevant stakeholders have access to these tools and associated policies. ● Develop robust and reliable biological indicators, and monitoring/assessment protocols for SB. ● Raise social awareness on SB loss and recovery; threats to SB including agricultural intensification and best practices for SB assessment; and management and monitoring for all land management activities.	● Guarantee soil health for all ecosystem vitality & human well-being. ● Support agriculture for sustainability, productivity, and resource use efficiency. ● Support farmers to reduce vulnerability by diminishing production costs, increasing yields and strengthening their capacity to design and implement SSM practices. ● SB can significantly contribute to tackle environmental problems. ● The knowledge that soil is alive expands the possibilities for human-soil relationships. ● SB must be considered a natural capital asset from which ecosystem services are produced.

Theme	Challenges/Gaps	Specific actions	*Cross-cutting actions	*Cross-cutting scopes
Threats to soil biodiversity	● It is crucial to envision land-use change and management as a trigger for other threats to SB. ● There are knowledge gaps in urban SB. ● Lack of knowledge of contaminant concentrations in soils and exposure thresholds for SB. ● Lack of understanding on interactive effects among multiple global change drivers on SB. ● Poor understanding of the role and impacts of threats to SB in selected ecoregions and global region. ● Inability to adequately map the importance of threats to SB at the global level.	● Consider SB and ecosystem services in land use planning. ● Foster activities to promote the practical application of SB, and integrate it into broader policy agendas for food security, ecosystem restoration, climate change adaptation and mitigation, and sustainable development. ● Promote sustainable planning management of urbanized environments and urban soil rehabilitation. ● Assessment of vulnerable species and landscapes to prioritize their protection. ● Minimise the drivers of SB loss and promote the improvement of soil health. ● Inclusion of SB into the risk assessment of agro-inputs. ● Regular assessment of soil contaminants and ecotoxicological test experiments with different target species. ● Perform detailed threat assessments on SB at various scales and/or for various taxa. ● Perform a regional and global synthesis of the threats to SB, using georeferenced and spatially relevant data. ● Promote Red-listing of endangered SB species at the national and global level.	● Advocate for the implementation of SSM under the VGSSM at national level. ● Implement the use/management and conservation of soil biodiversity as nature-based solutions. ● Promote ecosystem-based approaches that conserve, restore and avoid soil degradation and biodiversity loss. ● Develop partnerships that support multi-disciplinary approaches, foster synergies and ensure a multi-stakeholder perspective regarding SSM and SB. ● Implement the combined use of traditional knowledge, novel technologies and innovation and ensure that all relevant stakeholders have access to these tools and associated policies. ● Develop robust and reliable biological indicators, and monitoring/assessment protocols for SB. ● Raise social awareness on SB loss and recovery; threats to SB including agricultural intensification and best practices for SB assessment; and management and monitoring for all land management activities.	● Guarantee soil health for all ecosystem vitality & human well-being. ● Support agriculture for sustainability, productivity, and resource use efficiency. ● Support farmers to reduce vulnerability by diminishing production costs, increasing yields and strengthening their capacity to design and implement SSM practices. ● SB can significantly contribute to tackle environmental problems. ● The knowledge that soil is alive expands the possibilities for human-soil relationships. ● SB must be considered a natural capital asset from which ecosystem services are produced.

Theme	Challenges/Gaps	Specific actions	*Cross-cutting actions	*Cross-cutting scopes
Responses and opportunities	● Increase research on the field-scale performance of microbial inoculants and entomopathogenic nematodes as biological control of insect pests. ● Insufficient knowledge of the role of direct and indirect management of micro, meso and macrofauna in soil functioning and ecosystem service delivery. ● The portfolio of solutions to environmental problems is currently microbial-based; micro, meso and macrofauna are almost never included.	● Promote the prevention, suppression and control of pathogens and invasive species. ● Invest on targeted research on soil-borne diseases and promote integrated pest management. ● Privilege the development of whole community microbial inoculants over single microbial isolates. ● Implement nature-based solutions towards the micro, meso and macrofauna, not only in microbes.	● Advocate for the implementation of SSM under the VGSSM at national level. ● Implement the use/management and conservation of soil biodiversity as nature-based solutions. ● Promote ecosystem-based approaches that conserve, restore and avoid soil degradation and biodiversity loss. ● Develop partnerships that support multi-disciplinary approaches, foster synergies and ensure a multi-stakeholder perspective regarding SSM and SB. ● Implement the combined use of traditional knowledge, novel technologies and innovation and ensure that all relevant stakeholders have access to these tools and associated policies. ● Develop robust and reliable biological indicators, and monitoring/assessment protocols for SB. ● Raise social awareness on SB loss and recovery; threats to SB including agricultural intensification and best practices for SB assessment; and management and monitoring for all land management activities.	● Guarantee soil health for all ecosystem vitality & human well-being. ● Support agriculture for sustainability, productivity, and resource use efficiency. ● Support farmers to reduce vulnerability by diminishing production costs, increasing yields and strengthening their capacity to design and implement SSM practices. ● SB can significantly contribute to tackle environmental problems. ● The knowledge that soil is alive expands the possibilities for human-soil relationships. ● SB must be considered a natural capital asset from which ecosystem services are produced.

*This content applies to all themes addressed in table 1.
SB: soil biodiversity; N: nitrogen; SOC: soil organic carbon; SSM: sustainable soil management.

© Andy Murray

ANNEX I
COUNTRY RESPONSES TO THE SOIL BIODIVERSITY SURVEY

Disclaimer
The responses reflect the views of respondents in the country and are part of the GSP efforts to collect information on soil biodiversity at the national level. The GSP Secretariat has slightly modified the responses for harmonisation purposes.

1 | ASIA

1.1 | ASSESSMENT OF SOIL BIODIVERSITY

1.1.1 | CONTRIBUTIONS OF SOIL BIODIVERSITY TO ECOSYSTEMS SERVICES

Bhutan is applying more natural farming practices, if not organic then with very minimal use of chemical inputs. However, Bhutan does not have the capacity to study soil biodiversity and hence the information provided is largely based on the assumption that the clean environment, including the soil environment, is conducive for soil biodiversity to thrive and provide ecological functions.

China reported that soil harbours the greatest biodiversity on the planet, and soil biodiversity plays a significant role in maintaining carbon dynamics and nutrient cycling in terrestrial ecosystems. The ecosystem services provided by soil biodiversity played very important roles in China over the past ten years, especially reflected in soil formation, nutrient cycling, control of soil erosion, regulation of water supply and quality, pest and disease management and climate regulation (Fu, Zou and Coleman, 2009; Zhao *et al.*, 2011; Liu, Duan and Yu, 2013; Xu *et al.*, 2017; Lu *et al.*, 2018)[1].

In **Nepal**, drinking water in most part of the country is either from natural springs or from river. Therefore, soil is one of the important natural water purifiers. This might be one of the most important ecosystem services that this country is getting from soil resource (ICIMOD, 2020).

In **Thailand**, government agencies contributed to research and development and emphasized on soil microbial activities related to nutrient transformation in the soil and the conservation, restoration and protection of soil biodiversity, to reduce the rate of natural habitat loss. Moreover, several agencies encouraged land users to utilize organic matter incorporated in the soil. That is not only for soil improvement but also for the development of soil ecosystem services.

---

1   http://www.biodiversity-science.net/CN/1005-0094/home.shtml - http://www.csss.org.cn/

1.1.2 | NATIONAL ASSESSMENTS

In **Bhutan**, above-ground biodiversity has been thoroughly studied and documented. However, soil biodiversity remains unknown and yet to be studied.

China has assessed the status and trends of soil biodiversity through different ways, including a comprehensive assessment of the status and trends, scientific knowledge, innovations and practices of farmers, indigenous and traditional knowledge and maps[2].

Japan has the eDNA Project to develop soil biodiversity analysis system with environmental DNA[3] and implemented innovations and practices of farmers through the NIAES-NARO Symposium and Workshops[4].

Thailand has established the Action Plan of Biodiversity Management (2017-2021). It consists of four main strategies to integrate the value and management of biodiversity by participating at all stakeholder levels: conservation, restoration and protection of biodiversity, to reduce the rate of natural habitat loss, implement the rehabilitation of degraded ecosystems, and establish guidelines for supervising the use of genetically modified organisms; protect rights in national biodiversity and genetic resources to increase and share benefits from biodiversity in concept of the green economy; develop the knowledge and database system on biodiversity to be an international standard, through establishment of museums and genetic banks of plants, animals and microbes; and to promote and develop knowledge database on biodiversity in application and transfer technology related to sustainable biodiversity utilization. However, soil biodiversity is a part of this action plan of biodiversity management.

1.1.3 | PRACTICAL APPLICATIONS OF SOIL BIODIVERSITY

In **Bhutan**, some trials have been conducted on biological nitrogen fixation and therefore some information is available on this. However, the lack of capacity to conduct soil microbiology studies limits more sophisticated studies. Cropping practices such as intercropping with leguminous crops, and growing leguminous crops for green manuring, are encouraged and therefore the role of microbes in nitrogen fixing and soil fertility improvement has definitely contributed positively to food production. Furthermore, the use of effective microorganisms in composting to enhance the rate of the composting process also has contributed to food production using composts in the field to improve soil fertility.

---

2   http://soil.geodata.cn/; http://www.soilinfo.cn/map/
3   http://www.naro.affrc.go.jp/archive/niaes/project/edna/edna_en/index.html
4   http://www.naro.affrc.go.jp/english/laboratory/niaes/symposium/index.html

In **China**, soil beneficial microorganisms contribute directly (i.e. biological N_2 fixation, P solubilization and phytohormone production) or indirectly (i.e. antimicrobial compounds biosynthesis and elicitation of induced systemic resistance) to crop improvement and fertilizers efficiency. Microbial-based bioformulations have been developed fast in China and its application played important roles in improvement of soil quality and reducing soilborne diseases (Li *et al.*, 2009; Bai *et al.*, 2010). The use of soil biodiversity, especially soil microbial diversity, has contributed significantly to food production and nutrition in China (Li and Zhang, 2008; Xiaoping *et al.*, 2005)[5].

Japan reported on soilborne disease control, soil fertility management, GHG mitigation technology: Nutrient cycling (Bao, *et al.*, 2013), atmospheric composition and climate regulation including floods (Akiyama *et al.*, 2016), carbon sequestration and climate change including floods (Ocubo *et al.*, 2015), pest and disease regulation (Takahashi *et al.*, 2018).

Nepal is fostering the biological nitrogen fixing potentialities of the symbiotic and asymbiotic nitrogen fixing bacteria, endemic to Nepal. Phylogenetic diversity and symbiotic functioning in mungbean (*Vigna radiata* L. Wilczek), *Bradyrhizobia* from contrast agro-ecological regions of Nepal, and genetic diversity of native soybean *Bradyrhizobia* from different topographical regions along the southern slopes of the Himalayan Mountains in Nepal (Gartaula *et al.*, 2016).

In **Thailand**, there are studies of beneficial soil microorganism diversity on agriculture system. The introduction of soil-improvement crops, such as bean, pea and others clover legume can increase nitrogen. Products of the Land Development Department[6] include the use of *Trichoderma* spp. for soilborne disease control in agricultural systems; landscape support, land use planning, economic crops zoning of Thailand in provincial (local)/regional/national level; and introduction of land use in Thailand by following the soil's potential.

Thailand has several kinds of microbial products to utilize for soil improvement, crop productivity and biological pest control in agricultural activities. Almost all the microorganisms in such products are isolated from the soil, the soil and plant association, and plants, which consisted of bacteria, actinomycetes and fungi, and in the case of biofertilizer consisted of *Rhizobium*, Plant Growth Promoting Rhizobacteria (PGPR), Mycorrhiza and Phosphate Solubilizing Microorganism. In general, the government sector has encouraged land users to use microbial products as a way of reducing chemical fertilizer, pesticide and fungicide. At the same time, microbial and biofertilizer products would improve soil fertility and productivity, conserve the environment, and enhance safety and organic food production and sustainable agriculture. The initial stage of research and development is carried out by government agencies and universities and now plenty of private sector entities are established in Thailand for the production of several kind of microbial and biofertilizer products.

---

5   Also see: http://www.most.gov.cn/gnwkjdt/200604/t20060405_30186.htm (in Chinese)

6   Thailand's Land Development Department: www.ldd.go.th

1.1.4 | MAJOR PRACTICES NEGATIVELY IMPACTING SOIL BIODIVERSITY

Bhutan reported that agriculture practices are largely natural with minimum application of chemical inputs. Forest coverage is about 70 percent and there is minimum negative impact on soil.

In **China**, the overuse of chemical control has the most negative impact on soil biodiversity. Beyond that, monoculture, overgrazing and overuse of fertilizers are also serious agents on soil biodiversity.

Within the agricultural production system in **Thailand**, some farmers use hazardous chemical substances such as pesticides, herbicides, insecticides, disease protection and elimination substances. These have a residual effect if accumulated in soil and water, which may directly and indirectly affect soil microorganisms, small insects, earthworms, crabs and so on. Some agricultural practices have accelerated diseases of soil organic matter, which directly affects microorganisms and small living organisms in soil surface.

1.1.5 | INVASIVE ALIEN SPECIES (IAS)

In **China**, the agricultural production and biodiversity have been threatened, for example by fall webworm and giant African land snail. And from last year, a new IAS, fall armyworm (*Spodoptera frugiperda*) invaded China and resulted in devastating damage (Silver, 2019; Wescott, 2019).

Nepal reported the *Fusarium, Pythium* and *Rhizoctonia* as the major exotic soilborne diseases. Clubroot is a serious indigenous soilborne disease that affects brassica crops. Clubroot is caused by *Plasmodiophora brassicae*[7].

1.1.6 | MONITORING SOIL BIODIVERSITY

Bhutan has a good soil information system, but without the soil biodiversity information due to the lack of technical expertise on soil biodiversity.

---

7   https://ag.umass.edu/cafe/news/curbing-spread-of-clubroot-disease-in-nepal

China[8] has developed the Chinese Biodiversity Monitoring and Research Network (SinoBON), the Chinese Soil Microbial Diversity Monitoring and Research Network and the Chinese Soil Fauna Monitoring and Research Network. China has also developed a China Soil Database and conducts a national soil survey every a few years[9]. A huge number of methods exist to investigate soil organisms activity, biomass, biodiversity, toxic effects or bioaccumulation, but for reasons of cost efficiency, inventory quality and inventory repeatability, often only a few of them can be selected as indicator parameters, such as microbial biomass, microbial genetic diversity, potential C and N mineralization, abundance and diversity of nematodes, mites, enchytraeids, earthworms, anaerobic N ammonification, or nitrification.

Thailand reported some applications comprising Land Development Department Mobile Application programme and AgriMap Online programme that relate to land use planning or zoning that can help farmers to use their land, taking into consideration the soil or land potential[10].

Thailand reported that the biodiversity of soil organisms is important for ecosystem services in the soil and needs to be studied and monitored because such organisms are directly involved in circulation of nutrients in the soil. Currently, research on soil organisms is less prominent than any other research activity, due to its complex circumstances, environment and usually specific equipment. In Thailand, the most studied soil organisms are insects such as termites, ants, microbes (bacteria and fungi) and soil mycorrhizal fungi. These organisms are the most studied because they are very diverse and affect the soil ecosystem. However, research activity on soil biodiversity is necessary to understand soil properties, ecological characteristics, soil environment, land cover and more. The study of agricultural biodiversity is directly related to the conservation and sustainable use of biodiversity resources.

The Land Development Department has a research project that studies the biodiversity of soil organisms, such as the survey and study of useful soil microbial diversity in forest areas, organic farming areas in the northeast, central and southern regions of Thailand. The study of organic matter and biocharcoal application on change of biomass, microbial activities and communities, nutrient content transformation and development of plant root system under different soil management.

---

8   http://www.biodiversity-science.net/fileup/PDF/w2015-025-3.pdf
9   http://vdb3.soil.csdb.cn/
10 Thailand's Land Development Department: www.ldd.go.th

1.1.7 | INDICATORS USED TO EVALUATE SOIL BIODIVERSITY

In **China**, a huge number of methods exist to investigate soil organism activity, biomass, biodiversity, toxic effects or bioaccumulation, but for reasons of cost efficiency, inventory quality and inventory repeatability, often only a few of them can be selected as indicator parameters, such as microbial biomass, microbial genetic diversity, potential carbon and nitrogen mineralization, abundance and diversity of nematodes, mites, enchytraeids, earthworms, anaerobic nitrogen ammonification and nitrification.

Thailand reported several indicators used to evaluate soil biodiversity. The Living Planet Index (LPI) is an assessment of biodiversity by collecting data on various species of vertebrates to calculate the average population change during a specified period. The index is derived from the collection of scientific data on population of vertebrates, mammals, birds, fish, amphibians and reptiles. In general, biodiversity in soil surface in both animals and microbes will be used as an indicator of soil fertility and productivity including the abundance of ecosystems.

1.2 | RESEARCH, CAPACITY DEVELOPMENT AND AWARENESS RAISING

Bhutan has an Organic Flagship programme and is planning on setting up a soil microbiology laboratory to start with soil microbiology studies. Awareness is being raised on the importance of conserving or improving soil biodiversity through the use of fewer chemicals and more organic sources of plant nutrients. The country is also referring more to the Voluntary Guidelines for Sustainable Soil Management for the promotion of sustainable soil fertility management.

China has initiated many programmes via an integrated portfolio of large-scale programmes. The following review summarized and commented the investment on related programmes over the past twenty years (Bryan *et al.*, 2018). The Chinese government and academic organizations have developed educational programmes and held domestic and international meetings toward the conservation and sustainable use of soil biodiversity (e.g., 2nd Global soil biodiversity conference in 2017 -footnote and link to website- [11].

---

[11] The 2nd Global Soil Biodiversity Conference (GSBC2) was held in Nanjing, China, 15 - 19 October 2017: https://www.gbif.org/event/2Jf3Jo9N5SgyaeGAK46UCE/2nd-global-soil-biodiversity-conference-gsbc2 Other meetings organized (websites in Chinese) : http://www.edu.cn/rd/meeting/201010/t20101014_529564.shtml; http://wap.cnki.net/huiyi-TRXH201010001.html; http://www.rcees.ac.cn/xshd/xshd/201510/t20151015_4439631.html; http://www.ibcas.ac.cn/xueshu/Academic_report/201504/t20150403_4331964.html

In 2010, the Chinese Ministry of Environmental Protection (MEP) released draft Provisional Rules for the Environmental Management of Contaminated Sites. China has introduced robust practices for the risk-based management of contaminated sites, including definitions of contaminated sites, clear responsibility for environmental management and local standards that are steering national regulation[12].

India has a network project on Soil Biodiversity and Biofertilizers and a National Bureau of Agriculturally Important Microorganisms (NBAIM). Also, they are exploring and conserving cultivable microorganisms from different soil types, tillage practices and cropping systems of India. The country also has an integrated nutrient management and a National Mission on Soil Health Card.

In ***Nepal***, there are no soil biodiversity research programmes conducted so far. The Department of Agriculture launched the Sustainable Soil Management Program (SSMP) with the support from HELVETAS, but the programme was phased out. Now, Sustainable Soil Management Practices adopted in Nepal particularly focus on improved use of Farm Yard Manure (FYM), utilizing best uses of animal urine for soil fertility improvement and making biological pesticides. Nepal has Slope Agriculture Land Technology (SALT) for erosion control in sloping land. There are plans to develop a new programme to implement the Voluntary Guidelines for Sustainable Soil Management.

Pakistan reported on the Biodiversity Action Plan Pakistan[13]. Educational Institutes are playing a major role, and Life Sciences Departments are giving education for BS, MS and PhD students.

Thailand has given priority to research on microbial diversity. Researchers have carried out research projects on microbial diversity for a long time by supporting research funds from two organizations, the National Science and Technology Development Agency (NSTDA), and the Thailand Research Fund (TRF), whose main objectives are to encourage research activities and capacity building for researchers in soil biodiversity.

Soil biodiversity research projects in Thailand are carried out by various universities and government agencies from which the outcome is knowledge and strengthening researchers and technicians. Biodiversity of soil microbes usually is carried out for agricultural purposes in different regions of Thailand. Moreover, research activities to identify and classify of soil microbial community are also conducted, as well as soil microbial activities related to nutrient transformation in soil and microbial products for agricultural, Industry and energy purposes. However, several biodiversity resources are used for effective microbes in fixing nitrogen, dissolving, phosphating and production of plant growth regulators, bio fertilizer and antagonistic microorganisms in plant disease control to reduce the use of chemical fertilizers and agricultural substances. The use of freshwater algae that produces polysaccharides helps to improve soil structure.

---

12   Ministry of Environmental Protection of P.R. China (2010) Provisional Rules for the Environmental Management of Contaminated Sites – Draft for Comments https://link.springer.com/article/10.1007/s10980-018-0706-0

13   Available at: https://www.cbd.int/doc/world/pk/pk-nbsap-01-en.pdf

Conservation and rehabilitation of soil conditions, and research on the use of algae in oil production. There was also research to select local microbes for utilization and promotion of organic farming.

1.3 | MAINSTREAMING: POLICIES, PROGRAMMES, REGULATIONS AND GOVERNMENTAL FRAMEWORKS

In **Bhutan**, the Government expressed support for the Organic Agriculture Flagship programme in the Twelfth FYP of the Country. The objective is to go organic as much as possible in the country's agriculture practices. The flagship programme will support establishment of organic input manufacturing plants, laboratories and capacity development, both in terms of technical and human resource and linking and collaborating with organizations with similar objectives. The Country's Organic Flagship programme promotes organic agriculture, thereby supporting soil biodiversity in various ways for example, promotion of organic inputs, capacity development of both human and institutional to ensure conservation and sustainable use of soil biodiversity directly or indirectly. The country's objective to remain carbon neutral or even become carbon negative also recognizes the importance of sustainable soil and land management.

China reported the work of the Ministry of Ecology and Environment of the People's Republic of China and the Ministry of Agriculture and Rural Affairs of the People's Republic of China. The country has developed the Environmental Protection Law of the People's Republic of China; China Biodiversity Conservation Strategy and Action Plan (2011-2030)[14]; the Chinese Biodiversity Monitoring and Research Network (SinoBON), the Chinese Soil Microbial Diversity Monitoring and Research Network and the Chinese Soil Fauna Monitoring and Research Network.

India reported that the conservation and sustainable use of soil biodiversity is under the National Biodiversity Authority (NBA), a statutory autonomous body under the Ministry of Environment, Forests and Climate Change, established in 2003 by the Government of India to implement the provisions under the Biological Diversity Act 2002, after India signed the Convention on Biological Diversity (CBD) in 1992[15]. The Biological Diversity Act 2002 covers conservation, use of biological resources and associated knowledge occurring in India for commercial or research purposes or for the purposes of biosurvey and bioutilization. It provides a framework for access to biological resources and sharing the benefits arising out of such access and use. The National Biodiversity Authority (NBA) is a Statutory Body and it performs facilitative, regulatory and advisory functions for the

---

14 Available in Chinese : http://www.mee.gov.cn/gkml/hbb/bwj/201009/t20100921_194841.htm
15 http://nbaindia.org/content/22/2/1/aboutnba.html

Government of India on issues of conservation, sustainable use of biological resources and fair and equitable sharing of benefits arising out of the use of biological resources.

Nepal has policies related to climate change and environment in agriculture and food security (Thakur, 2017), but there are no policy regulations specific to soil biodiversity yet.

Pakistan has recently started the One Billion Tree Program for soil and biodiversity conservation.

Sri Lanka reported the Environment Act and the Soil Fauna and Flora Act.

Thailand has biodiversity strategies and a national plan to ensure soil and water conservation for sustainable use of soil resources. This includes the organic agriculture policy and the integrated action plan for organic agriculture[16]. Thailand has an integrated master plan in biodiversity resources for 2015-2021 that is used as a tool for driving its strategy of conservation and restoration of biodiversity resources. These resources conserve, restore and protect of species and genetic ecosystems of plants, agriculture, livestock, aquatic animals, including wild and native species, microbes and other species that are valuable in economic, cultural society and ecosystems, with a biodiversity action plan for 2017-2021. In case of production of microbial products for agricultural purposes, the regulation of the Fertilizer Act 2518 has specifications to control such microbial products in kind and amount of microorganisms, quality and effectiveness of microbial activities.

1.4 | ANALYSIS OF THE MAIN GAPS, BARRIERS AND OPPORTUNITIES IN THE CONSERVATION AND SUSTAINABLE USE OF SOIL BIODIVERSITY

Bhutan has a national biodiversity strategy and action plan which also enshrines soil biodiversity but as of now, there is a lack of expertise in conducting the studies.

In **China**, a soil biota database is currently being constructed to store observational data for public inquiry and analysis which will be taken into account in designing strategies to protect soil biodiversity and properly utilize soil resources. On a scale of four, especially "overly theoretical approach and lack of applicability" and "policy and institutional constraints" are two serious barriers to implementation of soil biodiversity management strategies.

---

16 Also see: www.ldd.go.th, www.moac.go.th

Pakistan reported that they plan to have: i) government, Educational Institutes and Forest Department collaboration; i) training of local uneducated communities about the importance of soil biodiversity and its conservation; and ii) availability of research funds.

On 12 June 1992, ***Thailand*** ratified the Convention on Biological Diversity at the United Nations Conference on Environment and Development or the Earth Summit. The Convention on Biological Diversity is an international agreement on the environment in the global community which needs and develop international cooperation between several agencies in both of government, public and private sectors in conservation of biodiversity and sustainable use of species and genetic ecosystems with three main objectives; i) to conserve biodiversity, ii) to sustainably utilize biodiversity components, and iii) to share the benefits in equality and just use of genetic resources.

In addition, Thailand has laws that facilitate the implementation of the obligations of the Convention. The main laws of Thailand related to biodiversity are the National Park Act 1961, the National Forest Act 1964, the Reserve and Protection Wildlife Act 1992, Plant Species Act 1975 and 1992, as well as the Science and Technology Development Act 1991, and the Environmental Quality Act 1992. The national policy and a national plan is a master plan for integrated management of biodiversity 2015-2021, including regulations and guidelines for specific government agencies, such as guidelines for biosafety for experiments in genetic engineering and biotechnology.

2 | EUROPE AND EURASIA

2.1 | ASSESSMENT OF SOIL BIODIVERSITY

2.1.1 | CONTRIBUTIONS OF SOIL BIODIVERSITY TO ECOSYSTEMS SERVICES

While the ecosystem services provided by soil biodiversity are highly relevant, ***France*** pointed out that the ranking ecosystem services is subjective and also depends on the context and the area. It is a subjective ranking as there are still unexplored areas of research/knowledge on biodiversity, soils and ecosystem services. Nonetheless, France has strong scientific competence in soil biodiversity (inventories, spatio-temporal distributions, ecological functions and services).

In addition to the ecosystem services listed in the survey, France has suggested others:

- Organic matter composition and degradation;
- Genetic resources for industry such as the agrofood industry, pharmaceutical industry (e.g. antibiotics, pollution treatment industries, biorefineries) and agriculture (agroecology);
- Enhancement of forest productivity;
- Cultural (education, research, inspiration);
- Pollination regulation (as many wild bees and bumblebees live in the soil).

Several of the listed ecosystem services are related to organic matter (which is either the trophic resource for soil organisms or their own production). As any biomass produced can be turned into organic matter, there is no "waste biomass" for soil ecosystems. Therefore, France suggests to replace "waste" by "urban waste in innocuous conditions".

Soil biodiversity provides ecosystem services as part of the soil ecosystem (with geographical, physical, chemical and biological properties). For example, the storage of carbon and the emission of greenhouse gases depend on soil organisms, but also on clay content and pH; the regulation of water supply or water quality depends on the physical properties of the soil.

The Ministry for Agriculture and Alimentation has produced a document on indicators for the organic and biologic state of agricultural soils (*Tour d'horizon des indicateurs relatifs à l'état organique et biologique des sols*). For each indicator, there is a description on what is observed or measured, how to proceed, who uses the indicator, why, what is the pertinent scale for using it, the status of the method and the results (uncertainty). The advantages and disadvantages of each indicator are listed (Ministère de l'agriculture et de l'alimentation, 2017).

In *Italy*, the National biodiversity framework has been implemented following the National Strategy for Biodiversity (NSB). The Ministry of Environment, Land and Sea has suggested, through the development of the NSB, several lines of action in respect of environmentally friendly agricultural policies for the management and conservation of biodiversity.

The objective of environmental protection is oriented towards the European "Common Agricultural Policy". This tool was adopted by the State-Regions Permanent Conference on 7 October 2010 in order to ensure a true integration of the country's development objectives and the protection of its biodiversity.

The Italian NBS is organized around three key themes: biodiversity and ecosystem services; biodiversity and climate change; and biodiversity and economic policies. Following the activity carried out on the NBS, the Ministry of Agriculture and Forestry, with the active collaboration of the regions, prepared the finalized National Plan for Agricultural Biodiversity (NPAB) to elaborate guidelines for the conservation and characterization of plant, animal and microbial genetic resources for food and agriculture. On 24 July 2012, a decree of the Minister of Agriculture, Food and Forestry was published on the official adoption of national guidelines for the protection of plant and animal biodiversity for food and agriculture, as well as food-related microbial and soil genetic resources (Ministero delle politiche agricole, alimentari e forestali, 2013). A modernization of the guidelines by a technical working group is currently underway.

After the publication of the NSB, which focused on soil biodiversity monitoring, a very important activity of the Italian Research Institution was carried out on suitable analytical methods to measure soil biodiversity. Practical protocols for characterizing and protecting soil biodiversity are now available.

The **Netherlands** reported that quantitative data on soil biodiversity and soil functions have been collected in the Netherlands soil monitoring network. The quantification of the contribution to soil ecosystem services is still under development and needs further empirical validation.

In the **Republic of Moldova**, soil invertebrates and soil microorganisms indices are the global indicators of soil quality. The diversity of invertebrates and microorganisms is one of the most important evaluation criteria of soil ecosystems, i.e. their resistance to different forms of degradation. In some ecosystems, the local diversity of soil fauna and microorganisms may be more important than the diversity of different groups of above-ground plants or animals. Excessive reduction of soil biodiversity, especially the loss of key species and/or species with unique functions, can have cascading ecological effects, leading to long-term deterioration of soil fertility and the loss of agricultural productive capacity.

Soil biodiversity can also have indirect effects on the soil's function as a carbon sink or source. Ecologically, soil biota is responsible for the regulation of several essential soil functions. Soil organisms provide a wide range of essential services for the sustainable functioning of all ecosystems by acting as regulators of the dynamics of soil organic matter, carbon sequestration and greenhouse gas emissions; by modifying soil physical structure and water regime; by enhancing the amount and efficiency of nutrient acquisition by vegetation and by enhancing plant health.

The main role of the soil biota is related to the mineralization of organic matter and the conservation of resources that have been formed within the limits of the ecosystem. In balanced ecosystems, the processes of microbial decomposition of organic matter and its synthesis are closely linked to plant growth, which ensures the stable existence of undisturbed ecosystems for long periods of time. In degraded ecosystems, the equilibrium is disturbed and mineralization processes are predominant. The stable deviation of biota indices from the equilibrium state of parameter values, either increasing or decreasing, indicates essential ecological changes or the destruction of the soil ecosystem.

Changes in biological properties may indicate the likely risk of soil degradation as a result of human activity. In this respect, the use of soil bio-indication as an integrated monitoring tool for soil degradation might be a possible solution. Nevertheless, the functions and services provided by the soil biota in agricultural ecosystems are poorly recognized in the ecological management of soils in the Republic of Moldova. Managers need to take into account recommendations on the use and management of soil biota for the long-term conservation and sustainable productivity of terrestrial ecosystems.

Spain reported that soil biodiversity has great potential for ecological restoration, pest control and improvement of plant nutritional quality when properly managed. However, soil biodiversity has not been fully taken into account for ecosystem management or restoration at present. Only organic agriculture is starting to include soil life health as a priority.

The **United Kingdom of Great Britain** and **Northern Ireland (UK)** has attempted to answer the question based on the current importance of soil biodiversity rather than its actual important. For soil formation, soil biological activity is one of the factors, but not the only one, and will always be important. Soil formation as a service is undervalued in England, but it can be relatively rapid in some circumstances. Nutrient cycling should be a fundamental service of soil biota, but in England most of the nutrient input come from inorganic bagged fertilizers (input of bagged N was 2.4 times compared to biologically fixed N in 2012).

Compared to other countries, soil erosion is less of a problem in lowland England due to rapid soil formation, but its impacts (sedimentation and eutrophication of water bodies) are of greater concern. However, at present, soil biology is rarely seen as a method of reducing erosion and more effort are being made to try to stop already mobile soil by using buffer strips. The most severe erosion occurs in upland peats, where activities such as drainage or burning are key factors and where soil biological activity is not likely to help.

Atmospheric regulation is strongly driven by soil biology, not only due to the development of recalcitrant soil C through soil biological processes, but also through the formation of nitrous oxide by soil microbes. Furthermore, the impact of soil biological activity on runoff is a strong mediator of flooding and its precedent causes of soil structural collapse and compaction. Flooding remains one of the key issues with soil degradation in the United Kingdom, although increased exposure to damage by flood due to development pressure accounts for increased flood damage, rather than increased flooding.

Pollutant degradation is scored lower because only some pollutants are amenable to remediation with soil biological activity and not all land is polluted. Recycling of waste biomass is scored highly because this is more or less entirely carried out by soil biota, whether *in situ* in the soil, or in composting facilities. Where waste nutrients are recycled (that is, not those entering watercourses or groundwater and flowing to the sea) these generally flow through sewage treatment plants and waste sewage sludge is predominantly disposed of on land for decomposition by soil biota.

Although not currently recognized, soil biota are likely to be of fundamental importance to the restoration of degraded land and ecosystems, in that they represent an integral part of those ecosystems to be restored - it will be impossible to restore the ecosystem without their soil biota. Natural nitrogen fixation (excepting that caused by lightning) is entirely carried out by soil biota (including those associated with plant roots).

There is huge potential for soil biota to contribute to pest and disease regulation, but this potential has not yet been realized. The mid-range score given indicates the likely unrecognized importance of soil biota currently, rather than an indication that it is moderately widely recognized.

Regarding "landscape support and biodiversity", there are two points: i) in landscape, the soil biota in the English countryside has little opportunity to express its influence, given that so much of the countryside is intensively farmed and the vegetation reflects that which is planted and supported by fertilizers, rather than a reflection of a comprehensive above-ground/below-ground ecosystem and ii) because the lack of supply of organic matter to ecosystems in intensive arable land, associated with increasing agricultural efficiency diverting more primary productivity towards human consumption, is likely to have had the general effect of reducing biomass and diversity of all wildlife - as seen with the decline in farmland birds and declining soil organic matter in intensively cultivated areas in England.

Farmers are now beginning to appreciate the role of soil biota in supporting productivity and increasingly are turning to no till and cover cropping approaches which cause less damage to the soil biota. Research and development is underway to embed this approach in agriculture nationally in England (AHDB BBSRO Soil Biology and Soil Health Programme). Advice on soils is delivered to farmers in water quality-priority areas through Catchment Sensitive farming in England. In Wales, assessments of earthworm numbers and diversity are now included in standard soil assessments for nutrient management planning (Farming Connect Service) and courses are now regularly provided for farmers including assessment of soil biological parameters.

There is little evidence of a link between soil biological activity and the nutritive value of human food. The natural hazard regulation does not, presumably, include flooding (already mentioned above). In England this is the greatest natural hazard and highly affected by soil biota. Although locally, fire, landslip and dust storms can occur these are not key natural hazards here. Human health scores low here because, although several key medicines are derived from soil biota, little effort is currently made to explore and develop these currently. There is no systematic approach in this respect for soils in the United Kingdom. In contrast, soilborne diseases are of relatively low importance and it is even possible that low contact with soil is related to higher incidence of allergies and intolerances. However, well-being plays a fundamental role in human health, especially in mental health and the roles that soil plays in delivering a healthy, enjoyable and fulfilling environment should not be underestimated. For this reason, soil organic matter has been included as a headline indicator for well-being in the National Indicators for Wales' Well-being Objectives, due to its role in underpinning resilience and function of natural ecosystems.

2.1.2 | NATIONAL ASSESSMENTS

Finland's soil fauna was considered in the Finnish IUCN assessments of threatened species in 2019[17]. For the assessment, various taxonomic expert groups provided species list for many soil mesofauna groups such as oribatid mites, mesostigmatid mites and collembolans, but these were not evaluated due the lack of information regarding the criteria. Most soil macrofauna, such as myriapods, earthworms, spiders, beetles, ants, pseudoscorpionids, harvestmen and woodlice were listed and assessed (Niemi, Karppinen and Uusitalo, 1997).

Earthworm communities in natural and arable habitats have been surveyed at regional scale. Finland has produced a great amount of scientific knowledge linked to soil biodiversity. More than 300 scientific soil biodiversity related projects have been funded by the Academy of Finland since 2010 (Terhivuo and Valovirta, 1978; Nieminen *et al.*, 2011). Distribution maps for Finnish oribatid mites were published in 1997. Additionally, distribution maps for spiders of Finland are available online since 2013[18].

Germany has reported that the nation-wide monitoring programme on biodiversity including soil biodiversity is currently under development and will be established within the next years. MonViA (Nationales Monitoring der biologischen Vielfalt in Agrarlandschaften)[19] is coordinated by the Thünen-Institute of Biodiversity in Braunschweig.

Assessments of soil biodiversity (effects) in agrarian systems are primarily carried out in Germany by the Thünen-Institute of Biodiversity in Braunschweig. As an example, a related meta-analysis for Germany found that reduced tillage enhances earthworm abundance and biomass in organic farming (Moos, Schrader, and Paulsen, 2017). The federal state of Baden-Württemberg analyses and publishes trends in soil biodiversity assessed in their permanent Forest Observation Sites[20].

---

17 Red list of Finnish species available at: https://www.environment.fi/redlist
18 Spider distribution maps: http://biolcoll.utu.fi/arach/aran2013/aranmaps.htm
19 https://www.bmel.de/SharedDocs/Pressemitteilungen/2019/070-Kickoff_Biodiversitaet_Monitoring.html
20 Also see: LUBW (2007, 2011) Collembolen an Wald- Dauerbeobachtungsflächen in Baden-Württemberg

Germany has published relevant material such as a new online soil-zoological data warehouse (Burkhardt, 2014)[21]. So far, more than 250 000 data sets on soil fauna from more than 13 000 sites are compiled in this database. Other examples are the state of knowledge of earthworm communities in German soils as a basis for biological soil quality assessment; the state of knowledge of enchytraeid communities in German soils as a basis for biological soil quality assessment (Jänsch et al., 2013); and a review with a focus on German data on tillage-induced changes in the functional diversity of soil biota (van Capelle, Schrader, and Brunotte, 2012). More recently, Germany has begun projects funded by the German Environmental Agency (UBA). Among these is "Development of reference values for soil organisms for soils in Germany", aiming to map soil biodiversity in Germany.

Italy, at the regional level, has monitored soil microbial biodiversity, according to the Guidelines for the conservation and characterization of plant, animal and microbial genetic resources for food and agriculture.

The Lazio Region has financed a monitoring programme for the BFI (Biologial fertility Index) to assess the degree of biological fertility of soil correlated with different soil agricultural production (Renzi et al., 2017). The Piemonte Region developed (according to BSI) a map of biological fertility of soil of one of the most important districts of the region for wine production (Petrella et al., 2011). The Pavia Province in the Lombardia Region was interested in a monitoring programme, carried out by the JRC, and in which were applied several biological indicators ranging from BFI to QBS to earthworms (Pompili et al., 2006; Beone et al., 2015). The Friuli-Venezia-Giulia Region and the Umbria Region are interested in the Biological Fertility measure in the Regional Development Rural Plan. Several training programmes to protect soil biodiversity by Research Institutions were developed (for example the summer school of SiPe).

The **Republic of Moldova** has partially developed an assessment system with indicators, criteria, statistical parameters and scale of the soil biota.

Spain reported that there is some important work done in taxonomy and distribution of some soil invertebrates, but an official assessment at the national scale is still missing[22].

Ukraine reported that the country is improving the National Digital Map of Ukraine on Soil Carbon as an integral part of the Global Soil Organic Carbon Map (GSOC Map)[23].

In the **United Kingdom of Great Britain** and **Northern Ireland**, the Countryside Survey (CS) 1998 and 2007 evaluated soil biological diversity using PLFAs, tRFLP for microbes and mesofauna were identified to broad (sub-class to order level) groups, over 436 sites

---

21   Also see: http://portal.edaphobase.org

22   Most available data in this sense have been published by the CSIC in the "Fauna Ibérica" series of books (http://www.fauna-iberica.mncn.csic.es/publicaciones/fivol.php)

23   Maklyuk, O.I. (Starchenko) et al., Conception of organic agriculture (soil-agrochemical supporting). Eds S.A. Baliuk and O.I. Makliuk. Kharkiv: Smugasta typografia, 2015 https://minagro.gov.ua/ua/news/v-ukraini-zbilshilas-kilkist-operatoriv-organichnogo-rinku-olga-trofimtseva Atlas of Soil Suitability of Ukraine for Organic Farming (Zonal Aspects) / Kharkiv, 2015. - 36 p

across Great Britain. In Wales, since 2013, there has been approximately 300 squares assessed (including the Countryside survey), to assess agri-environment impacts under the GMEP programme, which included some of the CS sites, but which also includes metabarcoding for bacteria, fungi and mesofauna.

In England, the Long Term Monitoring Network is a small network of 37 National Nature Reserves assessed for soil properties including PLFAs, tRFLP and mesofauna and some of the mesofauna samples have been subject to metabarcoding. It is unknown whether soil biological indicators will continue to be measured in future monitoring programmes.

The status of soil biodiversity in Wales, in terms of the presence or absence of organisms and habitat associations, has data to support it, although metabarcoding data will not necessarily be linked to known species or assemblages. However, understanding of detailed mesofauna communities is lacking in the CS 2007 data, although species data for oribatid mites is available for 1998. In Wales, comparisons can be made between CS and more recent Glastir monitoring for molecular approaches and for broad groups of mesofauna. In England, this is not available.

There has been no formal assessment of the state of scientific knowledge on soil biodiversity in the United Kingdom of Great Britain and Northern Ireland. However, there are some initiatives such as:

- Innovations and practices of farmers are reviewed in the NECR100 report linked above and as part of the AHDB BBSRO Soil Biology and Soil health Programme;

- The UK Soil Observatory[24] provides maps showing soil biological parameters based on the countryside survey and extrapolated using land cover mapping. This approach is based on limited data;

- AHDB BBSRO Soil biology and soil health programme[25];

- NECR 100 Managing Soil Biota report[26];

- Glastir (GMEP) portal[27];

- Natural England long Term Monitoring Network[28].

---

24 UK Soil Observatory available at: http://www.ukso.org
25 https://cereals.ahdb.org.uk/publications/2017/august/14/soil-biology-and-soil-health-partnership.aspx
26 http://publications.naturalengland.org.uk/publication/2748107
27 https://gmep.wales/
28 https://data.gov.uk/dataset/12ad05d1-a21a-4855-8545-4812db5f2cfd/long-term-monitoring-network-ltmn-soil-chemistry-and-biology-baseline

2.1.3 | PRACTICAL APPLICATIONS OF SOIL BIODIVERSITY

In **Finland**, soil biodiversity has been used for commercial microbial inoculates for legume root nodulation[29] and enhancement for tree seedling growth[30]; biological disease control products for the control of fungal plant pathogens in horticulture[31] and forestry[32]. No practical procedures are done, but few experimental scientific studies have been conducted, for example, Setälä and Huhta (1991) showed that the presence of soil fauna increase birch (Betula pendula) growth and nutrition in laboratory conditions (Setala and Huhta, 1991). Five similar studies have been conducted with trees but not with food plants.

In **France**, the Ministry for Agriculture and Food developed an agroecology project that includes soil biodiversity considerations, including an attempt to quantify the value of soil biodiversity and associated ecosystem services under the project (Pascual *et al.*, 2015). The main practical applications of soil biodiversity are nutrient recycling; agriculture and forestry productivity; landscaping; pest and disease regulation; regulation of water quality; soil and water remediation; carbon sequestration nitrogen and phosphorus cycle regulation; diagnosis of soil quality through indicators based on the diversity (taxonomic and functional) of soils; and the use of reference systems (type of soil, modes of use) (*Horrigue et al.*, 2016); biostimulants and biofertilizers. Agricultural productivity has increased, according to studies published by INRA or Agricultural chambers, through the use of inoculant for symbiosis.

Germany is maintaining or restoring self-regulation processes such as turnover of C, N, and P compounds otherwise depending on intensive use of agrochemicals (Lentendu *et al.*, 2014). Soil biodiversity is also contributing to multifunctionality of ecosystems (*Soliveres et al.*, 2016) and to stabilizing soils (soil aggregation) (Lehmann, Zheng, and Rillig, 2017). Increased biodiversity leads to more redundancy of functional traits covered by the soil biome. Hence the stability of soil functions is deemed to be correlated with biodiversity (e.g. Griffith *et al.*, 2004). This corresponds to the general theory of ecosystems, but detailed analyses are scarce. The following actions are related to soil biology in general, but not necessarily to soil biodiversity: pollutant degradation and soil remediation are applied in practice; application of soil biology for optimizing agricultural productivity and soilborne disease control are matter of active research (*Plaas et al.*, 2019)[33].

---

29 https://naturcom.fi/
30 http://verdera.fi/en/products/horticulture/gliomix/
31 http://verdera.fi/en/products/horticulture/mycostop/; http://verdera.fi/en/products/horticulture/prestop/
32 http://verdera.fi/en/products/forestry/rotstop/
33 In this article, German economic data were used exemplarily for the valuation of soil biodiversity with respect to soil-borne disease control. Also see for example, www.bonares.de

In **Germany**, there are still insufficient data regarding agriculture improvement, and it is subject to ongoing research (see *Rillig et al.*, 2018)[34]. However, it is broadly accepted that soil biodiversity is an important factor for soil health and thus for yield and product quality. Further, it has been shown that increasing soil biodiversity supports the suppression of pathogens. Moreover, intensification of land-use was shown to reduce (soil) biodiversity (Gossner *et al.*, 2016).

Italy reported the use of soil biodiversity on soil biomonitoring (IBS-bf and QBS-ar) (*Caoduro et al.*, 2014).

In the **Netherlands**, soil biodiversity drives all major soil ecosystem services mainly by trophic interactions in food webs (who eats whom). The services thus rely on functioning communities of soil organisms exposed to environmental changes and are essentially based on conversions of organic matter and energy between trophic levels, such as, for instance, primary decomposers (bacteria and fungi), microbivores and predators. Thus, the key components of the soil food web contribute to a healthy soil which reduces the need for external inputs of chemicals and fossil energy, reduces losses and emissions to the environment and is the core of sustainable management of land for agricultural production (Griffiths *et al.,* 2018). This applies to all agricultural (and forestry) soils including intensively managed soils. Replacing external inputs by improved use of functional soil biodiversity is the practical application in more sustainable agriculture and forestry. This includes biological nitrogen fixation, decomposition, carbon sequestration, nutrient mineralization and retention, soil aggregate formation and soilborne disease control (Koopmans *et al.*, 2006; Faber *et al.*, 2009; Korthals *et al.*, 2014; Bloem, Koopmans and Schils, 2017; Koopmans and Bloem, 2018; *Schouten et al.*, 2018)[35]. Enhanced use of soil biodiversity has contributed positively to food production and nutrition, and enhancing soil biodiversity can maintain production with lower external inputs.

The Republic of Moldova is promoting the recovery of the soil biota by green manuring; monitoring pesticides; contaminated soils; monitoring long term stored municipal sewage sludge and wastes amended soil; and assessment of soils infected with nematodes and soil fatigue prevention (Senikovskaya, Bogdevich, and Marinesku, 2008; Bogdevich and Senikovskaya, 2011; Senicovscaia *et al.*, 2012; Poiras *et al.*, 2013; Volosciuc and Josu, 2013; Senicovscaia, 2014; Senicovscaia, 2015).

Moldova is promoting the application of no-tillage practices with the introduction of green fertilizers. This practice in the southern zone of the Republic of Moldova improves the conditions for functioning of biota in the ordinary chernozem at the level of high values of parameters. As a result, yields of subsequent main crops increased by around 20 percent. For the application of grass cultivation, the use of the mixture of ryegrass and lucerne during three to five years led to the growth of the number of invertebrates

---

34 See for example: www.bonares.de; https://www.thuenen.de/en/cross-institutional-projects/soilman-ecological-and-economic-relevance-of-soil-biodiversity-in-agricultural-systems/).

35 Also see https://www.beterbodembeheer.nl/nl/beterbodembeheer.htm

and Lumbricidae family by 2.5 and 3.0 times, and their biomass by 1.6 to 2.0 times in comparison with the control plot. The annual population growth of earthworms reaches 14.9 ex/m^2. The microbial biomass accumulates in the soil in amounts of 132.5 kg/ha annually. The application of grass cultivation improved the invertebrates' diversity in the leached chernozem after five years. The grass-cultivated soil is characterized by a greater diversity of invertebrates. As a result, yields of subsequent main crops productivity has visibly increased (Leah and Cerbari, 2013; Senicovscaia, 2013; Wiesmeier *et al.*, 2015; Senicovscaia, 2018).

In **Spain**, the use of *Mycorrhiza* has improved plant performance and stress-resistance in agriculture (Martín-Robles *et al.*, 2018); to the use of *Mycorrhiza* for phytoremediation of polluted soils (Kohler *et al.*, 2015); soil bacteria as biofertilizers in agriculture (González-Andrés and James, 2016). Incipient work is being carried out on biological determinants of soil suppressiveness for biocontrol of plant root diseases (González-Andrés and James, 2016; Pastrana *et al.*, 2016; Arjona-Girona and López-Herrera, 2018) and soil erosion control in arid lands (Maestre *et al.*, 2006).

Ukraine has established the development and production of domestic microbial preparations with nitrogen-fixing, phosphate-mobilizing, growth-stimulating properties, biopesticides, aimed at replenishing soils with biological nitrogen, phosphorus and plant protection by biological methods. Farms widely implemented principles for production of organic agricultural products. Scientific monitoring is carried out to assess soil degradation on agricultural land and quality changes under biologization of agriculture.

Scientifically based agrarian technologies are implemented with reduced mechanical load on soil (zero tillage); optimizing the structure of agricultural lands and creating conditions for the sustainable functioning of soil microbial populations in the formation of agrobiocenoses and restoration of land, withdrawn from active use; implementation of soil protective crop rotation, including expansion of areas under perennial grasses; increase in the supply of organic matter, application of post-harvest residues; and implementation of effective technologies. The irrigation system on agricultural lands is also being restored and the number of livestock farms is increasing, which contributes to the growth of organic fertilizers in the soil[36].

Due in part to many years of research work on scientific support for organic farming and the formation of the legislative base, the total number of organic farm operators and the transition period has increased significantly to 426 (294 agroholdings, 381 173 ha total area of agricultural land, including 289 551 ha of organic land). Today in Ukraine there is an increase in the number of certified producers of organic products: 4 producers of organic dairy products, 1 of meat products, 15 of cereals, 11 of oils, and 3 producers of

---

36   O.I. Maklyuk (Starchenko) *et al.*, The strategy of balanced usage, reproduction and management of Ukrainian soil resources (Chapter 5.8 Optimization of soil biological condition). Kyiv: Agrarian Science, 2012. Biological Nitrogen Fixation: [monograph: in four volumes] / S.Ya. Kots, V.V. Patyka et. al.- K/:Logos, 2014. – 412 p. – Bibliography: p.314-385 Theory and practice of soil protective monitoring \ ed.by M.M. Miroshnychenko\ Kharkiv: FOP Broviv O.V. 2016. – 384 p.

organic spices[37].

In the **United Kingdom of Great Britain** and **Northern Ireland**, soil biological activity has huge importance for the delivery of the following services: production of food, fuel and fibre (nutrient cycling and fixation, water supply, soil structure formation, pest and disease control); climate regulation (carbon storage, methane production and consumption, nitrous oxide fluxes); flood risk mitigation (rebuilding and maintaining soil structure, water storage); water supply and quality (crop water holding capacity, biological treatment of water pollutants, mycorrhizal supply to crop roots); biodiversity support (soils as part of all terrestrial ecosystems; note that soils also represent a vast biodiversity resource in themselves); climate change adaptation (both in support of natural habitats, improving, linking and expanding and in terms of soil resource use for agriculture); waste/resource recycling (for example, sewage, crop residues, green waste); decontamination (organic pollutants, metals); and regulating air quality (generation of ammonia emissions, NOx gases, particulate matter). Soil biodiversity has a less important potential application for cultural services (e.g. it can be damaging to archaeological remains), but remains an underused resource for scientific exploration, education, communication and as a cultural wildlife phenomenon. Soil genetic diversity has huge potential to yield compounds of medicinal and industrial value but is largely unexplored.

Up until the discovery of the Haber process, soil biodiversity provided all the nutritional requirements for the entire world, including England and Wales. The potential for soil biology to deliver agricultural benefits under improved management has been reviewed by the United Kingdom[38]. The importance of soil biology to global ecosystem processes was explored[39] and the importance of soil biodiversity in supporting global biodiversity was reviewed as well (Bardgett and van der Putten, 2014).

2.1.4 | MAJOR PRACTICES NEGATIVELY IMPACTING SOIL BIODIVERSITY

In **Finland**, 50 percent of agricultural land is under cereal production with frequent tillage and low diversity in rotation, which results in less diverse earthworm communities. The effects of various practices have not been investigated comprehensively, but a few recent studies showed that forest clear felling harms red wood ant nest mounds and reduces the species richness of ant associates (beetles and oribatid mites).

---

37 https://agropolit.com/news/10552-v-ukrayini-zbilshilasya-kilkist-virobnikiv-organichnoyi-silskogospodarskoyi-produktsiyi Atlas of Soil Suitability of Ukraine for Organic Farming (Zonal Aspects) / Kharkiv, 2015. - 36 p.
38 http://publications.naturalengland.org.uk/publication/2748107
39 https://global.oup.com/academic/product/the-biology-of-soil-9780198525035?cc=gb&lang=en&

The **Netherlands** reported that, in general, functional soil biodiversity decreases with land use intensity and can be enhanced by reducing inputs and soil disturbance. Intensive practices are part of modern society and the challenge is to optimize the use of biodiversity and ecosystem services (Rutgers *et al.*, 2010).

In the **United Kingdom of Great Britain** and **Northern Ireland**, impacts due to overuse of pesticides are likely but are not well studied. There has generally been little land use change in the last ten years, but major changes in land use (increased intensification of agriculture) has resulted in historic impacts on soil biodiversity, although the nature of these impacts is largely unknown. This is supported by ongoing high rates of nutrient and agrochemical use. Practices leading to soil degradation include over cultivation and lack of organic matter inputs, along with traffickig of wet soils leading to compaction[40].

2.1.5 | INVASIVE ALIEN SPECIES (IAS)

Flatworms *Platydemus manokwari* (New Guinea flatworm) and *Obama nungara* were recently introduced in **France**. Some taxonomists regard it as invasive, but there are still controversies with soil ecologic scientists. Several alien invasive soilborne pathogens (for example, nematodes and fungal pathogens) are introduced in France mainly via potted plants or seeds, which cause important damage to crops or forest trees, and may have high economic impacts.

In **Italy**, the Durham slug (*Arion lusitanicus*) is a pest that has caused severe damage throughout Europe. Due to the rich soil biodiversity, this species is not considered a threat in Italy, as it is controlled by the natural soil food web.

The **United Kingdom of Great Britain** and **Northern Ireland** reported a range of antipodean flatworms known to have invaded the country, and these may have impacted earthworms and other soil fauna, but the scale of the impact is unknown, due to the lack of previous monitoring of earthworms. The spread of these is monitored by FERA through the Great Britain non-native species secretariat. Many smaller soil dwelling organisms are likely to have been introduced (e.g. a large number of *Symphypleona* springtails have recently been recorded in the country that appear to be new to science), but the provenance and impact of these is unknown[41].

---

40   https://environment.data.gov.uk/water-quality/view/landing
41   Also see: http://www.nonnativespecies.org/home/index.cfm (Selected soil species are: 1. Australian Flatworm (Australoplana sanguinea: https://secure.fera.defra.gov.uk/nonnativespecies/downloadDocument.cfm?id=349; 2. Kontikia flatworms (Kontikia ventrolineata and Kontikia Andersoni: http://www.nonnativespecies.org//download-Document.cfm?id=147; and 3. New Zealand Flatworm (Arthurdendyus triangulates: https://secure.fera.defra.gov.uk/nonnativespecies/downloadDocument.cfm?id=348)

2.1.6 | MONITORING SOIL BIODIVERSITY

At the European level, many efforts have been focused on assessing soil abiotic properties for creating maps such as the European Soil Database[42]. For agricultural lands more specifically, the LUCAS (Land Use/Cover Area frame statistical Survey) soil survey[43] addresses the main topsoil properties in 23 Member States of the European Union. The most recent survey (LUCAS 2018) assesses also soil biodiversity in 1 000 sampling points by DNA metabarcoding methods for several groups of organisms (Bacteria and *Archaea*, Fungi, Eukaryotes, nematodes, arthropods and earthworms). All these efforts can serve as a baseline and previous experience for future large scale (global) studies assessing soil biodiversity in agricultural lands in various climatic conditions and soil types. These studies are essential to create global datasets and maps for monitoring the status of agricultural soils and taking appropriate management measures at appropriate scales.

Finland does not have a separate soil information system, but the Finnish Biodiversity Information Facility (FinBIF) integrates a wide array of biodiversity research infrastructure approaches under the same umbrella, including soil biodiversity information. These include large-scale and multi-technology digitization of natural history collections; building a national DNA barcode reference library and linking it to species occurrence data; citizen science platforms enabling recording, managing and sharing of observation data; management and sharing of restricted data among authorities; community-driven species identification support; an e-learning environment for species identification; and IUCN Red Listing (Schulman, Juslén and Lahti, 2019).

Finland has expert groups on different organism group working under the supervision of A Steering Group for Evaluation and Monitoring of threatened species (LAUHA) led by the Ministry of Environment to gather observations of some soil fauna groups. However, there are no systematic monitoring schemes. Soil fauna in Finland is observed by, for example, expert groups for arachnids, expert group for beetles and expert group for crustaceans. The main duty of the expert groups is the IUCN threat assessment of Finnish species at ten-year intervals. The taxonomic knowledge has been scattered in different databases, but the Finnish Biodiversity Information Facility is unifying the information for example, by providing valid scientific names and distribution maps. The scientific collections material is preserved mainly in the Finnish Museum of Natural History (Luomus), which is part of the University of Helsinki and in the collections of the University of Turku[44].

France reported information systems at different levels: within the scientific interest group "Soil" (in French, GIS Sol – groupement d'intérêt scientifique "Sol"), the soil monitoring network (RMQS – réseau de mesure de la qualité des sols) looks at soil

---

[42] https://esdac.jrc.ec.europa.eu/content/european-soil-database-v20-vector-and-attribute-data
[43] https://esdac.jrc.ec.europa.eu/projects/lucas
[44] https://www.ymparisto.fi/en-S/Nature/Species/Species_protection_work. Also see the_species_expert_groups_in_Finland: https://luomus.fi/en/collections

biodiversity bacteria at the national level (ECOMIC-RMQS) and global soil biodiversity for Brittany (RMQS-BIODIV)[45].

Within the RMQS network, several maps have been produced at the national level (soil bacterial richness; soil molecular biomass; abundance of the most dominant bacterial and archeal phyla) (Dequiedt *et al.*, 2011; Terrat *et al.*, 2017; Karimi *et al.*, 2019) and at the regional level, that is, in Brittany (microbial biomass, bacterial biomass, nematodes abundance, plant-feeding nematodes, collembolan abundance, acari abundance, earthworms abundance and total macrofauna abundance) (Cluzeau *et al.*, 2015).

The first campaign has been achieved and a second campaign will cover the period from 2016 to 2027. A global soil biodiversity network is under construction and will complement the current network to monitor other biological compartments. Sampling procedures will be tested in 2020 and 2021 before the routine implementation. There are soil biodiversity indicators within the National Observatory of Biodiversity (in French, ONB – *Observatoire National de la Biodiversité*) for earthworms and bacteria. Participatory observatory of earthworms (in French, OPVT – *Observatoire Participatif des Vers de Terre*), coordinated by the University of Rennes for professionals (e.g. farmers) and gardeners. The agricultural observatory of biodiversity (in French, OAB – Observatoire agricole de la biodiversité) is a participatory observatory which includes observatory of earthworms by volunteer farmers, with the OPTV protocole.

In **Germany**, the German Environment Agency (Umweltbundesamt (UBA))[46] in Dessau-Roßlau stores and handles data of more than 800 long-term soil monitoring sites (*Bodendauerbeobachtungsflächen*) all over Germany beginning in the early 1980s. Soil biodiversity data is considered in many but not all sites (Schilli *et al.*, 2011)[47]. Soil biodiversity is optional in this programme and not all federal states provide information to this topic.

Most of the information on soil biodiversity from permanent monitoring sites have been transferred to this database run by the Senckenberg Museum of Natural History Görlitz (Burkhardt *et al.*, 2014). So far, more than 250 000 data sets on soil fauna from more than 13 000 sites are compiled in this database[48]. They develop maps of various scales on soil types and on different soil properties. A national soil information system, *FachInformationsSystem Bodenkunde* (FISBo BGR) is maintained by the Federal Institute of Geosciences and Resources[49]. However, this primarily comprises abiotic data. There are some Federal State Agencies that also include soil biotic information in

---

45 http://www.gissol.fr/le-gis/programmes/rmqs-34
46 https://www.umweltbundesamt.de
47 https://www.umweltbundesamt.de/publikationen/auswertung-veraenderungen-des-bodenzustands-fuer-0
    For the Soil monitoring program see https://www.umweltbundesamt.de/en/topics/soil-agriculture/soil-protection/soil-observation-assessment
48 See website of BGR (Bundesanstalt für Geowissenschaften und Rohstoffe = Federal Institute for Geosciences and Natural Resources) in Hannover: https://www.bgr.bund.de
49 BGR; see https://www.bgr.bund.de/DE/Themen/Boden/Produkte/produktkatalog_node.html

their information systems. The BonaRes Data Repository[50] also includes soil biological information.

The report *Soil Monitoring, Installation and Operation of Soil Monitoring Sites* was prepared by a working group on soil protection issues of the federal states and the federal government (LABO) (Rosenkranz *et al.*, 2000).

The assessment of soil biodiversity in long-term monitoring regimes is carried out by some federal states in the nation-wide Soil Permanent Observation Sites[51]. A few federal states include soil biodiversity in their long-term monitoring programmes. An example is the Forest Permanent Observation Sites (Wald Beobachtung) in Baden-Württemberg[52]. Due to the installation history of permanent monitoring sites by the federal states, there is no common "German" date for the beginning of the programme. For example, Bavaria and Baden-Wuerttemberg started 1985 but the new federal states (after reunification in 1990) installed the monitoring sites in 1992.

An assessment of available information on soil biodiversity was made in a project (Römbke *et al.*, 2012). The German Environmental Specimen Bank (ESB) also estimates terrestrial specimens (earthworms, soil) to permit load observations within both a spatial and a time framework but on a limited number of sites only[53]. Sampling areas, methods, analyses and sampling periods are available[54]. Within the Priority Program "Biodiversity Exploratories" of the German Science Foundation (DFG), diversity of soil organisms is analysed every three years on 150 grassland and 150 forest plots under variable land use intensity[55].

The national monitoring of the biological diversity in agricultural landscapes was launched March 2019. The Thünen Institute of Climate Smart Agriculture[56] carries out the Federal Soil Inventory (BZE-LW), while the Thünen Institute of Forest Ecosystems carries out the Forest Soil Inventory (BZE-Wald) in forest sites. Both inventories are beginning to include soil microbial parameters in its assessment protocols. A nation-wide monitoring programme on biodiversity including soil biodiversity is currently under development and will be established within the next years: MonViA (*Nationales Monitoring der biologischen Vielfalt in Agrarlandschaften*)[57]; a website is under construction. MonViA is coordinated by the Thünen-Institute of Biodiversity in Braunschweig.

Italy has proposed a National Monitoring Network of Soil Biodiversity and Degradation. Soil in Italy is a neglected medium and any investigation on its biota conflicts with the economics and policies currently adopted. The proposal has not been followed by

---

50 https://datenzentrum.bonares.de/research-data.php
51 https://www.umweltbundesamt.de/themen/boden-landwirtschaft/boden-schuetzen/boden-beobachten-bewerten
52 https://www.lubw.baden-wuerttemberg.de/klimawandel-und-anpassung/bodenleben
53 https://www.umweltprobenbank.de/en/documents
54 https://www.umweltprobenbank.de/en/documents/profiles/specimen_types/10033
55 https://www.biodiversity-exploratories.de/
56 https://www.thuenen.de/en/ak/projects/agricultural-soil-inventory-bze-lw
57 More information under https://www.bmel.de/SharedDocs/Pressemitteilungen/2019/070-Kickoff_Biodiversitaet_Monitoring.html

concrete actions (Floccia and Jacomini, 2012). Only soil loss is monitored on a regular basis. Most updated information is available on the government website.

The **Netherlands** reported on the Biological Indicator for Soil Quality within the Netherlands Soil Monitoring Network (Bloem and Breure, 2003; Bloem *et al.*, 2005; Rutgers *et al.*, 2008; Rutgers *et al.*, 2009). However, the National monitoring terminated in 2014 (capacity and expertise has been reduced or lost). Recently, a more limited set of indicators has been defined and will be further developed for practical application. The ambition is to include organic matter (total and labile), bacterial and fungal biomass, nematode diversity and number and earthworm number and diversity (Hanegraaf *et al.*, 2019).

The introduction of Soil Quality Monitoring in the **Republic of Moldova** in 2005-2008 was provided by the Decision of the Parliament no 415-XV of 24. 11. 2003, part II, point 2, action "a". The Parliament of the Republic of Moldova, concerned about the state of the environment and the soil cover, approved in 2003 the National Action Plan in the field of human rights for 2004-2008. Objective 2 of this Plan provided the population with quality, non-degraded, ecologically pure soils. To achieve this objective, the activity "Introduction of soil quality monitoring" was envisaged and executed by the Agency of Land and Cadastre Relations.

From 2005 to 2008, soil monitoring researches (including the monitoring of soil biota) were carried out to create the initial data base on the quality status of the main types and subtypes of regional soils. On typical arable land (horizontal surfaces), about 30 key polygons were located, within which the values of the main properties of the soils were appreciated. For some subtypes of zonal soils, key polygons were also located on the well-known lands, which allowed assessing the degree of degradation of the properties of the arable soils under the influence of anthropogenic factors, as well as the speed of remediation of these properties under the influence of the restored natural vegetation.

The research work related to soil biodiversity were carried out in the framework of projects: *National project No. 14 "Introduction of Soil Quality Monitoring"*, compartment "Creating a system for monitoring soil biota" with the financial support of the Agency of Land Relations and Cadastre and the Republican State Association for Soil Protection in 2005-2008; *National institutional project "Evaluation of the state and resistance of soil invertebrates and microorganisms aiming to reduce the degree of degradation and fertility conservation"* (State Registration No. 06-407- 035A) in 2006-2010; *National institutional project No. 11.817.04.33A "Evaluation of the quality status of soils, elaboration and testing of technologies to stop the degradation and increase of fertility through the modernization and extension of land improvements"* in 2011-2014.

The monograph, guidelines and current articles present data obtained during the first determination of soil properties in the network located at key monitoring sites, as well as some results with the monitoring aspect obtained during pedobiological studies conducted by the Nicolae Dimo Institute of Pedology, Agricultural Chemistry and Soil Protection in over the past few years.

In **Moldova**, soil information systems are available in the N. Dimo Institute of Pedology, Agricultural Chemistry and Soil Protection (Kishinev). The Institute developed a geoinformation system of soil quality for precision agriculture and it contains some information about the state of soil biota. A database of soil microbial biomass, enzyme activities, numbers of soil microorganisms and the state of edaphic fauna is kept in the Laboratory of Pedology. The results of research on the state of soil biota in Moldova are published in numerous articles, reports, monographs and guidelines. Some indicators have been monitored since 1958. In accordance with the concept of the Soil Register of the Republic of Moldova Information System, approved by Government Decision No. 1001 of 10 December 2014, the Land Management Project Institute has an information system on the state of the soil cover (arable soils).

There is no single national system of environmental information and an accessible information resource that includes data on soil and soil biodiversity, which complicates the process of finding information in the Republic of Moldova. In recent years, some environmental information systems have been developed in order to comply with national legislation and compliance with international obligations. Most of them, in particular the database on plant biodiversity, terrestrial fauna and biodiversity in aquatic ecosystems, were created in the framework of national and international projects. Most of them were created in the framework of international projects developed with donor support. This has led to a rapid increase in the number of data systems based on different, mostly commercial software, which are incompatible.

Ukraine is developing work to improve the National Digital Map of Ukraine on soil carbon as an integral part of the Global Soil Organic Carbon Map (GSOC Map). In 2019, the Ukrainian Soil Information Centre was established on the basis of the National Scientific Centre Institute for Soil Science and Agrochemistry Research O.N. Sokolovsky. The Centre set the following tasks: accumulation, processing and dissemination of data on the state and quality of soil in Ukraine, including by biological parameters; providing full up-to-date information on the state and quality of the soils of scientific institutions, public entities, state authorities, territorial communities, public organizations and the population of Ukraine; for exchange of information on soils with domestic and international organizations; and to carry out cross-sectoral research and to create information products on soil state and quality with a view to developing a national information infrastructure and its integrating with the relevant world infrastructure.

With the initiative of the Global Soil Partnership (GSP) and the support of the United Nations Food and Agriculture Organization (FAO), the Ukrainian Centre developed the first version of the GSOC Map for the 0-30 soil layer as one tool for achieving the Sustainable Development Goal 15 (protect, restore and promote sustainable use of terrestrial ecosystems, sustainably manage forests, combat desertification and halt and reverse land degradation and halt biodiversity loss). The work is underway to improve the National Digital Map of Ukraine on soil carbon as an integral part of the Global Soil

Organic Carbon Map (GSOC Map) for the 1 km x 1 km grid (FAO and ITPS, 2019)[58].

Within the framework of long-term field studies, scientific soil-monitoring is carried out at the stationary experiments of NSC ISSAR and other research institutions in order to assess soil degradation on agricultural land and changes in quality by the biologicalization of agriculture. The purposes of these observations determine the list of physical, agrochemical and biological parameters that determine the quality and fertility of soils. According to the National Program *Soil Resources: Forecast of Development, Sustainable Use and Management* (2016-2020), scientists of the Soil Microbiology Laboratory at NSC ISSAR have developed a system of biodiagnostic indicators and methodology for soil biological status to support soil biodiversity in Ukraine. The obtained experimental material is formed into blocks of biological indicators for the information centre, which is opened on the basis of the NSC ISSAR. In the Soil Resources Research Program of the NSC ISSAR in 2019-2020, the scientists started research projects on the development of measures for the prevention of environmental pollution by nitrates and pesticide residues in agricultural production, harmonization of methods of diagnosis and assessment of soil contamination, adopted in Ukraine and the world, improvement of monitoring methods in monitoring areas in the monitoring zone enterprises, roads, livestock complexes and storehouses for storage of agrochemicals.

Phytosanitary monitoring is a system of observation and control of the spread, density, intensity of development and danger of harmful organisms. Visual methods are based on direct examination and calculation of pests and plant organs damaged them and the intensity of their disease. Specialists of signalling and forecasts carry out detailed records systematically during the vegetation of plants not less than every ten days in the test areas of the selected fields. Depending on the time of conducting there are distinguished: autumn, spring (control) and vegetation (periodic) soil excavations, and depth ranges of shallow (up to 10 cm), normal (up to 45 cm to 50 cm) and deep (65 cm and deeper). Spring control excavations are carried out after thawing the soil, when it is scattered, in order to establish changes in the state (mortality) of pests during the wintering period and their density with the method of autumn surveys at least 10 percent of the areas surveyed during the autumn. Vegetative excavations are carried out during the growing season of crops to determine the density of soil pests and damaged plants by them. As a rule, these excavations are small - up to 20 cm - and the accounting pits are placed so that the row of plants is in their centre. The method of soil excavation also determines the number of pests that overwinter in the soil and damage the root system of perennial crops (hops, gardens, vineyards).

The **United Kingdom of Great Britain** and **Northern Ireland**'s past efforts include the Countryside Survey. Natural England continues to run the Long Term Monitoring Network over 37 NNRs, and Glastir GMEP has assessed 300 plots as described above and will be succeeded by the ERAMMP programme, which may continue to measure soil biological parameters. The NERC UK-SCAPE programme plans to establish a network

---

58   Also: http://www.issar.com.ua/uk/news/ukrayinskomu-gruntovomu-informaciynomu-centru-buty

of soil and vegetation monitoring plots where soil biota and their interactions with soil carbon will be explored (Cosby, 2018). UKSO displays the Countryside Survey soil biological information. The United Kingdom has a national soil information system (LANDIS) run by the National Soil Resources Institute at Cranfield University, but this doesn't contain any soil biological information (Cranfield University, 2020).

2.1.7 | INDICATORS USED TO EVALUATE SOIL BIODIVERSITY

Finland has indicators of soil biodiversity revealed by scientific studies for many of the themes listed, but not much for practical implications. However, the estimation of earthworm abundance and biodiversity is incorporated in the arable soil quality assessment tool developed for farmers ("*Peltomaan laatutesti*"), tests for soil capability to degrade hydrocarbons and toxicity of contaminated sites (Bionautit, 2020).

France reported that regarding ecotoxicology and soil pollution, the country is managing polluted sites, the use of organic wastes in agriculture and fertilizers, soil improvers and pesticides commercial authorization procedures. The genetic diversity has been catalogued in maps, monitoring and atlas. Soil fertility management is observed in forest (mainly when biomass is collected for bioenergy) and in the use of organic wastes in agriculture. Development of an operational diagnosis of soil biological quality (AgrInnov continued with REVA project). This was based on participatory science; France has linked research with a network of 250 farmers throughout the country. This project has developed and transferred training and dashboard indicators of soil biological quality directly to farmers. This is an operational project demonstration which has fulfilled its ambitions (more than 98 percent of farmers have been satisfied and about 60 percent have started to change their practices on the basis of a soil biology diagnosis). Others relevant initiatives were reported: jardibiodiv (ScaraB'Obs, 2015; Réseau-Agriville, 2018), the AgroEco-Sol Project (financement PIA, resp. AUREA, resp. Inra L. Ranjard) for the transfer of technology and expertise to agricultural development actors in order to develop a soil microbiology analysis chain with indicators.

Germany reported that there are no nationally implemented indicators for evaluating soil biodiversity related to the respective services/threats. However, there are numerous debates about how such indicators (not just for evaluating soil biodiversity but other soil functions as well) could be defined. At the Federal States level, some of those measures are in place but these are not standardized or harmonized.

The **Republic of Moldova** has a partially developed an assessment system with indicators, criteria, statistical parameters and scale of the soil biota.

In the **United Kingdom of Great Britain** and **Northern Ireland**, earthworms have been suggested as an indicator for the England Chemicals Strategy. The GMEP and National Indicators for Wales include soil carbon. Soil carbon is likely to be included in future indicators for England. The Environment Agency uses indicator thresholds for metal contamination to manage (among other things) waste to land applications, to protect soil biological function. Water quality parameters are measured in a network across the United Kingdom but are not explicitly linked to soil biological functions (UK Environment Agency, 2017, 2020).

2.2 | RESEARCH, CAPACITY DEVELOPMENT AND AWARENESS RAISING

Finland has developed the Finnish Biodiversity Information Facility, the national roadmap of research infrastructures that receives currently funding from several sources for the development of infrastructure. As part of the infrastructure funding for example, the Finnish Barcode of Life project aims to DNA barcode also the Finnish soil fauna during the following years. Finland participates in:

i. COST action European Soil-Biology Data Warehouse for Soil Protection (EUdaphobase)[59];

ii. MULTA (Baltic Sea Action Group, 2019a), a multi-benefit solution to climate-smart agriculture;

iii. TWINWIN, a nessling-funded project launched in spring 2019, aims to find out how plant and soil biodiversity affect the ability of the field to sequester carbon. Scientists design biodiversity-based practices to accelerate soil carbon sequestration and produce a computational model that considers impact of biodiversity on the carbon cycle (Baltic Sea Action Group, 2019b) in the field;

iv. Organic Farming Information site;

v. Enhancement of biodiversity in commercial forests[60];

vi. Soil biodiversity in the improvement of arable soil quality in the advisory activity of the main national advisory organization ProAgria (Alakukku *et al.*, 2017).

In **France**, soil and biodiversity have been included into the programme of the general streams of secondary education (Lycée). The *Agence de l'Environnement et de la Maîtrise de l'Énergie* (ADEME) supported awareness raising on soil biodiversity. In coordination

---

59 https://www.cost.eu/actions/CA18237/#tabs
60 https://www.metsakeskus.fi/monimetsa-hanke

with the Ministry in charge of the environment, a brochure and a card game called "Seven Families: The hidden life of the soil" were developed as part of the GESSOL research programme (the game has also been translated into English). Other activities include:

- Co-financing of the "Biodiversity atlas of soil bacteria";
- Co-financing of the "Biodiversity atlas of soil fungi" (in progress);
- Co-financing of the edition of the book "Planet collemboles";
- Co-edition of the book "Bioaccumulation, bioamplification of pollutants in the terrestrial fauna".

Participatory observatory programmes enable also awareness and training of farmers. The Experimentation and Monitoring Network for Agricultural Innovation (REVA - Réseau d'Expérimentation et de Veille à l'innovation Agricole) takes over from AGRINNOV the research project to train farmers, with the aim of changing their farming systems towards environmental and economic sustainability. REVA is locally funded by Water Agencies, ADEME, departmental and regional agriculture chambers, regions and farmers. A "rev-urban" is being set up based on the REVA model: an ANR Bis project has been accepted (CNRS, INRA). It aims to set up a participatory science approach to soil biology in urban and peri-urban areas. Some cities are involved in the project in connection with agricultural systems and a showroom on soil biodiversity with SVT high school classrooms was inaugurated by the Minister of Agriculture and the European Commission (400 visitors) during the First Global Soil Biodiversity Conference in Dijon (2014).

Two research projects had been funded under the GESSOL programme of the Ministry in charge of the environment: the project GENOSOIL on the extraction of DNA from soils and its analysis and one project on the measurement of total soil macrofauna. ADEME coordinated and co-funded a research programme called "bioindicators of soil quality" between 2002 and 2012: 80 bioindicators developed and tested during the first phase of the programme, 23 bioindicators selected among the 80 ones during the second phase and applied to 13 experimental sites, with three main topics (contamination with atmospheric deposits, polluted sites and spreading of organic wastes in agriculture) (Pérès *et al.*, 2011 ; Pauget *et al.*, 2013).

This programme made it possible to fund, in addition to the development of indicators, the start-up or implementation of soil biodiversity monitoring (ECOMIC-RMQS and MetaTAXOMIC-RMQS at the national level, RMQS-BIODIV at the level of Brittany which was co-funded by ADEME), to launch a participatory observatory of earthworms (OPVT), the definition of soil biodiversity indicators for the ONB (National Observatory of Biodiversity), the standardization of protocols on soil biology quality (AFNOR, CEN, ISO).

The *Agence de l'Environnement et de la Maîtrise de l'Énergie* funds research on the evaluation of impacts of polluted sites on ecosystems (including soil organisms) and on the development of soil bioindicators to assess polluted sites. ADEME also finances innovation investments to enable the industrialization of soil biological diagnostic tools (e.g. AGROECOSOL for agriculture) and the emergence of a service offering on soil quality bioindicators (ELISOL Environnement, VALORHIZ, GENOSCREEN). The AgroEco-SOl project (PIA funding, AUREA or Inra L. Ranjard) aims to transfer technology and expertise to agricultural development actors to develop a soil microbiology analysis chain, with indicators, adapted BDD.

The French Ministry of Agriculture funded also several research projects under the CASDAR programme (for example, AGRINNOV and MICROBIOTERRE). The agricultural observatory of biodiversity (Observatoire agricole de la biodiversité – OAB). The French Ministry of Agriculture funded a network about biodiversity and agriculture) ((Ministère de l'agriculture, de l'agroalimentaire et de la forêt, 2016) with research organisms and different schools of agriculture. The French research agency (ANR) funds also projects on soil biodiversity: "Agrosystems and functionnal soil biodiversity (SOFIA); Microbio-geography at French scale France by application of molecular tools to the RMQS (ECOMIC-RMQS); multi-scale evaluation of soil ecosystemic services within agroecosystems (SOILSERV); urban gardens and sustainable cities (JASSUR) (a part of the project concerned soil biodiversity). Some European research projects were coordinated by French researchers (for example, EcoFinders).

The French biodiversity agency (AFB) recently engaged a research programme "Soil biodiversity and agro-ecology". In addition, research has its own devices, long-term ones, some of them are integrated into the research infrastructure (IR). IR ANAEE, managed by CNRS and INRA (AnaEE, 2020) commits to most of these sites an observation of soil biodiversity (experimental plots, annual monitoring and effect of agricultural systems and practices on soil biology). Pioneering work is also done at landscape scales (IR OZCAR). INRA also has experimental facilities that have included soil biodiversity. For example, the CA-SYS Agroecology Platform at the landscape scale, which includes monitoring of soil biodiversity, according to agricultural systems (INRAE, 2020a). Soil biodiversity has also been monitoring in Lusignan. Many other programmes not specific to soil biodiversity have brought results, such as ANR (eg SYSTERRA, AgroBioSphere and CES) and EcoPhyto.

Germany presented some of the current research programmes: BonaRes (Federal Ministry of Education and Research, 2020a) (soil as a sustainable resource for the bioeconomy), FInAL (FNR, 2020) (Facilitating insects in agricultural landscapes through integration of renewable resources into cultivation systems – a scientifically supported pilot and demonstration project in landscape laboratories).

Germany's Federal Agency for Nature Conservation (*Bundesamt für Naturschutz, or BfN*) currently supervises the following research activities in the context of soil biodiversity: "*Ausarbeitung naturschutzfachlicher Leitplanken für die klimaschutzmotivierte Wiedervernässung von Niedermoorböden bei angepasster (nasser) landwirtschaftlicher Nutzung zur Maximierung der Synergien zwischen Klima- und Biodiversitätsschutz*" (FKZ 3518 81 0500). Ernst-Moritz-Arndt-Universität Greifswald, Humboldt-Universität Berlin und Hochschule für nachhaltige Entwicklung Eberswalde. "*Direkte und indirekte Auswirkungen von Düngung und Pflanzenschutzmaßnahmen auf die Biodiversität in Agrarlandschaften*" (FKZ 3518 84 0800). RWTH Aachen.

The recent "Research Initiative for the Conservation of Biodiversity (*Forschungsinitiative zum Erhalt der Artenvielfalt - Eine FONA-Leitinitiative*)" of the German Federal Government (BMBF) (Federal Ministry of Education and Research, 2020b) has one focus explicitly on soil biodiversity, its functions, drivers and anthropogenic influences.

Germany has a programme to support organic farming (UmweltBundesamt, 2018; Federal Ministry of Education and Research, 2020c). Again, this response relates to soil biological processes in general, not necessarily soil biodiversity: BMBF funding initiative "Soil as a sustainable resource for the bioeconomy – BonaRes" (Federal Ministry of Education and Research, 2020a), BMBF funding initiative "Plant roots and soil ecosystems: Importance of the rhizosphere for the bioeconomy" (starting 2020), DFG Priority Programme 2089 Rhizosphere Spatiotemporal Organisation - a Key to Rhizosphere Functions (Helmholtz Centre for Environmental Research -UFZ, 2017), and Biodiversity Exploratories (The Biodiversity Exploratories, 2020).

Germany reported examples of capacity development such as soil awareness raising, including soil biodiversity as a topic in the German environment agency with different projects (German Environment Agency Umweltbundesamt – UBA, 2014). Furthermore, the Commission for soil protection at the German Environment Agency (Umweltbundesamt), KBU, has some publications and activities in this topic (German Environment Agency Umweltbundesamt – UBA, 2020). Awareness raising activities are, for example, conducted in the framework of the BonaRes funding initiative. They will be intensified in the framework of the Scientific Year (Wissenschaftsjahr) 2020 "Bioeconomy", including through a Citizen Science action on soil health. The Senckenberg Museum of Natural History in Görlitz has developed school materials and maintained travelling exhibitions on soil and soil biodiversity for decades, which have been shown widely in Germany and throughout Europe. The most recent exhibition is the "Thin skin of our earth" ("Die Dünne Haut der Erde") (Senckenberg – Leibniz Institution for Biodiversity and Earth System Research, 2020).

Other permanent exhibitions on soil biodiversity exist for the general public, in the Museum am Schölerberg in Osnabrück (Museum am Schölerberg, 2020) and in the Hainich National Park Centre in Thiemsburg (Nationale Naturlandschaft, 2020). Information on sustainable soil management, including aspects of soil biodiversity, can be found in various information provided by the BLE. In addition, the Federal Ministry of Food and Agriculture provides information on various soil topics (German Federal Ministry of Food and Agriculture, 2020). Other relevant initiatives are the BonaRes (BONARES, 2020) funding initiative and Soil Taproot initiative within the German

Centre for integrative Biodiversity Research (iDiv) Halle – Jena – Leipzig (German Center for Integrative Biodiversity Research Halle-Jena-Leipzig – iDiv, 2020). Some events:

i. One prominent event each year in December is the announcement of "Boden des Jahres" ("Soil of the Year") by Kuratorium Boden des Jahres in Berlin (German Soil Science Society - DBG, 2020). Another annual event is a conference on the occasion of the World Soil Day (5 December) organized by the KBU and Umweltbundesamt –2019 with the topic soil biodiversity (German Environment Agency, KBU Conference "Biodiversity - everything is related to everything", 2019).

ii. International Thünen Symposium on Soil Metagenomics, every three years, taking place in December for three days at the Thünen Institute headquarters in Braunschweig (Thünen Institute, 2019).

iii. Sustainable soil management is inherent in the code of good agricultural practice. Support programmes in execution of the Common Agricultural Policy (GAP) are carried out for sustainable soil management too.

In *Italy*, at the regional level, several initiatives have been promoted to implement the development of sustainable soil management practices. Only sectoral policies (on pollinators, fungi, microbial diversity and agriculture) attempted to address the topic, sometimes in a very transdisciplinary sense, but with a lack of integration with national policies.

The Italian National Rural Development Network has proposed several biodiversity indicators linked with ecological ecosystem services: carbon sink in the soil and organic matter, sustainable use of fertilizers, sustainable use of pesticides, minimum tillage, cover crops, nutrient recycling, natural buffer zone and others.

All of these indicators were considered in the Development Rural Plans at regional level, with a positive trend in the preservation of the quality of the environment. Besides, the Ministry of Agriculture, Food and Forestry established the diffusion of Voluntary Guidelines for Sustainable Soil Management of the FAO through the Italian Development Rural Plan Network.

The Italian Society for Soil Sciences (SISS) has established a working group on soil biological monitoring through microarthropods (QBS-ar), which has organised three workshops on the topic and is divided into eight subgroups (with approximately 60 participants) (Società Italiana della Scienza del Suolo, 2020). The Italian Soil Science Society (SISS) from 2006 established the school of biodiversity and bioindication of soil to spread the knowledge on biodiversity of soil and its importance on sustainable soil management. Practical training for students, local administrators, farmers, and other stakeholders are organized. The programme changes every year and is attended by an average of 50 to 60 students per year. The main topics discussed are: analytical methods and soil sampling, soil ecosystem and biodiversity, soil pollution, soil forestry, biodiversity in agri-food sector, extreme soil, soil urbanization, biodiversity in the Mediterranean environment, biodiversity and sustainable use of soil.

Another important activity carried out by SISS, under the auspices of the Ministry of Agriculture, Food and Forestry, is the publication of a library series on analytical methods for soil characterization. In particular, they have published three volumes correlated with the characterization of soil biodiversity: microbial methods, biochemical methods and molecular methods (Nannipieri and Picci, 2002; Benedetti and Gianfreda, 2004; Mocali, 2010).

It is important to underline that Italy coordinates the European Microbial Research Infrastructure (MIRRI) platform on microbial collection, in which are considered soil microorganisms and food microorganisms. The Ministry of Agriculture, Food and Forestry financed the CREA to improve microbial genetic resource conservation for food and agriculture. In many cases, the collections are work collections. The conservation of soil microorganisms is based on ecosystem conservation that considers, according to Guidelines for the conservation and characterization of plant, animal and microbial genetic resources for food and agriculture, the relationship among microorganisms, plants and environment.

Italy coordinated the COST action 831, financed by the European Union with the participation of 16 European countries, on soil biotechnology for monitoring, conservation and restoration of biological fertility of soil (European Cooperation in Science & Technology - COST, 2020). At the end of the project, a handbook was published on microbial assessment on soil quality; it established a good start for a proposal of procedures to evaluate soil biodiversity.

Despite the crucial role performed by fungi within ecosystem processes, these organisms are still neglected in nature conservation plans and in the assessment of habitat protection priorities. In this perspective, ISPRA continues activities on mycological biodiversity started in 2003 by the former Agency for the Protection of the Environment and for Technical Services (APAT). The fields of interest were expanded and in 2007 the "Fungi Special Project" was established to provide unconventional operational tools to assess environmental quality and to promote the use of fungi as biological indicators. This special project was divided into 16 Research Topics, given the extent and complexity of the research subject. In 2010, a Fungi Special Project synoptic overview of its activities and results was illustrated within the national biodiversity conservation conference organized by ISPRA. At the end of 2011, after an intensive four-year (2008-2011) seminar activity carried out on a monthly basis, the need emerged to have a multidisciplinary ecosystem approach to scientific research on fungi, given the considerable diversity of the various physiographic areas of Italy. Thus, to endorse the results obtained by the various "Operating Units", the Project took part into the "Technical Table for the establishment of the National Network for Soil Biodiversity and Land Degradation Monitoring" (ReMo Programme).

To implement such activities, since 2012 the "Centres of Excellence for the study of soil biodiversity components" were established within the ISPRA "Fungi Special Project". Since 2012, thanks to the collaboration and intense activities of all the local operational structures (Operating Units and Centres of Excellence), the "Fungi Special Project" succeeded in constantly implementing the mapping and census data into the "ISPRA Mycological Diversity Information System" and in annually preparing several unedited editorial publications such as ISPRA-SNPA Manuals and Guidelines.

In Italy, some projects (Istituto Superiore per la Protezione e la Ricerca Ambientale – ISPRA, 2020) have attempted to address the specific problem with partial or local answers. This might seem a reductionist approach, while the great diversity of Italian soils and the extreme complexity of its biodiversity require often a limited approach to be effective, releasing specific projects on a widespread range of situations and lacking a national framework to coordinate, address and report them.

The **Netherlands** reported on research programmes or initiatives such as the PPS Beter Bodembeheer, a public private partnership. Researchers, companies and authorities cooperate to enhance knowledge, dissemination and application of relations between soil management, soil services and soil parameters/indicators (Beter Bodembeheer, 2020). The open bodem index (open soil index) is linked to PPS Beter bodembeheer and includes coalition of stakeholders (e.g. bank, investors and drinking water companies) (Open Bodemindex, 2020). The Delta Plan for Biodiversity Recovery (NIOO, 2019) and the Centre for Soil Ecology (The Centre for Soil Ecology – CSE, 2010). Regarding capacity development and awareness raising activities, the Netherlands has the Bodemacademie to Promote sustainable soil management (Bodem Academie, 2020), education programmes for schools (Globe Nederland, 2020) and the Bodemdierendag (National soil fauna day) (Bodemdierendagen, 2020).

Portugal is adopting the Guidelines and also promoted a seminar on Soil Sustainable Management Guidelines (Portuguese Partnership for Soil, 2018). Universities and soil laboratories have implemented capacity and skills to study aspects related with soil biodiversity (soil enzimology, soil microbial biomass, soil respiration).

In the **Republic of Moldova**, soil biodiversity research programs and projects (as part of a comprehensive study of soil) carried out by the group of soil biology (N. Dimo Institute of Pedology, Agricultural Chemistry and Soil Protection, Republic of Moldova) aim to support the development and implementation of sustainable soil management practices are as follows:

i. Complex programme of recovery of degraded lands and increase of soil fertility (2004-2005).

ii. Project "Elimination of acute risks from Obsolete Pesticide in Moldova" (department "Fitoremediation sites polluted by obsolete pesticides") with the support of MilieuKontakt, the Netherlands (2007 – 2008).

iii. National institutional project No 06-407-035A "Evaluation of the state and resistance of soil invertebrates and microorganisms aiming to reduce the degree of degradation and fertility conservation" (2006-2010).

iv. National institutional project No 11.817.04.33A "Evaluation of the quality status of soils, elaboration and testing of technologies to stop the degradation and increase of fertility through the modernization and extension of land improvements" (2011-2014).

v. IFAD international project No BES-036/14 RFSADP "Determination of soil physical, chemical and biological properties of demonstration plots in the implementation of conservative agriculture" with the support of Consolidated Unit of Programs Implementation IFAD (2014).

vi. IFAD international project No BES-017/15 RFSADP "Study of soil physical chemical and biological properties of demonstration lots in the implementation of Conservative Agriculture" with the support of Consolidated Unit of Programs Implementation IFAD (2015).

vii. IFAD international project No B&S-025/16 RFSADP "Study of soil physical chemical and biological properties of demonstration lots in the implementation of Conservative Agriculture" with the support of Consolidated Unit of Programs Implementation IFAD (2016).

viii. National project for young scientists No 15.819.05.09A "Interaction of microorganisms with the organic substance and the structural aggregates of the soil used for arable land and phytoamelioration" with the support of Academy of Sciences of Moldova and Ministry of Agriculture and Food Industry from Moldova (2015-2016).

ix. National institutional project No 1405177a "Preventive remediation of properties of the degraded arable layer in the southern Moldavian chernozems for the implementation of the conservative soil tillage system - No-till and / or Mini-till with subsoiling" (2015-2018).

In the framework of these projects, guidelines were issued for students and graduate students studying microbiology, soil science and ecology. In addition, many articles have been published and reports were presented at universities in the Republic of Moldova and abroad. The country partially promotes the adoption of sustainable soil management to avoid impairing key soil functions by land users as per the World Soil Charter and the Voluntary Guidelines for Sustainable Soil Management.

Spain reported that there are some general programmes, but not centred on soil biodiversity (Ministerio para la Transición Ecológica y el Reto Demográfico de España, 2020). In addition, there are initiatives for erosion minimization and inventory of polluted soils contemplated by different administrations. The Spanish Science Ministry

(MICINN) is currently supporting the thematic network ECOSOIL (soil biodiversity as an essential element in ecosystem functioning and sustainable use of the natural resources) in the framework of the Spanish National Program for the Promotion of Scientific and Technical Research of Excellence Dynamisation actions "Networks of Excellence" (CGL2017-90635-REDT) (2018-2020).

In **Ukraine**, scientists of NSC ISSAR, based on the results of long-term research, developed the National Soil Conservation Program of Ukraine; the Atlas of the suitability of soil of Ukraine for organic farming (zonal aspects); the Concept of organic agriculture; the monograph "Rational use of soil resources and reproduction of soil fertility: organizational, economic, environmental and regulatory aspects"; and the comprehensive assessment of biological state of soil microbial system and its transformation in modern farming systems.

The scientists continue to work on the National Program of the Academy of Agrarian Sciences "Soil resources: forecast of development, sustainable use and management" (2016-2020), which includes the task of improving soil biodiagnosis for biodiversity support and the NAAS Program 3 (Organic Agricultural Production (2016-2020) with the task of selection of new strains of nitrogen-fixing and phosphate-mobilizing bacteria that are promising for the creation of new, highly efficient biological preparations for organic farming.

The Ukrainian Soil Information Center is being created on the basis of the NSC ISSAR for the purpose of integration into the global soil-information network of data exchange, including biological soil indicators. Scientific institutions of the Agrarian academy of sciences are actively involved in the preparation of integrated programmes of innovation and investment development of rural territories and rural communities for scientific support of rational use of land resources, increase of soil fertility, conservation of biodiversity[61].

The NSC ISSAR scientists participated in various projects: the project of the Global Ecological Fund "Integrated management of natural resources of degraded lands of the steppe and Forest-Steppe zones of Ukraine"; the European Commission's project on data collection for mapping of the Danube basin: JRC/IPR/2017/D.3/0002/OC; "Collection of soil data in SOTER format from 14 Danube Strategy countries, at scale 1: 250 000" EC; in the preparation of project proposals for EU Framework Program on research and innovation "Horizon 2020". In cooperation with FAO, the Ukrainian version of the Voluntary Guidelines for Sustainable Soil Management was prepared.

---

61 National Soil Conservation Program of Ukraine, ed. by S.A. Baliuk, V.V. Medvedev, M.M. Miroshnychenko. Kharkiv. 2015. - 59 p. Rational use of soil resources and reproduction of soil fertility: organizational, economic, environmental and regulatory aspects / Monograph ed. by Baliuk, AESU Corresponding member A.V.Kucher. – Kharkiv, 2015. – 357-367 p. Atlas of Soil Suitability of Ukraine for Organic Farming (Zonal Aspects) / Kharkiv, 2015. - 36 p. Comprehensive assessment of the biological state of the microbial soil system and its transformation in modern farming systems / O. I. Maklyuk (Starchenko), O.Ye. Naydenova / Bulletin of agrarian science, special issue, October 2016.- p. 59-64

The United Kingdom of Great Britain and **Northern Ireland** reported on the AHDB BBSRO Soil Biology and Soil Health Programme (Agriculture and Horticulture Development Board – AHDB, 2017), which aims to develop, test and communicate methods for assessing and managing soil biological function for agriculture. The Soil Security Programme is funded by the Natural Environment Research Council and is exploring microbial grassland communities and mycorrhizal impacts of hedgerows as refuge (Soil Security Programme, 2020).

Some knowledge transfer activities have occurred and more are planned for the AHDB programme, in order to roll out soil biological indicators for farmers. Natural England has been providing training in soil biology (earthworm and mesofauna identification) on a very small scale to attempt to engender expertise to support biological recording and conservation. Farming Connect in Wales provides training on some soil biological indicators and Catchment Sensitive Farming in England provide some opportunities to engage with soil biology-themed events. Non-government groups such as the Earthworm Society of Britain (The Earthworm Society of Britain, 2020) provide training and guidance on earthworm recording and events are organized by other taxonomic groups such as the British Mycological Society (British Mycological Society, 2020) and the Association of British Fungus Groups (The Fungus Conservation Trust, 2015). The Soil Biology Special Interest Group of the AAB organizes events for knowledge exchange. The Field Studies Council (Field Studies Council – FSC, 2020) delivers some training of relevance to soil biology.

Some regulation to protect soils exists (such as farming rules for water), but these are often aimed at improving water quality issues. Application of materials to land are regulated to prevent dangerous levels of pollutants. Measures to encourage sustainable soil management include the following: In Wales, the Glastir agri-environment scheme aims to deliver sustainable soil management outcomes, including better nutrient management planning, less erosion, less runoff and to protect existing soil carbon stores or encourage sequestration (e.g. through habitat management such as restoration of degraded peatlands). In England, Countryside Stewardship options include several aimed at water quality protection, (such as buffer strips), but also some in-field options which may help prevent soil degradation, such as cover cropping, under sowing or legume-rich mixes. Options are also available to restore degraded peatlands through hydrological restoration and revegetation. Measures to protect and manage semi-natural and low intensity habitats will provide conservation of soil organisms associated with these. None of the options on offer are explicitly designed to improve soil biological function or to protect or enhance soil biodiversity.

2.3 | MAINSTREAMING: POLICIES, PROGRAMMES, REGULATIONS AND GOVERNMENTAL FRAMEWORKS

Finland has a National strategy on mainstreaming environmental work that includes soil biodiversity. The strategy has five objectives: focus on the mainstreaming of environmental issues across society; the introduction of new participants in the work to advance environmental causes; a decision-making process based on robust research data; and Finland's responsibility, as a member of the international community, for the global environment. The strategy also outlines policies linking the Sámi community's traditional knowledge to the protection of biodiversity.

In ***France***, producers are asked for commercial authorization procedures for pesticides, fertilizers and soil improvers and ecotoxicity tests on soil organisms (earthworms, *collembolla*, arbuscular mycorrhiza fungi) (Agence nationale de sécurité sanitaire de l'alimentation, de l'environnement et du travail – Anses, 2020). The French Ministry of Agriculture has promoted for the last five years conservation agriculture, through an agro-environmental measure (called MAEC "*semis direct sous couvert végétal*") and through supporting farmers association to create a rural network for the improvement and dissemination of soil conservation systems among local communities of farmers in France. This farmer to farmer approach has proven to be the most effective to mobilize farmers by shaping proposals fitting their own personal needs and capacity. Additionally, France has the Agroecology policy and the 4 per 1 000 initiative.

In ***Germany***, sustainable soil management is inherent in the code of good agricultural practice. Support programmes in execution of the Common Agricultural Policy (GAP) are carried out for sustainable soil management as well. Some initiatives are the BonaRes funding initiative (Bonares, 2020) and the Soil Taproot initiative within the German Centre for integrative Biodiversity Research (iDiv) Halle (Deutsches Zentrum für integrative Biodiversitätsforschung – iDiv, 2020) – Jena – Leipzig.

Table 6.3.1. Instruments addressing soil biodiversity in Germany.

Instrument	Date	Extent of implementation	Links to global frameworks	Addressing soil biodiversity
Soil protection act (or law)	1998	Implemented	National law	Natural soil functions (Art 2); Soil information system (Art 20) The German Federal Soil Protection Law (Bundes-Bodenschutzgesetz) explicitly mentions protection of soil biodiversity as a primary goal (§ 2), but this has not yet been implemented in the Federal Soil Protection and Contaminated Sites Ordinance of 1999.
Soil protection ordinance	1999	Implemented	National ordinance	In regard to soil precautionary values
National Strategy on Biological Diversity	2007	Implemented with Action Plan	Transposes Convention on Biodiversity (CBD)	Chapter. B 2.5 (Soil use). Protection, sustainable use and social aspects of the conservation of biodiversity. The strategy is linked to the German National Sustainability Strategy, the Biodiversity Strategy of the EU and the resolutions of the Convention on Biological Diversity (CBD). Addresses soil biology indirectly via reduction of nutrient input, pesticides, pollutants.
Climate Action Plan 2050 – Germany's long-term emission development strategy	2016	Implementation ongoing	Paris Climate Agreement	Includes measures on increasing soil carbon thereby contributes to soil biodiversity.
German Sustainable Development Strategy	2016 renewed 2018	Implemented	Agenda 2030, SDGs	Chapter: Protect, restore and promote sustainable use of terrestrial ecosystems, sustainably managed forests, combat desertification and halt and reverse land degradation and halt biodiversity loss

Further instruments include the following: the National Research Strategy on BioEconomy 2030 (2010) for sustainable agricultural production/soil management in accordance with climate protection, raw material supply, water availability und conservation of biodiversity; HighTech Strategy 2025 for conservation of biodiversity is a mission of the strategy, for example development of innovative technologies and methods to improve the monitoring of biodiversity, improving the systemic understanding of causes, dynamics and consequences of biodiversity changes; *Nationaler Aktionsplan zur nachhaltigen Anwendung von Pflanzenschutzmitteln* (2013), which is the national plan of action for sustainable use of pesticides; *Aktionsprogramm Insektenschutz* (2019), which is a plan of action for insect protection. Also, there is an indirect integration by sustainable soil management and good agricultural practice (for example, by adoption of the Common Agricultural policy (CAP) in Germany Art 6 AgrarZahlVerpflV and the code of good agricultural practice).

In **Italy**, the National Biodiversity Strategy (Ministry for the Environment, Land and Sea Protection, 2010) foresaw the need to establish a national programme to monitor soil biodiversity (page 58, objective 4-xiii). A National Plan on Agricultural Biodiversity (Ministry for the Agricultural, Alimentary, Forestry Policies and Tourism, 2008) provides regulation for agricultural biodiversity. The Fungi Special Project established a monitoring network of Operational Centres throughout the country and is addressing several topics related to soil biodiversity. Political innovation on environmental issues is seldom followed by actions, particularly if financial programmes are not covered by a national law. Only for fungi and selected indicators, a bulk of work has been provided without any financial investment to close the knowledge gap and to standardize activities and reporting (Ministero dell'Ambiente e della Tutela del Territorio e del Mare, 2010; Ministero delle Politiche Agricole Alimentari e Forestali, 2008). There is a relevant role for soil biodiversity in the country's policies.

The **Netherlands** reported on the Environment and Planning Act and policies include the *Bodemstrategie* (Soil strategy, Ministry of Agriculture, Nature and Food Quality, 2018) ((Government of the Netherlands, 2018); *Kamerbrief Bodemstrategie* 2019 (Letter to the parliament 2019 on Soil strategy) (Government of the Netherlands, 2018); Vision Ministry of Agriculture, nature and food quality (Government of the Netherlands, 2018); Letter to Parliament on nature-based agriculture (Government of the Netherlands, 2017); climate deal (Government of the Netherlands, 2019); and Dutch Soil Platform (formulates policy questions and coordinates strategic and applied research) (Dutch Soil Platform, 2020).

In **Portugal**, there are some agri-environmental measures under CAP-EU policies. There is no integration at this moment, but it is possible in the future.

In the **Republic of Moldova**, all national plans, strategies and programmes are aimed at the management and protection of plants and terrestrial fauna. In the Republic of Moldova, there are no programmes, normative acts or any other legal documents that contain a link to soil biodiversity. Specialists on issues related to biodiversity in soils hope that strategies and programmes for the conservation of protected habitats for plants and animals automatically apply to soil biota, since many species of terrestrial fauna have stages of larvae living in the soil and many plant diseases caused by phytopathogenic microorganisms are transmitted through the soil (FAO, 2000; Lozan, 2008; Government of the Republic of Moldova, 2014, 2018; Agency Moldsilva, 2020). The country is trying to integrate soil biodiversity into food security, environment, forestry, agriculture, protected areas, land management, climate change and family farmers.

Spain has the law 42/2007 "*Patrimonio Natural y de la Biodiversidad*", but it is not specific for soil.

Ukraine has the National report Sustainable Development Goals: Ukraine, which includes the conservation, restoration and sustainable use of terrestrial and internal freshwater ecosystems. The National Plan of Action to combat land degradation and desertification was adopted in 2016, where the leading measures are the development of a legislative framework for soil conservation and fertility protection, standards for soil quality, development of the draft Concept of environmental reform of the state environment. To date, according to the Article 4 of the Law of Ukraine, "On priority directions of innovation activity in Ukraine", the strategic focus is basic research on sustainable environmental management. The new Law on Organic Production N 2496-VIII was approved, which regulates relations in the field of organic production and the legislative basis for the further development of production of certified organic products (Government of Ukraine, 2019).

Since 2011, the National Academy of Agrarian Sciences has implemented a research programme "Organic production of agricultural products", and research institutions have participated in the development of the provisions of the Law "On the basic principles and requirements for organic production, circulation and labelling of organic production". The National Scientific Center Institute for Soil Science and Agrochemistry Research, named after O.N. Sokolovsky, conducts long-term research of soil resources within the framework of the academic programme Soil Resources: Forecast of Development, Sustainable Use and Management, where an important task is soil biodiagnosis and ways to support biodiversity (Government of Ukraine, 2012, 2019) [62].

The Land Code of Ukraine from 25.10. 2001 N° 2768 (hereinafter - LCU) ;

- The Law "On the Protection of Land" from 19.06.2003 N° 962;

- The Law "On State Control over Use and Protection of Land" from 19.06.2003 N°963;

- The Law "On State Land Cadastre" from 07.07.2011 N°3613;

- The Law "On Amendments to Certain Legislative Acts of Ukraine on the Conservation of Soil Fertility" from 04.06.2009 N°1443;

---

[62] Цілі Сталого Розвитку: Україна Sustainable Development Goals: Ukraine. 2015. https://menr.gov.ua/files/docs/%D0%9D%D0%B0%D1%86%D1%96%D0%BE%D0%B-D%D0%B0%D0%BB%D1%8C%D0%BD%D0%B0%20%D0%B4%D0%BE%D0%B-F%D0%BE%D0%B2%D1%96%D0%B4%D1%8C%20%D0%A6%D0%A1%D0%A0%20%D0%A3%D0%BA%D1%80%D0%B0%D1%97%D0%BD%D0%B8_%D0%BB%D0%B8%D0%BF%D0%B5%D0%BD%D1%8C%20 2017%20ukr.pdf

- The Law "On Land Valuation" from 11.12.2003, N°1378;

- Other laws adopted in accordance with them the legal acts, in particular: Cabinet of Ministers of Ukraine Regulations: "On the Monitoring of Soils on Agricultural Lands" from 26.02.2004, "On the State System of Environmental Monitoring" from 30.03.1998, "On the State Technological Center for the Protection of Soil Fertility" from 04.08.2000, Ministry of Agrarian Policy Order from 30.11.2003 "On the Agrochemical Passport of the Field", President Decree "On the Continuous Agrochemical Certification of Agricultural Land" from 02.12.2005.

In 2002, Ukraine joined the UN Convention to Combat Desertification (UNCCD) and committed itself to implementing its provisions, in particular to:

- Identify the natural and anthropogenic factors contributing to desertification and land degradation, as well as to propose measures that contribute to mitigating their negative effects;

- Formulate a long-term state policy, programme and action plan for addressing desertification and land degradation;

- Improve the state of affected agro ecosystems by changing land use and reducing the degree of land degradation;

- Introduce sustainable methods of managing agrarian resources and agriculture;

- Develop sustainable irrigation systems.

The concept of Combating Land Degradation and Desertification was approved to resolve the tasks set by the Cabinet of Ministers of Ukraine from October 22, 2014, No. 1024-p. After that, by the order of the Cabinet of Ministers of Ukraine dated March 30, 2016, No. 271-p, the National Plan of Action for Combating Land Degradation and Desertification was approved (UNCCD, 2018). This Plan of Action provided enhancing the effectiveness of state policy, rational use and protection of soils, strengthening and improving the coordination of activities of authorized state bodies and ensuring the implementation of planned activities. Implementation of the provisions of the Concept and the Plan of Action is complicated by the lack of a competent national body for the formulation and implementation of state policy in the field of rational land use and protection of soils, achieving a neutral level of their degradation.

In the **United Kingdom of Great Britain** and **Northern Ireland**, the Welsh Government's Natural Resources Policy, required by the 2016 Wales Environment Act, sets out priorities for natural resource management in Wales; it states under "we will continue to coordinate and embed best practice for our soil resources. This will be informed by monitoring trends in ... the functional importance of soil biodiversity...".

In England, there is extensive commitment to improving sustainable management of soils in the Government's 25 Year Plan and the agriculture bill, but no specific mention of soil biodiversity in any current government policy, programme, or enabling framework. Nature conservation in England and Wales is driven by the development of priority species and habitats on the basis of rarity, distinctiveness or threat, and these include many soil-dwelling organisms. However, comparatively few soil organisms have sufficient data to enable an assessment of their conservation status and so are vastly underrepresented in the priority species and habitat lists.

Where priority species are identified, conservation may be enacted through protection of sites, management planning, agri-environment measures, or other protection measures. The United Kingdom provides support for organic farming which is likely to support greater biological function in soils (Welsh Government, 2020; UK Government, 2020). Soil biodiversity has been recognized in Welsh policy (NRP, which cuts across all policy sectors) but is not explicitly recognized in English policy.

2.4 | ANALYSIS OF THE MAIN GAPS, BARRIERS AND OPPORTUNITIES IN THE CONSERVATION AND SUSTAINABLE USE OF SOIL BIODIVERSITY

In *France*, the French National Biodiversity Plan (2018-2022), includes several actions that concern (directly and indirectly) soil biodiversity: limit the consumption of natural, agricultural and forest areas to achieve the goal of "zero net land take" and economical soil management (actions 6 to 13); make agriculture an ally of biodiversity and accelerate the agroecological transition (action 21); act to promote soil conservation in agriculture (action 49); and develop a plan for strengthening soil biodiversity research (action 50) (Ministère de la Transition écologique, 2019).

Germany is monitoring the impacts on soil biodiversity (for example, pesticides, organic pollutants); monitoring climate change effects on soil biodiversity; nation-wide monitoring of insect fauna ((The Federal Agency for Nature Conservation – BfN, 2020); BMU (Federal Ministry of the Environment, "National Biodiversity Strategy" ((Federal Ministry for the Environment, Nature Conservation and Nuclear Safety, 2007); BMEL: "Ackerbaustrategie" (Strategy on farming, forthcoming). There is a lack of monitoring of soil biodiversity along a defined grid. A good example to follow is the French Atlas of Soil Bacterial Biodiversity (INRAE, 2020b).

In *Italy*, the National Biodiversity Strategy was designed to implement activities including on soil biodiversity in term of ecosystems, habitats, number of species and association of species, one of the richest in Europe. However, no Action Plan has been developed by Italian Governments in the last decade and only a few initiatives are working on a voluntary basis, such as the Fungi Special Project at ISPRA and QBS-Working Group at SISS. The situation is hopefully subject to change as soon as the European Union will address this topic in a specific Directive. The difficulty for monitoring biodiversity of soil for a long time related to the analytical methods to determine it. The progress carried out at a research and technology level, along with numerous initiatives about the spreading the knowledge of soil biodiversity, have raised awareness; consequently the monitoring of soil biodiversity is considered in several environmental programmes. In this decade, important progress has been made on soil biodiversity.

The crucial point is the diffusion of the Global Soil Partnership (GSP) activities. Since 2016, Italy has given itself a mirror organization with respect to the GSP and therefore the activities have been conducted within the Five Pillars. The alliance was promoted in numerous events, starting with schools, but also with farmers and professional organizations, and was recalled whenever there was a conference on the ground. At the national level, every year a schedule of activities is prepared trying to follow the addresses of the GSP and the ESP.

A permanent working group on the GSP has been set up at the Italian Society of Soil Science (see web page of SISS). At the end of 2018, the Ministry of Agriculture, Food, Forestry and Tourism funded a project for the establishment of a Soil HUB that started from the Italian Soil Partnership with the dual purpose of allowing an easier participation in the activities of the GSP and the ESP and to participate to the EJP Soil.

The Republic of Moldova is planning to have: i) Description of soil biota in conditions of the natural and agricultural ecosystem in view of assessing degrees of vulnerability and a new round of research on soil microorganisms by molecular genetic methods; ii) the development of methods and technologies for assuring recovery of soil biota; iii) the development of an information system of soil biodiversity for the national standard soil quality; iv) modernization of the soil biology educational institutions, including equipping them with modern equipment and technical facilities; v) organization of training programmes for soil microbiology and zoology professionals; vi) publication of training and information materials on soil biodiversity; and vii) increasing the social significance of soil biodiversity and ecosystem services through practical conferences and round-tables with farmers and local people.

Spain reported that there is a lack of effort in promoting the necessary shift from the use of conventional physical and chemical indicators. Sometimes soil is still seen as a substrate for plant growth and as a place to stock any kind of waste. There is growing concern in the agricultural sectors about the imminent renewal of the EC common agricultural policy that clearly focuses soil protection through ecological intensification (organic management) of the agricultural soils. It is expected to help to protect biodiversity in agricultural soils.

In the **United Kingdom of Great Britain** and **Northern Ireland**, Wales now has a more complete picture of its soil biodiversity thanks to the Glastir Monitoring and Evaluation Programme. England's understanding is based almost entirely on the countryside survey data. There remains a widespread lack of integration of soils into government policy and delivery of environmental objectives and soil biodiversity is particularly poorly represented. This is perhaps due to a perception that little is known about soil biodiversity, or that its exploration is prohibitively complicated or expensive. These issues have largely been overcome in recent years. There are pressure groups, such as the Sustainable Soils Alliance, which aim to promote more sustainable soil management and policy, but these also tend to ignore soil biodiversity.

In the United Kingdom, while soils and soil health are explicitly embedded into government agriculture and environmental policy and this relies on the sustained functioning of soil biology, there is no arrangement to embed the conservation of soil biodiversity itself. A global convention is needed which would require signatories to take action to fully understand their soil biodiversity and identify and address threats to its survival and function. Even the most developed countries in the world remain unaware of the diversity of life in their soils and by tolerating this ignorance, there is a risk of losing unexplored biodiversity and the ecosystem functions it provides, which maintain human life.

3 | LATIN AMERICA AND THE CARIBBEAN (LAC)

3.1 | ASSESSMENT OF SOIL BIODIVERSITY

3.1.1 | CONTRIBUTIONS OF SOIL BIODIVERSITY TO ECOSYSTEMS SERVICES

Brazil reported the growing importance of soil erosion and land degradation mitigation in efforts to recover degraded lands and reintegrate them into the production process, depending on soil characteristics and the agricultural suitability of the land. Ecosystem services can leverage the delivery of these services by mitigating burning and enabling the expansion of agriculture and livestock production without deforestation (zero deforestation) (de Freitas and Landers, 2014, Landers *et al.*, 2013, Embrapa, 2016).

Brazil has evolved from a food-insecure country in the early 1970s to one of the world's largest food producers and exporters. Production has increased steadily, and productivity gains have fostered a significant land-saving effect. However, wide variations in landscape, soil, climate, and plant diversity pose challenges for the application of soil health principles to enhance management practices. These challenges have been overcome through the application of conservation agriculture, a holistic integrated farming system that improves soil functioning and, consequently, soil health, crop growth and yield. Soil health encompasses aspects related to soil biodiversity, but it is a broader concept (it also includes aspects related to soil chemistry and physics), as it relates to the ability of a specific soil to function to provide biological productivity and other important environmental services, such as carbon sequestration, water storage and human, plant and animal health (EMBRAPA, 2019).

In **Chile**, the importance of soil microbiota in the provision of soil ecosystem service is just beginning to be recognised. Thus, the benefits of soil microbes are starting to be integrated into practices such as crop production and land restoration. In national reports on the state of soils, the soil biota aspect is hardly raised.

Colombia reported on the actions carried out by the Ministry of Environment and Sustainable Development in the framework of the Policy for the Sustainable Management of the soil (*Política para la Gestión Sostenible del suelo*–PGSS) where an integral conceptual model with an ecosystem vision was adopted, recognizing ecosystem functions and services related to biodiversity, climate change, water and food security, among others.

Actions were also carried out by the research institutes of the National Environmental System (*Institutos de investigación del Sistema Nacional Ambiental* - SINA). In the case of IDEAM, a monitoring and follow-up programme exists on the state of soils related to degradation, including erosion, salinization, loss of organic matter and carbon, deforestation monitoring and land cover, among others.

Mexico reported that soil is not yet perceived as an important natural resource by society in general. Soil biodiversity is perhaps the least valued subject compared to erosion or fertility. However, several research groups and projects are working on the subject. The National Information System on Biodiversity (in Spanish SNIB) includes 160 000 records of more than 4 700 species present in the soil (CONABIO, 2018; Negrete-Yankelevich and Barois-Boullard, 2012).

Peru has begun a process of raising awareness of the importance of soil biodiversity to reflect a better quality of products from family farming and agricultural products for export.

Suriname has a Centre for Agricultural Research. The country is aware of the close correlation between the activities of living organisms in the different soil layers and the chemical and physical properties of the soil that result in ecosystem services. Microorganisms, micro-, meso- and macrofauna are active in the soil in close contact with each other, the mineral and organic soil phases, the soil water and soil air phases and the residues, roots or other soil parts of native or cultivated vegetation. The reciprocal influence of the individual contacts can have positive or negative consequences on fertility values and soil health parameters. These values will be balanced under steady climatic conditions and with native cover of flora and fauna above ground. Disturbances of this balance are due to changes in the chemical and physical characteristics of the soil caused by direct and indirect consequences, above and below the soil surface, of ongoing climate change as well as human activities or soil management. Disturbed chemical and physical soil characteristics lead to changes in soil biodiversity, which in turn lead to changes in soil fertility and soil and thus in ecosystem services.

In **Venezuela (Bolivarian Republic of)**, soil scientists, aware of the functions of microorganisms in nutrient cycles and crop production, have started to study soil biota in various specific regional and local projects and in a wide range of soils and climatic conditions, with different agricultural uses and management. In the 1980s, issues related to biological nitrogen fixation by *Rhizobia* and mycorrhizal symbiosis, as well as the diversity of certain species of bacteria, fungi and algae were addressed, but with the use of cultural laboratory methods.

With the growing global understanding of the importance of soil biota in soil functions within ecosystems, in the 1990s, soil scientists expanded their studies on AM fungi and phosphate-solubilizing bacteria and their association with certain crops, as well as on the role of microbial biomass in the decomposition of organic matter in agricultural soils. With the beginning of the new millennium, some studies were carried out on microbial indicators such as microbial biomass and activities in soils from different geographical locations under different agricultural management practices, in relation to the chemical and physical properties of soils. Research on invertebrates and the identification of AM fungi in soils of savanna and forest ecosystems has also been conducted. A few laboratory studies were also carried on soils with artificial addition of municipal and industrial residual sludge. Interest in the use of biofertilizers and biopesticides in agricultural soils has increased in recent years. However, very little research has been conducted on soil biodiversity using biomolecular techniques. Although work has been carried out on various topics of soil biology, the biological properties of soils are not taken into account in routine laboratory soil analyses.

As a result of extension work, farmers are more aware of the importance of organic matter in the soil and its benefits for crop production, but they have little knowledge about soil biodiversity itself. Farmers also have the intuition that intensive agriculture can affect soil productivity, but they continue to use agrochemicals in an unsound way.

3.1.2 | NATIONAL ASSESSMENTS

Antigua and Barbuda is developing case studies with local groups to show their contributions to the different Aichi Biodiversity Targets and to empower local communities to advocate for the protection of Antigua and Barbuda's genetic resources and conservation areas and other local knowledge related to local groups. A particular focus is on local vetiver grass and its use for soil stabilization. This is a new technology whose application addresses soil and water challenges, including extreme weather conditions, and has potential for marine and coastal communities. These include landslides, soil erosion, infrastructure protection, soil and water conservation and sedimentation.

With the support of FAO/GSP, as part of the GSOC MAP process, **Chile** has developed a map of soil organic carbon, which will soon be linked to a biodiversity-related map. Chile has expressed some interest in integrating soil biodiversity as a parameter for assessing the impact of best soil practices, including those aimed at recovering the productive potential of degraded agricultural soils.

In ***Colombia***, activities to integrate soil management are carried out through agriculture extension programmes led by the Ministry of Agriculture and its local agencies. In addition to the Ministry of Agriculture, the Ministry of Environment and Sustainable Development has implemented actions related to the Sustainable Soil Management Policy (PGSS). In addition, Agrosavia has implemented research projects and IDEAM soil monitoring programmes.

Costa Rica reported on the *Rhizobium* distribution maps developed in 1992.

Mexico has produced a good amount of scientific knowledge, but soil biodiversity still needs to be taken into account in public policies that contribute to the conservation and sustainable use of soil biodiversity. Furthermore, there is indigenous and traditional soil knowledge catalogued in Mexico, which has at least 25 million indigenous people.

Suriname has developed two regional reports on the Guyagrofor project: "Development of Sustainable Agroforestry Systems Based on Indigenous" and "Maroon Knowledge in the Guyana Shield Region". There is also a research report on a local species of earthworm in Suriname (Ramnarain, Ansari and Ori, 2016).

In ***Venezuela (Bolivarian Republic of)***, the diversity of mites (*Acari: Prostigmata, Mesotigmata, Astigmata*) associated with leaf litter from plant formations (Vásquez, Sanchez & Valera, 2007) and soil macrofauna in silvopastoral systems with legumes and grasses (Medina *et al.,* 2011) were studied. Similarly, termite and oligochaete activities in savanna soils (López-Hernández, 2003). Two studies were conducted to describe the microbial crust and its biological and physical influence on soils in the arid region of Quíbor, Lara state (Toledo, 2006; Ospina, 2019, unpublished Doctoral Thesis).

3.1.3 | PRACTICAL APPLICATIONS OF SOIL BIODIVERSITY

In **Antigua and Barbuda**, soil biodiversity has allowed farmers to assess the levels of nitrogen, phosphorus, and potassium in the soil. The increasing amount of fertilizers added to the soil and the amount of pesticides used indicate that natural soil restoration should be a priority. Recycling and supporting the reuse of organic waste into compost is beneficial for soils and increases their water retention and storage capacity. Current efforts are aimed at creating income-generating opportunities and sustainable employment at the pilot stage. Results are being achieved but need to be scaled up.

Brazil reported that biological control and enhancement of soil biodiversity as integrated pest management (e.g. insects, nematodes, diseases and weeds) is very advanced. There is strong evidence of the benefits of soil biodiversity for food production, especially in areas where Conservation Agriculture (Zero-Tillage Conservation Agriculture) has been adopted. Not plowing, multi-annual crop rotation and cover crops, combined with traffic control, enhance soil biodiversity, increase the efficiency of fertilizers and control soil degradation and erosion (Landers *et al.*, 2013).

In Brazil, biological nitrogen fixation in soybeans is responsible for an annual economy of USD 13 billion. Nitrogen fertilization in not used in Brazil's soybean fields and therefore, from an environmental point of view, there is no leaching of NO_3 and the production of greenhouse gases is lower. More recently, strains of *Azospirilum brasiliense* have also been selected for crops such as corn, wheat and rice, and even for co-inoculation with *Bradyrhizobium* in soybean.

In 2019, Brazil launched a new technology called soil bioanalysis. This is a very simple and straightforward approach to assess soil health, based on determining the activity of two soil enzymes: b-glucosidase and arylsulfatase. Brazil has the presence of biological components in its routine soil analyses. The Brazilian market for biological control agents for soilborne diseases has grown considerably over the last decade in response environmental change and social acknowledgement of the need for increased sustainability (Mendes *et al.* 2015).

In **Chile** there is evidence of the importance of soil biodiversity for wine production (Viers *et al.*, 2013). There are also studies on the use of mycorrhizae to remediate metal-polluted soils and enhance plant growth (Cáceres and Kalinhoff, 2014).

Colombia reported that the main activities to integrate soil management are through agricultural extension programmes led by the Ministry of Agriculture and its local agencies, as well as the Ministry of Environment and Sustainable Development.

Cuba has an established programme for the production and development of bioproducts (biofertilizers, biostimulants and biopesticides) where soil biodiversity plays a fundamental role.

In *Jamaica*, the main practical applications of soil biodiversity are in the agriculture and waste management industries. Soil macro- and micro-organisms are often used in science-based fields such as composting and biological nitrogen fixation.

For *Mexico*, the work done on conservation tillage in the FIRA experimental field in Villa Diego, Guanajuato, has shown that it is possible to increase organic matter and with it soil biodiversity, thereby increasing the benefits while reducing the use of fertilizers. Healthy soils require communities of suitable organisms. In compacted or damaged soils with no biota, such as those that have been pastured or cultivated without rest and care, water drains quickly to the surface and plants or humans cannot benefit from it (Negrete *et al.*, 2012). Decreasing agrodiversity in traditional systems can have a negative impact on the ability of the mycorrhiza community to colonize maize roots, as well as on the availability of phosphorus, which is often the most limiting nutrient in tropical soils (Negrete *et al.*, 2012). Bacteria, fungi, earthworms, centipedes and collembolans recycle corpses and organic matter and make new nutrients available to plants. Centipede and millipede species (*Myriapoda: Chilopoda and Diplopoda*) as well as bacteria, fungi, earthworms and springtails live and feed on the soil surface and predate the lower fauna. They fragment leaf litter and dead animals, promoting decomposition, soil fertility and thus the productivity of the systems (Negrete *et al.*, 2012).

Another study analysed farm-level data for two agroecosystems with contrasting objectives in central Mexico: one aimed at staple crop production for self-subsistence and local markets, the other at cash crop production for export markets. Bivariate and multivariate trade-offs were analysed for different crop management strategies (conventional, organic, traditional, crop rotation) and their underlying socioeconomic drivers. There was a clear trade-off between crop yield and soil quality in self-subsistence systems. However, the other expected trade-offs between crop yields and soil quality have not always occurred, probably due to the overall soil quality of the region and the profile of most farms, which use low to medium inputs. The trade-offs depended heavily on farm-specific agricultural practices; organic, traditional and rotational management systems generally had lower trade-offs between yield and soil quality, pest control and biodiversity than conventional management systems. Factors perceived by farmers to be determining factors included rising prices for cash crops, higher input costs and extreme climatic events (e.g. drought, hail, frost). Farmers did not identify the regulation of soil quality, water quality, soil erosion, pests or pollinators as important constraints. Although acceptable yields can be maintained independently of key regulating and supporting services according to these perceptions, current levels of soil erosion and nutrient runoff are likely to have significant negative effects at the watershed scale. The sustainability of both agroecosystems could be substantially increased by promoting alternative practices to maintain biodiversity, soil quality and soil retention. Arbuscular mycorrhizal root colonization and soil P availability are positively related to agrodiversity in Mexican-maize polycultures (González-Esquivel *et al.*, 2015; González-Esquivel *et al.*, 2015; Negrete-Yanelevih and Barois Boullard, 2012).

Various soil conservation and restoration actions have been carried out by Mexico's National Forestry Commission, which contribute to controlling erosion and reducing surface runoff that directly affect the conservation and increase of soil biodiversity (e.g. construction of cultivation terraces, strip cultivation, natural or artificial soil cover, crop rotation, plot design, use of minimum tillage techniques, grazing control, slope protection and rectification of gullies and channels) (INECC, 2007). The Ministry of Agriculture has promoted conservation tillage and other agricultural land management actions that affect the conservation of organic matter and the preservation of soil biodiversity (for example, *Estrategia Nacional y Programa Nacional de Manejo Sustentable de Tierras*) (Semanart, 2008a).

However, a programme focusing on the conservation and increase of soil biodiversity has not been implemented. Knowledge of soil biodiversity has merited the attention and research of institutions such as INECOL, UNAM, IPN (and many others) where researchers have focused on documenting the species that inhabit the soils (e.g. mycorrhizal fungi associated with the roots of 80 percent of plant species) and how they provide nutrients that are indirectly accessible to them (Negrete-Yankelevich *et al.*, 2013).

Panama reported on the use of soil biodiversity to control soil-borne diseases (Ministerio de Desarrollo Agropecuario, 2020).

Among the practices implemented in **Peru**, the biological fixation of nitrogen by legumes is the most widely used in crop rotation practices. The National Institute of Agricultural Research (INIA) plans to work on related topics such as the characterization of *Phytophthora infestans* and *Ralstonia solanacearum* populations in three agroecological regions of Peru, and the strengthening of INIA's capacity for the continuous monitoring of the main potato pathogens (Chipana *et al.*, 2017; INIA, 2020).

Peru also reported that organic agriculture has been proven to enable sustainable use of soil and also improves the remuneration of producers through a better market price for the products (Terra Nuova, 2018). INIA, in the framework of its current projects, has developed a taxonomic and functional identification of native microorganisms potentially beneficial for the cultivation of potatoes present in the agricultural soils of the Huancavelica and Huánuco region (INIA, 2020).

Suriname has indicated that the application of soil biodiversity is specifically aimed at biological nitrogen fixation. It is common to include leguminous crops in the rotation cropping system: *Arachis hypogea* (peanut), *Glicine max* (soybean) and inoculum with *Rhizobium* spp. In the cassava field gene banks, intercropping includes rows and sticks to mark the corner of *Gliricida sepium* (for N-fixation, inter cropping, mulch, compost, fodder). *Gliricida* has also been introduced into the traditional slash-and-burn or subsistence cultivation system of tribal communities to enhance soil fertility and during the transition to permanent cropping systems.

Uruguay reported on the practical application of soil biodiversity in nitrogen fixation and for pest and disease control.

In ***Venezuela (Bolivarian Republic of)***, soil biodiversity is a component of a complex and dynamic metabolic system whose reduction through manipulation will inevitably have unpredictable results, therefore this compartment must be preserved and protected in as many scenarios as possible. Essential activities such as agriculture and mining must consider that sound management is one that maintains and promotes the presence of communities adapted to the environment as such. For this reason, within the framework of sustainable agriculture, the Venezuelan State has established an alternative for agricultural development, such as the production of microorganisms with biofertilizer characteristics, including nitrogen-fixing bacteria and phosphorus-solubilizing bacteria.

In addition there are other groups of microorganisms with pest and disease control characteristics, which are also being produced in order to develop a plan for the reduction of agrotoxins by these bio-inputs. This production is carried out under the guidance of the *Instituto de Salud Agrícola Integral* attached to the *Ministerio del Poder Popular de para la* Agricultura and has an infrastructure of 15 laboratories of biocontrollers (entomophages and entomophogens) and 11 laboratories of biofertilizers (nitrogen fixers and solubilizers of match) (Martínez *et al.*, 2006).

Soil biodiversity has been used in agroecological practices, on a small scale and in multi-species plots (multiple land use). In this sense, there are studies that show an increase in crop yields with the use of microbial biofertilizers as mentioned in the publication: "Effect of bacterial Biofertilizers on the growth of a corn cultivar in two Venezuelan contrasting soils". The results indicate that the native strains used have the potential to stimulate plant growth through both nitrogen fixation and phosphorus solubilization (López *et al.*, 2008).

Likewise, other authors have studied poorly fertile savanna soils and applied conservation practices using green manure sowing and chemical fertilization with phosphoric rock. For the management of soil biodiversity, arbuscular mycorrhizal fungi and solubilizing bacteria of calcium and phosphate were used as indicators. The results showed that a mixed combination of organic and inorganic fertilization can reflect the effect of microorganisms and will lead to a sustainable management of agroecosystems (Toro and López, 2008). Furthermore, in the ecosystems involved, Cuenca *et al.* (2002) demonstrated that the use of native shrubs (*Clusia pusilla*) from La Gran Sabana as "nurse" plants to promote the establishment of arbuscular mycorrhizal fungi (AMF) seeds and propagules around them could be a valid alternative to restart plant succession in areas that have been severely degraded.

In addition, it can be noted that in Venezuela, institutional and interdisciplinary actions are being carried out to resize and reorient conventional management practices based on high inputs, including high rates of inorganic fertilizer application in the country's main agrosystems, which has led to the degradation of all soil components. In this sense, a study was conducted on soils with different agricultural uses such as corn, sugar cane, milky, grass and natural forest.

The results indicated that in the cases managed with low organic matter content or intensive management, the nitrogen-fixing bacteria were better developed and in those with higher phosphorus content, either through organic or inorganic fertilization, a better development of phosphorus-solubilizing bacteria. It was also verified that the excessive use of machinery and agrochemicals decreased soil quality, affecting the development of microorganisms in the soil planted with corn (Padron *et al.*, 2012).

3.1.4 | MAJOR PRACTICES NEGATIVELY IMPACTING SOIL BIODIVERSITY

Antigua and Barbuda reported that high levels of toxic and hazardous substances that seep into the soil and ground water from unseparated buried waste also contaminate soil and water. This contamination has a significant impact on human health and the environment.

In ***Venezuela (Bolivarian Republic of)***, some researchers have reported intensively managed agricultural soils, negative effects on microbial biomass and activity due to conventional tillage (Rivero and Garcia, 2014) and the use of monoculture with pesticides and high doses of fertilizers (Aciego and Chacin, 2014). Little research has been carried out on biodiversity in contaminated soils. Under controlled conditions with soils of low fertility, the incorporation of raw and composted brewery sludge increased CO_2 evolution and decreased dehydrogenase activity (Aciego *et al.*, 1999) and the incorporation of herbicide mixes decreased cyanobacteria populations (Ojeda *et al.*, 1997).

3.1.5 | INVASIVE ALIEN SPECIES (IAS)

Chile reported that the lack of natural predators for beavers (*Castor Canadensis*) in the subpolar forests of southern Chile is of concern as they destroy native flora, threatening the soils underneath (Graells, Corcoran and Aravena, 2015).

Colombia reported microorganisms that invade many crops, such as *Fusarium* sp., *Phytoptora* sp. and nematodes. These are the species reported so far, but other species and microorganisms affect soil biodiversity. The floriculture sector has suffered and according to the 2009 DANE report (Census of Flower Production Farms), the main diseases in this sector were *Botrytis* with 26.2 percent, Dusty Mildeo with 22.2 percent, Villoso Mildeo with 16.2 percent and other percentages for *Fusarium* fungus and rust. Many of these phytopathogens cause damage to plants of different species and stages of development, which can survive permanently in the soil to a depth of 3 metres or more and also affect plants intended for consumption. Floriculture occupies the second line of cultivation after coffee, with the departments of Antioquia and Cundinamarca being responsible for the planting and production of more than 7 290 hectares, the soils under these crops suffer substantial biological deterioration that reduces their use, economic value and productivity.

Mexico has a list of 45 exotic species of different taxonomic groups that are associated with soil, 34 of which are already present in Mexico; the others are on a watch list. Two of these species have already been assessed through a risk analysis and are considered invasive for Mexico.

Nicaragua reported that *Azadirachta indica* as an invasive species that may have affected soil biodiversity since 1980.

Peru is working on a project called INIA to work on topics related to the study of *Phytophthora infestans* and *Ralstonia solanacearum* in three agroecology regions of Peru, as well as on the strength of capacity development and continuous monitoring of the main pathogens of potatoes (INIA, 2020).

In **Venezuela (Bolivarian Republic of)** reported the case of the Giant African Snail (*Achatina fulica*), introduced in 1990 as a pet. This pest has already spread to the entire Latin American continent, where it has been detected in Peru, Colombia, Venezuela and many Caribbean islands, including Cuba. Antigua and Barbuda has reported that the Giant African Snail is having a major impact on the island's agricultural sector, and that inputs *Sargassum* seaweed are having multiple negative impacts.

3.1.6 | MONITORING SOIL BIODIVERSITY

In **Antigua and Barbuda**, the soil department of the Ministry of Agriculture is no longer active due to resource constraints. Since then, projects have been financed by donors and partners. All ongoing and upcoming surveys and activities, including soil testing and soil analysis, are carried out by the Department of Analytical Services. A project has been submitted to extend local analytical services and capacity to the Minamata Convention Specific International Programme on the Caribbean Mercury Monitoring Network. This project includes soil and biodiversity data analyses (ABN, 2019).

Brazil has the Environmental Information System[63] of the Biota Research Program on Biodiversity Characterization, Conservation, Restoration and Sustainable Use[64] and the Brazilian Soil Information System, SISolos (EMBRAPA, 2020a). Brazil has developed a National Program for Soil Survey and Interpretation, named PRONASOLOS. Over the next three decades, this national effort will address the lack of information on soil and water resources through the largest pedological survey and land use interpretation.

In ***Colombia***, IGAC is the national authority responsible for compiling basic soil information[65]. There are studies on soil biodiversity at the local scale, but there is not specific monitoring programme or a database on soil biodiversity components. Colombia has a sustainable land management policy and a soil degradation monitoring programme. Within this programme, soil biodiversity is covered and planned by IDEAM, but the programme has not been implemented yet. There are some studies on soil biodiversity at the local scale, but there is no specific monitoring programme or database on soil biodiversity components.

Mexico monitors soil organic carbon content. Some studies present the spatial distribution of soil organic carbon at the country level (Segura-Castruita *et al.*, 2005). Mexico does not yet have a specific monitoring programme for soil biodiversity, but the country has one of the world's largest databases on soil biodiversity. INEGI has identified soil types at the national level, which is a valuable information that could be used to select priority areas for soil biodiversity monitoring. In addition, the National System of Biodiversity (in Spanish SNIB) includes information on the occurrence of more than 4 700 soil species. This system compiles general information on biodiversity, which includes soil species[66].

Panama has a soil laboratory as part of the *Instituto de Investigación Agropecuaria de Panamá* (IDIAP)[67].

In ***Peru***, the National Institute of Agricultural Research (in Spanish INIA) of the Ministry of Agriculture and Irrigation (MINAGRI), within the Department of Genetic Resources, includes the Area of Genetic Resources of Microorganisms and Associated Biodiversity among its lines of research. It should also be mentioned that in the 1980s there was an ecological surveillance project in the Central Huallaga (jungle), which lasted two years (INIA, 2020).

In ***Suriname***, the Soil Mapping Service produced a soil map in the 1970s that is still used today. The map contains the various soil types in the northern half of Suriname. The Center for Agricultural Research in Suriname (CELOS) produced a national soil map with remote sensing data containing both the northern and southern parts of the country. Detailed soil information can often be found in various reports, for example the CELOS

---

63 http://sinbiota.biota.org.br/
64 www.biota.org.br
65 https://www.igac.gov.co/es/contenido/areas-estrategicas/agrologia/laboratorio-nacional-de-suelos
66 http://www.snib.mx/
67 www.idiap.gob.pa

Multiple Landscape Analysis for the Carolina area (2012) and Pikin Slee (2018). Most of the reports are only available in hard copy.

Uruguay reported that there is no national soil information system in the country.

3.1.7 | INDICATORS USED TO EVALUATE SOIL BIODIVERSITY

Antigua and Barbuda indicated that some work is being done, but that it needs to be upscaled and linked to the country's realities.

Chile has an interest in integrating the evaluation of soil biological conditions within the framework of the national programme for the restoration of agricultural soils (SIRD).

Colombia has studies on soil degradation due to erosion and salinization, from which links with soil biodiversity loss can be established.

In ***Venezuela (Bolivarian Republic of)***, environmental indicators play a very important role in the development of environmental statistics, which can be methodologically compared at the international level, as the methodological approach of the Economic Commission for Latin America and the Caribbean (ECLAC) has been used. For studies and evaluations of environmental systems (aquatic, air and soil) in Venezuela, several research articles have been developed in which environmental indicators used to assess soil biodiversity are methodologically demonstrated, particularly with regard to genetic diversity, soil carbon sequestration, soil erosion, soil fertility management and soil pollution.

In Venezuela researchers have studied mesofauna diversity as mentioned above, as well as AM fungi diversity in selected ecological systems (Lovera and Cuenca, 2007). The effects of different soil managements on microbial properties were investigated using certain soil physiological methods as indicators of possible changes in the microbial community, such as basal respiration, metabolic quotients like specific respiration quotient (qCO_2) abind the microbial C/N ratio, substrate induced respiration (SIR), and soil enzymes activities (García and Rivero, 2012; Mendoza *et al.*, 2012; Armado *et al.*, 2009; Hernandez *et al.*, 2009; Gómez and Paolini, 2003). Very little research has been carried out on biomolecular analysis: for example, in a Vertisol of the Guárico plateau, the bacteria *Streptomyces* sp. and *Arthrobacter* sp. were associated with the decomposition of soybean residues and *Pseudonocardia* sp. and *Saccharopolyspora* sp. to maize, using molecular techniques, and communities of the fungi *Penicillium* sp. and *Aspergillus* sp. for maize and Fusarium sp. and Mortierella for soybean were recognized (España *et al.*, 2011a; España *et al.*, 2011b).

However, in the country, studies of soil biodiversity using molecular methods are limited due to the lack of knowledge and laboratories equipped for PLFAs, DNA and RNA analysis.

3.2 | RESEARCH, CAPACITY DEVELOPMENT AND AWARENESS RAISING

Antigua and Barbuda reported that the Caribbean Agricultural Research and Development Institute (CARDI) has a small office that conducts soil research, but is experiencing financial limitations. Antigua and Barbuda is to open a campus of the University of the West Indies (UWI), which will generate knowledge and enable research to be carried out. UWI/CAP NET and the Global Water partnership have offered services on a voluntary basis by training Antigua and Barbuda's local groups in soil information assessment.

Brazil has promoted sustainable soil management through practices such as no tillage, agroforestry systems and integrated crop-livestock-forestry systems (EMBRAPA, 2020b; EMBRAPA, 2020c). Over the last 40 years, agriculture models in the Cerrado biome have evolved more rapidly towards sustainability, with the widespread adoptions of no-tillage (de Freitas and Landers, 2014) and more recently the adoption of integrated crop-livestock systems (Cordeiro *et al.*, 2015), which have led to increased interest in soil health assessments (Balota *et al.*, 2004; Hungria *et al.*, 2009; de Carvalho Mendes, de Sousa, & dos Reis Junior, 2015; Mendes, 2016).

"The Low Carbon Brazilian Program" was launched by the Brazilian government in 2010 to promote specific agricultural activities based on best agricultural management practices which involved six main themes (Magalhães & Lunas Lima, 2014):

- Restoration of degraded pastureland and promotion of livestock intensification based on carrying capacity;

- Expansion of the area under no-tillage and associated cropping systems with high and diverse biomass-C inputs;

- Adoption of integrated crop-livestock-forestry-systems;

- Promotion of biological N fixation;

- Establishment of plantations of commercial forests and forestation;

- Application and recycling of industrial and animal residues.

In 2012, a successful public-private partnership called Integrated Network of Crop-Livestock-Forest was established. The main objective of the network is to accelerate the large-scale adoption of crop-livestock-forest integration systems by rural producers in an effort to achieve the sustainable intensification of Brazilian agriculture.

In Brazil, the best management practices adopted by Zero Tillage (no-till) Conservation Agriculture[68] and in the integrated crop-livestock-forest management systems as a means of improving soil conditions in tropical areas of the world consider that:

i. Brazil has abundant sunlight and rainfall in summer and throughout the year in most parts of the country, allowing for two to multiple successive crop harvests during the year, depending on the region, soil, crop species and other factors. To achieve this, crop varieties and soil, water and plant management practices have been adapted to make the best use of the available cropping periods, which are seasonally distributed according to the availability of sunlight, temperature variation, water availability and market prices;

ii. Cropping in the Amazon Rainforest cannot be done in the same way as in other regions, due to its specific characteristics;

iii. Tropical conditions are different from those in temperate regions, requiring different metrics for assessing vulnerability and sustainability as well as soil health and quality (Landers *et al.*, 2013).

Chile is implementing a programme for the restoration of degraded soils (Programa de Recuperación de Suelos Degradados - SIRD).

In **Colombia**, the evaluation of the adoption of the technologies developed by AGROSAVIA can be found in the Social Balance 2017 (Agrosavia, 2017) and 2018 (Agrosavia, 2018) integrated fertilization fractionation strategy for plantain cultivation, inoculation with nitrogen fixing bacteria in soy, recommendations for conditioning savannah land for agricultural production. Some courses have been developed, such as the ICGEB International Course - Second International Course on Soil Microbial Ecology: "An integrated and functional view for its application in agriculture" (Universidad Nacional de Colombia, 2019). Since IDEAM, no programme has been generated so far, but there is interest in working on this theme.

Costa Rica reported that the Microbial Soil Laboratory of the University of Costa Rica has carried out research on nematode ecology.

Mexico has promoted the reduction of soil erosion through the *Comisión Nacional Forestal* (CONAFOR) and some practices of the Ministry of Agriculture and Rural Development to promote the increase of organic matter, as well as the sustainable use of

---

68   ZERO TILLAGE (NO-TILL) CONSERVATION AGRICULTURE (ZT/CA) was defined by Landers and others (2013) as the set of practices and technologies which attends the technological pillars of no-tillage, crop rotation (pluri-annual rotation of annual crops), permanent soil cover and traffic control

water (CONAFOR, 2020). Mexico has strategies for the conservation of agricultural soils (Ávalos and Fernández, 2017), a national land management programme (SEMANART, 2008b), and institutions such as the Institute of Ecology (INECOL, 2020), the *Centro de Investigación en Ciencias de Información Geoespacial* (CONACYT, 2020), and the National institute for Sustainable Soil (*Instituto Nacional Del Suelo Sustentable*) that are continually developing research programmes on soil biodiversity (SEGOB, 2020).

In **Panama**, attempts are being made to introduce new practices using beneficial microorganisms as pathogen antagonists and plant growth promoters. The Ministry of Agricultural Development is trying to raise farmers' awareness on the effects of erosion on their crops and soil impoverishment, in addition to its effect on the eutrophication of inland waters (Ministerio de Desarrollo Agropecuario, 2020).

In **Peru**, there are investments in family farming, organic crops, agroecology and associated crops. There are various organic producers' associations that promote the production of crops in a sustainable way. The ANPE is an association formed by families that produce on an agroecological basis (ANPE Perú, 2020). Peru promotes sustainable soil management through soil survey studies for crop management and potential soil use purposes. Some regulations include the Regulation of Land Classification for its Greater Use Capacity (SENACE, 2009), and the Regulation for the Execution of Land Surveys (SENACE, 2010).

Trinidad and Tobago reported on the development of biofertilizers and its report on land capability.

Uruguay has plans for responsible use and management of soils. With regard to research, the country has soil microbiology laboratories, the network "*Red Uruguaya de Biodiversidad de Suelos*" (RUBIOS), the *Instituto de Ecología y Ciencias Ambientales, Facultad de Ciencias and Universidad de la República*.

Venezuela (Bolivarian Republic of) has university courses and agroecological initiatives. In addition, through the *Ministerio del Poder Popular para la Ciencia y la Tecnología*, technological innovation is carried out with the aim of building solutions within the framework of a national policy to promote projects. In this way, those involved in the development and implementation of sustainable soil management practices have the opportunity to apply for financial support from the Venezuelan state. Some examples:

i. Creation of the Socialist School of Tropical Agriculture, which is dedicated to the development of a multidisciplinary teaching with an agroecological approach to the sustainable management of agroecosystems.

ii. The National Institute for Agricultural Research is implementing the national soil plan to support the sowing of strategic items in the country, with a goal of 3 million hectares by 2019 (INIA, 2017).

iii. National Network of Laboratories for the Production of Biofertilizers and Biocontrollers of the National Institute of Integral Agricultural Health (INSAI, 2020).

In Venezuela, several institutions are dedicated to sustainable soil conservation, including soil maintenance and research, such as the Venezuelan Soil Society, which aims to make its extensive experience in agro-environmental studies available to all those who actively participate in agriculture and who also feel the need to protect the environment. Thus, producers, students, researchers, authorities and others can now count on a support centre that seeks to collaborate in the development of an agriculture that is in harmony with the ecosystems. On the official website of the Venezuelan Soil Society, various research studies are available:

- A digital library that protects ancient research heritage and facilitates the diaspora of cutting-edge knowledge;

- An informative digital magazine that presents the latest news in the field of agri-environmental technology and interviews with important personalities of the area, as well as exhibition of new tools, free information available online and information on events;

- Training and methodologies applied to environmental studies through online courses, which include tools such as Geographic Information Systems (GIS), Remote Sensing, Digital Cartography, Zootechnics, among others (Campo Ambiente, 2020).

3.3 | MAINSTREAMING: POLICIES, PROGRAMMES, REGULATIONS AND GOVERNMENTAL FRAMEWORKS

Antigua and Barbuda has recently acceded to the International Treaty on Plant Genetic Resources for Food and Agriculture (ITPGRFA) and a national working group will be established to take this treaty forward in the interest and for the rights of farmers. This is an ongoing process, as the actions to be seen are at the level of local communities. One of the key points is that participation in several key inter-agency committees allows for exchanges and linkages that can lead to policy processes that cut across many sectors and conventions. Local education awareness efforts, supported by mainstreaming and sharing knowledge and information, reach the nation through written articles and planned events, including other knowledge management processes such as videos and case studies. Sharing information on the impact of chemicals used for fogging exercises and by farmers on pollinators reaches the general public. Antigua and Barbuda's national dialogue on the IPBES report made technicians very aware of the serious consequences for the country, so they have committed to using alternatives. A stocktaking exercise is underway to see what is in stock and can be phased out.

Brazil has the National Soil Survey and Interpretation Program - Pronasolos - Decree no. 9.414/2018 (Ministério da Agricultura, Pecuária e Abastecimento, 2018) and the National Implementation of the Aichi Biodiversity Targets - National Biodiversity Program (Ministério do Meio Ambiente, 2020). The country has partially integrated soil biodiversity into sectoral and/or cross-sectoral policies through the National Biodiversity Policy, the Alelo Portal (Embrapa, 2020d) - microbial culture collections, Species Link (Species Link Project, 2020), Sistema Brasileiro de informações sobre Biodiversidade (SiBBr, 2020).

Chile has developed, through the Ministry of Environment, a document on the state of biodiversity (Ministerio del Medio Ambiente, 2014) in the country, but which does not focus on soil biodiversity. The "Resolución Exenta N° 1.690" established the methodology for identifying abandoned soils containing contaminants, but it does not focus on soil biodiversity. In Chile, legislation on soil conservation and protection is vague and scarce. Thus, there is a lack of specific regulations relating to soil biodiversity.

Colombia has sustainable soil management policies: "*Política nacional para la gestión sostenible del suelo*" (MADS 2016) (Minambiente, 2016a); "*Programa nacional de monitoreo y seguimiento a la degradación de los suelos*" (IDEAM) (Government of Colombia, 2012). The importance of soil biodiversity is included in the the policy on sustainable soil management has been taken into account in order to improve the national nutrition plan. Colombia has also put in place policies in favour of family farmers and protected areas (Government of Colombia, 2017; Government of Colombia, 2018). The country has developed some plans and National Policy on: i) Biodiversity Action Plan (Minambiente, 2012); ii) National climate change policy (Minambiente, 2016b); iii) Sustainable land management policy; iv) Regional climate change plan for Orinoquía (Corporinoquia, 2015); v) National Restoration Plan (Minambiente, 2015) and the Action Plan to Combat desertification and drought.

Ecuador has the Law of Plan of Soil, Law Organic of Land and the Constitution of Republic of Ecuador of 2008, which includes several articles. Ecuador reported on the National Soil Plan and the creation of the Directorate of Soil within the Ministry of Agriculture and Livestock.

Mexico has programmes focused mainly on soil erosion control. In addition, there are many laws dealing with the importance of soil and soil conservation, including the following: General Law on Ecological Balance and Environmental Protection; General Law on Sustainable Forest Development; General Law on Sustainable Rural Development; General Law on Climate Change. However, the existence of a legal framework has not led to the creation of specific soil conservation programmes, other than the actions carried out by CONAFOR, and no programmes have focused on soil biodiversity conservation.

Panama reported that the Panama Canal Authority currently has policies in place to reduce the effects of extensive agriculture and livestock in the Panama Canal Basin (Canal de Panamá, 2019). There are also some initiatives at the level of non-governmental offices, mainly in the area of ecosystem conservation, but not specifically in the area of soil conservation.

Peru has policies for family farmers, organic agriculture, crop rotation and biological nitrogen fixation; forestry; Indigenous peoples; erosion control through *andenería*. Peru has carried out soil surveys for land classification according to increased land use capacity, agroecological zoning, forest zoning and economic ecological zoning. Climate change, mitigation and adaptation measures are covered by the *Ley Marco del Cambio Climático*. Furthermore, the country also has a programme to reduce soil degradation "*Programa Presupuestal PP0089 Disminución de la Degradación de los Suelos*", in which erosion, overgrazing and improper land use are assessed.

Trinidad and Tobago reported on the National Environmental Policy and Vision 2030 with regard to greater care and sensitivity to the environment. Trinidad and Tobago has endorsed policies such as the National Environmental Policy and the National Waste Recycling Policy.

Venezuela (Bolivarian Republic of) reported on the law "*Ley de Salud Agrícola Integral*" (31 July 2008 Ministerio del Poder Popular Para la Agricultura y tierras). Article 1 of this law states: "The present decree, which has the value and force of law, aims to guarantee the integral health of agriculture. For the purposes of this decree with rank, value and force of law, integral agricultural health is understood as the primary health of animals, vegetables, products and by-products of both origins, soil, water, air, people and the close relationship between each of them, incorporating the principles of agro-ecological science that promote food security and sovereignty and popular participation, through the formulation, execution and control of policies, plans and programmes for the prevention, control and eradication of pests and diseases". Article 2 states that it is necessary to "Regulate the use of active ingredients in agricultural, domestic, public and industrial health products, particularly when the State considers that there is an imminent threat to human health or the environment".

Among the obstacles identified is the need for other mechanisms to generate soil biodiversity statistics. The term "comprehensive agricultural health" breaks down the agroecosystem into components. Soil resources can be differentiated from the other components and, to a certain extent, they are currently framed by a national soil biodiversity conservation policy that need to be implemented.

Venezuela has created and implemented the Degree Training Program (PFG) in Agroecology, through the Bolivarian University of Venezuela, as an integrated, cross-sectorial and sectoral policy. The formation of these professionals allows and encourage the practical application of soil biodiversity (e.g. the production of organic fertilizers). These practices are extended to rural areas and cities through workshops such as the National Institute of Integral Agricultural Health (INSAI) and the Training and Innovation Foundation for Rural Development (CIARA). In addition, in 2016, the Ministry of Popular Power for Urban Agriculture was created to promote sustainable production in urban areas (Domené-Painenao, 2019; Castellanos, 2011; Correo del Orinoco, 2016).

In Venezuela, policies in favour of agroecological production at the farmer level are being promoted at the national level, suggesting a less aggressive use of soil and water for food production. These policies are protected by various legal instruments, such as the National Drought Strategy, the National Strategy for the Conservation of Biological Diversity 2010-2020 and the Seed Law. Desertification, land degradation and drought are problems that affect all regions of the world. Within the framework of the United Nations Convention to Combat Desertification (UNCCD), the *Programa de Acción Nacional de lucha contra la Desertificación y Mitigación de la Sequía* (PAN) is being implemented. This is the instrument that guides national public policies for soil conservation, contributing in a coordinated way to guiding aspects related to the eradication of poverty and social exclusion in the areas of health, education, food, recognition of the rights of indigenous peoples, housing, productive employment and land tenure, among others.

The National Drought Strategy (MINEC, 2019) contains a set of strategic contributions and guidelines that contribute to the design of a national drought policy in Venezuela. Its implementation represents an important contribution together with the National Strategy of Biological Diversity and a future National Plan for Adaptation to Climate Change envisaged in the Plan de la Patria 2019 and 2025; and the 2030 Agenda for Sustainable Development.

The programme for the establishment of National Goals for Land Degradation Neutrality (NDT) was developed within the framework of Sustainable Development Goal 15 and its objective 15.3.1 with the determination of three indicators: plant cover/land use, land productivity and soil carbon reserve. With this determination, the Venezuelan State defines the goals and measures to achieve the NDT by 2030, envisaging the continued recovery of degraded land over 7 910.5 km^2 (791 050.00 ha.), which represents 1.08 percent of the Venezuelan territory. This will be achieved through the formulation of transformation projects to improve productivity conditions and soil recovery.

The Gender Plan on desertification and drought is being formulated in synergy with the Convention on Biological Diversity and Climate Change. This proposal, which responds to women's vulnerability to environmental problems, is a consequence of gender inequality in the unequal distribution of management power, private or common property, means of production and capacity development. As a result, a guidance document has been prepared that aims to support women in both rural and urban communities, with the objective of contributing to their prominent participation in decision-making, obtaining economic benefits and positively influencing livelihoods, to the benefit of the Venezuelan population in general.

Through the National Seed Plan, led by the National Institute of Agricultural Research (INIA), seeds adapted to the tropical climate and Venezuelan environmental conditions are provided. One of the objectives of the National Seed Plan is to reduce the country's dependence on imported seeds and social control of investments, in order to achieve an increasingly high substitution rate of national seeds and thus guarantee agri-food sovereignty. The implementation of this Plan represents a comprehensive policy of financial support to small and medium producers, with flexible financing rates, as well as comprehensive support to the worker through the provision of seeds, fertilizers and timely payment for crops, managing to strengthen and guide the national productive apparatus on a daily basis.

The *Plan Nacional de Agricultura Familiar, Indígena, Campesina, Urbana y Periurbana* (AFICUP) aims to contribute to the development of family farming as a productive mode of sustainable management of the agroecosystem, framed in actions that make it possible to strengthen and rebuild networks of solidarity relationships within each community, and to help diagnose, plan, investigate and support integrated human development projects.

The programme *Todas las Manos a la Siembra de la mission Agro-Venuzuela* aims to stimulate the nation's agricultural production and support the country's farmers and producers. The contribution to the Mission is related to academic, productive and technical aspects.

The plans and programmes for environmental education and technological innovation indicate the importance of social awareness with regard to problems related to desertification and drought, as well as environmental issues. Since 2005, the benefits of the *Ley Orgánica de Ciencia, Tecnología e Innovación* (LOCTI) have resulted in increased financial support for science, technology and innovation projects. This has enabled Venezuela to experience a significant change, from 0.48 percent in 2006 to 1.78 percent today. This plan marks the beginning for the application of the participatory approach in the formulation of public policies on science and technology, in accordance with the provisions of the legal framework of Venezuela. The National Science, Technology and Innovation Plan (PNCTI) 2005-2030, lays the foundations for a series of approaches, establishing strategies that allow, through a prospective vision, the development of science and technology in the country, one of its strategic objectives being the dissemination and information, as a basis for promoting the responsibility and participation of the different actors.

3.4 | ANALYSIS OF THE MAIN GAPS, BARRIERS AND OPPORTUNITIES IN THE CONSERVATION AND SUSTAINABLE USE OF SOIL BIODIVERSITY

Antigua and Barbuda has just submitted its sixth national report to the CBD, which has benefited from a major contribution from local groups, including farmers, beekeeping cooperatives that are doing advocacy and outreach to educate and inform technicians as well as local citizens. It must be acknowledged that local groups are contributing to the achievements of the objectives of the NBSAP and the post-2020 processes. Their actions must be valued and recognized, as well as the resources provided for their effective participation, knowledge sharing and involvement in all processes.

Brazil indicated that it plans to invest in the characterization of soil microorganisms, to fully implement the National Biodiversity Program (EPAMB), as well as the PRONASOLOS (an inter-ministerial arrangement that includes soil biodiversity) and to improve best practices in the different integrated agricultural systems.

In **Chile**, the Ministries of Environment and Agriculture are aware of the lack of information on this topic and efforts are being made to address it. For example, the Ministry of Agriculture actively participates in the Global Soil Partnership. There is currently a project in Chile related to the governance of soil erosion that aims to make information available for policy makes in this field, and one topic to be taken into account is soil biodiversity.

Colombia has capacity and there are regulations and policies related to soil biodiversity. However, the inter-sectoral and inter-institutional articulation still needs to be improved. Soil and its activities fall under three different Ministries and sometimes activities and policies can be seen as fragmented. The main idea is to promote the implementation of soil policy at different scales to address the importance of soils and in particular the components of biodiversity contained in soils. It is necessary to encourage regional and national decision-makers to recognize the importance of soil management in order to provide technical, human and financial support in their agendas. It is important to raise awareness of soil biodiversity and its loss factors in order to include it in planning instruments and strategies for environmental and productive sector management.

Costa Rica has ongoing policies for other soil functions, but not necessarily related to soil biodiversity.

In **Mexico**, there are programmes to conserve and increase forest cover. Due to a lack of soil biodiversity indicators, no direct action is being taken. The benefits are indirect, resulting from other public policies such as ENBIOMEX to promote actions to rehabilitate degraded soils. Despite decades of government programmes for soil conservation practices, farmer adoption as a basic principle for soil conservation, remains low (Cotler and Cuevas, 2017).

Panama is working to develop and implement more sustainable agriculture, reducing the use of agrochemicals as much as possible and encouraging farmers to develop agroecological and environmentally friendly practices, incorporating new technologies that take into account available biotic resources.

Suriname reported on the lack of financial support and cooperation between research institutes at the national level.

Trinidad and Tobago reported on Vision 2030 - The National Development Strategy of Trinidad and Tobago 2016-2030 (Government of the Republic of Trinidad and Tobago, 2016).

In ***Venezuela (Bolivarian Republic of)***, academics and researchers are promoting conservationist practices for agricultural soil management (e.g. fertilizing management, adopting reduced or no-till with cover crop residues and green manure, improved crop rotation) that indirectly affect soil biodiversity. Universities and national research institutes have carried out research directly on soil biology, despite limited resources, but very little on soil biodiversity. Moreover, in order to have an inventory of the existing microbiomes and fauna types, it would be necessary to have personnel trained in the new tools for the genetic study of soil organisms, taxonomists, and well-equipped laboratories. There is a need to continue working on understanding the relationship between biodiversity and soil functions.

4 | NEAR EAST AND NORTH AFRICA (NENA)

4.1 | ASSESSMENT OF SOIL BIODIVERSITY

4.1.1 | CONTRIBUTIONS OF SOIL BIODIVERSITY TO ECOSYSTEMS SERVICES

Most countries in this region reported that despite the relevance of soil biodiversity to the country, very little has been done to enhance it due to a lack of resources.

4.1.2 | NATIONAL ASSESSMENTS

The **Syrian Arab Republic** reported that as for the present time, it is in the recovery stage of the crisis, the government takes care of the soil by issuing a law to prevent urbanization in agricultural areas, which is a strict law aimed at protecting the soil from urbanization as well as protecting the soil from erosion. A law was also issued to analyze the soil at a nominal price for farmers in order to conserve the soil and its fertility. Finally, a number of seminars, conferences and meetings have been held, aimed at protecting the soil.

4.1.3 | PRACTICAL APPLICATIONS OF SOIL BIODIVERSITY

In the Syrian Arab Republic, soil organisms have been used as biofertilizers, phosphate solubilizing bacteria, organic manure, compost and other products (GCSAR, 2017). The Ministry of Agriculture created an office for organic agriculture (Ministry of Agriculture and Agrarian Reform of Syrian Arab Republic, 2017). The local fertilizer factories have been active in the manufacture of organic fertilizers such as humic and amino acids in addition to the biological fertilizers. GCSAR has started studies in the field of biochars and updating the fertilizer recommendation with the aim of conserving the soil fertility and biological diversity.

The Ministry of Agriculture in Syria has also established a number of biological control centres in the governorates of Hama and Lattakia that produce *Trichoderma* fungi with the aim of supporting the biological diversity of soils and combatting diseases. The Ministry of Agriculture supported the conservation agriculture system in Syria in a number of governorates, such as Hama, Daraa and Aleppo, by collaboration of NGOs in order to conserve the soil and its biological diversity. The government took care of biogas by establishing some biogas digesters in the Syrian governorates and with the support of NGOs. The government focused on recycling organic waste and adding it to the soil in order to enrich it and increase its biological diversity. GCSAR also achieved several studies in the field of soil pollution by heavy metals and their treatment through organic fertilizers. In the recent period, governance focused on climate change and adapting to it in order to preserve soil and during the year 2019 studied the effect of microorganisms isolated from Syrian soil in addressing oil-contaminated soil and MOWW through a large number of bacteria and fungi isolated from Syrian soil, which had great efficiency in the addressing of these substances.

4.1.4 | MAJOR PRACTICES NEGATIVELY IMPACTING SOIL BIODIVERSITY

In the **Syrian Arab Republic**, a large number of negative practices affect the soil, including the overuse of chemical fertilizers and pesticides, the use of surface irrigation, over-cutting of forests and overgrazing in the Syrian Badia, especially during the crisis in Syria, soil stress through cultivation of stressful crops of the soil, the use of wastewater in agriculture, especially vegetables, increasing soil salinity in the Euphrates basin as a result of the destruction of irrigation systems and the use of saline water from wells.

Tunisia faces several threats to its biodiversity, including water pollution from raw sewage; limited natural freshwater resources; toxic and hazardous waste disposal; overgrazing; erosion and desertification.

4.1.5 | INDICATORS USED TO EVALUATE SOIL BIODIVERSITY

The **Syrian Arab Republic** is carrying out numerous studies in the field of carbon sequestration in the soil by supporting the conservation agriculture system, as well as having implemented several projects in the field of rainwater harvesting, with the aim of increasing green water in the soil and reducing soil erosion. The government is concerned with managing soil fertility by securing fertilizers and integrated fertilization. As a result of the crisis in Syria, they are interested in studies of soil pollution resulting from the use of wastewater, waste and oil pollution. The Ministry of Agriculture also established a number of biological control centres in Syria.

4.2 | RESEARCH, CAPACITY DEVELOPMENT AND AWARENESS RAISING

In **Algeria**, there are research initiatives in some universities and extension programmes of the Ministry of Agriculture.

Lebanon reported that no till or minimum tillage is promoted by the Ministry of Agriculture and supported by GIZ as well as some private initiatives in organic farming.

Oman reported on the cooperation with international organizations on soil salinity and usage of unconventional water resources.

The **Syrian Arab Republic** started to consider the Voluntary Guidelines for Sustainable Soil Management. GCSAR is undertaking a lot of research related to soil biodiversity such soil microbiology, N- fixation nitrogen, PSB, the effects of organic manure, green manure, organic agriculture, clean agriculture, conservation agriculture, monitoring of forest fires, sustainability of Albadia by combatting overgrazing, isolating of useful microorganisms to addressing soil pollution, and others.

4.3 | MAINSTREAMING: POLICIES, PROGRAMMES, REGULATIONS AND GOVERNMENTAL FRAMEWORKS

Algeria has integrated agriculture, food security and health sectors to fight the rats causing leishmaniasis to humankind and affecting cereal yields.

Lebanon reported that Law 444 from 2003 is still not implemented to protect soil and water from contamination and mismanagement.

Oman reported that the country has land fragmentation law and pesticide law, implemented by the Ministry of Environment and the Ministry of Agriculture respectively.

The **Syrian Arab Republic** has several soil biodiversity programmes related to the environment, climate change adaptation, land taxonomy, extension services, land and water conservation, organic agriculture, transfer to modern irrigation, drought control, soil analysis laboratories, capacity building and others.

4.4 | ANALYSIS OF THE MAIN GAPS, BARRIERS AND OPPORTUNITIES IN THE CONSERVATION AND SUSTAINABLE USE OF SOIL BIODIVERSITY

Algeria reported the lack of sectoral coordination at the national level and the lack of significant arrangements related to conservation and sustainable use of soil biodiversity beside circumstantial arrangements related to health issues.

The **Syrian Arab Republic** reported the lack of soil conservation regulations and laws.

Tunisia reported that forests, climate change and changes in land use appeared to cause significant changes in ecosystem structure and biodiversity (Hanafi and Jauffret, 2008). In particular, the vegetation ecology of Tunisia has undergone a severe decline over the course of a century under the long-term impact of humans and livestock. Tunisia's territory is in danger of desertification. In particular, the loss of biodiversity caused by overgrazing is a serious issue in southern Tunisia. Because of the extremely dry soil, only 1 to 10 percent of vegetation covers the ground, and the dry biomass of perennial plants is 100 kg/ha/y (Gamoun *et al.*, 2012).

Government capacity building and adequate budget allocations are key to effectively protect all threatened ecosystems in the country. General actions that are needed include: ensuring equitable and managed urban development; improving infrastructure; sustainably managing water resources; reorganizing and developing the agriculture sector; ensuring ecosystems services; developing cultural heritage and landscape; protecting terrestrial fauna and flora; defining the conditions for a rational use of natural resources; enhancing social and economic development; promoting the participation of local communities in natural resource management; protecting lagoon and marine ecosystems; protecting natural heritage from droughts; improving legal frameworks for environmental protection; and improving institutional organization and knowledge base.

5 | NORTH AMERICA

5.1 | ASSESSMENT OF SOIL BIODIVERSITY

5.1.1 | CONTRIBUTIONS OF SOIL BIODIVERSITY TO ECOSYSTEMS SERVICES

In **Canada**, soil invertebrates, including mesofauna (0.15 mm to 10 mm in length) and macrofauna (> 2 mm in length), play a key role in recycling nutrients in soil and other detrital substrates and in regulating plants, fungi, microbials and invertebrates as predators/consumers or prey in food networks. Mesofauna is dominated by mites (*Acari*) and springtails (Collembola), but also includes potworms (*Enchytraeidae*), symphylans, and pseudoscorpions. Macrofauna includes both invertebrates that live entirely in the soil or litter layer (for example, earthworms, some beetles and spiders) and those that complete part of their life cycle in the soil (for example, many fly and beetle larvae).

Soil-dwelling beetles, spiders and soil- and water-inhabiting mites have also been used as bioindicators of soil and freshwater health (Behan-Pelletier, 1999; Beaulieu and Weeks, 2007; Pearce and Venier, 2006; Walter and Proctor, 2013). However, due to their small size, invertebrates (and particularly microarthropods) are often overlooked by researchers other than taxonomists, obscuring the awareness of their impact on agriculture, forestry, wildlife, human health and ecosystem services.

Soil biodiversity also provides all kinds of ecosystem services such as water purification, nutrient cycling and resilience of above-ground diversity to extreme events, and therefore has an indirect impact on air quality. Soil biodiversity supports ecosystem productivity and above-ground diversity and therefore contributes to carbon sequestration both directly (humus) and indirectly (plant roots, timber).

Bacteria can improve and maintain soil structure (tilth). Good soil tilth originates from the aggregation of soil particles due to microbial excretions acting like glue (Lupwayi and Hamel, 2010). To improve soil biodiversity, Agriculture and Agri-Food Canada (AAFC) and producers are implementing beneficial management practices (for example, soil cover approaches) (Government of Canada, 2020c; Lupwayi and Hamel, 2010).

In Canada, there have been relatively few studies specifically assessing the contributions of soil fauna to ecosystem services; however, the processes are expected to be similar to homologous climates in North America and Eurasia (Wall, ed., 2012; Whalen and Sampedro, 2010). Some key references that outline the roles and contributions of soil invertebrates to ecosystem processes include:

5.1.1.1 Soil formation

Soil invertebrates contributing to soil structure in Canada include: *Acari, Enchytraeidae,* Collembola, *Diptera* larvae and *Diplopoda*, with the contribution of each group depending on the ecosystem type (Pawluk, 1985). Soil microarthropods, particularly the dominant group of oribatid mites, make a particularly important contribution to soil formation by breaking down litter, comminuting soil, modifying soil structure and increasing decomposition; however, there is little recent data on their impact on Canadian soils (Fox, 2003). The contribution of microarthropods to soil formation is also impacted by the above-ground vegetation community (Berg and Pawluk, 1984). Earthworms are a recent addition to the Canadian fauna; they contribute to soil formation but also decrease soil biodiversity and may disrupt ecological processes (Migge-Kleian, 2006).

With respect to soil fungal diversity, arbuscular mycorrhizal fungi (AMF) are an important group of fungi that are symbiotically associated with 72 percent of vascular plants. This fungal-plant association has been effective for at least the last 500 million years.

AMF colonize plant roots and exchange nutrients with their hosts through a specialized intracellular structure named arbuscules. AMF produce extra radical mycelium (hyphae) which forage the soil for nutrients. The contribution of AM hyphae to soil formation is twofold. First, they develop a dense network of hyphae (hyphosphere) in the soil, representing up to 30 percent of the total soil microbial biomass (Olsson *et al.*, 1995). The development of such a network stabilizes and aerates the soil and supports the development and propagation of bacteria at the surface of the hyphae (Cao *et al.*, 2018; Zhang *et al.*, 2018). Second, hyphae contain and release organic compounds, including glomalin which is an alkaline-soluble glycoprotein (Magurno *et al.*, 2019). The turnover of hyphae leads to the release of glomalin in the soil. The concentration of glomalin-related soil protein is positively correlated with aggregate stability (Rillig, 2004). Moreover, glomalin has a relatively slow turnover in the soil, which contributes to long-lasting effects on aggregation.

5.1.1.2. Nutrient cycling

Microarthropods contribute to nutrient cycling by feeding on fungi, bacteria and decaying matter and by dispersing these organisms (many of which are primary recyclers) in the soil profile (Wall, 2012; Whalen and Sampedro, 2010).

Arbuscular mycorrhizal fungi play a key roles in soil nutrient cycling through close cooperation with other soil microorganisms. Their hyphae can strongly increase the mineralization of native soil organic matter (Paterson *et al.*, 2016). The AMF hyphosphere is tightly associated with phosphate-solubilizing bacteria that are able to solubilize inorganic phosphorus, which is then transferred into the hyphae and exchanged to plants via arbuscules for plant-derived carbon. This plant photosynthate transfer into the soil in turn stimulates bacterial growth.

Soil bacterial biodiversity plays a significant role in recycling nutrients during decomposition, converting organic nutrients into inorganic forms required by plants.

5.1.1.3. Soil erosion control

Through the production of glomalin and a dense network of hyphae, AMF prevent soil erosion. It has been shown that the extraradical length of the hyphae of arbuscular mycorrhiza has a direct effect on reducing soil erosion due to surface water flow (Mardhiah *et al.*, 2016). Therefore, the presence of a wide variety of AMF in soil with various hyphal features contributes to limit soil erosion.

Soils with a good tilth due to specific microbes, as indicated above, are less prone to erosion. Best management practices, such as direct seeding, no till and similar low soil disturbance practices, improve soil biodiversity and minimize erosion (Government of Canada, 2020d).

5.1.1.4. Atmospheric composition and climate regulation, including floods

As an indirect influence of soil invertebrates, AMF contribute to climate regulation by sequestering carbon in the soil through the transfer of carbon from the plant to the soil and the hyphae turnover.

5.1.1.5. Carbon sequestration and climate change responses

Soil fauna is indirectly important for carbon sequestration by forming and maintaining soil structure and promoting plant growth and populations. No studies have examined the importance of soil fauna for climate regulation, specifically in Canada.

There is a close positive correlation between AMF hyphal abundance and soil aggregation and C and N sequestration (Wilson *et al.*, 2009).

5.1.1.6. Regulation of water supply and quality

There is an indirect link via soil maintenance. Many studies show that AMF enhance plan tolerance to drought stress; this has been demonstrated in the case of maize, for example. Recent studies show that AMF significantly modify the radial transport of water through the roots due to the post-translational regulation of aquaporin activity. Although AMF do not directly improve water quality, they improve plant vigor under saline stress (Al-Karaki, 2006).

5.1.1.7. Pollutant degradation and soil remediation

Arbuscular mycorrhizal fungi contribute to soil remediation through their partnership with plants (phytoremediation). The extra radical mycelium connected to the plant root system significantly increases the efficiency of heavy metal or other pollutant uptake due to the small size of hyphae (5 to 10 μm in diameter), which allows high-resolution prospecting of the soil matrix (Göhre and Paszkowski, 2006). Arbuscular mycorrhizal fungi improve plant uptake in soils polluted by heavy metals and polycyclic aromatic hydrocarbons (PAHs) (Wang and Yin, 2005; Alarcón *et al.*, 2008). High soil biodiversity is expected to accelerate soil remediation, as Alarcón *et al.* (2008) showed that the highest rate of crude oil degradation in soils was measured when AMF (*Rhizophagus intraradices*) were combined with bacteria (*Sphingomonas paucimobilis*).

Mites and other soil arthropods have been used as bioindicators of soil contamination by toxins (Princz *et al.*, 2010; Princz *et al.*, 2012).

Soil microorganisms play a pivotal role in the degradation of chemical pesticides, thus minimizing serious residual effects on the environment and/or public health.

5.1.1.8. Recycling of waste biomass, nutrients and water

Diptera larvae (for example, soldier flies) have the potential to play a large role in the recycling of food and agricultural waste (Nguyen *et al.*, 2015). A diverse arthropod community is essential for the decomposition of cattle wastes in prairie ecosystems (Floate, 2011). Oribatid mites and earthworms contribute to the decomposition of crop residues (for example, Broadbent and Tomlin, 1982; Tomlin *et al.*, 1995). Arbuscular mycorrhizal fungi are soil-symbiotic fungi and do not participate in the recycling of waste biomass. Nevertheless, their presence contributes to maintaining a high level of biodiversity of soil microorganisms, including saprophytic fungi and bacteria which are directly involved in waste biomass degradation.

5.1.1.9. Restoration of degraded lands and ecosystems

Soil mites have been used as bioindicators of mine tailings restoration in Canada (St. John *et al.*, 2002). Mesofauna has also been found to respond negatively to forest harvesting and can provide reliable indicators of forest ecosystem health and recovery (Battigelli *et al.*, 2004).

Arbuscular mycorrhizal fungi can contribute to restoring degraded lands and ecosystems through artificial inoculation (bioinoculants) if air-borne spores are ineffective in recolonizing degraded lands. Eroded or disturbed soils show a limited level of AMF diversity and colonization. Nevertheless, the use of native AMF species, early seral and consortia is recommended (Asmelash *et al.*, 2016). Eroded or disturbed soils provide a highly stressful environment for plants. Numerous studies show how AMF can alleviate both biotic and abiotic stresses. Therefore, it is very important to ensure that plants will be able to cling to common AMF networks in order to survive the harsh conditions of a degraded land/ecosystem.

5.1.1.10. Nitrogen fixation

Nitrogen cycling in agroecosystems depends on arbuscular mycorrhizal fungi. Glasshouse experiments show that AMF improve nitrogen fixation of common beans (Ibijbijen *et al.*, 1996). Arbuscular mycorrhizal fungi improve plant vigor root biomass and increase crop Nitrogen Use Efficiency (NUE). Plants with an extensive root system are more likely to support more nodules and thus fix more atmospheric nitrogen. The promotion of "AMF friendly" agriculture practices (such as, direct seeding, no tilling and mulch cover) increases AMF abundance and diversity and thus improves crop NUE while reducing the use of nitrogen fertilizers (Verzeaux *et al.*, 2017). Root hair infection and nodulation are salt sensitive. AMF alleviate plant salt stress and it has been shown in greenhouse experiments that the inoculation of chickpeas with AMF under saline stress improves nodulation via flavonoid production.

Bacteria, especially *Rhizobia*, play an important role in nitrogen fixation in leguminous plants, for example, soybeans. The benefits are both economic and environmental. Non-legume crops grown in association or rotation with legumes generally have lower N-fertilizer requirements (Top Crop Manager, 2018).

5.1.1.11. Pest and disease regulation

Predatory mites (including mesostigmatids and prostigmatids) and predatory insects (e.g. staphylinid and carabid beetles) are important predators of invertebrate pests, including plant-feeding pests that live entirely in the soil (e.g. plant-feeding nematodes, root-feeding mites) or that spend part of their life-cycle in the soil (e.g. spider mites, thrips, fungus gnats, cutworms, root aphids).

Predatory mites can also be important control agents of potentially disease causing flies in manure and agricultural waste products (Azevedo *et al.*, 2018). Entomopathogenic nematodes are also effective against insect pests and have been used for biological control in agricultural systems.

Soil symbiotic fungi such as AMF enhance yield and resistance to leaf pathogens in wheat, and this bioprotection is achieved through priming, i.e. a preliminary induction of defense mechanisms which includes elicitation of defense-responsive genes, accumulation of defensive compounds at the local and systemic levels, and modulation of defence-related hormones (Fiorilli *et al.*, 2018).

Soils with active microorganisms are known to be suppressive to pathogens and pests due to antagonism from predation, antimicrobial production and nutrient competition. As such, soil management systems that foster microbial biodiversity are generally more sustainable (Green House Canada, 2016).

5.1.1.12. Landscape support (effects on above-ground biodiversity in agricultural landscapes)

The impact of microarthropod diversity on above-ground biodiversity is probably strong, via the processes listed above (nutrient cycling, soil structure formation and predation on pest species); however, little research examines this impact. Conversely, above-ground activities (e.g. grazing or crop rotation) have been shown to have an effect on the below-ground microarthropod community (Clapperton *et al.*, 2002; Osler *et al.*, 2008).

The introduction of non-native earthworms has had a significant effect on above-ground biodiversity and community structure in Canadian ecosystems (Eisenhauer *et al.*, 2007) and has negatively effects on native soil fauna (Cameron *et al.*, 2013b).

5.1.1.13. Enhance agricultural productivity and economic profitability

Soil mesofauna communities enhance agriculture both directly by consuming pest species and indirectly by maintaining soil quality.

The direct contribution of AMF to yield improvement under field conditions is still controversial due to the many environmental parameters to be taken into account (Cavagnaro *et al.*, 2019; Ryan and Graham 2018), nevertheless it has been shown that AMF enhance the temporal stability of plant community productivity in grasslands (Yang *et al.*, 2014). Arbuscular mycorrhizal fungi alleviate both biotic and abiotic stress factors while improving plant access to nutrients. Therefore, their presence can reduce the amount of chemicals to achieve high yields, which has a positive effect on profitability. Complementarity between soil fungi plays a key role in plant productivity, as Wagg *et al.* (2011) showed that it improves plant productivity by up to 82 percent and 85 percent.

Biological nitrogen fixation by symbiotic soil bacteria (collectively known as *Rhizobia*) associated with crop legumes (e.g. soybeans, beans, peas, lentils, peanuts, clover, alfalfa), minimizes the cost of and dependence on chemical nitrogen fertilizers in agriculture, enhances soil fertility and environmental sustainability (air, soil water), including by reducing greenhouse gas emissions from intensive manufacture of nitrogen fertilizers. In addition, bacteria play a significant role in soil formation, nutrient cycling and disease regulation. As such, they will have direct and indirect positive effects on productivity and profitability.

5.1.1.14. Enhance nutritional quality of food

There is an indirect link via soil quality. Yield quality can be improved through AMF. Based on a meta-analysis, Lehmann *et al.* (2014) showed that AMF positively affects Zn concentration in various crop plant tissues. AM symbiosis can stimulate the synthesis of secondary metabolites and enhance the accumulation of antioxidant compounds in plant tissues (Seeram, 2008), for example in greenhouse-grown lettuce (Baslam *et al.*, 2011). Based on a field experiment conducted on a real industrial tomato farm, Bona *et al.* (2017) showed that field inoculation of two *Pseudomonas* strains and a mixed mycorrhizal inoculum improved sugar and vitamin concentrations in the tomato fruits. The AMF improved the concentration of citric acid, while the bacteria positively modulated the sugar production and sweetness of the tomatoes. Hart *et al.* (2015) found that mycorrhizal inoculation enhanced the concentrations of several minerals (N, P, Cu), carotenoids and certain flavour compounds, as well as the antioxidant capacity of tomato fruits. Best management practices that foster soil biodiversity and the development of these soil microorganisms can therefore enhance the nutritional quality of food.

5.1.1.15. Natural hazard regulation

High soil biodiversity involves the development of humus and soil and above-ground vegetation which buffers flash floods due to heavy storms. The support of plant cover helps to mitigate the impact of the heat island effect.

5.1.1.16. Human health

In Canada, particularly in Eastern Canada, tick-vectored Lyme disease is increasing rapidly (Clow *et al.*, 2016). Ticks spend much of their lives in soil and litter environments and are depredated by soil-dwelling predators such as spiders, ants and beetles; however, little experimental research has been conducted to assess the impact of natural ecosystem predators on tick populations (Burtis *et al.*, 2019). By improving food quality, AMF contribute indirectly to human health.

5.1.1.17. Unknown – what needs to be studied

Soil biodiversity and its ecological roles in soil are still poorly known, in particular because less than half of the soil fauna is currently described (Beaulieu *et al.*, 2019 and other references in Langor and Sheffield, 2019).

The United States of America reported that the ecosystem services provided by soil biodiversity are extensive and that, as a result of advances in understanding and the ability to conduct sensitive analyses, the understanding of soil biodiversity and its impacts on a number of other phenomena is being demonstrated worldwide, not just in the United States of America.

5.1.2 | NATIONAL ASSESSMENTS

In **Canada**, a great deal of effort has been made in recent years to document the country's soil fauna. In particular, a recent volume of ZooKeys (Langor and Sheffield, 2019) contains comprehensive species counts for most soil and litter-dwelling taxa in Canada. This extensive undertaking provided an update to the last comprehensive status report from 1979 (Danks, 1979).

In addition, several publications related to soil biodiversity have been published on a variety of topics, including soybeans inoculated with root zone soils of native Canadian legumes (Bromfield *et al.*, 2017) and the diversity of bacteria associated with corn roots inoculated with Canadian woodland soils (Tchagang *et al.*, 2018). Canada has also developed species distribution maps.

In the **United States of America**, entities within the USDA have conducted extensive soil biodiversity assessments, but there are no comprehensive assessments of soil biodiversity across landscapes or across the country. Entities in this country have developed soil biodiversity information, but the data cover multiple countries. Examples include the NSF NEON website (The National Ecological Observatory Network, 2020), the Earth Microbiome Project (Earth Microbiome Project, 2020) (partly funded by United States of America organisations, but also supported by other countries) and the Global Soil Biodiversity Initiative (GSBI) (GSBI, 2020) and their Atlas (GSBI, 2020). The GSBI is located in Boulder, Colorado, but works across nations.

5.1.3 | PRACTICAL APPLICATIONS OF SOIL BIODIVERSITY

Canada reported that, in addition to the ecosystem services already mentioned (soil formation, decomposition, and biological regulation of pest species), the main practical applications of soil mesofauna are as bioindicators in natural and agricultural ecosystems (e.g. Battigelli *et al.,* 2004; Behan-Pelletier 1999; Behan-Pelletier, 2003) and in contaminated soils (Princz *et al.,* 2010; Princz *et al.,* 2012). Soil-dwelling beetles (for example, *Carabidae, Staphylinidae*) are also important predators in agricultural systems (Goulet, 2003) and bioindicators in natural ecosystems (Klimaszewski *et al.,* 2018).

The main practical applications of soil biodiversity in Canada are all the ecosystem services we take for granted (soil fertility, clean water, clean air, ecosystems resilience, temperature/precipitation regulation).

The main practical applications of bacteria are nitrogen fixation and the control of soil borne diseases, especially fungal pathogens. In addition, by exploiting the biodiversity of symbiotic soil bacteria, ecologically adapted and highly efficient nitrogen fixing bacteria to be selected for use as elite commercial inoculants to improve the sustainable production of crop legumes. Selection of traits to enhance the efficiency of nitrogen fixing symbiosis (e.g. photosynthetic symbiotic bacteria that provide their own energy for nitrogen fixation).

In Canada, more field trials are needed to gather evidence on the effect of soil biodiversity on food production and nutrition in the country's different agroecosystems. The Living Labs Initiative of AAFC will be instrumental in this effort.

There is ample evidence that the use of soil biodiversity of nitrogen fixing bacteria has contributed positively to food production and nutrition in Canada. For example, the commercial inoculation of crop legumes grown with selected efficient bacteria has been shown to improve overall sustainable crop production and nutrition, while also benefiting the environment (Stagnari, Maggio, Galieni *et al.*, 2017).

The **United States of America** reported that soil biodiversity has a number of practical applications, ranging from biological nitrogen fixation, carbon cycling (including sequestration, but also CH_4 and CO_2 emissions), to the creation and stabilization of soil aggregates that improve water infiltration and reduce erosion (both wind and water), improving soil water quality and aquifer and surface water recharge, through the metabolism and/or mineralization of nutrients and exogenous organic chemicals (e.g. herbicides, pesticides, antimicrobial contaminants), respectively to more natural forms and in the formation of soil ecosystems that are stable, ecologically sustainable and regenerative and (possibly) resilient to climate change (due to a greater diversity of selection to operate).

5.1.4 | MAJOR PRACTICES NEGATIVELY IMPACTING SOIL BIODIVERSITY

In *Canada*, changes in land use, over-grazing and inappropriate soil management are the major practices that may have had a negative impact on soil biodiversity over the past decade. Tillage has negative effects on mite communities (Beaulieu and Weeks, 2007; Behan-Pelletier 2003). In addition, tillage has been found to reduce the number of earthworms, which can adversely affect soil fertility (Chan, 2001; Clapperton *et al.*, 1997), but is complicated by the invasive nature and negative effect of earthworms on other biodiversity (Migge-Kleian *et al.*, 2006). However, between 1991 and 2011, Canada has seen more farmland move away from traditional tillage practices towards non-tillage. Overgrazing led to an increase in the number of prostigmatid mites and a decrease in the number of oribatids and some mesostigmatid mites (Clapperton *et al.*, 2002). This study also indicated that more diverse communities, as evidenced by less grazed sites, contributed to higher primary productivity (Clapperton *et al.*, 2002). Urbanization has been shown to either reduce macro-invertebrate diversity or modify it through a greater contribution of non-native species and homogenization of fauna with the loss of specialist species (Kotze *et al.*, 2011; Magura *et al.*, 2009). Similar effects are also probably occurring in the mesofauna, but these have been less well studied.

With regard to the overuse of chemical control mechanisms (e.g. disease control agents, pesticides, herbicides, veterinary drugs), further studies are needed. However, there is evidence that some pesticides reduce ground beetle (*Carabidae*) populations through sublethal effects in agricultural systems (Goulet, 2003), which may in turn affect herbivorous pest species. The impact of pesticides on the mesofauna is highly taxon-dependent, with some taxa (e.g. oribatid mites) being more negatively affected, while others (e.g. some prostigmatid mites and astigmatid mites) are unaffected or are increasing in numbers (see Behan-Pelletier (2003) for a review of Canadian studies assessing the impact of agricultural regimes on *Acari* and Collembola). Few studies have been conducted in Canada on the effect of fertilizers on soil fauna, but the addition of fertilizer has been found to affect community abundance and composition in other regions (e.g. Bird *et al.*, 2004; Lindberg and Persson, 2004) and probably has a significant effect on Canadian soils as well.

Canada has a set of agri-environmental indicators, which include soil indicators. For instance, the Soil Erosion Indicator tracks the health of Canada's agricultural soils, which includes monitoring organic matter and fertility loss. This indicator tracked the soil cover associated with Canadian agricultural activities from 1981 to 2011. Soil cover also has implications for land productivity, crop yield and quality, as well as for broader environmental issues such as wildlife habitat and water and air quality. Soil cover has increased on agricultural land. This improvement can mainly be attributed to the implementation of beneficial management practices such as the reduction of summer fallow – a practice of leaving fields bare, as well as the shift to reduced tillage and no-till practices, reducing the amount of bare soil exposed to degradation. Other related indicators are the Soil Organic Matter Indicator (soil carbon content), the Soil Salinization Indicator and the Wildlife Habitat Capacity on Farmland Indicator. When combined with the Soil Cover Indicator, it provides a snapshot of the biodiversity potential of farmland in Canada (Government of Canada, 2020c; Lupwayi and Hamel, 2010).

The United States of America reported that poor land management has been shown to impact soil biodiversity in a number of ways and that in, some ways, these impacts have resulted in losses in ecosystem's productivity. For example, heavy metal contamination of the areas around Kellogg, Idaho, by defective smelters has greatly reduced the biological diversity and biomass of soils and killed local tree stands (USDA, 2019a; USDA, 2019c; USDA, 2019d).

5.1.5 | INVASIVE ALIEN SPECIES (IAS)

Over 5 percent of **Canada**'s species (over 2000 species) are non-native (Langor, 2019), many of which have the potential to become invasive pests. These include groups such as mites, which are poorly known but dominate in soils. Therefore, an unknown number of mite species may be or have the potential to become invasive species. Canada has many beetles whose larvae in the soil have greatly affected crops and horticulture: *Otiorynchus* root weevils (Black Vine Weevil, *Otiorhynchus sulcatus*; Strawberry root weevil, *Otiorhynchus ovatus*; Alfalfa Snout Beetle, *Otiorhynchus ligustici*) as well as other European weevils such as *Otiorhynchus*, *Polydrusus* and *Sitona* (Campbell *et al.*, 1989). There are also very costly *Elateridae* (Dusky Wireworm, *Agriotes obscurus*; Lined click beetle, *Agriotes lineatus*; *Agriotes sputator*); and scarab beetles (Japanese beetle, *Popillia japonica*; European Chafer, *Amphimallon majale*) (Klimaszewski *et al.*, 2012). All of these species probably arrived in Canada as soil larvae (some were intercepted as larvae in imported plants with soil). The impacts of these species on native biodiversity has not been studied.

For *Staphylinidae*, the impact of IAS is probably very high on soils in agricultural systems and forest fragments, but almost nothing has been demonstrated experimentally. It is likely that the top-down control of predators on other organisms is disrupted, which would have an impact on nutrient cycling and potentially on the physical characteristics of the soil. Many other rove beetles (many introduced species) feed directly on decaying organic matter as saprophages. There are at least 153 non-native species of rove beetles in Canada (Klimaszewski and Brunke, 2018).

There are at least 54 adventive species of Carabidae in Canada (Bousquet *et al.*, 2013), mainly of Palearctic origin (Klimawszeski *et al.*, 2012). Although they may dominate the carabid community in urban and agricultural landscapes and are likely to impact other predatory and non-predatory arthropods, little experimental evidence has shown significant effects.

As "ecosystem engineers", earthworms have a major impact on soil biodiversity in Canada (Cameron *et al.*, 2013). Canada has been almost entirely earthworm-free since the last glaciation; however, several earthworm species were introduced by European settlers. These are now widespread in Canada and continue to expand their range in earthworm-free regions (Eisenhauer *et al.*, 2007). While earthworms may have a transient beneficial impact on soil biodiversity, ultimately biodiversity decreases in areas where the soil fauna has not co-evolved with earthworms (Migge-Kleian *et al.*, 2006). By considerably accelerating litter decomposition and homogenizing the soil-litter layer, earthworms have been shown to completely eliminate the litter layer in some areas, affecting all aspects of soil community composition and ecological processes (Migge-Kleian *et al.*, 2006 and references therein).

5.1.6 | MONITORING SOIL BIODIVERSITY

In **Canada**, the Alberta Biodiversity Monitoring Institute (ABMI) (ABMI, 2014) conducts the only real long-term monitoring of soil biodiversity (particularly oribatid mites), along with other wildlife and plant monitoring activities. It has been run continuously since 2007, with sampling locations throughout the province of Alberta, in a wide variety of habitats with varying levels of disturbance and human activity. Canada has not yet implemented a soil information system; however it does have products such as *Acari* of Canada (Beaulieu *et al.*, 2019) and publications (revisions, diagnostic tools). In addition, Canada maintains the Canadian National Collection of Insects, Arachnids and Nematodes, with unpublished data held/associated with all 17 million specimens that can used/mined for assessing biodiversity, links with ecosystem services and the impacts of climate change. Specimen data is currently being compiled into a database, with many records available or soon to be available online. There is also a soil information system in Canada, but it does not include soil biodiversity as part of a national soil survey programme (Government of Canada, 2018).

The **United States of America** reported that the NSF NEON and the Earth Microbiome Project (both mentioned above) have monitoring activities related to soil biodiversity, but NEON is focused not just on soil biodiversity but on North American locations - that it, it is continental in scope. The Earth Microbiome Project is an international effort. The United States of America has developed an interagency strategic plan for microbiome research (MIWG, 2018) that includes overviews and needs for additional research, including on soil microbiomes and their relationships with other microbiomes and with environmental and human health.

5.1.7 | INDICATORS USED TO EVALUATE SOIL BIODIVERSITY

In **Canada**, Oribatid mites have been used as indicators for soil toxicity (Princz *et al.*, 2010; Princz *et al.*, 2012). With regard to genetic diversity, much research is currently being conducted on the use of metabarcoding as a means of rapidly obtaining measures of soil biodiversity across taxonomic groups (Schwarzfeld, forthcoming). These methods are still being validated but could eventually provide extremely useful indicators of soil health, soil biodiversity and soil genetic diversity. For soil fertility management, nematodes, Collembola and Acari have been used as indicators of soil health and fertility, but most often grouped in higher taxonomic categories rather than at the species level (Nesbitt and Adl, 2014).

For water management, water mites have strong potential as bioindicators of water quality (Smith *et al.*, 2010), but the links between water management and soil faunal diversity have yet to be studied. With regard to soil pollution, soil mite species (Princz *et al.*, 2010; Princz *et al.*, 2012) and mesofaunal communities (Battigelli, 2011; St. John *et al.*, 2002) have been used to assess the reclamation of contaminated soils. For biological control of pests and diseases, the diversity of soil predatory mites and insects represents potential indicators for the biocontrol of plant pests that spend time in the soil. However, these indicators still need to be developed further.

In the **United States of America**, scientific teams from the USDA Agricultural Research Service are investigating the topics listed above in more than 100 different projects conducted by more than 400 different scientists in more than 50 laboratories located around the United States (USDA, 2019a; USDA, 2019b; USDA, 2019c; USDA, 2019d).

5.2 | RESEARCH, CAPACITY DEVELOPMENT AND AWARENESS RAISING

Canada reported that the maintenance of AAFC's (Agriculture and Agri-Food Canada) extensive collections and taxonomic research performed at AAFC definitively supports the development and implementation of sustainable soil management practices. Taxonomy, which may sometimes appear to be an "old-fashioned way" of doing biology, will be critical in future soil biodiversity research programmes. The lack of taxonomists in many fields is a real concern because taxonomists are the only experts that can rigorously collect and identify living organisms in the environment and raise concerns about the collapse of biodiversity (Government of Canada, 2020b; Fox and MacDonald, 2003). The lack of taxonomists (due to the absence of a national institution responsible for training taxonomists) and the small number of organic farmers across the country (1.8 percent) indicate that Canada, like many other countries, is far behind what it should be doing to implement the conservation and sustainable use of soil biodiversity.

Canada is promoting soil and land management as part of the country's agriculture practices (Government of Canada, 2020a). There are several AAFC-funded research projects that harness bacterial diversity for use in sustainable agriculture (Tambong, Xu and Bromfield, 2017; Yu *et al.*, 2014; Tian *et al.,* 2009). Under the current CAD 3 billion framework covering the period 2018-2023, the Canadian Agricultural Partnership (the Partnership), AAFC and the provinces and territories will build on the progress of the previous Framework by supporting various soil and water conservation initiatives. Through the Partnership, provinces and territories design and manage the delivery of cost-shared environmental and climate change programming, including stewardship programmes that increase farmers' awareness of on-farm environmental risks and their management (e.g., Environmental Farm Plans) and the adoption of practices and technologies that reduce these risks, including maintaining and improving soil health. Environment and climate change programming under the Partnership will also be the vehicle for the agriculture and agri-food sector's contributions to achieving the greenhouse gas reduction and climate adaptation commitments of the Pan-Canadian Framework on Clean Growth and Climate Change (PCF).

The **United States of America** has a growing interest in soil health, and universities and NGOs across the country are fostering commitment to soil health at the academic and local level (UCDAVIS, 2020; USDA, 2016). The USDA ARS conducts research in several of its area of work related to natural resources and sustainable agricultural systems. This research focuses on developing an understanding of soil ecosystems and then developing tools and practices that have led to enhanced soil management practices. The USDA NRCS has a number of programmes that promote these practices (USDA, 2020a; California Department of Food and Agriculture, 2020; South Dakota Soil Health Coalition, 2020; The Soil Health Institute, 2020; The National Association of Conservation Districts, 2020; Soil Health Institute, 2020; USDA, 2020c; USDA, 2020b; USDA, 2019).

5.3 | MAINSTREAMING: POLICIES, PROGRAMMES, REGULATIONS AND GOVERNMENTAL FRAMEWORKS

The United States of America does not have a specific policy on soil biodiversity, but has numerous programmes and frameworks, and there is extensive interest in the United States of America in examining and understanding the links between soil biodiversity and human health; for example, the Soil Health - Human Health meetings (Soil Health Institute, 2018), aspects of the One Health Initiative (One Health Initiative, 2020) and the Rodale Institute (Rodale Institute, 2018). There are also a number of research and public domain articles (The Atlantic, 2013; Brevik and Burgess, 2014; PEW, 2019; Rodale Institute, 2018) and support has been given to research in these areas (PND, 2019). Although there is no national policy, the USDA has in recent years emphasized the importance of soil biodiversity management through publications, programmes and other concerted efforts, such as the USDA Natural Resources Conservation Service (NRCS) focus on soil health. In addition, many states in the United States of America have established soil health initiatives or enacted legislation recognizing the current state of soil degradation and the need to manage and improve soil health.

5.4 | ANALYSIS OF THE MAIN GAPS, BARRIERS AND OPPORTUNITIES IN THE CONSERVATION AND SUSTAINABLE USE OF SOIL BIODIVERSITY

Canada reported a great need for additional taxonomists to document and describe Canada's fauna, especially for under-studied groups such as the soil mesofauna, before more biodiversity is lost due to invasive species, anthropogenic changes (such as landscape modification and climate change). In addition, ecologists and taxonomists need to work together to better correlate species community composition with ecosystem processes and patterns. To date, information on these patterns is scarce and many of the mechanisms are poorly studied.

Canada is moving towards greater consideration of soil biodiversity, although this is a relatively new area of direct focus. Canada is promoting practices (e.g. improved fertilizer management, adoption of no-till and improved crop rotations) that indirectly address soil biodiversity. Links between work on soil microbiomes and soil inventories are slowly being established through connections between bioinformatics teams and CanSIS. This will allow an effective start to inventory the part of the soil health equation that relates to the soil microbiome. In the meanwhile, the 'next generation' of soil health indicators (biological diversity, nutrient cycling capacity, disease suppressive capacity) are still missing to address future challenges and opportunities arising from soil/subsoil constraints, microbial metagenomics, inoculant technologies, biofertilizers and climate change adaption and mitigation.

Agriculture and Agrifood Canada is considering recommendations for future policies and programmes, including continuing to invest in soil research, development and technology to better understand soil ecosystem functions so that soil health can be measured and monitored and exploited to produce food and protect the environment more effectively, and investing in the development of a wider range of soil biological health indicators.

In the **United States of America**, the strategic plan for microbiome research provides some arrangements for the conservation and sustainable use of soil biodiversity (MIWG, 2018). In addition, the USDA ARS National Programs conduct research to solve agricultural and national resources problems in the United States of America (and for other nations). The Soil and Air National Program focuses on understanding soil biodiversity and developing tools for land managers and producers and others to understand and then manage soils for greater sustainability, productivity, resilience and ecosystem services. The NRSAS group (USDA, 2020a) also conducts complimentary research to address broader natural resources conservation and the sustainability of soil biodiversity.

In the USA, the barriers to implementing better soil biodiversity management lie in understanding the status, phenomena and dynamics of soil ecosystems, as well as their trends and stability, under different management systems and in the face of changing climatic conditions. Resources are needed to conduct extensive research to develop comprehensive soil censuses that provide an understanding of which organisms are present, their features (number, qualities, metabolic roles), their interactions and functioning in a three-dimensional physical realm (with relevance at different spatial and matrix scales) and over time (also at different important scales). These investigations should examine a multitude of sites to study the impacts of cropping systems, management systems, climate change conditions, other ecosystems (other than agricultural ecosystems) in order to generate an understanding of the censuses and their compartmentalized and holistic roles, but also to develop regionally relevant management approaches to support the optimization of soil biodiversity for sustainable and even regenerative ecosystem development and to increase ecosystem services for our world.

6 | SOUTH WEST PACIFIC

6.1 | ASSESSMENT OF SOIL BIODIVERSITY

6.1.1 | CONTRIBUTIONS OF SOIL BIODIVERSITY TO ECOSYSTEMS SERVICES

Most countries in this region reported that despite the relevance of soil biodiversity to the country, very little has been done to enhance it due to a lack of resources.

6.1.2 | NATIONAL ASSESSMENTS

Fiji reported the need to conduct a comprehensive assessment throughout the country to further improve Fiji's soil health.

Tonga reported the lack of national-scale programmes for soil biodiversity.

6.1.3 | PRACTICAL APPLICATIONS OF SOIL BIODIVERSITY

In ***Fiji***, a study on the influence of Mucuna on selected soil properties and dalo yield was done in 2013, which confirmed that soil health and crop yields can be improved through a Mucuna fallow cropping system. A further study will be published later next year which reconfirms these findings (Lal, Guinto and Smith, 2013). Another publication recently also showed similar results at the Tutu Training Centre (Pacific Farmers, 2018). The Fijian Government is promoting the use of Mucuna fallow cropping for building soil organic matter and improving soil physical, biological and chemical properties. Recently, the Ministry of Agriculture has established seed production nurseries of Mucuna, which will be officially launched and distributed to farmers for improving their soil health. In addition, the government has continuously supported the use of agroforestry system to improve the country degraded soils.

Kiribati reported applications related to soil health and soilborne diseases.

Tonga and ***Samoa*** reported that the main practical applications of soil biodiversity in the large volcanic islands of the Pacific are organic waste biodegradation/nutrient recycling and disease control.

6.1.4 | MAJOR PRACTICES NEGATIVELY IMPACTING SOIL BIODIVERSITY

In ***Fiji***, mono-cropping of taro, sugarcane and kava have many negative effects. Taveuni Island was the main producer of export taro, but unsustainable farming practices such as mass deforestation and mono-cropping has led to soil degradation and lower yields.

6.1.5 | MONITORING SOIL BIODIVERSITY

Fiji reported the need for capacity development and the implementation of tools to continuously monitor soil biodiversity. Soil surveys had been conducted previously with a focus on chemical soil fertility rather than soil biological properties.

6.1.6 | INDICATORS USED TO EVALUATE SOIL BIODIVERSITY

Studies conducted in the 1980s in central ***Fiji*** showed that the annual soil loss through soil erosion was 60 to 70 t/ha. Farmers in Taveuni, Fiji, have also seen the benefits of using biological farming to improve their soil health and crop yields.

6.2 | RESEARCH, CAPACITY DEVELOPMENT AND AWARENESS RAISING

In **Fiji**, the Ministry of Agriculture through its Land Use Unit promotes sustainable soil management. As a result, there is some awareness, but more needs to be done not only to accelerate awareness of soil degradation, but also to implement ways to improve degraded soils. A recently published document shows some of the problems of soil degradation and how soils could be improved (Pacific Farmers, 2018).

Kiribati and **Tonga** reported that the University of Queensland has an integrated crop management project underway in several countries, funded by the Australian Centre for International Agricultural Research (ACIAR). Regarding the adoption of sustainable soil management, few countries in the Pacific islands are working on it due to lack of resources to make this a priority.

6.3 | MAINSTREAMING: POLICIES, PROGRAMMES, REGULATIONS AND GOVERNMENTAL FRAMEWORKS

Fiji reported that the management of soil health and sustainable agricultural practices are continuously promoted across all land use programmes. The Ministry of Agriculture through its Land Use Unit promotes sustainable soil management in Fiji.

Kiribati reported that environment policies include linkages and collaboration regarding the conservation and sustainable use of soil biodiversity.

Tonga reported that it has been signatory to the Convention on Biological Diversity, which could lead to greater awareness of the conservation and sustainable use of soil biodiversity.

6.4 | ANALYSIS OF THE MAIN GAPS, BARRIERS AND OPPORTUNITIES IN THE CONSERVATION AND SUSTAINABLE USE OF SOIL BIODIVERSITY

In *Fiji*, the Ministry of Agriculture's Strategic Development Plan focuses on sustainable agricultural practices, including soil and water resource management. The Ministry of Agriculture is intending to set up experiments on incorporating an agro-forestry system for sustainable kava production. However, staff need to be trained to scientifically measure soil biodiversity.

7 | SUB-SAHARAN AFRICA (SSA)

7.1 | ASSESSMENT OF SOIL BIODIVERSITY

7.1.1 | CONTRIBUTIONS OF SOIL BIODIVERSITY TO ECOSYSTEMS SERVICES

Gabon gave priority to forests and environmental protection in its strategic orientation document which is clearly established in the Emerging Gabon Strategic Plan (in French, the Plan Stratégique Gabon Emergent - PSGE). All actions must be carried out within the framework of respect for the environment and the protection of forests, with an ambition to diversify the economy, which relies mainly on agriculture and the forestry sectors.

7.1.2 | NATIONAL ASSESSMENTS

Cameroon has assessed the consequences of shade management on the taxonomic patterns and functional diversity of termites (*Blattodea: Termitidae*) in cocoa agroforestry systems.

Madasgascar has reported that the LRI (*Le Laboratoire des RadioIsotopes*) is doing research on soil biodiversity and productivity approaches.

7.1.3 | PRACTICAL APPLICATIONS OF SOIL BIODIVERSITY

In ***Madagascar***, soil biodiversity has been supporting biological fixation of carbon and carbon and nutrient cycling regulation, according to the REDD+ report. There also is evidence that increased use of soil biodiversity has contributed positively to food production and nutrition in this country.

7.1.4 | MAJOR PRACTICES NEGATIVELY IMPACTING SOIL BIODIVERSITY

In ***Madagascar***, the most important threats are the overuse of chemical control mechanisms (e.g. pesticides, herbicides, fungicides, nematicides, insecticides, veterinary drugs such as antibiotics and growth promoters). Also cited were practices leading to soil and water degradation and forest clearing and habitat fragmentation, due to lack of knowledge and poverty.

7.1.5 | INVASIVE ALIEN SPECIES (IAS)

In ***Cameroon***, previous work showed that cocoa farmers of southern Cameroon (sub-Saharan Africa) are aware of the threat due to termites in their farms. The damages of termites on cocoa trees have been reported to be more serious in nurseries and old trees. The preferred plants parts by termites were trunks in heavy shaded systems while roots were the most preferred in poorly shaded systems. Most of the control methods relied mainly on chemicals and have been reported to lack of efficiency. The study of termites' diversity revealed that about 65 termite species are found in cocoa agroforests of Cameroon. It has been demonstrated that shaded systems shelter endemic and undescribed termite species. The study also showed that shade trees removal induces a loss of termite biodiversity in cocoa farms which may lead to emergence of invasive termite pests. About ten termite species have been identified as pests of cocoa with a high diversity in shaded systems. Their feeding behaviour analysis showed that some species like *Microcerotermes* spp. are mainly encountered on above parts of the crop while others like *Microtermes* spp. preferred the roots system of the crop.

7.1.6 | MONITORING SOIL BIODIVERSITY

With the support of UNCCD and FAO, **Cabo Verde** has developed several maps and data on soil organic carbon, land productivity, soil erosion and land use and land cover across the country, within the framework for the implementation of the Land Degradation Neutrality Progamme.

The sampling approach for national historical land use area change (2000-2015) is based on the systematic 1 km x 1 km grid sampling and has been designed and conducted using the high and medium resolution image repository available through Google Earth, Bing Maps and Earth Engine Explorer and Code Editor as a visual assessment exercise. The estimation of the areas corresponding to land-use and land-use changes categories in the framework of this systematic sampling approach (based on the visual assessment of the nodes of a 1 km x 1 km national grid) was based on assessments of area proportions.

In order to maximize the synergies with the United National Framework Convention on Climate Change (UNFCCC), Cabo Verde has opted to report its Land Degradation status harmonized with the IPCC 2006 Guidelines, using its six Land Use Classes (Forest, Cropland, Grassland, Otherland, Settlement and Wetland). In addition, Cabo Verde has a database at the national level, with information about land area with improved land cover, stable land cover, degraded land cover, improved soil organic carbon, stable soil organic carbon, degraded soil organic carbon, increasing productivity, stable productivity, and declining productivity.

7.1.7 | INDICATORS USED TO EVALUATE SOIL BIODIVERSITY

Cabo Verde has some indicators described on the National Report of UNCCD.

Gabon reported that indicators have been put in place as part of a survey on the areas of carbon sequestration.

Madagascar has indicators to assess soil biodiversity in relation to genetic diversity, soil carbon sequestration and soil fertility management.

7.2 | RESEARCH, CAPACITY DEVELOPMENT AND AWARENESS RAISING

In **Eswatini**, conservation agriculture is a major action to promote soil health, apart from the contour ploughing that was introduced many decades ago.

In **Gabon**, some research activities on soil biodiversity support the development and the implementation of sustainable practices and soil management.

Madagascar promotes the adoption of sustainable soil management to avoid impairing key soil functions by land users as per the World Soil Charter and the Voluntary Guidelines for Sustainable Soil Management (project ProSol).

In **Nigeria**, there were some capacity development programmes in the 1980s and 1990s through the Agricultural Development Programmes funded by the World Bank in all Nigerian states.

7.3 | MAINSTREAMING: POLICIES, PROGRAMMES, REGULATIONS AND GOVERNMENTAL FRAMEWORKS

Eswatini reported that the country is still dealing with land policy and unfortunately there are no specific policies related to soil biodiversity.

Gabon reported the Law *007/2014 du 1er aout 2014 relative à la protection de l'environnement en Republique Gabonaise* and the following policies and strategies: *le plan stratégique Gabon Emergent; le code de l'environnement; le code forestier; la loi portant développement de l'Agriculture durable, le Plan National d'Affectation des Terres, Plan National Climat, Plan d'Action sur la Diversité Biologique.* Gabon has adopted the national strategy and the national plan for the management of the human-wildlife conflict. The aim is to reconcile wildlife conservation with the food security strategy and the management of conflicts between agricultural activities and elephant protection. Gabon has put in place a land use plan for a better identification of agricultural areas in Gabon. Gabon has 13 national parks, representing nearly 11 percent of its 267,667 km^2 territorial area, in order to preserve biodiversity, a vital heritage for the planet.

7.4 | ANALYSIS OF THE MAIN GAPS, BARRIERS AND OPPORTUNITIES IN THE CONSERVATION AND SUSTAINABLE USE OF SOIL BIODIVERSITY

Eswatini reported that the issue of healthy soils is part of its agenda to deal with food security, but it remains not fully explored.

Gabon has reported that there is a general lack of understanding on how soil biodiversity influences ecosystem functions and services towards general biodiversity. The country has developed the following strategic plans and laws: *Le plan stratégique Gabon Emergent, Le code de l'environnement, Le code forestier, La loi portant développement de l'Agriculture durable.* At the level of agricultural planning, the administration in charge of the country's environmental policy and also the focal point to the Convention on Biological Diversity are strongly involved. This administration conducts environmental impact studies for any agricultural project and ensures that the commitments made by the country to preserve biodiversity are respected.

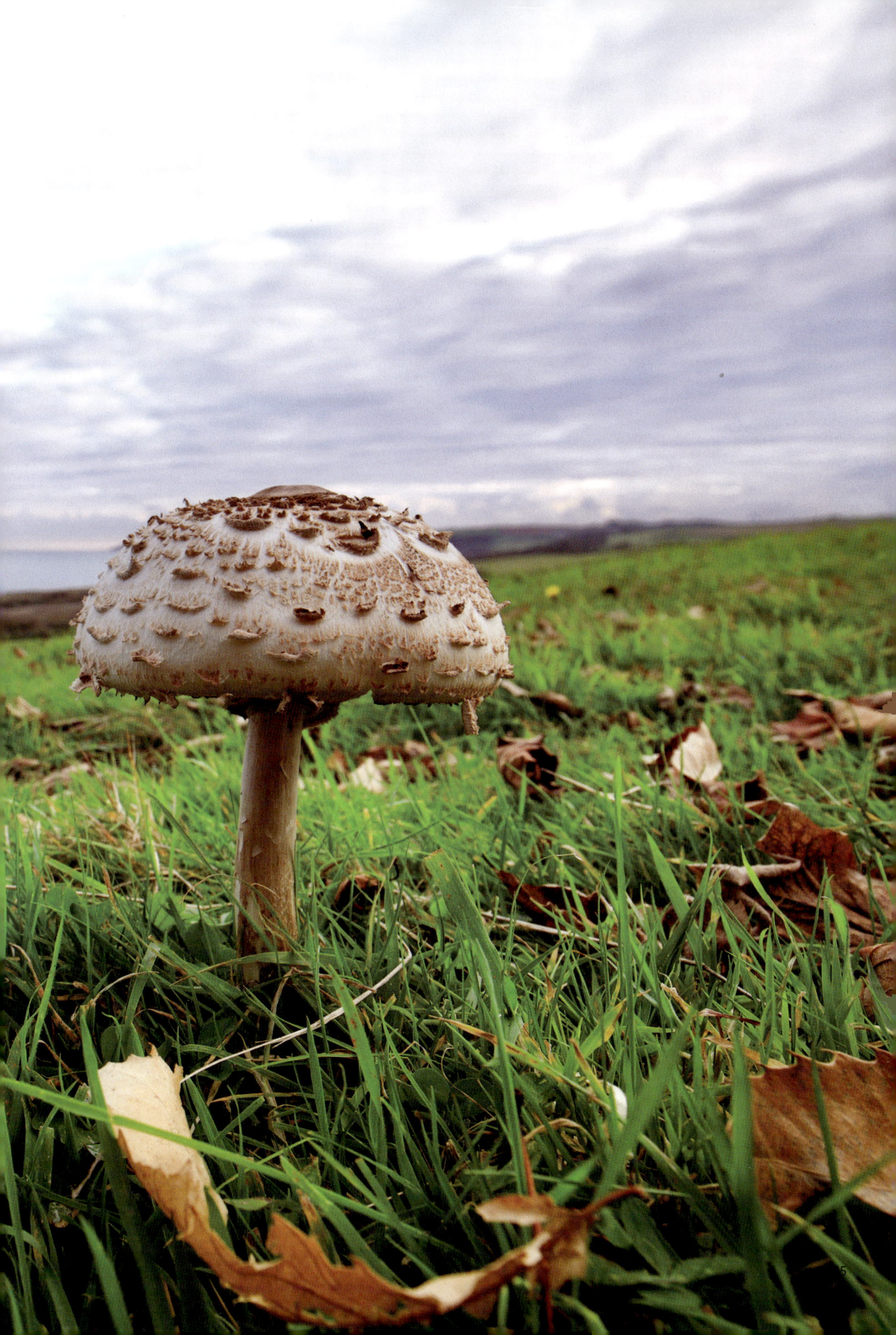

©FAO/Matteo Sala

ANNEX II
NATIONAL SURVEY ON STATUS OF SOIL BIODIVERSITY: KNOWLEDGE, CHALLENGES AND OPPORTUNITIES

Soils constitute one of the largest reservoirs of biodiversity on Earth and soil organisms are the source of key ecological functions and services that support agriculture, including soil conservation, water cycling, pest and diseases regulation, carbon sequestration and nitrogen fixation.

At the last UN Biodiversity Conference, held in Egypt in 2018, the Conference of the Parties (COP) to the Convention on Biological Diversity (CBD) invited the Food and Agriculture Organization of the United Nations (FAO) to prepare a report on the state of knowledge on soil biodiversity covering the current status, challenges and potentialities. Additionally, the COP requested the Secretariat of the CBD, in consultation with FAO under the aegis of the Global Soil Partnership (GSP) as well as other interested partners, to review the implementation of the International Initiative for the Conservation and Sustainable Use of Soil Biodiversity.

This survey is a first step in this process. The aim is to collect information at country level on the status of soil biodiversity, better understand concerns and threats to soil biodiversity, compile relevant policies, regulations or frameworks that have been implemented, and catalogue current soil biodiversity management and use efforts. The scope will be to make it available for consideration by the Subsidiary Body on Scientific, Technical and Technological Advice at a meeting held prior to CBD COP 15 (2020).

INSTRUCTIONS

This online survey consists of 16 questions and is divided into five sections: (I) General information; (II) Assessment; (III) Research, capacity building and awareness raising; (IV) Mainstreaming (policies, regulations and governmental frameworks); and (V) Gap analysis and opportunities.

Please note that in this survey, the term "COUNTRY" refers to the country for which you are answering the questions, not necessarily your country of origin.

IMPORTANT To be able to save what you have entered and edit the answers later on, you need to go to the end of the survey (PAGE 5) and SUBMIT the questionnaire. If you don't submit it, the responses will not be recorded. After submission, you will receive an email with the link to edit your survey, if necessary. Also, a copy of your responses will be emailed and you will have the possibility to print it.

DEADLINE Please complete the online survey, to the best of your ability and in line with the appropriate line Ministries and with relevant institutions at national level. Any follow up questions or comments can be directed to Ms. Monica Kobayashi (monica.kobayashi@fao.org) with copy to the Global Soil Partnership Secretariat (GSP-Secretariat@fao.org) by 8 September 2019.

Email address*:	

General Information

Full name (surname, first name) *:	
Country (for which you are answering the questions, not necessarily your country of origin)*:	
Ministry/Organization/University/Others (which you are representing):	

| ASSESSMENT

2.1 How do you perceive the importance of ecosystem services provided by soil biodiversity to this country over the past 10 years (1 being the lowest important and 5 being the highest). Please provide additional information to support the answers in the comments area below. Soil formation

- g. Soil formation
- h. Nutrient cycling
- i. Control soil erosion
- j. Atmospheric composition and climate regulation including floods
- k. Carbon sequestration and climate change responses
- l. Regulation of water supply and quality
- m. Pollutant degradation and soil remediation
- n. Recycling of waste biomass, nutrients and water
- o. Restoration of degraded lands and ecosystems
- p. Nitrogen fixation
- q. Pest and disease regulation
- r. Landscape support (effects on above-ground biodiversity in agricultural landscapes)
- s. Enhance agricultural productivity and economic profitability
- t. Enhance nutritional quality of food
- u. Natural hazard regulation
- v. Human health
- w. None
- x. Unknown
- y. Others (please specify):

Additional comments: Please link any relevant publications, documents or websites used in compiling these data.

2.2 What are the main practical applications of soil biodiversity (for instance, biological nitrogen fixation, monitoring the antimicrobials residues and/or antimicrobial resistance, soilborne disease control) in this country? Please attach any relevant publications, documents or websites used in compiling these data.

2.3 Is there any available evidence that the enhanced use of soil biodiversity has contributed positively to food production and nutrition in this country? Please link or attach any relevant publications, documents or websites used in compiling these data.

2.4 Please select all assessments linked to soil biodiversity that this country has developed:

    a. Comprehensive assessment of the status and trends
    b. Scientific knowledge
    c. Innovations and practices of farmers
    d. Indigenous and traditional knowledge
    e. Maps
    f. None
    g. Unknown
    h. Others (please specify):

Additional comments: Please link any relevant publications, documents or websites used in compiling these data.

2.5 Has this country developed and maintained a national soil information system including soil biodiversity information as part of a national soil survey programme? Please link any relevant publications, documents or websites used in compiling these data.

2.6 Does this country have monitoring activities related to soil biodiversity? Where possible provide information on the components of soil biodiversity that are monitored (monitoring programme name, monitoring activities, description, date of initiation, objectives, scale, indicators, sampling scheme, frequency of sampling, scientific name of organisms/species/varieties monitored, lessons learned, gaps, trends, priorities)

Additional comments: Please link any relevant publications, documents or websites used in compiling these data.

2.7 Are there any implemented indicators for evaluating soil biodiversity related to:

    a. Ecotoxicology
    b. Genetic diversity
    c. Soil carbon sequestration
    d. Soil erosion
    e. Soil fertility management
    f. Water management
    g. Soil pollution
    h. Biological control of pests and diseases

i. Antimicrobial resistance

j. None

k. Unknown

l. Others (please specify):

Additional comments: Please link any relevant publications, documents or websites used in compiling these data.

2.8 What are the major practices in this country that might have negatively impacted soil biodiversity in the last 10 years (1 being the lowest impact and 5 being the highest)? Please specify in the comments area below how they are measured, if applicable.

a. Over-use of chemical control mechanisms (for example, disease control agents, pesticides, herbicides, veterinary drugs)

b. Monoculture

c. Inappropriate soil management

d. Inappropriate water management

e. Practices leading to soil and water degradation

f. Over-grazing

g. Over-exploitation

h. Changes in land use

i. Forest clearing and habitat fragmentation

j. Over-use of fertilizers or other external inputs

k. None

l. Unknown

m. Others (please specify):

Additional comments: Please link any relevant publications, documents or websites used in compiling these data.

2.9 List any soil related invasive alien species (IAS) or soilborne disease that have had a significant effect on biodiversity in this country (scientific name of the IAS, biome and ecosystem services affected, production systems involved, how the IAS was introduced and when, measurement/monitoring procedure)

Additional comments: Please link any relevant publications, documents or websites used in compiling these data.

II | RESEARCH, CAPACITY DEVELOPMENT AND AWARENESS RAISING

3.1 Please describe any soil biodiversity research programmes or initiatives that exist in this country to support the development and implementation of sustainable soil management practices. Please link any relevant publications, documents or websites used in compiling these data.

3.2 Has this country implemented capacity development, training, extension or interdisciplinary educational programmes or awareness raising activities that target the conservation and sustainable use of soil biodiversity? Please link any relevant publications, documents or websites used in compiling these data.

3.3 Does this country promote the adoption of sustainable soil management to avoid impairing key soil functions by land users as per the World Soil Charter and the Voluntary Guidelines for Sustainable Soil Management (For more details: http://www.fao.org/3/a-i4965e.pdf and http://www.fao.org/3/a-bl813e.pdf)? Please link any relevant publications, documents or websites used in compiling these data.

III | MAINSTREAMING: POLICIES, REGULATIONS AND GOVERNMENTAL FRAMEWORKS

4.1 List any policies, programmes, regulation, enabling frameworks or any legal instrument that contain reference to soil biodiversity (name of the instrument, date of establishment, ministry, extent of implementation, links to global frameworks and agreements, indication of how exactly soil biodiversity is addressed, obstacles in developing or implementing the instrument, how they were identified, lessons learned)

Additional comments: Please link any relevant publications, documents or websites used in compiling these data.

4.2 Has this country integrated soil biodiversity into sectoral and/or cross-sectoral policies (such as those on health, food security, environment, forestry, agriculture, protected areas, land management, climate change, family farmers, Indigenous Peoples and Local Communities)? Please describe existing linkages and collaboration regarding the conservation and sustainable use of soil biodiversity.

IV | ANALYSIS OF GAPS AND OPPORTUNITIES

5.1 On a scale of 1-5 (1 being the lowest and 5 being the highest), how do you perceive the following options as barriers to implement a better soil biodiversity management strategies in this country?

    a. Lack of information and knowledge

    b. Lack of capacity and resource limitations

    c. Policy and institutional constraints

    d. Overly theoretical approach and lack of applicability

    e. Lack of research at national level

    f. None

    g. Unknown

    h. Others (please specify):

Additional comments: Please link any relevant publications, documents or websites used in compiling these data.

5.2 What arrangements are planned to ensure that the conservation and sustainable use of soil biodiversity are (directly or indirectly) taken into account in national planning and sectoral policy development (for example, National Biodiversity Strategies and Action Plan, national agricultural planning, national health planning)? Please describe the key priorities.

Additional comments: Please link any relevant publications, documents or websites used in compiling these data.

©FAO/Matteo Sala

ANNEX III
LIST OF COUNTRIES THAT RESPONDED TO THE SURVEY

- Algeria
- Antigua and Barbuda
- Azerbaijan
- Belgium
- Bhutan
- Brazil
- Cabo Verde
- Cameroon
- Canada
- Chile
- China
- Colombia
- Costa Rica
- Cuba
- Ecuador
- Estonia
- Eswatini
- Fiji
- Finland
- France
- Gabon
- Germany
- India
- Italy
- Jamaica
- Japan
- Kiribati
- Lebanon
- Lesotho
- Madagascar
- Mexico
- Nepal
- Netherlands
- Nicaragua
- Nigeria
- North Macedonia
- Oman
- Pakistan
- Panama
- Peru
- Portugal
- Republic of Moldova
- Spain
- Sri Lanka
- Sudan
- Suriname
- Syrian Arab Republic
- Thailand
- Tonga
- Trinidad and Tobago
- Tunisia
- Ukraine
- United Kingdom of Great Britain and Northern Ireland
- United States of America
- Uruguay
- Venezuela (Bolivarian Republic of)
- Zambia

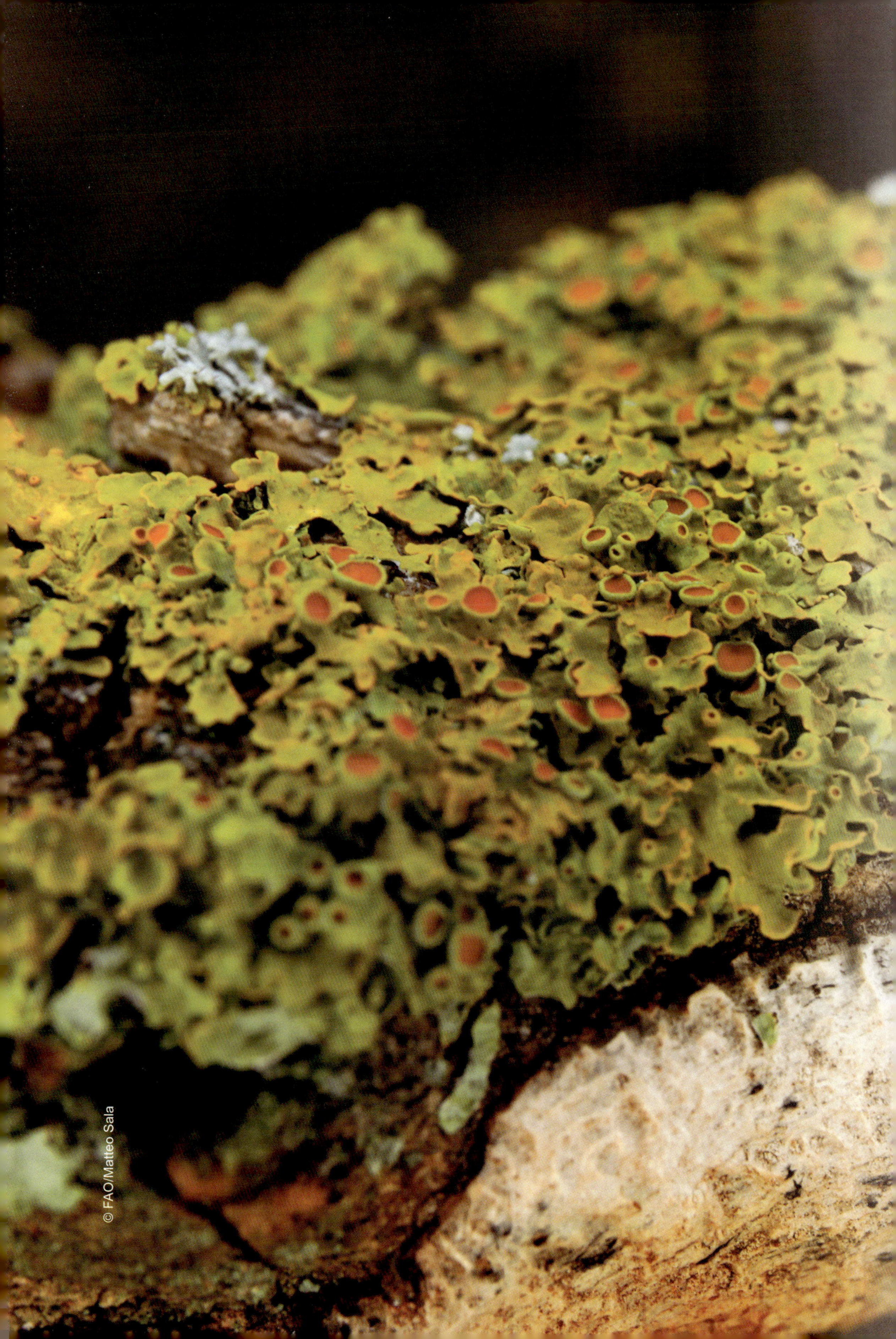

REFERENCES

A'Bear A.D., Johnson S.N. & Jones T.H. 2014. Putting the 'upstairs–downstairs' into ecosystem service: What can aboveground–belowground ecology tell us? *Biological Control*, 75: 97–107. doi: 10.1016/j.biocontrol.2013.10.004

A'Bear, A.D., Boddy, L., Raspotnig, G., & Jones, T.H. 2010. Non-trophic effects of oribatid mites on cord-forming basidiomycetes in soil microcosms. *Ecological Entomology*, 35: 477–484.

Abate, N.A., Wingfield, M.J., Slippers, B. & Hurley, B.P. 2017. Commercialisation of entomopathogenic nematodes: should import regulations be revised? *Biocontrol Science and Technology*, 27: 149-168.

Abatenh, E., Gizaw, B., Tsegaya, Z. & Wassie, M. 2017. Application of microorganisms in bioremediation-review. *Journal of Environmental Microbiology*, 1: 2–9.

Abba, A.M., Zufiaurre, E., Codesido, M. & Bilenca, D.N. 2015. Burrowing activity by armadillos in agroecosystems of central Argentina: Biogeography, land use, and rainfall effects. *Agriculture, Ecosystems and Environment*, 200: 54-56.

Abinandan, S., Subashchandrabose, S.R., Venkateswarlu, K. & Megharaj, M. 2019. Soil microalgae and cyanobacteria: the biotechnological potential in the maintenance of soil fertility and health. *Critical Reviews in Biotechnology*, 39(8): 981–998. https://doi.org/10.1080/07388551.2019.1654972

Abu-Bakar, N. & Ibrahim, N. 2013. Indigenous microorganisms production and the effect on composting process. *AIP Conference Proceedings*, 1571: 283; https://doi.org/10.1063/1.4858669

Aciego, J., Rojas, J., & Rivero, C. 1999. Effect of industrial residues (raw and composted) on soil biological properties. *Venesuelos* 7: 17- 26.

Aciego-Pietri, J. & Chacin, E. 2014. Propiedades biológicas de un suelo agrícola bajo uso intensivo de agroquímicos en Quibor, Venezuela. http://www.ucv.ve/uploads/media/Memorias_Jornadas_de_Investigaci%C3%B3n_2014.pdf

Ackerman, I. L., Teixeira, W. G., Riha, S. J., Lehmann, J., & Fernandes, E. C. M. 2007. The impact of mound-building termites on surface soil properties in a secondary forest of Central Amazonia. *Applied Soil Ecology*, 37: 267–276.

Addiscott, T. M. 2011. Emergence or self-organization?: Look to the soil population. *Communicative and integrative biology*, 4(4): 469–470. DOI:10.4161/cib.4.4.15547.

Adhikari, B. & Nadella, K. 2011. Ecological economics of soil erosion: A review of the current state of knowledge. *Annals of the New York Academy of Sciences*, 1219(1): 134-152.

Adhikari, K. & Hartemink, A.E. 2016. Linking soils to ecosystem services – a global review. *Geoderma*, 262: 101-111.

Agamennone, V., van Straalen, J., Brouwer, A., de Boer, T.E., Hensbergen, P.J., Zaagman, N., Braster, M., van Straalen, N.M., Roelofs, D. & Janssens, T.K.S. 2019. Genome annotation and antimicrobial properties of Bacillus toyonensis VU-DES13, isolated from the Folsomia candida gut. *Entomologia Experimentalis et Applicata*, 167: 269–285.

Agapito-Tenfen, S. Z., Okoli, A.S., Bernstein, M.J., Wikmark, O.G. & Myhr, A.I. 2018. Revisiting risk governance of GM plants: The need to consider new and emerging gene-editing techniques. *Frontiers in Plant Science*, 871: 1–16.

Agence nationale de sécurité sanitaire de l'alimentation, de l'environnement et du travail – Anses. 2020. [online]. [Cited 2 October 2020]. https://www.anses.fr/fr

Agency Moldsilva. 2020. Legislative and normative acts. [online]. [Cited 2 October 2020]. http://moldsilva.gov.md/pageview.php?l=ro&idc=262&t=/Legislatia/Acte-legislative-si-normative&

Aggarwal, A., Kadian, N., Tanwar, A., Yadav, A. & Gupta, K. K. 2011. Role of arbuscular mycorrhizal fungi (AMF) in global sustainable development. *Journal of Applied and Natural Science*, 3 (2): 340–351.

Ågren, G.I., Wetterstedt, J.M. & Billberger, M.F. 2012. Nutrient limitation on terrestrial plant growth–modeling the interaction between nitrogen and phosphorus. *New Phytologist*, 194, 953-960.

Agriculture and Horticulture Development Board – AHDB. 2017. [online]. [Cited 2 October 2020]. https://cereals.ahdb.org.uk/publications/2017/august/14/soil-biology-and-soil-health-partnership.aspx

Agronet. 2020. Peltomaan laatutesti. Agronet.fi [online]. [Cited 2 October 2020]. https://www.proagria.fi/www/peltomaan_laatutesti/index.php

Agrosavia. 2017. Balance Social 2017. [online]. [Cited 5 October 2020]. https://repository.agrosavia.co/bitstream/handle/20.500.12324/12011/110038_67755.pdf?sequence=1&isAllowed=y

Agrosavia. 2018. Balance Social 2018. [online]. [Cited 5 October 2020]. https://www.agrosavia.co/media/3123/balance-social-2018.pdf

Akhter, K., Ghous, T., Andleeb, S., Nasim, F.H., Ejaz, S., Abdin, Z., Khan, B.A. & Ahmed, M.N. 2017. Bioaccumulation of heavy metals by metal-resistant bacteria isolated from Tagetes minuta rhizosphere, growing in soil adjoining automobile workshops. *Pak. J. Zool*, 49(5): 1841-1846.

Aksoy, E., Louwagie, G., Gardi, C., Gregor, M., Schröder, C. & Löhnertz, M. 2017. Assessing soil biodiversity potentials in Europe. *Science of the Total Environment*, 589: 236-249.

Akiyama, H., Hoshino, Y.T., Itakura, M., Shimomura, Y., Wang, Y., Yamamoto, A., Tago, K., Nakajima, Y., Minamisawa, K. & Hayatsu, M. 2016. Mitigation of soil N_2O emission by inoculation with a mixed culture of indigenous Bradyrhizobium diazoefficiens. Scientific Reports, 6(1): 32869. https://doi.org/10.1038/srep32869

Al-Karaki, G. 2006. Nursery inoculation of tomato with arbuscular mycorrhizal fungi and subsequent performance under irrigation with saline water. *Scientia Horticulturae*, 109 (1): 1-7

Alakukku, L.E., Yli-Halla, M.J., Äijö, H., Mattila, T. & Peltonen, S., eds. 2017. Peltojen kunnostus. ProAgria maaseutukeskusten liitto. (also available at https://researchportal.helsinki.fi/en/publications/peltojen-kunnostus).

Alam, S. & Seth, R.K. 2014. Comparative Study on Effect of Chemical and BioFertilizer on Growth, Development and Yield Production of Paddy crop (Oryza sativa). Bhargava Agricultural Botany Laboratory, Department of Botany, University of Allahabad-211002 U.P. (India) International Journal of Science and Research (IJSR) ISSN (Online): 2319-7064. http://www.ijsr.net/

Alarcón, A., Davies Jr., F.T., Autenrieth, R.L., & Zuberer, D.A. 2008. Arbuscular Mycorrhiza and Petroleum-Degrading Microorganisms Enhance Phytoremediation of Petroleum-Contaminated Soil. *International Journal of Phytoremediation*, 10 (4): 251-263

Alberta Biodiversity Monitoring Institute-ABMI. 2014. The ABMI is a Leader in Biodiversity Monitoring. [online]. [Cited 5 October 2020]. https://abmi.ca/home.html

Aleklett, K. & Hart, M. 2013. The root microbiota—a fingerprint in the soil? *Plant Soil*, 370: 671–686.

Alele, P. O., Sheil, D., Surget-Groba, Y., Lingling, S. & Cannon, C. H. 2014. How does conversion of natural tropical rainforest ecosystems affect soil bacterial and fungal communities in the Nile river watershed of Uganda?. *PloS one*, 9(8): e104818.

Alig, R.J., Plantinga, A.J., Ahn, S. & Kline, J.D. 2003. *Land use changes involving forestry in the United States: 1952 to 1997, with projections to 2050.* Gen. Tech. Rep. PNW-GTR-587. Portland, OR. U.S. Department of Agriculture, Forest Service, Pacific Northwest Research Station.

Allen, D. C., Cardinale, B. J. & Wynn-Thompson, T. 2016. Plant biodiversity effects in reducing fluvial erosion are limited to low species richness. *Ecology*, 97: 17–24.

Alloway, B. J. 2008. *Zinc in Soils and Crop Nutrition.* International Zinc Association and International Fertilizer Industry Association, Brussels, Belgium, and Paris, France.

https://www.fertilizer.org/images/Library_Downloads/2008_IZA_IFA_ZincInSoils.pdf

Alori, E.T. & Babalola, O.O. 2018. Microbial inoculants for improving crop quality and human health in Africa. *Frontiers in Microbiology,* 9: 2213. doi: 10.3389/fmicb.2018.02213

Alvarez, R. 2005. A review of nitrogen fertilizer and conservation tillage effects on soil organic carbon storage. *Soil Use and Management*, 21: 38-52.

Aminov, R.I. 2010. A brief history of the antibiotic era: lessons learned and challenges for the future. *Frontiers in microbiology*, 1: 134. doi: 10.3389/fmicb.2010.00134

Amossé, J., Bettarel, Y., Bouvier, C., Bouvier, T., Tran Duc, T., Doan Thu, T. & Jouquet, P. 2013. The flows of nitrogen, bacteria and viruses from the soil to water compartments are influenced by earthworm activity and organic fertilization (compost vs. vermicompost). *Soil Biology and Biochemistry,* 66: 197–203.

Ampt, E.A., van Ruijven, J., Raaijmakers, J.M., Termorshuizen, A.J. & Mommer, L. 2018. Linking ecology and plant pathology to unravel the importance of soil-borne fungal pathogens in species-rich grasslands. *European Journal of Plant Pathology,* 154: 141-156.

Amundson, R., Guo, Y. & Gong, P. 2003. Soil diversity and land use in the United States. *Ecosystems,* 6: 470–482. https://doi.org/10.1007/s10021-002-0160-2.

AnaEE. 2020. AnaEE France - Accueil. In: Analysis and Experimentation on Ecosystems - France [online]. [Cited 2 October 2020]. https://www.anaee-france.fr/

Andersen, T., Elser, J.J. & Hessen, D.O. 2004. Stoichiometry and population dynamics. *Ecology Letters,* 7(9): 884–900.

Anderson, J.A., Ellsworth, P.C., Faria, J.C., Head, G.P., Owen, M.D.K., Pilcher, C.D., Shelton, A.M. & Meissle, M. 2019. Genetically engineered crops: Importance of diversified integrated pest management for agricultural sustainability. *Frontiers in Bioengineering and Biotechnology* 7: 1–14.

Andre, H. M., Ducarme, X. & Lebrun, P. 2002. Soil biodiversity: myth, reality or conning? *Oikos,* 96(1): 3–24.

Andrés, P. 1999. Ecological risks of the use of sewage sludge as fertilizer in soil restoration: effects on the soil microarthropod populations. *Land Degradation and Development* 10(1): 67–77. doi: 10.1002/(SICI)1099-145X(199901/02)10:1<67::AID-LDR322>3.0.CO;2-H

Andriuzzi, W.S., Franco, A.L., Ankrom, K.E., Cui, S., de Tomasel, C.M., Guan, P., Gherardi, L.A., Sala, O.E. & Wall, D.H. 2020. Body size structure of soil fauna along geographic and temporal gradients of precipitation in grasslands. *Soil Biology and Biochemistry,* 140: 107638.

ANPE Perú. 2020. [online]. [Cited 2 October 2020]. https://www.anpeperu.org/quienes-somos/presentacion

Antigua Breaking News (ABN). 2019. Minister Samantha Marshall addressing opening of Drought and Risk Management Training. In: Antigua Breaking News [Online]. [Cited 2 October 2020]. https://268today.com/local-news/video-minister-samantha-marshall-addresses-opening-of-drought-and-risk-management-training/

Antle, J.M. & Stoorvogel, J.J., 2006. Predicting the Supply of Ecosystem Services from Agriculture. *American Journal of Agricultural Economics,* 88: 1174-1180.

Arjona-Girona, I. & López-Herrera, C.J. 2018. Study of a new biocontrol fungal agent for avocado white root rot. Biological Control, 117: 6–12. https://doi.org/10.1016/j.biocontrol.2017.08.018

Armado, A., Contreras, F. García, P., & Paolini, J. 2009, Correlación de actividades enzimáticas con la respiración basal en sueloscacaoteros del occidente venezolano. https://www.academia.edu/4874028/Correlaci%C3%B3n_de_actividades_enzim%C3%A1ticas_con_la_respiraci%C3%B3n_basal_en_suelos_cacaoteros_del_occidente_venezolano

Arana, M.D., Martinez, G.A., Oggero, A.J., Natale, E.S. & Morrone, J.J. 2017. Map and shapefile of the biogeographic provinces of Argentina. *Zootaxa*, 4341(3): 420-422.

Armesto, J.J., Arroyo, M.T.K & Hinojosa, L.F. 2007. Chapter 11: The Mediterranean Environment of Central Chile. In Veblen, T., Young, K. & Orme, A., eds. *The Physical Geography of South America*. Oxford, U.K., Oxford University Press.

Armstrong, K.F. & Ball, S.L. 2005. DNA barcodes for biosecurity: Invasive species identification. *Philosophical Transactions of the Royal Society B: Biological Sciences* 360: 1813–1823.

Aronson, M.F., Nilon, C.H., Lepczyk, C.A., Parker, T.S., Warren, P.S., Cilliers, S.S., *et al.* 2016. Hierarchical filters determine community assembly of urban species pools. *Ecology*, 97: 2952-2963.

Arribas, P., Andújar, C., Hopkins, K., Shepherd, M. & Vogler, A.P. 2016. Metabarcoding and mitochondrial metagenomics of endogean arthropods to unveil the mesofauna of the soil. *Methods in Ecology and Evolution*, 7: 1071-1081.

Arthur, K.E., Kelez, S., Larsen, T., Choy, C.A. & Popp, B.N. 2014. Tracing the biosynthetic source of essential amino acids in marine turtles using $\delta 13C$ fingerprints. *Ecology*, 95: 1285-1293.

Aseman-Bashiz E., Asgharnia, H., Akbari, H., Iranshahi, L. & Mostafaii G. 2014. Bioremediation of the Soils Contaminated with Cadmium and Chromium, by the Earthworm Eisenia fetida. *Anuário do Instituto de Geociências* UFRJ. 37(2): 216-222. 10.11137/2014_2_216_222

Ashelford, K.E., Day, M.J. & Fry, J.C. 2003. Elevated abundance of bacteriophage infecting bacteria in soil. *Applied and Environmental Microbiology*, 69: 285–289.

Ashton, L.A., Griffiths, H.M., Parr, C.L., Evans, T.A., Didham, R.K., Hasan F., Teh, Y.A., Tin, H.S., Vairappan, C.S., & Eggleton, P. 2019. Termites Mitigate the Effects of Drought in Tropical Rainforest. *Science*, 363: 174–77.

Askary, T.H. & Abd-Elgawad, M.M.M. 2017. Beneficial nematodes in agroecosystems: a global perspective. In Abd-Elgawad, M.M.M., Askary, T.H. & Coupland, J. eds. *Biocontrol agents: entomopathogenic and slug parasitic nematodes*, pp. 3-25. CAB International, Wallingford, UK.

Asmelash, F., Bekele, T., & Birhane, E. 2016. The Potential Role of Arbuscular Mycorrhizal Fungi in the Restoration of Degraded Lands. *Frontiers in Microbiology*, 7: 1095

Atkinson, G., Bateman, I. & Mourato, S. 2012. Recent advances in the valuation of ecosystem services and biodiversity. *Oxford Review of Economic Policy*, 28: 22-47.

Atlas, R.M. & Philp, J. 2005. *Bioremediation: Applied Microbial Solutions for Real-World Environmental Cleanup.* ASM Press, Washington, D.C.

Atlas of Soil Suitability of Ukraine for Organic Farming (Zonal Aspects). 2015. Kharki, 2015–36 p.

Attanandana, T., Yost, R. & Verapattananirund, P. 2007. Empowering farmer leaders to acquire and practice site-specific nutrient management technology. *Journal of Sustainable Agriculture,* 30(1): doi: 10.1300/J064v30n01_08.

Attwood, G. T., Wakelin, S.A., Leahy, S.C., Rowe, S., Clarke, S., Chapman, D.F., Muirhead, R. & Jacobs, J.M.E. 2019. Applications of the Soil, Plant and Rumen Microbiomes in Pastoral Agriculture. *Frontiers in Nutrition,* 6:1–17.

Augé, R.M., Stodola, A.J.W, Tims J.E. & Saxton, A.M. 2001. Moisture retention properties of a mycorrhizal soil. *Plant Soil,* 230: 87-97.

Australian Academy of Science 2019. *Investigations of the causes of mass fish kills in the Menindee Region NSW over the summer of 2018-2019.* https://www.science.org.au/supporting-science/science-policy-and-sector-analysis/reports-and-publications/fish-kills-report

Australian Bureau of Statistics. 2016. *Land management and farming in Australia.* ABS, Canberra. https://www.abs.gov.au/ausstats/abs@.nsf/mf/4627.0

Australian Government 2017. *National Landcare Program: Twenty Million trees Program.* http://www.nrm.gov.au/national/20-million-trees

Averill, C., Turner, B.L. & Finzi, A.C. 2014. Mycorrhiza-mediated competition between plants and decomposers drives soil carbon storage. *Nature,* 505(7484): 543.

Ayers, R.S. & Westcot, W.D. 1976. *Water quality for agriculture.* FAO, Rome, Italy.

Azcón-Aguilar, C. & Barea, J.M. 2015. Nutrient cycling in the mycorrhizosphere. *Journal of soil science and plant nutrition,* 15(2): 372-396.

Azevedo, L.H, Ferreira, M.P., Castilho, R. de C., Cancado, P.H.D. & Moraes, G. J. de. 2018. Potential of Macrocheles species (Acari: Mesostigmata: Macrochelidae) as control agents of harmful flies (Diptera) and biology of Macrocheles embersoni Azevedo, Castilho and Berto on Stomoxys calcitrans (L.) and Musca domestica L. (Diptera: Muscidae). *Biological Control,* 123: 1-28.

Azziz, G., Giménez, M., Romero, H., Valdespino-Castillo, P.M., Falcón, L.I., Ruberto, L.A.M., MacCormack, W.P. & Batista, S. 2019. Detection of presumed genes encoding beta-lactamases by sequence based screening of metagenomes derived from Antarctic microbial mats. *Frontiers of Environment Science & Engineering,* 13: article 44. https://doi.org/10.1007/s11783-019-1128-1

Badri, D.V. & Vivanco, J.M. 2009. Regulation and function of root exudates. *Plant Cell and Environment,* 32(6): 666-681.

Baer, S. G., Bach, E.M., Meyer, C.K., Du Preez, C.C. & Six, J. 2015. Belowground Ecosystem Recovery During Grassland Restoration: South African Highveld Compared to US Tallgrass Prairie. *Ecosystems,* 18: 390–403.

Baert, J. M., Eisenhauer, N., Janssen, C.R. & De Laender, F. 2018. Biodiversity effects on ecosystem functioning respond unimodally to environmental stress. *Ecology Letters, 21(8)*: 1191-99

Bagyaraj, D.J., Thilagar, G., Ravisha, C., Kushalappa, C.G., Krishnamurthy, K.N. & Vaast, P. 2015. Below ground microbial diversity as influenced by coffee agroforestry systems in the Western Ghats, India. *Agriculture, Ecosystems and Environment*, 202: 198-202.

Bahram, M., Hildebrand, F., Forslund, S.K., Anderson, J.L., Soudzilovskaia, N.A, Bodegom, P.M., Bengtsson-Palme, J., et. al. 2018. Structure and function of the global topsoil microbiome. *Nature*, 560: 233-237. https://doi.org/10.1038/s41586-018-0386-6

Bai, Y., Wu, J., Clark, C.M., Naeem, S., Pan, Q., Huang, J., Zhang, L. & Han, X. 2010. Tradeoffs and thresholds in the effects of nitrogen addition on biodiversity and ecosystem functioning: evidence from inner Mongolia Grasslands. Global Change Biology, 16(1): 358–372. https://doi.org/10.1111/j.1365-2486.2009.01950.x

Bai, E., Li, S., Xu, W., Li, W., Dai, W. & Jiang, P. 2013. A meta-analysis of experimental warming effects on terrestrial nitrogen pools and dynamics. *New Phytol*, 199: 441–451. doi:10.1111/nph.12252

Bailey, V.L., Hicks Pries, C. & Lajtha, K. 2019. What do we know about soil carbon destabilization? *Environmental Resarch Letter* 14: 083004. https://doi.org/10.1088/1748-9326/ab2c11

Bailey, A. P., Rehman, T., Park, J., Keatinge, J. D. H. & Tranter, R. B., 1999. Towards a method for the economic evaluation of environmental indicators for UK integrated arable farming systems. *Agriculture, Ecosystems and Environment*, 72: 145-158.

Bajwa, M., Josan, A., Hira, G. & Singh, N. 1986. Effect of sustained saline irrigation on soil salinity and crop yields. *Irrigation Science,* 7: 27-35.

Baliuk, S.A., Medvedev, V.V., Miroshnychenko, M.M. 2015. Kharkiv. National Soil Conservation Program of Ukraine, 2015. - 59 p.

Baliuk, S.A. 2015. Rational use of soil resources and reproduction of soil fertility: organizational, economic, environmental and regulatory aspects. Monograph ed. by Baliuk, AESU Corresponding member A.V.Kucher. – Kharkiv, 2015. – 357-367 p.

Balmford, A., and Bond, W. 2005. Trends in the state of nature and their implications for human well-being. *Ecology Letters*, 8(11): 1218-1234.

Balota, E. L., Kanashiro, M., Colozzi Filho, A., Andrade, D. S., & Dick, R. P. 2004. Soil enzyme activities under long-term tillage and crop rotation systems in subtropical agro-ecosystems. *Brazilian Journal of Microbiology*, 35(4), 300-306.

Baltic Sea Action Group. 2019a. Carbon Action - STN MULTA Research Consortium. In: Carbon Action [online]. [Cited 2 October 2020]. https://carbonaction.org/en-stn-multa/

Baltic Sea Action Group. 2019b. TWINWIN Project. In: Carbon Action [online]. [Cited 2 October 2020]. https://carbonaction.org/twinwin-project/

Bampa, F., Daly, K., Fealy, R., O'Sullivan, L., Schulte, R.P.O., Gutzler, C., Wall, D. & Creamer, R. E. 2016. *Soil Status and Protection.* Report n° 179, Environmental Protection Agency.

Bang, H. S., Lee, J.-H., Kwon, O.S., Na, Y.E., Jang, Y.S. & Kim, W.H. 2005. Effects of paracoprid dung beetles (Coleoptera: Scarabaeidae) on the growth of pasture herbage and on the underlying soil. *Applied Soil Ecology*, 29: 165–171.

Banwart S.A., Nikolaidis N.P., Zhu Y.G., Peacock C.L. & Sparks D.L. 2019. Soil Functions: Connecting Earth's Critical Zone. *Annual Review of Earth and Planetary Sciences,* 47: 333-359.

Banwart, S. 2011. Save our soils. *Nature,* 474: 151-152.

Bao, Z., Matsushita, Y., Morimoto, S., Hoshino, Y.T., Suzuki, C., Nagaoka, K., Takenaka, M., Murakami, H., Kuroyanagi, Y., Urashima, Y., Sekiguchi, H., Kushida, A., Toyota, K., Saito, M. & Tsushima, S. 2013. Decrease in fungal biodiversity along an available phosphorous gradient in arable Andosol soils in Japan. Canadian Journal of Microbiology, 59(6): 368–373. https://doi.org/10.1139/cjm-2012-0612

Barber, N. A. Lamagdeleine-Dent, K.A., Willand, J.E., Jones, H.P., and McCravy, K.W. 2017. Species and functional trait re-assembly of ground beetle communities in restored grasslands. *Biodiversity and Conservation,* 26: 3481–3498.

Bardgett, R.D. 2005. *The Biology of Soil: A Community and Ecosystem Approach.* Oxford University Press, New York.

Bardgett, R.D. 2016. *Earth Matters: How soil underlies civilization*. Oxford University Press, London, UK.

Bardgett, R.D. & Wardle, D.A. 2010. *Aboveground-belowground linkages. Biotic interactions, ecosystem processes and global change.* Oxford University Press, Oxford, UK.

Bardgett, R.D. & Van der Putten, W.H. 2014. Belowground biodiversity and ecosystem functioning. *Nature*, 515: 505–511. doi: 10.1038/nature13855

Bardgett, R.D., Manning, P., Morriën, E. & De Vries, F.T. 2013. Hierarchical responses of plant–soil interactions to climate change: consequences for the global carbon cycle. *Journal of Ecology*, 101(2): 334-343.

Bardgett, R.D., Mommer, L. & De Vries, F.T. 2014. Going underground: root traits as drivers of ecosystem processes. *Trends in Ecology and Evolution*, 29(12): 692-699.

Bargués Tobella, A., Reese, H., Almaw, A., Bayala, A., Malmer, A., Laudon, H. & Ilstedt, U. 2014. The effect of trees on preferential flow and soil infiltrability inan agroforestry parkland in semiarid Burkina Faso. *Water Resources Research,* 50: 3342–3354.

Barnard, J. 2000. *Oregon's monster mushroom is world's biggest living thing.* The Independent on Sunday 6 Aug 2000, p 8.

Bar-On, Y.M., Phillips, R. & Milo, R. 2018. The biomass distribution on Earth. *Proceedings of the National Academy of Sciences of the United States of America*, 115: 6506–6511. https://doi.org/10.1073/pnas.1711842115

Barrett, J.E., Virginia, R.A., Wall, D.H. & Adams, B.J. 2008. Decline in a dominant invertebrate species contributes to altered carbon cycling in a low-diversity soil ecosystem. *Global Change Biology*, 14: 1734-1744.

Barrios, E. 2007. Soil biota, ecosystem services and land productivity. *Ecological Economics*, 64(2): 269-285. doi:10.1016/j.ecolecon.2007.03.004.

Barros, E., Pashanasi, B., Constantino, R. & Lavelle, P. 2002. Effects of Land-Use System on the Soil Macrofauna in Western Brazilian Amazonia. *Biology and Fertility of Soils*, 35(5): 338–47. https://doi.org/10.1007/s00374-002-0479-z.

Barrios, E., Sileshi, G.W., Shepherd, K. & Sinclair, F. 2012. Agroforestry and soil health: Linking trees, soil biota and ecosystem services. In D.H Wall, R.D. Bardgett, V. Behan-Pelletier, J.E. Herrick, T.H. Jones, K. Ritz, J. Six, D.R. Strong, W. van der Putten, eds. *Soil Ecology and Ecosystem Services*, pp. 315-330. Oxford, UK, Oxford University Press

Barrios, E., Valencia, V., Jonsson, M., Brauman, A., Hairiah, K., Mortimer, P. & Okubo, S. 2018. Contribution of trees to the conservation of biodiversity and ecosystem services in agricultural landscapes. *International Journal of Biodiversity Science, Ecosystem Services and Management*, 14(1): 1-16.

Barrios, M. 2007. Soil biota, ecosystem services and land productivity. *Ecological Economics*, 64: 269-285.

Barrios-Garcia, M. & Ballari, S. 2008. Impact of wild boar (Sus scrofa) in its introduced and native range: a review. *Biological Invasions*, 14: 2283–2300

Barros, E., Grimaldi, M., Sarrazin, M., Chauvel, A., Mitja, D., Desjardins, T. & Lavelle, P. 2004. Soil Physical Degradation and Changes in Macrofaunal Communities in Central Amazon. *Applied Soil Ecology*, 26(2): 157–68. https://doi.org/10.1016/j.apsoil.2003.10.012.

Barry, K.E., van Ruijven, J., Mommer, L., Bai, Y., Beierkuhnlein, C., Buchmann, N., de Kroon, H., Ebeling, A., Eisenhauer, N. & Guimaraes-Steinicke, C. 2019. Limited evidence for spatial resource partitioning across temperate grassland biodiversity experiments. *Ecology*, 101(1): e02905.

Bart, S., Amossé, J., Lowe, C.N., Mougin, C., Péry, A.R.R. & Pelosi, C. 2018. Aporrectodea caliginosa, a relevant earthworm species for a posteriori pesticide risk assessment: current knowledge and recommendations for culture and experimental design. *Environmental Science and Pollution Research*, 25: 33867-33881.

Bart, S., Barraud, A., Amossé, J., Péry, A.R.R., Mougin, C. & Pelosi, C. 2019. Understanding the life cycle of the earthworm Aporrectodea caliginosa: new data in an energy-based model. *Pedobologia*, 77: article no 1500592.

Bartkowski, B. 2017. Are diverse ecosystems more valuable? Economic value of biodiversity as result of uncertainty and spatial interactions in ecosystem service provision. *Ecosystem Service*, 24: 50-57.

Bartkowski, B., Hansjürgens, B., Möckel, S. & Bartke, S. 2018. Institutional Economics of Agricultural Soil Ecosystem Services. *Sustainability*, 10: 2447.

Bartley, R., Bainbridge, Z.T., Lewis, S.E., Kroon, F.J., Wilkinson, S.N., Brodie, J.E. & Silburn, D.M. 2014. Relating sediment impacts on coral reefs to watershed sources, processes and management: A review. *Science of the Total Environment*, 468: 1138-1153.

Bastias, B.A., Huang, Z.Q., Blumfield, T., Xu, Z. & Cairney, J.W.G. 2006. Influence of repeated prescribed burning on the soil fungal community in an eastern Australian wet sclerophyll forest. *Soil Biology and Biochemistry*, 38: 3492-3501

Batra, L. & Manna, M. C. 1997. Dehydrogenase activity and microbial biomass carbon in salt affected soils of semiarid and arid regions. *Arid Land Research and Management*, 11: 295–303.

Baslam, M., Garmendia, I., & Goicoechea, N. 2011. Arbuscular Mycorrhizal Fungi (AMF) Improved Growth and Nutritional Quality of Greenhouse-Grown Lettuce. *J. Agric. Food Chem.*, 59: 5504–5515

Battigelli, J. P. 2011. Exploring the world beneath your feet – soil mesofauna as potential biological indicators of success in reclaimed soils. Proceedings - Tailings and Mine Waste 2011, Vancouver, BC.

Battigelli J.P., Spence, J.R., Langor, D.W., & Berch, S.M. 2004. Short-term impact of forest soil compaction and organic matter removal on soil mesofauna density and oribatid mite diversity. *Canadian Journal of Forest Research*, 34: 1136–1149.

Baumgärtner, S. & Strunz, S. 2014. The economic insurance value of ecosystem resilience. *Ecological Economics*, 101: 21-32.

Baveye, P.C., Baveye, J. & Gowdy, J. 2016. Soil "Ecosystem" Services and Natural Capital: Critical Appraisal of Research on Uncertain Ground. *Frontiers in Environmental Science*, 4.

Baxter, C., Rowan, J.S., McKenzie, B.M. & Neilson, R. 2013 Understanding soil erosion impacts in temperate agroecosystems: bridging the gap between geomorphology and soil ecology using nematodes as a model organism. *Biogeosciences*, 10: 7133-7145.

Beaulieu, F. & Weeks, A.R. 2007. Free-living mesostigmatic mites in Australia: their roles in biological control and bioindication. *Australian Journal of Experimental Agriculture*, 47: 460-478.

Beaulieu, F., Knee, W., Nowell, V., Schwarzfeld, M., Lindo, Z., Behan-Pelletier, V.M., Lumley, L. *et al.* 2019. Acari of Canada. *Zookeys*, 819: 77-168.

Bebber, D.P., Ramotowski, M.A.T. & Gurr, S.J. 2013. Crop pests and pathogens move polewards in a warming world. *Nature Climate Change*, 3: 985–988.

Behan-Pelletier, V. 1999. Oribatid mite biodiversity in agroecosystems: role as bioindicators. *Agriculture, Ecosystems and Environment,* 74: 411–423.

Behan-Pelletier, V. M. 2003. Acari and Collembola biodiversity in Canadian agricultural soils. *Canadian Journal of Soil Science,* 83:279-288.

Beketov, M.A. & Liess, M. 2012. Ecotoxicology and macroecology-time for integration. *Environmental Pollution,* 162: 247-254.

Bell, T.H., Yergeau, E., Maynard, C., Juck, D., Whyte, L.G. & Greer C.W. 2013. Predictable bacterial composition and hydrocarbon degradation in Arctic soils following diesel and nutrient disturbance. *The ISME journal,* 7: 1200–1210.

Bellard, C., Bertelsmeier, C., Leadley, P., Thuiller, W. & Courchamp, F. 2012 Impacts of climate change on the future of biodiversity. *Ecology Letters,* 15: 365-377.

Bender, S.F., Wagg, C. & van der Heijden, M.G. 2016. An underground revolution: biodiversity and soil ecological engineering for agricultural sustainability. *Trends in Eecology and Evolution,* 31(6): 440-452. doi: 10.1016/j.tree.2016.02.016

Benedetti, A., & Gianfreda, L. 2004. *Metodi di analisi Biochimica del suolo.* Franco Angeli Roma, Italia.

Bengtson, P., Barker, J. & Grayston, S.J. 2012. Evidence of a strong coupling between root exudation, C and N availability, and stimulated SOM decomposition caused by rhizosphere priming effects. Ecology and Evolution 2(8): 1843-1852.

Bengtsson-Palme, J., & Larsson, D.G.J. 2015. Antibiotic resistance genes in the environment: Prioritizing risks. Nature Reviews Microbiology 13:396.

Benintende, S. 2010. Calidad de inoculantes comerciales para el cultivo de soja en la Argentina: concentración de rizobios viables y presencia de contaminantes (Quality of commercial inoculants for soybean crop in Argentina: concentration of viable rhizobia and presence of contaminants). Revista Argentina de Microbiología 42: 129-132 Conab (Companhia Nacional de Abastecimento), 2019- https://www.conab.gov.br/info-agro/safras/graos, accessed October, 10th 2019.

Beone, G.M., Cenci, R.M., Guidotti, L., Sena, F. & Umlauf, G. 2015. Indagine conoscitiva della qualità e dello stato di salute dei suoli lombardi (Progetto Soil). JRC Technical Reports. ISBN: 978-92-79-47101-8.

Berendse, F., van Ruijven, J., Jongejans, E. & Keesstra, S. D. 2015. Loss of plant species diversity reduces soil erosion resistance of embankments that are crucial for the safety of human societies in low-lying areas. *Ecosystems,* 18, 881–888.

Berg, G. & Smalla, K. 2009. Plant species and soil type cooperatively shape the structure and function of microbial communities in the rhizosphere. *Fems Microbiology Ecology* 68(1): 1-13.

Berg, G., Eberl, L. & Hartmann, A. 2005. Minireview The rhizosphere as a reservoir for opportunistic human pathogenic bacteria. *Environmental Microbiology,* 7:1673–1685.

Berg, N. W. & Pawluk, S. 1984. Soil mesofaunal studies under different vegetative regimes in north central Alberta. *Canadian Journal of Soil Science,* 64: 209-223

Bergstrom, D. M., Sharman, A., Shaw, J.D., Houghton, M., Janion-Scheepers, C., Achurch, H. & Terauds, A. 2018. Detection and eradication of a non-native Collembola incursion in a hydroponics facility in East Antarctica. *Biological Invasions*, 20:293–298.

Berhe, A.A., Barnes, R.T., Six, J. & Marín-Spiotta, E. 2018. Role of Soil Erosion in Biogeochemical Cycling of Essential Elements: Carbon, Nitrogen, and Phosphorus. *Annu. Rev. Earth Planet. Sci.,* 46, 521–548. doi:10.1146/annurev-earth-082517-010018

Bernard L., Chapuis-Lardy, L., Razafimbelo, T., Razafindrakoto, M., Pablo, A.L., Legname, E., Poulain, J., *et al.* 2012. Endogeic earthworms shape bacterial functional communities and affect organic matter mineralization in a tropical soil. *ISME J.,* 6 : 213–222.

Berninger, T., González López, O., Bejerano, A., Preininger, C. & Sessitsch, A. 2018. Maintenance and assessment of cell viability in formulation of non-sporulating bacterial inoculants. *Microb. Biotechnol.* 11:277 – 301.

Berruti, A., Lumini, E. & Bianciotto, V. 2017. AMF components from a microbial inoculum fail to colonize roots and lack soil persistence in an arable maize field. *Symbiosis*, 72: 73–80. doi: 10.1007/s13199-016-0442-7

Beter Bodembeheer. 2020. Soil Management research program. [online]. [Cited 2 October 2020]. https://www.beterbodembeheer.nl/nl/beterbodembeheer.htm

Bever, J.D., Platt, T.G. & Morton, E.R. 2012. Microbial population and community dynamics on plant roots and their feedbacks on plant communities. *Annual review of microbiology*, 66: 265.

Bever, J.D, Dickie, I.A., Facelli, E., Facelli, J.M., Klironomos, J., Moora, M., Rillig, M.C., Stock, W.D., Tibbett, M. & Zobel, M. 2010. Rooting theories of plant community ecology in microbial interactions. *Trends in Ecology and Evolution*, 25: 468–478.

Beylich, A., Oberholzer, H.; Schrader, S., Hoper, H. & Wilke, B. 2010. Evaluation of Soil Compaction Effects on Soil Biota and Soil Biological Processes in Soils. *Soil and Tillage Research,* 109(2): 133-143.

Bezemer, T.M. & van Dam, N.M. 2005. Linking aboveground and belowground interactions via induced plant defenses. *Trends in Ecology and Evolution,* 20: 617–624. doi: 10.1016/j.tree.2005.08.006 .

Bezemer, T.M., Fountain, M.T., Barea, J.M., *et al.* 2010. Divergent composition but similar function of soil food webs of individual plants: plant species and community effects. *Ecology*, 91: 3027–3036.

Bhadauria, T. 2016. Soil fauna as influenced by land use in Indo-gangetic plains. In K.G. Saxena & K.S. Rao, eds. *Soil Biodiversity: Inventory, Functions and Management,* pp. 263-284. Dehra Dun, India, Bishen Singh Mahendra Pal Singh.

Bhadauria, T., Kumar, P., Kumar, R., Maikhuri, R.K. & Saxena, K.G. 2012. Earthworm populations in a traditional village landscape in Central Himalaya, India. *Applied Soil Ecology*, 53: 83-93.

Bhakta, J.N., Ohnishi, K., Munekage, Y., Iwasaki, K. & Wei, M.Q. 2012. Characterization of lactic acid bacteria-based probiotics as potential heavy metal sorbents. *Journal of applied microbiology*, 112(6): 1193-120.

Bhatti, A.A., Haq, S. & Bhat, R.A. 2017. Actinomycetes benefaction role in soil and plant health. *Microbial Pathogenesis*, 111: 458-467.

Bhusal, D.R., Tsiafouli, M.A. & Sgardelis, S.P. 2015. Temperature-based bioclimatic parameters can predict nematode metabolic footprints. *Oecologia*, 179: 187-199. doi: 10.1007/s00442-015-3316-4

Bickel, S., Chen, X., Papritz, A. & Or, D. 2019. A hierarchy of environmental covariates control the global biogeography of soil bacterial richness. *Scientific Reports*, 9: 12129 | https://doi.org/10.1038/s41598-019-48571-w

Bin Jang, H., Bolduc, B., Zablocki, O., Kuhn, J.H., Roux, S., Adriaenssens, E.M., *et al.* 2019. Taxonomic assignment of uncultivated prokaryotic virus genomes is enabled by gene-sharing networks. *Nature Biotechnology*, 37: 632-639.

Bionautit. 2020. Maaperä. In: Bionautit. [online]. [Cited 2 October 2020]. http://www.bionautit.fi/maapera/

Bird, S.B., Coulson, R.N., & Fisher, R.F. 2004. Changes in soil and litter arthropod abundance following tree harvesting and site preparation in a loblolly pine (Pinus taeda L.) plantation. *Forest Ecology and Management*, 202: 195-208.

Bird, S.B., Herrick, J.E., Wander, M.M. & Wright, S.F. 2002. Spatial heterogeneity of aggregate stability and soil carbon in semi-arid rangeland. *Environmental Pollution*, 116: 445-455.

Birhane, E., Thomas, W., Frank, K., Stercka, J. & Bongers, F. 2010. Arbuscular mycorrhizal associations in Boswellia papyrifera (frankincense-tree) dominated dry deciduous woodlands of Northern Ethiopia. *Forest Ecology and Management*, 260: 2160–2169.

Birhane, E., Aregawi, K. & Giday, K. 2017. Changes in arbuscular mycorrhiza fungi spore density and root colonization of woody plants in response to exclosure age and slope position in the highlands of Tigray, Northern Ethiopia. *Journal of Arid Environments*, 142: 1-10 https://doi.org/10.1016/j.jaridenv.2017.03.002

Birnbaum, C., Hopkins, A.J.M, Fontaine, J.B. & Enright, N.J. 2019. Soil fungal responses to experimental warming and drying in a Mediterranean shrubland. *Science of the Total Environment*, 683: 524-536.

Birtel, J., Walser, J.C., Pichon, S., Bürgmann, H. & Matthews, B. 2015. Estimating bacterial diversity for ecological studies: methods, metrics, and assumptions. *PloS One*, 10(4): p.e0125356.

Birzele, L.T., Depner, M., Ege, M.J., Engel, M., Kublik, S., Bernau, C., Loss, *et al.* 2017. Environmental and mucosal microbiota and their role in childhood asthma. *Allergy: European Journal of Allergy and Clinical Immunology*, 72: 109–119.

Bissett, A., Fitzgerald, A., Meintjes, T., Mele, P.M., Reith, F., Dennis, P.G., Breed, M.F., *et al.* 2016. Introducing BASE: The Biomes of Australian Soil Environments soil microbial diversity database. *GigaScience*, 5: Article number: 21

Bittman, S., Sheppard, S.C., Poon, D. & Hunt, D.E. 2019. How efficient is modern peri-urban nitrogen cycling: A case study. *Journal of Environmental Management*, 244: 462–471. doi:https://doi.org/10.1016/j.jenvman.2019.05.054

Blackwell, M. 2011. The fungi: 1, 2, 3 ... 5.1 million species? *American Journal of Botany*, 98: 426–438.

Blackwell, M. & Vega, F.E. 2018. Lives within lives: Hidden fungal biodiversity and the importance of conservation. *Fungal Ecology*, 35: 127–134.

Blackwood C.B. & Buyer J.S. 2004. Soil microbial communities associated with Bt and non-Bt corn in three soils. *Journal of Environmental Quality*, 33(3): 832-6. DOI: 10.2134/jeq2004.0832.

Blair, N., Faulkner, R.D., Till, A.R., Körschens, M. & Schulz, E. 2006. Long-term management impacts on soil C, N and physical fertility: Part II: Bad Lauchstadt static and extreme FYM experiments. *Soil and Tillage Research*, 91: 39-47.

Blanchart, A., Séré, G., Cherel, J., Warot, G., Stas, M., Noël, C.J., Morel, J.L. & Schwartz, C. 2018. Towards an operational methodology to optimize ecosystem services provided by urban soils. *Landscape and Urban Planning*, 176: 1-9.

Blanchart, E., Ratsiatosika, O., Raveloson, H., Razafimbelo, T., Razafindrakoto, M., Sester, M., Becquer, T., Bernard, L. & Trap, J. 2019. Nitrogen supply reduces the earthworm-silicon control on rice blast disease in a Ferralsol. *Applied Soil Ecology*, 145:

Blanchart, E., Albrecht, A., Brown, G., Decaens, T., Duboisset, A., Lavelle, P., Mariani, L. & Roose, E. 2004. Effects of tropical endogeic earthworms on soil erosion. *Agriculture, Ecosystems & Environment*, 104: 303–315.

Blankinship, J.C., Niklaus, P.A. & Hungate, B.A. 2011. A meta-analysis of responses of soil biota to global change. *Oecologia*, 165(3): 553-565.

Bloem, J. & Breure, A.M. 2003. Chapter 8 Microbial indicators. Trace Metals and other Contaminants in the Environment, pp. 259–282. Elsevier. (also available at https://linkinghub.elsevier.com/retrieve/pii/S0927521503801388).

Bloem, J., Koopmans, C. & Schils, R. 2017. Effect van mest op de biologische bodemkwaliteit in de Zeeuwse akkerbouw. https://doi.org/10.18174/425171

Bloem, J., Schouten, A.J., Sørensen, S.J., Rutgers, M., Werf, A.K. van der & Breure, A.M. 2005. Monitoring and evaluating soil quality. Microbiological methods for assessing soil quality, pp. 23–49. CABI. (also available at https://research.wur.nl/en/publications/monitoring-and-evaluating-soil-quality)

Blouin, M., Hodson, M.E., Delgado, E.A., Baker, G., Brussaard, L., Butt, K.R., Dai, J., Dendooven, L., Pérès, G., Tondoh, J.E. & Cluzeau, D. 2013. A review of earthworm impact on soil function and ecosystem services. *European Journal of Soil Science*, 64(2): 161-182.

Blum, W.E., Zechmeister-Boltenstern, S., & Keiblinger, K.M. 2019. Does Soil Contribute to the Human Gut Microbiome?. *Microorganisms*, 7(9): 287.

Bobbink, R., Hicks, K., Galloway, J., Spranger, T., Alkemade, R., Ashmore, M. & De Vries, W. 2010. Global assessment of nitrogen deposition effects on terrestrial plant diversity: A synthesis. *Ecological Applications*, 20(1): 30–59.

Bodem Academie. 2020. Ecosystem services. [online]. [Cited 2 October 2020]. http://bodemacademie.nl/bodemecosysteemdiensten/

Bodemdierendagen. 2020. Benthic animals live in healthy soil. [online]. [Cited 2 October 2020]. https://bodemdierendagen.nl/

Boettcher, S.E. & Kalisz, P.J. 1991. Single-tree influence on earthworms in forest soils of Eastern Kentukcy. *Soil Science Society of America Journal*, 55(3): 862-865.

Bogdevich, O. & Senikovkaya, I. 2011. Assessment of soil microbiology by bioremediation of POPS contaminated soils. Paper presented at 11th International HCH and Pesticides Forum., 7 September 2011, Gabala, Azerbaijan.

Boivin, A. & Poulsen, V. 2017. Environmental risk assessment of pesticides: state of the art and prospective improvement from science. *Environmental Science and Pollution Research*, 24: 6889-6894.

Bolan, N.S., Adriano, D.C. & Curtin, D. 2003. Soil acidification and liming interactions with nutrient and heavy metal transformation and bioavailability. *Advances in Agronomy*, 78: 215–272.

Bolduc, B., Jang, H.B., Doulcier, G., You, Z.-Q., Roux, S. & Sullivan, M.B. 2017. vConTACT: an iVirus tool to classify double-stranded DNA viruses that infect Archaea and Bacteria. *PeerJ*, 5: e3243.

Bolourian, A. & Mojtahedi, Z. 2018. Streptomyces, shared microbiome member of soil and gut, as 'old friends' against colon cancer. *FEMS microbiology ecology*, 94(8): fiy120.

Bommarco, R., Kleijn, D. & Potts, S.G. 2013. Ecological intensification: harnessing ecosystem services for food security. *Trends in Ecology and Evolution*, 28: 230-238.

Bommarco, R., Miranda, F., Bylund, H. & Björkman, C. 2011. Insecticides Suppress Natural Enemies and Increase Pest Damage in Cabbage. *Journal of Economic Entomology*, 104: 782-791.

Bona, E., Cantamessa, S., Massas, N., Manassero, P., Marsano, F., Copetta, A., Lingua, G. *et al.* 2017. Arbuscular Mycorrhizal Fungi and Plant Growth-Promoting Pseudomonads Improve Yield, Quality and Nutritional Value of Tomato: A Field Study. *Mycorrhiza*, 27:1–11.

Bonachela, J. A., Pringle, R. M., Sheffer, E., Coverdale, T.C., Guyton, J.A., Caylor, K.K., Levin, S.A & Tarnita, C.E. 2015. Termite mounds can increase the robustness of dryland ecosystems to climatic change. *Science*, 347: 651–655.

BONARES. 2020. Initiative of the German Federal Ministry for Education and Research (BMBF). [online]. [Cited 2 October 2020]. www.bonares.de

Bongers, T. 1990. The maturity indiex: an ecological measure of environmental disturbance based on nematode species composition. *Oecologia*, 83: 14-19.

Borchhardt, N., Baum, C., Mikhailyuk, T. & Karsten, U. 2017. Biological Soil Crusts of Arctic Svalbard—Water Availability as Potential Controlling Factor for Microalgal Biodiversity. *Frontiers in Microbiology*, 8: 1485. https://doi.org/10.3389/fmicb.2017.01485

Borrelli, P., Van Oost, K., Meusburger, K., Alewell, C., Lugato, E., & Panagos, P. 2018. A step towards a holistic assessment of soil degradation in Europe: Coupling on-site erosion with sediment transfer and carbon fluxes. *Environmental research*, 161: 291-298.

Borrelli, P., Robinson, D.A., Fleischer, L.R., Lugato, E., Ballabio, C., Alewell, C. & Bagarello, V. 2017. An assessment of the global impact of 21st century land use change on soil erosion. *Nature communications*, 8(1): 2013.

Bossuyt, H., Six, J. & Hendrix, P.F. 2006. Interactive effects of functionally different earthworm species on aggregation and incorporation and decomposition of newly added residue carbon. *Geoderma*, 130: 14–25.

Bottinelli, N., Jouquet, P., Capowiez, Y., Podwojewski, P., Grimaldi, M. & Peng, X. 2015. Why is the influence of soil macrofauna on soil structure only considered by soil ecologists? *Soil and Tillage Research*, 146: 118–124.

Boudreau, M.A. 2013. Diseases in intercropping systems. *Annual Review of Phytopathology*, 51: 499–519. doi: 10.1146/annurev-phyto-082712-102246

Bourguibignon, T., Lo, N., Šobotník, J., Ho, S.Y.W., Iqbal, N., Coissac, E., Lee., M., Jendryka, M.M., *et al.* 2017. Mitochondrial Phylogenomics Resolves the Global Spread of Higher Termites, Ecosystem Engineers of the Tropics. *Molecular Biology and Evolution*, 34: 589-597.

Bousquet, Y., Bouchard, P., Davies, A.E., & Sykes, S.S. 2013. Checklist of beetles (Coleoptera) of Canada and Alaska. Second edition. *ZooKeys*, 360: 1-44

Bouwman, L., Goldewijk, K.K., Van Der Hoek, K.W., Beusen, A.H.W., Van Vuuren, D.P., Willems, J., Rufino, M.C., Stehfest, E. 2013. Exploring global changes in nitrogen and phosphorus cycles in agriculture induced by livestock production over the 1900–2050 period. *Proceedings of the National Acadamy of Sciences of the United States of America*, 110: 20882 LP – 20887. doi:10.1073/pnas.1012878108

Bowles, T.M., Jackson, L.E., Loeher, M. & Cavagnaro, T.R. 2017. Ecological intensification and arbuscular mycorrhizas: a meta-analysis of tillage and cover crop effects. *Journal of Applied Ecology*, 54: 1785–1793.

Bradáčová, K., Florea, A.S., Bar-Tal, A., Minz, D., Yermiyahu, U., Shawahna, R., Kraut-Cohen, *et al.* 2019. Microbial consortia versus single-strain inoculants: an advantage in PGPR-assisted tomato production? *Agronomy*, 9(2): 105. DOI: https://doi.org/10.3390/agronomy9020105

Bradford, M.A., Jones, T.H., Bardgett, R.D., Black, H.I., Boag, B., Bonkowski, M., Cook, R., *et al.* 2002. Impacts of soil faunal community composition on model grassland ecosystems. *Science*, 298(5593), 615-618.

Bradshaw, C.J.A. 2012. Little left to lose: deforestation and forest degradation in Australia since European colonization. *Journal of Plant Ecology*, (5)1: 109-120.

Brady, M., Sahrbacher, C., Kellermann, K. & Happe, K. 2012. An agent-based approach to modeling impacts of agricultural policy on land use, biodiversity and ecosystem services. *Landscape Ecology*, 27: 1363-1381.

Brady, M., Kellermann, K., Sahrbacher, C., Jelinek, L. & Lobianco, A. 2009. Impacts of Decoupled Agricultural Support on Farm Structure, Biodiversity and Landscape Mosaic: Some EU Results. *Journal of Agricultural Economics*, 60: 563-585.

Brady, V.M., Hristov, J., Wilhelmsson, F. & Hedlund, K. 2019. Roadmap for Valuing Soil Ecosystem Services to Inform Multi-Level Decision-Making in Agriculture. *Sustainability*, 11(19): 5285.

Brauman, A. 2000. Effect of gut transit and mound deposit on soil organic matter transformations in the soil feeding termite: a review. *European Journal of Soil Biology*, 36: 117-125.

Braun, P., G. Grass, A. Aceti, L. Serrecchia, A. Affuso, E. Georgi, B. Northoff, *et al.* 2015. Microevolution of Anthrax from a Young Ancestor (M.A.Y.A.) Suggests a Soil-Borne Life Cycle of Bacillus anthracis. *PLoS One*, 10(8): 1–23.

Bretagnolle, V. & Gaba, S. 2015. Weeds for Bees? A Review. *Agronomy for Sustainable Development.* 35: 891–909. https://doi.org/10.1007/s13593-015-0302-5.

Breure, A.M., De Deyn, G.B., Dominati, E., Eglin, T., Hedlund, K., Van Orshoven, J. & Posthuma, L. 2012. Ecosystem services: a useful concept for soil policy making! *Current Opinion in Environmental Sustainability*, 4: 578-585.

Brevik, E. C. & Burgess, L. C. 2014. The Influence of Soils on Human Health. Nature Education Knowledge 5(12):1. [online]. [Cited 5 October 2020]. https://www.nature.com/scitable/knowledge/library/the-influence-of-soils-on-human-health-127878980/

Brevik, E.C. & Sauer, T.J. 2015. The past, present, and future of soils and human health studies. *Soil*, 1: 35–46.

Breznak, J.A. 2000. Ecology of prokaryotic Microbes in the guts of wood-and litter feeding Termites. In Y. Abe, D.E. Bignell & T. Higashi, eds. *Termites: Evolution, Sociality, Symbioses, Ecology,* pp. 235-242. Dordrecht, Netherlands, Kluwer Academic Press.

Briones, M.I.J. 2018. The Serendipitous Value of Soil Fauna in Ecosystem Functioning: The Unexplained Explained. *Frontiers in Environmental Science*, 6: 149.

British Mycological Society. 2020. [online]. [Cited 2 October 2020]. https://www.britmycolsoc.org.uk/

Broadbent, A. B. & Tomlin, A.D. 1982. Comparison of two methods for assessing the effects of carbofuran on soil animal decomposers in cornfields. *Environmental Entomology,* 11: 1036-1042.

Brodie, G. 2018. The use of physics in Weed control. In K. Jabran & B.S. Chauhan, eds. *Non-Chemical weed control*, pp. 33-59. Academic Press.

Brooker, R.W., Bennett, A.E., Cong, W.-F., *et al.* 2015. Improving intercropping: a synthesis of research in agronomy, plant physiology and ecology. *New Phytologist,* 206: 107–117. doi: 10.1111/nph.13132

Brookes, G. & Barfoot, P. 2014. Key global economic and environmental impacts of genetically modified (GM) crop use 1996-2012. *GM Crops Food,* 5(2):149-160.

Brouwer. C., Goffeau, A. & Heibloem, M. 1985. *Irrigation Water Management: Training Manual No. 1 - Introduction to Irrigation.* Rome, FAO. http://www.fao.org/3/r4082e/r4082e00.htm.

Brown, J., Scholtz, C.H., Janeau, J.L., Grellier, S. & Podwojewski, P. 2010. Dung beetles (Coleoptera: Scarabaeidae) can improve soil hydrological properties. *Applied Soil Ecology*, 46: 9–16.

Bruckner, A., Schuster, R., Smit, T. & Heethoff, M. 2018. Imprinted or innate food preferences in the model mite Archegozetes longisetosus (Actinotrichida Ordbatida, Trhypochthoniidae). *Pedobiologia,* 66:74–80.

Brum, J.R., Ignacio-Espinoza, J.C., Roux, S., Doulcier, G., Acinas, S.G., Alberti, A., *et al.* 2015. Patterns and ecological drivers of ocean viral communities. *Science*, 348: 1261498.

Brune, A. 2011. Microbial Symbioses in the Digestive Tract of Lower Termites. In E. Rosenberg & U. Gophna, eds. *Beneficial Microorganisms in Multicellular Life Forms,* pp. 3-25. Berlin, Germany, Springer-Verlag.

Bruns, M.A. & Byrne, L.B. 2004. Scale model of a soil aggregate and associated organisms: A teaching tool for soil ecology. *Journal of Natural Resources and Life Science Education*, 33: 85-91.

Brussaard, L., De Ruiter, P.C. & Brown, G.G. 2007. Soil biodiversity for agricultural sustainability. *Agriculture, Ecosystems and Environment*, 121: 233-244.

Bryan, B.A., Gao, L., Ye, Y., Sun, X., Connor, J.D., Crossman, N.D., Stafford-Smith, M., Wu, J., He, C., Yu, D., Liu, Z., Li, A., Huang, Q., Ren, H., Deng, X., Zheng, H., Niu, J., Han, G. & Hou, X. 2018. China's response to a national land-system sustainability emergency. Nature, 559(7713): 193–204. https://doi.org/10.1038/s41586-018-0280-2

Bryan, J.E., Shearman, P.L., Asner, G.P, Knapp, D.E, Aoro, G. & Lokes, B. 2013. Extreme Differences in Forest Degradation in Borneo: Comparing Practices in Sarawak, Sabah, and Brunei. *PLoS ONE*, 8(7): e69679 https://doi.org/10.1371/journal.pone.0069679.

Brzeszcz, J. & Kaszycki, P. 2018. Aerobic bacteria degrading both n-alkanes and aromatic hydrocarbons: an undervalued strategy for metabolic diversity and flexibility. *Biodegradation*, 29: 359-407.

Büdel, B., Dulić, T., Darienko, T., Rybalka, N. & Friedl, T. 2016. Cyanobacteria and Algae of Biological Soil Crusts. *In* B. Weber, B. Büdel & J. Belnap, Hrsg. *Biological Soil Crusts: An Organizing Principle in Drylands*, S. 55–80. Ecological Studies. Cham, Springer International Publishing. (verfügbar unter http://link.springer.com/10.1007/978-3-319-30214-0_4).

Bugiel, L.N, Livingstone, S.W., Isaac, M.E., Fulthorpe, R.R. & Martin, A.R. 2019. Impacts of invasive plant species on soil biodiversity: a case study of dog-strangling vine (Vincetoxicum rossicum) in a Canadian National Park. *Canadian Journal of Soil Science*, 98:716-723.

Buhl, J., Gautrais, J., Solé, R.V., Kuntz, P., Valverde, S., Deneubourg, J.L. & Theraulaz, G. 2004. Efficiency and robustness in ant networks of galleries. *European Physical Journal B*, 42: 123–129.

Bulgarelli, D., Rott, M., Schlaeppi, K., *et al.* 2012. Revealing structure and assembly cues for Arabidopsis root-inhabiting bacterial microbiota. *Nature*, 488: 91–95. doi: 10.1038/nature11336

Bulgarelli, D., Schlaeppi, K., Stijn Spaepen, S., van Themaat, E.V.L. & Schulze-Lefert, P. 2013. Structure and functions of the bacterial microbiota of plants. *Annual Review of Plant Biology*, 64: 807–838.

Bullock, C., Kretsch, C. & Candon, E. 2008. *The Economic and Social Aspects of Biodiversity - Benefits and Costs of Biodiversity in Ireland.* Dublin, Ireland. The Stationary Office, Government of Ireland. 195pp.

Bulot, A., Dutoit, T., Renucci, M. & Provost, E. 2014. A new transplantation protocol for harvester ant queens Messor barbarus (Hymenoptera: Formicidae) to improve the restoration of species-rich plant communities. *Myrmecological News*, 20: 43–52.

Buol, S. 1994. Saprolite-Regolith Taxonomy- An Approximation. In D.L. Cremeens, R.B. Brown & J.H. Huddleston, eds. *Whole Regolith Pedology,* pp. 119-132. Soil Science Society of America Special Publication nº 34. Madison, U.S.A. Soil Science Society of America.

Burgess, T.I., McDougall, K.L., Scott, P.M., Hardy, G.E.StJ., & Garnas, J. 2019. Predictors of Phytophthora diversity and community composition in natural areas across diverse Australian ecoregions. *Ecography,* 42: 594-607.

Buringh, P. 1978. Food production potential of the world, In R. Sinha, ed. *The World Food Problem: Consensus and Conflict*, pp. 477-485. Oxford, UK. Pergamon Press.

Burke, J.L., Maerz, J.C., Milanovich, J.R., Fisk, M.C., Gandhi, K.J.K. 2011. Invasion by Exotic Earthworms Alters Biodiversity and Communities of Litter- and Soil-dwelling Oribatid Mites. *Diversity,* 3: 155-175.

Burkhardt, U., Russell, D.J., Decker, P., Döhler, M., Höfer, H., Lesch, S., Rick, S., Römbke, J., Trog, C., Vorwald, J., Wurst, E. & Xylander, W.E.R. 2014. The Edaphobase project of GBIF-Germany—A new online soil-zoological data warehouse. Applied Soil Ecology, 83: 3–12. https://doi.org/10.1016/j.apsoil.2014.03.021

Burns, A., Gleadow, R., Cliff, J., Zacarias, A. & Cavagnaro, T. 2010. Cassava: The drought, war and famine crop in a changing world. *Sustainability*, 2: 3572–3607.

Burri, K., Gromke, C. & Graf, F. 2013. Mycorrhizal fungi protect the soil from soil erosion: A wind tunnel study. *Land Degradation and Development*, 24: 385–392.

Burrowes, B., Harper, D.R., Anderson, J., McConville, M. & Enright, M.C. 2011. Bacteriophage therapy: potential uses in the control of antibiotic-resistant pathogens. *Expert review of anti-infective therapy,* 9(9): 775-785.

Burtis, J.C., Yavitt, J.B., Fahey, T.J., & Ostfeld, R.S. 2019. Ticks as soil-dwelling arthropods: an intersection between disease and soil ecology. Journal of Medical Entomology, tjz116. DOI: 10.1093/jme/tjz116

Butt, T.M., Coates, C.J., Dubovskiy, I.M. & Ratcliffe, N.A. 2016. Entomopathogenic fungi: new insights into host–pathogen interactions. *Advances in genetics*, 94: 307-364

Byrne, L.B. 2007. Habitat structure: a fundamental concept and framework for urban soil ecology. *Urban Ecosystems*, 10: 255-274.

Byrne, L.B. 2016. Overcoming Challenges of the Soil Education Gap: Part 2. Beneath our Feet: The Blog of the Global Soil Biodiversity Initiative. https://www.globalsoilbiodiversity.org/blog-beneath-our-feet/2018/10/2/overcoming-challenges-of-the-soil-education-gap-part-2

Byrne, L.B. 2020. Socioecological Soil Restoration in Urban Cultural Landscapes. In J.A. Stanturf and M.A. Callaham, eds. *Soils and Landscape Restoration*. Hoboken, USA. Wiley Publishers.

Byrne, L.B., Thiet, R.K., and Chaudhary, V.B. 2016. Pedagogy for the pedosphere. *Frontiers in Ecology and the Environment,* 14: 238-240.

Cáceres, A. and Kalinhoff, C. 2014. Efecto de la perturbación producida por la extracción de arena sobre las micorrizas arbusculares (MA) en un bosque seco tropical.

Cáceres, T., Megharaj, M. & Naidu, R. 2008. Toxicity and transformation of fenamiphos and its metabolites by two micro algae *Pseudokirchneriella subcapitata* and *Chlorococcum* sp. *Science of The Total Environment*, 398(1–3): 53–59. https://doi.org/10.1016/j.scitotenv.2008.03.022

Cakmak, I. 2002. Plant nutrition research: Priorities to meet human needs for food in sustainable ways. *Plant and Soil*, 247: 3–24.

California Department of Food and Agriculture (cdfa). 2020. Healthy Soils Program. [online]. [Cited 5 October 2020]. https://www.cdfa.ca.gov/oefi/healthysoils/

Callaham, M.A., Jr., González, G.G., Hale, C.M., Heneghan, L., Lachnicht, S.L. & Zou. X. 2006. Policy and management responses to earthworm invasions in North America. *Biological Invasions*, 8: 1317-1329.

Cameron, E. K., M. Vilà, and M. Cabeza. 2016. Global meta-analysis of the impacts of terrestrial invertebrate invaders on species, communities and ecosystems. *Global Ecology and Biogeography*, 25: 596–606.

Cameron, E. K., Proctor, H.C., & Bayne, E.M. 2013. Effects of an ecosystem engineer on belowground movement of microarthropods. *PLoS ONE*, 8(4), e62796.

Cameron, E. K., Knysh, K. M., Proctor, H. C., & Bayne, E. M. 2013b. Influence of two exotic earthworm species with different foraging strategies on abundance and composition of boreal microarthropods. *Soil Biology and Biochemistry*, 57, 334-340.

Cameron, E.K., Martins, I.S., Lavelle, P., Mathieu, J., Tedersoo, L., Gottschall, F. & Guerra, C.A. 2018. Global gaps in soil biodiversity data. *Nature Ecology and Evolution*, 2: 1042-1043. doi: 10.1038/s41559-018-0573-8.

Cameron, E. K., Martins, I. S., Lavelle, P., Mathieu, J., Tedersoo, L., Bahram, M., Siebert, J, *et al.* 2019. Global mismatches in aboveground and belowground biodiversity. *Conservation Biology*, 33: 1187–1192. https://doi.org/10.1111/cobi.13311

Cammeraat, L.H., Willott, S.J., Compton, S.G. & Incoll, L.D. 2002. The effects of ants' nests on the physical, chemical and hydrological properties of a rangeland soil in semi-arid Spain. *Geoderma*, 105: 1–20.

Campbell, J. M., Sarazin, M. J., & Lyons, D. B. 1989. Canadian beetles (Coleoptera) injurious to crops, ornamentals, stored products, and buildings. *Agriculture Canada, Research branch, Publication 1826*. 491 pp.

Campbell, B.M., Hansen, J., Rioux, J., Stirling, C.M., Twomlow, S. & Wollenberg, E.L. 2018. Urgent action to combat climate change and its impacts (SDG 13): transforming agriculture and food systems. *Current Opinion in Environmental Sustainability*, 34: 13–20.

Campo Ambiente. 2020. Tecnologías al servicio del ambiente y la agricultura . [online]. [Cited 6 October 2020]. https://campoambiente.com/

Canal de Panamá. 2019. Canal de Panamá. [online]. [Cited 5 October 2020]. www.micanaldepanama.com

Canfora, L., Bacci, G., Pinzari, F., Papa, G.L., Dazzi, C. & Benedetti, A. 2014. Salinity and bacterial diversity: to what extent does the concentration of salt affect the bacterial community in a saline soil? *PLoS One*, 9(9): p.e106662.

Cannon, R.J.C. 1998. The implications of predicted climate change for insect pests in the UK, with emphasis on non-indigenous species. *Global Change Biology*, 4: 785–796.

Cao, Z., Han, X., Hu, C., Chen, J., Zhang, D. & Steinberger, Y. 2011. Changes in the abundance and structure of a soil mite (Acari) community under long-term organic and chemical fertilizer treatments. *Applied Soil Ecology*, 49: 131-138.

Cao, J., Wang, C., Dou, Z., Liu, M., & Ji, D. 2018. Hyphospheric impacts of earthworms and arbuscular mycorrhizal fungus on soil bacterial community to promote oxytetracycline degradation. *Journal of Hazardous Materials*, 341: 346–354.

Caoduro, G., Battiston, R., Giachino, P.M., Guidolin, L. & Lazzarin, G. 2014. Gli indici di biodiversità per la valutazione della qualità di aria, acqua e suolo della certificazione "Biodiversity Friend" in aree temperate. Biodiversity Journal, 5(1): 69–86.

Cape, J.N., Fowler, D. & Davison, A. 2003. Ecological effects of sulfur dioxide, fluorides, and minor air pollutants: recent trends and research needs. *Environment International*, 29: 201-211.

van Capelle, C., Schrader, S. & Brunotte, J. 2012. Tillage-induced changes in the functional diversity of soil biota – A review with a focus on German data. European Journal of Soil Biology, 50: 165–181. https://doi.org/10.1016/j.ejsobi.2012.02.005

Capowiez, Y., Pierret, A., Daniel, O., Monestiez, P. & Kretzschmar, A. 1998. 3D skeleton reconstructions of natural earthworm burrow systems using CAT scan images of soil cores. *Biology and Fertility of Soils*, 27: 51–59.

Capowiez Y., Samartino S., Cadoux S., Bouchant P., Richard G. & Boizard H. 2012. Role of earthworms in regenerating soil structure after compaction in reduced tillage systems. *Soil Biology and Biochemistry*, 55: 93-103.

Caravaca, F., Masciandaro, G. & Ceccanti, B. 2002. Land use in relation to soil chemical and biochemical properties in a semiarid Mediterranean environment. *Soils and Tillage Research*, 68: 23-30. doi:10.1016/ S0167-1987(02)00080-6.

Cardinale, B.J., Duffy, J.E., Gonzalez, A., Hooper, D.U., Perrings, C., Venail, P., Narwani, *et al.* 2012. Biodiversity loss and its impact on humanity. *Nature*. 486(7401): 59-67.

Cardoso, I.M. & Kuyper, T.W. 2006. Mycorrhizas and tropical soil fertility. *Agriculture, Ecosystems and Environment*, 116: 72–84.

Carini, P., Delgado-Baquerizo, M., Hinckley, E.-L.S., Holland-Moritz, H., Brewer, T.E., Rue, G., Vanderburgh, C., McKnight, D. & Fierer, N. 2018. Unraveling the effects of spatial variability and relic DNA on the temporal dynamics of soil microbial communities (preprint). *BioRxiv*. https://doi.org/10.1101/402438

Carlos, V., Sánchez, C., & Valera, N. 2007. Mite diversity (Acari: Prostigmata, Mesotigmata, Astigmata) associated to soil litter from two vegetation zones at the University Park UCLA, Venezuela. *Iheringia. Serie Zoologia* 97 https://doi.org/10.1590/S0073-47212007000400017

Caron, D.A., Worden, A.Z., Countway, P.D., Demir, E. & Heidelberg, K.B. 2008. Protists are microbes too: a perspective. *ISME Journal*, 3: 4-12.

Carrillo, Y., Bradford, M.A., Jordan, C.F. & Molina, M. 2011. Soil fauna alter the effects of litter composition on nitrogen cycling in a mineral soil. *Soil Biology and Biochemistry*, 43: 1440–1449.

Carroll, J.J. & Viglierchio, D.R. 1981. On the transport of nematodes by the wind. *Journal of Nematology*, 13: 476-483.

Caruso, T., Schaefer, I., Monson, F. & Keith, A.M. 2019. Oribatid mites show how climate and latitudinal gradients in organic matter can drive large-scale biodiversity patterns of soil communities. *Journal of Biogeography*, 46(3): 611-620 DOI: https://doi.org/10.1111/jbi.13501.

Carvalho, T.S.C. de, Jesus, E. da C., Barlow, J., Gardner, T.A., Soares, I.C., Tiedje, J.M. & Moreira, F.M. de S. 2016. Land Use Intensification in the Humid Tropics Increased Both Alpha and Beta Diversity of Soil Bacteria. *Ecology*, 97(10): 2760–71.

Casalla, R. & Korb, J. 2019. Termite diversity in Neotropical dry forests of Colombia and the potential role of rainfall in structuring termite diversity. *Biotropica*, 51(2): DOI: 10.1111/btp.12626

Cassman, K.G., Dobermann, A., Walters, D.T. & Yang, H. 2003. Meeting Cereal Demand While Protecting Natural Resources and Improving Environmental Quality. *Annual Review of Environmental Resources*, 28: 315-358.

Castellanos M. Valecillos U, F. 2011. Diseño de Modulo de Lombricultura. [ebook] Trujillo: Universidad de Los Angeles. [online]. [Cited 5 October 2020]. http://bdigital.ula.ve/storage/pdftesis/pregrado/tde_arquivos/34/TDE-2012-09-23T06:16:29Z-1765/Publico/castellanosfreda_valecillosmaria.pdf

Cavagnaro, T. R. 2015. Biologically Regulated Nutrient Supply Systems: Compost and Arbuscular Mycorrhizas—A Review. *Advances in Agronomy*, 129: 293–321.

Cavagnaro et al. 2019. is Rillig, M.C., Aguilar-Trigueros, C.A., Camenzind, T., Cavagnaro, T.R., Degrune, F., Hohmann, P. et al. 2019. Why farmers should manage the arbuscular mycorrhizal symbiosis. *New Phytologist*, 222 (3): https://doi.org/10.1111/nph.15602

Cavicchioli, R., Ripple, W.J., Timmis, K.N., Azam, F., Bakken, L.R., Baylis, M., Behrenfeld, M.J. et al. 2019. Scientists' warning to humanity: microorganisms and climate change. *Nature Reviews Microbiology*, 17: 569-586. doi: 10.1038/s41579-019-0222-5.

CBD. 1992. Convention on Biological Diversity. [Online]. [Cited 5 October 2020]. https://www.cbd.int/doc/legal/cbd-en.pdf.

Čerevková, A., Bobul'ská, L., Miklisová, D. & Renčo, M. 2019. A case study of soil food web components affected by Fallopia japonica (Polygonaceae) in three natural habitats in Central Europe. *The Journal of Nematology*, 51: DOI: 10.21307/jofnem-2019-042

Cerri, C.E.P., Paustian, K., Bernoux, M., Victoria, R.L., Melillo, J.M. & Cerri, C.C. 2004. Modeling Changes in Soil Organic Matter in Amazon Forest to Pasture Conversion with the Century Model. *Global Change Biology*, 10(5): 815–32. https://doi.org/10.1111/j.1365-2486.2004.00759.x.

César Terrer, C., Vicca, S., Hungate, B.A., Phillips, R.P., Prentice, I.C. 2016. Mycorrhizal association as a primary control of the CO2 fertilization effect. *Science*, 353, 72-74. DOI: 10.1126/science.aaf4610

Chabrier, C. & Quénéhervé, P. 2008. Preventing nematodes from spreading: a case study with Radopholus similis (Cobb) Thorne in a banana field. *Crop Protection*, 27: 1237-1243.

Chamberlain, P.M., McNamara, N.P., Chaplow, J., Stott, A.W. & Black, H.I.J. 2006. Translocation of surface litter carbon into soil by Collembola. *Soil Biology and Biocemistry*, 38: 2655–2664. doi:https://doi.org/10.1016/j.soilbio.2006.03.021

Chamizo, S., Mugnai, G., Rossi, F., Certini, G. & De Philippis, R. 2018. Cyanobacteria Inoculation Improves Soil Stability and Fertility on Different Textured Soils: Gaining Insights for Applicability in Soil Restoration. *Frontiers in Environmental Science*, 6: doi: 10.3389/fenvs.2018.00049

Chan, K. Y. 2001. An Overview Of Some Tillage Impacts On Earthworm Population Abundance And Diversity: Implications For Functioning In Soils. *Soil And Tillage Research*, 57: 179-191.

Chandler, M., See, L., Copas, K., Bonde, A.M.Z., López, B.C., Danielsen, F., Legind, J.K., et al. 2017. Contribution of citizen science towards international biodiversity monitoring. *Biological Conservation*, 213(b): 280-294. doi: 10.1016/j.biocon.2016.09.004.

Chandrajith, R., Chaturangani, D., Abeykoon, S., Barth, J.A.C., van Geldern, R., Edirisinghe, E.A.N.V. & Dissanayake, C.B. 2014. Quantification of groundwater-seawater interaction in a coastal sandy aquifer system: a study from Panama, Sri Lanka. *Environmental Earth Sciences*, 72: 867-877.

Chaparro, J.M., Sheflin, A.M., Manter, D.K. & Vivanco, J.M. 2012. Manipulating the soil microbiome to increase soil health and plant fertility. *Biology and Fertility of Soils*, 48: 489–499. doi: 10.1007/s00374-012-0691-4

Charyulu D.K & Biswas S. 2010. *Organic input production and marketing in India – Efficiency, issues and policies.* Ahmedabad, India. Allied Publishers.

Chauvel, A., Grimaldi, M., Barros, E., Blanchart, E., Desjardins, T., Sarrazin, M. & Lavelle, P. 1999. Pasture Damage by an Amazonian Earthworm. *Nature*, 398: 32–33.

Cheik, S., Bottinelli, N., Minh, T.T., Doan, T.T. & Jouquet, P. 2019. Quantification of three dimensional characteristics of macrofauna macropores and their effects on soil hydraulic conductivity in northern Vietnam. *Frontiers in Environmental Science*, 7: 1–10.

Chen, Y.S., Yanagida, F. & Shinohara, T. 2005. Isolation and identification of lactic acid bacteria from soil using an enrichment procedure. *Letters in applied microbiology*, 40(3): 195-200.

Chhabra, R. 1996. *Soil salinity and water quality*. Rotterdam, Netherlands. CRCPress/Balkema.

Chipana, V., Clavijo, C., Medina, P., & Castillo, D. 2017. Inoculación de vainita (Phaseolus vulgaris L.) con diferentes concentraciones de Rhizobium etli y su influencia sobre el rendimiento del cultivo. Ecología Aplicada, 16(2), 91-98. https://dx.doi.org/10.21704/rea.v16i2.1012

Choi, S.-W. 2015. Bottom-up impact of soils on the network of soil, plants, and moths (Lepidoptera) in a South Korean temperate forest. *Canadian Entomologist,* 147(4): 405-418.

Choudhary, O.P., Josan, A.S., Bajwa, M.S. & Kapur, M.L. 2004. Effect of sustained sodic and saline-sodic irrigation and application of gypsum and farmyard manure on yield and quality of sugarcane under semi-arid conditions. *Field Crops Research*, 87: 103-116.

Chown, S.L., Slabber, S., McGeoch, M.A., Janion, C. & Leinaas, H.P. 2007. Phenotypic plasticity mediates climate change responses among invasive and indigenous arthropods. *Proceedings of the Royal Society B: Biological Sciences,* 274: 2531–2537.

Chrzan, A. 2016. Monitoring bioconcentration of potentially toxic trace elements in soils trophic chains. *Environmental Earth Sciences,* 75: 1–8.

Ciobanu, M., Eisenhauer, N., Stoica, I.A. & Cesarz, S. 2019. Natura 2000 priority and non-priority habitats do not differ in soil nematode diversity. *Applied soil ecology,* 135: 166-173.

Clapperton, M. J., Kanashiro, D.A., & Behan-Pelletier, V.M. 2002. Changes in abundance and diversity of microarthropods associated with Fescue Prairie grazing regimes. *Pedobiologia,* 46: 496-511.

Clapperton, M. J., Miller. J.J., Larney, F.J., & Lindwall, C.W. 1997. Earthworm populations as affected by long-term tillage practices in southern Alberta, Canada. *Soil Biol. Biochem.* 29: 631-633.

Clarholm, M. 1985. Interactions of bacteria, protozoa and plants leading to mineralization of soil nitrogen. *Soil Biology and Biochemistry,* 17: 181-187.

Classen, A.T., Sundqvist, M.K., Henning, J.A., Newman, G.S., Moore, J.A.M., Cregger, M.A., Moorhead, L.C. & Patterson, C.M. 2015. Direct and indirect effects of climate change on soil microbial and soil microbial-plant interactions: What lies ahead? *Ecosphere,* 6: article 130.

Clavel, J., Julliard, R. & Devictor, V. 2011. Worldwide Decline of Specialist Species: Toward a Global Functional Homogenization? *Frontiers in Ecology and the Environment*, 9(4): https://doi.org/10.1890/080216.

Clay, N.A., Lucas, J., Kaspari, M. & Kay, A.D. 2013. Manna from heaven: refuse from an arboreal ant links aboveground and belowground processes in a lowland tropical forest. *Ecosphere*, 4: 1e15.

Clemmensen, K.E., Bahr, A., Ovaskainen, O., Dahlberg, A., Ekblad, A., Wallander, H., Stenlid, J., Finlay, R.D., Wardle, D.A. & Lindahl, B.D. 2013. Roots and associated fungi drive long-term carbon sequestration in boreal forest. *Science,* 339(6127): 1615-1618.

Cleveland, C.C., & Liptzin, D. 2007. C:N:P stoichiometry in soil: is there a "Redfield ratio" for the microbial biomass? *Biogeochemistry*, 85: 235–252.

Clow, K. M., Ogden, N.H., Lindsay, L. R., Michel, P., Pearl, D. J., & Jardine, M. 2016. Distribution Of Ticks And The Risk Of Lyme Disease And Other Tick-Borne Pathogens Of Public Health Significance In Ontario, Canada. *Vector-Borne And Zoonotic Diseases*, 16(4): 215-222.

Cluzeau, D., Guernion, M., Chaussod, R., Martin-Laurent, F., Villenave, C., Cortet, J., Ruiz-Camacho, N., Pernin, C., Mateille, T., Philippot, L., Bellido, A., Rougé, L., Arrouays, D., Bispo, A. & Pérès, G. 2012. Integration of biodiversity in soil quality monitoring: Baselines for microbial and soil fauna parameters for different land-use types. European Journal of Soil Biology, 49: 63–72. https://doi.org/10.1016/j.ejsobi.2011.11.003

Coggan, N.V., Hayward, M.W. & Gibb, H. 2018. A global database and "state of the field" review of research into ecosystem engineering by land animals. *Journal of Animal Ecology*, 87: 974–994.

Colavecchio, A., Cadieux, B., Lo, A. & Goodridge, L.D. 2017. Bacteriophages contribute to the spread of antibiotic resistance genes among foodborne pathogens of the Enterobacteriaceae family–a review. *Frontiers in microbiology*, 8: 1108.

Cole, L., Buckland, S.M. & Bardgett, R.D. 2005. Relating microarthropod community structure and diversity to soil fertility manipulations in temperate grassland. *Soil Biology and Biochemistry*, 37: 1707-1717.

Cole, R.J., Holl, K.D., Zahawi, R.A., Wickey, P. & Townsend, A.R. 2016. Leaf litter arthropod responses to tropical forest restoration. *Ecology and Evolution*, 6: 5158–5168.

Coleman, D.C., Callaham, M.A., Crossley, D.A. 2018. *Fundamentals of Soil Ecology (third edition)*. London, UK. Academic Press, Elsevier Inc.

Coleman, D.C. 2008. From peds to paradoxes: Linkages between soil biota and their influences on ecological processes. *Soil Biology and Biochemistry*, 40: 271–289.

Coleman, D.C., Whitman, W.B. 2005. Linking species richness, biodiversity and ecosystem functioning in soil systems. *Pedobiologia*, 49: 479–497

Colman, B.P., Schimel, J.P. 2013. Drivers of microbial respiration and net N mineralization at the continental scale. *Soil Biology and Biochemistry*, 60: 65–76. doi:https://doi.org/10.1016/j.soilbio.2013.01.003

Colombo, F., Macdonald, C.A., Jeffries, T.C., Powell, J.R. & Singh, B.K. 2016. Impact of forest management practices on soil bacterial diversity and consequences for soil processes. *Soil Biology and Biochemistry*, 94: 200-210.

Comisión Nacional para el Conocimiento y Uso de la Biodiversidad (CONABIO). 2020. Sistema Nacional de Información sobre Biodiversidad de México. In: Comisión Nacional para el Conocimiento y Uso de la Biodiversidad [Online]. [Cited 2 October 2020]. http://www.snib.mx/

Compant, S., Samad, A., Faist, H. & Sessitsch, A. 2019. A review on the plant microbiome: Ecology, functions, and emerging trends in microbial application. *Journal of Advanced Research*, 19: 29-37.

CONABIO. 2018. Sistema Nacional de Información sobre Biodiversidad. Registros de ejemplares. [online]. [Cited 2 October 2020]. http://www.snib.mx/

CONACYT. 2020. Centro de Investigación en Ciencias de Información Geoespacial. 2020. [online]. [Cited 5 October 2020]. https://www.centrogeo.org.mx

CONAFOR. 2020. [online]. [Cited 5 October 2020]. https://www.gob.mx/conafor

Conant, R.T., Cerri, C.E.P., Osborne, B.B. & Paustian, K. 2017. Grassland management impacts on soil carbon stocks: A new synthesis. *Ecological Applications*, 27: 662–668.

Cong, R.-G., Termansen, M. & Brady, M.V. 2015. Managing soil natural capital: a prudent strategy for adapting to future risks. *Annals of Operations Research*, 255: 439-463.

Cong, R-G., Hedlund, K., Andersson, H. & Brady, M. 2014. Managing soil natural capital: An effective strategy for mitigating future agricultural risks? *Agricultural Systems*, 129: 30-39.

Convention on Biological Diversity. 2019. *Nagoya Protocol on Access to Genetic Resources and the Fair and Equitable Sharing of Benefits Arising from their Utilization to the Convention on Biological Diversity.* https://www.cbd.int/abs/ (retrieved 30 Sept 2019).

Cook, M., Molto, E. & Anderson, C. 1989. Fluorochrome labelling in Roman period skeletons from Dakhleh Oasis, Egypt. *American Journal of Physical Anthropology*, 80(2): 137-143. https://eurekamag.com/research/007/355/007355691.php

Coq, S., Barthès, B.G., Oliver, R., Rabary, B. & Blanchart, E. 2007. Earthworm activity affects soil aggregation and soil organic matter dynamics according to the quality and localization of crop residues – An experimental study (Madagascar). *Soil Biology and Biochemistry*, 39: 2119-2128.

Cordeiro, L. A. M., Balbino, L. C., Galerani, P. R., Domit, L. A., Silva, P. C., Kluthcouski, J., Vilela, L., Marchão, R. L., Skorupa, L. A., & Wruck, F. J. 2015. Transferência de tecnologias para adoção da estratégia de integração lavoura-pecuária-floresta. In Cordeiro, L. A. M., Vilela, L., Kluthcouski, J. & Marchão, R. L., eds. *Integração Lavoura-Pecuária-Floresta: o produtor pergunta, a Embrapa responde*. Brasília, DF: Embrapa, pp. 377-393.

Cork, S., Eadie, L., Mele, P., Price, R. & Yule, D. 2012. *The relationships between land management practices and soil condition and the quality of ecosystem services delivered from agricultural land in Australia.* Report prepared by Kiri-ganai Research for the Department of Agriculture, Fisheries and Forestry, Canberra.

Corkidi, L., Allen, E.B., Merhaut, D., Allen, M.F., Downer, J., Bohn, J. & Evans, M. 2004. Assessing the Infectivity of Commercial Mycorrhizal Inoculants in Plant Nursery Conditions. *Journal of Environmental Horticulture*, 22: 149–154. doi: 10.24266/0738-2898-22.3.149

Corporinoquia. 2015. Boletín Informativo Noviembre 2015. [Online]. [Cited 2 October 2020]. https://www.corporinoquia.gov.co/index.php/multimedia/segundo-boletin-informativo-virtual/156-segundo-boletin-informativo-del-2015.html

Correo del Orinoco. 2016. Para garantizar la soberanía alimentaria | Creados ministerios para la Producción Agrícola y Tierras, de Pesca y Acuicultura, y de Agricultura Urbana [online]. [Cited 5 October 2020]. http://www.correodelorinoco.gob.ve/creados-ministerios-para-produccion-agricola-y-tierras-pesca-y-acuicultura-y-agricultura-urbana/

Cosby, J. 2018. SOC-D: Soil Organic Carbon Dynamics. In: UK Centre for Ecology & Hydrology [online]. [Cited 2 October 2020]. https://www.ceh.ac.uk/uk-scape/soc-d-soil-organic-carbon-dynamics

Costa, E. M., Lima, W., Oliveira-Longatti, S.M. & De Souza, F.M. 2015. Phosphate-Solubilising Bacteria Enhance Oryza Sativa Growth and nutrient accumulation in an oxisol fertilized with rock phosphate. *Ecological Engineering*, 83: 380-385

Costa, O.Y., Raaijmakers, J.M. & Kuramae, E.E. 2018. Microbial extracellular polymeric substances: ecological function and impact on soil aggregation. *Frontiers in Microbiology*, 9: 1636.

Costantini, E. A. C. L'Abate, G. 2009. The Soil Cultural Heritage of Italy: Geodatabase, Maps, and Pedodiversity Evaluation. *Quaternary International*, 209(1–2): 142–153. https://doi.org/10.1016/j.quaint.2009.02.028

Costantini, E.A.C. & L'Abate, G. 2016. Beyond the concept of dominant soil: Preserving pedodiversity in upscaling soil maps.". *Geoderma*, 271: 243–253. https://doi.org/10.1016/j.geoderma.2015.11.024.

Costantini, E.A.C., Castaldini, M., Diago, M.P., Giffard, B., Lagomarsino, A., Schroers, H.-J., Priori, S., *et al.* 2018. Effects of soil erosion on agro-ecosystem services and soil functions: A multidisciplinary study in nineteen organically farmed European and Turkish vineyards. *Journal of Environmental Management*, 223: 614–624.

Costanza, R., de Groot, R., Sutton, P., van der Ploeg, S., Anderson, S.J., Kubiszewski, I., Farber, S. & Turner, R.K. 2014. Changes in the global value of ecosystem services. *Global Environmental Change*, 26: 152–158

Costanza, R. & Daly, H. E. 1992. Natural Capital and Sustainable Development. *Conservation Biology*, 6: 37-46.

Costanza, R., d'Arge, R., de Groot, R., Farber, S., Grasso, M., Hannon, B., Limburg, K., *et al.* 1997. The value of the world's ecosystems services and natural capital. *Nature*, 387: 253-260.

Cotler Ávalos, H. and Cuevas Fernández, M.L.. 2017. Estrategias de conservación de suelos en agroecosistemas de México. Fundación Río Arronte. Espacios Naturales. México. https://www.centrogeo.org.mx/stories/archivos/users/hcotler/Cotler_y_Cuevas-_Estrategias-de-conservacion-de-suelos-en-agroecosistemas-de-mexico.pdf

Coulis M., Bernard L., Gérard F., Hinsinger P., Plassard C., Villeneuve M. & Blanchart E. 2014. Endogeic earthworms modify soil phosphorus, plant growth and interactions in a legume-cereal intercrop. *Plant and Soil*, 379: 149-160.

Coulson, S.J., Hodkinson, I.D., Webb, N.R. & Harrison, J.A. 2002. Survival of terrestrial soil-dwelling arthropods on and in seawater: implications for trans-oceanic dispersal. *Functional Ecology*, 16: 353-356.

Cowie, A. L., Orr, B.J., Castillo Sanchez, V.M., Chasek, P., Crossman, N.D., Erlewein, A., Louwagie, G., *et al.* 2018. Land in balance: The scientific conceptual framework for Land Degradation Neutrality. *Environmental Science and Policy*, 79: 25–35.

Coyle, D.R., Nagendra, U.J., Taylor, M.K., Campbell, J.H., Cunard, C.E., Joslin, A.H., Mundepi, A., Phillips, C.A. & Callaham, M.A. 2017. Soil fauna responses to natural disturbances, invasive species, and global climate change: Current state of the science and a call to action. *Soil Biology and Biochemistry*, 110: 116–133.

Cranfield University. 2020. LandIS - Land Information System - Homepage Soil Portal [online]. [Cited 2 October 2020]. http://www.landis.org.uk/

Craven, D., Thakur, M.P., Cameron, E.K., Frelich, L.E., Beauséjour, R., Blair, R.B., Blossey, B., *et al.* 2017. The unseen invaders: introduced earthworms as drivers of change in plant communities in North American forests (a meta-analysis). *Global Change Biology*, 23(3): 1065-1074.

Crawford, J. W., Deacon, L., Grinev, D., Harris, J.A., Ritz, K., Singh, B.K. & Young, I. 2012. Microbial diversity affects self-organization of the soil–microbe system with consequences for function. *Journal of the Royal Society interface*, 9(71): 1302-1310. DOI: https://doi.org/10.1098/rsif.2011.0679

Crawford, J.W., Sleemant, B.D. & Young, I.M. 1993. On the relation between number-size distributions and the fractal dimension of aggregates. *European Journal of Soil Science*, 44(4): 555-565. https://doi.org/10.1111/j.1365-2389.1993.tb02321.x.

Creed, I.F., Duinker, P.N., Serran, J. & Steenberg, J.W.N. 2019. Managing risks to Canada's boreal zone: transdisciplinary thinking in pursuit of sustainability. *Environmental Reviews,* 27: 407- 418 DOI 10.1139/er-2018-0070

Cristescu, R. H., Frère, C. & Banks, P.B. 2012. A review of fauna in mine rehabilitation in Australia: Current state and future directions. *Biological Conservation*, 149: 60–72.

Crofts, T.S., Gasparrini, A.J. & Dantas, G. 2017. Next-generation approaches to understand and combat the antibiotic resistome. *Nature Reviews Microbiology,* 15(7): 422. doi:10.1038/nrmicro.2017.28

Crotty, F.V. 2011. *Elucidating the relative importance of the bacterial and fungal feeding channels within the soil food web under differing land managements.* PhD thesis. University of Plymouth.

Crouzet, O., Consentino, L., Pétraud, J.-P., Marrauld, C., Aguer, J.-P., Bureau, S., Le Bourvellec, C., Touloumet, L. & Bérard, A. 2019. Soil Photosynthetic Microbial Communities Mediate Aggregate Stability: Influence of Cropping Systems and Herbicide Use in an Agricultural Soil. *Frontiers in Microbiology*, 10: 1319. https://doi.org/10.3389/fmicb.2019.01319

Crow, S.E., Filley, T.R., McCormick, M., Szlávecz, K., Stott, D.E., Gamblin, D. & Conyers, G. 2009. Earthworms, stand age, and species composition interact to influence particulate organic matter chemistry during forest succession. *Biogeochemistry*, 92: 61–82.

Crowther, T.W., Boddy, L. & Jones, T.H. 2012. Functional and ecological consequences of saprotrophic fungus–grazer interactions. *The ISME Journal*, 6: 1992–2001.

Crowther, T.W., Stanton, D.W., Thomas, S.M., A'Bear, A.D., Hiscox, J., Jones, T. H., Boddy, L., et al. 2013. Top-down control of soil fungal community composition by a globally distributed keystone consumer. *Ecology,* 94(11): 2518-2528.

Crowther, T.W., Van Den Hoogen, J., Wan, J., Mayes, M.A., Keiser, A.D., Mo, L., Maynard, D.S., et al. 2019. The global soil community and its influence on biogeochemistry. *Science,* 365(6455): eaav0550.

Crowther, T.W., Maynard, D.S., Leff, J.W., Oldfield, E.E., McCulley, R.L., Fierer, N. & Bradford, M.A. 2014. Predicting the Responsiveness of Soil Biodiversity to Deforestation: A Cross-Biome Study. *Global Change Biology*, 20: 2983–94. https://doi.org/10.1111/gcb.12565.

Cuenca, G., De Andrade, Z. & Meneses, E. 2001. The presence of aluminum in arbuscular mycorrhizas of clusia multiflora exposed to increased acidity. *Plant And Soil,* 231: 233–241 Https://Doi.Org/10.1023/A:1010335013335

Cuenca, G., De Andrade, Z., Milagros, L., Farardo, L., Meneses, E., Marquez, M., & Machuca, R. 2002. El uso de arbustos nativos micorrizados para la rehabilitación de áreas degradadas de La Gran Sabana, Estado Bolívar, Venezuela. *Interciencia*, 27: 165- 172.http://ve.scielo.org/scielo.php?script=sci_abstract&pid=S0378-18442002000400003&lng=es&nrm=iso&tlng=es

Cui, X., Zhang, Y., Gao, J., Peng, F. & Gao, P. 2018. Long-term combined application of manure and chemical fertilizer sustained higher nutrient status and rhizospheric bacterial diversity in reddish paddy soil of Central South China. Scientific Reports, 8(1): 16554. https://doi.org/10.1038/s41598-018-34685-0

Curry, J.P. 1994. *Grassland Invertebrates*. London, UK. Chapman and Hall.

Curry, J.P. & Schmidt, O. 2007. The feeding ecology of earthworms – A review. *Pedobiologia*, 50: 463–477.

da C Jesus, E., Marsh, T.L., Tiedje, J.M. & de S Moreira, F.M. 2009. Changes in land use alter the structure of bacterial communities in Western Amazon soils. *The ISME journal*, 3(9): 1004.

da Silva, R.T., Fleskens, L., van Delden, H. & van der Ploeg, M. 2018. Incorporating soil ecosystem services into urban planning: status, challenges and opportunities. *Landscape Ecology*, 33: 1087-1102.

Dagar, J.C., Singh, G. & Singh, N.T. 2001. Evaluation of forest and fruit trees used for rehabilitation of semiarid alkali-sodic soils in India. *Arid Land Research and Management*, 15(2): 115-133.

Daily, G.C., Polasky, S., Goldstein, J., Kareiva, P.M., Mooney, H.A., Pejchar, L., Ricketts, T.H., Salzman, J. & Shallenberger, R. 2009. Ecosystem services in decision making: time to deliver. *Frontiers in Ecology and the Environment,* 7: 21-28.

Dandeniya, W.S. & Attanayake, R.N. 2016. *A comparison of soil fungal communities of dry-zone and wet-zone forests using a metagenomic approach.* Proceedings of the 1st Annual International Conference of Bioscience and Biotechnology. 12th- 14th January. Colombo, Sri Lanka.

Dang, Y.P., Moody, P.W., Bell, M.J., Seymour, N.P., Dalal, R.C., Freebairn, D.M. & Walker, S.R. 2015. Strategic tillage in no-till farming systems in Australia's northern grains-growing regions: II. *Implications for agronomy, soil and environment, Soil and Tillage Research,* 152: 115–123. doi.org/10.1016/ j.still.2014.12.013

Danks, H. V. 1979. Canada and its insect fauna. Memoirs of the Entomological Society of Canada. Ottawa, Canada.

Danovaro, R., Corinaldesi, C., Dell'Anno, A., Fuhrman, J.A., Middelburg, J.J. & Noble, R.T., *et al.* 2011. Marine viruses and global climate change. *FEMS Microbiology Reviews,* 35: 993-1034.

Dantas, G., Sommer, M.O., Oluwasegun, R.D. & Church, G.M. 2008. Bacteria subsisting on antibiotics. *Science,* 320(5872): 100-103.

Darbas, T., Laredo, L., Bonnett, G. & Baker, G. 2013. *Environmental stocktake of the northern grains production region.* CSIRO Sustainable Agriculture Flagship and Grains Research and Development Corporation, Canberra.

Das, S.C., Mandal, B. & Mandal, L.N. 1991. Effect of growth and subsequent decomposition of blue-green algae on the transformation of iron and manganese in submerged soils. *Plant and Soil,* 138(1): 75–84. https://doi.org/10.1007/BF00011810

Davison, J., Moora, M., Öpik, M., Adholeya, A., Ainsaar, L., Bà, A., Burla, S., *et al.* 2015. Global assessment of arbuscular mycorrhizal fungus diversity reveals very low endemism. *Science,* 349(6251): 970-973. https://doi.org/10.1126/science.aab1161

Dawes, T. Z. 2010. Reestablishment of ecological functioning by mulching and termite invasion in a degraded soil in an Australian savanna. *Soil Biology and Biochemistry,* 42: 1825–1834.

D'costa, V.M., King, C.E., Kalan, L., Morar, M., Sung, W.W., Schwarz, C., Froese, D., *et al.* 2011. Antibiotic resistance is ancient. *Nature,* 477(7365): 457-461.

de Araujo, A.S.F., Mendes, L. W. Lemos, L. N. Antunes, J. E. L. Beserra, J. E. A. de Lyra, M. d. C. C. P. Figueiredo, M. d. V. B. 2018. Protist species richness and soil microbiome complexity increase towards climax vegetation in the Brazilian Cerrado. *Communications Biology,* 1: 135.

de Boer, W., Folman, L.B., Summerbell, R.C., Boddy, L. 2005. Living in a fungal world: impact of fungi on soil bacterial niche development. *Fems Microbiology Reviews*, 29: 795-811.

de Carvalho Mendes, I., de Sousa, D. M. G., & dos Reis Junior, F. B. 2017. Bioindicadores de qualidade de solo: dos laboratórios de pesquisa para o campo. *Cadernos de Ciência & Tecnologia*, 32(1/2), 191-209.

de Freitas, P.L., & Landers, J.N. 2014. The Transformation of Agriculture in Brazil Through Development and Adoption of Zero Tillage Conservation Agriculture. *International Soil and Water Conservation Research*, 2. ISSN 2095-6339,

De Deyn, G.B., Cornelissen, J.H. & Bardgett, R.D. 2008. Plant functional traits and soil carbon sequestration in contrasting biomes. *Ecology Letters*, 11(5): 516-531.

De Deyn, G.B., Raaijmakers, C.E., Zoomer, H.R., Berg, M.P., de Ruiter, P.C., Verhoef, H.A., Bezemer, T.M. & van der Putten, W.H. 2003. Soil invertebrate fauna enhances grassland succession and diversity. *Nature*, 422(6933): 711.

de Freitas P.L. and Landers, J.N. 2014. The transformation of agriculture in Brazil through development and adoption of Zero Tillage Conservation Agriculture [online]. [Cited 5 October 2020]. https://ainfo.cnptia.embrapa.br/digital/bitstream/item/118124/1/ISWCR-Pedro-Freitas.pdf

de Graaff, M. A., Hornslein, N., Throop, H., Kardol, P. & van Diepen, L.T. 2019. Effects of agricultural intensification on soil biodiversity and implications for ecosystem functioning: A meta-analysis. In Sparks, D.L. ed. *Advances in Agronomy*, pp. 719-743. Cambridge, USA. Academic Press Inc.

de Graaff, M.A., Adkins, J., Kardol, P. & Throop, H.L. 2015. A meta-analysis of soil biodiversity impacts on the carbon cycle. *Soil*, 1: 257-271.

de Groot, G.A., Jagers op Akkerhuis, G.A.J.M., Dimmers, W.J., Charrier, X. & Faber, J.H. 2016. Biomass and diversity of soil mite functional groups respond to extensification of land management, potentially affecting soil ecosystem services. *Frontiers in Environmental Science*, 4: article 15. https://doi.org/10.3389/fenvs.2016.00015

De Ruiter, P.C., Neutel, A.-M. & Moore, J.C. 1995. Energetics, patterns of interaction strengths, and stability in real ecosystems. *Science*, 269: 1257–1260. doi: 10.1126/science.269.5228.1257

De Ruiter, P.C., Neutel, A.-M. & Moore, J.C. 1994. Modelling food webs and nutrient cycling in agro-ecosystems. *Trends in Ecology & Evolution*, 9: 378–383.

de Souza, O. & Cancello, E.M. 2010. Termites and ecosystem function. International Commission on Tropical Biology and Natural Resources, Encyclopedia of Life Support Systems (EOLSS), 14 pp.

De Vries, F.T. & Shade, A., 2013. Controls on soil microbial community stability under climate change. *Frontiers in Microbiology*, 4: 265. doi:10.1371/journal.pone.0080522.

de Vries, F.T., Manning, P., Tallowin, J.R., Mortimer, S.R., Pilgrim, E.S., Harrison, K.A., Hobbs, P.J., Quirk, H., Shipley, B. & Cornelissen, J.H. 2012. Abiotic drivers and plant traits explain landscape-scale patterns in soil microbial communities. *Ecology Letters*, 15(11): 1230-1239.

de Vries, F.T., Thébault, E., Liiri, M., Birkhofer, K., Tsiafouli, M.A., Bjørnlund, L., Bracht, H., *et al.* 2013. Soil food web properties explain ecosystem services across European land use systems. *Proceedings of the National Academy of Sciences*, 110: 14296–14301.

De Vries, F.T., Liiri, M.E., Bjørnlund, L., Bowker, M.A., Christensen, S., Setälä, H.M. & Bardgett, R.D. 2012. Land Use Alters the Resistance and Resilience of Soil Food Webs to Drought. *Nature Climate Change*, 2: 276–280. https://doi.org/10.1038/nclimate1368.

Decaëns, T. 2008. Priorities for conservation of soil animals. *CAB Reviews: Perspectives in Agriculture, Veterinary Science, Nutrition and Natural Resources*, 3:

Decaëns, T. 2010. Macroecological patterns in soil communities. *Global Ecology and Biogeography*, 19: 287-302. https://doi.org/j.1466-8238.2009.00517.x

Decaëns, T., Jiménez, J.J., Gioia, C., Measey, G.J., Lavelle, P. 2006. The values of soil animals for conservation biology. *European Journal of Soil Biology*, 42 S23–S38. https://doi:10.1016/j.ejsobi.2006.07.001.

Decagon Devices Inc. 2006. Measuring Specific Surface of Soil with the WP4. ICT International. Decagon Devices. Application Note. http://ictinternational.com/content/uploads/2015/02/MeasuringSpecificSurfaceofSoilwiththeWP4.pdf.

Deharveng, L. 2004. Recent advances in Collembola systematics. *Pedobiologia (Jena)*, 48: 415–433. doi:10.1016/j.pedobi.2004.08.001

Del Toro, I., Ribbons, R.R. & Ellison, A.M. 2015. Ant-mediated ecosystem functions on a warmer planet: effects on soil movement, decomposition and nutrient cycling. *Journal of Animal Ecology* 84(5): 1233–1241. doi:10.1111/1365-2656.12367

Del Toro, I., Ribbons, R.R. & Pelini, S.L. 2012. The little things that run the world revisited: a review of ant-mediated ecosystem services and disservices (Hymenoptera: Formicidae). Myrmecological News 17: 133–146.

Delgado-Baquerizo, M. & Eldridge, D.J. 2019. Cross-Biome Drivers of Soil Bacterial Alpha Diversity on a Worldwide Scale. *Ecosystems 22*, pp.1-12.

Delgado-Baquerizo, M., Eldridge, D.J., Ochoa, V., Gozalo, B., Singh, B.K., Maestre, F.T. 2017. Soil microbial communities drive the resistance of ecosystem multifunctionality to global change in drylands across the globe. *Ecology Letters*, 20:1295-1305.

Delgado-Baquerizo, M., Bardgett, R.D., Vitousek, P.M., Maestre, F.T., Williams, M.A., Eldridge, D.J., Lambers, H., Neuhauser, S., Gallardo, A., García-Velázquez, L. & Sala, O.E. 2019. Changes in belowground biodiversity during ecosystem development. *Proceedings of the National Academy of Sciences*, 116(14), pp.6891-6896.

Delgado-Baquerizo, M., Giaramida, L., Reich, P.B., Khachane, A.N., Hamonts, K., Edwards, C., Singh, B.K., *et al.* 2016. Lack of functional redundancy in the relationship between microbial diversity and ecosystem functioning. *Journal of Ecology*, 104(4): 936-946.

Delgado-Baquerizo, M., Maestre, F.T., Reich, P.B., Trivedi, P., Osanai, Y., Liu, Y.-R., Hamonts, K., Jeffries, T.C. & Singh, B.K. 2016. Carbon content and climate variability drive global soil bacterial diversity patterns. *Ecological Monographs,* 86:373-390. https://doi.org/10.1002/ecm.1216

Delgado-Baquerizo, M., Maestre, F. T., Reich, P. B., Jeffries, T. C., Gaitan, J. J., Encinar, D., Singh, B.K., *et al.* 2016. Microbial diversity drives multifunctionality in terrestrial ecosystems. *Nature communications,* 7: 10541.

Delgado-Baquerizo, M., Oliverio, A.M., Brewer, T.E., Benavent-González, A., Eldridge, D.J., Bardgett, R.D., Maestre, F.T., Singh, B.K. & Fierer, N. 2018. A global atlas of the dominant bacteria found in soil. *Science,* 359:320-325.

Delgado-Bacquerizo, M., Powell, J.R., Hamonts, K., Reith, F., Mele, P., Brown, M.V., Dennis, P.G., *et al.* 2017. Circular linkages between soil biodiversity, fertility and plant productivity are limited to topsoil at the continental scale. *New Phytologist,* 215: 1186–1196 doi.org/10.1111/nph.14634

Delgado-Bacquerizo, M., Reith, F., Dennis, P.G., Hamonts, K., Powell, J.R., Young, A., Singh, B.K. & Bissett, A. 2018. Ecological drivers of soil microbial diversity and soil biological networks in the Southern Hemisphere. *Ecology,* 99, 583-596. doi.org/10.1002/ecy.2137

Demnerová, K., M. Mackova, V. Speváková, K. Beranova, L. Kochánková, P. Lovecká, E. Ryslavá & T. Macek. 2005. Two approaches to biological decontamination of ground-water and soil polluted by aromatics-characterization of microbial populations. *International Microbiology,* 8: 205–211.

Deng, L., Krauss, S., Feichtmayer, J., Hofmann, R., Arndt, H. & Griebler, C. 2014. Grazing of heterotrophic flagellates on viruses is driven by feeding behaviour. *Environmental Microbiology Reports,* 6: 325-330.

Deng, L., Shangguan, Z.-P., Wu, G.-L. & Chang, X.-F. 2017. Effects of grazing exclusion on carbon sequestration in China's grassland. *Earth-Science Reviews,* 173: 84–95.

Dequiedt, S., Saby, N.P.A., Lelievre, M., Jolivet, C., Thioulouse, J., Toutain, B., Arrouays, D., Bispo, A., Lemanceau, P. & Ranjard, L. 2011. Biogeographical patterns of soil molecular microbial biomass as influenced by soil characteristics and management. *Global Ecology and Biogeography,* 20(4): 641–652. https://doi.org/10.1111/j.1466-8238.2010.00628.x

Derpsch., R.D., Friedrich, T., Kassam, A. & Hongwen, L. 2010. Current status of adoption of no-till farming in the world and some of its main benefits. *International Journal of Agricultural and Biological Engineering,* 3: 1-25.

Desiré, T.V., Fosah, M.R., Desiré, M.H. & Fotso. 2018. Effect of indigenous and effective microorganism fertilizers on soil microorganisms and yield of Irish potato in Bambili, Cameroon. *African Journal of Microbiology Research,* 12(15): 345-353 DOI: 10.5897/AJMR2017.8601

Destoumieux-Garzón, D. 2018. The One Health Concept : 10 Years Old and a Long Road Ahead. *Frontiers in Veterinary Science,* 5: 1–13.

Di Pietro, A., Gut-Rella, M., Pachlatko, J.P. & Schwinn, F.J. 1992. Role of antibiotics produced by Chaetomium globosum in biocontrol of Pythium ultimum, a causal agent of damping-off. *Phytopathology,* 82(2): 131-135.

Di, H.J., Cameron, K.C., Shen, J.-P., Winefield, C.S., O'Callaghan, M.O., Bowatte, S. & He, J.-Z. 2010. Ammonia-oxidizing bacteria and archaea grow under contrasting soil nitrogen conditions. *FEMS Microbiology Ecology,* 72(3): 386-394

Diamond, S.E., Penick, C.A., Pelini, S.L., Ellison, A.M., Gotelli, N.J., Sanders, N.J. & Dunn, R. 2013. Using physiology to predict the responses of ants to climatic warming. *Integrative and Comparative Biology,* 53: 965–974. doi:10.1093/icb/ict085.

Diaz, S., Pascal, U., Stenseke, M., Martin-Lopez, B., Watson, R.T., Molnar, Z., Hill, R., Chan, K.M., Baste, I.A., Brauman, K.A., et al 2018. Assessing nature's contribution to people. *Science,* 359: 270-272.

Dibog, L., Eggleton, P., Norgrove, L., Bignell, D.E. & Hauser, S. 1999. Impacts of canopy cover on soil termite assemblages in an agrisilvicultural system in southern Cameroon. *Bulletin of Entomological Research,* 89: 125–132.

Dixit, R., Wasiullah, Malaviya, D., Pandiyan, K., Singh, U.B., Sahu, A., Shukla, R., *et al.* 2015. Bioremediation of Heavy Metals from Soil and Aquatic Environment: An Overview of Principles and Criteria of Fundamental Processes. *Sustainability,* 7: 2189-2212.

Djigal, D., Baudoin, E., Philippot, L., Brauman, A. & Villenave, C. 2010. Shifts in size, genetic structure and activity of the soil denitrifier community by nematode grazing. *European Journal of Soil Biology,* 46: 112-118.

Djuideu, T.C.L. 2017. *Perception et étude des communautés de termites des systèmes agroforestiers cacaoyers du Sud-Cameroun (Master).* Département de Biologie et Physioogie Animales. Yaoundé, Cameroon. Université de Yaoundé 1.

Doan, T.T., Ngo, P.T., Rumpel, C., Van Nguyen, B. & Jouquet, P. 2013. Interactions between compost, vermicompost and earthworms influence plant growth and yield: A one-year greenhouse experiment. *Scientia Horticulturae,* 160: 148–154.

Dobermann, A. & Cassman, K.G. 2002. Plant nutrient management for enhanced productivity in intensive grain production systems of the United States and Asia. *Plant and Soil*, 247: 153-175.

Dobrovolsky, G.V. & Nikitin, E.D. 2009. *Red Book of Soils of Russia: The Objects of the Red Book and Inventory of the Most Valuable Soils.* Moscow, Russia. MAX Press.

Domené-Painenao Cruces F. Herrera, O. 2019. LA AGROECOLOGÍA EN VENEZUELA: TENSIONES ENTRE EL RENTISMO PETROLERO Y LA SOBERANÍA AGROALIMENTARIA. [ebook] Maracay: Programa de Formación de Grado en Agroecología, Universidad Bolivariana de Venezuela, Maracay, Venezuela, pp.56, 57, 58, 59, 60, 61. [online]. [Cited 5 October 2020] http://ecopoliticavenezuela.org/biblioteca/textos/Agroecologia%20en%20Venezuela.pdf

Dominati, E., Patterson, M. & Mackay, A. 2010. A framework for classifying and quantifying the natural capital and ecosystem services of soils. *Ecological Economics,* 69: 1858-1868.

Dominati, E., Mackay, A., Green, S. & Patterson, M. 2014. A soil change-based methodology for the quantification and valuation of ecosystem services from agro-ecosystems: a case study of pastoral agriculture in New Zealand. *Ecological Economics,* 100: 119–129.

Domínguez, J., Bohlen, P.J. & Parmelee, R.W. 2004. Earthworms increase nitrogen leaching to greater soil depths in row crop agroecosystems. *Ecosystems,* 7: 672–685.

Dong, H. & Lu, A. 2012. Mineral–microbe interactions and implications for remediation. *Elements,* 8: 95-100.

Donhauser, J. & Frey, B. 2018. Alpine soil microbial ecology in a changing world. *FEMS microbiology ecology,* 94(9): fiy099.

Donoso, D.A. 2017. Tropical ant communities are in long-term equilibrium. *Ecological Indicators,* 83: 515–523. doi:10.1016/j.ecolind.2017.03.022

Dorador, C., Vila, I., Witzel, K.P. & Imhoff, J.F. 2013. Bacterial and archaeal diversity in high altitude wetlands of the Chilean Altiplano. *Fundamental and Applied Limnology,* 182(2): 135–159.

Dore, A.J., Hallsworth, S., McDonald, A.G., Werner, M., Kryza, M., Abbot, J., Nemitz, E., *et al.* 2014. Quantifying missing annual emission sources of heavy metals in the United Kingdom with an atmospheric transport model. *Science of the Total Environment,* 479: 171-180.

Downing JA, and McCauley E. 1992. The nitrogen: phosphorus relationships in lakes. *Limnol Oceanography,* 37: 936-945.

Du, H., Qu, C., Liu, J., Chen, W., Cai, P., Shi, Z., *et al.* 2017. Molecular investigation on the binding of Cd (II) by the binary mixtures of montmorillonite with two bacterial species. *Environmental Pollution,* 229: 871-878. doi: 10.1016/j.envpol.2017.07.052

Duangurai, T., Indrawattana, N. & Pumirat, P. 2018. Burkholderia pseudomallei Adaptation for Survival in Stressful Conditions. *Biomedical Research Institute Review,* 05272018:

Duboise, S.M., Moore, B.E., Sorber CA & Sagik, B.P. 1979. Viruses in Soil Systems. *CRC Critical Reviews in Microbiology,* 7: 245-285.

Dudal, R., Nachtergaele, F.O. & Purnell, M.F. 2002. Human factor of soil formation. *Proceedings of the 17th Congress of Soil Science, Bangkok 2002.*

Duran, P., Thiergart, T., Garrido-Oter, R., Agler, M., Kemen, E., Shulze-Lefert, P. & Hacquard, S. 2018. Microbial interkingdom interactions in roots promote Arabidopsis survival. *Cell,* 175(4): 973-983 doi: https://doi.org/10.1101/354167

Dutch Soil Platform (DSP). 2020. [online]. [Cited 2 October 2020]. https://www.bodemambities.nl/dsp

Dyk, A., Leckie, D., Tinis, S. & Ortlepp, S. 2015. *Canada's National Deforestation Monitoring System*. Natural Resources Canada, Information report BC-X-439. Victoria, Canada. Canadian Forest Service Pacific Forestry Centre.

Earth Microbiome Project. 2020. [online]. [Cited 5 October 2020]. http://www.earthmicrobiome.org/

Easter, M., Paustian, K., Killian, K., Williams Feng, S.T., Al-Adamat, R., Batjes, N.H., Bernoux, M., *et al.* 2007. The GEFSOC soil carbon modelling system: A tool for conducting regional-scale soil carbon inventories and assessing the impacts of land use change on soil carbon. *Agriculture, Ecosystems and Environment,* 122:13–25. https://doi.org/10.1016/j.agee.2007.01.004

Eastwood, D.C., Floudas, D., Binder, M., Majcherczyk, A., Schneider, P., Aerts, A., Asiegbu, F.O., *et al.* 2011. The plant cell wall–decomposing machinery underlies the functional diversity of forest fungi. *Science,* 333(6043): 762-765.

Eckert, M., Gaigher, R., Pryke, J.S. & Samways, M.J. 2019. Rapid recovery of soil arthropod assemblages after exotic plantation tree removal from hydromorphic soils in a grassland-timber production mosaic. *Restoration Ecology,* 27(6): DOI: 10.1111/rec.12991

Edaphobase. 2020. Edaphobase Data - Query Portal. In: Edaphobase [online]. [Cited 1 October 2020]. https://portal.edaphobase.org/

Eden S.P. Bromfield, Sylvie Cloutier, James T. Tambong, Thu Van Tran Thi. 2017. Corrigendum to "Soybeans inoculated with root zone soils of Canadian native legumes harbour diverse and novel *Bradyrhizobium* spp. that possess agricultural potential". Systematic and Applied Microbiology, Volume 40, Issue 8, December 2017, Pages 517. DOI https://doi.org/10.1016/j.syapm.2017.07.007

Edigi, E., Delgado-Baquerizo, M., Plett, J.M., Wang, J., Eldridge, D.J., Bardgett, R.D., Maestre, R.D. & Singh, B.K. 2019. A few Ascomycota taxa dominate soil fungal communities worldwide. *Nature Communications,* 10: 2369 https://doi.org/10.1038/s41467-019-10373-z

Egerer, M., Ossola, A. & Lin, B.B. 2018. Creating socioecological novelty in urban agroecosystems from the ground up. *BioScience,* 68: 25-34.

Eggleton, P. 2000. Global patterns of termite diversity. In T. Abe, D.E. Bignell & M. Higashi, eds. *Termites: Evolution, Sociality, Symbioses and Ecology*, pp. 25-51. Dordrecht, Netherlands. Springer.

Eglin, T., Ciais, P., Piao, S.L., Barre, P., Bellassen, V., Cadule, P., Chenu, C., *et al.* 2010. Historical and future perspectives of global soil carbon response to climate and land-use changes. *Tellus B,* 62: 700-718. https://doi.org/10.1111/j.1600-0889.2010.00499.x

Ehlers, W. 1975. Observations on earthworm channels and infiltration on tilled and untilled loess soil. *Soil Science,* 119: 242–249.

Eilers, K. G. Debenport. Anderson & S. Fierer, N. 2012. Digging deeper to find unique microbial communities: The strong effect of depth on the structure of bacterial and archaeal communities in soil. *Soil Biology and Biochemistry*, 50: 58–65. DOI: https://doi.org/10.1016/j.soilbio.2012.03.011

Eisenhauer, N. 2010. The action of an animal ecosystem engineer: identification of the main mechanisms of earthworm impacts on soil microarthropods. *Pedobiologia,* 53(6): 343-352.

Eisenhauer, N. & Guerra, C.A. 2019. Global maps of soil nematode worms. *Nature,* 572: 187-188.

Eisenhauer, N., Bonn, A. & Guerra, C.A. 2019. Recognizing the quiet extinction of invertebrates. *Nature Communications,* 10: Article 50. https://doi.org/10.1038/s41467-018-07916-1.

Eisenhauer, N., P. B. Reich & F. Isbell. 2012b. Decomposer diversity and identity influence plant diversity effects on ecosystem functioning. *Ecology,* 93: 2227–2240.

Eisenhauer, N., Hörsch, V., Moeser, J. & Scheu, S. 2010. Synergistic effects of microbial and animal decomposers on plant and herbivore performance. *Basic and Applied Ecology,* 11(1): 23-34.

Eisenhauer, N., Partsch, S., Parkinson, D., & Scheu, S. 2007. Invasion of a deciduous forest by earthworms: Changes in soil chemistry, microflora, microarthropods and vegetation. *Soil Biology and Biochemistry*, 39: 1099-1110.

Eisenhauer, N., Vogel, A., Jensen, B. & Scheu, S. 2018. Decomposer diversity increases biomass production and shifts aboveground-belowground biomass allocation of common wheat. *Scientific reports,* 8(1): 17894.

Eisenhauer, N., Cesarz, S., Koller, R., Worm, K. & Reich, P. B. (2012a). Global change belowground: impacts of elevated CO_2, nitrogen, and summer drought on soil food webs and biodiversity. *Global Change Biology,* 18(2): 435-447.

Eisenhauer, N., Herrmann, S., Hines, J., Buscot, F., Siebert, J. & Thakur, M.P. 2018. The dark side of animal phenology. *Trends in Ecology and Evolution*, 33: 898-901.

Eisenhauer, N., Dobies, T., Cesarz, S., Hobbie, S. E., Meyer, R. J., Worm, K. & Reich, P. B. (2013). Plant diversity effects on soil food webs are stronger than those of elevated CO_2 and N deposition in a long-term grassland experiment. *Proceedings of the National Academy of Sciences,* 110(17): 6889-6894.

Ekelund F., Rønn, R. & Christensen, S. 2001. Distribution with depth of protozoa, bacteria and fungi in soil profiles from three Danish forest sites. *Soil Biology and Biochemistry,* 33(4-5): 475-481. https://doi.org/10.1016/s0038-0717(00)00188-7.

Ekström, G. & Ekbom, B. 2011. Pest Control in Agro-ecosystems: An Ecological Approach. *Critical Reviews in Plant Sciences,* 30: 74-94.

El Fantroussi, S. & Agathos, S.N. 2005. Is bioaugmentation a feasible strategy for pollutant removal and site remediation? *Current Opinion in Microbiology,* 8: 268–275.

Eldridge, D.J. 1993. Effect of ants on sandy soils in semi-arid eastern australia: Local distribution of nest entrances and their effect on infiltration of water. *Australian Journal of Soil Research*, 31: 509–518.

Eldridge, D.J. 2011. The resource coupling role of animal foraging pits in semi-arid woodlands. *Ecohydrology*, 4: 623-630.

Eldridge, D.J. & Rath, D. 2002. Hip holes: kangaroo (Macropus spp.) resting sites modify the physical and chemical environment of woodland soils. *Australian Ecology*, 27: 527-536.

Eldridge D.J., Koen T.B., Killgore A., Huang N. & Whitford W.G. (2012) Animal foraging as a mechanism for sediment movement and soil nutrient development: evidence from the semi-arid Australian woodlands and the Chihuahuan Desert. *Geomorphology*, 157: 131–141

Eldridge, D.J., Woodhouse, J.N., Curlevski, N.J.A., Hayward, M., Brown, M.V. & Neilan, B.A. 2015. Soil-foraging animals alter the composition and co-occurrence of microbial communities in a desert shrubland. *ISME Journal*, 9: 2671-2681.

Eldridge, D.J., Delgado-Baquerizo, M., Travers, S.K., Val, J., Oliver, I., Hamonts, K. & Singh, B.K. 2017. Competition drives the response of soil microbial diversity to increased grazing by vertebrate herbivores. *Ecology*, 98: 1922-1931.

Elser, J.J., Dobberfuhl, D.R., MacKay, N.A. & Schampel, J.H. 1996. Organism size, life history, and N:P stoichiometry. *Bioscience*, 46: 674–684.

Elser, J.J., Okie, J., Lee, Z. & Souza, V. 2018. The Effect of Nutrients and N:P Ratio on Microbial Communities: Testing the Growth Rate Hypothesis and Its Extensions in Lagunita Pond (Churince). In F. García-Oliva, J.J. Else & V. Souza, eds. *Ecosystem Ecology and Geochemistry of Cuatro Cienegas. How to Survive in an Extremely Oligotrophic Site*, pp. 31-41. Cham. Springer.

Elser, J.J., Bracken, M.E.S., Cleland, E.E., Gruner, D.S., Harpole, W.S., Hillebrand, H., Ngai, J.T., Seabloom, E.W., Shurin, J.B. & Smith, J.E. 2007. Global analysis of nitrogen and phosphorus limitation of primary producers in freshwater, marine and terrestrial ecosystems. *Ecology Letters*, 10: 1135– 1142.

Elser, J.J., Sterner, R.W., Gorokhova, E., Fagan, W.F., Markow, T.A., Cotner, J.B., Harrison, J.F., Hobbie, S.E., Odell, G.M. & Weider, L.J. 2000. Biological stoichiometry from genes to ecosystems. *Ecology Letters*, 3: 540–550.

Emam, T. 2016. Local soil, but not commercial AMF inoculum, increases native and non-native grass growth at a mine restoration site. *Restoration Ecology*, 24: 35–44. doi: 10.1111/rec.12287

EMBRAPA. 2016. Número Temático O solo como fator de integração entre os componentes ambientais e a produção agropecuária. [online]. [Cited 2 October 2020]. https://seer.sct.embrapa.br/index.php/pab/article/view/22572/13571

EMBRAPA. 2019. FAO e Fundação BB reconhecem como boas práticas contribuições da Embrapa Florestas. [online]. [Cited 2 October 2020]. https://www.embrapa.br/busca-de-noticias/-/noticia/46095358/fao-e-fundacao-bb-reconhecem-como-boas-praticas-contribuicoes-da-embrapa-florestas

EMBRAPA. 2020a. Sistema de Informação de solos brasileiros. In: EMBRAPA [Online]. [Cited 2 October 2020]. https://www.sisolos.cnptia.embrapa.br/

EMBRAPA. 2020b. EMBRAPA Solos. In: EMBRAPA [Online]. [Cited 2 October 2020]. https://www.embrapa.br/solos

EMBRAPA. 2020c. PronaSolos. In: EMBRAPA [Online]. [Cited 2 October 2020]. https://www.embrapa.br/pronasolos

EMBRAPA. 2020d. Embrapa Recursos Genéticos e Biotecnologia. [online]. [Cited 5 October 2020]. https://www.embrapa.br/alelo

Emerson, J.B. 2019. Soil Viruses: A New Hope. *mSystems,* 4: e00120-19.

Emerson, J.B., Roux, S., Brum, J.R., Bolduc, B., Woodcroft, B.J., Jang, H.B., *et al.* 2018. Host-linked soil viral ecology along a permafrost thaw gradient. *Nature Microbiology*, 3: 870-880.

Erb, M., Lenk, C., Degenhardt, J., Turlings, T.C. 2009. The underestimated role of roots in defense against leaf attackers. *Trends in Plant Science*, 14(12): 653-659.

ERAMMP. 2020. Overview of ERAMMP. In: *Environnement & rural affairs monitoring and modelling programme* [online]. [Cited 2 October 2020]. https://erammp.wales/en

España, M., Rasche, F., Kandeller, E., Brune, T., Rodríguez, B., Bending, G. & Cadisch, G. 2011a. Assessing the effect of organic residue quality on active decomposing Fungi in a tropical vertisol using 15N-DNA stable isotope probing. *Fungal Ecology*, 4: 115-119.

España, M., Rasche, F., Kandeller, E., Brune, T., Rodríguez, B., Bending, G. & Cadisch, G. 2011b. Identification of active bacteria involved in decomposition of complex maize and soybean residues in a tropical Vertisol using 15N-DNA stable isotope probing. *Pedobiologia*, 54(3):187-193

Estrada, G.A., Baldani, V.L.D., De Oliveira, D.M., Urquiaga, S. & Baldani, J.I. 2013. Selection of phosphate-solubilizing diazotrophic Herbaspirillum and Burkholderia strains and their effect on rice crop yield and nutrient uptake. *Plant Soil.* 369, 115-129.

Ettema, C.H., Wardle, D.A. 2002. Spatial soil ecology. *Trends in Ecology and Evolution,* 17(4): 177–183.

Ettl, H. & Gärtner, G. 2014. *Syllabus der Boden-, Luft- und Flechtenalgen*. Berlin, Heidelberg, Springer Berlin Heidelberg. (verfügbar unter http://link.springer.com/10.1007/978-3-642-39462-1).

European Commission Joint Research Center & Global Soil Biodiversity Initiative. 2016. *Global Soil Biodiversity Atlas.* Luxembourg. https://esdac.jrc.ec.europa.eu/content/global-soil-biodiversity-atlas

European Commission. 2006. Impact Assessment of the Thematic Strategy on Soil Protection, SEC. https://ec.europa.eu/environment/archives/soil/pdf/SEC_2006_620.pdf

European Cooperation in Science & Technology (COST). 2020. Biotechnology of soil: monitoring, conservation and remediation. [online]. [Cited 2 October 2020]. http://cost.eu/cost-action/biotechnology-of-soil-monitoring-conservation-and-remediation

Euteneuer P., Wagentristl H., Steinkellner S., Scheibreithner C. & Zaller J.G. 2019. Earthworms affect decomposition of soil-borne plant pathogen Sclerotinia sclerotiorum in a cover crop field experiment. *Applied Soil Ecology,* 138: 88-93.

Eydallin G, Ryall B, Maharjan R, Ferenci T. 2014. The nature of laboratory domestication changes in freshly isolated Escherichia coli strains. *Environmental Microbiology,* 16: 813–828. doi: 10.1111/1462-2920.12208

Faber, J.H., Akkerhuis, G.A.J.M.J. op, Bloem, J., Lahr, J., Diemont, W.H. & Braat, L.C. 2009. Ecosysteemdiensten en bodembeheer : maatregelen ter verbetering van biologische bodemkwaliteit. *Wageningen, Alterra.* No. 1813. (also available at https://library.wur.nl/WebQuery/wurpubs/377468).

Falkinham, J.O., Wall, T.E., Tanner, J.R., Tawaha, K., Alali, F.Q., Li, C. & Oberlies, N.H. 2009. Proliferation of antibiotic-producing bacteria and concomitant antibiotic production as the basis for the antibiotic activity of Jordan's red soils. *Applied Environmental Microbiology,* 75(9): 2735-2741.

Fanin, N., Gundale, M.J., Farrell, M., Ciobanu, M., Baldock, J.A., Nilsson, M.C., Wardle, D.A., *et al.* 2018. Consistent effects of biodiversity loss on multifunctionality across contrasting ecosystems. *Nature ecology and evolution,* 2(2): 269.

FAO. 2000. Technology and information transfer: improving capability to fight defoliating insects in the Republic of Moldova. [online]. [Cited 2 October 2020]. http://www.fao.org/3/y5507e/y5507e07.htm (FAO, 2000)

FAO. 2011. *How does international price volatility affect domestic economies and food security?* [online]. The state of food insecurity in the world 2011. http://www.fao.org/3/i2330e/i2330e.pdf

FAO. 2014. *Building a common vision for sustainable food and agriculture – Principles and approaches* [online]. http://www.fao.org/3/a-i3940e.pdf

FAO. 2015. *Status of the World's Soil Resources: Technical Summary.* Rome, Italy. FAO. p. 94. (http://www.fao.org/3/a-i5126e.pdf)

FAO. *Pesticide registration Toolkit* [online]. http://www.fao.org/pesticide-registration-toolkit/tool/page/pret/limites-maximales-de-rsidus

FAO. *Why soil matters?* [online]. Soils Portal. [09/26/2019] http://www.fao.org/soils-portal/about/en/

FAO. *What is a Soil?* [online]. Soils Portal. [09/26/2019] http://www.fao.org/soils-portal/about/all-definitions/en/

FAO & ITPS. 2015. *Status of the World's Soil Resources– Main Report.* [online] Food and Agriculture Organization of the United Nations and Intergovernmental Technical Panel on Soils. Rome, Italy.

FAO & ITPS. 2017. *Global Soil Organic Carbon Map- version I.* [online]. Rome, Italy. http://54.229.242.119/GSOCmap/

FAO & ITPS. 2019. Global Soil Organic Carbon (GSOC) Map [online]. [Cited 7 September 2020]. http://www.fao.org/global-soil-partnership/pillars-action/4-information-and-data-new/global-soil-organic-carbon-gsoc-map

Fargione, J., Bassett, S., Boucher, T., Bridgham, S., Conant, R.T., Cook-Patton, S., Ellis, P.W., *et al.* 2018. Natural Climate Solutions for the United States. *Science Advances*, 4(11): 1–15.

Farzaneh, M., Vierheilig, H., Löss A. & Kaul, H.P. 2011. Arbuscular mycorrhiza enhances nutrient uptake in chickpea. *Plant Soil and Environment,* 57(10): 465–470

Faulkner, K.T., Robertson, M.P., Rouget, M. & Wilson, J.R.U. 2016. Understanding and managing the introduction pathways of alien taxa: South Africa as a case study. *Biological Invasions*, 18: 73–87.

Fausto, C., Mininni, A.N., Sofo, A., Crecchio, C., Scagliola, M., Dichio, B. & Xiloyannis, C. 2018. Olive orchard microbiome: characterisation of bacterial communities in soil-plant compartments and their comparison between sustainable and conventional soil management systems. *Plant Ecology and Diversity,* 11: 597–610. doi: 10.1080/17550874.2019.1596172

Federal Ministry of Education and Research. 2020a. BonaRes [online]. [Cited 2 October 2020]. https://www.bonares.de/

Federal Ministry of Education and Research. 2020b. Research Initiative for the Conservation of Biodiversity - A FONA Flagship Initiative. In: Research for Sustainable Development – FONA [online]. [Cited 2 October 2020]. https://www.fona.de/en/measures/funding-measures/preservation-of-biodiversity.php

Federal Ministry of Education and Research. 2020c. Bundesprogramm [online]. [Cited 2 October 2020]. https://www.bundesprogramm.de/

Federal Ministry for the Environment, Nature Conservation and Nuclear Safety. 2007. National Strategy on Biological Diversity. [online]. [Cited 2 October 2020]. https://www.bmu.de/fileadmin/bmu-import/files/english/pdf/application/pdf/broschuere_biolog_vielfalt_strategie_en_bf.pdf

Fee, E. 2019. Implementing the Paris Climate Agreement: Risks and Opportunities for Sustainable Land Use. In H. Ginzky, E. Dooley, I. L. Heuser, E. Kasimbazi, T. Markus, & T. Qin, eds. *Yearbook of Soil Law and Policy 2018*, pp. 249–270. Cham, Switzerland. Springer.

Felson, A.J., Bradford, M.A. & Terway, T.M. 2013. Promoting earth stewardship through urban design experiments. *Frontiers in Ecology and the Environment,* 11: 362-367.

Fenner, K., Canonica, S., Wackett, L.P. & Elsner, M. 2013. Evaluating pesticide degradation in the environment: blind spots and emerging opportunities. *Science,* 341(6147): 752-758.

FERA. No date. Pesticide Usage Surveys; https://secure.fera.defra.gov.uk/pusstats/. https://secure.fera.defra.gov.uk/pusstats/.

Ferlian, O., Eisenhauer, N., Aguirrebengoa, M., Camara, M., Ramirez-Rojas, I., Santos, F., Tanalgo, K. and Thakur, M.P. 2018. Invasive earthworms erode soil biodiversity: A meta-analysis. *Journal of Animal Ecology,* 87(1): 162-172.

Ferris, H. 2010. Contribution of nematodes to the structure and function of the soil food web. *The Journal of Nematology,* 42: 63-67.

Field Studies Council (FSC). 2020. [online]. [Cited 2 October 2020] https://www.field-studies-council.org/

Fierer, N. 2017. Embracing the unknown: disentangling the complexities of the soil microbiome. *Nature Reviews Microbiology,* 15: 579.

Fierer, N. & Jackson, R.B. 2006. The diversity and biogeography of soil bacterial communities. *Proceedings of the National Academy of Sciences,* 103: 626-631. https://doi.org/10.1073/pnas.0507535103

Fierer, N., Schimel J.P. & Holden, P.A. 2003. Variations in microbial community composition through two soil depth profiles. *Soil Biology and Biochemistry,* 35(1): 167-176. https://doi.org/10.1016/S0038-0717(02)00251-1

Fierer, N., Bradford, M. A., & Jackson, R. B. (2007). Toward an ecological classification of soil bacteria. *Ecology,* 88(6): 1354-1364.

Fierer, N., Strickland, M.S., Liptzin, D., Bradford, M.A. & Cleveland, C.C. 2009. Global patterns in belowground communities. *Ecology Letters,* 12: 1238–1249.

Figuerola, E.L., Guerrero, L.D., Türkowsky, D., Wall, L.G. & Erijman, L. 2015. Crop monoculture rather than agriculture reduces the spatial turnover of soil bacterial communities at a regional scale. *Environmental Microbiology,* 17(3), 678-688.

Filser, J. 2002. The role of Collembola in carbon and nitrogen cycling in soil: Proceedings of the Xth international Colloquium on Apterygota, České Budějovice 2000: Apterygota at the Beginning of the Third Millennium. *Pedobiologia (Jena),* 46: 234–245. doi:https://doi.org/10.1078/0031-4056-00130

Finke, D.L. & Denno, R.F. 2005. Predator diversity and the functioning of ecosystems: the role of intraguild predation in dampening trophic cascades. *Ecology Letters,* 8: 1299-1306.

Finlay, B.J. 2002. Global dispersal of free-living microbial eukaryote species. *Science,* 296(5570): 1061-1063.

Fiorilli, V., Catoni, M., Miozzi, L., Novero, M., Accotto, G.P., & Lanfranco, L. 2009. Global and cell-type gene expression profiles in tomato plants colonized by an arbuscular mycorrhizal fungus. *New Phytol.,* 184(4):975-987. doi:10.1111/j.1469-8137.2009.03031.x

Fiorilli, V., Vannini, C., Ortolani, F., Garcia-Seco, D., Chiapello, M., Novero, M., Domingo, G., Terzi, V., Morcia, C., Bagnaresi, P., Moulin, L, Bracale, M., & Bonfante, P. 2018. Omics approaches revealed how arbuscular mycorrhizal symbiosis enhances yield and resistance to leaf pathogen in wheat. *Scientific reports*, 8(1), 1-18.

Fischer, G., Shah, M., van Velthuizen H. & Nachtergaele, F.O. 2002. Global Agro-ecological Assessment for Agriculture in the 21st century: Methodology and Results. FAO/IIASA. Laxenburg, IIASA and Rome, FAO. 155 pp. (also available at http://pure.iiasa.ac.at/id/eprint/6667/1/RR-02-002.pdf)

Fischer, C., Tischer, J., Roscher, C., Eisenhauer, N., Ravenek, J., Gleixner, G., Attinger, S., *et al.* 2015. Plant species diversity affects infiltration capacity in an experimental grassland through changes in soil properties. *Plant and Soil,* 397(1-2): 1-16.

Floate, K. D. 2011. Arthropods in cattle dung on Canada's grasslands. In K. D. Floate, ed. *Arthropods of Canadian Grasslands*, Vol. 2, Biological Survey of Canada, pp. 71-88.

Floccia, F. & Jacomini, C. 2012. Programma RE MO. Rete nazionale monitoraggio della biodiversità e del degrado dei suoli. Quaderni – Natura e biodiversità No. 4. ISPRA Ambiente. (also available at https://www.isprambiente.gov.it/files/pubblicazioni/quaderni/natura-e-biodiversita/files/Quaderno_42012_ReMo.pdf).

FNCA. 2011. *Forum for nuclear cooperation in Asia (FNCA) Overview of FNCA Biofertilizer project 2010. Issue No. 9 March 2011.* FNCA Biofertilizer Newsletter. https://www.fnca.mext.go.jp/english/bf/news_img/nl09.pdf

FNR. 2020. FNR: Fachagentur Nachwachsende Rohstoffe. In: FNR [online]. [Cited 2 October 2020]. https://www.fnr.de/projektfoerderung/ausgewaehlte-projekte/projekte/news/foerderung-von-insekten-in-agrarlandschaften-durch-integrierte-anbausysteme-mit-nachwachsenden-rohsto/?tx_news_pi1%5Bcontroller%5D=News&tx_news_pi1%5Baction%5D=detail&cHash=2a2a99f21382fcdab6a5eaac44fdf6ef

Foley, J.A., Ramankutty, N., Brauman, K.A., Cassidy, E.S., Gerber, J.S., Johnston, M., Mueller, N.D., *et al.* 2011b. Solutions for a cultivated planet. *Nature,* 478: 337-342.

Forest, I., Reich, P.B., Tilman, D., Hobbie, S.E., Stephen Polasky, S. & Binder, S. 2013. Nutrient enrichment, biodiversity loss, and consequent declines in ecosystem productivity. *Proceedings of the National Acadamy of Sciences of the United States of America, 2013 Jul 16*, 110(29): 11911–11916. doi: 10.1073/pnas.1310880110

Fornara, D.A. & Tilman, D. 2008. Plant functional composition influences rates of soil carbon and nitrogen accumulation. *Journal of Ecology,* 96(2): 314-322.

Forsberg, K.J., Patel, S., Gibson, M.K., Lauber, C.L., Knight, R., Fierer, N. & Dantas, G. 2014. Bacterial phylogeny structures soil resistomes across habitats. *Nature,* 509(7502): 612.

Forst, S. & Clarke, D. 2002. Bacteria-nematode symbiosis. In R. Gaugler, ed. *Entomopathogenic nematodlogy*, pp. 57-78. Wallingford. CABI.

Foster, S. & Custodio, E. 2019. Groundwater Resources and Intensive Agriculture in Europe – Can Regulatory Agencies Cope with the Threat to Sustainability? *Water Resources Management,* 33: 2139–51. https://doi.org/10.1007/s11269-019-02235-6.

Fowler, D., Coyle, M., Skiba, U., Sutton, M.A., Cape, J.N., Reis, S., Voss, M, *et al.* 2013. The global nitrogen cycle in the twenty-first century. *Philosophical Transactions of the Royal Society of London. Series B, Biological Sciences,* 368(1621): 20130164. doi: 10.1098/rstb.2013.0164

Fox, C. A. 2003. Characterizing soil biota in Canadian agroecosystems: state of knowledge in relation to soil organic matter. *Canadian Journal of Soil Science,* 83: 245-257.

Fox, C. A. and MacDonald, K. B. 2003. Challenges related to soil biodiversity research in agroecosystems – Issues within the context of scale of observation. *Can. J. Soil Sci.* 83: 231–244

Frampton, R.A., Pitman, A.R. & Fineran, P.C. 2012. Advances in bacteriophage-mediated control of plant pathogens. *International Journal of Microbiology,* 2012: Article ID 326452 doi:10.1155/2012/326452

França, F.M., Frazão, F.S., Korasaki, V., Louzada, J. & Barlow, J. 2017. Identifying Thresholds of Logging Intensity on Dung Beetle Communities to Improve the Sustainable Management of Amazonian Tropical Forests. *Biological Conservation,* 216: 115–22. https://doi.org/10.1016/j.biocon.2017.10.014.

Francis, CA. 2005. Crop Rotations. In D. Hilel, ed. *Encyclopedia of Soils in the Environment.* New York, USA. Academic Press.

Franco, A.A. 2009. Fixação biológica de nitrogênio na cultura da soja no Brasil:Uma lição para o futuro. (Biological nitrogen fixation on soybean crop in Brazil: A lesson for the future.) Boletim Informativo da Sociedade Brasileira de Ciência do Solo (Brazilian Soil Science Society Newsletter) Jan. Apr: 23-24

Franco, A.L.C., Sobral, B.W., Silva, A.L.C. & Wall, D.H. 2019. Amazonian Deforestation and Soil Biodiversity. *Conservation Biology,* 33: 590–600. https://doi.org/10.1111/cobi.13234.

Franco, A.L.C., Bartz, M.L.C., Cherubin, M.R., Baretta, D., Cerri, C.E.P., Feigl, B.J., Wall, D.H., Davies, C.A. & Cerri, C.C. 2016. Loss of Soil (Macro)Fauna Due to the Expansion of Brazilian Sugarcane Acreage. *The Science of the Total Environment,* 563–564: 160–68. https://doi.org/10.1016/j.scitotenv.2016.04.116.

Franken, O., Huizinga, M., Ellers, J. & Berg, M.P. 2018. Heated communities: large inter- and intraspecific variation in heat tolerance across trophic levels of a soil arthropod community. *Oecologia,* 186: 311–322.

Fred, E.B., Baldwin, I.L. & McCoy, E. 1932. Root nodule bacteria and leguminous plants. *University of Wisconsin Studies in Science, No. 5 Chapter 13: Natural and artificial inoculation*: 229-256.

Frelich, L.E., Blossey, B., Cameron, E.K., Dávalos, A., Eisenhauer, N., Fahey, T., Ferlian, O., *et al.* 2019. Side-swiped: ecological cascades emanating from earthworm invasions. *Frontiers in Ecology and the Environment,* 17(9): 502–510.

Fritsch, C., Coeurdassier, M., Gimbert, F., Crini, N., Scheifler R. & De Vaufleury, A. 2011. Investigations of responses to metal pollution in land snail populations (Cantareus aspersus and Cepaea nemoralis) from a smelter-impacted area. *Ecotoxicology,* 20: 739–759.

Frost, P.C., Evans-White, M.A., Finkel, Z.V., Jensen, T.C. & Matzek, V. 2005. Are you what you eat? Physiological constraints on organismal stoichiometry in an elementally imbalanced world. *Oikos,* 109(1):18–28.

Fu, S., Zou, X. & Coleman, D. 2009. Highlights and perspectives of soil biology and ecology research in China. *Soil Biology and Biochemistry,* 41(5): 868–876. https://doi.org/10.1016/j.soilbio.2008.10.014

Fujimaki, R., Sato, Y., Okai, N. & Kaneko, N. 2010. The train millipede (Parafontaria laminata) mediates soil aggregation and N dynamics in a Japanese larch forest. *Geoderma,* 159: 216–220.

Fulekar, M.H. 2017. Microbial degradation of petrochemical waste-polycyclic aromatic hydrocarbons. *Bioresources and Bioprocessing,* 4(28): 1-16. doi: 10.1186/s40643-017-0158-4.

Gagelidze, N.A., Amiranashvili, L.L., Sadunishvili, T.A., Kvesitadze, G.I., Urushadze, T.F. & Kvrivishvili, T.O. 2018. Bacterial composition of different types of soils of Georgia. *Annals of Agrarian Science,* 16(1): 17-21. https://doi.org/10.1016/j.aasci.2017.08.006.

Gallagher, R.V., Randall, R.P. & Leishman, M.R. 2015. Trait differences between naturalized and invasive plant species independent of residence time and phylogeny. *Conservation Biology,* 29: 360–369. doi: 10.1111/cobi.12399.

Gamoun, M., Ouled Belgacem, A., Hanchi, B., Neffati, M., & Gillet, F. 2012. Impact of grazing on the floristic diversity of arid rangelands in South Tunisia. *Revue d'Écologie (Terre Vie),* 67:271-282.

Gao, Z., Karlsson, I., Geisen, S., Kowalchuk, G. & Jousset, A. 2019. Protists: Puppet Masters of the Rhizosphere Microbiome. *Trends in Plant Science,* 24: 165-176.

Garbeva, P., Van Veen, J.A. & van Elsas, J.D. 2004. Microbial diversity in soil: Selection of Microbial Populations by Plant and Soil Type and Implications for Disease Suppressiveness. *Annual Review of Phytopathology,* 42(1): 243-70. https://doi.org/10.1146/annurev.phyto.42.012604.135455

García, A., & Rivero, C. 2012. Efecto de residuos de maíz sobre la actividad enzimática con diferentes sistemas de labranza. http://saber.ucv.ve/ojs/index.php/rev_venes/article/view/4566

García A. & Rivero, C. 2014. Efecto de la incorporación de residuos orgánicos sobre la respiración edáfica basal previamente sometidos a diferentes tipos de labranza. http://www.ucv.ve/uploads/media/Memorias_Jornadas_de_Investigaci%C3%B3n_2014.pdf

García-Gutierrez Báez, C., San José Martínez, F. & Caniego, J. 2017. A protocol for fractal studies on porosity of porous media: High quality soil porosity images. *Journal of Earth Science,* 28: 888. https://doi.org/10.1007/s12583-017-0777-x

Garcia-Montiel, D.C., Neill, C., Melillo, J., Thomas, S., Steudler, P.A. & Cerri, C.C. 2000. Soil Phosphorus Transformations Following Forest Clearing for Pasture in the Brazilian Amazon. *Soil Science Society of America Journal*, 64(5): 1792. https://doi.org/10.2136/sssaj2000.6451792x.

García-Orenes, F., V. Arcenegui, V., Chrenková, K., Mataix-Solera, J., Moltó, J., Jara-Navarro, A.B. & Torres, M.P. 2017. Effects of salvage logging on soil properties and vegetation recovery in a fire-affected Mediterranean forest: A two-year monitoring research. *Total of the Science Environmental*, 586: 1057-1065.

García-Palacios, P., Maestre, F.T., Kattge, J. & Wall, D.H. 2013. *Ecology Letters*, 16: 1045–1053.

Gardi, C., Jeffery, S. & Saltelli, A. 2013. An estimate of potential threats levels to soil biodiversity in EU. *Global Change Biology*, 19: 1538–1548. doi: 10.1111/gcb.12159

Garnier-Sillam E. & Tessier, D. 1991. Rôle des termites sur le spectre poral des sol forestiers tropicaux; cas de Thoracotermes macrothorax (Sjöstedt) et de Macrotermes mûlleri (Sjöstedt). *Insectes sociaux*, 38: 397-412.

Gartaula, H., Patel, K., Johnson, D., Devkota, R., Khadka, K. & Chaudhary, P. 2017. From food security to food wellbeing: examining food security through the lens of food wellbeing in Nepal's rapidly changing agrarian landscape. Agriculture and Human Values, 34(3): 573–589. https://doi.org/10.1007/s10460-016-9740-1

Gaur, A. & Adholeya, A. 2004. Prospects of arbuscular mycorrhizal fungi in phytoremediation of heavy metal contaminated soils. *Current Science*, 86(4): 528-534.

Gaur, N., Flora, G., Yadav, M. & Tiwari, A. 2014. A review with recent advancements on bioremediation-based abolition of heavy metals. *Environmental Science Process Impacts*, 16: 180-193

Ge, T., Wu, X., Chen, X., Yuan, H., Zou, Z., Li, B., Zhou, P., Liu, S., Tong, C., Brookes, P. & Wu, J. 2013. Microbial phototrophic fixation of atmospheric CO_2 in China subtropical upland and paddy soils. *Geochimica et Cosmochimica Acta*, 113: 70–78. https://doi.org/10.1016/j.gca.2013.03.020

Gebremikael, M.T., Steel, H., Buchan, D., Bert, W. & De Neve, S. 2016. Nematodes enhance plant growth and nutrient uptake under C and N-rich conditions. *Scientific Reports*, 6: 32862.

Geisen, S., Mitchell, E.A.D., Wilkinson, D.M., Adl, S., Bonkowski, M., Brown, M.W., Fiore-Donno, A.M., et al. 2017. Soil protistology rebooted: 30 fundamental questions to start with. *Soil Biology and Biochemistry*, 111: 94-103.

Geisen, S., Mitchell, E.A.D., Adl, S., Bonkowski, M., Dunthorn, M., Ekelund, F., et al. 2018. Soil protists: a fertile frontier in soil biology research. *FEMS Microbiology Reviews*, 42: 293-323.

Geisen, S., Briones, M.J.I., Gan, H., Behan-Pelletier, V.M., Friman, V.-P., de Groot, G.A., Hannula, S.E., *et al.* 2019. A methodological framework to embrace soil biodiversity. *Soil Biology and Biochemistry,* 136: 107536.

Geisen, S., Wall, D.H. & van der Putten, W.H. 2019. Challenges and opportunities for soil biodiversity in the Anthropocene. *Current Biology,* 29: R1036-R1044.

General Commission for Scientific Agricultural Research - GCSAR. 2017. [online]. [Cited 5 October 2020]. http://gcsar.gov.sy

George, P.B.L. & Lindo, Z. 2015a. Application of body size spectra to nematode trait-index analyses. *Soil Biology and Biochemistry,* 84: 15-20.

George, P.B.L. & Lindo, Z. 2015b. Congruence of community structure between taxonomic identification and T-RFLP analyses in free-living soil nematodes. *Pedobiologia,* 58: 113-117.

George, P.B., Lallias, D., Creer, S., Seaton, F.M., Kenny, J.G., Eccles, R.M., Griffiths, R.I., *et al.* 2019. Divergent national-scale trends of microbial and animal biodiversity revealed across diverse temperate soil ecosystems. *Nature Communications,* 10(1): 1107.

George, P.B.L., Keith, A.M., Creer, S., Barrett, G.L., Lebron, I., Emmett, B.A., Robinson, D.A. & Jones, D.L. 2017. Evaluation of mesofauna communities as soil quality indicators in a national-level monitoring programme. *Soil Biology and Biochemistry,* 115: 537-546.

Gerasimova, M.I., Bogdanova, M.D. & Nikitin, E.D. 2014. Geographical and genetic aspects of the Red Book of Russian Soils. *Moscow Univeristy Soil Science Bulletin,* 69: 49-54. https://doi.org/10.3103/S0147687414020045.

Gergócs, V. & L. Hufnagel. 2009. *Applied Ecology and Environmental Research,* 7: 79–98.

Gerlach, J., Samways, L. & Pryke, M.J. 2013. Terrestrial invertebrates as bioindicators: an overview of available taxonomic groups. *Journal of Insect Conservation,* 17: 831.

German Center for Integrative Biodiversity Research Halle-Jena-Leipzig – iDiv. 2020. [online]. [Cited 2 October 2020]. https://www.idiv.de/de/index.html

German Environment Agency (Umweltbundesamt – UBA). 2014. Learning about and experiencing soil. [online]. [Cited 2 October 2020]. https://www.umweltbundesamt.de/en/topics/soil-agriculture/learning-about-experiencing-soil

German Environment Agency (Umweltbundesamt – UBA). 2019. KBU Conference "Biodiversity - everything is related to everything". [online]. [Cited 2 October 2020]. https://www.umweltbundesamt.de/service/termine/kbu-tagung-biodiversitaet-alles-haengt-allem

German Environment Agency (Umweltbundesamt – UBA). 2020. Soil Protection Commission. [online]. [Cited 2 October 2020]. https://www.umweltbundesamt.de/themen/boden-landwirtschaft/kommissionen-beiraete/kommission-bodenschutz-0#textpart-1

German Federal Ministry of Food and Agriculture. 2020. Ground. [online]. [Cited 2 October 2020]. https://www.bmel.de/DE/Landwirtschaft/Pflanzenbau/Boden/Boden_node.html

German Soil Science Society – DBG. 2020. World Soil Day. [online]. [Cited 2 October 2020]. https://www.dbges.de/de/boden-des-jahres

Gerson U., R. L. Smiley, R.L. & Ochoa, R. 2003. *Mites (Acari) for Pest Control*. Oxford, UK. Blackwell Science. 539 pp.

Gerz, M., Guillermo Bueno, C., Ozinga, W.A., Zobel, M. & Moora, M. 2018. Niche differentiation and expansion of plant species are associated with mycorrhizal symbiosis. *Journal of Ecology*, 106(1): 254-264.

Ghabrial, S.A., Castón, J.R., Jiang, D., Nibert, M.L. & Suzuki, N. 2015. 50-plus years of fungal viruses. *Virology*, 479-480: 356-368.

Ghodbane, R. & M. Drancourt. 2013. Non-human sources of Mycobacterium tuberculosis. *Tuberculosis*, 93: 589–595.

Ghosh, D., Roy, K., Williamson, K.E., Srinivasiah, S., Wommack, K.E. & Radosevich, M. 2009. Acyl-Homoserine Lactones Can Induce Virus Production in Lysogenic Bacteria: an Alternative Paradigm for Prophage Induction. *Applied and Environmental Microbiology*, 75: 7142.

Ghosh, D., Roy, K., Williamson, K.E., White, D.C., Wommack, K.E., Sublette, K.L. & Radosevich, M. 2008. Prevalence of Lysogeny among Soil Bacteria and Presence of 16S rRNA and trzN Genes in Viral-Community DNA. *Applied and Environmental Microbiology*, 74: 495.

Ghoulam, C., Foursy, A. & Fares, K. 2002. Effects of salt stress on growth, inorganic ions and proline accumulation in relation to osmotic adjustment in five sugar beet cultivars. *Environmental and Experimental Botany*, 47: 39– 50.

Gianinazzi, S., Gollotte, A., Binet, M.-N., van Tuinen, D., Redecker, D. & Wipf, D. 2010. Agroecology: the key role of arbuscular mycorrhizas in ecosystem services. *Mycorrhiza*, 20: 519–530. doi: 10.1007/s00572-010-0333-3

Giardino, J. & Houser, C., eds. 2015. *Principles and Dynamics of the Critical Zone*. Amsterdam, Netherlands. Elsevier. 674 pp.

Gibb, H., Sanders, N.J., Dunn, R.R., Arnan, X., Vasconcelos, H.L., Donoso, D.A., Andersen, A.N., et al. 2018. Habitat disturbance selects against both small and large species across varying climates. *Ecography*, 41(7): 1184–1193. doi:10.1111/ecog.03244

Gibb, H., Sanders, N.J., Dunn, R.R., Watson, S., Photakis, M., Abril, S., Andersen, A.N., et al. 2015. Climate mediates the effects of disturbance on ant assemblage structure. *Proceedings of Royal Society London B*, 282: 20150418. doi:10.1098/rspb.2015.0418

Gibson, L., Lee, T.M., Koh, L.P., Brook, B.W., Gardner, T.A., Barlow, J., Peres, C.A. et al., 2011. Primary Forests Are Irreplaceable for Sustaining Tropical Biodiversity. *Nature*, 478: 378–81. https://doi.org/10.1038/nature10425.

Giles, M., Morley, N., Baggs, E.M. & Daniell, T.J. 2012. Soil nitrate reducing processes: drivers, mechanisms for spatial variation, and significance for nitrous oxide production. *Frontiers in Microbiology*, 3: DOI: 407. 10.3389/fmicb.2012.00407.

Giller, K.E. 2001. Nitrogen Fixation in Tropical Cropping Systems, 2nd Edition. Wallingford, UK. CAB International.

Gillespie, I. & Philp, J. 2013. Bioremediation, an environmental remediation technology for the bioeconomy. *Trends in Biotechnology,* 31(6): 329-332 doi: 10.1016/j.tibtech.2013.01.015.

Gilyarov, M.S. 1949. *Characteristic Features of Soil as a Habitat and Its Significance in the Evolution of Insects*. Moscow, Russia. Nauka.

Gisin, H. 1943. Ökologie und Lebensgemeinschaften der Collembolen im schweizerischen Exkursionsgebiet Basels. Geneva, Switzerland. A. Kundig. 94pp.

Glaser, A., ed. 2012. *America's Grasslands Conference: Status, Threats, and Opportunities.* Proceedings of the 1st Biennial Conference on the Conservation of America's Grasslands. August 15- 17, 2011, Sioux Falls, SD. Washington, DC, National Wildlife Federation & Brookings, SD, South Dakota State University.

Glick, P.A. 1939. *The distribution of insects, spiders and mites in the air.* Technical Bulletin No. 673. Washington, DC. U.S. Department of Agriculture. 147 pp.

Global Harvest Choice. 2010. *Agro-ecological Zones of Sub-Saharan Africa*. Washington, DC, International Food Policy Research Institute & St. Paul, MN, University of Minnesota.

Globe Nederland. 2020. The living soil. [online]. [Cited 2 October 2020]. https://globenederland.nl/onderzoeksproject/bodem/

GMEP. 2020. *Glastir Monitoring and Evaluation Programme* [online]. [Cited 2 October 2020]. https://gmep.wales/

Godfrey, S.A.C. & Marshall, J.W. 2002. Soil on imported shipping containers provides a source of new Pseudomonad biodiversity into New Zealand. *New Zealand Journal of Crop and Horticultural Science,* 30: 19–27.

Göhre, V. & Paszkowski, U. 2006. Contribution of the Arbuscular Mycorrhizal Symbiosis to Heavy Metal Phytoremediation. Planta, 223 (6): 1115-1122

Gómez, Y. & Paolini, J. 2003. Actividad microbiana en suelos de sabanas de los Llanos Orientalesde Venezuela convertidas en pasturas. *Revista de Biología Tropical* 54(2): 273-285. https://www.academia.edu/4874023/Actividad_microbiana_en_suelos_de_sabanas_de_los_Llanos_Orientales_de_Venezuela_convertidas_en_pasturas

Gongalsky K.B. & Zaitsev A.S. 2016. The Role of Spatial Heterogeneity of the Environment in Soil Fauna Recovery after Fires. *Doklady Earth Sciences,* 471(2): 1265–1268. DOI: 10.1134/S1028334X16120035.

Gongalsky, K.B., Malmström, A., Zaitsev, A.S., Shakhab, S.V., Bengtsson, J. & Persson, T. 2012. Do burned areas recover from inside? An experiment with soil fauna in a heterogeneous landscape. *Applied Soil Ecology,* 59: 73-86

González, J.M. & Suttle, C. 1993. Grazing by marine nanoflagellates on viruses and virus-sized particles: Ingestion and digestion. *Marine Ecology-progress Series,* 94: 1-10.

González-Andrés, F. & James, E., eds. 2016. *Biological Nitrogen Fixation and Beneficial Plant-Microbe Interaction.* Springer International Publishing. (also available at https://www.springer.com/gp/book/9783319325262).

González-Esquivel, C., Gavito, M.E., Astier, M., Cadena-Salgado, M., del Val, E., Villamil-Echeverri, L., Merlín-Uribe, Y. Balvanera, P. 2015. Ecosystem service trade-offs, perceived drivers and sustainability in contrasting agroecosystems in Central Mexico. *Ecology and Society.* 20(1): 38.

Gopal, M., Gupta, A. 2016. Microbiome Selection Could Spur Next-Generation Plant Breeding Strategies. *Front Microbiol,* 7: doi: 10.3389/fmicb.2016.01971

Gossner, M.M., Lewinsohn, T.M., Kahl, T., Grassein, F., Boch, S., Prati, D., Arndt, H, *et al.* 2016. Land-use intensification causes multitrophic homogenization of grassland communities. *Nature,* 540(7632): 266.

Gould, I.J., Quinton, J.N., Weigelt, A., De Deyn, G.B. & Bardgett, R.D. 2016. Plant diversity and root traits benefit physical properties key to soil function in grasslands. *Ecology Letters,* 19(9): 1140-1149.

Goulet, H. 2003. Biodiversity of ground beetles (Coleoptera: Carabidae) in Canadian agricultural soils. *Can. J. Soil Sci.,* 83: 259–264.

Government of Canada. 2018. Canadian Soil Information Service. [online]. [Cited 5 October 2020]. http://sis.agr.gc.ca/cansis/

Government of Canada. 2020a. Soil and land. [online]. [Cited 5 October 2020]. http://www.agr.gc.ca/eng/science-and-innovation/agricultural-practices/soil-and-land/?id=1370345323701

Government of Canada. 2020b. Government of Canada Publications. [online]. [Cited 5 October 2020]. http://publications.gc.ca/collections/Collection/A42-70-2-1997E.pdf

Government of Canada. 2020c. Soil Cover Indicator. [online]. [Cited 5 October 2020]. http://www.agr.gc.ca/eng/science-and-innovation/agricultural-practices/soil-and-land/soil-cover-indicator/?id=1462489641309#c

Government of Canada. 2020d. Soil management. [online]. [Cited 5 October 2020]. http://www.agr.gc.ca/eng/science-and-innovation/agricultural-practices/soil-and-land/soil-management/?id=1370346218601

Government of Colombia. 2017. Ministerio de Agricultura y Desarollo Rural. Resolución numero 000464 de 2017. [online]. [Cited 5 October 2020]. https://www.minagricultura.gov.co/Normatividad/Resoluciones/Resoluci%C3%B3n%20No%20000464%20de%202017.pdf

Government of Colombia. 2012. PROGRAMA NACIONAL DE MONITOREO Y SEGUIMIENTO DE LA DEGRADACIÓN DE SUELOS Y TIERRAS EN COLOMBIA [online]. [Cited 5 October 2020]. http://www.ideam.gov.co/documents/11769/153422/Adicionalmente+consulte_C2-C-RE-S%C3%A1nchez%2C+R.pdf/d5a21efa-18e7-486c-b925-80560ae91b3c

Government of Colombia. 2018. Ley 1930 de 2018. [online]. [Cited 5 October 2020]. https://www.funcionpublica.gov.co/eva/gestornormativo/norma.php?i=87764

Government of the Republic of Moldova. 2014. DECISION No. 301 of 24.04.2014 on the approval of the Environmental Strategy for 2014-2023 and the Action Plan for its implementation[online]. [Cited 2 October 2020]. http://lex.justice.md/viewdoc.php?action=view&view=doc&id=352740&lang=1

Government of the Republic of Moldova. 2018. DECISION OF THE GOVERNMENT OF THE REPUBLIC OF MOLDOVA - On the approval of the Strategy on biological diversity of the Republic of Moldova for 2015-2020 and the Action Plan for its implementation. [online]. [Cited 2 October 2020]. http://base.spinform.ru/show_doc.fwx?rgn=76072

Government of the Netherlands. 2017. Letter to Parliament on nature-based agriculture. [online]. [Cited 2 October 2020]. https://www.government.nl/documents/parliamentary-documents/2017/07/10/letter-to-parliament-on-nature-based-agriculture

Government of the Netherlands. 2018. Vision Ministry of Agriculture, Nature and Food Quality. [online]. [Cited 2 October 2020]. https://www.government.nl/documents/policy-notes/2018/11/19/vision-ministry-of-agriculture-nature-and-food-quality---english

Government of the Netherlands. 2018. Letter to Parliament - Soil strategy. [online]. [Cited 2 October 2020]. https://www.rijksoverheid.nl/ministeries/ministerie-van-landbouw-natuur-en-voedselkwaliteit/documenten/kamerstukken/2018/05/23/kamerbrief---bodemstrategie

Government of the Netherlands. 2019. Climate deal makes halving carbon emissions feasible and affordable. [online]. [Cited 2 October 2020]. https://www.government.nl/latest/news/2019/06/28/climate-deal-makes-halving-carbon-emissions-feasible-and-affordable

Government of the Republic of Trinidad and Tobago. 2016. Vision 2030. [online]. [Cited 5 October 2020]. https://www.planning.gov.tt/sites/default/files/Vision%202030-%20The%20National%20Development%20Strategy%20of%20Trinidad%20and%20Tobago%202016-2030.pdf

Government of Ukraine. 2012. LAW OF UKRAINE About priority directions of innovative activity in Ukraine. [online]. [Cited 2 October 2020]. https://zakon.rada.gov.ua/laws/show/3715-17/ed20110908

Government of Ukraine. 2019. LAW OF UKRAINE On the basic principles and requirements for organic production, circulation and labeling of organic products. [online]. [Cited 2 October 2020]. https://zakon.rada.gov.ua/laws/show/2496-19

Government of Ukraine. 2019. LAW OF UKRAINE On the basic principles and requirements for organic production, circulation and labeling of organic products. [online]. [Cited 2 October 2020]. https://zakon.rada.gov.ua/laws/show/2496-19

Graells, G., Corcoran, D. & Aravena, J. C. 2015. Invasion of North American beaver (Castor canadensis) in the province of Magallanes, Southern Chile: comparison between dating sites through interviews with the local community and dendrochronology. *Revista chilena de historia natural*, 88(1): 3.

Graham, E.B., Knelman, J.E., Schindlbacher, A., Siciliano, S., Breulmann, M., Yannarell, A., Beman, J.M., *et al.* 2016. Microbes as engines of ecosystem function: when does community structure enhance predictions of ecosystem processes? *Frontiers in Microbiology,* 7: 214.

Graham, R.D. & Welch, R.M. 1997. A strategy for breeding staple-food crops with high micronutrient density. In R. Fischer, P.W.F. Labbe, M.R. Cockell, K.A. Gibson, eds. *Trace Elements in Man and Animals 9,* pp. 447-450. Ottawa, Canada. National Research Council Canada.

Granli, T. & Bøckman, O.C. 1994. Nitrous oxide from agriculture. *Norwegian Journal of Agricultural Sciences Supplement,* 12: 7–128.

Grassé, P.-P. 1982. *Termitologia. Tome II. Anatomie, Physiologie, Reproduction des termites.* Paris, France. Fondation Singer-Polignac. 676 p.

Grassé, P.-P. 1986. *Termitologia, Tome III : Comportement, Socialité, Écologie, Évolution, Systematique.* Paris, France. Fondation Singer-Polignac.

Graves, A.R., Morris, J., Deeks, L.K., Rickson, R.J., Kibblewhite, J.A., Farewell, T.S. & Truckle, I. 2015. The total costs of soil degradation in England and Wales. *Ecological Economics,* 119: 399-413.

Greaver, T.L., Sullivan, T.J., Herrick, J.D., Barber, M.C., Baron, J.S., Cosby, B.J. & Novak, K.J. 2012. Ecological effects of nitrogen and sulfur air pollution in the US: what do we know? *Frontiers in Ecology and the Environment,* 10(7): 365–372. doi: 10.1890/110049

Green House Canada. 2016. Growing Interest in Biopesticides. [online]. [Cited 5 October 2020]. https://www.greenhousecanada.com/inputs/biocontrols/growing-interest-in-biopesticides-31051

Greenslade, P.J.N. 1985. Pterygote insects and the soil: their diversity, their effects on soil and the problem of species identification. *Quaestiones entomology,* 21: 571-585.

Gren, I.-M. 1995. The Value of Investing in Wetlands for Nitrogen Abatement. *European Review of Agricultural Economics,* 22: 157-172.

Grêt-Regamey, A., Sirén, E., Brunner, S.H. & Weibel, B. 2017. Review of decision support tools to operationalize the ecosystem services concept. *Ecosystem Services,* 26: 306–315.

Griffin, D.W. 2007. Atmospheric movement of microorganisms in clouds of desert dust and implications for human health. *Clinical Microbiology Reviews,* 20: 459-477.

Griffiths, B.S. & L. Philippot. 2013. Insights into the resistance and resilience of the soil microbial community. *FEMS Microbiology Reviews,* 37: 112–129.

Griffiths, B.S., de Groot, G.A, Laros, I., Stone, D. & Geisen, S. 2018. The need for standardisation: Exemplified by a description of the diversity, community structure and ecological indices of soil nematodes. *Ecological Indicators,* 87: 43-46.

Griffiths, B.S., Ritz, K., Bardgett, R.D., Cook, R., Christensen, S., Ekelund, F., Sørensen, S.J., *et al.* 2000. Ecosystem Response of Pasture Soil Communities to Fumigation-Induced Microbial Diversity Reductions: An Examination of the Biodiversity - Ecosystem Function Relationship. *Oikos*, 90: 279–294.

Griffiths, B.S., Kuan, H.L., Ritz, K., Glover, L.A., McCaig, A.E. & Fenwick, C. 2004. The Relationship between Microbial Community Structure and Functional Stability, Tested Experimentally in an Upland Pasture Soil. *Microbial Ecology*, 47(1): 104–113. https://doi.org/10.1007/s00248-002-2043-7

Griffiths, B.S., Ritz, K., Wheatley, R., Kuan, H. L., Boag, B., Christensen, S., Bloem, J., *et al.* 2001. An examination of the biodiversity–ecosystem function relationship in arable soil microbial communities. *Soil Biology and Biochemistry,* 33(12-13): 1713-1722.

Griffiths, R.I., Thomson, B.C., Plassart, P., Gweon, H.S., Stone, D., Creamer, R.E., Lemanceau, P. & Bailey, M.J. 2016. Mapping and validating predictions of soil bacterial biodiversity using European and national scale datasets. *Applied Soil Ecology,* 97: 61–68. http://dx.doi.org/10.1016/j.apsoil.2015.06.018

Griggs, D., Stafford-Smith, M., Gaffney, O., Rockström, J., Öhman, M.C., Shyamsundar, P., Steffen, W., Glaser, G., Kanie, N. & Noble, I. 2013. Policy: sustainable development goals for people and planet. *Nature,* 495: 305-307.

Griscom, B.W., Adams, J., Ellis, P.W., Houghton, R.A., Lomax, G., Miteva, D.A., Schlesinger, W.H., *et al.* 2017. Natural climate solutions. *Proceedings of the National Academy of Sciences,* 114: 11645–11650.

Grossbard, E. 1952. Antibiotic production by fungi on organic manures and in soil. *Microbiology,* 6(3-4): 295-310.

GSBI. 2016. GLOBAL SOIL BIODIVERSITY ATLAS. VIEW AND DOWNLOAD ATLAS CHAPTERS. [online]. [Cited 5 October 2020]. https://www.globalsoilbiodiversity.org/atlas-introduction

GSBI. 2020. Welcome to the Global Soil Biodiversity Initiative. [online]. [Cited 5 October 2020]. https://www.globalsoilbiodiversity.org/

Guadagnini, A., San José Martínez, F. & Pachepsky, Y. 2013. Scaling in Soil and Other Complex Porous Media. *Vadose Zone Journal,* 12(3): DOI: 10.2136/vzj2008.0127

Guéi, A.M., Baidai, Y., Tondoh, E.J. & Huising, H. 2012. Functional attributes: Compacting versus decompacting earthworms and influence on soil structure. *Current Zoology,* 58: 556-565. DOI: 10.1093/czoolo/58.4.556.

Guerra, C.A., Heintz-Buschart, A., Sikorski, J., Chatzinotas, A., Guerrero-Ramírez, N., Cesarz, S., Beaumelle, L., *et al.* 2019. Blind spots in global soil biodiversity and ecosystem function research. *bioRxiv,* 774356: doi: 10.1101/774356.

Guerrero-Ramírez, N.R. & Eisenhauer, N. 2017. Trophic and non-trophic interactions influence the mechanisms underlying biodiversity–ecosystem functioning relationships under different abiotic conditions. *Oikos,* 126(12): 1748-1759.

Guerrero-Ramírez, N.R., Craven, D., Reich, P.B., Ewel, J.J., Isbell, F., Koricheva, J., Parrotta, J.A., *et al.* 2017. Diversity-dependent temporal divergence of ecosystem functioning in experimental ecosystems. *Nature Ecology and Evolution,* 1(11): 1639-1642.

Guidi, L., Chaffron, S., Bittner, L., Eveillard, D., Larhlimi, A., Roux S., *et al.* 2016. Plankton networks driving carbon export in the oligotrophic ocean. *Nature,* 532: 465-470.

Guilland, C., Maron, P. A., Damas, O. & Ranjard, L. 2018. Biodiversity of urban soils for sustainable cities. *Environmental Chemistry Letters,* 16: 1267-1282.

Guo, D., Fan, Z., Lu, S., Ma, Y., Nie, X., Tong, F. & Peng, X. 2019. Changes in rhizosphere bacterial communities during remediation of heavy metal-accumulating plants around the Xikuangshan mine in southern China. *Scientific Reports,* 9: 1–11.

Guo, S., Xiong, W., Xu, H., Hang, X., Liu, H., Xun, W., Li, R., & Shen, Q. 2018. Continuous application of different fertilizers induces distinct bulk and rhizosphere soil protist communities. *European Journal of Soil Biology,* 88: 8-14.

Guo, X., Liu, N., Li, X., Ding, Y., Shang, F., Gao, Y., Ruan, J., Huang, Y. 2015. Red soils harbor diverse culturable actinomycetes that are promising sources of novel secondary metabolites. *Applied Environmental Microbiology,* 81: 3086 –3103. doi:10.1128/AEM.03859-14.

Gutterman, Y. 1987. Dynamics of porcupine (Hystrix indica Kerr) diggings: their role in the survival and renewal of geophytes and hemicryptophytes in the Negev Desert highlands. *Israel Journal of Botany,* 36: 133–143

Haddaway, N.R., Hedlund, K., Jackson, L.E., Kätterer, T., Lugato, E., Thomsen, I.K., Jørgensen, H.B. & Isberg, P.-E. 2017. How does tillage intensity affect soil organic carbon? A systematic review. *Environmental Evidence,* 6: article 30.

Hagvar, S. 2016. From Litter to Humus in a Norwegian Spruce Forest: Long-Term Studies on the Decomposition of Needles and Cones. *Forests,* 7(9): 186.

Haichar, F.Z., Santaella, C., Heulin, T. & Achouak, W. 2014. Root exudates mediated interactions belowground. *Soil Biology and Biochemistry,* 77: 69-80. doi: 10.1016/j.soilbio.2014.06.017

Hailemariam, M., Birhane, E., Asfaw, Z. & Zewdie, S. 2013. Arbuscular mycorrhizal association of indigenous agroforestry tree species and their infective potential with maize in the rift valley, Ethiopia. *Agroforest Systems,* 87: 1261–1272, DOI 10.1007/s10457-013-9634-9

Haines-Young, R.H. & Potschin, M. 2010. The links between biodiversity, ecosystem services and human well-being. In D. Raffaelli & C. Frid, eds. *Ecosystem Ecology: A New Synthesis. BES Ecological Reviews Series,* pp. 110–139. Cambridge, UK. Cambridge University Press.

Hall, B.G. & Barlow, M. 2004. Evolution of the serine β-lactamases: past, present and future. *Drug Resistance Updates,* 7(2): 111-123.

Hamza, M.A. & Anderson, W.K. 2005. Soil compaction in cropping systems: a review of the nature, causes and possible solutions. *Soil and Tillage Research,* 82: 121-145.

Hanafi, A. & Jauffret, S. 2008. Are long-term vegetation dynamics useful in monitoring and assessing desertification processes in the arid steppe, southern Tunisia. *Journal of Arid Environments*, 72 : 557-572.

Hanegraaf, M., Elsen, E. van den, Haan, J. de & Visser, S. 2019. Bodemkwaliteitsbeoordeling van landbouwgronden in Nederland - indicatorset en systematiek, versie 1.0. (795). https://doi.org/10.18174/498307

Hansel, C.M, Fendorf, S., Jardine, P.M. & Francis, C.A. 2008. Changes in bacterial and archaeal community structure and functional diversity along a geochemically variable soil profile. *Applied Environmental Microbiology*, 74: 1620–1633. DOI: 10.1128/AEM.01787-07.

Happe, K., Kellermann, K. & Balmann, A. 2006. Agent-based Analysis of Agricultural Policies: an Illustration of the Agricultural Policy Simulator AgriPoliS, its Adaptation and Behavior. *Ecology and Society*, 11: 49. http //www.ecologyandsociety.org/vol11/iss41/art49/

Haque, S.A. 2006. Salinity problems and crop production in coastal regions of Bangladesh. *Pakistan Journal of Botany*, 38: 1359-1365.

Hardham, A.R. & Blackman, L.M. 2018. Phytophthora cinnamomi. *Molecular Plant Pathology*, 19: 260–285.

Hardoim, P.R., van Overbeek, L., Berg, G., Pirttilä, A.M., Compant, S., Campisano, A., Döring, M. & Sessitsch, A. 2015. The hidden world within plants: ecological and evolutionary considerations for defining functioning of microbial endophytes. *Microbiology and Molecular Biology Review*, 79: 293-320.

Harms, H., Schlosser, D. & Wick, L.Y. 2011. Untapped potential : exploiting fungi in bioremediation of hazardous chemicals. *Nature Reviews Microbiology*, 9: 177-192.

Harmsen, J. 2007. Measuring bioavailability: from a scientific approach to standard methods. *Journal of Environmental Quality*, 36: 1420-1428.

Harris, C.A., Scott, A.P., Johnson, A.C., Panter, G.H., Sheahan, D., Roberts, M. & Sumpter, J.P. 2014. Principles of sound ecotoxicology. *Environmental Science and Technology*, 48: 3100-3111.

Hart, M., Ehret, D.L., Krunmbein, A., Leung, C., Murch, S., Turi, C. & Franken, P. 2015. Inoculation With Arbuscular Mycorrhizal Fungi Improves the Nutritional Value of Tomatoes. *Mycorrhiza*, 25 : *359–376*. doi: 10.1007/s00572-014-0617-0

Hartemink, A.E., Balks, M.R., Chen, Z.S., Drohan, P., Field, D.J., Krasilnikov, P., *et al.* 2014. The joy of teaching soil science. *Geoderma*, 217: 1-9.

Hartmann, A., Holzinger, A., Ganzera, M. & Karsten, U. 2016. Prasiolin, a new UV-sunscreen compound in the terrestrial green macroalga *Prasiola calophylla* (Carmichael ex Greville) Kützing (Trebouxiophyceae, Chlorophyta). *Planta*, 243(1): 161–169. https://doi.org/10.1007/s00425-015-2396-z

Hartmann, M., Howes, C.G., VanInsberghe, D., Yu, H., Bachar, D., Christen, R., Mohn, W.W., *et al.* 2012. Significant and persistent impact of timber harvesting on soil microbial communities in Northern coniferous forests. *The ISME journal*, 6(12): 2199.

Hartmann, M., Niklaus, P.A., Zimmermann, S., Schmutz, S., Kremer, J., Abarenkov, K., Frey, B., *et al.* 2014. Resistance and resilience of the forest soil microbiome to logging-associated compaction. *The ISME journal*, 8(1): 226.

Hassan S.E.D., Boon, E., St-Arnaud, M. & Hijri, M. 2011. Molecular biodiversity of arbuscular mycorrhizal fungi in trace metal-polluted soils. *Molecular Ecology*, 20: 3469–3483

Hassanin, A., Johnston, A.E., Thomas, G.O. & Jones, K.C. 2005. Time trends of atmospheric PBDEs inferred from archived UK herbage. *Environmental Science and Technology*, 39: 2436-2441.

Hättenschwiler, S. & Gasser, P. 2005. Soil animals alter plant litter diversity effects on decomposition. *Proceedings of the National Academy of Sciences*, 102(5): 1519-1524.

Hättenschwiler, S., Tiunov, A.V. & Scheu, S. 2005. Biodiversity and litter decomposition in terrestrial ecosystems. *Annual Review of Ecology, Evolution, and Syststematics*, 36: 191-218.

Havlin, J., Balster, N., Chapman, S., Ferris, D., Thompson, T. & Smith, T. 2010. Trends in soil science education and employment. *Soil Science Society of America Journal*, 74: 1429-1432.

Hawksworth, D.L. & Lücking, R. 2017. Fungal diversity revisited: 2.2 to 3.8 million species. *Microbiology spectrum*, 5(4): doi: 10.1128/microbiolspec.FUNK-0052-2016.

Hayes, F., Spurgeon, D.J., Lofts, S. & Jones, L. 2018. Evidence-based logic chains demonstrate multiple impacts of trace metals on ecosystem services. *Journal of Environmental Management*, 223: 150-164.

Heal, G. 2000. *Nature and the market place: Capturing the value of ecosystem services.* Washington, USA. Island Press.

Hector A. & Bagchi, R. 2007. Biodiversity and ecosystem multifunctionality. *Nature*, 448: 188-190.

Hedlund, K. 2012. *Soil as natural capital: Agricultural production, soil fertility and farmers economy.* SOILSERVICE Policy Brief. Lund University. 4 pp.

Hedlund, K., Griffiths, B., Christensen, S., Scheu, S., Setälä, H., Tscharntke, T. & Verhoef, H. 2004. Trophic interactions in changing landscapes: Responses of soil food webs. *Basic and Applied Ecology*, 5: 495-503. doi: 10.1016/j.baae.2004.09.002

Heemsbergen, D.A., Berg, M.P., Loreau, M., Van Hal, J.R., Faber, J.H. & Verhoef, H.A. 2004. Biodiversity effects on soil processes explained by interspecific functional dissimilarity. *Science*, 306(5698): 1019-1020.

Helmholtz Centre for Environmental Research (UFZ). 2017. Rhizosphere Spatiotemporal Organisation – a Key to Rhizosphere Functions" (SPP 2089) [online]. [Cited 2 October 2020]. https://www.ufz.de/spp-rhizosphere/

Hendrix, P.F., Mueller, B.R., Bruce, R.R., Langdale, G.W. & Parmelee, R.W. 1992. Abundance and distribution of earthworms in relation to landscape factors on the Georgia Piedmont, USA. *Soil Biology and Biochemistry,* 24: 1357–1361.

Hendrix, P.F., Callaham, M.A., Drake, J.M., Huang, C.-Y., James, S.W., Snyder, B.A. & Zhang, W. 2008. Pandora's contained bait: The global problem of introduced earthworms. *Annual Review of Ecology, Evolution, and Systematics,* 39: 593-613.

Hernández-Becerra, N., Tapia-Torres, Y., Beltrán-Paz, O., Blaz, J., Souza, V. & Garcia-Oliva, F. 2016. Agricultural land-use change in a Mexican oligotrophic desert depletes ecosystem stability. *PeerJ,* 4: e2365

Hernández, R.M., Castro, I., Lozano, Z., Toro, M., Bravo, C., Torres, A., Ojeda, A., *et al.* 2009. Actividad microbiana en suelos de sabanas cultivados con maíz asociado a pastos, leguminosas y diferentes fuentes de fósforo. XVIII Congreso Venezolano de la Ciencia del Suelo. Santa Bárbara, estado Zulia, Venezuela.

Heywood, V.H. 1995. *Global biodiversity assessment.* United Nations Environment Programme. Cambridge, UK. Cambridge University Press.

Hickman, Z.A. & Reid, B.J. 2008. Earthworm assisted bioremediation of organic contaminants. *Environment International,* 34: 1072–1081.

Higa, T. 1996. Effective microorganisms–Their role in Kyusei Nature Farming. In J.F. Parr, *et al.*, eds. *Proceedings of the 3rd International Nature Farming Conference,* pp. 20-23. Washington, USA. United States Department of Agriculture.

Hiiesalu, I., Pärtel, M., Davison, J., Gerhold, P., Metsis, M., Moora, M., Öpik, M., Vasar, M., Zobel, M. & Wilson, S.D. 2014. Species richness of arbuscular mycorrhizal fungi: associations with grassland plant richness and biomass. *New Phytologist,* 203(1): 233-244.

Hinchliff, C.E., Smith, S.E., Allman, J.F., Burleigh, J.G., Chaudhary, R., Coghill, L.M., Crandall, *et al.* 2015. Synthesis of phylogeny and taxonomy into a comprehensive tree of life. *Proceedings of the National Academy of Sciences,* 112: 12764–12769.

Hirano, T. & Tamae, K. 2011. Earthworms and Soil Pollutants. *Sensors,* 11: 11157-11167.

Hirsch, P.R. 2019. Microrganisms cycling soil nutrients. In J.D. van Elsas, J.T. Trevors, A. Soares Rosado & P. Nannipieri, eds. *Modern Soil Microbiology III,* pp. 179-194. Boca Raton, USA. CRC Press.

Hobbie, S.E., Reich, P.B., Oleksyn, J., Ogdahl, M., Zytkowiak, R., Hale, C. & Karolewski, P. 2006. Tree species effects on decomposition and forest floor dynamics in a common garden. *Ecology,* 87: 2288-2297.

Hobbs, R.J., Higgs, E. & Harris, J.A. 2009. Novel ecosystems: implications for conservation and restoration. *Trends in Ecology and Evolution,* 24: 599-605.

Hoffmann, L. 1989. Algae of terrestrial habitats. *The Botanical Review,* 55(2): 77–105. https://doi.org/10.1007/BF02858529

Hooper, D.U., Adair, E.C., Cardinale, B.J., Byrnes, J.E., Hungate, B.A., Matulich, K.L., Gonzalez, A., Duffy, J.E., Gamfeldt, L. & O'Connor, M.I. 2012. A global synthesis reveals biodiversity loss as a major driver of ecosystem change. *Nature,* 486: 105-108. https://doi.org/10.1038/nature11118

Hoogen, J., Geisen, S., Routh, D., Ferris, H., Traunspurger, W., Wardle, D.A., de Goede, R.G.M., et al. 2019. Soil nematode abundance and functional group composition at a global scale. *Nature,* 572: 194–198 doi:10.1038/s41586-019-1418-6.

Hopkin, A.S.P. & Martin, M.H. 1982. The Distribution of Zinc, Cadmium, Lead and Copper within the Woodlouse Oniscus asellus (Crustacea, Isopoda). *Oecologia,* 54: 227–232.

Hopkin, S.P. 2007. *A Key to the Collembola (Springtails) of Britain and Ireland.* Telford, UK. Field Studies Council publications. 252 pp.

Horrigue, W., Dequiedt, S., Chemidlin Prévost-Bouré, N., Jolivet, C., Saby, N.P.A., Arrouays, D., Bispo, A., Maron, P.-A. & Ranjard, L. 2016. Predictive model of soil molecular microbial biomass. *Ecological Indicators,* 64: 203–211. https://doi.org/10.1016/j.ecolind.2015.12.004

Horton, A.A., Walton, A., Spurgeon, D.J., Lahive, E. & Svendsen, C. 2017. Microplastics in freshwater and terrestrial environments: Evaluating the current understanding to identify the knowledge gaps and future research priorities. *Science of the Total Environment* 586: 127-141.

Howard-Varona, C., Hargreaves, K.R., Abedon, S.T. & Sullivan, M.B. 2017. Lysogeny in nature: mechanisms, impact and ecology of temperate phages. *The ISME Journal,* 11: 1511.

Hoy M.A. 2011. *Agricultural Acarology: Introduction to Integrated Mite Management.* Boca Raton, USA. CRC Press, Taylor and Francis Groups. 410 pp.

Hu, L., Xia, M., Lin, X., Xu, C., Li, W., Wang, J., Zeng, R. & Song, Y. 2018. Earthworm gut bacteria increase silicon bioavailability and acquisition by maize. *Soil Biology and Biochemistry,* 125: 215-221.

Huang, L.-F., Song, L.-X., Xia, X.-J., Mao, W.-H., Shi, K., Zhou, Y.-H. & Yu, J.-Q. 2013. Plant-Soil Feedbacks and Soil Sickness: From Mechanisms to Application in Agriculture. *Journal of Chemical Ecology,* 39(2): 232–242. https://doi.org/10.1007/s10886-013-0244-9

Humphrey, G.S. & Mitchell, P.B. 1983. A preliminary assessment of the role of bioturbation and rain wash on sandstone hillslopes in the sydney basin. In R.W. Young & G.C. Nansan, eds. *Aspects of Australian sandstone landscape,* pp. 66-79. Australian and New Zealand Geomorphology group.

Hungria, M., Franchini, J. C., Brandao-Junior, O., Kaschuk, G., & Souza, R. A. 2009. Soil microbial activity and crop sustainability in a long-term experiment with three soil-tillage and two crop-rotation systems. *Applied Soil Ecology,* 42(3), 288-296.

Hunt, H.W., Coleman, D.C. & Ingham, E.R. 1987. The detrital food-web in a short-grass prairie. *Biology and Fertility of Soils,* 3: 57–68

Husain, Q., Husain, M. & Kulshrestha, Y. 2009. Remediation and treatment of organopollutants mediated by peroxidases: a review. *Critical Reviews in Biotechnology,* 29(2): 94–119.

Ibáñez, J.J. 2018. Diversity of soils. In: *Oxford Bibliographies. Geography* [online]. Oxford, UK. Oxford University Press. https://www.oxfordbibliographies.com/view/document/obo-9780199874002/obo-9780199874002-0104.xml

Ibáñez, J.J. & Bockheim, J. 2013. *Pedodiversity*. Boca Raton, USA. C.R.C. Press. 256pp.

Ibáñez, J.J. & Boixadera, J. 2002. The search for a new paradigm in pedology: a driving force for new approaches to soil classification. In: E. Micheli, F. Nachtergaele, R.J.A. Jones & L. Montanarella, eds. Soil Classification 2001, pp. 93-110. EU JRC, Hungarian Soil. Sci. Soc., FAO, EU, Italy. EUR - Scientific and Technical Research Reports.

Ibáñez, J.J. Jiménez-Ballesta, R. & García-Álvarez, A. 1990. Soil Landscapes and drainage basins in Mediterranean mountain areas. *Catena,* 17(6): 573-583. https://doi.org/10.1016/0341-8162(90)90031-8.

Ibáñez, J.J. Krasilnikov, P., & Saldaña, A. 2012. Archive and refugia of soil organisms: applying a pedodiversity framework for the conservation of biological and non-biological heritages. *Journal of Applied Ecology,* 49 (6): 1267–1277. https://doi.org/10.1111/j.1365-2664.2012.02213.x.

Ibáñez, J.J., De-Alba, S., Bermúdez, F.F. & García-Álvarez, A. 1995. Pedodiversity: concepts and measures. *Catena,* 24: 215-232. https://doi.org/10.1016/0341-8162(95)00028-Q.

Ibáñez, J.J. De-Alba, S. Lobo, A. & Zucarello, V. 1998. Pedodiversity and global soil patterns at coarser scales (with Discussion). *Geoderma,* 83: 171-192. https://doi.org/10.1016/S0016-7061(97)00147-X

Ibáñez, J.J., García Álvarez, A. Saldaña, & A. Recatalá, L. 2003. Scientific rationality, quantitative criteria and practical implications in the design of soil reserves networks: their role in soil biodiversity and soil quality studies. In: M.C. Lobo & J.J. Ibáñez, eds. *Preserving Soil Quality and Soil Biodiversity; The Role of Surrogate Indicators,* pp. 191-274. Zaragoza, Spain. IMIA-CSIC.

Ibijbijen, J., Urquiaga, S., Ismali, M., Alves, B.J.R., & Boddey, R.M. 1996. Effect of arbuscular mycorrhizal fungi on growth, mineral nutrition and nitrogen fixation of three varieties of common beans (*Phaseolus vulgaris*). *New Phytologist*, 134 (2): 353-360

ICIMOD. 2020. ICIMOD. In: *ICIMOD* [online]. [Cited 1 October 2020]. https://www.icimod.org/

IFOAM. 2019. The world of organic agriculture. Statistics and emerging trends 2019. Research institute of organic agriculture. https://shop.fibl.org/chen/mwdownloads/download/link/id/1202/

INECC. 2007. Parte II- Recursos Naturales. [online]. [Cited 1 October 2020]. http://www2.inecc.gob.mx/publicaciones2/libros/16/parte2.html

INECOL. 2020. INECOL. In: INECOL [online]. [Cited 5 October 2020]. https://www.inecol.mx

Ingham, E.R. 2009. Chapter 4: Soil Fungus. In *Soil Biology Primer,* pp. 22-23. Ankeny, USA. Soil and Water Conservation Society.

Innocenti, G. & Sabatini, M.A. 2018. Collembola and plant pathogenic, antagonistic and arbuscular mycorrhizal fungi: a review. *Bulletin of Insectology,* 71: 71–76.

INPE. 2017. *Monitoramento Da Floresta Amazónica Brasileira Por Satélite* [online]. Projeto PRODES. http://www.obt.inpe.br/OBT/assuntos/programas/amazonia/prodes

INRAE. 2020a. Plateforme CA-SYS - Accueil. In: *INRAE* [online]. [Cited 2 October 2020]. https://www6.inrae.fr/plateforme-casys/

INRAE. 2020b. INRAE. In: INRAE [online]. [Cited 2 October 2020]. http://www.inra.fr/Chercheurs-etudiants/Agroecologie/Toutes-les-actualites/Atlas-francais-des-bacteries-du-sol

Instituto Geográfico Agustín Codazzi (IGAC). 2020. Laboratorio Nacional de Suelos. In: Instituto Geográfico Agustín Codazzi [Online]. [Cited 2 October 2020]. https://www.igac.gov.co/es/contenido/areas-estrategicas/agrologia/laboratorio-nacional-de-suelos

Instituto Nacional de Investigaciones Agricolas (INIA). 2017. INIA active plan nacional de suelos en apoyo a la siembra de rubros estratégicos para el país [online]. [Cited 6 October 2020] http://www.inia.gob.ve/index.php/informacion/reportajes/1036-inia-activa-su-plan-nacional-de-suelos-en-apoyo-a-la-siembra-de-rubros-estrategicos-para-el-pais-2

Instituto Nacional de Innovación Agraria (INIA). 2020. Área de Recursos Genéticos de Microorganismos y Biodiversidad Asociada. In: INIA [Online]. [Cited 2 October 2020]. https://drgb.inia.gob.pe/area-de-recursos-geneticos-de-microorganismos-y-biodiversidad-asociada/

Instituto Nacional de Salud Agricola Integral (INSAI). 2020. Biofertilizantes [online]. [Cited 2 October 2020]. http://www.insai.gob.ve/?page_id=175

Instituto Superiore per la Protezione e la Ricerca Ambientale – ISPRA. 2020. Dare priorità agli interventi di riuso o riorganizzazione rispetto a nuovi impegni del suolo In: ISPRA. [online]. [Cited 2 October 2020]. http://www.sinanet.isprambiente.it/gelso/vocabolari/territorio-e-paesaggio/dare-priorita-agli-interventi-di-riuso-o-riorganizzazione-rispetto-a-nuovi-impegni-del-suolo

IPBES. 2018a. The IPBES assessment report on land degradation and restoration. Bonn, Germany. Secretariat of the Intergovernmental Science-Policy Platform on Biodiversity and Ecosystem Services. 744 pp.

IPBES. 2018b. *Summary for policymakers of the assessment report on land degradation and restoration of the Intergovernmental Science- Policy Platform on Biodiversity and Ecosystem Services.* Bonn, Germany. IPBES secretariat.

IPCC. 2006. IPCC Guidelines for National Greenhouse Gas Inventories Volume 4: Agriculture, Forestry and other Land Use. Hayama, Japan. The Institute for Global Environmental Strategies.

IPPC. 2007. *Movement of soil and growing media in association with plants in international trade.* Rome, Italy. IPPC.

IPCC. 2014. *Climate Change 2014: Synthesis Report.* Contribution of Working Groups I, II and III to the Fifth Assessment Report of the Intergovernmental Panel on Climate Change. Geneva, Switzerland. IPCC. 151 pp.

IPCC. 2019. *Climate change and land: Summary for Policymakers.* IPCC Special Report on Climate Change, Desertification, Land Degradation, Sustainable Land Management, Food Security, and Greenhouse gas fluxes in Terrestrial Ecosystems. Cheltenham and Camberley, UK. Edward Elgar Publishing.

Isaac, P., Martínez, F.L., Bourguignon, N., Sánchez, L.A. & Ferrero, M.A. 2015. Improved PAHs removal performance by a defined bacterial consortium of indigenous Pseudomonas and actinobacteria from Patagonia, Argentina. *International Biodeterioration & Biodegradation,* 101: 23-31. doi: 10.1016/j.ibiod.2015.03.014

Isbell, F., Adler, P R., Eisenhauer, N., Fornara, D., Kimmel, K., Kremen, C., Scherer-Lorenzen, M, *et al.* 2017b. Benefits of increasing plant diversity in sustainable agroecosystems. *Journal of Ecology,* 105(4): 871-879.

Isbell, F., Craven, D., Connolly, J., Loreau, M., Schmid, B., Beierkuhnlein, C., Ebeling, A., *et al.* 2015. Biodiversity increases the resistance of ecosystem productivity to climate extremes. *Nature,* 526(7574): 574.

Isbell, F., Gonzalez, A., Loreau, M., Cowles, J., Diaz, S., Hector, A., Turnbull, L.A., *et al.* 2017a. Linking the influence and dependence of people on biodiversity across scales. *Nature,* 546(7656): 65.

IUCN Environmental Law Programme. 2018. *The Year in Review 2018.* IUCN Environmental Law Centre & IUCN World Commission on Environmental Law. 16 pp. https://www.iucn.org/sites/dev/files/content/documents/iucn_elp_year_in_review_2018.pdf

Jackson, R.B., Lajtha, K., Crow, S.E., Hugelius, G., Kramer, M.G. & Piñeiro, G. 2017. The Ecology of Soil Carbon: Pools, Vulnerabilities, and Biotic and Abiotic Controls. *Annual Review of Ecology, Evolution, and Systematics,* 48: 419–445.

James, A.I., Eldridge, D.J. & Hill, B.M. 2009. Foraging animals create fertile patches in an Australian desert shrubland. *Ecography,* 32: 723-732.

Jandl, R., Rodeghiero, M., Martinez, C., Cotrufo, M.F., Bampa, F., van Wesemael, B., Harrison, R.B., *et al.* 2014. Current status, uncertainty and future needs in soil organic carbon monitoring. *Science of the Total Environment,* 468–469: 376–383. https://doi.org/10.1016/j.scitotenv.2013.08.026

Jangid, K., Williams, M.A., Franzluebbers, A.J., Sanderlin, J.S., Reeves, J.H., Jenkins, M.B., Endale, D.M., Coleman, D.C. & Whitman, W.B. 2008. Relative impacts of land-use, management intensity and fertilization upon soil microbial community structure in agricultural systems. *Soil Biology and Biochemistry,* 40(11): 2843-2853. doi:10.1016/j.soilbio.2008.07.030.

Janion, C., Leinaas, H.P., Terblanche, J.S. & Chown, S.L. 2010. Trait means and reaction norms: The consequences of climate change/invasion interactions at the organism level. *Evolutionary Ecology,* 24: 1365–1380.

Jankowski, K., Schindler, D.E. & Horner-Devine, M.C. 2014. Resource availability and spatial heterogeneity control bacterial community response to nutrient enrichment in lakes. *PLoS One,* 9: p.e86991.

Jänsch, S., Steffens, L., Höfer, H., Horak, F., Roß-Nickoll, M., Russell, D., Burkhardt, U., Toschki, A. & Römbke, J. 2013. State of knowledge of earthworm communities in German soils as a basis for biological soil quality assessment. : 19.

Jeasen, C.E., Percich, J.A. & Graham, P.H. 2002. Integrated management strategies of bean root rot with Bacillus subtillis and Rhizobium in Minnesota. *Field Crops Research,* 74: 107-115.

Jeffery, S., Gardi, C., Jones, A., Montanarella, L., Marmo, L., Miko, L., Ritz, K., Peres, G., Rǿmbke, J. & van der Putten, W.H., eds. 2010. *European Atlas of Soil Biodiversity.* Luxembourg. European Commission, Publications Office of the European Union. 128 pp.

Jeffery, M. AC, AO(Mil), CVO, MC (Retd). 2017. *Restore the Soil: Prosper the Nation.* A report submitted to the Australian Prime Minister December 2017.

Jeffery, S., Jones, A., Gardi, C., Montanarella, L., Marmo, L., Miko, L., Ritz, K., Peres, G., Römbke, J. & van der Putten, W., eds. 2010. *European Atlas of Soil Biodiversity.* Luxembourg. Publications Office of the European Union.

Jelinski, N.A., Moorberg, C.J., Ransom, M.D. & Bell, J.C. 2019. A Survey of Introductory Soil Science Courses and Curricula in the United States. *Natural Sciences Education,* 48: 1-13.

Jenkinson, D.S. 1990. The turnover of organic carbon and nitrogen in soil. *Philosophical Transactions of the Royal Society London B,* 329: 361-388. https://doi.org/10.1098/rstb.1990.0177

Jing, X., Sanders, N.J., Shi, Y., Chu, H., Classen, A.T., Zhao, K., Chen, L., Shi, Y., Jiang, Y. & He, J-.S. 2015. The links between ecosystem multifunctionality and above- and belowground biodiversity are mediated by climate. *Nature Communication,* 6: 8159.

Jo, I., Fridley, J.D. & Frank, D.A. 2017. Invasive plants accelerate nitrogen cycling: evidence from experimental woody monocultures. *Journal of Ecology,* 105: 1105–1110. doi:10.1111/1365-2745.12732

Johnson, M.J., Lee, K.Y. & Scow, K.M. 2003. DNA fingerprinting reveals links among agricultural crops, soil properties, and the composition of soil microbial communities. *Geoderma,* 114: 279–303.

Johnston, A.S.A., Hodson, M.E., Thorbek, P., Alvarez, T. & Sibly, R.M. 2014. An energy budget agent-based model of earthworm populations and its application to study the effects of pesticides. *Ecological Modelling,* 280: 5-17.

Joimel, S., Schwartz, C., Hedde, M., Kiyota, S., Krogh, P. H., Nahmani, J., *et al.* 2017. Urban and industrial land uses have a higher soil biological quality than expected from physicochemical quality. *Science of The Total Environment,* 584: 614-621.

Joner, E.J., Briones, R. & Leyval, C. 2000. Metal-binding capacity of arbuscular mycorrhizal mycelium. *Plant Soil,* 226: 227–234.

Jones, A., Stolboyov, V., Rusco, E., Gentile, A.R., Gardi, C., Marechal, B. & Montanarella, L. 2009. Climate change in Europe. 2. Impact on soil. A review. *Agronomy for Sustainable Development,* 29: 423-432. https://doi.org/10.1051/agro:2008067

Jones, F.G.W. & Thomasson, A.J. 1976. Bulk density as an indicator of pore space in soils usable by nematodes. *Nematologica,* 22: 133-137.

Jones, J.B., Jackson, L.E., Balogh, B., Obradovic, A., Iriarte, F.B. & Momol, M.T. 2007. Bacteriophages for plant disease control. *Annual Review of Phytopathology,* 45: 245-262.

Jónsson, J.O.G. & Davíðsdóttir, B. 2016. Classification and valuation of soil ecosystem services. *Agricultural Systems,* 145: 24-38.

Joosten, H., Tapio-Biström, M-.L. & Tol, S., eds. 2012. *Peatlands – guidance for climate change mitigation by conservation, rehabilitation and sustainable use. Second edition.* Mitigation of Climate Change in Agriculture Series 5. Rome, Italy. FAO.

Jouquet, E., Bloquel, E., Doan, T.T., Ricoy, M., Orange, D., Rumpel, C. & Duc, T.T. 2011. Do Compost and Vermicompost Improve Macronutrient Retention and Plant Growth in Degraded Tropical Soils? *Compost Science & Utilization,* 19: 15–24.

Jouquet, P., Dauber, J., Lagerlöf, J., Lavelle, P. & Lepage, M. 2006. Soil invertebrates as ecosystem engineers: Intended and accidental effects on soil and feedback loops. *Applied Soil Ecology,* 32: 153–164.

Jouquet, P., Janeau, J.L., Pisano, A., Sy, H.T., Orange, D., Minh, L.T.N. & Valentin, C. 2012. Influence of earthworms and termites on runoff and erosion in a tropical steep slope fallow in Vietnam: A rainfall simulation experiment. *Applied Soil Ecology,* 61: 161–168.

Jouquet, P., Plumere, T., Thu, T.D., Rumpel, C., Duc, T.T. & Orange, D. 2010. The rehabilitation of tropical soils using compost and vermicompost is affected by the presence of endogeic earthworms. *Applied Soil Ecology,* 46: 125–133.

Jung, W.J., An, K.N., Jin, Y.L., Park, R.D., Lim, K.T., Kim, K.Y. & Kim, T.H. 2003. Biological control of damping-off caused by Rhizoctonia solani using chitinase-producing Paenibacillus illinoisensis KJA-424. *Soil Biology and Biochemistry,* 36: 1261-1264.

Jusselme, M.D., Poly, F., Miambi, E., Mora, P., Blouin, M., Pando, A. & Rouland-Lefèvre, C. 2012. Effect of earthworms on plant Lantana camara Pb-uptake and on bacterial communities in root-adhering soil. *Science of the Total Environment,* 416: 200–207.

Kaiser, D., Lepage, M., Konaté, S. & Linsenmair, K.E. 2017. Ecosystem services of termites (Blattoidea: Termitoidae) in the traditional soil restoration and cropping system Zaï in northern Burkina Faso (West Africa). *Agriculture, Ecosystems and Environment,* 236: 198–211.

Kaiser, K., Wemheuer, B., Korolkow, V., Wemheuer, F., Nacke, H., Schöning, I., Daniel, R., *et al.* 2016. Driving forces of soil bacterial community structure, diversity, and function in temperate grasslands and forests. *Scientific Reports,* 6: 33696.

Kakumanu, M.L. & Williams, M.A. 2014. Osmolyte dynamics and microbial communities vary in response to osmotic more than matric water deficit gradients in two soils. *Soil Biology and Biochemistry,* 79: 14-24.

Kalingan, M., Rajagopal, S. & Venkatachalam, R. 2016. Effect of Metal Stress due to Strontium and The Mechanisms of Tolerating it by Amaranthus caudatus L. *Biochemistry & Physiology,* 5(3): 10000207.

Kamble, P.N., Gaikwad, V.B., Kuchekar, S.R. & Bååth, E. 2014. Microbial growth, biomass, community structure and nutrient limitation in high pH and salinity soils from Pravaranagar (India). *European Journal of Soil Biology,* 65: 87-95.

Kampichler, C. & Bruckner, A. 2010. The role of microarthropods in terrestrial decomposition: a meta-analysis of 40 years of litterbag studies. *Biological Reviews of the Cambridge Philosophical Society,* 84: 375–389. doi:10.1111/j.1469-185X.2009.00078.x

Kaneda, S. & Kaneko, N. 2011. Influence of Collembola on nitrogen mineralization varies with soil moisture content. *Soil Science and Plant Nutrition,* 57: 40–49. doi:10.1080/00380768.2010.551107

Kardol, P. & Wardle, D.A. 2010. How understanding aboveground-belowground linkages can assist restoration ecology. *Trends in Ecology and Evolution,* 25: 670–679. doi: 10.1016/j.tree.2010.09.001

Kardol, P., Bezemer, T.M. & Van der Putten, W.H. 2009. Soil organism and plant introductions in restoration of species-rich grassland communities. *Restoration Ecology,* 17: 258–269. doi: 10.1111/j.1526-100X.2007.00351.x

Kardol, P., Cregger, M.A., Campany, C.E. & Classen, A.T. 2010. Soil ecosystem functioning under climate change: plant species and community effects. *Ecology,* 91(3): 767-781.

Kardol, P., Reynolds, W.N., Norby, R.J. & Classen, A.T. 2011. Climate change effects on soil microarthropod abundance and community structure. *Applied Soil Ecology,* 47(1): 37-44.

Kardol, P., Throop, H.L., Adkins, J. & de Graaff, M.A. 2016. A hierarchical framework for studying the role of biodiversity in soil food web processes and ecosystem services. *Soil Biology and Biochemistry,* 102: 33-36.

Kareiva, P., Tallis, H., Ricketts, T.H., Daily, G.C. & Polasky, S., eds. 2011. *Natural Capital: Theory and Practice of Mapping Ecosystem Services.* Oxford, UK. Oxford University Press.

Karimi, B., Dequiedt, S., Terrat, S., Jolivet, C., Arrouays, D., Wincker, P., Cruaud, C., Bispo, A., Chemidlin Prévost-Bouré, N. & Ranjard, L. 2019. Biogeography of Soil Bacterial Networks along a Gradient of Cropping Intensity. *Scientific Reports,* 9(1): 3812. https://doi.org/10.1038/s41598-019-40422-y

Karsten, M., Addison, P., Van Vuuren, B.J. & Terblanche, J.S. 2016. Investigating population differentiation in a major African agricultural pest: Evidence from geometric morphometrics and connectivity suggests high invasion potential. *Molecular Ecology*, 25: 3019–3032.

Kaspari, M. 2019. In a globally warming world, insects act locally to manipulate their own microclimate. *PNAS*, 116(12): 5220-5222. doi:10.1073/pnas.1901972116

Kaspari, M., Clay, N.A., Donoso, D.A. & Yanoviak, S.P. 2014. Sodium fertilization increases termites and enhances decomposition in an Amazonian forest. *Ecology*, 95(4): 795–800.

Kaspari, M., Bujan, J., Roeder, K.A., de Beurs, K. & Weiser, M.D. 2019. Species Energy and Thermal Performance Theory predict 20- year changes in ant community abundance and richness. *Ecology*, 100(12):

Kearney, S.G., Carwardine, J., Reside, A.E., Fisher, D.O., Maron, M., Doherty, T.S., Legge, S., et al. 2018. The threats to Australia's imperilled species and implications for a national conservation response. *Pacific Conservation Biology*, 25: 231-244. doi.org/10.1071/PC18024

Keiluweit, M., Bougoure, J.J., Nico, P.S., Pett-Ridge, J., Weber, P.K. & Kleber, M. 2015. Mineral protection of soil carbon counteracted by root exudates. *Nature Climate Change*, 5(6): 588-595.

Keith, A.M., Schmidt, O. & McMahon, B.J. 2016. Soil stewardship as a nexus between Ecosystem Services and One Health. *Ecosystem Services*, 17: 40–42.

Keith, A.M., Griffiths, R.I., Henrys, P.A., Hughes, S., Lebron, I., Maskell, L.C., Ogle, S.M., et al. 2015. Monitoring Soil Natural Capital and Ecosystem Services by Using Large-Scale Survey Data. In M. Stromberger, D. Lindbo & N. Comerford, eds. *Soil Ecosystem Services:* Advances in Agricultural Systems Modeling. ASA-CSSA-SSSA.

Kendrick, J., Ribbons, R.R., Classen, A.T. & Ellison, A.M. 2015. Changes in canopy structure and ant assemblages affect soil ecosystem variables as a foundation species declines. *Ecosphere*, 6(5): Article 77. doi:10.1890/ES14-00447.1

Kerfahi, D., Tripathi, B.M., Dong, K., Go, R. & Adams, J.M. 2016. Rainforest conversion to rubber plantation may not result in lower soil diversity of bacteria, fungi, and nematodes. *Microbial Ecology*, 72(2): 359-371.

Keuskamp, J.A., Dingemans, B.J.J., Lehtinen, T., Sarneel, J.M. & Hefting, M.M. 2013. Tea Bag Index: A novel approach to collect uniform decomposition data across ecosystems. *Methods in Ecology and Evolution*, 4(11): doi: 10.1111/2041-210X.12097.

Khush, G.S. 1999. Green revolution: Preparing for the 21st century. *Genome*, 42: 646–655.

Kibblewhite, M.G., Ritz, K. & Swift, M.J. 2008. Soil health in agricultural systems. *Philosophical transactions of the Royal Society of London. Series B, Biological sciences*, 363: 685–701. doi: 10.1098/rstb.2007.2178

Kimber, A. & Eggleton, P. 2018. Strong but taxon-specific responses of termites and wood-nesting ants to forest regeneration in Borneo. *Biotropica*, 50(2): 266-273.

Kimura, M., Jia, Z-.J., Nakayama, N., Asakawa, S. 2008. Ecology of viruses in soils: Past, present and future perspectives. *Soil Science and Plant Nutrition,* 54: 1-32.

Kinell, G., Söderqvist, T., Elmgren, R., Walve, J. & Franzén, F. 2012. *Cost-Benefit Analysis in a Framework of Stakeholder Involvment and Integrated Costal Zone Modeling.* Cere working paper 2012:1. Umeå, Sweden. Centre for Environmental and Resource Economics. 31 pp.

King, J.R. 2016. Where do eusocial insects fit into soil food webs? *Soil Biology and Biochemistry,* 102: 55–62. doi:10.1016/j.soilbio.2016.07.019

Kionte, K. & Fitch, D.H.A. 2013. Nematodes. *Current Biology,* 23: R862-R864.

Kirchmann, H., Bergström, L., Kätterer, T., Mattsson, L. & Gesslein, S. 2007. Comparison of Long-Term Organic and Conventional Crop–Livestock Systems on a Previously Nutrient-Depleted Soil in Sweden. *Agronomy Journal,* 99: 960-972.

Kitagami, Y., Tanikawa, T., Mizoguchi, T. & Matsuda, Y. 2018. Nematode communities in pine forests are shaped by environmental filtering of habitat conditions. *Journal of Forestry Research,* 23: 346-353

Klarner, B., Maraun, M. & Scheu, S. 2013. Trophic diversity and niche partitioning in a species rich predator guild – Natural variations in stable isotope ratios (13C/12C, 15N/14N) of mesostigmatid mites (Acari, Mesostigmata) from Central European beech forests. *Soil Biology and Biochemistry,* 57: 327-333.

Kleian, S. 2006. The influence of invasive earthworms on indigenous fauna in ecosystems previously uninhabited by earthworms. *Biological Invasions*, 8: 1275-1285.

Kleijn, D., Rundlöf, M., Scheper, J., Smith, H.G. & Tscharntke, T. 2011. Does conservation on farmland contribute to halting the biodiversity decline? *Trends in Ecology and Evolution,* 26: 474-481.

Klimaszewski, J., & Brunke, A.J. 2018. Canada's adventive rove beetle (Coleoptera, Staphylinidae) fauna In: Biology of Rove Beetles (Staphylinidae). In O. Betz, U. Irmler, & J. Klomaszewski eds. Biology of Rove Beetles (Staphylinidae), pp. 65-79. Springer International Publishing.

Klimaszewski, J., Bourdon, C. & Pelletier, G. 2018. Synopsis of adventive species of Coleoptera (Insecta) recorded from Canada. Part 4: Superfamilies Scarabaeoidea, Scirtoidea, Buprestoidea, Byrrhoidea, Elateroidea, Derodontoidea, Bostrichoidea, and Cleroidea. *Pensoft Series Faunistica No. 104*. Pensoft, Sofia, Bulgaria. 215 pp.

Klimaszewski, J., Brunke, A.J., Work, T.T., & Venier, L. 2018. Rove Beetles (Coleoptera, Staphylinidae) as Bioindicators of Change in Boreal Forests and Their Biological Control Services in Agroecosystems: Canadian Case Studies. In O. Betz, U. Irmler, & J. Klomaszewski eds. **Biology of Rove Beetles (Staphylinidae)**, pp. 161-181. Springer International Publishing.

Klimaszewski, J., Langor, D., Batista, R., & Dorval, J-A. 2012. Synopsis of adventive species of Coleoptera (Insecta) recorded from Canada. Part 1: Carabidae. *Pensoft Series Faunistica No. 103*. Pensoft Publishers, Sofia, Bulgaria. 96 pp.

Klironomos, J., Zobel, M., Tibbett, M., Stock, W.D., Rillig, M.C., Parrent, J.L., Moora, M., et al. 2011. Forces that structure plant communities: quantifying the importance of the mycorrhizal symbiosis. *New Phytologist,* 189: 366–370.

Köchy, M., Don, A., van der Molen, M.K. & Freibauer, A. 2015. Global distribution of soil organic carbon – Part 2: Certainty of changes related to land use and climate. *Soil,* 1: 367-380. doi:10.5194/soil-1-367-2015

Kohler, J., Caravaca, F., Azcón, R., Díaz, G. & Roldán, A. 2015. The combination of compost addition and arbuscular mycorrhizal inoculation produced positive and synergistic effects on the phytomanagement of a semiarid mine tailing. *Science of The Total Environment,* 514: 42–48. https://doi.org/10.1016/j.scitotenv.2015.01.085

Kooijman, S.A.L.M. 2010. *Dynamic energy budget theory for metabolic organization.* Cambridge, UK. Cambridge University Press. 514 pp.

Koopmans, C. & Bloem, J. 2018. Soil quality effects of compost and manure in arable cropping: Results from using soil improvers for 17 yearsin the MAC trial. *Bunnik, the Netherlands, Louis Bolk Institute.* pp. 40.

Koopmans, C.J., Smeding, F.W., Rutgers, M., Bloem, J. & Eekeren, N.J.M. van. 2006. Biodiversiteit en bodembeheer in de landbouw. Driebergen, Louis Bolk Institut. (also available at https://research.wur.nl/en/publications/biodiversiteit-en-bodembeheer-in-de-landbouw).

Kortenkamp, A. & Faust, M. 2018. Regulate to reduce chemical mixture risk. *Science,* 361: 224–226.

Korthals, G.W., Thoden, T.C., van den Berg, W. & Visser, J.H.M. 2014. Long-term effects of eight soil health treatments to control plant-parasitic nematodes and Verticillium dahliae in agro-ecosystems. *Applied Soil Ecology,* 76: 112–123. https://doi.org/10.1016/j.apsoil.2013.12.016

Kosola, K.R., Eissenstat, D.M. & Graham, J.H. 1995. Root demography of mature citrus trees: the influence of Phytophthora nicotianae. *Plant and Soil,* 171(2): 283-288.

Kotze, J., Venn, S., Niemelä, J., & Spence, J. 2011. Effects Of Urbanization On The Ecology And Evolution Of Arthropods. In J. Niemelä, J.H. Breuste, T.Elmqvist , G. Guntenspergen, P. James, & N.E. Mcintyre, Eds. *Urban Ecology: Patterns, Processes And Applications. Oxford Online Scholarship. doi:10.1093/Acprof:Oso/9780199563562.003.0019*

Krishna, K., Grimaldi, D.A., Krishna, V. & Engel, M.S. 2013. Treatise on the Isoptera of the world. *Bulletin of the American Museum of Natural History,* 377.

Krivtsov, V., Illian, J.B., Liddell, K., Garside, A., Bezginova, T., Salmond, R., Thompson, J., et al. 2003. Some aspects of complex interactions involving soil mesofauna: analysis of the results from a Scottish woodland. *Ecological Modelling,* 170: 441–452.

Kroeger, M.E., Delmont, T., Eren, A.M., Meyer, K.M., Guo, J., Khan, K. & Tiedje, J.M., et al. 2018. New biological insights into how deforestation in amazonia affects soil microbial communities using metagenomics and metagenome-assembled genomes. *Frontiers in microbiology,* 9: 1635.

Kuffner, M., Puschenreiter, M., Wieshammer, G., Gorfer, M. & Sessitsch, A. 2008. Rhizosphere bacteria affect growth and metal uptake of heavy metal accumulating willows. *Plant Soil,* 304(1-2): 35-44. doi: 10.1007/s11104-007-9517-9

Kuiper, I., de Deyn, G.B., Thakur, M.P., & van Groenigen, J.W. 2013. Soil invertebrate fauna affect N_2O emissions from soil. *Global Change Biology,* 19: 2814–2825.

Kumar, A., Bisht, B., Joshi, V & Dhewa, T. 2011. Review on Bioremediation of Polluted Environment : A Management Tool. *International Journal of Environmental Sciences,* 1: 1079–1093.

Kumar, B.L., & Gopal, D.V.R.S. 2015. Effective role of indigenous microorganisms for sustainable environment. *3 Biotech,* 5(6), 867–876. https://doi.org/10.1007/s13205-015-0293-6

Kun, M.E. 2015. Do orbatid mites enhance fungal growth in Austrocedrus chilensis leaf litter? *Systematic and Applied Acarology,* 20:171–176.

Kuperman, R.G., Checkai, R.T., Simini, M., Sunahara, G.I. & Hawari, J. 2018. Energetic contaminants inhibit plant litter decomposition in soil. *Ecotoxicology and Environmental Safety,* 153: 32-39.

Kuperman, R.G., Minyard, M.L., Checkai, R.T., Sunahara, G.I., Rocheleau, S., Dodard, S.G., Paquet, L. & Hawari, J. 2017. Inhibition of soil microbial activity by nitrogen-based energetic materials. *Environmental Toxicology and Chemistry,* 36: 2981-2990.

Kuzyakov, Y. 2010. Priming effects: interactions between living and dead organic matter. *Soil Biology and Biochemistry,* 42(9): 1363-1371.

Laban, P., Metternicht, G. & Davies, J. 2018. *Soil biodiversity and soil organic carbon: keeping drylands alive.* Gland, Switzerland. International Union for Conservation of Nature and Natural Resources. 36 pp.

Ladeiro, B. 2012. Saline Agriculture in the 21st Century: Using Salt Contaminated Resources to Cope Food Requirements. *Journal of Botany,* 2012: Article 310705.

Lal, R. 2001. Soil degradation by erosion. *Land Degradation and Development,* 12: 519–539.

Lal, R. 2004. Soil carbon sequestration impacts on global climate change and food security. *Science,* 304(5677): 1623-1627. DOI: 10.1126/science.1097396.

Lal, R. 2010a. Beyond Copenhagen: mitigating climate change and achieving food security through soil carbon sequestration. *Food Security,* 2: 169-177.

Lal, R. 2010b. Managing Soils and Ecosystems for Mitigating Anthropogenic Carbon Emissions and Advancing Global Food Security. *Bioscience,* 60: 708-721.

Lal, R. 2015. Restoring soil quality to mitigate soil degradation. *Sustainability,* 7: 5875-5895. https://doi.org/10.3390/su7055875

Lal, R. 2019. Accelerated soil erosion as a source of atmospheric CO2. *Soil and Tillage Research,* 188: 35–40. https://doi.org/10.1016/j.still.2018.02.001

Lal, R., Horn, R. & Kosaki, T. 2018. *Soil and sustainable development goals*. Stuttgart, Germany. Catena-Schweizerbart. 213 pp.

Lal, R., Griffin, M., Apt, J., Lave, L. & Morgan, M.G. 2004. Managing Soil Carbon. *Science*, 304: 393.

Lal R., Guinto D.F. & Smith M. 2013. Influence of mucuna fallow crop on selected soil properties, weed suppression and taro yields in Taveuni, Fiji. In: *XXIX International Horticultural Congress on Horticulture: Sustaining Lives, Livelihoods and Landscapes* (IHC2014): 1118 (pp. 89-94). DOI 10.17660/ActaHortic.2016.1118.13

Lambers, H., Raven, J.A., Shaver, G.R. & Smith, S.E. 2008. Plant nutrient-acquisition strategies change with soil age. *Trends in Ecology and Evolution*, 23(2): 95-103.

Landers, J. N., Rass, G., Freitas, P. D., Basch, G., González Sanchez, E., Tabaglio, V., Kassan, A., Derpsch, R., Friedrich, T. & Giupponi, L. 2013. Effects of Zero Tillage (No-Till) Conservation Agriculture on soil physical and biological properties and their contributions to sustainability. Geophysical Research Abstracts. Vol. 15.

Landrigan, P.J., Fuller, R., Hu, H., Caravanos, J., Cropper, M.L., Hanrahan, D., Sandilya, K., Chiles, T.C., Kumar, P. & Suk, W.A. 2018. Pollution and Global Health – An Agenda for Prevention. *Environmental Health Perspectives*, 126: 3–8.

Landrigan, P.J., Fuller, R., Acosta, N.J.R., Adeyi, O., Arnold, R., Basu, N., Balde, A.B., *et al.* 2018. The Lancet Commission on pollution and health. *Lancet*, 391: 462-512.

Lange, L. 2014. The importance of fungi and mycology for addressing major global challenges. *IMA Fungus*, 5: 463–471.

Langewald, J., Mitchell, J.D.N.K. & Kooyman, C. 2003. Microbial control of termites in Africa. In P. Neuenschwander, C. Borgemeister, Langewald J, eds. *Microbial Control in IPM Systems in Africa*, 227-242. Cotonou, Benin. International Institute of Tropical Agriculture.

Langor, D. W. & Sheffield, C. S. 2019. The Biota of Canada: terrestrial arthropods. *Zookeys*, 819: 1-4.

Lapola, D.M, Martinelli, L.A., Peres, C.A., Ometto, J.P.H.B., Ferreira, M.E., Nobre, C.A., Aguiar, A.D.P., *et al.* 2014. Pervasive Transition of the Brazilian Land-Use System. *Nature Climate Change*, 4: 27–35. https://doi.org/10.1038/NCLIMATE2056.

Larsen, T., Taylor, D.L., Leigh, M.B. & O'Brien, D.M. 2009. Stable isotope fingerprinting: a novel method for identifying plant, fungal, or bacterial origins of amino acids. *Ecology*, 90: 3526-3535.

Lauber, C.L., Hamady, M., Knight, R. & Fierer, N. 2009. Pyrosequencing-based assessment of soil pH as a predictor of soil bacterial community structure at the continental scale. *Applied Environmental Microbiology*, 75: 5111–5120.

Laudicina, V.A., Novara, A., Barbera, V., Egli, M. & Badalucco, L. 2015. Longterm tillage and cropping system effects on chemical and biochemical characteristics of soil organic matter in a Mediterranean environment. *Land Degradation and Development*, 26: 45–53. doi:10.1002/ldr.2293.

Laurance, W.F., Sayer, J. & Cassman, K.G. 2014. Agricultural Expansion and Its Impacts on Tropical Nature. *Trends in Ecology and Evolution,* 29(2): 107-116. https://doi.org/10.1016/j.tree.2013.12.001.

Lavelle, P. 1997. Faunal activities and soil processes: adaptive strategies that determine ecosystem function. *Advances in ecological research,* 27: 93-132.

Lavelle, P. & Martin, A. 1992. Small-scale and large-scale effects of endogeic earthworms on soil organic matter dynamics in soils of the humid tropics. *Soil Biology and Biochemistry,* 24: 1491-1498.

Lavelle P. & Spain A.V. 2006. *Soil ecology, 2nd edition.* Amsterdam, Netherlands. Kluwer Scientific Publications.

Lavelle P., Decaëns T., Aubert M., Barot S., Blouin M., Bureau F., Margerie P., Mora P. & Rossi J.P. 2006. Soil invertebrates and ecosystem services. *European Journal of Soil Biology,* 42: S3–S15.

Lavelle, P., Bignell, D., Lepage, M., Wolters, V., Roger, P., Ineson, P., Heal, O.W & Dhillion, S. 1997. Soil function in a changing world: the role of invertebrate ecosystem engineers. *European Journal of Soil Biology,* 33: 159–193.

Lavelle, P., Dugdale, R., Scholes, R., Berhe, A.A., Carpenter, E., Codispoti, L., Izac, A-.M., *et al.* 2005. Nutrient cycling. In R. Hassan, R. Scholes & N. Ash, eds. *Ecosystems and human well-being: current state and trends, vol 1.* Findings of the condition and trends working group of the Millennium Ecosystem Assessment. Washington, USA. Island Press.

Lavelle, P., Rodriguez, N., Arguello, O., Bernal, J., Botero, C., Chaparro, P., Gómez, Y., *et al.* 2014. Soil ecosystem services and land use in the rapidly changing Orinoco River Basin of Colombia. *Agriculture, Ecosystem and Environment,* 185: 106–117.

Lavelle P., Spain A.V., Blouin M., Brown G.G., Decaens, T., Grimaldi M., Jimenez J.J., *et al.* 2016. Ecosystem engineers in a self-organized soil: a review of concepts and future research questions. *Soil Science,* 181: 91-109.

Lazzaro, L., Mazza, G., d'Errico, G., Fabiani, A., Guiliani, C., Inghilesi, A.F., Lagomarsino, A., *et al.* 2018. How ecosystems change following invasion by Robinia pseudoacacia: Insights from soil chemical properties and soil microbial, nematode, microarthropod and plant communities. *Science of the Total Environment,* 622-623: 1509-1518. doi: 10.1016/j.scitotenv.2017.10.017.

Leah, T. & Cerbari, V. 2013. Degradation of ordinary chernozem in the south of Moldova and phytotechnical measures to remedy of their fertility. *Scientific Papers. Series A. Agonomy*, LVI: 56–61.

Lehmann, A., Zheng, W. & Rillig, M.C. 2017. Soil biota contributions to soil aggregation. *Nature Ecology and Evolution,* 1(12): 1828.

Lehmann, A., Stavros, D., Veresoglou, E., Leifheit, F. & Rillig, M.C. 2014. Arbuscular mycorrhizal influence on zinc nutrition in crop plants – A meta-analysis. *Soil Biology and Biochemistry,* 69: 123-131. ISSN 0038-0717.

Lehmitz, R., Russell, D., Hohberg, K., Christian, A. & Xylander, W.E. 2011. Wind dispersal of oribatid mites as a mode of migration. *Pedobiologia,* 54: 201-207.

Lemanceau, P., Maron, P-.A., Mazurier, S. Mougel, C., Pivato, B., Plassart, P., Ranjard, L., Revellin, C., Tardy, V. & Wipf, D. 2015. Understanding and managing soil biodiversity: a major challenge in agroecology. *Agronomy for Sustainable Development,* 35: 67–81.

Lenoir, L., Persson, T., Bengtsson, J., Wallander, H. & Wiren, A. 2007. Bottom-up or top-down control in forest soil microcosms? Effects of soil fauna on fungal biomass and C/N mineralization. *Biology and Fertility of Soils,* 43: 281–294.

Lentendu, G., Wubet, T., Chatzinotas, A., Wilhelm, C., Buscot, F. & Schlegel, M. 2014. Effects of long-term differential fertilization on eukaryotic microbial communities in an arable soil: a multiple barcoding approach. *Molecular Ecology,* 23(13): 3341–3355. https://doi.org/10.1111/mec.12819

Léonard, J. & Rajot, J.L. 2001. Influence of termites on runoff and infiltration: Quantification and analysis. *Geoderma,* 104: 17–40.

Li, S. & Wu, F. 2018. Diversity and co-occurrence patterns of soil bacterial and fungal communities in seven intercropping systems. *Frontiers in Microbiology,* 9: 1521

Li, W., Ciais, P., Guenet, B., Shushi, P., Chang, J., Chaplot, V., Khudyaev, S., *et al.* 2018. Temporal response of soil organic carbon after grassland-related land-use change. *Global Change Biology,* 24: 4731-4736. https://doi.org/10.1111/gcb.14328

Li, F.-R., Liu, J.-L., Ren, W. & Liu, L.-L. 2018. Land-use change alters patterns of soil biodiversity in arid lands of northwestern China. *Plant and Soil,* 428(1-2): 371-388.

Li, G., Sun, G.X., Ren, Y., Luo, X.S. & Zhu, Y.G. 2018. Urban soil and human health: a review. *European Journal of Soil Science,* 69: 196–215.

Li, Y.S., Robin, P., Cluzeau, D., Bouché, M., Qiu, J.P., Laplanche, A., Hassouna, M., Morand, P., Dappelo, C. & Callarec, J. 2008. Vermifiltration as a stage in reuse of swine wastewater: Monitoring methodology on an experimental farm. *Ecological Engineering,* 32: 301–309.

Li, Z., Tian, D., Wang, B., Wang, J., Wang, S., Chen, H.Y., Xu, X., Wang, C., He, N. & Niu, S. 2019. Microbes drive global soil nitrogen mineralization and availability. *Global Change Biology,* 25: 1078-1088.

Li, Z. & Zhang, H. 2008. Application of Microbial Fertilizers in Sustainable Agriculture. *Journal of Crop Production,* 3(1): 337–347. https://doi.org/10.1300/J144v03n01_28

Liang, Z., Elsgaard, L., Nicolaisen, M.H., Lyhne-Kjærbye, A., & Olesen, J.E. 2018. Carbon mineralization and microbial activity in agricultural topsoil and subsoil as regulated by root nitrogen and recalcitrant carbon concentrations. *Plant and Soil,* 433: 65-82.

Lichner, L., Hallett, P.D., Drongová, Z., Czachor, H., Kovacik, L., Mataix-Solera, J. & Homolák, M. 2013. Algae influence the hydrophysical parameters of a sandy soil. *CATENA,* 108: 58–68. https://doi.org/10.1016/j.catena.2012.02.016

Liese, R., Leuschner, C., & Meier, I.C. 2019. The effect of drought and season on root life span in temperate arbuscular mycorrhizal and ectomycorrhizal tree species. *Journal of Ecology*, 107(5): 2226-2239.

Lin, F., Penton, C.R., Ruan, Y., Shen, Z., Xue, C., Li, R. & Shen, Q. 2017. Inducing the rhizosphere microbiome by biofertilizer application to suppress banana Fusarium wilt disease. *Soil Biology and Biochemistry*, 104: 39-48

Lindahl, B.D., Nilsson, R.H., Tedersoo, L., Abarenkov, K., Carlsen, T., Kjøller, R., Kõljalg, U., *et al.* 2013. Fungal community analysis by high-throughput sequencing of amplified markers–a user's guide. *New Phytologist*, 199(1): 288-299.

Lindbo, D.L., Kozlowski, D.A. & Robinson, C. 2012. *Know Soil, Know Life.* Madison, USA. Soil Science Society of America.

Lindberg, N. & Persson, T. 2004. Effects of long-term nutrient fertilisation and irrigation on the microarthropod community in a boreal norway spruce stand. *Forest Ecology And Management*, 188: 125-135.

Liu, Y., Duan, M. & Yu, Z. 2013. Agricultural landscapes and biodiversity in China. *Agriculture, Ecosystems & Environment*, 166: 46–54. https://doi.org/10.1016/j.agee.2011.05.009

Liu, C., Wang, L., Song, X., Chang, Q., Frank, D.A., Wang, D., Li, J., Lin, H. & Du, F. 2018. Towards a mechanistic understanding of the effect that different species of large grazers have on grassland soil N availability. *Journal of Ecology*, 106: 357-366.

Liu, J.-L., Ren, W., Zhao, W.-Z. & Li, F.-R. 2018. Cropping systems alter the biodiversity of ground- and soil-dwelling herbivorous and predatory arthropods in a desert agroecosystem: Implications for pest biocontrol. *Agriculture, Ecosystems and Environment*, 266: 109-121.

Liu, T., Guo, R., Ran, W., Whalen, J.K. & Li, H. 2015. Body size is a sensitive trait-based indicator of soil nematode community response to fertilization in rice and wheat agroecosystems. *Soil Biology and Biochemistry*, 88: 275-281.

Liu, Y., Wang, L., He, R., Chen, Y., Xu, Z., Tan, B., Zhang, L., *et al.* 2019. Higher soil fauna abundance accelerates litter carbon release across an alpine forest-tundra ecotone. *Scientific Reports*, 9: 10561, doi: 10.1038/s41598-019-47072-0.

Liu, Z., Klümper, U., Shi, L., Ye, L. & Li, M. 2019. From pig breeding environment to subsequently produced pork: Comparative analysis of antibiotic resistance genes and bacterial community composition. *Frontiers in Microbiology*, 10: 1–12.

Llewellyn, R.S. & d'Emden, D.H. 2010. *Adoption of no-till cropping practices in Australian grain growing regions*, pp. 1–31. Kingston, Australia. Grains Research and Development Corporation.

Lo Papa, G., Palermo, V. & Dazzi, C. 2011. Is land-use change a cause of loss of pedodiversity? The case of the Mazzarrone study area, Sicily. *Geomorphology*, 135: 332–342. https://doi.org/10.1016/j.geomorph.2011.02.015

Lobe, J.W., Callaham, M.A., Jr., Hendrix, P.F. & Hanula, J.L. 2014. Removal of an invasive shrub (Chinese privet: Ligustrum sinense Lour) reduces exotic earthworm abundance and promotes recovery of native North American earthworms. *Applied Soil Ecology*, 83: 133-139.

Lobell, D.B., Burke, M.B., Tebaldi, C., Mastrandrea, M.D., Falcon, W.P. & Naylor, R.L. 2008. Prioritizing Climate Change Adaptation Needs for Food Security in 2030. *Science*, 319: 607–610.

Lobry de Bruyn, L.A., Conacher, A.J. 1990. The role of termites and ants in soil modification: a review. *Australian Journal of Soil Research*, 28: 55–93.

Loeppmann, S., Forbush, K., Cheng, W. & Pausch, J. 2019 Subsoil biogeochemical properties induce shifts in carbon allocation pattern and soil C dynamics in wheat. *Plant and Soil*, 442: 369-383.

López M., Martínez V.M., Brossard F.M., Bolívar A., Alfonso N., Amelia A.A. & Pereira A.H. 2008. Efecto de biofertilizantes bacterianos sobre el crecimiento de un cultivar de maíz en dos suelos contrastantes venezolanos. *Agronomía Tropical*, 58(4): 391-401.

López-Hernández, D. 2003. Soil activity of macrofauna (Termites and Oligochaeta) in savanna soils. *Venesuelos*, 15 – 25. 1http://saber.ucv.ve/ojs/index.php/rev_venes/article/view/961/890

López-Hernández, D., Brossard, M., & Fardeau, J. 2006 Effect of different termite feeding groups on P sorption and P availability in African and South American savannas. *Biol Fertil Soils*, 42: 207–214. https://doi.org/10.1007/s00374-005-0017-x

Jordán López, A., García Moreno, J., Gordillo Rivero, Á.J., Martínez Zavala, L.M. & Cerdá García, A. 2014. Organic carbon, water repellency and soil stability to slaking under different crops and managements: a case study at aggregate and intra-aggregate scales. *Soil Discuss*, (1): 295-325.

Lorenz, K. & Lal, R. 2009. Biogeochemical C and N cycles in urban soils. *Environment International*, 35: 1–8. doi: https://doi.org/10.1016/j.envint.2008.05.006

Lorenz, K., Lal, R. & Ehlers, K. 2019. Soil organic carbon stocks as an indicator for monitoring land ans soil degradation in relation to United Nations' Sustainable Development Goals. *Land Degradation and Development*, 30: 824-838. https://doi.org/10.1002/ldr.3270

Louca, S., Mazel, F., Doebeli, M. & Parfrey, L.W. 2019. A census-based estimate of Earth's bacterial and archaeal diversity. *PLoS Biology*, 17: p.e3000106.

Lovera, M. & Cuenca, G. 2007. Diversidad de hongos micorrízicos arbusculares (hma) y potencial micorrízico del suelo de una sabana natural y una sabana perturbada de la gran sabana, Venezuela. Interciencia, 32: http://ve.scielo.org/cgi-bin/wxis.exe/iah/

Lowe, S., Browne, M., Boudjelas, S. & De Pooter, M. 2004. *100 of the World's Worst Invasive Alien Species- A selection from the Global Invasive Species Database.* Auckland, New Zealand. Invasive Species Specialist Group.

Lowenfels, J. & Lewis, W. 2010. *Teaming with Microbes: A Gardener's Guide to the Soil Food Web*. Portland, USA. Timber Press.

Lozan, A. 2008. Biosafety Concerns in the Republic of Moldova: opportunities and challenges. Chisinau, 2008. [online]. [Cited 5 October 2020] http://www.biosafety.md/public/214/en/Concerns2.pdf

Lu, F., Hu, H., Sun, W., Zhu, J., Liu, G., Zhou, W., Zhang, Q., Shi, P., Liu, X., Wu, X., Zhang, L., Wei, X., Dai, L., Zhang, K., Sun, Y., Xue, S., Zhang, W., Xiong, D., Deng, L., Liu, B., Zhou, L., Zhang, C., Zheng, X., Cao, J., Huang, Y., He, N., Zhou, G., Bai, Y., Xie, Z., Tang, Z., Wu, B., Fang, J., Liu, G. & Yu, G. 2018. Effects of national ecological restoration projects on carbon sequestration in China from 2001 to 2010. *Proceedings of the National Academy of Sciences*, 115(16): 4039. https://doi.org/10.1073/pnas.1700294115

Lubbers I.M., Pulleman M.M. & Van Groenigen J.W. 2017. Can earthworms simultaneously enhance decomposition and stabilization of plant residue carbon? *Soil Biology and Biochemistry*, 105: 12-24.

Lugato, E., Paustian, K., Panagos, P., Jones, A. & Borrelli, P. 2016. Quantifying the erosion effect on current carbon budget of European agricultural soils at high spatial resolution. *Global Change Biology*, 22: 1976-1984. https://doi.org/10.1111/gcb.13198

Luke, S.H., Fayle, T.M., Eggleton, P., Turner, E.C. & Davies, R.G. 2014. Functional Structure of Ant and Termite Assemblages in Old Growth Forest, Logged Forest and Oil Palm Plantation in Malaysian Borneo. *Biodiversity and Conservation*, 23(11): 2817–32. https://doi.org/10.1007/s10531-014-0750-2.

Luna, L., Pastorelli, R., Bastida, F., Hernández, T., García, C., Miralles, I. & Solé-Benet, A. 2016. The combination of quarry restoration strategies in semiarid climate induces different responses in biochemical and microbiological soil properties. *Applied Soil Ecology*, 107: 33–47. doi: 10.1016/j.apsoil.2016.05.006

Lundberg, D.S., Lebeis, S.L., Paredes, S.H., Yourstone, S., Gehring, J., Malfatti, S., Tremblay, J., et al. 2012. Defining the core Arabidopsis thaliana root microbiome. *Nature*, 488: 86–90. doi: 10.1038/nature11237

Luo, Y., Ahlström, A., Allison, S.D., Batjes, N.H., Brovkin, V., Carvalhais, N., Zhou, T., et al. 2016. Toward more realistic projections of soil carbon dynamics by Earth system models. *Global Biogeochemical Cycles*, 30: 40–56.

Luo, Z., Wang, E. & Sun, O.J. 2010. Soil carbon change and its responses to agricultural practices in Australian agro-ecosystems: A review and synthesis. *Geoderma*, 155: 211-223.

Lupwayi, N. & Hamel, C. 2010. Soil Biology of the Canadian Prairies. *Prairie Soils & Crops J.* 3: 16-24.

Lupwayi, N.Z., Larney, F.J., Blackshaw, R.E., Kanashiro, D.A., Pearson, D.C. & Petri, R.M. 2017. Pyrosequencing reveals profiles of soil bacterial communities after 12 years of conservation management on irrigated crop rotations. *Applied Soil Ecology*, 121: 65-73.

Lv, Y., Wang, C., Jia, Y., Wang, W., Ma, X., Du, J. & Tian, X. 2014. Effects of sulfuric, nitric, and mixed acid rain on litter decomposition, soil microbial biomass, and enzyme activities in subtropical forests of China. *Applied Soil Ecology,* 79: 1–9. doi: https://doi.org/10.1016/j.apsoil.2013.12.002

Lynch, J.P. 2007. Turner review no. 14. Roots of the second green revolution. *Australian Journal of Botany,* 55: 493–512.

Lyngwi, N.A. & Joshi, S.R. 2015. Traditional Sacred Groves, an ethnic strategy for conservation of microbial diversity. *Indian Journal of Traditional Knowledge,* 14: 474-480.

Ma, H.-K., Pineda, A., Van der Wurff, A. & Bezemer, T.M. 2018. Synergistic and antagonistic effects of mixing monospecific soils on plant-soil feedbacks. *Plant and Soil,* 429: 271-279.

Ma, C. & Yin, X. 2019. Responses of soil invertebrates to different types in the Changbai Mountains of China. *Journal of Forestry Research,* 24(3): 153-161.

Maaß, S., Caruso, T. & Rillig, M.C. 2015. Functional role of microarthropods in soil aggregation. *Pedobiologia (Jena),* 58: 59–63. doi:https://doi.org/10.1016/j.pedobi.2015.03.001

Macías, F. & Camps Arbestain, M. 2010. Soil carbon sequestration in a changing global environment. *Mitigation and Adaptation Strategies for Global Change,* 15: 511–529. doi: 10.1007/s11027-010-9231-4

MacKenzie, N.J., Hairsine P.B., Gregory, L.J., Austin, J., Baldock, J.A., Webb, M.J., Mewett, J., Cresswell, H.P., Welti, N. & Thomas M. 2017. *Priorities for improving soil condition across Australia's agricultural landscapes.* Report prepared for the Australian Government Department of Agriculture and Water Resources. Canberra, Australia. Commonwealth Scientific and Industrial Research Organisation. 129 pp.

Madigan, M.T., Martinko, J.M., Bender, K.S., Buckley, D.H. & Stahl, D.A. 2015. *Brock biology of microorganisms (14th edition).* Boston, USA. Pearson.

Madureira, L., Rambonilaza, T. & Karpinski, I. 2007. Review of methods and evidence for economic valuation of agricultural non-commodity outputs and suggestions to facilitate its application to broader decisional contexts. *Agriculture, Ecosystems and Environment,* 120: 5-20.

Maestre, F.T., Delgado-Baquerizo, M., Jeffries, T.C., Eldridge, D.J., Ochoa, V., Gozalo, B., Quero, J.L., et al. 2015. Increasing aridity reduces soil microbial diversity and abundance in global dryland. *Proceedings of the National Academy of Sciences,* 112: 15684-15689. www.pnas.org/cgi/doi/10.1073/pnas.1516684112

Maestre, F.T., Martín, N., Díez, B., López-Poma, R., Santos, F., Luque, I. & Cortina, J. 2006. Watering, Fertilization, and Slurry Inoculation Promote Recovery of Biological Crust Function in Degraded Soils. *Microbial Ecology,* 52(3): 365–377. https://doi.org/10.1007/s00248-006-9017-0

Maestre, F.T., Bowker, M.A., Cantón, Y., Castillo-Monroy, A.P., Cortina, J., Escolar, C., Escudero, A., Lazaro, R. & Martinez, I. 2011. Ecology and functional roles of biological soil crusts in semi-arid ecosystems of Spain. *Journal of Arid Environments,* 75: 1282-1291.

Maestre, F.T., Quero, J.L., Gotelli, N.J., Escudero, A., Ochoa, V., Delgado-Baquerizo, M., García-Palacios, P., *et al.* 2012. Plant species richness and ecosystem multifunctionality in global drylands. *Science,* 335(6065): 214-218.

Magalhães, M. M., & Lunas Lima, D. 2014. Low-Carbon Agriculture in Brazil: The Environmental and Trade Impact of Current Farm Policies. International Centre for Trade and Sustainable Development, Geneva, Sitzerland. Issue Paper No. 54.

Magura, T., Lövei, G.L., & Tóthmérész, B. 2009. Does urbanization decrease diversity in ground beetle (Carabidae) assemblages? *Global Ecology and Biogeography*, 19: 16-26.

Magurno, F., Posta, M., & Wozniak, K. 2019. Glomalin gene as molecular marker for functional diversity of arbuscular mycorrhizal fungi in soil. *Biology and Fertility of Soil*, 55: 411-417

Mahé, F., de Vargas, C., Bass, D., Czech, L., Stamatakis, A., Lara, E., Singer, D., *et al.* 2017. Parasites dominate hyperdiverse soil protist communities in Neotropical rainforests. *Nature Ecology and Evolution,* 1: 0091.

Majer, J.D. 1989. Long term recolonization of fauna in reclaimed land. In J.D. Majer, ed. *Animals in primary succession, the role of fauna in reclaimed land,* 143-174. Cambridge, UK. Cambridge University press.

Maklyuk, O. I., Naydenova, O.Ye. 2016. Comprehensive assessment of the biological state of the microbial soil system and its transformation in modern farming systems. Bulletin of agrarian science, special issue, October 2016. p. 59-64.

Mallen-Cooper, M., Nakagawa, S. & Eldridge, D.J. 2019. Global meta-analysis of soil-disturbing vertebrates reveals strong effects on ecosystem patterns and processes. *Global Ecology and Biogeography,* 28: 661-679.

Malthus, T. 1798. *Essay on the principle of population.* London, UK. Reeves and Turner.

Mancini, M. 2019. Agricoltura Organica e Rigenerativa. Florence, Italy. Terra Nuova Edizioni.

Mando, A., Brussaard, L. & Stroosnijder, L. 1999. Termite- and mulch-mediated rehabilitation of vegetation on crusted soil in West Africa. *Restoration Ecology,* 7: 33–41.

Mangan, S.A., Schnitzer, S.A., Herre, E.A., Mack, K.M.L., Valencia, M.C., Sanchez, E. & Bever, J.D. 2010. Negative plant-soil feedback predicts tree-species relative abundance in a tropical forest. *Nature,* 466: 752–755. doi: 10.1038/nature09273

Mangkoedihardjo, S. 2006. Biodegradability improvement of industrial wastewater using hyacinth. *Journal of Applied Science,* 6: 1409–1414.

Manter, D.K., Delgado, J.A. & Moore-Kucera, J. 2018. Integrated soil health management: a framework for soil conservation and regeneration. Pages 69–87 In D. Reicosky, ed. *Managing soil health for sustainable agriculture Volume 1: Fundamentals,* pp. 69-87. Cambridge, UK. Burleigh Dodds Science Publishing.

Maraun, M., Visser, S. & Scheu, S. 1998. Oribatid mites enhance the recovery of the microbial community after a strong disturbance. *Applied Soil Ecology,* 9: 175–181.

Marcotullio, P.J., Braimoh, A.K. & Onishi, T. 2008. The impact of urbanization on soils. In A.K. Braimoh & P.L.G. Vlek, eds. *Land Use And Soil Resources,* pp. 201-250. Stockholm, Sweden. Springer.

Mardhiah, U., Caruso, T., Gurnell, A. & Rillig, M.C. 2016. Arbuscular mycorrhizal fungal hyphae reduce soil erosion by surface Water flow in a greenhouse experiment. *Applied soil Ecology,* 99: 137-140

Marichal, R., Mathieu, J., Couteaux, M.-M. Mora, P., Roy, J. & Lavelle, P. 2011. Earthworm and microbe response to litter and soils of tropical forest plantations with contrasting C:N:P stoichiometric ratios. *Soil Biology and Biochemistry,* doi:10.1016/j.soilbio.2011.04.001

Maron, P.A., Sarr, A., Kaisermann, A., Lévêque, J., Mathieu, O., Guigue, J., Chabbi, A., *et al.* 2018. High microbial diversity promotes soil ecosystem functioning. *Applied Environmental Microbiology,* 84(9): e02738-17.

Marschner, H. 1995. *Mineral Nutrition of Higher Plants.* London, UK. Academic Press.

Marsh, P. & Wellington, E.M.H. 1994. Phage-host interactions in soil. *FEMS Microbiology Ecology,* 15: 99-107.

Martín-Robles, N., Lehmann, A., Seco, E., Aroca, R., Rillig, M.C. & Milla, R. 2018. Impacts of domestication on the arbuscular mycorrhizal symbiosis of 27 crop species. *New Phytologist,* 218(1): 322–334. https://doi.org/10.1111/nph.14962

Martinez, J.L. 2009. Environmental pollution by antibiotics and by antibiotic resistance determinants. *Environmental Pollution,* 157(11): 2893-2902.

Martínez V.R., López, M., Brossard F.M., Tejeda G.G.; Pereira A.,H., Parra Z.C., Rodriguez S., J. & Alba, A. 2006. Procedimientos para el estudio y fabricación de Biofertilizantes Bacterianos. Maracay; Venezuela. Instituto Nacional de Investigaciones Agrícolas. 88p.

Martinez-Beltran, J. & Manzur, K.L. 2005. Overview of salinity problems in the world and FAO strategies to address the problem. *In International salinity forum-managing saline soils and water: science, technology and social issues,* pp. 311-314. Riverside, USA. USDA-ARS Salinity Laboratory.

Martins, S.A., Schurt, D.A., Seabra, S.S., Martins, S.J., Ramalho, M.A.P., Moreira, F.M.S., Silva, J.C.P., Silva, J.A.G. & Medeiros, F.H.V. 2018. Common bean (Phaseolus vulgaris L.) growth promotion and biocontrol by rhizobacteria under Rhizoctonia solani suppressive and conducive soils. *Applied Soil Ecology,* 127: 129-135.

Matscheko, N., Lundstedt, S., Svensson, L., Harju, J. & Tysklind, M. 2002. Accumulation and elimination of 16 polycyclic aromatic compounds in the earthworm (Eisenia fetida). *Environmental toxicology and chemistry,* 21(8): 1724-1729.

Matz, C., & Kjelleberg, S. 2005. Off the hook-how bacteria survive protozoan grazing. *Trends in Microbiology,* 13: 302-307.

Mavi, M.S. & Marschner, P. 2013. Salinity affects the response of soil microbial activity and biomass to addition of carbon and nitrogen. *Soil Research,* 51(1): 68-75.

Mavi, M.S., Sanderman, J., Chittleborough, D.J., Cox, J.W. & Marschner, P. 2012. Sorption of dissolved organic matter in salt-affected soils: Effect of salinity, sodicity and texture. *Science of the Total Environment,* 435: 337-344.

Mayer, P. & Holmstrup, M. 2008. Passive dosing of soil invertebrates with polycyclic aromatic hydrocarbons: Limited chemical activity explains toxicity cutoff. *Environmental Science and Technology,* 42: 7516-7521.

McConnell, K.E. & Bockstael, N.E. 2005. Valuing the Environment as a Factor of Production. In K.-G. Mäler & J.R. Vincent, eds. *Handbook of Environmental Economics,* Vol. 2, Ch. 14. Amsterdam, Netherlands. North-Holland.

McCormack, M.L., Adams, T.S., Smithwick, E.A.H. & Eissenstat, D.M. 2012. Predicting fine root lifespan from plant functional traits in temperate trees. *New Phytologist,* 195(4): 823-831.

McCormack, M.L., Dickie, I.A., Eissenstat, D.M., Fahey, T.J., Fernandez, C.W., Guo, D., Helmisaari, H.S., Hobbie, E.A., Iversen, C.M. & Jackson, R.B. 2015. Redefining fine roots improves understanding of below-ground contributions to terrestrial biosphere processes. *New Phytologist,* 207(3): 505-518.

McElroy, M.T. & Donoso, D.A. 2019. Ant Morphology Mediates Diet Preference in a Neotropical Toad. *Copeia,* 107(3): 430–438. doi: 10.1643/CH-18-162

McGuire, K.L., Henkel, T.W. & Granzow de la Cerda, I. 2008. Dual mycorrhizal colonization of forest-dominating tropical trees and the mycorrhizal status of non-dominant tree and liana species. *Mycorrhiza,* 18: 217. https://doi.org/10.1007/s00572-008-0170-9.

McManus, P.S., Stockwell, V.O., Sundin, G.W. & Jones, A.L. 2002. Antibiotic use in plant agriculture. *Annual review of phytopathology,* 40(1): 443-465.

McNeill, M.R., Phillips, C.B., Robinson, A.P., Aalders, L., Richards, N., Young, S., Dowsett, C., James, T. & Bell, N. 2017. Defining the biosecurity risk posed by transported soil: Effects of storage time and environmental exposure on survival of soil biota. *NeoBiota,* 32: 65–88.

McNeill, M.R., Phillips, C.B., Bell, N.L & Proffitt, J.R. 2006. Potential spread of pests in New Zealand through commercial transport of nursery plants. *New Zealand Plant Protection,* 59: 75–79.

McNeill, M.R., Phillips, C., Young, S., Shah, F., Aalders, L., Bell, N., Gerard, E. & Littlejohn, R. 2011. Transportation of nonindigenous species via soil on international aircraft passengers' footwear. *Biological Invasions,* 13: 2799–2815.

MEA. 2005. *Millennium Ecosystem Assessment.* Ecosystems and human well-being: scenarios. Washington, USA. Island Press.

Medina-Sauza, R.M., Álvarez-Jiménez, M., Delhal, A., Reverchon, F., Blouin, M., Guerrero-Analco, J.A., Cerdán, C.R., Guevara, R., Villain, L. & Barois I. 2019. Earthworms building up soil microbiota, a review. *Frontiers in Environmental Science,* 7: 1-20.

Meena, S.S., Sharma, R.S., Gupta, P., Karmakar, S. & Aggarwal, K.K. 2016. Isolation and identification of Bacillus megaterium YB3 from an effluent contaminated site efficiently degrades pyrene. *Journal of Basic Microbiology*, 56(4): 369-78. doi: 10.1002/jobm.201500533

Medina, M.G., García, D.E., Moratinos, P., Clavero, T. & Iglesias, J.M. 2011. Macrofauna edáfica en sistemas silvopastoriles con *Morus alba*, Leucaena leucophala y pastos. *Zootecnia Tropical*, 29: 301-311. http://ve.scielo.org/scielo.php?script=sci_arttext&pid=S0798-72692011000300006

Megharaj, M., Kantachote, D., Singleton, I. & Naidu, R. 2000. Effects of long-term contamination of DDT on soil microflora with special reference to soil algae and algal transformation of DDT. *Environmental Pollution*, 109(1): 35–42. https://doi.org/10.1016/S0269-7491(99)00231-6

Meier, S., Borie, F., Bolan, N., Cornejo, P. 2012. Phytoremediation of Metal-Polluted Soils by Arbuscular Mycorrhizal Fungi. *Critical Reviews in Environmental Science and Technology*, 42(7): 741-775, DOI: 10.1080/10643389.2010.528518

Meisner, A., Jacquiod, S., Snoek, L.B., ten Hooven, F.C. & van der Putten, W.H. 2018. Drought legacy effects on the composition of soil fungal and prokaryote communities. *Frontiers in Microbiology*, 9: 294.

Meloni, D., Oliva, M., Ruiz, H. & Martinez, C. 2001. Contribution of proline and inorganic solutes to osmotic adjustment in cotton under salt stress. *Journal of Plant Nutrition*, 24: 599–612.

Mendes, I.C.. 2016. Indicadores biológicos de qualidade de solo em sistemas de plantio direto no Brasil: estado atual e perspectivas futuras. In: Moreira, F.M.S., Kasuya, M.C.M. (Eds.), *Fertilidade e biologia do solo: integração e tecnologia para todos*. Sociedade Brasileira de Ciência de Solo, Viçosa, pp. 297-322.

Mendes IC, Kappes C, Ono FB, Sousa DMG, Reis-Junior FB, Lopes AAC, Semler TD, Zancanaro L. 2015. Qualidade biológica do solo: por que e como avaliar. Rondonópolis: Boletim de Pesquisa da Fundação MT; 2017. p. 98-105. Portilho IIR, Scorza Júnior RP, Salton JC, Mendes IM, Mercante FM. Persistência de inseticidas e parâmetros microbiológicos em solo sob sistemas de manejo. Cienc Rural. 2015;45:22-8

Mendes, R., Kruijt, M., de Bruijn, I., Dekkers, E., van der Voort, M., Schneider, J.H.M., Piceno, Y.M., *et al.* 2011. Deciphering the rhizosphere microbiome for disease-suppressive bacteria. *Science*, 332: 1097-1100.

Mendoza, B., Florentino, A., Hernández-Hernández R., Aciego, J., & Torres, D. 2012. Atributos biológicos del suelo con aplicación de abono orgánico y soluciones salinas. XIX Congreso Latinoamericano y XXIII Congreso Argentino de la Ciencia del Suelo

Menge, D.N.L., Lichstein, J.W. & Ángeles-Pérez, G. 2014. Nitrogen fixation strategies can explain the latitudinal shift in nitrogen-fixing tree abundance. *Ecology*, 95: 2236–2245. https://doi.org/10.1890/13-2124.1

Menz, F.C. & Seip, H.M. 2004. Acid rain in Europe and the United States: an update. *Environmental Science and Policy,* 7(4): 253–265. doi: https://doi.org/10.1016/j.envsci.2004.05.005

Mergeay, M., Monchy, S., Vallaeys, T., Auquier, V., Benotmane, A., Bertin, P., Taghavi, S., Dunn, J., van der Lelie, D. & Wattiez, R. 2003. Ralstonia metallidurans, a bacterium specifically adapted to toxic metals: towards a catalogue of metal-responsive genes. *FEMS Microbiology Reviews,* 27: 385–410.

Merloti, L.F., Mendes, L.W., Pedrinho, A., de Souza, L.F., Ferrari, B.M. & Tsai, S.M. 2019. Forest-to-agriculture conversion in Amazon drives soil microbial communities and N-cycle. *Soil Biology and Biochemistry,* 137: 107567.

Metternicht, G.I. & Zinck, J.A. 2003. Remote sensing of soil salinity: potentials and constraints. *Remote sensing of the environment,* 85: 1 - 20.

Meyer, K.M., Klein, A.M., Rodrigues, J.L., Nüsslein, K., Tringe, S.G., Mirza, B.S. & Bohannan, B.J. 2017. Conversion of Amazon rainforest to agriculture alters community traits of methane-cycling organisms. *Molecular ecology,* 26(6): 1547-1556.

Microbiome Interagency Working Group – MIWG. 2018. Interagency Strategic Plan for Microbiome Research FY 2018-2022. [online]. [Cited 5 October 2020]. https://science.osti.gov/-/media/ber/pdf/workshop-reports/Interagency_Microbiome_Strategic_Plan_FY2018-2022.pdf?la=en&hash=5FEC6B8ABBB2B58BC7DBD5076C93639D1F5E351A

Middleton, E.L., Richardson, S., Koziol, L., Palmer, C.E., Yermakov, Z., Henning, J.A., Schultz, P.A. & Bever, J.D. 2015. Locally adapted arbuscular mycorrhizal fungi improve vigor and resistance to herbivory of native prairie plant species. *Ecosphere,* 6: 276. doi: 10.1890/ES15-00152.1

Migge-Kleian, S. 2006. The influence of invasive earthworms on indigenous fauna in ecosystems previously uninhabited by earthworms. *Biological Invasions,* 8: 1275-1285.

Mikutta, R., Mikutta, C., Kalbitz, K., Scheel, T., Kaiser, K. & Jahn, R. 2007. Biodegradation of forest floor organic matter bound to minerals via different binding mechanisms. *Geochimica Et Cosmochimica Acta,* 71: 2569-2590.

Miller, J.J., Battigelli, J.P., Beasley, B.W. & Drury, C.F. 2017. Response of Soil Mesofauna to Long-Term Application of Feedlot Manure on Irrigated Cropland. *Journal of Environmental Quality,* 46: 185-192.

Mills, J.G., Weinstein, P., Gellie, N.J.C., Weyrich, L.S., Lowe, A.J. & Breed, M.F. 2017. Urban habitat restoration provides a human health benefit through microbiome rewilding: the Microbiome Rewilding Hypothesis. *Restoration Ecology,* 25: 866–872. doi: 10.1111/rec.12610

Milner, A.M. & Boyd, I.L. 2017. Toward pesticidovigilance Can lessons from pharmaceutical monitoring help to improve pesticide regulation? *Science,* 357. 1232-1234.

Minambiente. 2012. Política nacional para la gestión integral de la biodiversidad y sus servicios ecosistémicos. In: Minambiente. [online]. [Cited 6 October 2020]. http://www.humboldt.org.co/images/pdf/PNGIBSE_espa%C3%B1ol_web.pdf

Minambiente. 2015. Plan Nacional de Restauraciòn. In: Minambiante. [online]. [Cited 5 October 2020]. https://www.minambiente.gov.co/images/BosquesBiodiversidadyServiciosEcosistemicos/pdf/Ordenaci%C3%B3n-y-Manejo-de-Bosques/PLAN_NACIONAL_DE_RESTAURACI%C3%93N_2.pdf

Minambiente. 2016a. Política para la gestión sostenible del suelo. In: Minambiante. [online]. [Cited 5 October 2020]. http://www.minambiente.gov.co/images/AsuntosambientalesySectorialyUrbana/pdf/suelo/Pol%C3%ADtica_para_la_gesti%C3%B3n_sostenible_del_suelo_FINAL.pdf

Minambiente. 2016b. Política Nacional de Cambio Climático. In: Minambiante. [online]. [Cited 5 October 2020]. http://www.minambiente.gov.co/images/cambioclimatico/pdf/Politica_Nacional_de_Cambio_Climatico_-_PNCC_/PNCC_Politicas_Publicas_LIBRO_Final_Web_01.pdf

MINEC. 2019. MINEC innova la estrategia nacional de la sequía. [online]. [Cited 5 October 2020]. http://www.minec.gob.ve/minec-innova-la-estrategia-nacional-de-la-sequia/

Minor, M.A. & Norton, R.A. 2004. Effects of soil amendments on assemblages of soil mites (Acari: Oribatida, Mesostigmata) in short-rotation willow plantings in central New York. *Canadian Journal of Forest Research*, 34: 1417–1425.

Ministério da Agricultura, Pecuária e Abastecimento. 2018. DECRETO Nº 9.414, DE 19 DE JUNHO DE 2018. Institui o Programa Nacional de Levantamento e Interpretação de Solos do Brasil. [online]. [Cited 5 October 2020]. http://www.planalto.gov.br/ccivil_03/_ato2015-2018/2018/decreto/D9414.htm

Ministère de l'agriculture, de l'agroalimentaire et de la forèt. 2016. *RMT Biodiversité et Agriculture* [online]. [Cited 2 October 2020]. http://www.rmt-biodiversite-agriculture.fr/moodle/

Ministère de l'agriculture et de l'alimentation. 2017. *Tour d'horizon des indicateurs relatifs à l'état organique et biologique des sols* [online]. [Cited 2 October 2020]. https://agriculture.gouv.fr/tour-dhorizon-des-indicateurs-relatifs-letat-organique-et-biologique-des-sols

Ministerio de Desarrollo Agropecuario. 2020. Ministerio de Desarrollo Agropecuario. [online]. [Cited 2 October 2020]. http://mida.gob.pa

Ministerio del Medio Ambiente. 2014. QUINTO INFORME NACIONAL DE BIODIVERSIDAD DE CHILE. [online]. [Cited 5 October 2020]. https://mma.gob.cl/wp-content/uploads/2017/08/Libro_Convenio_sobre_diversidad_Biologica.pdf

Ministério do Meio Ambiente. 2020. Biodiversidade. [online]. [Cited 5 October 2020]. https://www.mma.gov.br/biodiversidade/convenção-da-diversidade-biologica/estrategia-e-plano-de-ação-nacionais-para-a-biodiversidade-epanb

Ministerio para la Transición Ecológica y el Reto Demográfico de España. 2020. Conservación de la biodiversidad en España. [online]. [Cited 2 October 2020]. https://www.miteco.gob.es/es/biodiversidad/temas/conservacion-de-la-biodiversidad/conservacion-de-la-biodiversidad-en-espana/default.aspx

Ministero dell'Ambiente e della Tutela del Territorio e del Mare. 2010. La Strategia Nazionale per la Biodiversità. [online]. [Cited 2 October 2020]. https://www.minambiente.it/sites/default/files/archivio/allegati/biodiversita/Strategia_Nazionale_per_la_Biodiversita.pdf

Ministère de la Transition écologique. 2019. Plan biodiversité. [online]. [Cited 2 October 2020]. https://www.ecologique-solidaire.gouv.fr/plan-biodiversite

Ministero delle politiche agricole, alimentari e forestali. 2013. *Biodiversitá - Linee guida per la conservazione e la caratterizzazione della biodiversità vegetale, animale e microbica di interesse per l'agricoltura*. INEA edition. Roma. (also available at https://www.reterurale.it/flex/cm/pages/ServeBLOB.php/L/IT/IDPagina/9580).

Ministero delle politiche agricole, alimentari e forestali. 2008. Plano Nazionale Sulla Biodiversità di Interesse Agricolo [online]. [Cited 2 October 2020]. http://www.isprambiente.gov.it/files/biodiversita/20080313_SR_Piano_nazionale_biodiversita_agricoltura.pdf

Ministry of Agriculture and Agrarian Reform of Syrian Arab Republic. 2017. Agriculture Journal Issue 59 [online]. [Cited 5 October 2020]. http://moaar.gov.sy/main/

Mitter, B., Pfaffenbichler, N., Flavell, R., Compant, S., Antonielli, L., Petric, A., Berninger, T., *et al.* 2017. A new approach to modify plant microbiomes and traits by introducing beneficial bacteria at flowering into progeny seeds. *Frontiers in Microbiology,* 8: 11. https://doi.org/10.3389/fmicb.2017.00011

Miura, F., Nakamoto, T., Kaneda, S., Okano, S., Nakajima, M. & Murakami, T. 2008. Dynamics of soil biota at different depths under two contrasting tillage practices. *Soil Biology and Biochemistry,* 40(2): 406-414.

MoEF. 2011. *Bioremediation, its applications to contaminated sites in India - A state of the art report by Ministry of Environment and Forests*. New Delhi, India. Ministry off Environment and Forests, Government of India.

Mocali, S. 2010. *Metodi di analisi molecolare del suolo*. Franco Angeli, Roma, Italia.

Mohan Kumar, B.M. 2016. Homegardens as harbingers of belowground diversity in the humid tropics. In K.G. Saxena & K.S. Rao, eds. *Soil Biodiversity: Inventory, Functions and Management,* pp. 373-382. Dehra Dun, India. Bishen Singh Mahendra Pal Singh.

Montanarella, L. 2007. Trends in Land Degradation in Europe. In M.K. Sivakumar & N. Ndiang'ui, eds. *Climate and Land Degradation,* pp. 83-104. Heidelberg, Germany. Springer.

Montero, E. 2005. Rényi dimensions analysis of soil particle-size distributions. *Ecological Modelling,* 182 (3–4): 305-315. https://doi.org/10.1016/j.ecolmodel.2004.04.007.

Moore, J.C. & Hunt, H.W. 1988. Resource compartmentation and the stability of real ecosystems. *Nature,* 333: 261–263

Moore, J.C., Berlow, E.L., Coleman, D.C., de Ruiter, P.C., Don, Q., Hastings, A. & Collins Johnson, N. 2004. Detritus, trophic dynamics and biodiversity. *Ecology Letters,* 7: 584–600. doi: 10.1111/j.1461-0248.2004.00606.x

Moore, J.C., McCann, K. & de Ruiter, P.C. 2005. Modeling trophic pathways, nutrient cycling, and dynamic stability in soils. *Pedobiologia,* 49: 499-510.

Moore, J.C. & de Ruiter, P.C. 2012. *Energetic Food Webs.* Oxford, UK. Oxford University Press.

Moore, J.C. & de Ruiter, P.C. 2000. Invertebrates in detrital food webs along gradients of productivity. In D.C. Coleman & P.F. Hendix, eds. *Invertebrates as Webmasters in Ecosystems,* pp. 161-184. Oxford, USA. CABI Publishing.

Moos, J.H., Schrader, S. & Paulsen, H.M. 2018. Reduced tillage enhances earthworm abundance and biomass in organic farming: A meta-analysis. *Landbauforschung - applied agricultural and forestry research*(67(3–4)): 123–128. https://doi.org/10.3220/LBF1512114926000

Mora, C., Tittensor, D.P., Adl, S., Simpson, A.G. & Worm, B. 2011. How many species are there on Earth and in the ocean? *PLoS Biology,* 9: p.e1001127.

Moradi, J., Vicentini, F., Šimáčková, H., Pižl, V. Tajovský, K., Stary, J. & Frouz, J. 2018. An investigation into the long-term effect of soil transplant in bare spoil heaps on survival and migration of soil meso and macrofauna. *Ecological Engineering,* 110: 158–164. doi: 10.1016/j.ecoleng.2017.11.012

Morand, P., Robin, P., Pourcher, A.M., Oudart, D., Fievet, S., Luth, D., Cluzeau, D., Picot, B. & Landrain, B. 2011. Design of an integrated piggery system with recycled water, biomass production and water purification by vermiculture, macrophyte ponds and constructed wetlands. *Water Science and Technology,* 63: 1314–1320.

Morari, F., Lugato, E., Berti, A. & Giardini, L. 2006. Long-term effects of recommended management practices on soil carbon changes and sequestration in north-eastern Italy. *Soil Use and Management,* 22: 71-81.

Morgan, J.A.W., Bending, G.D. & White, P.J. 2005. Biological costs and benefits to plant-microbe interactions in the rhizosphere. *Journal of Experimental Botany,* 56: 1729-1739

Morillo, E. & Villaverde, J. 2017. Advanced technologies for the remediation of pesticide-contaminated soils. *Science of The Total Environment,* 586: 576–597.

Morriën, E., Hannula, S.E., Snoek, L.B., Helmsing, N.R., Zweers, H., de Hollander, M., Soto, R.L *et al.* 2017. Soil networks become more connected and take up more carbon as nature restoration progresses. *Nature Communications,* 8: 14349.

Mouginot, C., Kawamura, R., Matulich, K.L., Berlemont, R., Allison, S.D., Amend, A.S. & Martiny, A.C. 2014. Elemental stoichiometry of Fungi and Bacteria strains from grassland leaf litter. *Soil Biology and Biochemistry,* 76: 278-285.

Mueller, K.E., Tilman, D., Fornara, D.A. & Hobbie, S.E. 2013. Root depth distribution and the diversity-productivity relationship in a long-term grassland experiment. *Ecology*, 94(4): 787-793.

Mueller, K.E., Eisenhauer, N., Reich, P.B., Hobbie, S.E., Chadwick, O.A., Chorover, J., Dobies, T., et al. 2016. Light, earthworms, and soil resources as predictors of diversity of 10 soil invertebrate groups across monocultures of 14 tree species. *Soil Biology and Biochemistry*, 92: 184-198.

Mueller, N.D., Gerber, J.S., Johnston, M., Ray, D.K., Ramankutty, N. & Foley, J.A. 2012. Closing yield gaps through nutrient and water management. *Nature*, 490: 254-257.

Mueller, R.C., Rodrigues, J.L.M., Usslein, K.N. & Brendan, J.M. Bohannan. 2016. Land Use Change in the Amazon Rain Forest Favours Generalist Fungi. *Functional Ecology*, 30: 1845–53. https://doi.org/10.1111/1365-2435.12651.

Mueller, U.G. & Sachs, J.L. 2015. Engineering microbiomes to improve plant and animal health. *Trends in Microbiology*, 23: 606–617. doi: 10.1016/j.tim.2015.07.009

Muleta, D., Assefa, F., Nemomissa, S. & Granhall, U. 2007. Composition of coffee shade tree species and density of indigenous arbuscular mycorrhizal fungi (AMF) spores in Bonga natural coffee forest, southwestern Ethiopia. *Forest Ecology and Management*, 241: 145–154.

Müller, D.B., Vogel, C., Bai, Y. & Vorholt, J.A. 2016. The Plant Microbiota: Systems-Level Insights and Perspectives. *Annual Review of Genetics*, 50: 211–234. doi: 10.1146/annurev-genet-120215-034952

Mus, F., Crook, M.B., Garcia, K., Costas, A.G., Geddes, B.A., Kouri, E.D., Paramasivan, P., et al. 2016. Symbiotic nitrogen fixation and the challenges to its extension to nonlegumes. *Applied and Environmental Microbiology*, 82: 3698-3710.

Museum am Schölerberg. 2020. The universe under our feet. [online]. [Cited 2 October 2020]. https://www.museum-am-schoelerberg.de/das-erwartet-sie/dauerausstellungen/unterwelten.html

Myers, B., Webster, K.L., Mclaughlin, J.W. & Basiliko, N. 2012. Microbial activity across a boreal peatland nutrient gradient: the role of fungi and bacteria. *Wetlands Ecology and Management*, 20: 77-88.

Nagy, L.G., Toth, R., Kiss, E., Slot, J., Gacser, A. & Kovacs, G.M. 2017. Six Key Traits of Fungi: Their Evolutionary Origins and Genetic Bases. *Microbiology Spectrum*, 5(4): doi: 10.1128/microbiolspec.FUNK-0036-2016.

Nash, L. 1993. Water quality and health. In Gleick, P., ed. *Water in crisis: a guide to the world's fresh water resources*, pp. 25-39. Oxford, UK. Oxford University Press.

National Research Council. 2001. *Basic Research Opportunities in Earth Science*. Washington, USA. The National Academies Press. doi:10.17226/9981.

National Soil Conservation Program of Ukraine. 2015. ed. by S.A. Baliuk, V.V. Medvedev, M.M. Miroshnychenko. Kharkiv. 2015. - 59 p. Rational use of soil resources and reproduction of soil fertility: organizational, economic, environmental and regulatory aspects / Monograph ed. by Baliuk, AESU Corresponding member A.V.Kucher. – Kharkiv, 2015. – 357-367 p. Atlas of Soil Suitability of Ukraine for Organic Farming (Zonal Aspects) / Kharkiv, 2015. - 36 p. Comprehensive assessment of the biological state of the microbial soil system and its transformation in modern farming systems / O. I. Maklyuk (Starchenko), O.Ye. Naydenova / Bulletin of agrarian science, special issue, October 2016.- p. 59-64

Natural England. 2018. Long-term Monitoring Network (LTMN) - Soil Chemistry and Biology Baseline. In: *Find open data* [online]. [Cited 2 October 2020]. https://data.gov.uk/dataset/12ad05d1-a21a-4855-8545-4812db5f2cfd/long-term-monitoring-network-ltmn-soil-chemistry-and-biology-baseline

Nature Editorial. 2019. The value of biodiversity is not the same as its price. *Nature,* 573: 463-464. [doi: 10.1038/d41586-019-02882-0]

Nawaz, M.F., Bourrié, G. & Trolard, F. 2013. Soil compaction impact and modelling. *Agronomy for Sustainable Development,* 33(2): 291-309.

Negrete-Yankelevich, S., **& Barois-Boullard, I.** 2012. Bajo tus pies: La vida en el suelo. CONABIO. *Biodiversitas*, 105: 6-9.

Negrete-Yankelevich, S., Maldonado-Mendoza, I. E., Lázaro-Castellanos, J. O., Sangabriel-Conde, W., & Martínez-Álvarez, J. C. 2013. Arbuscular mycorrhizal root colonization and soil P availability are positively related to agrodiversity in Mexican maize polycultures. Biology and Fertility of Soils, 201-212. https://doi.org/10.1007/s00374-012-0710-5

Neill, C., Piccolo, M.C., Cerri, C.C., Steudler, P.A., Melillo, J.M & Brito, M. 1997. Net Nitrogen Mineralization and Net Nitrification Rates in Soils Following Deforestation for Pasture across the Southwestern Brazilian Amazon Basin Landscape. *Oecologia,* 110(2): 243–52. https://doi.org/10.1007/s004420050157.

Nélieu, S., Delarue, G., Ollivier, E., Awad, P., Fraillon, F. & Pelosi, C. 2016. Evaluation of epoxiconazole bioavailability in soil to the earthworm Aporrectodea icterica. *Environmental science and pollution research international,* 23: 2977-2986.

Nelson, M.L., Dinardo, A., Hochberg, J. & Armelagos, G.J. 2010. Brief communication: mass spectroscopic characterization of tetracycline in the skeletal remains of an ancient population from Sudanese Nubia 350–550 CE. *American journal of physical anthropology,* 143(1): 151-154.

Nesbitt, J. E. & Adl, S.M. 2014. Differences in soil quality indicators between organic and sustainably managed potato fields in Eastern Canada. Ecological Indicators, 37: 119-130.

Nesme, J. & Simonet, P. 2015. The soil resistome: a critical review on antibiotic resistance origins, ecology and dissemination potential in telluric bacteria. *Environmental microbiology,* 17(4): 913-930.

Nesme J., Cécillon, S., Delmont, T.O., Monier, J.M., Vogel, T.M. & Simonet, P. 2014. Large-Scale metagenomic-based study of antibiotic resistance in the environment. *Current Biology,* 19: 1096-100.

Neuenkamp, L., Prober, S.M., Price, J.N., Zobel, M. & Standish, R.J. 2018. Benefits of mycorrhizal inoculation to ecological restoration depend on plant functional type, restoration context and time. *Fungal Ecology*, 40: 140-149. doi: 10.1016/j.funeco.2018.05.004

Newton, L.H. & Dillingham, C.K. 1994. *Classic cases in environmental ethics.* Belmont, USA. Wadsworth Publishing Co.

Newton, P.C.D., Clark, H., Bell, C.C., Glasgow, E.M., Tate, K.R., Ross, D.J., Yeates, G.W. & Saggar, S. 1995. Plant growth and soil processes in temperate grassland communities at elevated CO_2. *Journal of Biogeography*, 22: 235e240.

Ng, E.L., Patti, A.F., Rose, M.T., Schefe, C.R., Smernik, R.J. & Cavagnaro, T.R. 2015. Do organic inputs alter resistance and resilience of soil microbial community to drying? *Soil Biology and Biochemistry*, 81: 58–66.

Nguyen, T. T., Tomberlin, J.K., & Vanlaerhoven, S. 2015. Ability of black soldier fly (Diptera: Stratiomyidae) larvae to recycle food waste. *Environmental Entomology*, 44: 406-410.

Nichols, E., Spector, S., Louzada, J., Larsen, T., Amezquita, S. & Favila, M.E. 2008. Ecological functions and ecosystem services provided by Scarabaeinae dung beetles. *Biological Conservation*, 141: 1461–1474.

Nichols, R.V., Vollmers, C., Newsom, L.A., Wang, Y., Heintzman, P.D., Leighton, M., Green, R.E. & Shapiro, B. 2018. Minimizing polymerase biases in metabarcoding. *Molecular Ecology Resources*, 18: 927–939.

Nielsen, U.N., Wall, D.H. & Six, J. 2015. Soil biodiversity and the environment. *Annual Review of Environment and Resources*, 40: 63-90.

Nielsen, U.N., Ayres, E., Wall, D.H. & Bardgett, R.D. 2011. Soil biodiversity and carbon cycling: a review and synthesis of studies examining diversity–function relationships. *European Journal of Soil Science*, 62(1): 105-116.

Nielsen, U.N., Ayres, E., Wall, D.H., Li, G., Bardgett, R.D., Wu, T. & Garey, J.R. 2014. Global-scale patterns of assemblage structure of soil nematodes in relation to climate and ecosystem properties. *Global Ecology and Biogeography*, 23: 968-978. https://doi.org/10.1111/geb.12177

Nieminen, M., Ketoja, E., Mikola, J., Terhivuo, J., Sirén, T. & Nuutinen, V. 2011. Local land use effects and regional environmental limits on earthworm communities in Finnish arable landscapes. *Ecological Applications*, 21(8): 3162–3177. https://doi.org/10.1890/10-1801.1

Nijkamp, P., Vindigni, G. & Nunes, P.A.L.D. 2008. Economic valuation of biodiversity: A comparative study. *Ecological Economics*, 67: 217-231.

Niklasson, M. & Granström, A. 2000. Numbers and sizes of fires: long-term spatially explicit fire history in a Swedish boreal landscape. *Ecology*, 81: 1484-1499

Nkem, J.N., Lobry De Bruyn, L.A. Grant, C.D. & Hulugalle, N.R. 2000. The impact of ant bioturbation and foraging activities on surrounding soil properties. *Pedobiologia*, 44: 609–621.

Nkem, J.N., Wall, D.H., Virginia, R.A., Barrett, J.E., Broos, E.J., Porazinska, D.L. & Adams, B.J. 2006. Wind dispersal of soil invertebrates in the McMurdo Dry Valleys, Antarctica. *Polar Biology*, 29: 346-352.

Noirot, C. & Alliot, H. 1947. *La lutte contre les termites.* Paris, France. Masson and Cie. 98 pp.

Nordén, J., Penttilä, R., Siitonen, J., Tomppo, E. & Ovaskainen, O. 2013. Specialist Species of Wood-Inhabiting Fungi Struggle While Generalists Thrive in Fragmented Boreal Forests. *Journal of Ecology*, 101(3): 701–12. https://doi.org/10.1111/1365-2745.12085.

Novara, A., Gristina, L., Saladino, S.S., Santoro, A. & Cerdà, A. 2011. Soil erosion assessment on tillage and alternative soil managements in a Sicilian vineyard. *Soil and Tillage Research*, 117: 140–147. doi:10.1016/j.still. 2011.09.007.

Nunan, N., Wu, K., Young, I.M., Crawford, J.W. & Ritz, K. 2003. Spatial distribution of bacterial communities and their relationships with the micro-architecture of soil. *FEMS Microbiology Ecology*, 44(2): 203-215. 10.1016/S0168-6496(03)00027-8

Nunez, D., Nahuelhual, L. & Oyarzun, C. 2006. Forests and water: The value of native temperate forests in supplying water for human consumption. *Ecological Economics*, 58: 606-616.

O'Connor, M.I., Piehler M.F., Leech D.M., Anton, A. & Bruno, J.F. 2009. Warming and resource availability shift food web structure and metabolism. *PLoS Biology*, 7: e1000178. doi:10.1371/journal.

O'Neill, J. 2016. *Tracking drug-resistant infections globally: final report and recommendations.* Rome, Italy. Food and Agriculture Organization of the United Nations. [Accessed 3rd September 2019]. https://amr-review.org/sites/default/files/160525_Final%20paper_with%20cover.pdf

Oades, J.M. 1984. Soil Organic-Matter and Structural Stability - Mechanisms and Implications for Management. *Plant and Soil*, 76: 319-337.

Oades, J.M. 1988. The Retention of Organic-Matter in Soils. *Biogeochemistry*, 5: 35-70.

Ochoa-Hueso, R., Rocha, I., Stevens, C.J., Manrique, E. & Luciañez, M.J. 2014. Simulated nitrogen deposition affects soil fauna from a semiarid Mediterranean ecosystem in central Spain. *Biology and Fertility of Soils*, 50(1): 191–196.

Ochoa-Hueso, R., Collins, L.S., Delgado-Baquerizo, M., Hamonts, K., Pockman, W.T., Sinsabaugh L.R. & Power S.A. 2018. Drought consistently alters the composition of soil fungal and bacterial communities in grasslands from two continents. *Global Change Biology*, 24(7): 2818–2827. doi: 10.1111/gcb.14113

Ohlson, K. 2014. *The Soil Will Save Us: How Scientists, Farmers, and Foodies Are Healing the Soil to Save the Planet.* Emmaus, USA. Rodale Books.

Ohta, T., Niwa, S., Agetsuma, N. & Hiura, T. 2014. Calcium concentration in leaf litter alters the community composition of soil invertebrates in warm-temperate forests. *Pedobiologia*, 57: 257-262.

Ohtani, K. & Shimizu, T. 2016. Regulation of Toxin Production in Clostridium perfringens. Toxins, 8(7): 1–14.

Ojeda A., Rojas, J. & J. Aciego, J. 1997. Effect of different herbicides mixes on the cyanobacteria population dynamic in a soil of El Sombrero, Guárico State. http://saber.ucv.ve/ojs/index.php/rev_venes/article/view/1076/1005

Okubo, T., Liu, D., Tsurumaru, H., Ikeda, S., Asakawa, S., Tokida, T., Tago, K., Hayatsu, M., Aoki, N., Ishimaru, K., Ujiie, K., Usui, Y., Nakamura, H., Sakai, H., Hayashi, K., Hasegawa, T. & Minamisawa, K. 2015. Elevated atmospheric CO_2 levels affect community structure of rice root-associated bacteria. *Frontiers in Microbiology*, 6. https://doi.org/10.3389/fmicb.2015.00136

Oleson, K.W., Lawrence, D.M., Bonan, G.B., Flanner, M.G., Kluzek, E., Lawrence, P.J., Levin, S., Swenson, S.C. & Thornton, B. 2010. *Technical Description of Version 4.0 of the Community Land Model.* Boulder, USA. National Center for Atmospheric Research.

Olsson, P.A., Bååth, E., Jakobsen, I. & Söderström, B. 1995. The use of phospholipid and neutral lipid fatty acids to estimate biomass of arbuscular mycorrhizal fungi in soil. *Mycological Research*, 99: 623-629.

Oliverio, A.M., Huijie, G., Wickings, K. & Fierer, N. 2018. A DNA metabarcoding approach to characterize soil arthropod communities. *Soil Biology and Biochemistry,* 125: 37-43.

Olley, J., Brooks, A., Spencer, J., Pietsch, T. & Borombovits, D. 2013. Subsoil erosion dominates the supply of fine sediment to rivers draining into Princess Charlotte Bay, Australia. *Journal of Environmental Radioactivity,* 124: 121-129.

One Health Initiative. 2020. One Health Initiative will unite human and veterinary medicine. [online]. [Cited 5 October 2020]. http://www.onehealthinitiative.com/

Open Bodemindex. 2020. Simple and affordable insight into soil quality and soil improvement. [online]. [Cited 5 October 2020]. https://openbodemindex.nl/

Öpik, M., Zobel, M., Cantero, J.J., Davison, J., Facelli, J.M., Hiiesalu, I., Jairus, T., *et al.* 2013. Global sampling of plant roots expands the described molecular diversity of arbuscular mycorrhizal fungi. *Mycorrhiza,* 23(5): 411-430.

Orgiazzi, A. & Panagos, P. 2018. Soil biodiversity and soil erosion: It is time to get married. *Global Ecology and Biogeography,* 27: 1155–1167.

Orgiazzi, A., Dunbar, M.B., Panagos, P., Jones, A. & Fernández-Ugalde, O. 2015. Soil biodiversity and DNA barcodes: opportunities and challenges. *Soil Biology and Biochemistry,* 80: 244–250.

Orgiazzi, A., Ballabio, C., Panagos, P. Jones, A. & Fernández-Ugalde, O. 2017. LUCAS Soil, the largest expandable soil dataset for Europe: a review. European Journal of Soil Science. 69(1): 140–53.

Orgiazzi, A., Panagos, P., Yigini, Y., Dunbar, M.B., Gardi, C., Montanarella, L. & Ballabio, C. 2016. A Knowledge-Based Approach to Estimating the Magnitude and Spatial Patterns of Potential Threats to Soil Biodiversity. *Science of the Total Environment*, 545–546: 11–20. https://doi.org/10.1016/j.scitotenv.2015.12.092.

Orgiazzi, A., Bardgett, R.D., Barrios, E., Behan-Pelletier, V., Briones, M.J.I., Chotte, J-L., De Deyn, G.B., *et al.* 2016. *Global Soil Biodiversity Atlas.* Luxembourg. Publications Office of the European Union. 176 pp.

Orr, C.H., James, A., Leifert, C., Cooper, J.M. & Cummings, S.P. 2011. Diversity and activity of free-living nitrogen-fixing bacteria and total bacteria in organic and conventionally managed soils. *Applied and Environmental Microbiology*, 77: 911-919.

Ortega-Gonzalez, D.K., Martinez-Gonzalez, G., Flores, C.M., Zaragoza, D., Cancino-Diaz, J.C., Cruz-Maya, J.A., *et al.* 2015. Amycolatopsis sp. Poz14 isolated from oil-contaminated soil degrades polycyclic aromatic hydrocarbons. *International Biodeteriation and Biodegradation*, 99: 650-173. doi: 10.1016/j.ibiod.2015.01.008

Osler, G.H.R. & Sommerkorn, M. 2007. Toward a complete soil C and N cycle: incorporated the soil fauna. *Ecology*, 88: 1611-1621.

Osler, G. H. R., Harrison, L., Kanashiro, D.K., & Clapperton, J. 2008. Soil microarthropod assemblages under different arable crop rotations in Alberta, Canada. *Applied Soil Ecology* 38: 71-78.

Ossola, A. & Livesley, S.J. 2016. Drivers of soil heterogeneity in the urban landscape. In R.A. Francis, J.D.A. Millington & M.A. Chadwick eds. Urban Landscape Ecology, pp. 37-59. London, UK. Routledge.

Ostfeld, R.S. & Keesing, F. 2012. Effects of Host Diversity on Infectious Disease. *Annual Review of Ecology, Evolution, and Systematics*, 43: 157-182.

Otero, X.L., De La Peña-Lastra, S., Pérez-Alberti, A., Ferreira, T.O. & Huerta-Diaz, M.A. 2018. Seabird colonies as important global drivers in the nitrogen and phosphorus cycles. *Nature Communications*, 9: doi:10.1038/s41467-017-02446-8

Otte, M.J. & Chilonda, P. 2002. *Cattle and small ruminant production in sub-Saharan Africa: A systematic review.* Rome, Italy. FAO.

Otto, C.C. & Haydel, S.E. 2013. Microbicidal clays: composition, activity, mechanism of action, and therapeutic applications. In Méndez-Vilas, ed. *Microbial Pathogens and Strategies for Combating them: Science, Technology and Education*, pp. 1169-1180. Badajoz, Spain. Formatex Research Center.

Pachepsky, Y.A., Crawford, J.W. & Rawls, W.J. 2000. Fractals in Soil Science. *Developments in Soil Science, V. 27.* Elsevier Science, Amsterdam.

Pachepsky, Y.A., Giménez, D., Crawford, J.W. & Rawls, W.J. 2000. Conventional and fractal geometry in soil science. *Developments in Soil Science*, 27: 7-18. https://doi.org/10.1016/S0166-2481(00)80003-3.

Pacific Farmers. 2018. Review of the Tutu Rural Training Centre Courses. [online]. [Cited 2 October 2020]. https://pacificfarmers.com/resource/a-review-of-the-tutu-rural-training-centre-courses/

Paez-Espino, D., Eloe-Fadrosh, E.A., Pavlopoulos, G.A., Thomas, A.D., Huntemann, M., Mikhailova, N., et al. 2016. Uncovering Earth's virome. *Nature*, 536: 425-430.

Pal, B., Singh, C. & Singh, H. 1984. Barley yield under saline water cultivation. *Plant and Soil*, 81: 221–228.

Palková, Z. 2004. Multicellular microorganisms: laboratory versus nature. *EMBO Reports*, 5: 470–476. doi: 10.1038/sj.embor.7400145

Pan, G., Smith, P. & Pan, W. 2009. The role of soil organic matter in maintaining the productivity and yield stability of cereals in China. *Agriculture, Ecosystems and Environment*, 129: 344-348.

Panagos, P., Borrelli, P., Poesen, J., Ballabio, C., Lugato, E., Meusburger, K., Montanarella, L. & Alewell, C. 2015. The new assessment of soil loss by water erosion in Europe. *Environmental Science and Policy*, 54: 438-447. DOI: 10.1016/j.envsci.2015.08.012

Panagos P., Borrelli P., Meusburger K., Yu B., Klik A., Lim K.J., Yang J.E, et al. 2017. Global rainfall erosivity assessment based on high-temporal resolution rainfall records. *Scientific Reports*, 7: 4175. DOI: 10.1038/s41598-017-04282-8.

Pande, M. & Tarafdar, J.F. 2004. Arbuscular Mycorrhizal fungal diversity in Neem based Agroforestry Systems in Rajistan. *Applied Soil Ecology*, 26: 233-241.

Padron, L., Torres Rodriguez, D. G., Contreras Olmos, J., López, M. & Colmenares, C. 2012. Aislamientos de cepas fijadoras de nitrógeno y solubilizadoras de fósforo en un suelo alfisol venezolano. R*evista mexicana de ciencias agrícolas*, 3(2): 285-297.

Panke-Buisse, K., Poole, A.C., Goodrich, J.K., Ley, R.E., Kao-Kniffin, J. 2015. Selection on soil microbiomes reveals reproducible impacts on plant function. *The ISME Journal*, 9: 980–989. doi: 10.1038/ismej.2014.196

Pankhurst, C.E., Yu, S., Hawke, B.G. & Harch, B.D. 2001. Capacity of fatty acid profiles and substrate utilization patterns to describe differences in soil microbial communities associated with increased salinity or alkalinity at three locations in South Australia. *Biology and Fertility of Soils*, 33: 204-217.

Pankratova, E.M. 2006. Functioning of cyanobacteria in soil ecosystems. *Eurasian Soil Science*, 39(S1): S118–S127. https://doi.org/10.1134/S1064229306130199

Paoletti, M.G., Dufour, D.L., Cerda, H., Torres, F., Pizzoferrato, L. & Pimentel, D. 2000. The importance of leaf- and litter-feeding invertebrates as sources of animal protein for the Amazonian Amerindians. *Proceedings of the Royal Society B: Biological Sciences*, 267: 2247–2252.

Pardeshi, M. & Prusty, B.A.K. 2010. Termites as ecosystem engineers and potentials for soil restoration. *Current Science*, 99: 11.

Paris Agreement. 2015. *Paris Agreement under the United Nations Framework Convention on Climate Change.* Paris, France. 27 pp.

Park, H. & DuPonte, M.W. 2008. How to cultivate indigenous microorganisms. Honolulu, USA. University of Hawaii. 6 p.

Park, I.W., Hooper, J., Flegal, J.M., & Jenerette, G.D. 2018. Impacts of climate, disturbance and topography on distribution of herbaceous cover in Southern California chaparral: Insights from a remote-sensing method. *Diversity and Distribution,* 24: 497- 508 DI 10.1111/ddi.12693

Parnell, J.J., Berka, R., Young, H.A., Sturino, J.M., Kang, Y., Barnhart, D.M. & DiLeo, M.V. 2016. From the Lab to the Farm: An Industrial Perspective of Plant Beneficial Microorganisms. *Frontiers in Plant Science,* 7: 1110. doi: 10.3389/fpls.2016.01110

Parton, W.J., Scurlock, J.M.O. & Ojima, D.S. 1993. Observations and modeling of biomass and soil organic matter dynamics for the grassland biome worldwide. *Global Biogeochemical Cycles,* 7: 785-809. https://doi.org/10.1029/93GB02042

Pascual, U., Termansen, M., Hedlund, K., Brussaard, L., Faber, J.H., Foudi, S., Lemanceau, P. & Jorgensen, S.L. 2015. On the value of soil biodiversity and ecosystem services. *Ecosystem Services,* 15: 11-18.

Pastrana, A.M., Basallote-Ureba, M.J., Aguado, A., Akdi, K. & Capote, N. 2016. Biological control of strawberry soil-borne pathogens Macrophomina phaseolina and Fusarium solani, using Trichoderma asperellum and Bacillus spp. *Phytopathologia Mediterranea,* 55(1): 109–120.

Paterson, E., Sim, A., Davidson, J., & Daniell, T.J. 2016. Arbuscular mycorrhizal hyphae promote priming of native soil organic matter mineralisation. *Plant and soil,* 408 (1-2): 243-254.

Pauget, B., Gimbert, F., Coeurdassier, M., Crini, N., Pérès, G., Faure, O., Douay, F., Hitmi, A., Beguiristain, T., Alaphilippe, A., Guernion, M., Houot, S., Legras, M., Vian, J.-F., Hedde, M., Bispo, A., Grand, C. & de Vaufleury, A. 2013. Ranking field site management priorities according to their metal transfer to snails. *Ecological Indicators,* 29: 445–454. https://doi.org/10.1016/j.ecolind.2013.01.012

Paul, E.A. 2016. The nature and dynamics of soil organic matter: Plant inputs, microbial transformations, and organic matter stabilization. *Soil Biology and Biochemistry,* 98: 109–126.

Paul, E.A., Paustian, K., Elliott, E.T. & Cole, C.V. 1997. *Soil Organic Matter in Temperate Agroecosystems: Long-Term Experiments in North America.* Boca Raton, USA. CRC Press.

Paula, F.S., Rodrigues, J.L., Zhou, J., Wu, L., Mueller, R.C., Mirza, B.S., Pellizari, V.H., *et al.* 2014. Land use change alters functional gene diversity, composition and abundance in Amazon forest soil microbial communities. *Molecular ecology,* 23(12): 2988-2999.

Paustian, K., Lehmann, J., Ogle, S., Reay, D., Robertson, G.P. & Smith, P. 2016. Climate-smart soils. *Nature,* 532: 49–57.

Pavao-Zuckerman, M. & Pouyat, R.V. 2017. The effects of urban expansion on soil health and ecosystem services: An overview. In C. Gardi, ed. *Urban Expansion, Land Cover and Soil Ecosystem Services,* pp. 123-145. London, UK. Routledge.

Pawluk, S. 1985. Soil micromorphology and soil fauna: problems and importance. *Quaestiones Entomologicae* 21: 473-496.

Pearce, J.L. & Venier, L.A. 2006. The use of ground beetles (Coleoptera: Carabidae) and spiders (Araneae) as bioindicators of sustainable forest management: A review. *Ecological Indicators*, 6: 780-793.

Peay, K.G., Kennedy, P.G. & Talbot, J.M. 2016. Dimensions of biodiversity in the Earth mycobiome. *Nature Reviews Microbiology*, 14: 434–447.

Pecl, G.T., Araújo, M.B., Bell, J.D., Blanchard, J., Bonebrake, T.C., Chen, I.C., & Clark, T.D. 2017. Biodiversity redistribution under climate change: Impacts on ecosystems and human well-being. *Science*, 355(6332): eaai9214

Pelosi, C., Joimel, S. & Makowski, D. 2013. Searching for a more sensitive earthworm species to be used in pesticide homologation tests - A meta-analysis. *Chemosphere*, 90. 895-900.

Peña-Neira, S. 2018. Climate and Biological Diversity: How Should the Effects of Climate Change on Biological Diversity Be Legally Addressed in International and Comparative Law and Solutions? In W. Leal Filo & J. Barbir, eds. *Handbook on Climate Change and Biodiversity*, pp. 325-335. Cham, Switzerland. Springer.

Pena-Neira, S., Guzman, D. & Montero, P. 2005. La diversidad biológica y genética, relaciones jurídicas con la conservación, uso sustentable y utilización de los recursos que se encuentran en el suelo. *Medio Ambiente and Derecho: Revista electronica de derecho ambiental*, 12-13: 1576-3196. https://dialnet.unirioja.es/servlet/articulo?codigo=1371928]

Pennell, K.D. 2016. Specific surface area. In S.A. Elias, ed. *Reference Manual in Earth Systems and Environmental Sciences*. Oxford, UK. Elsevier. DOI: 10.1016/B978-0-12-409548-9.09583-X.

Pepoyan, A., Balayan, M., Manvelyan, A., Galstyan, L., Pepoyan, S., Petrosyan, S., Tsaturyan, V., Kamiya, S., Torok, T. & Chikindas, M. 2018. Probiotic Lactobacillus acidophilus strain INMIA 9602 Er 317/402 administration reduces the Numbers of Candida albicans and abundance of enterobacteria in the gut microbiota of Familial Mediterranean fever patients. *Frontiers in Immunology*, 9: 1–11.

Pérès, G., Vandenbulcke, F., Guernion, M., Hedde, M., Beguiristain, T., Douay, F., Houot, S., Piron, D., Richard, A., Bispo, A., Grand, C., Galsomies, L. & Cluzeau, D. 2011. Earthworm indicators as tools for soil monitoring, characterization and risk assessment. An example from the national Bioindicator programme (France). *Pedobiologia*, 54: S77–S87. https://doi.org/10.1016/j.pedobi.2011.09.015

Perna, A., Jost, C., Couturier, E., Valverde, S., Douady, S. & Theraulaz, G. 2008. The structure of gallery networks in the nests of termite Cubitermes spp. revealed by X-ray tomography. *Naturwissenschaften*, 95: 877–884.

Persson, J., Fink, P., Goto, A., Hood, J.M., Jonas, J., & Kato, S. 2010. To be or not to be what you eat: regulation of stoichiometric homeostasis among autotrophs and heterotrophs. *Oikos*, 119: 741–751.

Persson, T. & Wiren, A. 1995. Nitrogen mineralisation and potential nitrification at different depths in acid forest soils. *Plant and Soil,* 168: 55-65.

Peter, H., Ylla, I., Gudasz, C., Romani, A.M., Sabater, S. & Tranvik, L.J. 2011. Multifunctionality and diversity in bacterial biofilms. *PloS one,* 6(8): e23225.

Petermann, J.S., Fergus, A.J.F., Turnbull, L.A., & Schmid, B. 2008. Janzen-Connell effects are widespread and strong enough to maintain diversity in grasslands. *Ecology,* 89: 2399–2406

Petrella, F. 2011. Atlante pedologico del Piemonte. La fertilità biologica dei suoli. In A. Benedetti& R. Francaviglia, eds. *Biodiversità e pedodiversità: affinità e divergenze nell'areale italiano.* Libro dei riassunti Convegno celebrazioni giornata mondiale sul suolo pp. 63-72. Roma, Italia.

Petrzik, K., Vondrák, J., Kvíderová, J. & Lukavský, J. 2015. Platinum Anniversary: Virus and Lichen Alga Together More than 70 Years. *PLOS ONE,* 10(3): e0120768. https://doi.org/10.1371/journal.pone.0120768

PEW. 2019. Soil Health Can Combat Climate Change From the Ground Up. [online]. [Cited 5 October 2020]. https://www.pewtrusts.org/en/research-and-analysis/blogs/stateline/2019/08/23/soil-health-can-combat-climate-change-from-the-ground-up

Pey, B., Laporte, M. A., Nahmani, J., Auclerc, A., Capowiez, Y., Caro, G. & Joimel, S. 2014. A thesaurus for soil invertebrate trait-based approaches. *PloS One,* 9(10): e108985.

Philanthropy News Digest - PND. 2019. $10 Million Effort for Soil Health, Climate Change Mitigation Launched. [online]. [Cited 5 October 2020]. https://philanthropynewsdigest.org/news/10-million-effort-for-soil-health-climate-change-mitigation-launched

Philippot, L., Raaijmakers, J.M., Lemanceau, P., & van der Putten, W.H. 2013. Going back to the roots: the microbial ecology of the rhizosphere. *Nature Reviews Microbiology,* 11(11): 789-799.

Philippot, L., Spor, A., Hénault, C., Bru, D., Bizouard, F., Jones, C.M., Maron, P.A., *et al.* 2013. Loss in microbial diversity affects nitrogen cycling in soil. *The ISME journal,* 7(8): 1609.

Phillips, H.R.P., Cameron, E.K., Ferlian, O., Türke, M., Winter, M. & Eisenhauer, N. 2017. Red list of a black box. *Nature Ecology and Evolution,* 1: article 0103. DOI:10.1038/s41559-017-0103.

Phillips, H.R.P., Guerra, C.A., Bartz, M.L.C., Briones, M.J.I., Brown, G., Ferlian, O., Gongalsky, K.B., *et al.* 2019. Global distribution of earthworm diversity. *Science,* 366(6464): 480-485. https://doi.org/10.1101/587394

Phillips, H., Beaumelle, L., Tyndall, K., Burton, V., Cameron, E., Eisenhauer, N. & Ferlian, O. 2019. The effects of global change on soil faunal communities: a meta-analytic approach. *Research Ideas and Outcomes,* 5: e36427.

Phillips, J.D. 2001a. The relative importance of intrinsic and extrinsic factors in pedodiversity. *Annals of the Association of American Geographers,* 91: 609–621. https://doi.org/10.1111/0004-5608.00261.

Phillips, J.D. 2001b. Divergent evolution and spatial structure of soil landscape. variability. *Catena,* 43: 101–113. https://doi.org/10.1016/S0341-8162(00)00122-3

Phua, C.K.H., Wahid, A.N.A. & Rahim, A. 2011. Development of multifunctional bio fertilizer formulation from indigenous microorganisms and evaluation of their N2-fixing capabilities on chinese cabbage using 15 N tracer technique. *Pertanika Journal of Tropical Agricultural Science,* 35(3): 673–679.

Picci, G. & Nannipieri, P. 2002. *Metodi di analisi microbiologica del suolo.* MIPAF, Franco Angeli, Roma, Italia.

Pickett, S.T. & Cadenasso, M.L. 2009. Altered resources, disturbance, and heterogeneity: a framework for comparing urban and non-urban soils. *Urban Ecosystems,* 12: 23-44.

Pieper, S. & Weigmann, G. 2008. Interactions between isopods and collembolans modulate the mobilization and transport of nutrients from urban soils. *Applied Soil Ecology,* 39: 109–126. doi: https://doi.org/10.1016/j.apsoil.2007.11.012

Pimentel, D., Hepperly, P., Hanson, J., Douds, D. & Seidel, R. 2005. Environmental, Energetic, and Economic Comparisons of Organic and Conventional Farming Systems. *Bioscience,* 55: 573-582.

Pimentel, D. & Kounang, N. 1998. Ecology of soil erosion in ecosystems. *Ecosystems,* 1: 416-426.

Pimentel, D., Wilson, C., McCullum, C., Huang, R., Dwen, P., Flack, J., *et al.* 1997. Economic and Environmental Benefits of Biodiversity. *BioScience,* 47(11): 747-757. doi:10.2307/1313097

Pineda, A., Kaplan, I. & Bezemer, T.M. 2017. Steering soil microbiomes to suppress aboveground insect pests. *Trends in Plant Science,* 22: 770–778. doi: 10.1016/j.tplants.2017.07.002

Ping, C.L., Jastrow, J.D., Jorgenson, M.T., Michaelson, G.J. & Shur, Y.L. 2015. Permafrost soils and carbon cycling. *Soil,* 1: 147–171.

Piotrowska-Długosz, A. & Charzyński P. 2015. The impact of soil sealing on microbial biomass, enzymatic activity, and physicochemical properties in the Ekranic Technosols of Toruń (Poland). *Journal of Soils and Sediments,* 15: 47-59.

Pires, D.P., Boas, D.V., Sillankorva, S. & Azeredo, J. 2015. Phage therapy: a step forward in the treatment of Pseudomonas aeruginosa infections. *Journal of virology,* 89(15): 7449-7456.

Piron, D., Boizard, H., Heddadj, D., Pérès, G., Hallaire, V. & Cluzeau, D. 2017. Indicators of earthworm bioturbation to improve visual assessment of soil structure. *Soil and Tillage Research,* 173: 53-63.

Plaas, E., Meyer-Wolfarth, F., Banse, M., Bengtsson, J., Bergmann, H., Faber, J., Potthoff, M., Runge, T., Schrader, S. & Taylor, A. 2019. Towards valuation of biodiversity in agricultural soils: A case for earthworms. *Ecological Economics*, 159: 291–300. https://doi.org/10.1016/j.ecolecon.2019.02.003

Plante, A.F., Conant, R.T., Stewart, C.E., Paustian, K. & Six, J. 2006. Impact of soil texture on the distribution of soil organic matter in physical and chemical fractions. *Soil Science Society of America Journal*, 70: 287-296.

Platt, B.F., Kolb, D.J., Kunhardt, C.G., Milo, S.P. & New L.G. 2016. Burrowing through the literature: The impact of soil-disturbing vertebrates on physical and chemical properties of soil. *Soil Science*, 181: 175-191.

Poeplau, C., Don, A., Vesterdal, L., Leifeld, J., van Wesemaels, B., Schumacher, J. & Gensior, A. 2011. Temporal dynamics of soil organic carbon after land-use change in the temperate zone – carbon response functions as a model approach. *Global Change Biology*, 17: 2415–2427. https://doi.org/10.1111/j.1365-2486.2011.02408.x

Pointing, S.B. & Belnap, J. 2012. Microbial colonization and controls in dryland systems. *Nature Reviews Microbiology*, 10: 551.

Poiras, L., Iurcu-Straistraru, E., Bivol, A., Poiras, N., Toderas, I., Bugaciuc, M. & Boincean, B.P. 2014. Effects of Long-Term Fertility Management on the Soil Nematode Community and Cyst Nematode Heterodera schachtii Population in Experimental Sugar Beet Fields. *In* D. Dent, ed. *Soil as World Heritage*, pp. 37–43. Dordrecht, Springer Netherlands. (also available at https://doi.org/10.1007/978-94-007-6187-2_6).

Pokarzhevskii, A.D., van Straalen, N.M., Zaboev, D.P. & Zaitsev, A.S. 2003. Microbial links and element flows in nested detrital food-webs. *Pedobiologia*, 47: 213–224.

Pompili, L., Mellina, A. S., Benedetti, A., & Bloem, J. 2006. Microbial indicators for assessing biological fertility status of soils. In R. M. Cenci, & F. Sena eds., *Biodiversity-bioindication to evaluate soil health*, pp. 46-54. Scientific and Technical Research Series No. EUR 22245). Ispra, Italy.

Porre, R.J., van Groenign, J.W., De Deyn, G.B., de Goede, R.G.M. & Lubbers, I.M. 2016. Exploring the relationship between soil mesofauna, soil structure and N_2O emissions. *Soil Biology and Biochemistry*, 96: 55-64.

Portuguese Partnership for Soil. 2018. Seminar "Sustainable Soil Management Guidelines". [online]. [Cited 2 October 2020]. https://www.flipsnack.com/parportuguesasolo/new-flipbook.html

POST. 2011. *Ecosystem Service Valuation*. POSTNOTE Number 378. London, UK. Parliamentary Office of Science & Technology.

Postma-Blaauw, M.B., de Goede, R.G.M., Bloem, J., Faber, J.H. & Brussaard, L. 2010. Soil biota community structure and abundance under agricultural intensification and extensification. *Ecology*, 91: 460-473.

Postma-Blaauw, M.B., de Goede, R.G.M., Bloem, J., Faber, J.H. & Brussaard, L. 2012. Agricultural intensification and de-intensification differentially affect taxonomic diversity of predatory mites, earthworms, enchytraeids, nematodes and bacteria. *Applied Soil Ecology,* 57: 39–49.

Potapov, A.A., Semenina, E.E., Korotkevich, A.Y., Kuznetsova, N.A. & Tiunov, A.V. 2016. Connecting taxonomy and ecology: Trophic niches of collembolans as related to taxonomic identity and life forms. *Soil Biology and Biochemistry,* 101: 20–31. doi: https://doi.org/10.1016/j.soilbio.2016.07.002

Potapov, A.M., Korotkevich, A.Yu. & Tiunov, A.V. 2018. Non-vascular plants as a food source for litter-dwelling Collembola: Field evidence. *Pedobiologia,* 66: 11–17. https://doi.org/10.1016/j.pedobi.2017.12.005

Potschin-Young, M., Haines-Young, R., Görg, C., Heink, U., Jax, K. & Schleyer, C. 2018. Understanding the role of conceptual frameworks: Reading the ecosystem service cascade. *Ecosystem Services,* 29C: 428-440.

Potter, K.M. 2018. Large-scale patterns of forest fire occurrence in the conterminous United States, Alaska, and Hawaii, 2016. In K.M. Potter & B.L. Conkling, eds. *Forest Health Monitoring: national status, trends, and analysis 2017*, pp. 45–64. General Technical Report SRS-233. Asheville, USA. U.S. Department of Agriculture, Forest Service, Southern Research Station.

Potter, K.M., Paschke, J.L. & Zweifler, M. 2018. Large-scale patterns of insect and disease activity in the conterminous United States, Alaska and Hawai'i from the national insect and disease survey, 2016. In K.M. Potter & B.L. Conkling, eds. *Forest Health Monitoring: national status, trends, and analysis 2017*, pp. 23-44. General Technical Report SRS-233. Asheville, USA. U.S. Department of Agriculture, Forest Service, Southern Research Station.

Poudel, R., Jumpponen, A., Schlatter, D.C., Paulitz, T.C., McSpadden Gardener, B.B., Kinkel, L.L. & Garrett, K.A. 2016. Microbiome networks: a systems framework for identifying candidate microbial assemblages for disease management. *Phytopathology,* 106: 1083-1096.

Pouyat, R.V., Setälä, H., Szlavecz, K., Yesilonis, I. D., Cilliers, S., Hornung, E., *et al.* 2017. Introducing GLUSEEN: a new open access and experimental network in urban soil ecology. *Journal of Urban Ecology,* 3: jux002. doi.org/10.1093/jue/jux002

Powers, J.S., Corre, M.D., Twine, T.E. & Veldkamp, E. 2011. Geographic bias of field observations of soil carbon stocks with tropical land-use changes precludes spatial extrapolation. *Proceedings of the National Academy of Sciences,* 108: 6318-6322. https://doi.org/10.1073/pnas.1016774108

Pratama, A.A. & van Elsas, J.D. 2018. The 'neglected' soil virome: Potential role and impact. *Trends in Microbiology,* 26: 649-662.

Prather, C.M., Pelini, L., Laws, S., Rivest, A., Woltz, E., Bloch, M. & Del Toro, I. 2012. Invertebrates, ecosystem services and climate change. *Biological Reviews,* 88: 327–348. doi:10.1111/brv.12002

Pressler, Y., Moore, J.C. & Cotrufo, M.F. 2019. Belowground community responses to fire: meta-analysis reveals contrasting responses of soil microorganisms and mesofauna. *Oikos,* 128: 309–327. DOI: 10.1111/oi k .05738.

Pretty, J. 2008. Agricultural sustainability: concepts, principles and evidence. *Philosophical Transactions of the Royal Society,* 363: 447-465.

Přibyl, P. & Cepák, V. 2019. Screening for heterotrophy in microalgae of various taxonomic positions and potential of mixotrophy for production of high-value compounds. *Journal of Applied Phycology*, 31(3): 1555–1564. https://doi.org/10.1007/s10811-019-1738-9

Princz, J. I., Behan-Pelletier, V.M., Scroggins, R.P., & Siciliano, S.D. 2010. Oribatid mites in soil toxicity testing-the use of Oppia nitens (C.L. Koch) as a new test species. *Environ. Toxicol. Chem.* 29: 971-979.

Princz, J. I., Moody, M., Fraser, C., Vand der Vliet, L., Lemieux, H., Scroggins, R., & Siciliano, S.D. 2012. Evaluation of a new battery of toxicity tests for boreal forest soils: assessment of the impact of hydrocarbons and salts. *Environ. Toxicol. Chem.* 31: 766-777

Prober, S.M., Leff, J.W., Bates, S.T., Borer, E.T., Firn, J., Harpole, W.S., Lind, E.M., *et al.* 2015. Plant diversity predicts beta but not alpha diversity of soil microbes across grasslands worldwide. *Ecology letters,* 18(1): 85-95.

Prommer, J., Walker, T.W.N., Wanek, W., Braun, J., Zezula, D., Hu, Y., Hofhansl, F. & Richter, A. 2019. Increased microbial growth, biomass, and turnover drive soil organic carbon accumulation at higher plant diversity. *Global Change Biology,* 26(2): https://doi.org/10.1111/gcb.14777

Ptatscheck, C., Gansfort, B. & Traunspurger, W. 2018. The extent of wind-mediated dispersal of small metazoans, focusing nematodes. *Scientific Reports,* 8: 6814.

Qafoku, N.P. 2015. Climate-change effects on soils: Accelerated weathering, soil carbon, and elemental cycling. *Advances in Agronomy,* 131: 111-172. https://doi.org/10.1016/bs.agron.2014.12.002

Qin, Y., Druzhinina, I.S., Pan, X. & Yuan, Z. 2016. Microbially Mediated Plant Salt Tolerance and Microbiome-based Solutions for Saline Agriculture. *Biotechnology Advances,* 34: 1245–1259. doi: 10.1016/j.biotechadv.2016.08.005

Qin, Z., Xie, J.F., Quan, G.M., Zhang J.E., Mao, D.J. & Wang, J.X. 2018. Changes in the soil meso- and micro-fauna community under the impacts of exotic Ambrosia artemisiifolia. *Ecological Resources,* 34: 265-276.

Querner, P., Bruckner, A., Drapela, T., Moser, D., Zaller, J.G. & Frank, T. 2013. Landscape and site effects on Collembola diversity and abundance in winter oilseed rape fields in eastern Austria. *Agriculture, Ecosystems and Environment,* 164: 145-154.

Quinton, J.N., Govers, G., Van Oost, K. & Bardgett, R.D. 2010. The impact of agricultural soil erosion on biogeochemical cycling. *Natural Geoscience,* 3: 311–314. doi:10.1038/ngeo838

Raaijmakers, J.M. & Mazzola, M. 2016. Soil immune responses. *Science,* 352: 1392–1393. doi: 10.1126/science.aaf3252

Rabot, E., Wiesmeier, M., Schlüter, S. & Vogel, H.J. 2018. Soil Structure as an Indicator of Soil Functions: A Review. *Geoderma,* 314: 122–137. https://doi.org/10.1016/j.geoderma.2017.11.009.

Rajkumar, M., Sandhya, S., Prasad, M.N.V. & Freitas, H. 2012. Perspectives of plant-associated microbes in heavy metal phytoremediation. *Biotechnology Advances,* 30: 1562–1574.

Ram, A. & Aaron, Y. 2007. Negative and positive effects of topsoil biological crusts on water availability along a rainfall gradient in a sandy arid area. *Catena,* 70: 437–442.

Ramirez, K.S., Craine, J.M. & Fierer, N. 2012. Consistent effects of nitrogen amendments on soil microbial communities and processes across biomes. *Global Change Biology,* 18(6): 1918–1927. doi: 10.1111/j.1365-2486.2012.02639.x

Ramirez, K.S., Geisen, S., Morrien, E., Snoek, B.L. & van der Putten, W.H. 2018. Network analysis can advance above-belowground ecology. *Trends in Plant Science,* 9: 759-768

Ramirez, K.S., Leff, J.W., Barberán, A., Bates, S.T., Betley, J., Crowther, T.W., Kelly, E.F., *et al.* 2014. Biogeographic patterns in below-ground diversity in New York City's Central Park are similar to those observed globally. *Proceedings of the Royal Society B: Biological Sciences,* 28: 20141988.

Ramirez, K.S., Döring, M., Eisenhauer, N., Gardi, C., Ladau, J., Leff, J.W., Lentendu, G., *et al.* 2015. Toward a global platform for linking soil biodiversity data. *Frontiers in Ecology and Evolution,* 3: 91.

Ramnarain, Y.I., Ansari, A., Ori, L. 2016. RESEARCH REPORT ON ONE LOCAL EARTHWORM SPECIES IN SURINAME. *Journal of Biology and Nature.* 5(1): 26-30.

Rath, K.M. & Rousk J. 2015. Salt effects on the soil microbial decomposer community and their role in organic carbon cycling: A review. *Soil Biology and Biochemistry,* 81: 108-123.

Rath, K.M., Maheshwari, A. & Rousk, J. 2019. Linking microbial community structure to trait distributions and functions using salinity as an environmental filter. *mBio,* 10: e01607-19.

Rath, K.M., Fierer, N., Murphy, D.V. & Rousk, J. 2018. Linking bacterial community composition to soil salinity along environmental gradients. *The ISME Journal,* 13: 836-846. doi:10.1038/s41396-018-0313-8.

Ray, J.D., McIntyre, N.E., Wallace, M.C., Teaschner, A.P. & Schoenhals, M.G. 2016. Factors influencing burrowing owl abundance in prairie dog colonies on the Southern High Plains of Texas. *Journal of Raptor Research,* 50: 185-193.

Redfield, A.C. 1958. The biological control of chemical factors in the environment. *American Scientist,* 46: 205-221.

Reed, H.E. & Martiny, J.B. 2007. Testing the functional significance of microbial composition in natural communities. *FEMS microbiology ecology,* 62(2): 161-170.

Reese, A.T., Savage, A., Youngsteadt, E., McGuire, K.L., Koling, A., Watkins, O., *et al.* 2016. Urban stress is associated with variation in microbial species composition—but not richness—in Manhattan. *The ISME Journal,* 10: 751.

Regan, K., Stempfhuber, B., Schloter, M., Rasche, F., Boeddinghaus, R.S., Prati, D., Philippot, L., Kandeler, E. & Marhan, S. 2017. Temporal and spatial shifts in abundances and potential activities of nitrogen cycling communities in an unfertilized temperate grassland exhibit both variability and stability. *Soil Biology and Biochemistry,* 109: 214-226.

Reich, P.B., Tilman, D., Isbell, F., Mueller, K., Hobbie, S.E., Flynn, D.F. & Eisenhauer, N. 2012. Impacts of biodiversity loss escalate through time as redundancy fades. *Science,* 336(6081): 589-592.

Reichman, O.J. & Seabloom, E.W. 2002. The role of pocket gophers as subterranean ecosystem engineers. *Trends in Ecology and Evolution,* 17(1): 44–49.

Ren, X. M., Guo, S.J., Tian, W., Chen, Y., Han, H., Chen, E., Li, B.L., Li, Y.Y. & Chen, Z.J. 2019. Effects of plant growth-promoting bacteria (PGPB) inoculation on the growth, antioxidant activity, cu uptake, and bacterial community structure of rape (Brassica napusL.) grown in cu-contaminated agricultural soil. *Frontiers in Microbiology,* 10: 1–12.

Rendina, N., Nuzzaci, M., Sofo, A., Campiglia, P., Scopa, A., Sommella, E., Pepe, G., *et al.* 2019. Yield parameters and antioxidant compounds of tomato fruit: the role of plant defence inducers with or without Cucumber mosaic virus infection. *Journal of the Science of Food and Agriculture,* 99: 5541–5549. doi: 10.1002/jsfa.9818

Rengasamy, P. 2006. World salinization with emphasis on Australia. *Journal of Experimental Botany,* 57: 1017-1023.

Renker, C., Otto, P., Schneider, K., Zimdars, B., Maraun, M. & Buscot, F. 2005. *Microbial Ecology,* 50: 518–528.

Restrepo Rivera, J. & Hensel, J. 2007. *El ABC de la agricultura orgánica fostitos y panes de piedra.* Santiago de Cali, Colombia. Feriva S.A.

Renzi, G., Canfora, L., Salvati, L., & Benedetti, A. 2017. Validation of the soil Biological Fertility Index (BFI) using a multidimensional statistical approach: A country-scale exercise, *CATENA,* 149:294-299.

Réseau-Agriville. 2018. Jardibiodiv, l'observatoire de la biodiversité du sol dans les jardins urbains. In: *Réseau-Agriville* [online]. [Cited 2 October 2020]. https://reseau-agriville.com/jardibiodiv/

Richards, R. 1983. Should selection for yield in saline regions be made on saline or non-saline soils? *Euphytica,* 32: 431–438.

Richards, R., Dennett, C., Qualset, C., Epstein, E., Norlyn, J. & Winslow, M. 1987. Variation in yield of grain and biomass in wheat, barley, and triticale in a salt-affected field. *Field and Crops Research,* 15: 277-287.

Riding, R. 2000. Microbial carbonates: the geological record of calcified bacterial-algal mats and biofilms: *Microbial carbonates. Sedimentology*, 47: 179–214. https://doi.org/10.1046/j.1365-3091.2000.00003.x

Rietkerk, M., Ketner, P., Burger, J., Hoorens, B. & Olff, H. 2000. Multiscale soil and vegetation patchiness along a gradient of herbivore impact in a semi-arid grazing system in West Africa. *Plant Ecology*, 148: 207–224.

Riley, H. & Bakkegard, M. 2006. Declines of soil organic matter content under arable cropping in southeast Norway. *Acta Agriculturae Scandinavica, B*, 56: 217-223.

Rillig, M.C. 2004. Arbuscular mycorrhizae, glomalin, and soil aggregation. *Canadian Journal of Soil Science*, 84(4): 355-363

Rillig, M.C., Antonovics, J., Caruso, T., Lehmann, A., Powell, J.R., Veresoglou, S.D. & Vergruggen, E. 2015. Interchange of entire communities: microbial community coalescence. *Trends in Ecology and Evolution*, 30: 470–476. doi: 10.1016/j.tree.2015.06.004

Rillig, M.C. 2004. Arbuscular mycorrhizae, glomalin, and soil aggregation. *Canadian Journal of Soil Science*, 84(4): 355–363. https://doi.org/10.4141/S04-003

Rillig, M.C. & Mummey, D.L. 2006. Mycorrhizas and soil structure. *New Phytologist*, 171(1): 41-53.

Rillig, M.C., Tsang, A. & Roy, J. 2016. Microbial community coalescence for microbiome engineering. *Frontiers in Microbiology*, 7: 1967. doi: 10.3389/fmicb.2016.01967

Rillig, M.C., Lehmann, A., Lejmann, J., Camenzind, T. & Rauh, C. 2018. Soil biodiversity effects from field to fork. *Trends in Plant Science*, 23: 17-24.

Rillig, M.C., Aguilar-Trigueros, C.A., Bergmann, J., Verbruggen, E., Veresoglou, S.D. & Lehmann, A. 2015. Plant root and mycorrhizal fungal traits for understanding soil aggregation. *New Phytologist*, 205(4): 1385-1388.

Ripple, W.J., Wolf, C. & Newsome, T.M. 2019. World Scientists' Warning of a Climate Emergency. *Bioscience Magazine*, 2000: 1–20.

Robin, C., Piou, D., Feau, N., Douzon, G., Schenck, N. & Hansen, E.M. 2011. Root and aerial infections of Chamaecyparis lawsoniana by Phytophthora lateralis: A new threat for European countries. *Forest Pathology*, 41: 417–424.

Robinson, D.A., Lebron, I. & Vereecken, H. 2009. On the definition of the natural capital of soils: a framework for description, evaluation and monitoring. *Soil Science Society of America Journal*, 73: 1904–1911.

Robinson, D.A., Hockley, N., Cooper, D.M., Emmett, B.A., Keith, A.M., Lebron, I., Reynolds, B., *et al.* 2013. Natural capital and ecosystem services, developing an appropriate framework as a basis for valuation. *Soil Biology & Biochemistry*, 57: 1023-1033.

Robinson, D.A., Fraser, I., Dominati, E.J., Davíðsdóttir, B., Jónsson, J.O.G., Jones, L., Jones, S.B., et al. 2014. On the Value of Soil Resources in the Context of Natural Capital and Ecosystem Service Delivery. *Soil Science Society of America Journal,* 78: 685-700.

Robinson, R.A. & Sutherland, W.J. 2002. Post-war changes in arable farming and biodiversity in Great Britain. *Journal of Applied Ecology,* 39: 157–176.

Rocha, I., Ma, Y., Carvalho, M.F., Magalhaes, C., Janouskova, M., Vosatka, M., Freitas, H. & Oliveira, R.S. 2018. Seed coating with inocula of arbuscular mycorrhizal fungi and plant growth promoting rhizobacteria for nutritional enhancement of maize under different fertilization regimes. *Archives of Agronomy and Soil Science,* 65: 31-43

Rockström, J., Steffen, W., Noone, K., Persson, A., Stuart Chapin III, F., Lambin, E.F., Lenton, T.M., Scheffer, M., Folke, C., Schellnhuber, H.J., Nykvist, B., de Wit, C.A., Hugues, T.,van der Leeuw, S., Rodhe, H., Sorlin S., Snyder, P.K., Costanza, R., Svedin, U., Falkenmark, M., Karlberg, L., Corell, R.W., Fabry, V.J., Hansen, J., Walker, B., Liverman, D., Richardson, K., Crutzen, P., & Foley, J.A. 2009. A safe operating space for humanity. *Nature,* 461: 472-475.

Rodale Institute. 2018. 10 Ways to Connect Soil and Human health. [online]. [Cited 5 October 2020].https://rodaleinstitute.org/blog/10-ways-to-connect-soil-and-human-health/

Rodale Institute. 2018. 8 Ways to Fight Climate Change With Soil Health. [online]. [Cited 5 October 2020]. https://rodaleinstitute.org/blog/8-ways-to-fight-climate-change-with-soil-health/

Rodrigues, J.L.M., Pellizari, V.H., Mueller, R., Baek, K., Jesus, E. de C., Paula, F.S., Mirza, B., et al. 2013. Conversion of the Amazon Rainforest to Agriculture Results in Biotic Homogenization of Soil Bacterial Communities. *Proceedings of the National Academy of Sciences,* 110(3): 988–93. https://doi.org/10.1073/pnas.1220608110.

Rodriguez, H., Vessely, S., Shah, S. & Glick, B.R. 2008. Effect of a nickel-tolerant ACC deaminase-producing Pseudomonas strain on growth of non-transformed and transgenic canola plants. *Current Microbiology,* 57(2): 170-174. doi: 10.1007/s00284-008-9181-1

Roger, F., Bertilsson, S., Langenheder, S., Osman, O.A. & Gamfeldt, L. 2016. Effects of multiple dimensions of bacterial diversity on functioning, stability and multifunctionality. *Ecology,* 97(10): 2716-2728.

Roger-Estrade, J., Anger, C., Bertrand, M. & Richard, G. 2010. Tillage and soil ecology: Partners for sustainable agriculture. *Soils and Tillage Research,* 111(1): 33-40. doi:10.1016/j.still.2010.08.010.

Rogozhin, E., Sadykova, V., Baranova, A., Vasilchenko, A., Lushpa, V., Mineev, K., Georgieva, M., et al. 2018. A novel lipopeptaibol emericellipsin A with antimicrobial and antitumor activity produced by the extremophilic fungus Emericellopsis alkalina. *Molecules,* 23(11): 2785 - 2797.

Römbke, J., Jänsch, S., Roß-Nickoll, M., Toschki, A., Höfer, H., Horak, F., Russell, D., Burkhardt, U. & Schmitt, H. 2012. *Erfassung und Analyse des Bodenzustands im Hinblick auf die Umsetzung und Weiterentwicklung der Nationalen Biodiversitätsstrategie*. Dessau-Roßlau, Umweltbundesamt. (also available at https://www.umweltbundesamt.de/publikationen/erfassung-analyse-des-bodenzustands-im-hinblick-auf).

Romeis, J., Raybould, A., Bigler, F., Candolfi, M.P., Hellmich, R.L., Huesing, J.E. & Shelton, A.M. 2013. Deriving criteria to select arthropod species for laboratory tests to assess the ecological risks from cultivating arthropod-resistant genetically engineered crops. *Chemosphere*, 90: 901-909.

Romero, G.Q., Gonçalves-Souza, T., Vieira, C. & Koricheva, J. 2015. Ecosystem engineering effects on species diversity across ecosystems: a meta-analysis. *Biological Reviews*, 90: 877–890.

Roose, E., Kabore, V. & Guenat, C. 1999. Zai practice: A west african traditional rehabilitation system for semiarid degraded lands, a case study in burkina faso. *Arid Soil Research and Rehabilitation*, 13: 343–355.

Root-Bernstein, M. & Ebensperger, L.A. 2013. Meta-analysis of the effects of small mammal disturbances on species diversity, richness and plant biomass. *Australian Ecology*, 38: 289–299

Ros, M.B., Hiemstra T., van Groenigen, J.W., Chareesri, A. & Koopmans, G.F. 2017. Exploring the pathways of earthworm-induced phosphorus availability. *Geoderma*, 303: 99–109.

Rosenkranz, D., Bachmann, G., König, W. & Einsele, G., eds. 2000. *Handbuch Bodenschutz – ergänzbare Loseblattsammlung*.

Rosling, A., Landeweert, R., Lindahl, B.D., Larsson, K.-H., Kuyper, T.W., Taylor, A.F.S. & Finlay, R.D. 2003. Vertical Distribution of Ectomycorrhizal Fungal Taxa in a Podzol Soil Profile. *The New Phytologist*, 159(3): 775-783 https://www.jstor.org/stable/1514274

Roslund, M.I., Grönroos, M., Rantalainen, A.-L., Jumpponen, A., Romantschuk, M., Parajuli, A., Hyöty, H, Laitinen, O. & Sinkkonen, A. 2018. Half-lives of PAHs and temporal microbiota changes in commonly used urban landscaping materials. *PeerJ*, 6: e4508. doi:10.7717/peerj.4508

Rouland-Lefevre, C. 2000. Symbiosis with fungi. In Y. Abe, Bignell, D.E. & Higashi, T., eds. *Termites: Evolution, Sociality, Symbiosis, Ecology*, pp. 289-306. Dordrecht, Netherlands. Kluwer Academic Press.

Rousseau, L., Fonte, S.J., Téllez, O., Van Der Hoek, R. & Lavelle, P. 2013. Soil Macrofauna as Indicators of Soil Quality and Land Use Impacts in Smallholder Agroecosystems of Western Nicaragua. *Ecological Indicators*, 27: 71–82. https://doi.org/10.1016/j.ecolind.2012.11.020.

Roux, S., Brum, J.R., Dutilh, B.E., Sunagawa, S., Duhaime, M.B., Loy, A., et al. 2016. Ecogenomics and potential biogeochemical impacts of globally abundant ocean viruses. *Nature*, 537: 689-693.

Rowe, H.I., Brown, C.S. & Claassen, V.P. 2007. Comparisons of mycorrhizal responsiveness with field soil and commercial inoculum for six native montane species and Bromus tectorum. *Restoration Ecology*, 15(1): 44–52. doi: 10.1111/j.1526-100X.2006.00188.x

Rowe, E.C., Smart, S.M. & Emmett, B.A. 2014. Phosphorus availability explains patterns in a productivity indicator in temperate semi-natural vegetation. *Environmental Science-Processes and Impacts*, 16: 2156-2164.

Roy, S., Saxena, P. & Bano, R. 2012. Soil biota build-up under organic and inorganic fertilization in semi-arid central India. *Annals of Arid Zone*, 51(2): 91-98.

Ruf, A. & Beck, L. 2005. The use of predatory soil mites in ecological soil classification and assessment concepts, with perspectives for oribatid mites. *Ecotoxicology and Environmental Safety*, 62: 290-299.

Rundel, P.W., Villagran, P.E., Dillon, M.O., Roig-Juñent, S. & Debandi, G. 2007. Chapter 10: Arid and Semi-Arid Ecosystems. In T. Veblen, K. Young & A. Orme, eds. *The Physical Geography of South America*, pp. 158-183. Oxford, UK. Oxford University Press.

Rusek, J. 1998. Biodiversity of Collembola and their functional role in the ecosystem. *Biodiversity and Conservation*, 7: 1207-1219.

Rutgers, M., Mulder, C., Schouten, A., Bloem, J., Bogte, J., Breure, A., Brussaard, L., De Goede, R., Faber, J. & Keidel, H. 2008. Soil ecosystem profiling in the Netherlands with ten references for biological soil quality. *RIVM report 607604009*.

Rutgers, M., Jagers of Akkerhuis, G.A.J.M., Bloem, J., Schouten, A.J. & Breure, A.M. 2010. Priority areas in the Soil Framework Directive. The significance of soil biodiversity and ecosystem services. , p. 62. Bilthoven, The Netherlands. (also available at https://www.rivm.nl/bibliotheek/rapporten/607370002.pdf).

Rutgers, M., Orgiazzi, A., Gardi, C., Römbke, J., Jänsch, S., Keith, A.M., Neilson, R., *et al.* 2016. Mapping earthworm communities in Europe. *Applied Soil Ecology*, 97: 98-111. https://doi.org/10.1016/j.apsoil.2015.08.015

Rutgers, M., van Leeuwen, J.P., Vrebos, D., van Wijnen, H.J., Schouten, T. & de Goede, R.G.M. 2019. Mapping Soil Biodiversity in Europe and the Netherlands. *Soil Systems*, 3: 39.

Rutgers, M., Schouten, A.J., Bloem, J., Eekeren, N.V., Goede, R.G.M.D., Akkerhuis, G.A.J.M.J., Wal, A.V. der, Mulder, C., Brussaard, L. & Breure, A.M. 2009. Biological measurements in a nationwide soil monitoring network. *European Journal of Soil Science*, 60(5): 820–832. https://doi.org/10.1111/j.1365-2389.2009.01163.x

Ryan, M.H. & Graham, J.G. 2018. Little evidence that farmers should consider abundance or diversity of arbuscular mycorrhizal fungi when managing crops. *New Phytologist*, 220: 1092–1107

Ryan, S.F., Adamson, N.L., Aktipis, A., Andersen, L.K., Austin, R., Barnes, L., Beasley, M.R., *et al.* 2018. The role of citizen science in addressing grand challenges in food and agriculture research. *Proceedings of the Royal Society B: Biological Sciences*, 285: 20181977. doi: 10.1098/rspb.2018.1977.

Sala, O. E., Chapin, F. S., Armesto, J. J., Berlow, E., Bloomfield, J., Dirzo, R., Leemans, R., et al. 2000. Global biodiversity scenarios for the year 2100. *Science,* 287(5459): 1770-1774.

Salamon, J.-A., Alphei, J., Ruf, A., Scharfer, M., Scheu, S., Schneider, K., Sührig, A. & Maraun, M. 2006. Transitory dynamic effects in the soil invertebrate community in a temperate deciduous forest: effects of resource quality. *Soil Biology and Biochemistry,* 38: 209-221.

Salmon, S., Ponge, J.F., Gachet, S., Deharveng, L., Lefebvre, N. & Delabrosse, F. 2014. Linking species, traits and habitat characteristics of Collembola at European scale. *Soil Biology and Biochemistry,* 75: 73-85.

San José Martínez, F., Muñoz Ortega, F.J., Caniego Monreal, F.J., Kravchenko, A.N. & Wang, W. 2015. Soil aggregate geometry: Measurements and morphology. *Geoderma,* 237-238: 36-48. DOI: https://doi.org/10.1016/j.geoderma.2014.08.003

Sanabria, C., Lavelle, P. & Fonte, S.J. 2014. Ants as indicators of soil-based ecosystem services in agroecosystems of the Colombian Llanos. *Applied Soil Ecology,* 84: 24–30. doi:10.1016/j.apsoil.2014.07.001

Sanchez-Cañizares, C., Jorrín, B., Poole, P.S. & Tkacz, A. 2017. Understanding the holobiont: the interpendence of plants and their microbiome. *Current Opinion in Microbiology,* 38: 188-196.

Sánchez-Moreno, S., Nicola, N.L., Ferris, H. & Zalom, F.G. 2009. Effects of agricultural management on nematode–mite assemblages: Soil food web indices as predictors of mite community composition. *Applied Soil Ecology,* 41(1): 107-117.

Saporito, R.A., Donnelly, M.A., Norton, R.A., Garraffo, M., Spande, T.F. & Daly, J.W. 2007. Oribatid mites as a major dietary source for alkaloids in poison frogs. *Proceedings of the National Academy of Sciences of the USA,* 104: 8885–8890.

Sardinha, M., Muller, T., Schmeisky, H. & Joergensen, R.G. 2003. Microbial performance in soils along a salinity gradient under acidic conditions. *Applied Soil Ecology,* 23: 237-244.

Sartori, M., Philippidis, G., Ferrari, E., Borrelli, P., Lugato, E., Montanarella, L. & Panagos, P. 2019. A linkage between the biophysical and the economic: Assessing the global market impacts of soil erosion. *Land Use Policy,* 86: 299-312.

Setälä, H. & Huhta, V. 1991. Soil fauna increase Betula pendula growth: laboratory experiments with coniferous forest floor. Ecology 72: 665-671.

Scalenghe, R. & Marsan, R.A. 2009. The anthropogenic sealing of soils in urban areas. *Landscape and Urban Planning,* 90: 1-10.

ScaraB'Obs. 2015. *ScaraB'Obs – Étude & Conservation des Bousiers de France* [online]. [Cited 2 October 2020]. https://scarab-obs.fr/

Schaffelke, B., Collier, C., Kroon, F., Lough, J., McKenzie, L., Ronan, M., Uthicke, S. & Brodie, J. 2017. *Chapter 1: The condition of coastal and marine ecosystems of the Great Barrier Reef and their responses to water quality and disturbances.* Scientific Consensus Statement 2017: A synthesis of the science of land-based water quality impacts on the Great Barrier Reef. State of Queensland, 2017.

Scharlemann, J.P.W., Tanner, E.V.J., Hiederer, R. & Kapos, V. 2014. Global soil carbon: understanding and managing the largest terrestrial carbon pool. *Carbon Management,* 5: 81-91. https://doi.org/10.4155/cmt.13.77

Scherber, C., Eisenhauer, N., Weisser, W. W., Schmid, B., Voigt, W., Fischer, M., Beßler, H., et al. 2010. Bottom-up effects of plant diversity on multitrophic interactions in a biodiversity experiment. *Nature,* 468(7323): 553.

Scheu, S., Theenhaus, A. & Jones, T.H. 1999. Links between the detritivore and the herbivore system: Effects of earthworms and Collembola on plant growth and aphid development. *Oecologia,* 119: 541–551.

Scheu, S., Ruess, L. & Bonkowski, M. 2005. Interactions Between Microorganisms and Soil Micro- and Mesofauna. In Varma, A. & Buscot, F., eds. *Microorganisms in Soils: Roles in Genesis and Functions,* pp. 235-275. Berlin, Germany. Springer-Verlag.

Scheunemann, N., Digel, C., Scheu, S. & Butenschoen, O. 2015. Roots rather than shoot residues drive soil arthropod communities of arable fields. *Oecologia,* 179: 1135–1145.

Schilli, C., Rinklebe, J., Lischeid, G., Kaufmann-Boll, C. & Lazar, S. 2011. *Auswertung der Veränderungen des Bodenzustands für Boden-Dauerbeobachtungsflächen (BDF) UBA-Text 90/2011 Teil B: Datenauswertung und Weiterentwicklung.* Umweltbundesamt. (also available at https://www.umweltbundesamt.de/publikationen/auswertung-veraenderungen-des-bodenzustands-fuer-0).

Schlaeppi, K. & Bulgarelli, D. 2015. The plant microbiome at work. *Molecular Plant-Microbe Interactions,* 28: 212-217.

Schmidt, M.W.I., Torn, M.S., Abiven, S., Dittmar, T., Guggenberger, G., Janssens, I.A., Kleber, M., et al. 2011. Persistence of soil organic matter as an ecosystem property. *Nature,* 478: 49-56.

Schmidt, O., Dyckmans, J. & Schrader, S. 2016. Photoautotrophic microorganisms as a carbon source for temperate soil invertebrates. *Biology Letters,* 12(1): 20150646. https://doi.org/10.1098/rsbl.2015.0646

Schmidt, R., Gravuer, K., Bossange, A.V., Mitchell, J. & Scow, K. 2018. Long-term use of cover crops and no-till shift soil microbial community life strategies in agricultural soil. *PLoS One,* 13(2): e0192953.

Schmidt, S.N., Smith, K.E.C., Holmstrup, M. & Mayer, P. 2013. Uptake and toxicity of polycyclic aromatic hydrocarbons in terrestrial springtails studying bioconcentration kinetics and linking toxicity to chemical activity. *Environmental Toxicology and Chemistry,* 32: 361-369.

Schmitz, O.J. 2010. *Resolving ecosystem complexity.* Princeton, USA. Princeton University Press. 175 pp.

Schmitz, O.J., Hawlena, D. & Trussell, G.C. 2010. Predator control of ecosystem nutrient dynamics. *Ecology Letters,* 13: 1199-1209.

Schneider A.K. & Schröder B. 2012. Perspectives in modelling earthworm dynamics and their feedbacks with abiotic soil properties. *Applied Soil Ecology,* 58: 29-36.

Schneider, F.D., Scheu, S. & Brose, U. 2012. Body mass constraints on feeding rates determine the consequences of predator loss. *Ecology Letters,* 15: 436-443.

Schneider, K. & Maraun, M. 2005. Feeding preferences among dark pigmented fungal taxa ("Dematiacea") indicate limited trophic niche differentiation of oribatid mites (Oribatida, Acari). *Pedobiologia,* 49: 61–67.

Schneider, K., Migge, S., Norton, R.A., Scheu, S., Langel, R., Reineking, A. & Maraun, M. 2004. Trophic niche differentiation in soil microarthropods (Oribatida, Acari): evidence from stable isotope ratios (15N/14N). *Soil Biology and Biochemistry,* 36: 1769–1774.

Schnitzer, S.A., Klironomos, J.N., HilleRisLambers, J., Kinkel, L.L., Reich, P.B., Xiao, K., Rillig, M.C., *et al.* 2011. Soil microbes drive the classic plant diversity-productivity pattern. *Ecology,* 92: 296–303.

Schnürer, J., Clarholm, M. & Rosswall, T. 1985. Microbial biomass and activity in an agricultural soil with different organic matter contents. *Soil Biology and Biochemistry,* 17: 611-618.

Schoelz, J.E. & Stewart, L.R. 2018. The Role of Viruses in the Phytobiome. *Annual Review of Virology,* 5: 93-111.

Schouten, T., Bloem, J., Goede, R.G.M. de, Eekeren, N. van, Deru, J., Zanen, M., Sukkel, W., Balen, D.J.M. van, Korthals, G. & Rutgers, M. 2018. Veldexperimenten uitgelicht: Niet-kerende grondbewerking goed voor de bodembiodiversiteit? *Bodem*(3): 20–23.

Schuldt, A., Assmann, T., Brezzi, M., Buscot, F., Eichenberg, D., Gutknecht, J., Härdtle, W., *et al.* 2018. Biodiversity across trophic levels drives multifunctionality in highly diverse forests. *Nature communications,* 9(1): 2989.

Schulman, L., Juslén, A. & Lahti, K. 2019. FinBIF: An all-embracing, integrated, cross-sectoral biodiversity data infrastructure. *Biodiversity Information Science and Standards,* 3: e37253. https://doi.org/10.3897/biss.3.37253

Schulz-Bohm, K., Geisen, S., Wubs, E.R.J., Song, C., de Boer, W. & Garbeva, P. 2017. The prey's scent - Volatile organic compound mediated interactions between soil bacteria and their protist predators. *The ISME Journal,* 11: 817–820.

Schütz, K., Bonkowski, M. & Scheu, S. 2008. Effects of Collembola and fertilizers on plant performance (Triticum aestivum) and aphid reproduction (Rhopalosiphum padi). *Basic Applied Ecology,* 9: 182–188. doi: https://doi.org/10.1016/j.baae.2006.07.003

Schwarzmüller, F., Eisenhauer, N. & Brose, U. 2015. 'Trophic whales' as biotic buffers: weak interactions stabilize ecosystems against nutrient enrichment. *Journal of Animal Ecology,* 84(3): 680-691.

Scott, A., Mangan, S.A., Schnitzer, S.A., Herre, E.A., Mack, K.M.L., Valencia, M.C., Sanchez, E.I. & Bever, J.D. 2010. Negative plant–soil feedback predicts tree-species relative abundance in a tropical forest. *Nature,* 466: 752–755. https://doi.org/10.1038/nature09273

See, C.R., McCormack, M.L., Hobbie, S.E., Flores-Moreno, H., Silver, W.L. & Kennedy, P.G. 2019. Global patterns in fine root decomposition: climate, chemistry, mycorrhizal association and woodiness. *Ecology Letters,* 22(6): 946-953.

Seeram, N.P. 2008. Berry Fruits: Compositional Elements, Biochemical Activities, and the Impact of Their Intake on Human Health, Performance, and Disease. *J. Agric. Food Chem.*, 56: 627–629.

Segura-Castruita, M. A., Sánchez-Guzmán, P., Ortiz-Solorio, C. A. & del Carmen Gutiérrez-Castorena, M. 2005. Carbono orgánico de los suelos de México. *Terra Latinoamericana*, 23(1): 21-28.

SEMANART. 2008a. Plan de Acción de Combate a la Desertificación. [online]. [Cited 2 October 2020]. https://www.gob.mx/cms/uploads/attachment/file/31167/pnacdd.pdf

SEMANART. 2008b. Programa Nacional Manejo Sustentable de Tierras [online]. [Cited 5 October 2020]. https://www.gob.mx/cms/uploads/attachment/file/31167/pnacdd.pdf

Semchenko, M., Leff, J.W., Lozano, Y.M., Saar, S., Davison, J., Wilkinson, A., Jackson, B.G., *et al.* 2018. Fungal diversity regulates plant-soil feedbacks in temperate grassland. *Science Advances*, 4: eaau4578.

SENACE. 2009. DECRETO SUPREMO N° 017-2009-AG. [online]. [Cited 5 October 2020]. https://www.senace.gob.pe/download/senacenormativa/NAT-3-7-01-DS-017-2009-AG.pdf

SENACE. 2010. DECRETO SUPREMO N° 013-2010-AG. [online]. [Cited 5 October 2020]. https://www.senace.gob.pe/wp-content/uploads/2016/10/NAT-3-7-03-DS-013-2010-AG.pdf

Senckenberg – Leibniz Institution for Biodiversity and Earth System Research. 2020. [online]. [Cited 2 October 2020]. https://www.senckenberg.de/duennehautdererde

Senicovscaia, I. 2014. Evaluation of organic amendments effects on soil microorganisms and enzymatic activities of degraded chernozem. *Scientific papers. Series A. Agronomy*, LVII: 6.

Senicovscaia, I. 2015. Monitoring and recovery of the soil biota in conditions of the degradation processes intensification in the Republic of Moldova. *MPRA Paper*, p. MPRA Paper No. 69375. University Library of Munich, Germany. (also available at https://ideas.repec.org/p/pra/mprapa/69375.html).

Senicovscaia, I. 2018. Restoration of biota in the ordinary chernozem by green manuring in the southern zone of the Republic of Moldova. *Agronomie și agroecologie*, 52(1). (also available at https://ibn.idsi.md/ro/vizualizare_articol/89562).

Senicovscaia, I. undated. Conservation of invertebrates' biodiversity in soils of the Republic of Moldova. International symposium "Agrarian economy and rural development – realities and perspectives for Romania

Senicovscaia, I., Marinescu, C., Andrieş, S., Filipciuc, V., Boincean, B., Bulat, L., Burghelea, A., Botezatu, T. & Daniliuc, R. 2012. *Methodological Instructions on the Assessment and Increase of the Soil Biota Stability in Conditions of the Degradation Processes Intensification. (Rom)*. Chisinau, Pontos. (also available at https://www.ipaps.md/book/).

Senikovskaya, I., Bogdevich, O. & Marinesku, K. 2008. Microbiological characteristic of pesticide polluted soils. Paper presented at 9th International HCH and Pesticides Forum for Central and Eastern European, Caucasus and Central Asia countries, 2008, Chisinau.

Senthilkumar, S., Basso, B., Kravchenko, A.N. & Robertson, G.P. 2009. Contemporary Evidence Of Soil Carbon Loss In The U.S. Corn Belt. *Soil Science Society of America Journal*, 73: 2078-2086.

Seo, J.-S., Keum, Y.-S. & Li, Q.X. 2009. Bacterial Degradation of Aromatic Compounds. *International Journal of Environmental Research and Public Health*, 6: 278–309. doi: 10.3390/ijerph6010278

Seppey, C.V.W., Singer, D., Dumack, K., Fournier, B., Belbahri, L., Mitchell, E.A.D. & Lara, E. 2017. Distribution patterns of soil microbial eukaryotes suggests widespread algivory by phagotrophic protists as an alternative pathway for nutrient cycling. *Soil Biology and Biochemistry*, 112: 68-76.

Serna-Chavez, H.M., Fierer, N. & van Bodegom, P.M. 2013. Global drivers and patterns of microbial abundance in soil: Global patterns of soil microbial biomass. *Global Ecology and Biogeography*, 22: 1162–1172. https://doi.org/10.1111/geb.12070

Sessitsch, A., Brader, G., Pfaffenbichler, N., Gusenbauer, D. & Mitter, B. 2018. The contribution of plant microbiota to economy growth. *Microbiology and Biotechnology*, 11(5): doi:10.1111/1751-7915.13290

Setala, H. & Huhta, V. 1991. Soil Fauna Increase Betula Pendula Growth: Laboratory Experiments With Coniferous Forest Floor. *Ecology*, 72(2): 665–671. https://doi.org/10.2307/2937206

Setala, H., de Vries, F.T., Hemerik, L., Sgardelis, S.P., Brady, M.V., Tsiafouli, M.A., Bardgett, R.D., et al. 2013. Soil food web properties explain ecosystem services across European land use systems. *Proceedings of the National Academy of Sciences*, 110(35): 14296–14301. doi: 10.1073/pnas.1305198110

Setia, R., Marschner, P., Baldock, J., Chittleborough, D., Smith, P. & Smith, J. 2011. Salinity effects on carbon mineralization in soils of varying texture. *Soil Biology and Biochemistry*, 43: 1908-1916.

Shaffer, J.A. & DeLong, J.P. 2019. *The effects of management practices on grassland birds– An introduction to North American grasslands and the practices used to manage grasslands and grassland birds*. U.S. Geological Survey Professional Paper 1842. 63 p. https://doi.org/10.3133/pp1842A.

Shainberg, I. & Letey, J. 1984. Response of soils to sodic and saline conditions. *Hilgardia*, 52: 1-57.

Sheehan, C., Kirwan, L., Connolly, J. & Bolger, T. 2006. The effects of earthworm functional group diversity on nitrogen dynamics in soils. *Soil Biology and Biochemistry*, 38(9): 2629-2636.

Shen, Z., Ruan, Y., Chao, X., Zhang, J., Li, R. & Shen, Q. 2015. Rhizosphere microbial community manipulated by 2 years of consecutive biofertilizer application associated with banana Fusarium wilt disease suppression. *Biology and Fertility of Soils*, 51(5): 553-562.

Shen, Z., Xue, C., Taylor, P.W.J., Ou, Y., Wang, B., Zhao, Y., Ruan, Y., Li, R. & Shen, Q. 2018. Soil pre-fumigation could effectively improve the disease suppressiveness of biofertilizer to banana Fusarium wilt disease by reshaping the soil microbiome. *Biology and Fertility of Soils*, 54: 793-806.

Shen, Z., Xue, C., Penton, C.R., Thomashow, L.S., Zhang, N., Wang, B., Ruand, Y., Li, R. & Shen, Q. 2019. Suppression of banana Panama disease induced by soil microbiome reconstruction through an integrated agricultural strategy. *Soil Biology and Biochemistry*, 128: 164-174.

Shen, Z., Wang, B., Zhu, J., Hu, H., Tao, C., Ou, Y., Deng, X., Ling, N., Li, R. & Shen, Q. 2019. Lime and ammonium carbonate fumigation coupled with bio-organic fertilizer application steered banana rhizosphere to assemble a unique microbiome against Panama disease. *Microbial Biotechnology*, 12(3): 515-527.

Sheng, X.F., Xia, J.J., Jiang, C.Y., He, L.Y. & Qian, M. 2008. Characterization of heavy metal-resistant endophytic bacteria from rape (Brassica napus) roots and their potential in promoting the growth and lead accumulation of rape. *Environmental Pollution*, 156(3): 1164-70.

Shipitalo, M.J. & Protz, R. 1987. Comparison of morphology and porosity of a soil under conventional and zero tillage. *Canadian Journal of Soil Science*, 67: 445–456.

Shortle, J. & Horan, R.D. 2017. Nutrient Pollution: A Wicked Challenge for Economic Instruments. *Water Economics and Policy*, 03: 1650033.

Shtangeeva, I., Niemelä, M. & Perämäki, P. 2019. Effects of bromides of potassium and ammonium on some crops. *Journal of Plant Nutrition*, 42: 2209–2220.

Shtina, E.A. 1974. The principal directions of experimental investigations in soil algology with emphasis on the U.S.S.R. *Geoderma*, 12(1–2): 151–156. https://doi.org/10.1016/0016-7061(74)90047-0

Shuster, W.D., McDonald, L.P., McCartney, D.A., Parmelee, R., Studer, N. & Stinner, B. 2002. Nitrogen source and earthworm abundance affected runoff volume and nutrient loss in a tilled-corn agroecosystem. *Biology and Fertility of Soils*, 35: 320–327.

SiBBr. 2020. Sistema de Informação sobre a Biodiversidade Brasileira. [online]. [Cited 5 October 2020]. www.sibbr.gov.br

Siciliano, S.D., Palmer, A.S., Winsley, T., Lamb, E., Bissett, A., Brown, M.V., van Dorst J., et al. 2014. Soil fertility is associated with fungal and bacterial richness, whereas pH is associated with community composition in polar soil microbial communities. *Soil Biology and Biochemistry*, 78: 10-20. https://doi.org/10.1016/j.soilbio.2014.07.005

Siddiky, M.R.K., Kohler, J., Cosme, M. & Rillig, M.C. 2012. Soil biota effects on soil structure: Interactions between arbuscular mycorrhizal fungal mycelium and collembola. *Soil Biology and Biochemistry*, 50: 33–39. doi: https://doi.org/10.1016/j.soilbio.2012.03.001

Siebert, J., Eisenhauer, N., Poll, C., Marhan, S., Bonkowski, M., Hines, J., Koller, R., Reuss, L. & Thakur, M.P. 2019. Earthworms modulate the effects of climate warming on the taxon richness of soil meso- and macrofauna in an agricultural system. *Agriculture, Ecosystems and Environment*, 278: 72-80.

Siebert, J., Thakur, M.P., Reitz, T., Schadler, M., Schulz, E., Yin, R., Eisenhauer, N., *et al.* 2019. Extensive grassland-use sustains high levels of soil biological activity, but does not alleviate detrimental climate change effects. *Resilience in Complex Socioecological Systems*, 60: 25.

Sieriebriennikov, B., Ferris, H. & de Goede, R.G.M. 2014. NINJA: An automated calculation system for nematode-based biological monitoring. *European Journal of Soil Biology*, 61: 90-93.

Silva, R.A., Oliveira, C.M.G. & Inomoto, M.M. 2008. Fauna de Fitonematóides Em Áreas Preservadas e Cultivadas Da Floresta Amazônica No Estado de Mato Grosso. *Tropical Plant Pathology*, 33: 204–211.

Silver, A. 2019. Caterpillar's devastating march across China spurs hunt for native predator. *Nature*, 570(7761): 286–287. https://doi.org/10.1038/d41586-019-01867-3

Simberloff, D., Martin, J.-L., Genovesi, P., Maris, V., Wardle, D.A., Aronson, J., Courchamp, F., *et al.* 2013. Impacts of biological invasions: what's what and the way forward. *Trends in Ecology and Evolution*, 28: 58-66.

Simmons, B.L., Wall, D.H., Adams, B.J., Ayers, E., Barrett, J.E. & Virginia, R.A. 2009. Long-term experimental warming reduces soil nematode populations in the McMurdo Dry Valleys, Antarctica. *Soil Biology and Biochemistry*, 41: 2052-2060.

Singh, B.K., Quince, C., Macdonald, C.A., Khachane, A., Thomas, N., Al-Soud, W.A., Crooks, B., *et al.* 2014. Loss of microbial diversity in soils is coincident with reductions in some specialized functions. *Environmental microbiology*, 16(8): 2408-2420.

Singh, K. 2016. Microbial and enzyme activities of saline and sodic soils. *Land Degradation and Development*, 27: 706–718.

Sitch, S., Smith, B., Prentice, C., Arneth, A., Bondau, A., Cramer, W., Kaplans, J.O., *et al.* 2003. Evaluation of ecosystem dynamics, plant geography and terrestrial carbon cycling in the LPJ dynamic global vegetation model. *Global Change Biology*, 9: 161-185. https://doi.org/10.1046/j.1365-2486.2003.00569.x

Six, J., Frey, S.D., Thiet, R.K. & Batten, K.M. 2006. Bacterial and fungal contributions to carbon sequestration in agroecosystems. *Soil Science Society of America Journal*, 70(2): 555-569.

Six, J., Paustian, K., Elliott, E.T. & Combrink, C. 2000. Soil structure and organic matter: I. Distribution of aggregate-size classes and aggregate-associated carbon. *Soil Science Society of America Journal*, 64: 681-689.

Sizmur, T. and Hodson, M.E. 2009. Do earthworms impact metal mobility and availability in soil? - A review. *Environmental Pollution,* 157: 1981–1989.

Skubala, P. and Zaleski, T. 2012. Heavy metal sensitivity and bioconcentration in oribatid mites (Acari, Oribatida): Gradient study in meadow ecosystems. *Science of the Total Environment,* 414: 364–372.

Slabber, S., Worland, M.R., Leinaas, H.P. & Chown, S.L. 2007. Acclimation effects on thermal tolerances of springtails from sub-Antarctic Marion Island: Indigenous and invasive species. *Journal of Insect Physiology,* 53: 113–125.

Slade, E.M., Riutta, T., Roslin, T. & Tuomisto, H.L. 2016. The role of dung beetles in reducing greenhouse gas emissions from cattle farming. *Scientific Reports,* 6: 1–9.

Slizovskiy, I.B. & Kelsey, J.W. 2010. Soil sterilization affects aging-related sequestration and bioavailability of p,p'-DDE and anthracene to earthworms.

Smith, D.G., Martinelli, R., Besra, G.S., Illarionov, P.A., Szatmari, I., Brazda, P., Allen, M.A., *et al.* 2019. Identification and characterization of a novel anti-inflammatory lipid isolated from Mycobacterium vaccae, a soil-derived bacterium with immunoregulatory and stress resilience properties. *Psychopharmacology,* 236(5): 1653-1670.

Smith, I.M., Cook, D.R. & Smith, B.P. 2009. Water mites (Hydrachnida) and other arachnids. In: J.H. Thorp & A.P. Covich, eds. *Ecology and Classification of North American Freshwater Invertebrates*. Academic Press, San Diego, pp. 485–586.

Smith, P., House, J.I., Bustamante, M., Sobocká, J., Harper, R., Pan, G., West, P.C., *et al.* 2016. Global Change Pressures on Soils from Land Use and Management. *Global Change Biology,* 22(3): 1008–28. https://doi.org/10.1111/gcb.13068.

Smith, P., Soussana, J., Angers, D., Schipper, L., Chenu, C., Rasse, D.P., Batjes, N.H., *et al.* 2019. How to measure, report and verify soil carbon change to realize the potential of soil carbon sequestration for atmospheric greenhouse gas removal. *Global Change Biology,* 26(1): 1–23.

Smith, S.E. & Read, D.J. 2008. *Mycorrhizal Symbiosis, 3rd Edition.* New York, USA. Academic Press.

Snapp, S.S., Gentry, L.E. & Harwood, R. 2010. Management Intensity - Not Biodiversity - the Driver of Ecosystem Services in a Long-Term Row Crop Experiment. *Agriculture, Ecosystems and Environment,* 138: 242–48. https://doi.org/10.1016/j.agee.2010.05.005.

Snyder, B.A., Callaham, M.A., Jr. & Hendrix, P.F. 2011. Spatial variability of an invasive earthworm (Amynthas agrestis) population and potential impacts on soil characteristics and millipedes in the Great Smoky Mountains National Park, USA. *Biological Invasions,* 13: 349-358.

Società Italiana della Scienza del Suolo. 2019. Gruppo di Lavoro SISS Qualità biologica del suolo basata sui microartropodi (QBS-ar). [online]. [Cited 2 October 2020]. https://www.scienzadelsuolo.org/QBS-ar.php

SoE. 2016. *Australia State of Environment 2016.* Independent report to the Australian Government Minister for the Environment. Canberra, Australia. State of the Environment 2016 Committee.

SoE. 2018. *Australia State of Environment 2018.* Independent report to the Australian Government Minister for the Environment. Canberra, Australia. State of the Environment 2016 Committee.

Sofo, A., Ricciuti, P., Fausto, C., Mininni, A.N., Crecchio, C., Scagliola, M., Malerba, A.D., Xilioyannis, C. & Dichio, B. 2019. The metabolic and genetic diversity of soil bacterial communities depends on the soil management system and C/N dynamics: The case of sustainable and conventional olive groves. *Applied Soil Ecology,* 137: 21–28. doi: 10.1016/j.apsoil.2018.12.022

SOFR. 2018. *Australia's State of the Forests Report 2018.* Montreal Process Implementation Group for Australia and National Forest Inventory Steering Committee, ABARES. Canberra, Australia.

Soga, M. & Gaston, K.J. 2016. Extinction of experience: the loss of human–nature interactions. *Frontiers in Ecology and the Environment,* 14: 94-101.

Soka, G. & Ritchie, M. 2014. Arbuscular Mycorrhizal Symbiosis, Ecosystem Processes And Environmental Changes In Tropical Soils. *Applied Ecology and Environmental Research,* 13(1): 229-245.'

Soil Health Institute. 2018. Conference on Connections Between Soil Health and Human Health, October 16 -17, 2018. Recommendations and next steps. [online]. [Cited 5 October 2020]. https://soilhealthinstitute.org/humanhealthconference/

Soil Health Institute. 2020. Soil Health Educational Resources. [online]. [Cited 5 October 2020]. https://soilhealthinstitute.org/resources/soil-health-educational-resources/

Soil Science Society of China. 2020. *Soil Science Society of China* [online]. [Cited 1 October 2020]. http://www.csss.org.cn/

Soil Security Programme. 2020. Soil Security Programme. [online]. [Cited 2 October 2020]. https://www.soilsecurity.org/

Soliveres, S., van der Plas, F., Manning, P., Prati, D., Gossner, M.M., Renner, S.C., Alt, F., Arndt, H., Baumgartner, V., Binkenstein, J., Birkhofer, K., Blaser, S., Blüthgen, N., Boch, S., Böhm, S., Börschig, C., Buscot, F., Diekötter, T., Heinze, J., Hölzel, N., Jung, K., Klaus, V.H., Kleinebecker, T., Klemmer, S., Krauss, J., Lange, M., Morris, E.K., Müller, J., Oelmann, Y., Overmann, J., Pašalić, E., Rillig, M.C., Schaefer, H.M., Schloter, M., Schmitt, B., Schöning, I., Schrumpf, M., Sikorski, J., Socher, S.A., Solly, E.F., Sonnemann, I., Sorkau, E., Steckel, J., Steffan-Dewenter, I., Stempfhuber, B., Tschapka, M., Türke, M., Venter, P.C., Weiner, C.N., Weisser, W.W., Werner, M., Westphal, C., Wilcke, W., Wolters, V., Wubet, T., Wurst, S., Fischer, M. & Allan, E. 2016. Biodiversity at multiple trophic levels is needed for ecosystem multifunctionality. *Nature,* 536(7617): 456.

Somerville, K.D., Clarke, B., Keppel, G., McGill, C., Newby, Z.J., Wyse, S.V., James, S.A. & Offord, C.A. 2018. Savings rainforests in the South Pacific: challenges in ex situ conservation. *Australian Journal of Botany,* 65(8): 609-624.

Sommer, R. & Bossio, D. 2014. Dynamics and climate change mitigation potential of soil organic carbon sequestration. *Journal of Environmental Management,* 144: 83–87.

Song, J., Wan, S., Piao, S., Knapp, A.K., Classen, A.T., Vicca, S., Kardol, P., et al. 2019. A meta-analysis of 1,119 manipulative experiments on terrestrial carbon-cycling responses to global change. *Nature Ecology and Evolution,* 3: 1309-1320.

Song, Y.S., Li, X.W., Li, F. & Li, H.M. 2015. Influence of different types of surface on the diversity of soil fauna in Beijing Olympic Park. *Chinese Journal of Applied Ecology,* 26: 1130-1136.

Soomro, A., Mirjat, M., Oad, F., Soomro, H., Samo, M. & Oad, N. 2001. Effect of irrigation intervals on soil salinity and cotton yield. *OnLine Journal of Biological Sciences,* 1: 472–474.

Soong, J.L. & Nielsen, U.N. 2016. The role of microarthropods in emerging models of soil organic matter. *Soil Biology and Biochemistry,* 102: 37–39. doi:https://doi.org/10.1016/j.soilbio.2016.06.020

Soong, J.L., Vandegehuchte, M.L., Horton, A.J., Nielsen, U.N., Denef, K., Shaw, E.A., de Tomasel, C.M., Parton, W., Wall, D.H. & Cotrufo, M.F. 2016. Soil microarthropods support ecosystem productivity and soil C accrual: Evidence from a litter decomposition study in the tallgrass prairie. *Soil Biology and Biochemistry,* 92: 230–238. doi:https://doi.org/10.1016/j.soilbio.2015.10.014

South Dakota Soil Health Coalition. 2020. South Dakota Soil Health Coalition. [online]. [Cited 5 October 2020]. https://www.sdsoilhealthcoalition.org/

Species Link Project. 2020. Species Link Project [online]. [Cited 5 October 2020]. http://splink.cria.org.br/

Spurgeon, D.J., Rowland, P., Ainsworth, G., Rothery, P., Long, S. & Black, H.I.J. 2008. Geographical and pedological drivers of distribution and risks to soil fauna of seven metals (Cd, Cu, Cr, Ni, Pb, V and Zn) in British soils. *Environmental Pollution,* 153: 273-283.

Šrut, M., Menke, S., Höckner M. & Sommer, S. 2019. Earthworms and cadmium – Heavy metal resistant gut bacteria as indicators for heavy metal pollution in soils? *Ecotoxicology and Environmental Safety,* 171: 843–853.

St. John, M.G., Bagatto, G., Behan-Pelletier, V., Lindquist, E.E., Shorthouse, J.D., & Smith, I.M. 2002. Mite (Acari) colonization of vegetated mine tailings near Sudbury, Ontario, Canada. *Plant and Soil* 245: 295-305.

Staddon, P., Lindo, Z., Crittenden, P.D., Gilbert, F. & Gonzales, A. 2010. Connectivity, non-random extinction and ecosystem function in experimental metacommunities. *Ecology Letters,* 13: 543-552.

Stagnari, F., Maggio, A., Galieni, A. *et al.* 2017. Multiple benefits of legumes for agriculture sustainability: an overview. *Chem. Biol. Technol. Agric.* **4, 2. https://doi.org/10.1186/s40538-016-0085-1

Stanford, J., O'Brien, M. & Grange, J. 2008. Successful immunotherapy with Mycobacterium vaccae in the treatment of adenocarcinoma of the lung. *European journal of cancer,* 44: 224-227

Steffan, S.A. & Dharampal, P.S. 2018. Undead food-webs: integrating microbes into the food-chain. *Food Webs,* 18: e00111.

Steffan, S.A., Chikaraishi, Y., Currie, C.R., Horn, H., Gaines-Day, H.R., Pauli, J.N., Zalapa, J.E. & Ohkouchi, N. 2015. Microbes are trophic analogs of animals. *Proceedings of the National Academy of Sciences,* 112: 15119-15124.

Steffen, W., Richardson, K., Rockström, J., Cornell, S.E., Fetzer, I., Bennett, E.M., Biggs, R., Carpenter, S.R., de Vries, W., de Wit, C.A., Folke, C., Gerten, D., Heinke, J., Mace, G.M., Persson, L.M., Ramanathan, V., Reyers, B., & Sörlin, S. 2015. Planetary boundaries: Guiding human development on a changing planet. *Science* 347 (6223). DOI: 10.1126/science.1259855.

Steffen, W., Rockström, J., Richardson, K., Lenton, T.M., Folke, C., Liverman, D., Summerhayes, C.P., *et al.* 2018. Trajectories of the Earth System in the Anthropocene. *Proceedings of the National Academy of Sciences of the United States of America,* 115: 8252–8259.

Stegen, J.C., Bottos, E.M. & Jansson, J.K. 2018. A unified conceptual framework for prediction and control of microbiomes. *Current Opinions in Microbiology,* 44: 20-27.

Steidinger, B.S., Crowther, T.W., Liang, J., Van Nuland, M.E., Werner, G.D.A., Reich, P.B., Peay, K.G., *et al.* 2019. Climatic controls of decomposition drive the global biogeography of forest-tree symbioses. *Nature,* 569: 404-408. https://doi.org/10.1038/s41586-019-1128-0

Stein, A., Gerstner, K. & Kreft, H. 2014. Environmental heterogeneity as a universal driver of species richness across taxa, biomes and spatial scales. *Ecology Letters,* 17: 866-880.

Sterken, M.G., Snoek, L.B., Kammenga, J.E. & Andersen, E.C. 2015. The laboratory domestication of Caenorhabditis elegans. *Trends in Genetics,* 31: 224–231. doi: 10.1016/j.tig.2015.02.009

Sterner, R.W. & Elser, J.J. 2002. *Ecological stoichiometry: the biology of elements from molecules to the biosphere.* Princeton, USA. Princeton University Press.

Stevens, C.J., Dise, N.B., Mountford, J.O. & Gowing, D.J. 2004. Impact of nitrogen deposition on the richness of grasslands. *Science,* 303: 1876–1879.

Stevens, C.J., Thompson, K., Grime, J.P., Long, C.J., Gowing, D.J.G., 2010. Contribution of acidification and eutrophication to declines in species richness of calcifuge grasslands along a gradient of atmospheric nitrogen deposition. *Functional Ecology,* 24: 478-484.

Stevens, C.J., Dupr, C., Dorland, E., Gaudnik, C., Gowing, D.J.G., Bleeker, A. & Dise, N.B. 2011. The impact of nitrogen deposition on acid grasslands in the Atlantic region of Europe. *Environmental Pollution,* 159(10): doi: 10.1016/j.envpol.2010.11.026

Stevens, R.M. & Partington, D.L. 2013. Grapevine recovery from saline irrigation was incomplete after four seasons of non-saline irrigation. *Agricultural Water Management,* 122: 39-45.

Stirling, G.R. 2018. Biological control of plant-parasitic nematodes: An ecological perspective, a review of progress and opportunities for future research. In K. Davies & Y. Spiegel, eds.. *Biological Control of Plant-Parasitic Nematodes*, pp. 1- 38. CRC Press. United States of America.

Stone, D., Ritz, K., Griffiths, B.G., Orgiazzi, A. & Creamer, R.E. 2016. Selection of biological indicators appropriate for European soil monitoring. *Applied Soil Ecology,* 97: 12-22. doi: 10.1016/j.apsoil.2015.08.005

Stork, N.E. 2018. How many species of insects and other terrestrial arthropods are there on earth? *Annual Review of Entamology,* 63: 31–45.

Strickland, M.S. & Rousk, J. 2010. Considering fungal:bacterial dominance in soils - Methods, controls, and ecosystem implications. *Soil Biology and Biochemistry,* 42: 1385-1395.

Stuble, K.L., Patterson, C.M., Rodriguez-Cabal, M.A., Ribbons, R.R., Dunn, R.R. & Sanders, N.J. 2014. Ant-mediated seed dispersal in a warmed world. *PeerJ,* 2: e286. doi:10.7717/peerj.286

Suarez, A.V., Bolger, D.T. & Case, T.J. 1998. Effects of fragmentation and invasion on native ant communities in coastal southern California. *Ecology,* 79: 2041–2056.

Subke, J.A., Inglima, I. & Cotrufo, M.F. 2006. Trends and methodological impacts in soil CO_2 efflux partitioning: A metaanalytical review. *Global Change Biology,* 12(6): 921-943.

Subler, S., Baranski, C.M. & Edwards, C.A. 1997. Earthworm additions increased short-term nitrogen availability and leaching in two grain-crop agroecosystems. *Soil Biology and Biochemistry,* 29: 413–421.

Sukhdev, P., Wittmer, H., Schröter-Schlaack, C. & Nesshöver, C. 2010. *The Economics of Ecosystems and Biodiversity*. Mainstreaming the Economics of Nature: A synthesis of the approach, conclusions and recommendations of TEEB. United Nations Environment Program.

Sun, H., Terhonen, E., Koskinen, K., Paulin, L., Kasanen, R. & Asiegbu, F.O. 2014. Bacterial diversity and community structure along different peat soils in boreal forest. *Applied Soil Ecology,* 74: 37-45.

Sustr, V. & Simek, M. 2009. Methane release from millipedes and other soil invertebrates in Central Europe. *Soil Biology and Biochemistry,* 41: 1684–1688.

Suttle, C.A. 2007. Marine viruses - major players in the global ecosystem. *Nature Reviews Microbiology,* 5: 801-812.

Suttle, C.A. 2013. Viruses: unlocking the greatest biodiversity on Earth. *Genome,* 56: 542-544.

Swenson, W., Wilson, D.S. & Elias, R. 2000. Artificial ecosystem selection. *Proceedings of the National Academy of Sciences*, 97: 9110–9114. doi: 10.1073/pnas.150237597

Swift, M.J., Andren, O., Brussaard, L., Briones, M., Couteaux, M.M., Ekschmitt, K., Kjoller, A., Loiseau, P. & Smith, P. 1998. Global change, soil biodiversity, and nitrogen cycling in terrestrial ecosystems: three case studies. *Global Change Biology*, 4: 729 - 743.

Sylvia, D.M., Hartel, P.G. Fuhrmann, J.J. & Zuberer, D.A. 2005. *Principles and Applications of Soil Microbiology*. Upper Saddle River, USA. Pearson Prentice Hall.

Syphard, A.D., Brennan, T.J. & Keeley, J. 2019. Extent and drivers of vegetation type conversion in Southern California Chaparral. *Ecosphere*, 10(7): .DOI 10.1002/ecs2.2796

Szabolcs, I. 1989. *Salt-affected soils*. Boca Raton, USA. CRC Press.

Szoboszlay, M., Dohrmann, A.J., Poeplau, C., Don, A. & Tebbe, C.C. 2017. Impact of land-use change and soil organic carbon quality on microbial diversity in soils across Europe. *FEMS Microbiology Ecology*, 93(12): https://doi/10.1093/femsec/fix146

Takahashi, H., Matsushita, Y., Ito, T., Nakai, Y., Nanzyo, M., Kobayashi, T., Iwaishi, S., Hashimoto, T., Miyashita, S., Morikawa, T., Yoshida, S., Tsushima, S. & Ando, S. 2018. Comparative analysis of microbial diversity and bacterial seedling disease-suppressive activity in organic-farmed and standardized commercial conventional soils for rice nursery cultivation. *Journal of Phytopathology*, 166(4): 249–264. https://doi.org/10.1111/jph.12682

Tambong J.T., Xu, R. & Bromfield, E.S.P. 2017. Pseudomonas canadensis sp. nov., a biological control agent isolated from a field plot under long-term mineral fertilization Int J Syst Evol Microbiol 2017;67:889–895 DOI 10.1099/ijsem.0.001698

Tamburini, G., De Simone, S., Sigura, M., Boscutti, F. & Marini, L. 2016. Soil Management Shapes Ecosystem Service Provision and Trade-Offs in Agricultural Landscapes. *Proceedings of the Royal Society B: Biological Sciences*, 283(1837): 20161369. https://doi.org/10.1098/rspb.2016.1369.

Tang, C.Y., Fu, Q.S., Criddle, C.S. & Leckie, J.O. 2007. Effect of flux (transmembrane pressure) and membrane properties on fouling and rejection of reverse osmosis and nanofiltration membranes treating perfluorooctane sulfonate containing wastewater. *Environmental Science and Technology*, 41: 2008–2014.

Tano, Y., Yapi, A. & Kouassi, K.P. 2005. Diversité biologique et importance de termites (Isoptères) dans les écosystèmes de savane et de foret de Côte d'Ivoire. *Bioterre*, 5:

Tardiff, S.E & Stanford, J.A. 1998. Grizzly bear digging: effects on subalpine meadow plants in relation to mineral nitrogen availability. *Ecology*, 79: 2219-2228.

Tarrand, J.J., Krieg, N.R. & Döbereiner, J. 1978. A taxonomic study of the Spirillum lipoferum group, with descriptions of a new genus, Azospirillum gen. nov. and two species, Azospirillum lipoferum (Beijerinck) comb. nov. and Azospirillum brasilense sp. nov. *Canadian Journal of Microbiology*, 24: 967-980.

Tchagang, C.F., Xu, R., Overy, D., Blackwell, B., Chabot, D., Hubbard, K., Doumbou, C.L., Bromfield, E.S.P., Tambong, J.T. 2018. Diversity of bacteria associated with corn roots inoculated with Canadian woodland soils, and description of Pseudomonas aylmerense sp. nov. Doi https://doi.org/10.1016/j.heliyon.2018.e00761

Tchagang, C.F., Xu, R., Overy, D., Blackwell, B., Chabot, D., Hubbard, K., Doumbou, C.L., Bromfield, E.S.P., Tambong, J.T. 2018. Diversity of bacteria associated with corn roots inoculated with Canadian woodland soils, and description of Pseudomonas aylmerense sp. nov. Doi https://doi.org/10.1016/j.heliyon.2018.e00761

Tecon, R. & Or, D. 2017. Biophysical processes supporting the diversity of microbial life in soil. *FEMS Microbiology Review,* 41: 599-623.

Tedersoo, L. 2017. Biogeography of mycorrhizal symbiosis. Berlin, Germany. Springer.

Tedersoo, L., Bahram, M., Põlme, S., Kõljalg, U., Yorou, N.S., Wijesundera, R., Smith, M.E., *et al.* 2014. Global diversity and geography of soil fungi. *Science,* 346(6213): 1256688.

Teng, H., Rossel, R.A.V., Shi, Z., Behrens, T., Chappell, A. & Bui, E. 2016. Assimilating satellite imagery and visible-near infrared spectroscopy to model and map soil loss by water erosion in Australia. *Environmental Modelling and Software,* 77: 156-167.

Terhivuo, J. & Valovirta, I. 1978. Habitat spectra of lumbricidae (oligochaeta) in Finland. *Annales Zoologici Fennici,* 15(3): 202–209.

Terra Nuova. 2018. La producción orgánica nacional tiene más de 30 cadenas de cultivos. In: Terra Nova [Online]. [Cited 2 October 2020]. http://www.terranuova.org.pe/la-produccion-organica-nacional-mas-30-cadenas-cultivos/

Terrat, S., Horrigue, W., Dequietd, S., Saby, N.P.A., Lelièvre, M., Nowak, V., Tripied, J., Régnier, T., Jolivet, C., Arrouays, D., Wincker, P., Cruaud, C., Karimi, B., Bispo, A., Maron, P.A., Prévost-Bouré, N.C. & Ranjard, L. 2017. Mapping and predictive variations of soil bacterial richness across France. *PLOS ONE,* 12(10): e0186766. https://doi.org/10.1371/journal.pone.0186766

Teste, F.P., Kardol, P., Turner, B.L., Wardle, D.A., Zemunik, G., Renton, M. & Laliberté, E. 2017. Plant-soil feedback and the maintenance of diversity in Mediterranean-climate shrublands. *Science,* 355: 173–176. doi: 10.1126/science.aai8291

Thakali, S., Allen, H.E., Di Toro, D.M., Ponizovsky, A.A., Rooney, C.P., Zhao, F.J., McGrath, S.P., *et al.* 2006. Terrestrial biotic ligand model. 2. Application to Ni and Cu toxicities to plants, invertebrates, and microbes in soil. *Environmental Science and Technology,* 40: 7094-7100.

Thakur, S.B. 2017. Climate change related policy environment in agriculture and food security in Nepal. *Journal of Agriculture and Environment,* 18: 120–130. https://doi.org/10.3126/aej.v18i0.19897

Thakur, M.P., van Groenigen, J.W., Kuiper, I. & De Deyn, G.B. 2014. Interactions between microbial-feeding and predatory soil fauna trigger N2O emissions. *Soil Biology and Biochemistry,* 70: 256-262.

Thakur, M.P., van der Putten, W.H., Cobben, M.M.P., van Kleunen, M. & Geisen, S. 2019. Microbial invasions in terrestrial ecosystems. *Nature Reviews Microbiology,* 17: 621–631.

Thakur, M. P., Milcu, A., Manning, P., Niklaus, P. A., Roscher, C., Power, S., Guo, H., et al. 2015. Plant diversity drives soil microbial biomass carbon in grasslands irrespective of global environmental change factors. *Global Change Biology,* 21(11): 4076-4085.

Thakur, M.P., Reich, P.B., Hobbie, S.E., Stefanski, A., Rich, R., Rice, K.E., Eisenhauer, N., et al. 2018. Reduced feeding activity of soil detritivores under warmer and drier conditions. *Nature Climate Change,* 8(1): 75.

Thakur, M.P., Phillips, H.R.P., Brose, U., de Vries, F., Lavelle, P., Loreau, M., Mathieu, J., et al. 2019. Toward an integrative understanding of soil biodiversity. *Biological Reviews,* 95(2): 350-364.

Thakur, M.P., Del Real, I.M., Cesarz, S., Steinauer, K., Reich, P.B., Hobbie, S., Eisenhauer, N., et al. 2019. Soil microbial, nematode, and enzymatic responses to elevated CO_2, N fertilization, warming, and reduced precipitation. *Soil Biology and Biochemistry,* 135: 184-193.

Thakur, M.L. 2016. Diversity and distribution of termites of Western and North-Western Himalaya. In K.G. Saxena and K.S. Rao, eds. *Soil Biodiversity: Inventory, Functions and Management,* pp. 225-246. Dehradun, India. Bishen Singh Mahendra Pal Singh.

The Atlantic. 2013. Healthy Soil Microbes, Healthy People. [online]. [Cited 2 October 2020]. https://www.theatlantic.com/health/archive/2013/06/healthy-soil-microbes-healthy-people/276710/

The Biodiversity Exploratories. 2020. Exploratories for large-scale and long-term functional biodiversity research. [online]. [Cited 2 October 2020]. https://www.biodiversity-exploratories.de/

The Centre for Soil Ecology (CSE). 2010. CSE. [online]. [Cited 2 October 2020]. https://www.soilecology.eu/

The Earthworm Society of Britain (ESB). 2020. ESB. [online]. [Cited 2 October 2020]. https://www.earthwormsoc.org.uk/

The Federal Agency for Nature Conservation (BfN). 2020. Nationwide insect monitoring is being developed. [online]. [Cited 2 October 2020]. https://www.bfn.de/themen/biologische-vielfalt/nationale-strategie/projekt-des-monats/insektenmonitoring.html

The Fungus Conservation Trust. 2015. The Fungus Conservation Trust [online]. [Cited 2 October 2020]. http://www.abfg.org/

The National Association of Conservation Districts. 2020. Soil. [online]. [Cited 5 October 2020]. https://www.nacdnet.org/about-nacd/what-we-do/soil/

The National Ecological Observatory Network. 2020. Terrestrial Organisms. [online]. [Cited 5 October 2020]. https://www.neonscience.org/data-collection/terrestrial-organisms

The Netherlands Institute of Ecology (NIOO). 2019. Delta Plan for Biodiversity Recovery: working together for a richer Netherlands. [online]. [Cited 2 October 2020]. https://nioo.knaw.nl/en/news/delta-plan-biodiversity-recovery-working-together-richer-netherlands

The Soil Health Institute. 2020. About the Soil Health Institute. [online]. [Cited 5 October 2020]. https://soilhealthinstitute.org/about-us/

Thiele-Bruhn, S., Bloem, J., de Vries, F.T., Kalbitz, K. & Wagg, C. 2012. Linking soil biodiversity and agricultural soil management. *Current Opinion in Environmental Sustainability*, 4: 523–528. doi: 10.1016/j.cosust.2012.06.004

Thirkell, T.J., Charters, M.D., Elliott, A.J., Sait, S.M. & Field, K.J. 2017. Are mycorrhizal fungi our sustainable saviours? Considerations for achieving food security. *Journal of Ecology*, 105: 921–929.

Tomlin, A.D., Tu, C.M., & Miller, J.J. 1995. Response of earthworms and soil biota to agricultural practices in corn, soybean and cereal rotations. *Acta Zool. Fenn.*, 196: 195-199.

Thomloudi, E.-E., Tsalgatidou P.C., Douka D., Spantidos T.-N., Dimou, M., Venieraki A. & Katinakis P. 2019. Multistrain versus single-strain plant growth promoting microbial inoculants - The compatibility issue. *Hellenic Plant Protection Journal*, 12: 61-77. DOI 10.2478/hppj-2019-0007

Thomsen, I.K. & Christensen, B.T. 2004. Yields of wheat and soil carbon and nitrogen contents following long-term incorporation of barley straw and ryegrass catch crops. *Soil Use and Management*, 20: 432-438.

Thünen Institute. 2019. 4th Thünen Symposium on Soil Metagenomics - Understanding and Managing Soil Microbiomes. [online]. [Cited 2 October 2020]. www.soil-metagenomics.org

Tian, D. & Niu, S. 2015. A global analysis of soil acidification caused by nitrogen addition. *Environmental Research Letters*, 10(2). doi: 10.1088/1748-9326/10/2/024019

Tian, G., Pauls, K.P., Dong, Z., Reid, L.M., and Tian, L.-N. 2009. Colonization of the nitrogen-fixing bacterium Gluconacetobacter diazotrophicus in a large number of Canadian corn plants. Can. J. Plant Sci. 89(6), pp. 1009-1016. DOI: 10.4141/CJPS08040

Tiede, Y., Donoso, D.A., Bendix, J., Brandl, R. & Farwig, N. 2017. Ants as indicators of environmental change and ecosystem processes. *Ecological Indicators*, 83: 527–537. doi: 10.1016/j.ecolind.2017.01.029

Tilman, D., Wedin, D., and Knops, J. 1996. Productivity and sustainability influenced by biodiversity in grassland ecosystems. *Nature*, 379(6567): 718.

Tilman, D., Cassman, K.G., Matson, P.A., Naylor, R. & Polasky, S. 2002. Agricultural sustainability and intensive production practices. *Nature*, 418: 671-677.

Tilman, D., Reich, P.B., Knops, J., Wedin, D., Mielke, T. & Lehman, C. 2001. Diversity and productivity in a long-term grassland experiment. *Science*, 294(5543): 843-845.

Tirado, R. & Allsopp, M. 2012. *Phosphorus in agriculture: Problems and solutions.* Amsterdam, Netherlands. Greenpeace International.

Tiunov, A.V. & Scheu, S. 2005. Facilitative interactions rather than resource partitioning drive diversity-functioning relationships in laboratory fungal communities. *Ecology Letters,* 8(6): 618-625.

Transparency Market Research. 2017. *Global Biofertilizers Market: Prosperity of Organic Food Industry Augur Well for Future, observes TMR* [online]. https://www.transparencymarketresearch.com/pressrelease/biofertilizers-market.htm

Tohu, H., Peay, K.G., Yamamichi, M., Narisawa, K., Hiruma, K., Naito, K., Fukuda, S, *et al.* 2018. Core microbiomes for sustainable agroecosystems. *Nature Plants,* 4: 247-257. doi10.1038/s41477-018-0139-4

Toledo, V. 2006. *Caracterización de la costra microbiótica y su influencia biológica y física en suelos de la región árida de Quíbor, estado Lara.* Tesis doctoral. Universidad Central de Venezuela, Facultad de Agronomía. Maracay, Venezuela. https://scielo.conicyt.cl/scielo.php?script=sci_nlinks&ref=6636411&pid=S0716-078X201100010000100144&lng=es

Tondoh, E.J., Dimobe, K., Guéi, M.A., Adahe, L., Baidai, Y., Julien K. N'Dri, K.J., & Forkuor, G. 2019. Soil health changes over a 25-year chronosequence from forest to plantations in rubber tree (Hevea brasiliensis) landscapes in Southern Côte d'Ivoire: do earthworms play a role? Frontiers in Environmental Science 1-19: doi: 10.3389/fenvs.2019.00073

Top Crop Manager. 2018. Going native with nitrogen fixation. [online]. [Cited 5 October 2020]. https://www.topcropmanager.com/going-native-with-nitrogen-fixation-21468/

Toro, M., Bazó, I. & López, M. 2008. Micorrizas arbusculares y bacterias promotoras de crecimiento vegetal, biofertilizantes nativos de sistemas agrícolas bajo manejo conservacionista. *Agronomía Tropical,* 58(3): 215-221.

Torsvik, V., Goksoyr, J. & Daae, F.L. 1990. High Diversity in DNA of soil bacteria. *Applied and Environmental Microbiology,* 56: 782- 787.

Torsvik, V., Øvreås, L. & Thingstad, T.F. 2002. Prokaryotic diversity -magnitude, dynamics, and controlling factors. *Science,* 296(5570): 1064-1066.

Toyota, A., Kaneko, N. & Ito, M.T. 2006. Soil ecosystem engineering by the train millipede Parafontaria laminata in a Japanese larch forest. *Soil Biology and Biochemistry,* 38: 1840–1850.

Traoré, S., Bottinelli, N., Aroui, H., Harit, A. & Jouquet, P. 2019. Termite mounds impact soil hydrostructural properties in southern Indian tropical forests. *Pedobiologia,* 74: 1–6.

Treseder, K.K. 2004. A meta-analysis of mycorrhizal responses to nitrogen, phosphorus, and atmospheric CO_2 in field studies. *New Phytologist,* 164: 347-355.

Treseder, K.K. 2008. Nitrogen additions and microbial biomass: a meta-analysis of ecosystem studies. *Ecological Letters,* 11: 1111–1120.

Tripathi, G., Ram, S., Sharma, B.M. & Singh, G. 2009. Fauna-associated changes in soil biochemical properties beneath isolated trees in a desert pastureland of India and their importance in soil restoration. *Environmentalist*, 29(3): 318-329.

Trivedi, P., Anderson, I.C. & Singh, B.K. 2013. Microbial modulators of soil carbon storage: integrating genomic and metabolic knowledge for global prediction. *Trends in microbiology*, 21(12): 641-651.

Trivedi, P., Delgado-Baquerizo, M., Anderson, I.C. & Singh, B.K. 2016. Response of soil properties and microbial communities to agriculture: implications for primary productivity and soil health indicators. *Frontiers in Plant Science*, 7: 990.

Trivedi, C., Delgado-Baquerizo, M., Hamonts, K., Lai, K., Reich, P.B. & Singh, B.K. 2019. Losses in microbial functional diversity reduce the rate of key soil processes. *Soil Biology and Biochemistry*, 135: 267-274.

Trivedi, P., Delgado-Baquerizo, M., Trivedi, C., Hu, H., Anderson, I. C., Jeffries, T.C., Singh, B.K., *et al.* 2016. Microbial regulation of the soil carbon cycle: evidence from gene–enzyme relationships. *The ISME journal*, 10(11): 2593.

Trubl, G., Jang, H.B., Roux, S., Emerson, J.B., Solonenko, N., Vik, D.R., *et al.* 2018. Soil Viruses Are Underexplored Players in Ecosystem Carbon Processing. *mSystems*, 3: e00076-18.

Trubl, G., Solonenko, N., Chittick, L., Solonenko, S.A., Rich, V.I., & Sullivan, M.B. 2016. Optimization of viral resuspension methods for cabon-rich soils along a permafrost thaw gradient. *PeerJ*, 4: e1999.

Tscharntke, T., Bommarco, R., Clough, Y., Crist, T.O., Kleijn, D., Rand, T.A., Tylianakis, J.M., Nouhuys, S.v. & Vidal, S. 2007. Conservation biological control and enemy diversity on a landscape scale. *Biological Control*, 43: 294-309.

Tscharntke, T., Clough, Y., Bhagwat, S.A., Buchori, D., Faust, H., Hertel, D., Hölscher, D., *et al.* 2011. Multifunctional shade-tree management in tropical agroforestry landscapes–a review. *Journal of Applied Ecology*, 48(3): 619-629.

Tsiafouli, M.A., Thebault, E., Sgardelis, S.P., de Ruiter, P.C., van der Putten, W.H., Birkhofer, K., Hemerik, L., *et al.* 2015. Intensive agriculture reduces soil biodiversity across Europe. *Global Change Biology*, 21: 973-985.

Tsitsilas, A., Hoffmann, A.A., Weeks, A.R. & Umina, P.A. 2011. Impact of groundcover manipulations within windbreaks on mite pests and their natural enemies. *Australian Journal of Entomology*, 50: 37-47.

Turbé, A., De Toni, A., Benito, P., Lavelle, P., Lavelle, P., Camacho, N.R., van der Putten, W.H., Labouze, E., & Mudgal, S. 2010. *Soil Biodiversity: Functions, Threats and Tools for Policy Makers*. Final Report. Brussels, Belgium. European Commission. 250 pp.

Turk, M., Montiel, V., Žigon, D., Plemenitaš, A. & Ramos, J. 2007. Plasma membrane composition of Debaryomyces hansenii adapts to changes in pH and external salinity. *Microbiology*, 153: 3586-3592.

Turnbull, M.S., George, P.B.L. & Lindo, Z. 2014. Weighing in: size spectra as a standard tool in soil community analyses. *Soil Biology and Biochemistry,* 68: 366-372.

Tyrrell, C., Burgess, C.M., Brennan, F.P. & Walsh, F. 2019. Antibiotic resistance in grass and soil. *Biochemical Society Transactions,* 47: 477–486.

Udovic, M., Plavc, Z. & Lestan, D. 2007. The effect of earthworms on the fractionation, mobility and bioavailability of Pb, Zn and Cd before and after soil leaching with EDTA. *Chemosphere,* 70: 126–134.

UCDAVIS. 2020. Sustainable Agriculture Research & Education Program. [online]. [Cited 5 October 2020]. https://asi.ucdavis.edu/programs/ucsarep/research-initiatives/are/ecosystem/teaching-soil-health

UK Center for Ecology & Hydrology. 2020. Countryside Survey - measuring change in our countryside. In: *Countryside Survey* [online]. [Cited 2 October 2020]. https://countrysidesurvey.org.uk/

UK Environment Agency. 2017. Soil screening values for assessing ecological risk. In: *GOV.UK* [online]. [Cited 2 October 2020]. https://www.gov.uk/government/publications/soil-screening-values-for-assessing-ecological-risk

UK Environment Agency. 2020. *Water quality data archive.* [online]. [Cited 2 October 2020]. https://environment.data.gov.uk/water-quality/view/landing

UK Government. 2020. 25 Year Environment Plan. [online]. [Cited 2 October 2020]. https://www.gov.uk/government/publications/25-year-environment-plan

UmweltBundesamt. 2018. Organic farming. In: *Umweltbundesamt* [online]. [Cited 2 October 2020]. https://www.umweltbundesamt.de/en/topics/soil-agriculture/toward-ecofriendly-farming/organic-farming

United Nations Convention on Biological Diversity. 2014. *Pathways of introduction of invasive species, their prioritization and management.* Montreal, Canada.

UNCCD. 2018. Ukrainian Coordination Council to Combat Land Degradation and Desertification meets to discuss implementation of national action plan. [online]. [Cited 2 October 2020]. https://www.unccd.int/news-events/ukrainian-coordination-council-combat-land-degradation-and-desertification-meets

United Nations Environment. 2019. *Global environment outlook - GEO-6: Healthy planet, healthy people.* Cambridge, UK. Cambridge University Press.

UNFCC. 1992. *United Nations Framework Convention on Climate Change.* New York, USA. United Nations.

United Nations Forum on Forests. 2007. *Report of the seventh session (24 February 2006 and 16 to 27 April 2007).* Economic and Social Council. Supplement No. 22. New York, USA. United Nations.

United Nations, Department of Economic and Social Affairs, Population Division. 2019. *World Urbanization Prospects: The 2018 Revision (ST/ESA/SER.A/420).* New York, USA. United Nations.

United Nations. 1976. *Convention on wetlands of international importance especially as waterfowl habitat.* Paris, France. United Nations.

Universidad Nacional de Colombia. 2019. Second Soil Microbial Ecology International Course. [Online]. [Cited 5 October 2020]. http://somiecoc.unal.edu.co/home/

Upchurch, R., Chiu, C.Y., Everett, K., Dyszynski, G., Coleman, D.C. & Whitman, W.B. 2008. Differences in the composition and diversity of bacterial communities from agricultural and forest soils. *Soil Biology and Biochemistry,* 40(6): 1294-1305.

US Salinity Laboratory Staff. 1954. *Diagnosis and improvement of saline and alkali soils.* USDA Handbook No. 60. Washington, USA. US Government Printing Office.

USDA. 1954. No. 60. Diagnosis and improvement of saline and alkaline soils. Washington, USA. United States Department of Agriculture. 160 pp.

USDA. 2016. An Economic Perspective on Soil Health. [online]. [Cited 5 October 2020]. https://www.ers.usda.gov/amber-waves/2016/september/an-economic-perspective-on-soil-health/

USDA. 2019a. National Program 212: Soil and Air. [online]. [Cited 5 October 2020]. https://www.ars.usda.gov/natural-resources-and-sustainable-agricultural-systems/soil-and-air/

USDA. 2019b. National Program 103: Animal Health. [online]. [Cited 5 October 2020]. https://www.ars.usda.gov/animal-production-and-protection/animal-health/

USDA. 2019c. National Program 216: Sustainable Agricultural Systems Research. [online]. [Cited 5 October 2020]. https://www.ars.usda.gov/animal-production-and-protection/animal-health/

USDA. 2019d. National Program 211: Water Availability and Watershed Management. [online]. [Cited 5 October 2020]. https://www.ars.usda.gov/natural-resources-and-sustainable-agricultural-systems/water-availability-and-watershed-management/

USDA. 2020a. Natural Resources and Sustainable Agricultural Systems. https://www.ars.usda.gov/natural-resources-and-sustainable-agricultural-systems/

USDA. 2020b. Soil Health. [online]. [Cited 5 October 2020]. https://www.nrcs.usda.gov/wps/portal/nrcs/main/soils/health/

USDA. 2020c. Soil Education. [online]. [Cited 5 October 2020]. https://www.nrcs.usda.gov/wps/portal/nrcs/main/soils/edu/

Usman, K., Al-Ghouti, M.A. & Abu-Dieyeh, M.H. 2019. The assessment of cadmium, chromium, copper, and nickel tolerance and bioaccumulation by shrub plant Tetraena qataranse. *Scientific Reports,* 9: 1–11.

Vaclav, R. & Kaluz, S. 2014. The effect of herbivore faeces on the edaphic mite community: implications for tapeworm transmission. *Experimental and Applied Acarology*, 62: 377–390.

Valiente-Banuet, A., Aizen, M.A., Alcántara, J.M., Arroyo, J., Cocucci, A., Galetti, M., *et al.* 2015. Beyond species loss: the extinction of ecological interactions in a changing world. *Functional Ecology*, 29(3): 299-307.

van den Hoogen, J., Geisen, S., Routh, D., Ferris, H., Traunspurger, W., Wardle, D.A., de Goede, R.G.M., *et al.* 2019. Soil nematode abundance and functional group composition at a global scale. *Nature*, 572: 194–198. https://doi.org/10.1038/s41586-019-1418-6

van der Bij, A.U., Weijters, M.J., Bobbink, R., Harris, J.A., Pawlett, M., Ritz, K., Benetková, P., Moradi, J., Frouze, J., & van Diggelen, R. 2018. Facilitating ecosystem assembly: Plant-soil interactions as a restoration tool. *Biological Conservation*, 220: 272–279. doi: 10.1016/j.biocon.2018.02.010

Van der Heijden, M.G.A., Bardgett, R.D. & van Straalen, N.M. 2008. The unseen majority: soil microbes as drivers of plant diversity and productivity in terrestrial ecosystems. *Ecology Letters*, 11: 296–310. doi: 10.1111/j.1461-0248.2007.01139.x

Van Der Heijden, M.G., Martin, F.M., Selosse, M.A. & Sanders, I.R. 2015. Mycorrhizal ecology and evolution: the past, the present, and the future. *New Phytologist*, 205(4): 1406-1423.

Van Der Heijden, M.G., De Bruin, S., Luckerhoff, L., Van Logtestijn, R.S. & Schlaeppi, K. 2016. A widespread plant-fungal-bacterial symbiosis promotes plant biodiversity, plant nutrition and seedling recruitment. *The ISME journal*, 10(2): 389.

Van der Putten, W.H., Van Dijk, C., & Peters, B.A.M. 1993. Plant-specific soil-borne diseases contribute to succession in foredune vegetation. *Nature*, 362(6415): 53.

Van der Putten, W.H., Cook, R., Costa, S., Davies, K.G., Fargette, M., Freitas, H., Hol, W.H.G., *et al.* 2006. Nematode interactions in nature: models for sustainable control of nematode pests of crop plants? *Advances in Agronomy*, 89: 227-260.

Van der Putten, W.H., Bardgett, R.D., Bever, J.D., Bezemer, T.M., Casper, B.B., Fukami, T., Kardol, P., *et al.* 2013. Plant–soil feedbacks: the past, the present and future challenges. *Journal of Ecology*, 101: 265–276. doi: 10.1111/1365-2745.12054

van Elsas, J.D., Chiurazzi, M., Mallon, C.A., Elhottová, D., Krištůfek, V. & Falcão Salles, J. 2012. Microbial diversity determines the invasion of soil by a bacterial pathogen. *Proceedings of the National Academy of Sciences of the United States of America*, 109: 1159-1164.

van Elsas, J.D., Costa, R., Jansson, J., Sjöling, S., Bailey, M., Nalin, R., Vogel, T.M. & Van Overbeek, L. 2008. The metagenomics of disease-suppressive soils–experiences from the METACONTROL project. *Trends in Biotechnology*, 26(11): 591-601.

Van Goethem, M.W., Pierneef, R., Bezuidt, O.K., Van De Peer, Y., Cowan, D.A. & Makhalanyane, T.P. 2018. A reservoir of 'historical' antibiotic resistance genes in remote pristine Antarctic soils. *Microbiome*, 6(1): 40-52.

Van Groenigen, K.J., Qi, X., Osenberg, C.W., Luo, Y. & Hungate, B.A. 2014. Faster decomposition under increased atmospheric CO_2 limits soil carbon storage. *Science,* 344: 508-509. https://doi.org/10.1126/science.1249534

Van Hoesel, W., Tiefenbacher, A., König, N., Dorn, V.M., Hagenguth, J.F., Prah, U., Widhalm, T., *et al.* 2017. Single and combined effects of pesticide seed dressings and herbicides on earthworms, soil microorganisms, and litter decomposition. *Frontiers in Plant Science,* 8: 215.

van Leeuwen, J.P., Saby, N.P.A., Jones, A., Louwagie, G., Micheli, E., Rutgers, M. & Creamer, R.E. 2017. Gap assessment in current soil monitoring networks across Europe for measuring soil functions. *Environmental Research Letters,* 12(12): 124007. doi: 10.1088/1748-9326/aa9c5c

van Leeuwen, J.P., Creamer, R.E., Cluzeau, D., Debeljak, M., Gatti, F., Henriksen, C.B., *et al.* 2019. Modeling of soil functions for assessing soil quality: soil biodiversity and habitat provisioning. *Frontiers in Environmental Science,* 7: 113. doi: 10.3389/fenvs.2019.00113pbio.1000178

Vanlauwe, B., Six, J., Sanginga, N. & Adesina, A.A. 2015. Soil fertility decline at the base of rural poverty in sub-Saharan Africa. *Nature Plants,* 1: 15101. doi: 10.1038/nplants.2015.101

Vargas-Rojas, R., Cuevas-Corona, R., Yigini, Y., Tong, Y., Bazza, Z. & Wiese, L. 2019. Unlocking the Potential of Soil Organic Carbon: A Feasible Way Forward. Pages 373–395 in T. Ginzky, H. Dooley, E. Heuser, I.L. Kasimbazi, E. Markus, T. Qin, eds. *International Yearbook of Soil Law and Policy, 2018,* pp. 373-395. Cham, Switzerland. Springer.

Vasenev, V. & Kuzyakov, K. 2018. Urban soils as hot spots of anthropogenic carbon accumulation: Review of stocks, mechanisms and driving factors. *Land Degradation and Development,* 29(6): https://doi-org.inee.bib.cnrs.fr/10.1002/ldr.2944

Vasquez, C., Quiros De G., M., Aponte, O. & Sandoval, D. M. F. 2008. Primer reporte de Raoiella indica Hirst (Acari: Tenuipalpidae) en Sur América. *Neotrop. entomol.* [online]. 2008, 37:739-740.

Vasquez, C., Sanchez, C., & Valera, N. 2007. Mite diversity (Acari: Prostigmata, Mesotigmata, Astigmata) associated to soil litter from two vegetation zones at the University Park UCLA. IHERINGIA. SERIE ZOOLOGIA, 97(4): 466-471.

Vécrin, M.P. & Muller, S. 2003. Top-soil translocation as a technique in the re-creation of species-rich meadows. *Applied Vegetation Science,* 6(2): 271–278. doi: 10.1111/j.1654-109X.2003.tb00588.x

Veen, G.F., De Long, J.R., Kardol, P., Sundqvist, M.K., Snoek, L.B. & Wardle, D.A. 2017. Coordinated responses of soil communities to elevation in three subarctic vegetation types. *Oikos,* 126: 1586-1599.

Veen, G.F., Wubs, E.R.J., Bardgett, R., Barrios, E., Bradford, M., Carvalho, S., De Deyn, G., *et al.* 2019. Applying the aboveground-belowground interaction concept in agriculture: spatio-temporal scales matter. *Frontiers in Ecology and Evolution,* 7: 300.

Velki, M. & Ečimović, S. 2016. Important Issues in Ecotoxicological Investigations Using Earthworms. In P. de Voogt, ed. *Reviews of Environmental Contamination and Toxicology Volume 239.* Cham, Switzerland. Springer.

Velthof, G., Barot, S., Bloem, J., Butterbach-Bahl, K., de Vries, W., Kros, J., Lavelle, P., Olesen, J.E. & Oenem, O. 2011. *Nitrogen as a threat to European soil quality.* Cambridge, UK. Cambridge University Press.

Venkateswarlu, B., Wani, S.P. & Vineela, C. 2016. Soil microbial diversity in rainfed agroecosystems. In K.G. Saxena & K.S. Rao, eds. *Soil Biodiversity: Inventory, Functions and Management,* pp. 373-382. Dehradun, India. Bishen Singh Mahendra Pal Singh.

Veresoglou, S.D., Halley, J.M. & Rillig, M.C. 2015. Extinction risk of soil biota. *Nature Communications,* 6: 8862.

Verzeaux, J., Hirel, B., Dubois, F., Lea, P.J. & Tétu, T. 2017. Agricultural Practices to Improve Nitrogen Use Efficiency Through the Use of Arbuscular Mycorrhizae: Basic and Agronomic Aspects. *Plant Science,* 264: 48-56

Vessey, J.K. 2003. Plant growth promoting rhizobacteria as biofertilizers. *Plant Soil,* 255: 571–586.

Viers, J.H., Williams, J.N., Kimberly, A.N., Barbosa, O., Kotzé, I., Spence, L., Webb, L.B., Merenlender, A., Reynolds, M. 2013. Vinecology: pairing wine with nature. DOI https://doi.org/10.1111/conl.12011

Vitousek, P.M., Naylor, R., Crews, T., David, M.B., Drinkwater, L.E., Holland, E., Nziguheba, G, *et al.* 2009. Nutrient imbalances in agricultural development. *Science,* 324(5934): 1519-1520.

Vitti, A., Pellegrini, E., Nali, C., Lovelli, S., Sofo, A., Valerio, M., Scopa, A. & Nuzzaci, M. 2016. Trichoderma harzianum T-22 induces systemic resistance in tomato infected by Cucumber mosaic virus. *Frontiers in Plant Science,* 7: 1520

Vodyanitskii, Y.N. 2015. Organic matter of urban soils. *Eurasian Soil Science,* 48: 802-811. https://doi.org/10.1134/S1064229315080116

Volosciuc, L. & Josu, V. 2014. Ecological Agriculture to Mitigate Soil Fatigue. *In* D. Dent, ed. *Soil as World Heritage,* pp. 431–435. Dordrecht, Springer Netherlands. (also available at https://doi.org/10.1007/978-94-007-6187-2_45).

Von Hertzen, L. & Haahtela, T. 2006. Disconnection of man and the soil: reason for the asthma and atopy epidemic? *Journal of Allergy and Clinical Immunology,* 117: 334-344.

Von Hertzen, L., Hanski, I. & Haahtela, T. 2011. Natural immunity. *EMBO reports,* 12: 1089-1093.

Vonk, J.A., Breure, A.M. & Mulder, C. 2013. Environmentally-driven dissimilarity of trait-based indices of nematodes under different agricultural management and soil types. *Agriculture, Ecosystems and Environment,* 179: 133-138.

Vos, H.M.J., Koopmans, G.F., Beezemer, L., de Goede, R.G.M., Hiemstra, T. & van Groenigen J.W. 2019. Large variations in readily-available phosphorus in casts of eight earthworm species are linked to cast properties. *Soil Biology and Biochemistry,* 138: 107583.

Wagg, C., Jansa, J., Schmid, B. & van der Heijden, M.G.A. 2011. Belowground biodiversity effects of plant symbionts support aboveground productivity. Ecology Letters 14:1001–1009.

Wagg, C., Bender, S.F., Widmer, F. & van der Heijden, M.G.A. 2014. Soil biodiversity and soil community composition determine ecosystem multifunctionality. *Proceedings of the National Academy of Sciences,* 111(14), 5266-5270.

Wagg, C., Dudenhöffer, J.-H., Widmer, F. & van der Heijden, M.G.A. 2018. Linking diversity, synchrony and stability in soil microbial communities. *Functional Ecology,* 32: 1280–1292. doi.org/10.1111/1365-2435.13056

Wagg, C., Jansa, J., Schmid, B., & Van der Heijden, M.G.A. 2011. Belowground biodiversity effects of plant symbionts support aboveground productivity. *Ecology Letters*, 14 (10): 1001-1009

Wagg, C., Schlaeppi, K., Banerjee, S., Kuramae, E.E. & van der Heijden, M.G.A. 2019. Fungal-bacterial diversity and microbiome complexity predict ecosystem functioning. *Nature Communications,* 10: 4841.

Waldrop, M.P. & Creamer, C. 2019. Soil microbial communities and global change. In J.D. van Elsas, J.T. Trevors, A. Soares Rosado & P. Nannipieri, eds. *Modern Soil Microbiology III,* pp. 331-342. Boca Raton, USA. CRC Press.

Walker, X., Baltzer, J., Cumming, S., Day, N., Ebert, C., Goetz, S., Johnstone, J., *et al.* 2019. Increasing wildfires threaten historic carbon sink of boreal forest soils. Nature, 572: 520-523. DOI: 10.1038/s41586-019-1474-y

Wall, D. H. (Ed). 2012. *Soil Ecology and Ecosystem Services.* Oxford University Press, Oxford, UK.

Wall, D.H., Nielsen, U.N. & Six, J. 2015. Soil biodiversity and human health. *Nature,* 52: 69-76.

Wallwork, J.A. 1976. *The Distribution and Diversity of Soil Fauna.* London, UK. Academic Press.

Walter, D.E. & Proctor, H.C. 2013. *Mites: Ecology, Evolution and Behaviour: life at a microscale, 2nd edition.* Cham, Switzerland. Springer. 494 pp.

Walther, G.-R., Roques, A., Hulme, P.E., Sykes, M.T., Pyšek, P., Kühn, I., Zobel, M., *et al.* 2009. Alien species in a warmer world: risks and opportunities. *Trends in Ecology and Evolution,* 24: 686-693.

Wang, B. & Qiu, Y.L. 2006. Phylogenetic distribution and evolution of mycorrhizas in land plants. *Mycorrhiza,* 16: 299–363.

Wang, B., Li, R., Ruan, Y., Ou, Y., Zhao, Y. & Shen, Q. 2015. Pineapple-banana rotation reduced the amount of Fusarium oxysporum more than maize-banana rotation mainly through modulating fungal communities. *Soil Biology and Biochemistry*, 86: 77-86

Wang, D., Liu, H., Ma, J., Qu, J., Guan, J., Lu, N., Lu, Y. & Yuan, X. 2016. Recycling of hyper-accumulator: Synthesis of ZnO nanoparticles and photocatalytic degradation for dichlorophenol. *Journal of Alloys and Compounds*, 680: 500-505

Wang, F. Lin, X. & Yin, R. 2005. Heavy metal uptake by arbuscular mycorrhizas of *Elsholtzia splendens* and the potential for phytoremediation of contaminated soil. *Plant Soil*, 269: 225. https://doi.org/10.1007/s11104-004-0517-8

Wang, F., Wang, C., Liu, P., Lei, C., Hao, W., Gao, Y., Liu, Y.G. & Zhao, K. 2016. Enhanced rice blast resistance by CRISPR/ Cas9-Targeted mutagenesis of the ERF transcription factor gene OsERF922. *PLoS ONE*, 11: 1–18.

Wang, H., Liu, H., Wang, Y., Xu, W., Liu, A., Ma, Z., Mi, Z., Zhang, Z., Wang, S. & He, J.S. 2017. Warm- and cold- season grazing affect soil respiration differently in alpine grasslands. *Agriculture, Ecosystems and Environment*, 248: 136-143.

Wang, H., Zhang, Y., Chen, G., Hettenhausen, C., Liu, Z., Tian, K. & Xiao, D. 2018. Domestic pig uprooting emerges as an undesirable disturbance on vegetation and soil properties in a plateau wetland ecosystem. *Wetlands Ecology and Management*, 26: 509-523.

Wang, H.Y. & Ortiz-Bobea, A. 2019. Market-Driven Corn Monocropping in the U.S. Midwest. *Agriculture and Resource Economics Review*, 48: 274- 296. DI 10.1017/age.2019.4

Wang, K., Peng, C., Zhu, Q., Zhou, X., Wang, M., Zhang, K. & Wang, G. 2017. Modeling global soil carbon and soil microbial carbon by integrating microbial processes into the ecosystem process model TRIPLEX-GHG. *Journal of Advances in Modeling Earth Systems*, 9: 2368-2384.

Wang, K., Zhang, Y., Tang, Z., Shangguan, Z., Chang, F., Jia, F., Chen, Y., He, X., Shi, W. & Deng, L. 2019. Effects of grassland afforestation on structure and function of soil bacterial and fungal communities. *Science of the Total Environment*, 676: 396-406.

Wang, X., Wu, J. & Kumari, D. 2018. Composition and functional genes analysis of bacterial communities from urban parks of Shanghai, China and their role in ecosystem functionality. *Landscape and Urban Planning*, 177: 83-91.

Wardle, D. 2002. *Communities and Ecosystems*. Princeton, USA. Princeton Universiy Press.

Wardle, D.A., Bardgett, R.D., Klironomos, J.N., Setälä, H., van der Putten, W.H., Diana, H. & Wall, D.H. 2004. Ecological Linkages between Aboveground and Belowground Biota. *Science*, 304(5677): 1629-1633. DOI: 10.1126/science.1094875

Warith, M., Ferehner, R., & Fernández, L. 1992. Bioremediation of organic contaminated soil. *Hazardous Waste and Hazardous Materials*, 9: 137-147.

Watson, J.E.M., Jones, K.R., Fuller, R.A., Marco, M.D., Segan, D.B., Butchart, S.H.M., Allan, J.R., McDonald-Madden, E. & Venter, O. 2016. Persistent disparities between recent rates of habitat conversion and protection and implications for future global conservation targets. *Conservation Letters,* 9: 413-421. doi.org/10.1111/conl.12295

Wei, X., Shao, M., Gale, W. & Li, L. 2014. Global patterns of soil carbon losses due to the conversion of forests to agricultural land. *Scientific Reports,* 4: 4062. https://doi.org/10.1038/srep04062

Weil, R.R., Brady, N.C. 2017. The Nature and Properties of Soils. Edition: 15th. Publisher: Pearson Education. ISBN: 978-0133254488.

Weisser, W.W., Roscher, C., Meyer, S.T., Ebeling, A., Luo, G.J., Allan, E., Besser, H., Barnard, R.L., Buchmann, N., Buscot, F., *et al.* 2017. Biodiversity effects on ecosystem functioning in a 15-year grassland experiment: Patterns, mechanisms, and open questions. *Basic and Applied Ecology,* 23: 1-73.

Weitz, J.S., Stock, C.A., Wilhelm, S.W., Bourouiba, L., Coleman, M.L., Buchan, A., *et al.* 2015. A multitrophic model to quantify the effects of marine viruses on microbial food webs and ecosystem processes. *The ISME Journal,* 9: 1352.

Welemariam, M., Kebede, F., Bedadi, B. & Birhane, E. 2018. Effect of Community-based Soil and Water Conservation Practices on Soil Glomalin, Aggregate Size Distribution, Aggregate Stability and Aggregate Associated Organic Carbon in Northern Highlands of Ethiopia. *Agriculture and Food Security,* 7: 42. https://doi.org/10.1186/s40066-018-0193-1

Welsh Government. 2020. Natural resources policy. [online]. [Cited 2 October 2020]. https://gov.wales/natural-resources-policy

Welti, E.A.R., Sanders, N.J., de Beurs, K.M. & Kaspari, M. 2019. A distributed experiment demonstrates widespread sodium limitation in grassland food webs. *Ecology,* 100: e02600

Wepking, C., Badgley, B., Barrett, J.E., Knowlton, K.F., Lucas, J.M., Minick, K.J., Ray, P.P., Shawver, S.E. & Strickland, M.S. 2019. Prolonged exposure to manure from livestock-administered antibiotics decreases ecosystem carbon-use efficiency and alters nitrogen cycling. *Ecology Letters,* 22(12): 13390.

Werner, I. & Hitzfeld, B. 2012. 50 Years of Ecotoxicology since Silent Spring - A Review. *GAIA- Ecological Perspectives for Science and Society,* 21: 217-224.

Wertz, S., Degrange, V., Prosser, J. I., Poly, F., Commeaux, C., Guillaumaud, N. & Le Roux, X. 2007. Decline of soil microbial diversity does not influence the resistance and resilience of key soil microbial functional groups following a model disturbance. *Environmental Microbiology,* 9(9): 2211-2219.

Wescott, B. 2019. Destructive pest could spread to all of China's grain production in 12 months. *CNN,* May 2019. (Also available at https://www.cnn.com/2019/05/09/asia/china-armyworm-grain-intl/index.html).

Whalen, J.K. & Sampedro, L. 2010. *Soil Ecology and Management.* Cambridge University Press, Cambridge.

Whitford, W.G. & Kay, F.R. 1999. Biopedturbation by mammals in deserts: a review. *Journal of Arid Environments,* 41: 203-230.

Wichern, J., Wichern, F. & Joergensen, R.G. 2006. Impact of salinity on soil microbial communities and the decomposition of maize in acidic soils. *Geoderma,* 137: 100-108

Wickings, K. & Grandy, A.S. 2011. The oribatid mite Scheloribates moestus (Acari: Oribatida) alters litter chemistry and nutrient cycling during decomposition. *Soil Biology and Biochemistry,* 43(2): 351–358.

Wieder, W.R., Allison, S.D., Davidson, E.A., Georgiou, K., Hararuk, O., He, Y. & Xu, X. 2015. Explicitly representing soil microbial processes in Earth system models. *Global Biogeochemical Cycles,* 29: 1782–1800.

Wiesmeier, M., Lungu, M., Hübner, R. & Cerbari, V. 2015. Remediation of degraded arable steppe soils in Moldova using vetch as green manure. *Solid Earth,* 6(2): 609–620. https://doi.org/10.5194/se-6-609-2015

Wiesmeier, M., Urbanski, L., Hobley, E., Lang, B., von Lützow, M., Martin-Spiotta, E., *et al.* 2019. Soil organic carbon storage as a key function of soils - A review of drivers and indicators at various scales. *Geoderma,* 333: 149-162. https://doi.org/10.1016/j.geoderma.2018.07.026

Wiesmeier, M., Poeplau, C., Sierra, C.A., Maier, H., Frühauf, C., Hübner, R., Kühnel, A., *et al.* 2016. Projected loss of soil organic carbon in temperate agricultural soils in the 21 st century: Effects of climate change and carbon input trends. *Science Reports,* 6: 32525. https://doi.org/10.1038/srep32525

Wilhelm, S.W. & Suttle, C.A. 1999. Viruses and Nutrient Cycles in the Sea: Viruses play critical roles in the structure and function of aquatic food webs. *BioScience,* 49: 781-788.

Wilkinson, D.M. & Mitchell, E.A.D. 2010. Testate amoebae and nutrient cycling with particular reference to soils. *Geomicrobiology Journal,* 27: 520-533.

Williams K.J., Ford, A., Rosauer, D.F., De Silva, N., Mittermeier, R., Bruce, C., Larsen, F.W. & Margules, C. 2011. Forests of East Australia: The 35th Biodiversity Hotspot. In F. Zachos & J. Habel, eds. *Biodiversity Hotspots.* Berlin, Germany. Springer.

Williamson, K.E., Fuhrmann, J.J., Wommack, K.E. & Radosevich, M. 2017. Viruses in Soil Ecosystems: An Unknown Quantity Within an Unexplored Territory. *Annual Review of Virology,* 4: 201-219.

Williamson, K.E., Radosevich, M. & Wommack, K.E. 2005. Abundance and Diversity of Viruses in Six Delaware Soils. *Applied and Environmental Microbiology,* 71: 3119-3125.

Willis, K.J., ed. 2017. *State of the World's Plants 2017.* Report. London, UK. Royal Botanic Gardens, Kew.

Wilpiszeski, R.L., Aufrecht, J.A., Retterer, S.T., Sullivan, M.B., Graham, D.E., Pierce, E.M., *et al.* 2019. Soil Aggregate Microbial Communities: Towards Understanding Microbiome Interactions at Biologically Relevant Scales. *Applied and Environmental Microbiology,* 85: e00324-19.

Wilson, D. & Read, J. 2003. Kangaroo harvesters: fertilising the rangelands. *The Rangeland Journal*, 25: 47-55.

Wilson, G.W.T., Rice, C.W., Rillig, M.C., Springer, A., & Harnett, D.C. 2009. Soil Aggregation and Carbon Sequestration Are Tightly Correlated With the Abundance of Arbuscular Mycorrhizal Fungi: Results From Long-Term Field Experiments. *Ecology Letters*, 12 (5): 452-461

Wodika, B.R. & Baer, S.G. 2015. If we build it, will they colonize? A test of the field of dreams paradigm with soil macroinvertebrate communities. *Applied Soil Ecology*, 91: 80–89.

Wodika, B.R., Klopf, R.P. & Baer, S.G. 2014. Colonization and Recovery of Invertebrate Ecosystem Engineers during Prairie Restoration. *Restoration Ecology*, 22: 456–464.

Wollrab, S., Diehl, S. & De Roos, A.M. 2012. Simple rules describe bottom-up and top-down control in food webs with alternative energy pathways. *Ecology Letters*, 15: 935-946.

Wolters, V., Silver, W.L., Bignell, D.E., Coleman, D.C., Lavelle, P., van der Puttern, W.H., de Ruiter, P., *et al.* 2000. Effects of Global Changes on Above- and Belowground Biodiversity in Terrestrial Ecosystems: Implications for Ecosystem Functioning: We identify the basic types of interaction between vascular plants and soil biota; describe the sensitivity of each type to changes in species composition; and, within this framework, evaluate the potential consequences of global change drivers on ecosystem processes. *Bioscience*, 50: 1089–1098.

Wong, M.-K., Tsukamoto, J., Yusuyin, Y., Tanaka, S., Iwasaki, K. & Tan, N.P. 2016. Comparison of soil macro-invertebrate communities in Malaysian oil palm plantations with secondary forest from the viewpoint of litter decomposition. *Forest Ecology and Management*, 381: 63-73.

Wong, V.N.L., Dalal, R.C. & Greene R.S.B. 2009. Carbon dynamics of sodic and saline soils following gypsum and organic material additions: A laboratory incubation. *Applied Soil Ecology*, 41: 29-40.

Wong, V.N.L., Greene, R.S.B., Dalal, R.C. & Murphy, B.W. 2010. Soil carbon dynamics in saline and sodic soils: a review. *Soil Use and Management*, 26: 2-11.

World Bank. 2016. *World Development Indicators 2016* [online]. Washington, USA. World Bank. https://openknowledge.worldbank.org/handle/10986/23969.

Wright, S.F. & Upadhyaya, A. 1998. A survey of soils for aggregate stability and glomalin, a glycoprotein produced by hyphae of arbuscular mycorrhizal fungi. *Plant and Soil*, 198(1): 97-107.

Wu, J., Zhang, W., Shao, Y. & Fu, S. 2017. Plant-facilitated effects of exotic earthworm Pontoscolex corethrurus on the soil carbon and nitrogen dynamics and soil microbial community in a subtropical field ecosystem. *Ecology and Evolution*, 7(21): 8709–8718.

Wu, L., Sun, Q., Sugawara, H., Yang, S., Zhou, Y., McCluskey, K., Vasilenko, A., *et al.* 2013. Global catalogue of microorganisms (gcm): a comprehensive database and information retrieval, analysis, and visualization system for microbial resources. *BMC Genomics*, 14: 933.

Wu, T., Ayres, E., Bardgett, R.D., Wall, D.H., & Garey, J.R. 2011. Molecular study of worldwide distribution and diversity of soil animals. *Proceedings of the National Academy of Sciences,* 108: 17720-17725.

Wu, X., Cao, R., Wei, X., Xi, X., Shi, P., Eisenhauer, N. & Sun, S. 2017. Soil drainage facilitates earthworm invasion and subsequent carbon loss from peatland soil. *Journal of Applied Ecology,* 54(5): 1291-1300.

Wu, X., Jousset, A., Guo, S., Karlsson, I., Zhao, Q., Wu, H., Kowalchuk, G.A., Shen, Q., Li, R. & Geisen, S. 2018. Soil protist communities form a dynamic hub in the soil microbiome. *The ISME Journal,* 12: 634-638.

Wu, X., Guo, S., Jousset, A., Zhao, Q., Wu, H., Li, R., Kowalchuk, G.A. & Shen, Q. 2017. Bio-fertilizer application induces soil suppressiveness against Fusarium wilt disease by reshaping the soil microbiome. *Soil Biology and Biochemistry,* 114: 238-247

Wubet, T., Kottke, I., Teketay, D. & Oberwinkler, F. 2003. Mycorrhizal status of indigenous trees in dry Afromontane forests of Ethiopia. *Forest Ecology and Management,* 179: 387–399.

Wubs, E.R.J., Melchers, P.D. & Bezemer, T.M. 2018. Potential for synergy in soil inoculation for nature restoration by mixing inocula from different successional stages. *Plant Soil,* 433: 147–156. doi: 10.1007/s11104-018-3825-0

Wubs, E.R.J., Van der Putten, W.H., Bosch, M. & Bezemer, T.M. 2016. Soil inoculation steers restoration of terrestrial ecosystems. *Nature Plants,* 2: 16107. doi: 10.1038/NPLANTS.2016.107

Wubs, E.R.J., Van der Putten, W.H., Mortimer, S.R., Korthals, G.W., Duyts, H., Wagenaar, R. & Bezemer, T.M. 2019. Single introductions of soil biota and plants generate long-term legacies in soil and plant community assembly. *Ecology Letters,* 22: 1145–1151. doi: 10.1111/ele.13271

Wurzburger, N. & Clemmensen, K.E. 2018. From mycorrhizal fungal traits to ecosystem properties–and back again. *Journal of Ecology,* 106(2): 463-467.

Xiao Z., Jiang L., Chen X., Zhang Y., Defossez E., Hu F., Liu M. & Rasmann S. 2019. Earthworms suppress thrips attack on tomato plants by concomitantly modulating soil properties and plant chemistry. *Soil Biology and Biochemistry,* 130: 23-32.

Xiaoping, Z., Wenpo, W., Wei, L., Shuguo, G. & Dongjun, H. 2005. The Influences of Microbial Fertilizers on Nutrients Uptake of Pepper and Changes of Soil Nutrients. *Chinese Agricultural Science Bulletin.* (also available at http://en.cnki.com.cn/Article_en/CJFDTotal-ZNTB200505082.htm).

Xie, T., Liu, W., Anderson, B.D., Liu, X. & Gray, G.C. 2017. A system dynamics approach to understanding the One Health concept. *PLoS ONE* 12: 1–11. https://doi.org/10.1371/journal.pone.0184430

Xu, J., Ebada, S.S. & Proksch, P. 2010. Pestalotiopsis a highly creative genus: chemistry and bioactivity of secondary metabolites. *Fungal Diversity,* 44(1): 15-31.

Xu, W., Xiao, Y., Zhang, J., Yang, W., Zhang, L., Hull, V., Wang, Z., Zheng, H., Liu, J., Polasky, S., Jiang, L., Xiao, Y., Shi, X., Rao, E., Lu, F., Wang, X., Daily, G.C. & Ouyang, Z. 2017. Strengthening protected areas for biodiversity and ecosystem services in China. *Proceedings of the National Academy of Sciences*, 114(7): 1601. https://doi.org/10.1073/pnas.1620503114

Xu, X., Thornton, P.E. & Post, W.M. 2013. A global analysis of soil microbial biomass carbon, nitrogen and phosphorus in terrestrial ecosystems: Global soil microbial biomass C, N and P. *Global Ecology and Biogeography*, 22: 737–749. https://doi.org/10.1111/geb.12029

Xu, Y. & Zhou, N. 2017. Microbial remediation of aromatics-contaminated soil. *Frontiers of Environmental Science and Engineering*, 11(2): 1-9. doi: 10.1007/s11783-017-0894-x

Xuelei, Z., Chen, J., Tan, M. & Sun, Y. 2007. Assessing the Impact of Urban Sprawl on Soil Resources of Nanjing City Using Satellite Images and Digital Soil Databases. *Catena*, 69(1): 16–30. https://doi.org/10.1016/j.catena.2006.04.020.

Xuelei, Z., Jie, C., Ganlin, Z., Manzhy, T. & Ibáñez, J.J. 2003. Pedodiversity analysis in Hainan Island. *Journal of Geographical Sciences*, 13(2): 181-186. https://doi.org/10.1007/BF02837456.

Xun, W., Li, W., Xiong, W., Ren, Y., Liu, Y., Miao, Y., Xu, Z., Zhang, N., Shen, Q. & Zhang, R. 2019. Diversity-triggered deterministic bacterial assembly constrains community functions. *Nature Communications*, 23: 3833.

Yachi, S. & Loreau, M. 1999. Biodiversity and ecosystem productivity in a fluctuating environment: the insurance hypothesis. *Proceedings of the National Academy of Sciences of the United States of America*, 96: 1463–1468.

Yang, Y., Tilman, D., Furey, G. & Lehman, C. 2019. Soil carbon sequestration accelerated by restoration of grassland biodiversity. *Nature Communications*, 10: 718.

Yang, G., Liu, N., Lu, W., Wang, S., Kan, H., Zhang, Y., Xu, L., & Chen, Y. 2014. The interaction between arbuscular mycorrhizal fungi and soil phosphorus availability influences plant community productivity and ecosystem stability. *Journal of Ecology*, 2014, 102, 1072–1082

Yapi, A. 1991. *Biologie, Ecologie et metabolisme digestif de quelques espèces de termites humivores de savane.* Thèse de Doctorat Université. Abdijan, Ivory Coast. Faculté des Sciences et Techniques de l'Université d'Abidjan. 94 pp.

Yarza, P., Yilmaz, P., Pruesse, E., Glöckner, F.O., Ludwig, W., Schleifer, K.H., Whitman, W.B., Euzéby, J., Amann, R. & Rosselló-Móra, R. 2014. Uniting the classification of cultured and uncultured bacteria and archaea using 16S rRNA gene sequences. *Nature Reviews Microbiology*, 12: 635.

Yeates, G.W., Ferris, H., Moens, T. & van der Putten, W.H. 2009. The role of nematodes in ecosystems. In M.J. Wilson & T. Kakouli-Duarte, eds. *Nematodes as Environmental Indicators*, pp. 1-45. Wallingford, UK. CABI Publishing.

Yeung, A.C.Y., Paltsev, A., Daigle, A., Duinker, P.N. & Creed, I.F. 2019. Atmospheric change as a driver of change in the Canadian boreal zone. *Environmental Reviews,* 27: 346-376. DOI: 10.1139/er-2018-0055

Yimyam, N., Youpensuk, S., Rerkasem, B. & Rerkasem, K. 2016. The role of mycorrhizal fungi and Macaranga denticulata symbiosis in maintaining productivity of rice in shifting cultivation in Thailand. In K.G. Saxena & K.S. Rao, eds. *Soil Biodiversity: Inventory, Functions and Management,* pp. 365-372. Dehradun, India. Bishen Singh Mahendra Pal Singh.

Young, I.M. & Crawford, J.W. 2004. Interactions and Self-Organization in the Soil-Microbe Complex. *Science,* 304(5677): 1634-1637. DOI: 10.1126/science.1097394

Youngsang, C. 2016. Jadam Organic Farming. The way to ultra-low cost agriculture. En.jadam.kr.

Yu, X.-M., Cloutier, S., Tambong, J.T., and Bromfield, E.S.P. 2014. "Bradyrhizobium ottawaense sp. nov., a symbiotic nitrogen fixing bacterium from root-nodules of soybeans in Eastern Canada. Intern. J. Syst. and Evol. Microbiol., 64(Pt. 9), pp. 3202-3207. DOI: 10.1099/ijs.0.065540-0

Yu, H.L., He, N.P., Wang, Q.F., Zhu, J.X., Gao, Y., Zhang, Y.H., Jia, Y.L. & Yu, G.R. 2017. Development of atmospheric acid deposition in China from the 1990s to the 2010s. *Environmental Pollution,* 231: 182-190.

Yu, J., Xue, Z., He, X., Liu, C. & Steinberger, Y. 2017. Shifts in composition and diversity of arbuscular mycorrhizal fungi and glomalin contents during revegetation of desertified semiarid grassland. *Applied Soil Ecology,* 115: 60-67.

Yuan, H., Ge, T., Chen, C., O'Donnell, A.G. & Wu, J. 2012. Significant Role for Microbial Autotrophy in the Sequestration of Soil Carbon. *Applied and Environmental Microbiology,* 78(7): 2328–2336. https://doi.org/10.1128/AEM.06881-11

Zaitsev, A.S., Gongalsky, K.B., Korobushkin, D.I., Butenko, K.O., Gorshkova, I.A., Rakhleeva, A.A., Saifutdinov, R.A., Kostina, N.V., Shakhab, S.V. & Yazrikova, T.E. 2017. Reduced Functionality of Soil Food Webs in Burnt Boreal Forests: a Case Study in Central Russia. *Contemporary Problems of Ecology,* 10(3): 277–285. DOI: 10.1134/S199542551703012X

Zaitsev, A.S., Gongalsky, K.B., Malmström, A., Persson, T. & Bengtsson, J. 2016. Why are forest fires generally neglected in soil fauna research? A mini-review. *Applied Soil Ecology,* 98: 261-271. DOI: 10.1016/j.apsoil.2015.10.012.

Zaitsev, A.S. & van Straalen, N.M. 2001. Species diversity and metal accumulation in oribatid mites (Acari, Oribatida) of forests affected by a metallurgical plant. *Pedobiologia,* 45:467–479.

Zak, D.R., Pregitzer, K.S., Curtis, P.S., Teeri, J.A., Fogel, R. & Randlett, D.L. 1993. Elevated atmospheric CO_2 and feedback between carbon and nitrogen cycles. *Plant and Soil,* 151: 105-117.

Zak, J.C., Willig, M.R., Moorhead, D.L. & Wildman, H.G. 1994. Functional diversity of microbial communities: a quantitative approach. *Soil Biology and Biochemistry,* 26: 1101-1108.

Zaller, J.G., Heigl, F., Grabmaier, A., Lichtenegger, C., Piller, K. & Allabashi, R. 2011. Earthworm-mycorrhiza interactions can affect the diversity, structure and functioning of establishing model grassland communities. *PLoS ONE,* 6: e29293.

Zarraonaindia, I., Owens, S.M., Weisenhorn, P., West, K., Hampton-Marcell, J., Lax, S., Bokulich, N.A., *et al.* 2015. The soil microbiome influences grapevine-associated microbiota. *mBio,* 6: e02527-14, DOI: 10.1128/mBio.02527-14

Zechmeister-Boltenstern, S., Keiblinger, K.M., Mooshammer, M., Peñuelas, J., Richter, A., Sardans, J. & Wanek, W. 2015. The application of ecological stoichiometry to plant–microbial–soil organic matter transformations. *Ecological Monographs,* 85: 133–155. doi:10.1890/14-0777.1

Zehe, E., Blume, T. & Blöschl, G. 2010. The principle of "maximum energy dissipation": A novel thermodynamic perspective on rapid water flow in connected soil structures. *Philosophical Transactions of the Royal Society B: Biological Sciences,* 365: 1377–1386.

Zhang, J. & Elser, J.J. 2017. Carbon:nitrogen:phosphorus stoichiometry in fungi: a meta-analysis. *Frontiers in Microbiology,* 8: 1281.

Zhang, Y.-M. & Rock, C.O. 2008. Membrane lipid homeostasis in bacteria. *Nature Reviews Microbiology,* 6: 222.

Zhang, Z.-Q. 2013. Animal biodiversity: An update of classification and diversity in 2013. In Z.-Q. Zhang, ed. *Animal Biodiversity: An Outline of Higher-level Classification and Survey of Taxonomic Richness,* pp. 1-82. Auckland, New Zealand. Magnolia Press.

Zhang, H., Tomodai, H., Tabata, N., Miura, H., Namikoshi, M., Yamaguchi, Y., Masuma, R. & Omura, S. 2001. Cladospolide D, a new 12-membered macrolide antibiotic produced by Cladosporium sp. FT-0012. *The journal of Antibiotics,* 54(8): 635-641.

Zhang, K., Cheng, X., Shu, X., Liu, Y. & Zhang, Q. 2018. Linking soil bacterial and fungal communities to vegetation succession following agricultural abandonment. *Plant and Soil,* 431(1-2): 19-36.

Zhang, K., Shi, Y., Cui, X., Yue, P., Li, K., Liu, X., Tripathi, B.M. & Chu, H. 2019. Salinity Is a Key Determinant for Soil Microbial Communities in a Desert Ecosystem. *MSystems,* 4(1): e00225-18.

Zhang, L., Shi, N., Fan, J., Wang, F. George, T.S. & Feng, G. 2018. Arbuscular Mycorrhizal Fungi Stimulate Organic Phosphate Mobilization Associated With Changing Bacterial Community Structure Under Field Conditions. *Environmental Microbiology,* 20 (7):2639-2651

Zhang, X., Crawford, J.W. & Young, I.M. 2016. A Lattice Boltzmann model for simulating water flow at pore scale in unsaturated soils. *Journal of Hydrology,* 538: 152-160. https://doi.org/10.1016/j.jhydrol.2016.04.013

Zhang, Y.F., He, L.Y., Chen, Z.J., Wang, Q.Y., Qian, M. & Sheng, X.F. 2011. Characterization of ACC deaminase-producing endophytic bacteria isolated from copper-tolerant plants and their potential in promoting the growth and copper accumulation of Brassica napus. *Chemosphere*, 83(1): 57-62. doi: 10.1016/j.chemosphere.2011.01.041

Zhao, Q.-G., He, J.-Z., Yan, X.-Y., Zhang, B., Zhang, G.-L. & Cai, Z.-C. 2011. Progress in Significant Soil Science Fields of China over the Last Three Decades: A Review. *Pedosphere*, 21(1): 1–10. https://doi.org/10.1016/S1002-0160(10)60073-2

Zhao, J., Wang, F., Li, J., Zou, B., Wang, X., Li, Z. & Fu, S. 2014. Effects of experimental nitrogen and/or phosphorus additions on soil nematode communities in a secondary tropical forest. *Soil Biology and Biochemsity*, 75: 1-10.

Zhou, H., Wang, H., Huang, Y. & Fang, T. 2016. Characterization of pyrene degradation by halophilic Thalassospira sp. strain TSL5-1 isolated from the coastal soil of Yellow Sea, China. *International Journal of Biodeteriation and Biodegradation*, 107: 62-69. doi: 10.1016/j.ibiod.2015.10.022

Zhou, H., Zhang, D., Jiang, Z., Sun, P., Xiao, H., Yuxin, W. & Chen, J. 2019. Changes in the soil microbial communities of alpine steppe at Qinghai-Tibetan Plateau under different degradation levels. *Science of the Total Environment*, 651: 2281-2291.

Zhu, S., Vivanco, J.M. & Manter, D.K. 2016. Nitrogen fertilizer rate affects root exudation, the rhizosphere microbiome and nitrogen-use-efficiency of maize. *Applied Soil Ecology*, 107: 324–333. doi: 10.1016/j.apsoil.2016.07.009

Zhu, Z., Chang, L., Li, J., Liu, J., Feng, L. & Wu, D. 2018. Interactions between earthworms and mesofauna affect CO_2 and N_2O emissions from soils under long-term conservation tillage. *Geoderma*, 332: 153-160.

Zhu, D., Ke, X., Wu, L., Christie, P & Luo, Y. 2016. Biological transfer of dietary cadmium in relation to nitrogen transfer and 15N fractionation in a soil collembolan-predatory mite food chain. *Soil Biology and Biochemistry*, 101: 207-216.

Zhu, X., Chang, L., Liu, J., Zhou, M., Li, J., Gao, B. & Wu, D. 2016. Exploring the relationships between soil fauna, different tillage regimes and CO_2 and N_2O emissions from black soil in China. *Soil Biology and Biochemistry*, 103: 106-116.

Zieger, S.L., Ammerschubert, S., Polle, A. & Scheu, S. 2017. Root-derived carbon and nitrogen from beech and ash trees differentially fuel soil animal food webs of deciduous forests. *PLoS ONE*, 12(12): e0189502.

Zomer, R.J., Bossio, D.A., Sommer, R. & Verchot, L.V. 2017. Global Sequestration Potential of Increased Organic Carbon in Cropland Soils. *Scientific Reports*, 7: 15554.